LEAVING CERTIFICA

Real World PHYSICS

DAN O'REGAN

FOLENS

Editor
Margaret Burns

Design and Layout
Liz Murphy

Illustrations
Michael Phillips
Jean de Lemos

Picture Research
Sinéad Kelleher

ISBN 1 84131 391 2

Folens Publishers,
Hibernian Industrial Estate,
Greenhills Road,
Tallaght,
Dublin 24.

The author and Publisher would like to thank the following for permission to use material: Slidefile; Popperfoto; The Telegraph Colour Library; Kos Picture Source; Science and Society Picture Library; Genesis Space Photo Library; Science Photo Library; Eircell; The International Salvage Union; Oxford Scientific Films; Imperial War Museum; The J Allan Cash Photo library; Bruce Coleman Collection; Redferns Music Picture Library; Shout Picture Library; The Wellcome Trust Medical Photographic Library; Eaglemoss publications; The Ronald Grant Archive; Neil Bruce's Automobile Photolibrary; Image Select International; Saint Patrick's College Maynooth; Oxford University Press; Corbis Images Ltd, Radiological Protection Institute of Ireland. We would also like to thank the Department of Education and Science for questions reproduced.

CONTENTS

PREFACE

This textbook is specifically written for the new Leaving Certificate Physics Syllabus and presents the entire course for both Higher Level and Ordinary Level students. It gives a clear and accurate presentation of the requirements of the new syllabus, including both options.

Particular emphasis has been paid to the new STS material, with reference to many applications of Physics in the real world. Step-by-step instructions are given for mandatory experiments with experimental notes and warnings given where necessary. Problems and solutions follow each section as do exercises. Further completion and drawing-type questions, as well as more challenging problems, can be found in the accompanying Workbook. A 'Checklist' at the end of each chapter brings key points together for ease of revision.

Higher level material is clearly marked with a dotted line down the side; Option Chapters 32 and 33 are completely Higher Level and have no line. Higher Level students study Chapters 1–31 and either Chapter 32 or 33. Ordinary level students study the relevant parts of Chapters 1–11 and 14–31. They are not required to study Chapters 12, 13, 32 or 33.

I wish to thank each of the following without whose constant help, support and advice the production of this book would not have been possible; Mr John O'Connor, Ms Margaret Burns, Ms Liz Murphy and Ms Sinéad Kelleher of Folens; Mr Declan Kennedy; Dr Michael G Egan, Electrical Engineering Department, UCC; Mr John O'Regan, Avalon Computing Ltd; Mr Diarmuid Ó Mathúna, Principal, Rochestown College, Cork; Mr Peter Brennan; Mr Brian Scannell; Mr John Scannell. For providing endless support and encouragement I would also like to thank my parents Henry and Madeleine O'Regan; my parents in law, Maurice and Elizabeth Carroll; my wife, Jackie, and family Simon, Robert, Sophie and Lola.

Dan O'Regan
Rochestown College
Rochestown
Cork

Introduction

CHAPTER 1

PHYSICS: AN ANSWER TO MANY QUESTIONS

Scientists are very inquisitive people; they want to know how and why things happen. Physicists try to discover the laws that govern the physical world because these laws give us a better understanding of the world in which we live. Many questions that you may have about how the world works will be answered during this course. For instance, have you ever wondered why when you rub a pen in your hair the pen can pick up a piece of paper or why a spoon immersed in water appears to bend where it enters the water? Have you asked yourself any of these questions:

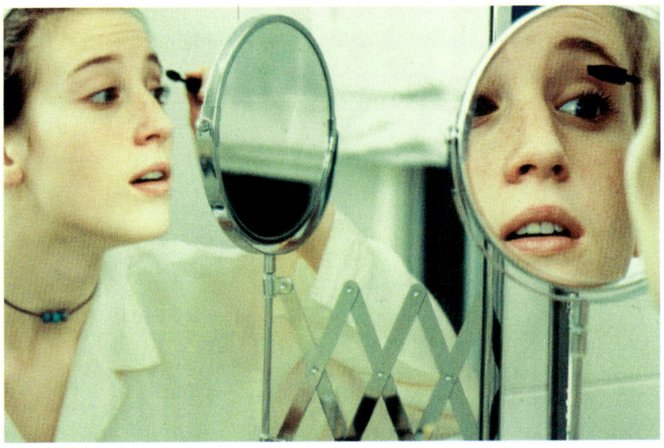

- Why do I see colours on my CD?
- How do we see reflections in a mirror?
- Why is there sometimes a rainbow when it's wet?
- How can we see inside a person's body with an X-ray machine?
- Why is a magnet attracted to Iron?
- How do the Gardaí's speed guns work?
- What is a mirage and why do they happen?

If you are curious enough to wonder about any of these things then you will enjoy studying Physics. It is a subject that comes to life as it is experienced first hand. As you proceed through the course you will use special scientific equipment and apply theories to see how and why certain things happen; all of the above questions and many more will be answered. *Real World Physics* helps you to develop an awareness of how Physics relates to everyday life; you will learn to appreciate the enormous impact that Physics has on modern society.

STUDYING PHYSICS

While studying physics:

- you must think like a scientist by being curious and looking for explanations,
- don't be afraid to ask questions and keep asking until you understand clearly the area of Physics being taught to you,
- try to think logically,
- take pride in doing the experiments and getting the correct results!

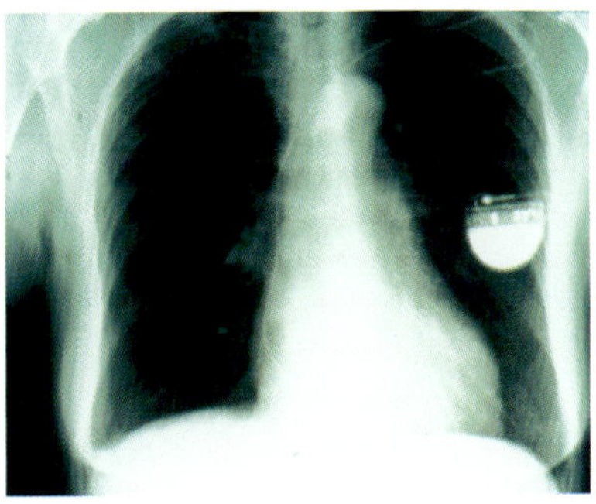

WHAT IS PHYSICS?

Physics is the branch of science that studies matter, energy and the relationship between them.

Areas in Physics You Will Study

In Leaving Certificate Physics you will study various topics in each of the following areas:

- Mechanics and Motion
- Heat
- Waves and Sound
- Light and Optics
- Electricity and Magnetism
- Atomic Physics and Electronics
- Nuclear and Particle Physics

Experiments in Physics

There are two main types of experiment in physics: those involving observation but not measurement, and those involving measurement.

OBSERVATION

If you have studied science before you probably performed or saw an experiment to show that a solid expands when heated. You may also have seen an experiment showing that a coil of wire carrying an electric current behaves like a bar magnet. In these experiments you *observed* but did not *measure* anything.

MEASUREMENT

A civil engineer might need to know by how much a bridge would expand during the hottest day in summer. He or she would, therefore, have to know by **how much** steel and concrete expand when heated. An electrical engineer who wanted to design an electromagnet to lift scrap iron would need to know what **size** electric current and coil would be needed to give a strong enough magnetic field to lift scrap. Some of the experiments in Leaving Certificate Physics that you will do in the laboratory involve **measurement**. This introduces the idea of a **physical quantity**. It is defined as follows:

PHYSICAL QUANTITY
Any property of matter that can be measured is called a ***physical quantity*** (or simply a quantity).

EXAMPLES OF QUANTITIES:

- time
- length
- area
- volume
- distance
- displacement
- speed
- velocity
- acceleration
- mass
- momentum
- force
- moment of a force
- work
- energy
- power
- temperature
- frequency
- sound intensity
- sound intensity level
- electric charge
- electric field strength
- capacitance
- electric current
- potential difference
- resistance
- magnetic flux
- magnetic flux density
- activity of a radioactive source

What Is Measuring?

MEASURING
When you measure a quantity you ***compare*** it with a standard amount of the same quantity. This standard amount is called a ***unit***.

- A measurement finds that the length of a piece of wire is 12 metres. This means that this piece of wire is 12 times longer than the metre. The metre is the unit of length.
- A measurement finds that the size of an electric current flowing in a wire is two amperes. This means that the current in this wire is twice as large as the ampere. The ampere is the unit of electric current.

The result of a measurement is always a number multiplied by a unit. This (the number and unit) is called the **magnitude** (or size) of the quantity being measured.

SI Units

In 1960 scientists agreed to use a particular system of units. They called this system the International System of Units. Any unit of this system is called an SI unit. SI units will be used throughout your physics course. Some examples of quantities and their SI units are shown in FIG. 1.1.

Quantity	Time	Displacement	Speed	Force	Resistance	Frequency
SI Unit	second	metre	metre per second	newton	ohm	hertz

Fig. 1.1

Basic Units

An independent unit could have been chosen for the measurement of each quantity. Different units would then not be related to each other in an easy way. To simplify matters, seven quantities, called basic quantities, have been chosen and the unit given to each is called a **basic unit** or a **base unit**. These seven base units are precisely defined. We need not worry about the exact definitions for the present. FIG. 1.2 shows the name and symbol of the five basic quantities and basic units on our course.

Basic quantity	Symbol	Name of SI base unit	Symbol
length	l or s	metre	m
time	t	second	s
mass	m	kilogram	kg
electric current	I	ampere	A
temperature	T	kelvin	K

Fig. 1.2

Derived Units

The unit of every other quantity is called a **derived unit** because it can be expressed as the product or quotient of one or more of the basic units. FIG. 1.3 shows some examples of quantities and their derived units.

The ***unit*** of any quantity can be written as a product or quotient of one or more base units.

Physical Quantity	Symbol	Name of SI Unit	Symbol
Area	A	square metre	m^2
Volume	V	cubic metre	m^3
Speed	v or u	metre per second	$m\ s^{-1}$ or m/s
Density	ρ	kilogram per cubic metre	$kg\ m^{-3}$ or kg/m^3

Fig. 1.3

Finding the Derived Unit of a Quantity

Problem 1: Find the SI unit of volume

Solution: Since: Volume = length × breadth × height

SI unit of volume = (unit of length)(unit of breadth)(unit of height) = (m)(m)(m)

= m^3 = the cubic metre

Problem 2: Find the SI unit of speed

Solution:

$$\text{Speed} = \frac{\text{distance}}{\text{time}} \Rightarrow \text{SI unit of speed} = \frac{\text{unit of distance}}{\text{unit of time}} = \frac{\text{m}}{\text{s}} = \text{m s}^{-1}$$

It is called the metre per second.*

* Recall from your maths that: $\frac{1}{x}$ is written as x^{-1}. Thus $\frac{\text{m}}{\text{s}} = \text{m}\left(\frac{1}{\text{s}}\right) = \text{m s}^{-1}$

Problem 3: Find the SI unit of density

Solution:

$$\text{Since density} = \frac{\text{mass}}{\text{volume}}$$

$$\text{SI unit of density} = \frac{\text{unit of mass}}{\text{unit of volume}} = \frac{\text{kilogram}}{\text{cubic metre}} = \text{kilogram per cubic metre}$$

$$= \text{kg m}^{-3} \text{ or kg/m}^3$$

Problem 4: When a force F acts on a body of mass m and causes that body to get an acceleration a you will see in Chapter 9 that: Force = mass x acceleration i.e. $F = ma$. If the unit of acceleration is the m s^{-2}, find the SI unit of force in terms of basic units.

Solution: $F = ma \Rightarrow$ Unit of Force = (unit of mass)(unit of acceleration) = (kg)(m s^{-2}) = kg m s^{-2}

This unit is usually called the **newton** (N) in honour of the great scientist and mathematician Isaac Newton.

Problem 5: When a force F moves through a distance s in the same direction as the force the Work W done is given by the formula: Work = Force x Distance, i.e. $W = Fs$. Find:

(a) the SI unit of work in terms of the newton.

(b) the SI unit of work in terms of basic units.

Solution:

(a) Work = Force x Distance $\Rightarrow$ SI unit of work = (unit of force)(unit of distance)
= newton metre.

This unit is usually called the **joule (J)** in honour of the scientist James Joule.
Thus 1 J = 1 N m

(b) From Problem 4: 1 N = 1 kg m s^{-2}

$\Rightarrow$ Unit of work = N m = (kg m s^{-2})(m) = kg m^2 s^{-2}

Thus the unit of work, the joule = kg m^2 s^{-2}

Sometimes if a unit is expressed in terms of base units it is complicated. It may then be given another name – usually that of a scientist who made some important discovery in the area of science where that unit often occurs. When this happens **the unit symbol is written with a capital letter but the name of the unit is not.** Fig. 1.4 shows some units of this kind that you will meet.

Quantity	Symbol	Unit	Symbol	Alternative Symbol	Symbol in terms of basic units
Force	F	newton	N	kg m s^{-2}	kg m s^{-2}
Pressure	P or p	pascal	Pa	N m^{-2}	kg m^{-1} s^{-2}
Work	W	joule	J	N m	kg m^2 s^{-2}
Energy	E	joule	J	N m	kg m^2 s^{-2}
Power	P	watt	W	J s^{-1}	kg m^2 s^{-3}
Frequency	f	hertz	Hz	s^{-1}	s^{-1}
Magnetic flux density	B	tesla	T		
Magnetic flux	Φ	weber	Wb		
Activity of a radioactive source	A	becquerel	Bq		

Fig. 1.4

USING SI UNITS

- When solving numerical problems in physics you should use only SI units. In any problem you do, **make sure that the value of each quantity is expressed in the correct SI unit before you start.**
- When doing calculations it is not necessary to write down units at each step of the argument. However **when a result is arrived at, it must be written with the correct unit (or unit symbol) after it. Marks will definitely be lost in an exam if you do not!**
- When writing down the magnitude of a quantity with the appropriate unit symbol, **a space is left between the number and the unit symbol.** For example, a length of 5 metres is written as 5 m, not 5m.
- When writing a unit in terms of basic units, a space is left between the symbol of each basic unit. For example, the unit of density is abbreviated kg m^{-3} not kgm^{-3} , the unit of speed is abbreviated m s^{-1} not ms^{-1}.

Fig. 1.5

MULTIPLES AND FRACTIONS OF THE STANDARD UNITS

Sometimes the standard SI units are too large or too small to be used easily. We therefore use the multiples or fractions of the standard units shown in FIG. 1.6. Note the following:

- The name of a unit has the prefix written before it, e.g. 10 kilometres, 5 millinewtons
- There is no space between the prefix and the symbol for the unit, e.g. 60 millimetres is written a 60 mm; 20 kilowatts is written as 20 kW.
- Only multiples of 10 to the power of 3 are usually used. The prefix 'centi' is normally only used with the metre. A table similar to FIG. 1.6 can be found on **page 5 of your maths tables**.

Problem 6: Express each of the following in standard SI units:

(i) 12 mm
(ii) 400 mJ
(iii) 7 MW

Solution:

(i) 12 mm = 12 x 10^{-3} m = 1.2 x 10^{-2} m
(ii) 400 mJ = 400 x 10^{-3} J = 0.4 J
(iii) 7 MW = 7 x 10^{6} W

Mutiple or Fraction	Prefix	Symbol
10^{9}	giga-	G
10^{6}	mega-	M
10^{3}	kilo-	k
10^{-2}	centi-	c
10^{-3}	milli-	m
10^{-6}	micro-	μ
10^{-9}	nano-	n
10^{-12}	pico-	p

Fig. 1.6

EXERCISE 1.1

1. Show that the SI unit of area is the square metre.
2. If a body of mass m kilograms is moving with a velocity of v metres per second its momentum P is defined by the equation:
   $P = mv$. Find the SI unit of momentum in terms of basic units.
3. When the velocity of a body changes its average acceleration a is given by:
   $$a = \frac{\text{change in velocity}}{\text{time taken for change}}$$
   Find the SI unit of acceleration.
4. What is the SI unit of density? Prove that your answer is correct.
5. Pressure P is defined as force per unit area, i.e. $P = F/A$. If the SI unit of force is the newton (N) find the SI unit of pressure in terms of the newton and the metre.
6. (a) How many square centimetres in one square metre?
   (b) How many cubic centimetres in one cubic metre?
   (c) How many grams in a kilogram?
7. Express each of the following in terms of standard SI units:
   (i) 5 cm^2 (ii) 40 cm^2 (iii) 1 cm^3 (iv) 456 cm^3 (v) 1000 000 000 cm^3.
8. Express each of the following measurements in standard SI units:
   (i) 105 km
   (ii) 57 mm
   (iii) 6.67×10^{-11} cm
   (iv) 6×10^{27} grams
   (v) 9 grams per cubic centimetre
   (vi) 100 km h^{-1}
   (vii) 5 nN
   (viii) 10 μW
   (ix) 5 Gm
9. The unit of acceleration a is the metre per second squared (m s^{-2}). Force is equal to mass by acceleration, i.e. $F = m\,a$. The unit of force is the newton. Express the newton in terms of basic units.
10. By definition work W is equal to force F multiplied by distance s travelled, i.e. $W = F\,s$. The unit of work is the joule (J). Express the joule in terms of the newton.
11. Use the results of the previous two questions to write the unit of work, the joule, in terms of basic units.
12. By definition power P is equal to work done divided by time taken, i.e. $P = \frac{W}{t}$. The unit of power is called the watt (W). Express the watt in terms of the joule and the second. Then express the watt in terms of base units.

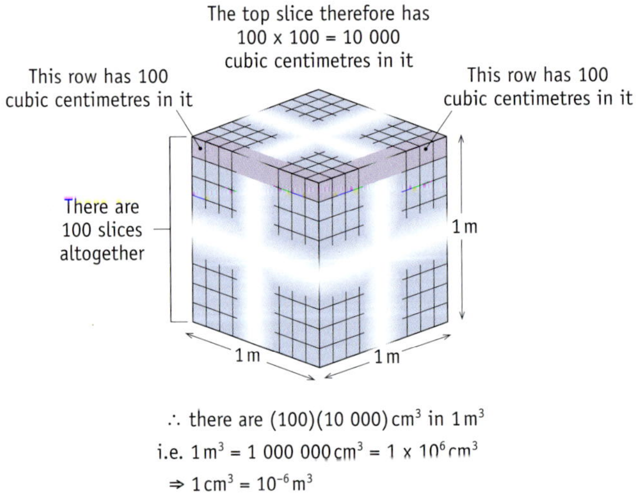

Fig. 1.7
How many cm^3 in 1 m^3?

Geometrical Optics

CHAPTER 2

Without Light You Cannot See

You may be in a room as you read this. If the lights in the room are switched off and the windows and doors blocked so that no light can enter, you will not be able to see anything in the room. To be able to see an object, light from that object must enter your eye. Since light is necessary for sight it is sometimes called 'visible' light. Note, however, that you do not see the light itself – you only see the object from which the light comes.

Some objects give out their own light. An object that gives out its own light is called a **self-luminous object**. The Sun, the stars, a fire, a bulb (that is switched on and operating) are examples of self-luminous objects. A **non-luminous object** is one which does not give out its own light. You can only see such an object when light from some other source bounces off the object and goes into your eye.

What Is Light?

Light is a form of energy that travels outwards from the source producing it at the incredibly fast speed of 300 000 000 metres per second. This is about 186 000 miles per second (no need to remember these figures).

LIGHT
Light is a form of energy that travels away from the source producing it at a speed of about 3×10^8 m s^{-1}.

Problem 1: The Sun is 1.5×10^{11} m from the Earth. How long does it take light to travel from the Sun to the Earth?

Solution: $$\text{Time taken} = \frac{\text{Distance}}{\text{Speed}} = \frac{1.5 \times 10^{11}}{3 \times 10^{8}} = 500 \text{ s} = 8.33 \text{ min}$$

Thus the light that left the Sun 8.33 minutes ago reaches the Earth now.

HOW DO WE KNOW THAT LIGHT IS A FORM OF ENERGY?

- Energy of some other form is always needed to produce light in any source of light, e.g. in a bulb, electrical energy is converted to light energy; in a fire, chemical energy is converted to light energy.
- Light can cause a photocell to produce electric current. Solar panels are large numbers of photocells together converting larger amounts of light energy to electrical energy (Fig. 2.1).

Fig. 2.1
In full summer sunlight this solar panel has a power output of 1000 watts.

Fig. 2.2
Crooke's radiometer converts light energy into kinetic energy.

- When light from the sun or from a lamp shines on a device called **Crooke's radiometer** (FIG. 2.2), the vanes rotate. Thus it converts light energy to kinetic energy.

LIGHT TRAVELS IN STRAIGHT LINES

Light usually travels in straight lines. You probably know this already if you have ever seen a beam of light from a laser or a searchlight or a car headlamp shining through dust or mist. The dust or mist shows up the path of the light. The sides of the light beam are seen to be straight lines (FIG. 2.3).

Fig. 2.3
Dust and moisture in the air show up the path of the light. The light travels in straight lines.

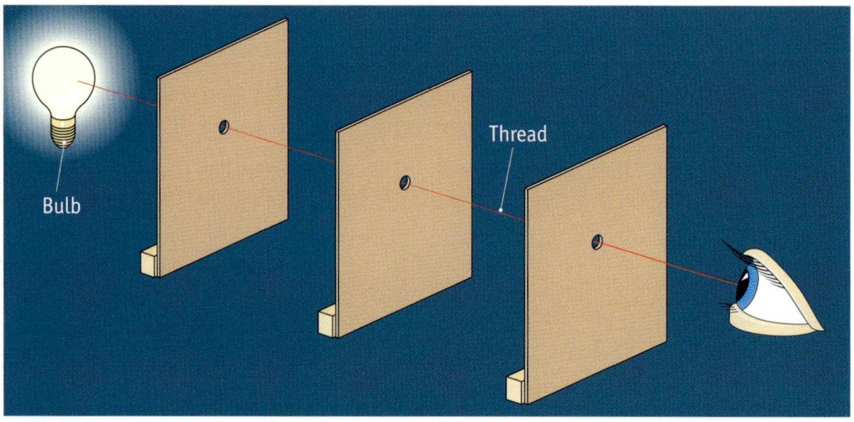

Fig. 2.4
Equipment used in an experiment to show that light travels in straight lines.

To show that light travels in straight lines, set up a bulb and three or four sheets of cardboard, each with a small hole in its centre as in FIG. 2.4. Pass a piece of thread through the holes. If you try to see the lighting bulb through the holes, you will find that you will not see the bulb unless the holes are lined up in a straight line, i.e. when the string is pulled tight. This shows that the light from the bulb cannot turn corners to pass through the holes if they are not in line.

A beam of light may be spreading out like the light from a torch. Such a beam is called a **diverging beam**. A beam of light may be getting narrower. Such a beam is called a **converging beam**. A beam of light may remain the same width. Such a beam is called a **parallel beam**. FIG. 2.5 shows a parallel, a converging and a diverging beam of light.

A straight line showing the direction in which light is travelling is called a **light ray**. Since a very narrow parallel beam is a straight line showing the direction in which light travels, it is a light ray.

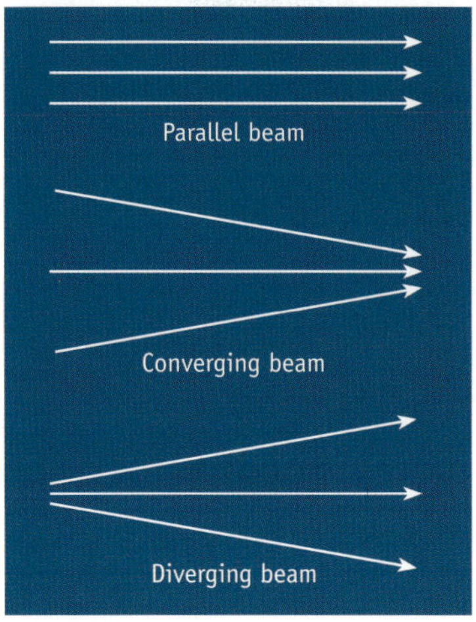

Fig. 2.5
A parallel, a converging and a diverging beam of light.

REFLECTION OF LIGHT

When light strikes the surface of an object some of the light is always absorbed into the surface and some always bounces back from the surface.

REFLECTION OF LIGHT
The bouncing of light off an object is called ***reflection***.

DIFFUSE REFLECTION

If the surface on which the light shines is rough, the reflected light is scattered in all directions from the surface (FIG. 2.6). This is called **diffuse reflection**. Most of the objects in the room that you are now in are reflecting light diffusely in all directions and can be seen by you from many positions as you move around the room.

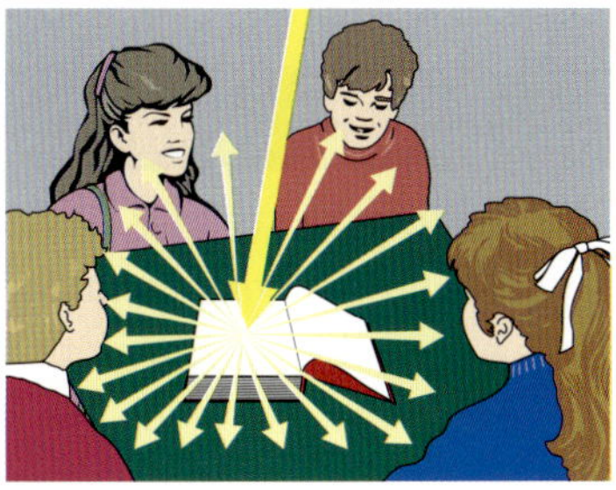

Fig. 2.6
Diffuse reflection. Each person can see the object.

REGULAR REFLECTION

If the surface of the object on which light shines is silvered and polished smooth – such a surface is called a **mirror** – the light falling on it is not scattered in all directions. Instead it is reflected off it as shown in FIG. 2.7.

Fig. 2.7
Regular reflection of light from a plane mirror.

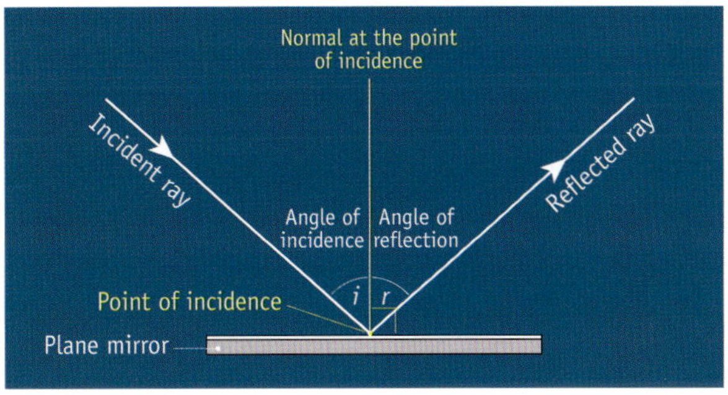

Fig. 2.8

This is called **regular reflection**. If the mirror is flat it is called a **plane mirror**. In FIG. 2.8:

- The ray of light falling on the mirror is called the **incident ray**,
- The line drawn perpendicular to the mirror where the ray strikes the mirror is called the **normal at the point of incidence**,
- The ray of light leaving the mirror is called the **reflected ray**,
- The angle between the incident ray and the normal is called the **angle of incidence**. Its symbol is i,
- The angle between the reflected ray and the normal is called the **angle of reflection**. Its symbol is r.

THE LAWS OF REFLECTION OF LIGHT

When light is reflected from a plane mirror the following laws hold.

LAWS OF REFLECTION OF LIGHT

Law 1: The incident ray, the normal at the point of incidence and the reflected ray all lie in the same plane.

Law 2: The angle of incidence is equal to the angle of reflection ($i = r$).

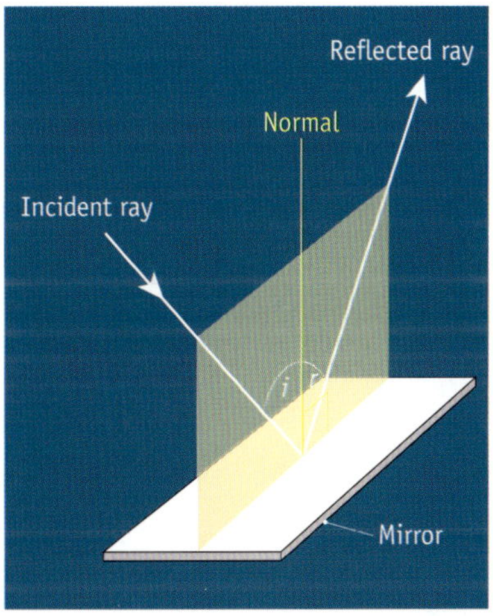

Fig. 2.9

Law 1: The incident ray, the normal and the reflected ray lie in the same plane.

Law 2: $\angle i = \angle r$

FIG. 2.9 shows the laws in diagrammatic form. Law 1 simply says that the reflected ray lies in the plane formed by the incident ray and the normal. The reflected ray is not deflected to the right or to the left.

EXPERIMENT

TO DEMONSTRATE THE LAWS OF REFLECTION.

- Set up the equipment shown in FIG. 2.7 (page 9).
- Adjust the ray box until it is producing a very narrow beam, i.e. a ray.
- Place the mirror on the drawing paper and mark its position with a pencil.
- Shine the ray onto the mirror. With a pencil mark the direction of the ray before it strikes the mirror and after it reflects from the mirror.
- Remove the mirror. Draw in the normal at the point of incidence, the incident ray and the reflected ray.
- With a protractor, measure the angle of incidence (*i*) and the angle of reflection (*r*). Record these values.
- Repeat the experiment for different values of the angle of incidence.

Result

- Within the limits of experimental error it will be found that **the angle of incidence is equal to the angle of reflection**, thus showing Law 2.
- **The reflected ray will be travelling parallel to the paper**, i.e. it will not be heading up from the paper or heading into the paper. It will be in the same plane as the incident ray and the normal, thus showing Law 1.

HOW AN IMAGE IS FORMED IN A PLANE MIRROR

FIG. 2.10 shows some of the rays from a point object striking a plane mirror. Each ray is reflected from the mirror obeying the laws of reflection. To the eye of an observer looking at the mirror, the object P appears to be at Q – i.e. behind the mirror. Q is called an image of P. The rays entering the eye appear to the eye to be coming from Q, even though no light rays actually pass through Q. We say the image at Q is formed by the apparent intersection of rays. Such an image is called a **virtual image**.

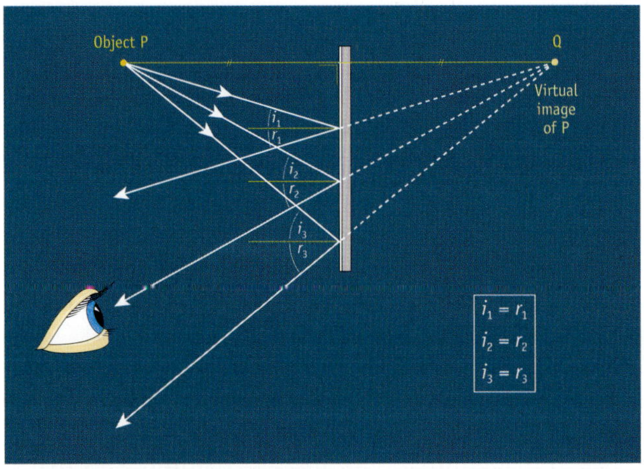

Fig. 2.10 How an image is formed in a plane mirror.

VIRTUAL IMAGE
A ***virtual image*** is an image formed by the apparent intersection of rays.

Experimentally it is found that a virtual image in a plane mirror is:

- On the perpendicular from the object to the mirror.
- The same distance behind the mirror as the object is in front of the mirror.

IMAGE OF AN EXTENDED OBJECT IN A PLANE MIRROR

In FIG. 2.11 each point on the hand has an image in the mirror. The result of this is a virtual image of the complete hand behind the mirror. The image is the same size as the object. Note that the image is different from the object in the sense that if the object is a right hand, the image seems to be a left hand. This is called **lateral inversion**. Thus writing appears strange in a mirror. The front of an ambulance has ‘ƎƆИAJUBMA’ written on it, so that the driver of a car ahead will see ‘AMBULANCE’ in the rear view mirror and hopefully let the ambulance pass.

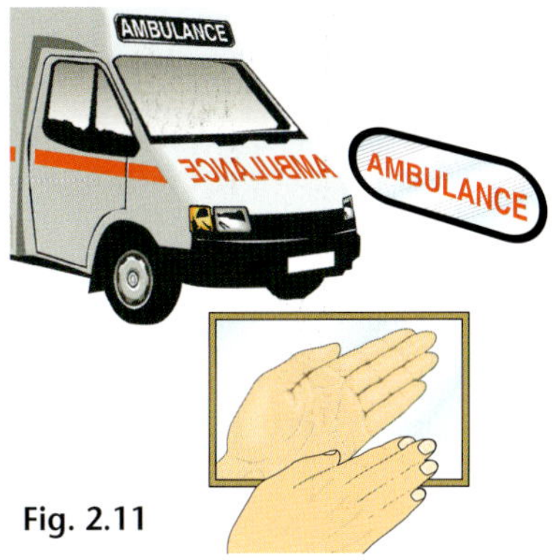

Fig. 2.11

USES OF PLANE MIRRORS

- A plane mirror can be used to see yourself. In FIG. 2.12 some of the light from the boy's hair reflects off the mirror and enters his eye. He is therefore able to see his hair in the mirror.
- The periscope. The periscope is a simple device that is based on the laws of reflection. It is made up of two plane mirrors (FIG. 2.13(A)). The light from the object you are looking at is reflected from each mirror and then into your eye, enabling you to see over or around an obstacle which would otherwise block your view (FIG. 2.13(B)).

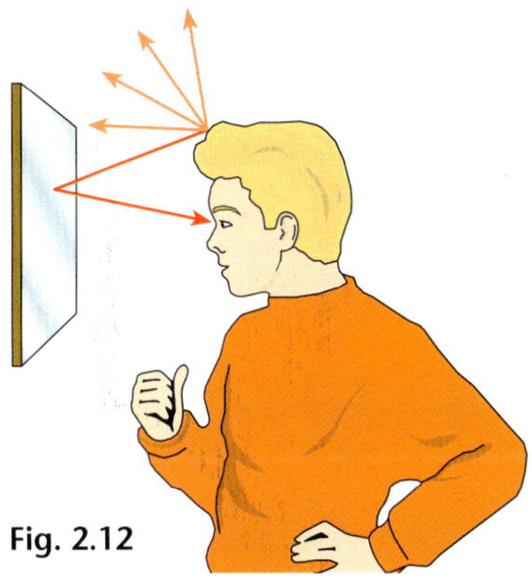
Fig. 2.12

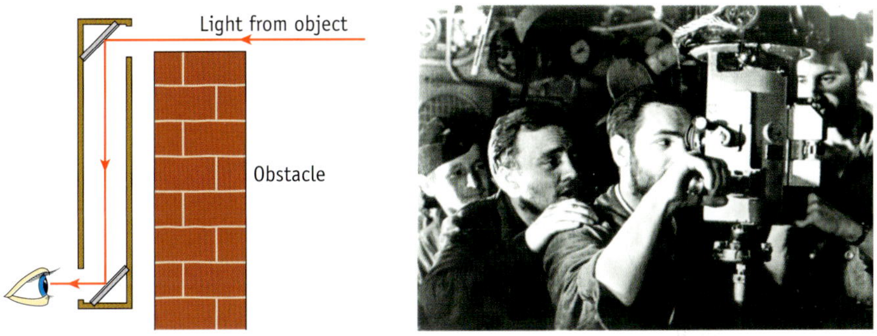

(A) (B)
Fig. 2.13

The apparent movement of one object relative to another due to the motion of the observer is called ***parallax***. The object that is farthest away appears to move with the observer.

THE METHOD OF NO PARALLAX FOR LOCATING AN IMAGE

Hold two pencils in front of you as shown in FIG. 2.14(A), so that the pencils appear to be in the same vertical line.

Without moving either of the pencils move your head to the right. Notice that one of the pencils – namely the one that is farthest away from you – also appears to move to the right relative to the other one. Move your head to the left and note again that the same pencil appears to move with you. This apparent relative motion of the pencils due to the motion of the observer is called **parallax**.

If you now place the pencils as in FIG. 2.14(B) so that their tips coincide you will observe no relative motion between them as you move your head to the right or to the left. They are said to be in a **state of no parallax**. Obviously if there is no parallax between things they must be in the same line and the same distance from the observer. We can use this fact to locate the image of an object in a plane mirror.

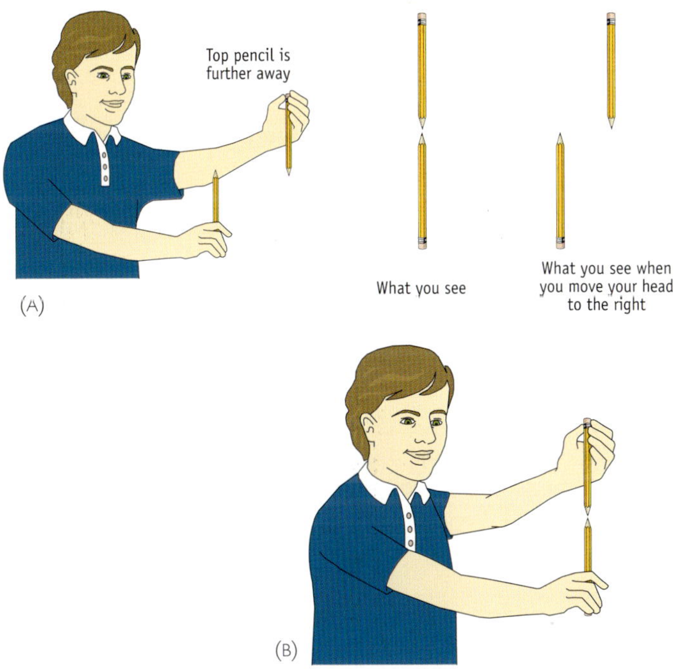

(A) (B)
Fig. 2.14

EXPERIMENT

TO LOCATE AN IMAGE IN A PLANE MIRROR BY THE METHOD OF NO PARALLAX.

Method

- Set up the equipment shown in FIG. 2.15. A virtual image of the object pin will be seen in the mirror.
- Adjust the height of the search pin so that its tip is just above the top of the mirror.
- Looking towards the mirror and ignoring the object pin you should see:
  (i) the image in the mirror,
  (ii) the search pin above the mirror.
- Adjust the search pin until it appears to be in line with the image.
- Move the search pin towards or away from the mirror until there is no parallax between the image and the search pin.The search pin then points out the position of the image.
- Using a metre stick, measure the distance (u) from the object to the silver of mirror. Measure the distance (v) from the search pin to the silver of the mirror (which is equal to the distance from the image to the mirror).

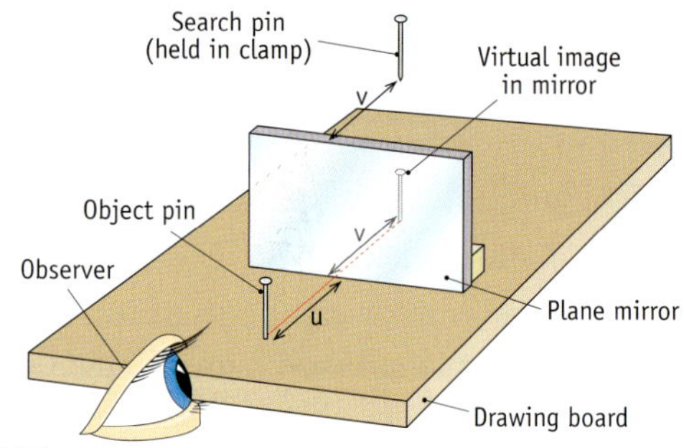

Fig. 2.15

Result

Within the limits of experimental error u will be equal to v, thus verifying that the image in a plane mirror is the same distance behind the mirror as the object in front of the mirror.

EXERCISE 2.1

1. The Moon is 380 000 000 m from the Earth. The speed of light is 300 000 000 m s^{-1}. How long does it take for light reflected from the Moon to reach the Earth?
2. In FIG. 2.16 name A, B, C, D, and E.

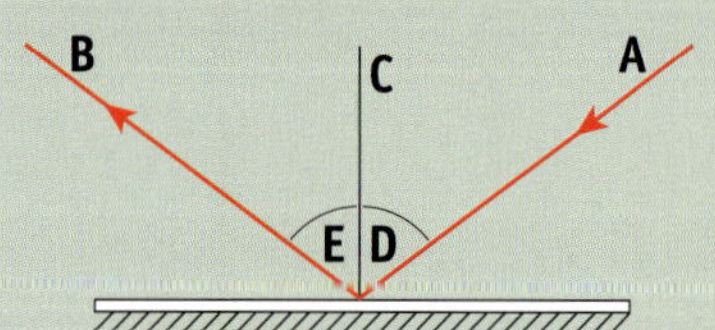

Fig. 2.16

3. FIG. 2.17 shows three rays of light leaving an object O and falling onto a plane mirror. Copy the diagram and draw in accurately each ray after reflection. Show the position of the image of O.

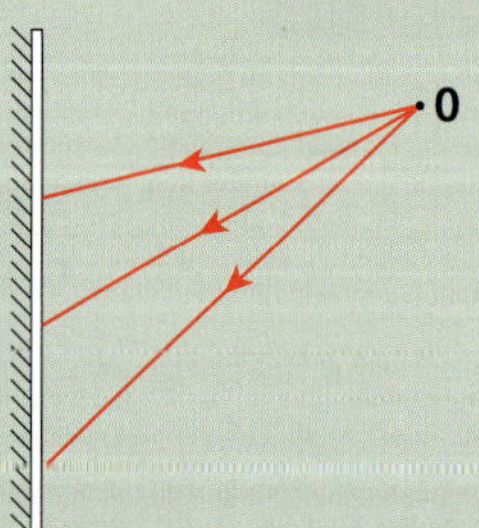

Fig. 2.17

4. A boy is standing 4 m from a plane mirror. How far and in what direction must he move so that he will be:
   (i) 4 m from his image,
   (ii) 1 m from his image?
5. A being from a distant galaxy is 2 m tall and has one small eye situated exactly at the top of its head. What is the least length of mirror that it needs in order to see itself fully in the mirror. Show with a ray diagram that it does not depend on how far the mirror is held from the alien being.
6. The nearest star is 39 700 000 000 000 km away. How many years does it take light to travel this distance?
   (speed of light = 3×10^8 m s^{-1})

7. A ray of light makes an angle of incidence of 30° on a plane mirror (Fig. 2.18). With the incident ray remaining fixed, the mirror is then rotated clockwise through 20° about the point *p*.
   (i) What is the new angle of incidence?
   (ii) What is the new angle of reflection?
   (iii) Through what angle did the reflected ray rotate?

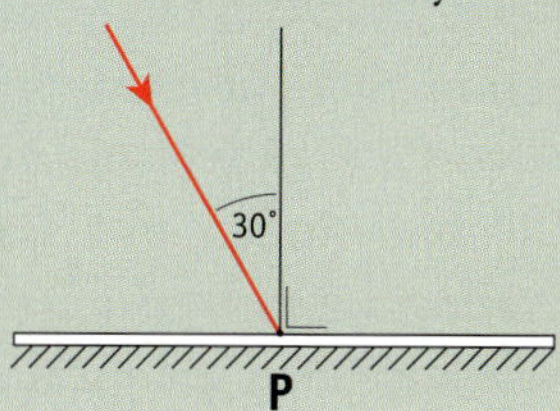

Fig. 2.18

8. A person moving straight towards a plane mirror is approaching his image at 4 m s^{-1}. With what speed is the person moving?

9. A woman who is 1.8 m tall, has her eyes situated 1.2 m from the ground when standing upright. She stands in front of a plane mirror and can just see herself fully in the mirror which is in a vertical plane. What is the length of the mirror?

10. Define each of the following: Self-luminous object; Non-luminous object; Converging beam; Diverging beam; Parallel beam; Reflection of light; Diffuse reflection; Regular reflection; Incident ray; Reflected ray; Normal at the point of incidence; Angle of incidence; Angle of reflection.

11. State the Laws of Reflection of Light.

12. What is a Virtual image?

CHAPTER CHECKLIST

- **Recall** the answers to questions 10, 11 and 12 above from memory.
- **Describe** an experiment to demonstrate the Laws of Reflection.
- **Draw** a ray diagram showing how an image is formed in a plane mirror.
- **Use** the method of no parallax to locate an image.

CHAPTER 3

Reflection of Light from Spherical Mirrors

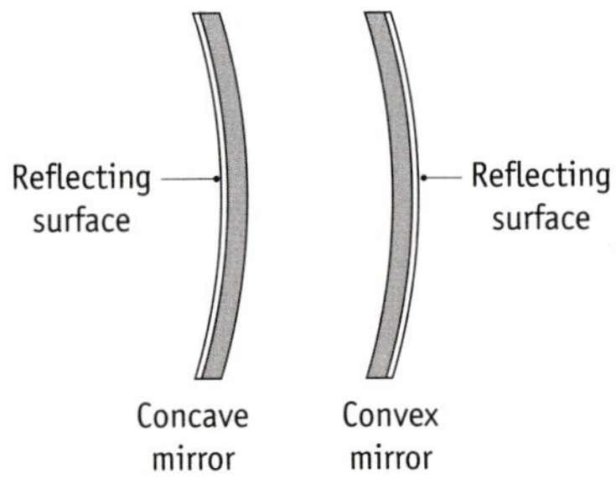

Fig. 3.1
Spherical Mirrors

CONCAVE AND CONVEX MIRRORS

In this chapter we shall look at the reflection of light from spherical mirrors. There are two types of spherical mirror; a concave and a convex (FIG. 3.1). From FIG. 3.1 you see that the reflecting surface of a **concave mirror** 'caves in' at the centre, whereas the reflecting surface of a **convex mirror** bulges out at the centre. These mirrors are called spherical mirrors because the glass (or other material) from which they are made is part of the surface of a sphere.

TERMS USED TO DESCRIBE SPHERICAL MIRRORS

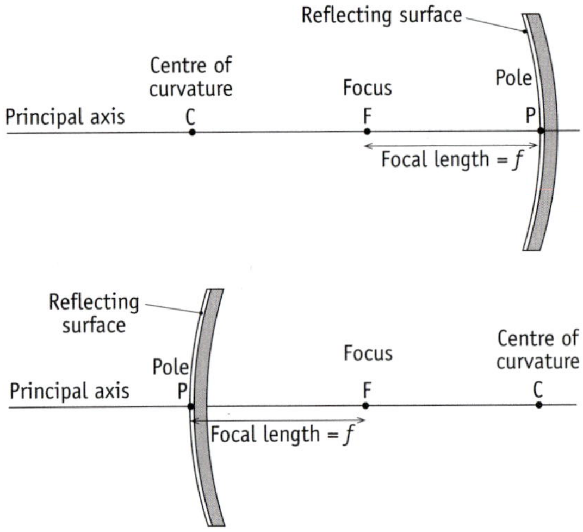

Fig. 3.2

- The centre of a spherical mirror is called the **pole** of the mirror. In FIG. 3.2, P is the pole.
- The centre of the sphere from which the mirror is made is called the **centre of curvature** of the mirror. In FIG. 3.2, C is the centre of curvature.
- The straight line joining the pole to the centre of curvature is called the **principal axis** of the mirror. We shall just call it the **axis**.
- The point half way between the centre of curvature and the pole is called the **focus** or **focal point** of the mirror. In FIG. 3.2, F is the focus.
- The distance from the focus to the pole is called the **focal length** of the mirror. It is normally symbolised by the letter *f*.

REFLECTION OF LIGHT FROM A CONCAVE MIRROR

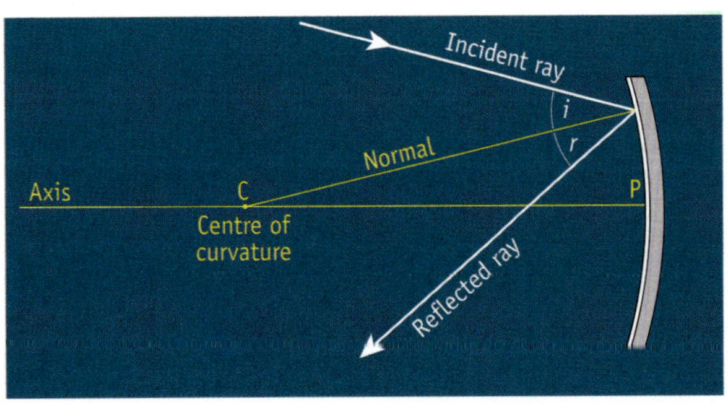

Fig. 3.3

The laws of reflection of light also apply to spherical mirrors. Suppose a ray of light strikes a concave mirror as in FIG. 3.3. Where does the ray go after reflection? To find out, draw the normal at the point of incidence. The normal for such a mirror is the line joining the point of incidence to the centre of curvature of the mirror. The angle of incidence *i* is the angle between the incident ray and the normal. The laws of reflection tell us that the angle of reflection *r* is the same size as *i*. Therefore the ray is reflected as shown. By this method any other ray falling on the mirror can be located after reflection.

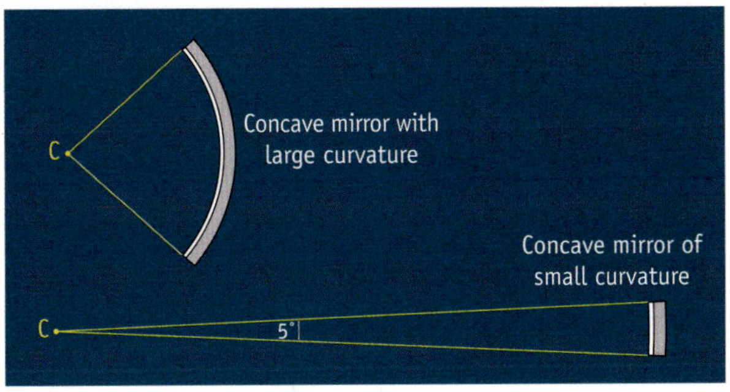

Fig. 3.4

This process of finding the reflected ray is time consuming. We shall only use the following results which are consequences of the laws of reflection.

A ray which strikes the pole is ***reflected at an equal angle*** with the axis.

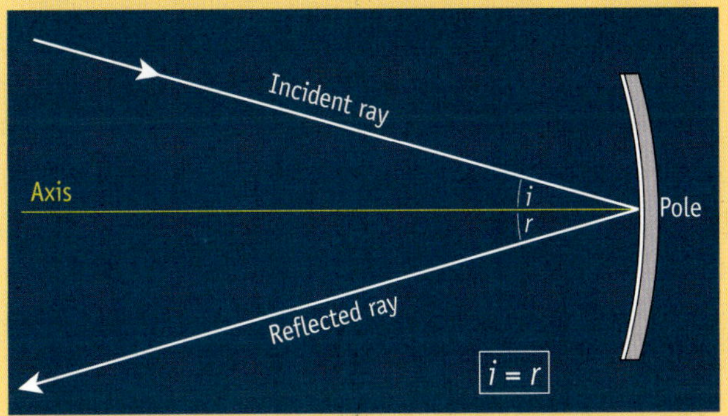

Fig. 3.5

A ray which passes through the centre of curvature and then strikes the mirror is ***reflected back along its own path.***

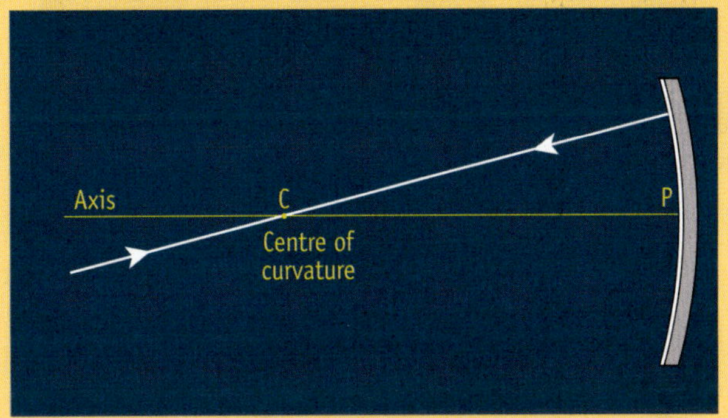

Fig. 3.6

If the mirror is not very curved (i.e. if the angle subtended by the mirror at the centre of curvature is less than about 5° (Fig. 3.4)), it can be proved that:

A ray which comes in parallel to the axis ***passes through the focus after reflection*** at the mirror.

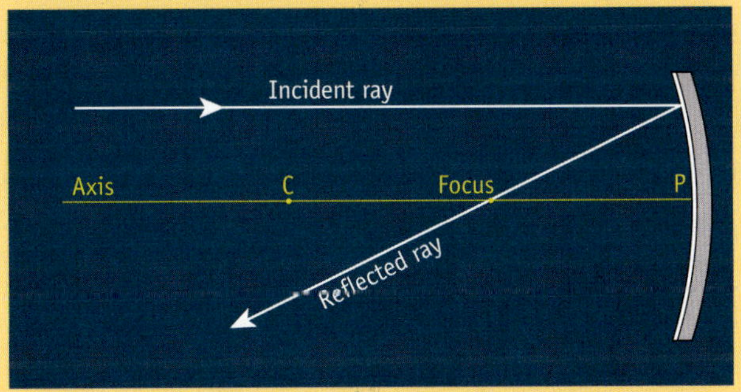

Fig. 3.7

A ray which passes through the focus and then strikes the mirror is ***reflected out parallel to the axis.***

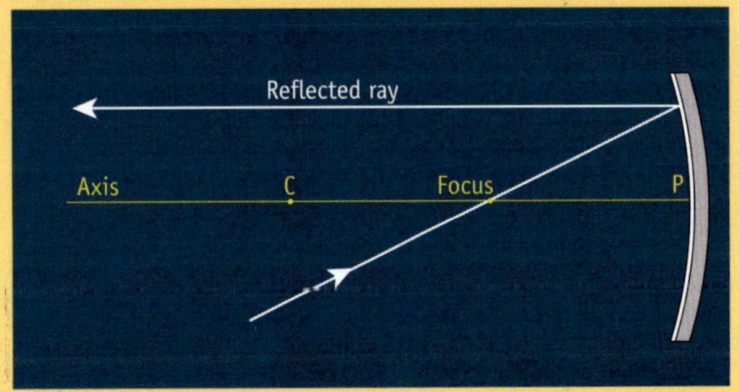

Fig. 3.8

You do not need to be able to prove these results. Simply remember them. They can easily be verified in the laboratory with a ray box and a concave mirror.

How an Image is Formed in a Concave Mirror

In Fig. 3.9 each point on the object is giving out light in all directions. Some of the light rays from the top of the object are shown striking the mirror and being reflected. Each of the reflected rays passes through the same point X after reflection. It can be shown that any other ray from the top of the object would also pass through point X after reflection. It follows that the eye (i.e. an observer) as shown will see an image of the tip of the object at X, since rays which were originally diverging from P are now diverging from X. By repeating this process for other points on the object you would see that there is an image formed of the whole object as in Fig. 3.9.

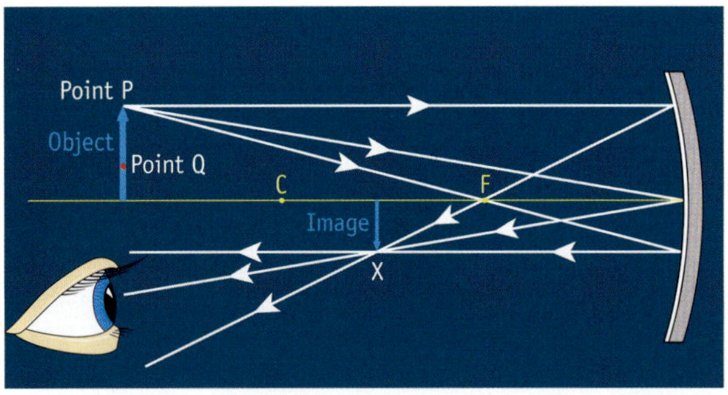

Fig. 3.9

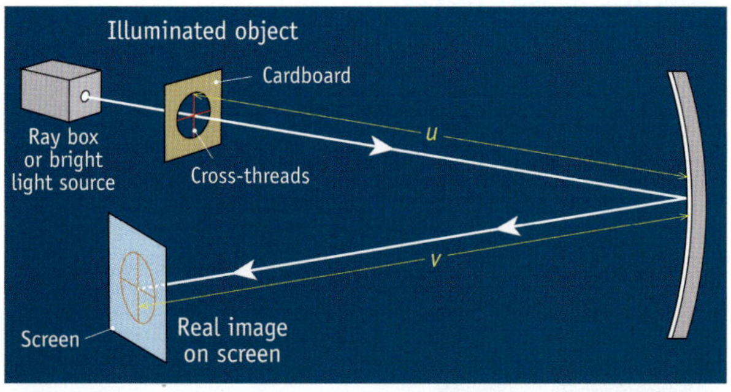

Fig. 3.10

This image is called a **real image** since it is caused by the **actual intersection of rays**. Note also that the image is a different size to the object. For a concave mirror the image will be a different size to the object unless the object is at the centre of curvature (Fig. 3.11(b)).

REAL IMAGE

A ***real image*** is an image formed by the actual intersection of light rays.

Such an image can be located on a screen or by the method of no parallax.

Properties of a Real Image

A real image can be seen on a screen – e.g. the image on a cinema screen is a real image. You can project a real image on a screen easily with the equipment shown in Fig. 3.10. The piece of cardboard has a hole cut in it and two pieces of thread stretched across the hole. These cross-threads are illuminated by the ray box and are used as an illuminated object. If the distance from the screen to the mirror is varied, a position will be found for which there is a clear image of the cross-threads formed on the screen. If such an image cannot be found, the mirror is too near the ray box and should be moved away from it. A real image may also be located by the method of no parallax outlined on page 12.

To find out where the image in a concave mirror is formed, we need not draw all the rays shown in Fig. 3.9. Any two rays will do. In Fig. 3.11 ray diagrams are drawn showing how to find the image when the object is outside *C*; at *C*; between *C* and *F*; at *F* and inside *F*.

Try to draw the relevant ray diagrams yourself before looking at Fig. 3.11. From the ray diagrams in Fig. 3.11 we see the following:

FOR A CONCAVE MIRROR:

If the object is ***outside the focus*** the image is ***real*** and located ***in front*** of the mirror.

If the object is ***inside or at the focus*** the image ***is virtual*** and is located ***behind*** the mirror.

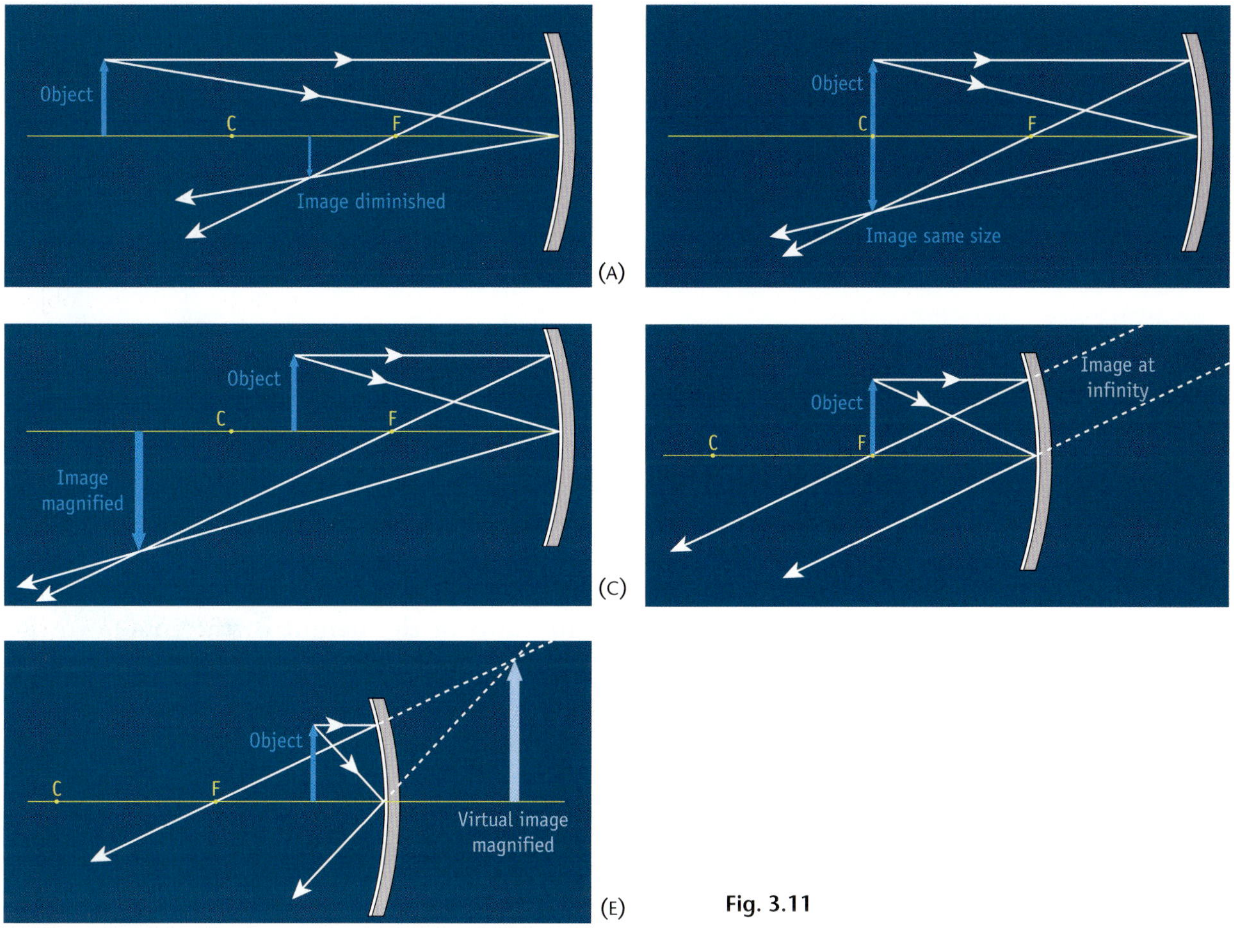

Fig. 3.11

VIRTUAL IMAGE

You saw on page 10 that a virtual image is an image formed by the apparent intersection of rays. Any image formed in a plane mirror is a virtual image. An image formed in a concave mirror is virtual if the object is inside or at the focus of the mirror.

> **VIRTUAL IMAGE**
> A ***virtual image*** is an image formed by the apparent intersection of rays.
>
> Such an image can never be formed on a screen. It can be located by the method of no parallax.

IMAGE OF A DISTANT OBJECT IN A CONCAVE MIRROR

If an object is a large distance from a concave mirror, the image is real and at the focus of the mirror. Suppose P is a point on a distant object (FIG. 3.12). P gives out light in all directions. Since P is far away, light from P arrives at the mirror as a parallel beam. Rays from another point Q on the distant object will also arrive as a parallel beam.

In FIG. 3.12, the beam from P is brought to a focus as shown. The beam from Q is brought to a focus at the focal point of the mirror. Thus a real image of the distant object is formed at the focal plane of the mirror. This can easily be shown in the laboratory as in FIG. 3.13.

> Light from any point on a distant object arrives as a ***beam of parallel light***.

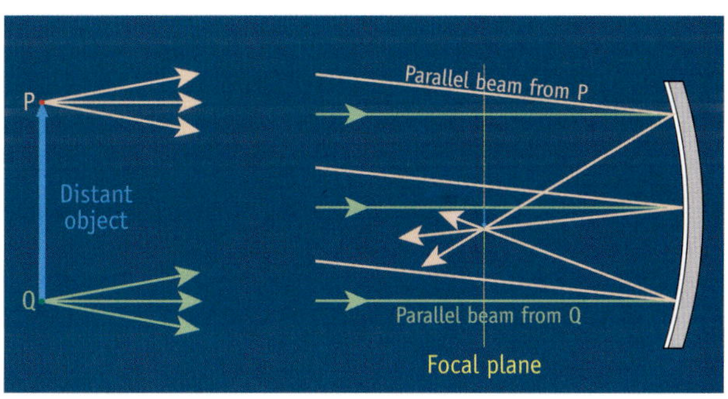

Fig. 3.12

Fig. 3.13
A concave mirror producing a real image of a distant window on a cardboard screen in the laboratory.

Formula for a Concave Mirror

The distance of the image from a concave mirror changes as the distance of the object from the mirror changes and these distances are only the same when the object is at the centre of curvature. A simple formula which you need not prove, relates these distances.

Suppose u is the **distance from the object to the mirror**,

Suppose v is the **distance from the image to the mirror**.

Suppose f is the **focal length**. Then:

For a Concave Mirror:

If the image is **real**: $\frac{1}{u} + \frac{1}{v} = \frac{1}{f}$ If the image is **virtual**: $\frac{1}{u} - \frac{1}{v} = \frac{1}{f}$

Magnification

Usually the height of the image is different from the height of the object. Suppose the height of the object is 4 cm and the height of the image is 20 cm. Then the image is five times bigger than the object. We say the magnification is five. **The ratio of the height of the image to the height of the object is called the magnification (m).**

It can be proved that the magnification produced by a concave mirror = image distance / object distance. You need not worry about the proof.

$$\text{Magnification} = \frac{\text{Height of image}}{\text{Height of object}} = \frac{\text{Image distance}}{\text{Object distance}} \quad \text{i.e. } m = \frac{v}{u}$$

Problem 1: An object is placed 30 cm in front of a concave mirror of focal length 20 cm. Find the position and nature of the image.

Solution: Here $u = 30$ and $f = 20$. v is to be found. Since the object is outside the focus, the image is real and the formula to be used is $1/u + 1/v = 1/f$. Even if we were not sure whether the image was real or virtual we can still use this formula. If v then turns out to be positive the image is real; if v turns out to be negative the image is virtual.

$$\frac{1}{u} + \frac{1}{v} = \frac{1}{f} \Rightarrow \frac{1}{30} + \frac{1}{v} = \frac{1}{20} \Rightarrow \frac{1}{v} = \frac{1}{20} - \frac{1}{30} \Rightarrow \frac{1}{v} = \frac{3-2}{60} = \frac{1}{60} \Rightarrow v = 60 \text{ cm}$$

Thus the image is real (its nature) and is located 60 cm in front of the mirror.

Problem 2: An object is placed 12 cm in front of a concave mirror of focal length 20 cm. Find the position, nature and magnification of the image.

Solution: The object is inside the focus therefore the image is virtual and the formula to use is $1/u - 1/v = 1/f$. However, let us assume we do not know whether the image is real or virtual and use the formula $1/u + 1/v = 1/f$.

$$\frac{1}{u}+\frac{1}{v}=\frac{1}{f} \Rightarrow \frac{1}{12}+\frac{1}{v}=\frac{1}{20} \Rightarrow \frac{1}{v}=\frac{1}{20}-\frac{1}{12} \Rightarrow \frac{1}{v}=\frac{3-5}{60}=\frac{-1}{30} \Rightarrow v=-30\text{cm}$$

A **negative value** tells us the **image is virtual**. It is located 30 cm **behind** the mirror.

$$\text{Magnification} \quad m = \frac{v}{u} = \frac{30}{12} = 2.5$$

Thus the image is 2.5 times as high as the object.

Problem 3: An object placed 15 cm in front of a concave mirror produces a real image which is 30 cm from the front of the mirror. Calculate the focal length of the mirror. If the object is 5 cm high what is the height of the image?

Solution: Here $u = 15$, $v = 30$ and f is unknown.

$$\frac{1}{u}+\frac{1}{v}=\frac{1}{f} \Rightarrow \frac{1}{15}+\frac{1}{30}=\frac{1}{f} \Rightarrow \frac{30+15}{(15)(30)}=\frac{1}{f} \Rightarrow \frac{45}{450}=\frac{1}{f} \Rightarrow f=\frac{450}{45}=10\text{ cm}$$

i.e. focal length of mirror = 10 cm

$$\text{Magnification} = \frac{\text{Height of image}}{\text{Height of object}} = \frac{v}{u}$$

$$\Rightarrow \frac{\text{Height of image}}{5} = \frac{30}{50} \Rightarrow \text{Height of image} = \frac{(5)(30)}{15} = 10\text{ cm}$$

Problem 4: An image is formed in a concave mirror of focal length 20 cm. The image is three times the size of the object. Where must the object be placed if:

(i) the image is real and

(ii) the image is virtual?

Solution: Magnification: $\frac{v}{u} = 3 \Rightarrow v = 3u$

Real image: $$\frac{1}{u}+\frac{1}{v}=\frac{1}{f} \Rightarrow \frac{1}{u}+\frac{1}{3u}=\frac{1}{f} \Rightarrow \frac{3+1}{3u}=\frac{1}{20}$$

$$\Rightarrow \frac{4}{3u}=\frac{1}{20} \Rightarrow u = 26.67\text{ cm}$$

Virtual Image: $$\frac{1}{u}-\frac{1}{v}=\frac{1}{f} \Rightarrow \frac{1}{u}-\frac{1}{3u}=\frac{1}{f} \Rightarrow \frac{3-1}{3u}=\frac{1}{20}$$

$$\Rightarrow \frac{2}{3u}=\frac{1}{20} \Rightarrow u = 13.33\text{ cm}$$

For a real image the object must be placed 26.67 cm from the mirror and for a virtual image the object must be 13.33 cm from the mirror.

Uses of Concave Mirrors

If a bulb is placed at the focal point of a concave mirror, light from the bulb which strikes the mirror is reflected out parallel to the axis (Fig. 3.14). Practical use of this fact is made in searchlights and some floodlights. If an object is placed inside the focus of a concave mirror the image is magnified and the right way up. A shaving or make-up mirror makes use of this fact (Fig. 3.15). A dentist can also see around a patients mouth with such a mirror. A concave mirror is also used in a slide projector to reflect light from the bulb which would otherwise be wasted. The mirror reflects the light back through the slide producing a brighter image.

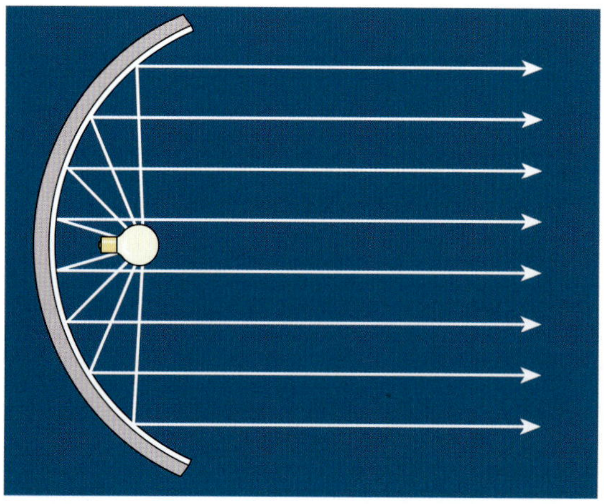

Fig. 3.14

Fig. 3.15

EXERCISE 3.1

1. An object is placed 30 cm in front of a concave mirror and produces a real image that is 50 cm from the mirror. Calculate the focal length of the mirror.
2. An object is placed 20 cm in front of a concave mirror and produces a virtual image that is 30 cm behind the mirror. Find the focal length of the mirror.
3. An object is placed 15 cm in front of a concave mirror of focal length 10 cm. At what distance from the mirror will its image be formed? What is the nature and magnification of the image? If the height of the object is 2 cm what is the height of the image?
4. An object is placed 10 cm in front of a concave mirror of focal length 20 cm. Find the position, nature and magnification of the image.
5. A concave mirror of focal length 40 cm forms a real image which is four times the height of the object. How far from the mirror is the object?
6. A concave mirror of focal length 20 cm forms an image which is three times the height of the object. Find the distance of the object from the mirror if:
   (i) the image is real,
   (ii) the image is virtual.
7. Find the two positions in which an object may be put so that an image which is twice the height of the object is formed in a concave mirror of focal length 50 cm.
8. How far in front of a concave mirror of focal length 100 cm must you stand in order to see an image of your face which is upright and three times magnified?
9. An object is at a very large distance from a concave mirror of focal length 40 cm. Find the distance from the image to the mirror. Is the image real or virtual?
10. Where must an object be put so that an image which is one third the height of the object is formed in a concave mirror of focal length 40 cm?

EXPERIMENT **LIGHT 1**

TO MEASURE THE FOCAL LENGTH OF A CONCAVE MIRROR.

Summary of Method

In this experiment you will place an illuminated cross-threads as object in front of a concave mirror and locate the image of the cross-threads on a white cardboard screen. You will measure the distance *u* from the cross-threads to the mirror and the distance *v* from the screen to the mirror.

Using the formula: $\frac{1}{u} + \frac{1}{v} = \frac{1}{f}$ the focal length *f* of the mirror can be found.

Equipment Needed

- A concave mirror and stand
- A retort stand and clamp
- An illuminated object. This can be made by cutting a small circular hole in a sheet of cardboard and sticking two pieces of thread across the hole with Sellotape. It is illuminated by light from a ray box.
- A ray box, a sheet of white cardboard and a metre stick

Method

1. Find an approximate value for the focal length of the mirror by focusing an image of a distant object on a sheet of paper (FIG. 3.13, PAGE 17). The distance from the image to the mirror is its focal length approximately. Measure and record this distance.
2. Set up the equipment as in FIG. 3.10 on page 16. The distance from the object to the mirror must be greater than the focal length found in Step 1; otherwise you will not get a real image.
3. Mount the sheet of cardboard on the retort stand. This is the screen.
4. Adjust the position of the screen until the image of the cross-threads is in sharpest focus on it.
5. With a metre stick measure:
   - the distance *u* from the cross-threads to the pole of the mirror,
   - the distance *v* from the real image on the screen to the pole of the mirror.
   Record these values.
6. Change the value of *u* and repeat Steps 4 and 5 at least four times.

Experimental Data and Results:

Approximate focal length found by focusing image of distant object on screen =

u / cm	v / cm	$\frac{1}{u}$	$\frac{1}{v}$	$\frac{1}{f} = \left(\frac{1}{u} + \frac{1}{v}\right)$	f / cm

Average value of focal length f =

Handling the Data

Method 1

Complete the Table and calculate the average value of *f*. This is the focal length of the mirror. When calculating $1/u + 1/v$ retain at least three significant figures in your calculations.

Method 2

On graph paper, plot a graph of $1/u$ against corresponding values of $1/v$. Draw the straight line that best fits the points and produce it to intersect both axes.

Read from the graph the value of $1/u$ where the graph cuts the $1/u$ axis and the value of $1/v$ where it cuts the $1/v$ axis. Get the average of these two values ($= 1/f$). Find the reciprocal of this result. It should agree closely with the value obtained in method 1 for the focal length.

Sources of Error

Errors arise when:

(i) deciding when the image on the screen is in sharpest focus,

(ii) measuring the distances *u* and *v* with the metre stick.

Obviously there is a larger error in *v* than in *u* since *u* only involves error in using the metre stick.

Questions

1. If you could not find any clear image of the cross-threads on the screen what is likely to be the problem? How would you rectify it?
2. Why is the value for *u* obtained more accurate than the value for *v*?
3. List two precautions that should be taken in measuring *u*.
4. Errors occur both in locating the position of the image and measuring its distance from the pole. State one precaution in each case you could take to reduce these errors.

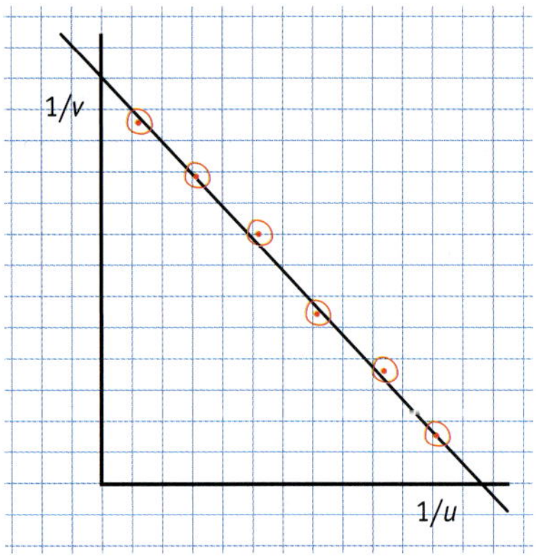

Fig. 3.16

Graph of $\frac{1}{u}$ against $\frac{1}{v}$.

Explanation of the Graph

The graph of $\frac{1}{u}$ against $\frac{1}{v}$ (on the y-axis) is a straight line (Fig. 3.16).

You know from your maths (co-ordinate geometry of the straight line) that any straight line cuts the x-axis when $y = 0$

$\therefore$ the line $\frac{1}{u} + \frac{1}{v} = \frac{1}{f}$ cuts the $\frac{1}{u}$ axis when $\frac{1}{v} = 0$

Giving: $\frac{1}{u} + 0 = \frac{1}{f} \Rightarrow$ The value of $\frac{1}{u}$ where line cuts $\frac{1}{u}$ axis is $\frac{1}{f}$

Similarly value of $\frac{1}{v}$ where line cuts $\frac{1}{v}$ axis is $\frac{1}{f}$

Reflection of Light From a Convex Mirror

The laws of reflection of light also apply to a convex mirror. Suppose a ray of light strikes a convex mirror as in Fig. 3.17. The normal at the point of incidence is the line joining the point of incidence to the centre of curvature of the mirror. The angle of incidence i is also shown in the diagram. The laws of reflection tell us that the angle of reflection r is the same size as i, ($i = r$). Therefore the ray is reflected as shown.

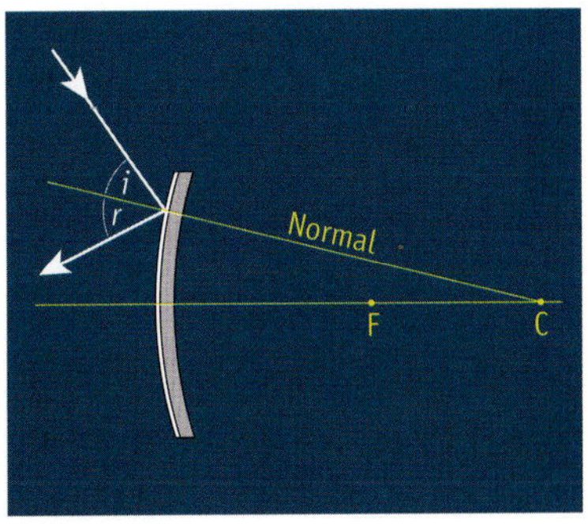

Fig. 3.17

By this method any other ray falling on the mirror can be located after reflection. As with the concave mirror, this process of finding the reflected ray is time consuming.

We shall only use the following results which follow from the Laws of Reflection. You do not need to be able to prove them but simply accept and remember them.

A ray which strikes the pole is reflected at an ***equal angle*** with the axis.

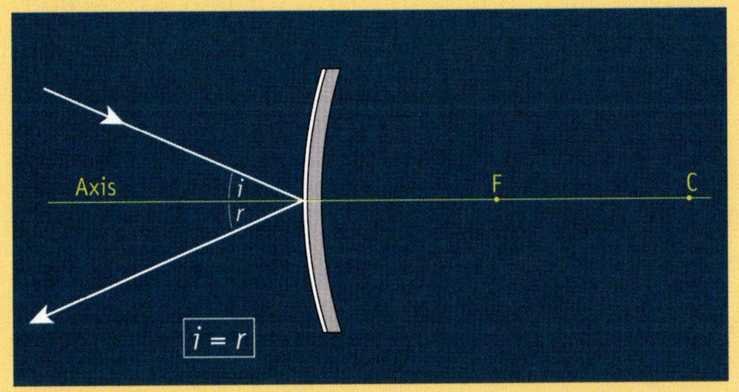

Fig. 3.18

A ray which is heading for the centre of curvature is ***reflected back along its own path.***

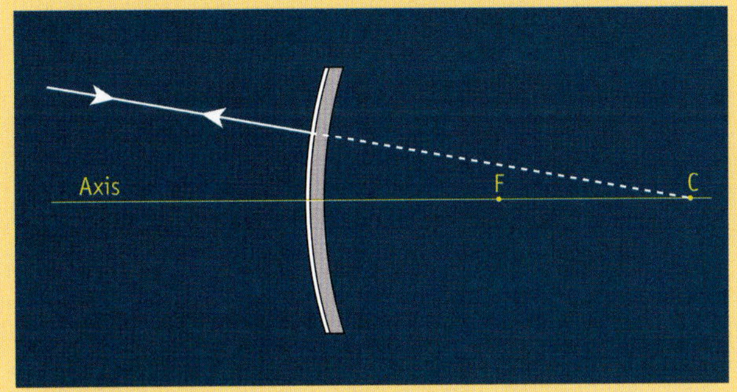

Fig. 3.19

If the mirror is not very curved it is found that:

A ray which enters parallel to the axis is ***reflected as if it came from the focus.***

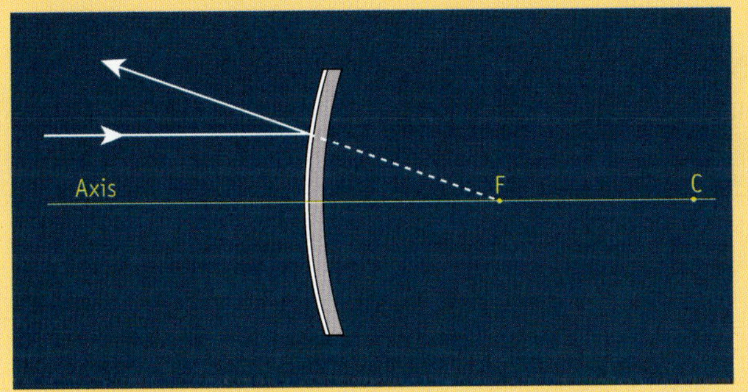

Fig. 3.20

A ray which is heading for the focus and strikes the mirror is ***reflected out parallel to the axis.***

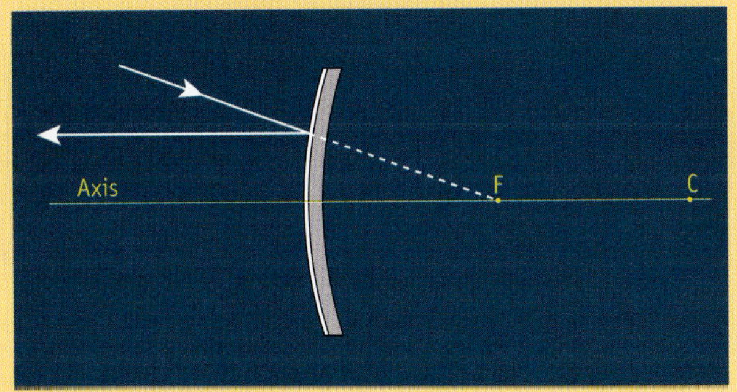

Fig. 3.21

The Formation of an Image by a Convex Mirror

Fig. 3.22 (page 24) shows an object in front of a convex mirror. Every point on the object is giving out light in all directions. We have drawn in some of the light rays from the top of the object and shown where these rays go after reflection. **What do we notice?** Each of the rays appears to diverge from the same point (X) after reflection. It can also be shown that any other ray from the top of the object would also appear to diverge from X after reflection. Thus an eye will see an **image** of the tip of the object at X, since rays which originally were diverging from P now appear to be diverging from X.

Likewise by drawing rays from any other point Q on the object, the image of that point can be found. By repeating this process for other points on the object we would see that there is an image formed of the whole object as shown in FIG. 3.22. This image is a **virtual image** since it is caused by the apparent intersection of rays.

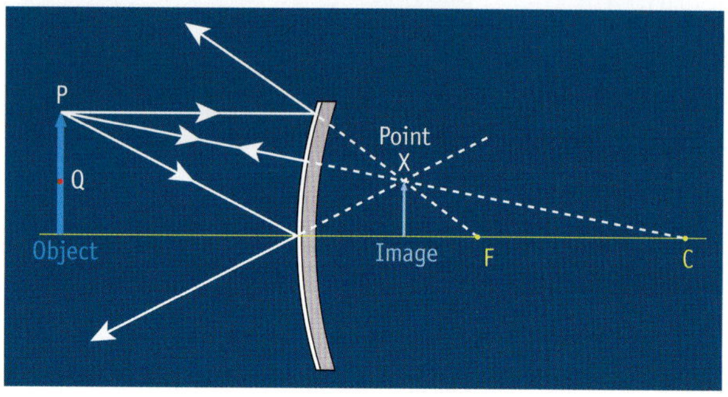

Fig. 3.22

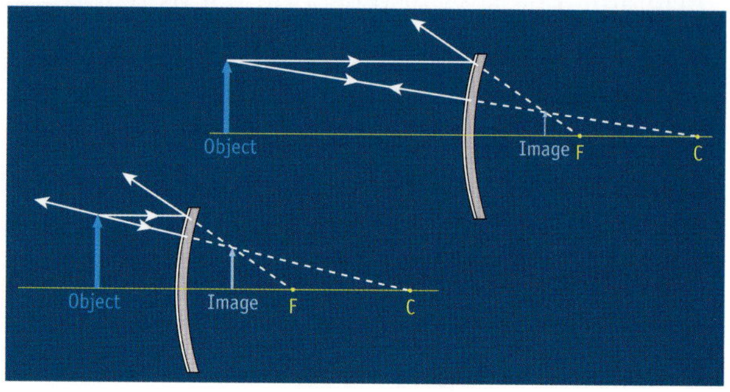

Fig. 3.23

To find out where the image in a convex mirror is formed, we need not draw all the rays shown in FIG. 3.22. Any two rays will do. If you locate the image for a number of different positions of the object (FIG. 3.23) you will see that:

FOR A CONVEX MIRROR:
The image is always ***virtual*** and located behind the mirror.

The image is always ***diminished***. The nearer the object is to the mirror the bigger the image.

FORMULA FOR A CONVEX MIRROR

A formula very similar to that for a concave mirror holds for a convex mirror. As before, let u be the distance from the **object** to the mirror, let v be the distance from the **image** to the mirror and let f be the **focal length** of the mirror. Then:

For a Convex Mirror:

$$\frac{1}{u} - \frac{1}{v} = -\frac{1}{f}$$ Note that $\frac{1}{v}$ and $\frac{1}{f}$ are always negative.

MAGNIFICATION

The ratio of the height of image to the height of the object is called the **magnification**. It can be proved that the magnification produced by a convex mirror = v/u.

$$\text{Magnification} = \frac{\text{Height of image}}{\text{Height of object}} = \frac{\text{Image distance}}{\text{Object distance}} \quad \text{i.e. } m = \frac{v}{u}$$

Uses of Convex Mirrors

Since the image formed in a convex mirror is **always diminished** the field of view of a convex mirror is bigger than that of a plane mirror (Fig. 3.24). Such mirrors are sometimes used in:

(i) the door mirror of a car,
(ii) at concealed entrances to give a view of oncoming traffic,
(iii) in shops to deter shoplifters (see Fig. 3.25),
(iv) in railway stations.

> The formulae for the concave and convex mirror can be summarised in one formula as follows:
>
> $$\frac{1}{u} + \frac{1}{v} = \frac{1}{f}$$
>
> where u is always +
>
> v is + for a real image
> v is – for a virtual image
>
> f is + for a concave mirror
> f is – for a convex mirror

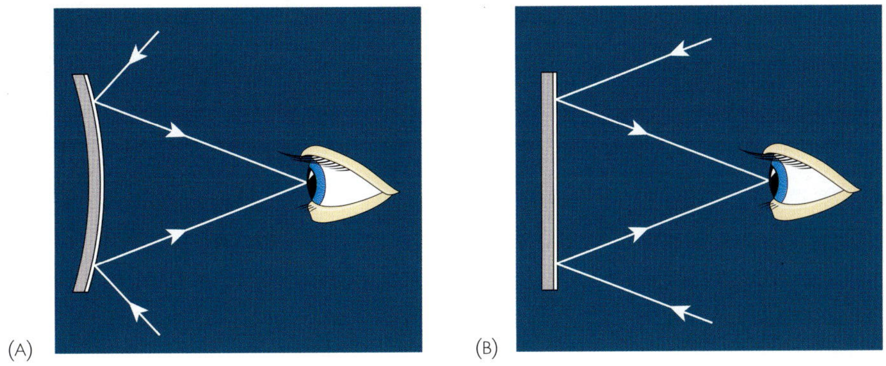

Fig. 3.24
The convex mirror gives a larger field of view than a plane mirror.

Problem 5: An object is placed 40 cm in front of a convex mirror of focal length 24 cm. At what distance from the mirror will the image be formed? Find the nature and magnification of the image.

Solution:
$$\frac{1}{u} - \frac{1}{v} = -\frac{1}{f} \Rightarrow \frac{1}{40} - \frac{1}{v} = -\frac{1}{24} \Rightarrow -\frac{1}{v} = -\frac{1}{24} - \frac{1}{40}$$

$$\Rightarrow -\frac{1}{v} = \frac{-5-3}{120} \Rightarrow 8v = 120 \Rightarrow v = 15 \text{ cm}$$

∴ The image is 15 cm from the mirror. The image is virtual since an image is always virtual in a convex mirror.

Magnification $= \frac{v}{u} = \frac{15}{40} = \frac{3}{8} \Rightarrow$ The image is ⅜ the size of the object.

Problem 6: A convex mirror of focal length 15 cm forms an image that is one third the size of the object. Find the positions of the object and the image.

Solution: Size of image = (⅓) Size of object ⇒ Magnification = ⅓

$$\Rightarrow \frac{v}{u} = \frac{1}{3} \Rightarrow 3v = u \text{ Substituting into: } \frac{1}{u} - \frac{1}{v} = -\frac{1}{f} \text{ gives } \frac{1}{3v} - \frac{1}{v} = -\frac{1}{15}$$

$$\Rightarrow \frac{1-3}{3v} = -\frac{1}{15} \Rightarrow -\frac{2}{3v} = -\frac{1}{15} \Rightarrow 3v = 30 \Rightarrow v = 10\text{cm}$$

$u = 3v \Rightarrow u = (3)(10) = 30$ cm.

Thus the image is 10 cm from the mirror and the object is 30 cm from the mirror.

EXERCISE 3.2

1. An object is placed 10 cm in front of a convex mirror of focal length 12 cm. Find the position, nature and magnification of the image.

Fig. 3.25

2. An object is placed 30 cm in front of a convex mirror of focal length 12 cm. Find the position, nature and magnification of the image.
3. A convex mirror of focal length 10 cm forms an image that is one quarter the size of the object. Find the positions of the object and the image.
4. A convex mirror forms an image which is 10 cm from the mirror. The height of the object is 5 cm and the height of the image is 2 cm. Find the focal length of the mirror.
5. A convex mirror forms an image which is 4 cm from the mirror. The height of the image is 4 cm and the height of the object is 6 cm. What is the focal length of the mirror?
6. Where must an object be placed so that an image one third the height of the object is formed in a convex mirror of focal length 20 cm?
7. Where must an object be put so that the height of the image formed in a convex mirror of focal length 40 cm is half the height of the object?
8. Where must an object be put so that the height of the image formed in a convex mirror is half the height of the object?

CHAPTER CHECKLIST

- **Say** what each of the following mean in the context of spherical mirrors: Pole, Centre of curvature, Axis, Focus, Focal length.
- **Define** each of the following: Real image, Virtual image, Magnification.
- **Draw** a ray diagram to locate the image formed in a concave or convex mirror no matter what the position of the object.
- **Recall** and use the formulae: $\frac{1}{u} + \frac{1}{v} = \frac{1}{f}$ and $m = \frac{v}{u}$ to solve problems.
- **Describe** and carry out an experiment to measure the focal length of a concave mirror.
- **List** practical uses of a concave and a convex mirror.

Refraction

CHAPTER 4

Refraction of Light

You saw on page 8 that light usually travels in straight lines. When a ray of light goes from one medium to another it may change direction on going into the second medium, i.e. the ray may bend (Fig. 4.1(a)). This bending is called the **refraction*** of light. Note that the ray does not bend (no refraction occurs) if it strikes the second medium at right angles (Fig. 4.1(b)).

REFRACTION
The bending of a ray of light when it goes from one medium to another is called ***refraction***.

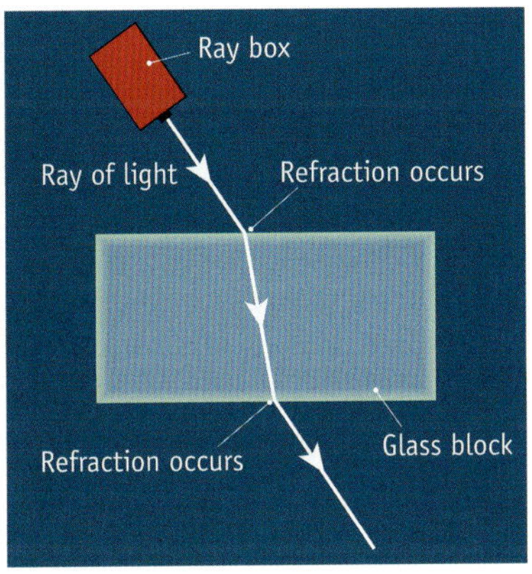

Fig. 4.1 (a)

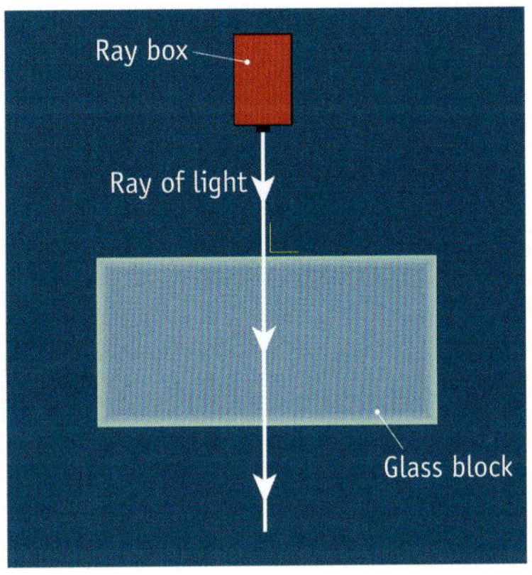

Fig. 4.1 (b)
No refraction occurs when the ray strikes the glass at right angles.

When a ray of light falls on a glass block (Fig. 4.2) or on another transparent medium, the following terms are used:

- The ray of light falling on the glass block is called the **incident ray**.
- The line drawn at right angles to the glass block where the ray strikes the block is called the **normal at the point of incidence**.
- The ray of light **in the glass block** is called the **refracted ray**.
- The angle between the incident ray and the normal is called the **angle of incidence (*i*)**.
- The angle between the refracted ray and normal is called the **angle of refraction (*r*)**.

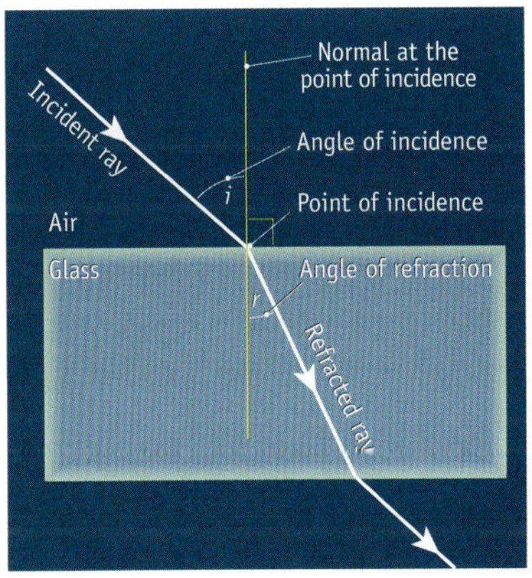

Fig. 4.2

* You will see in Chapter 16 that refraction will only occur if the speed of light in the second medium is different from the speed in the first. Refraction obviously also occurs when light enters or leaves a vacuum (other than at right angles).

EXPERIMENT

TO DEMONSTRATE REFRACTION.

- With the equipment in FIG. 4.3 adjust the ray box to produce a very narrow beam, i.e. a ray.
- Shine the ray onto one side of the glass block. Note how the ray bends on entering the glass and bends again on leaving the block. Note that the ray coming out of the glass block is parallel to the ray striking the block.
- Shine the ray onto the glass block so that it strikes the block at right angles. Note that the ray does not bend on entering the glass but simply passes straight through.
- Repeat the experiment for different angles of incidence. You will see that the angle of refraction increases as the angle of incidence increases (FIG. 4.4). If you measured the size of the angles you would see that they do not increase proportionally.

Fig. 4.3

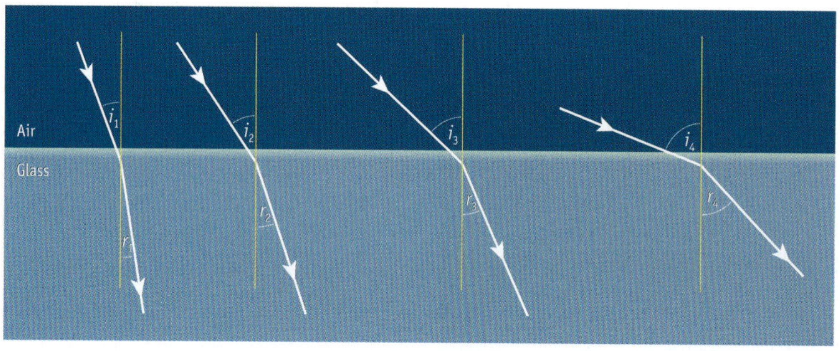

Fig. 4.4
As the angle of incidence increases so does the angle of refraction, but they do not increase proportionally.

In the last experiment you saw that when a ray of light goes from air into glass it is refracted (bent) towards the normal and when it goes from glass into air it is refracted away from the normal. In general:

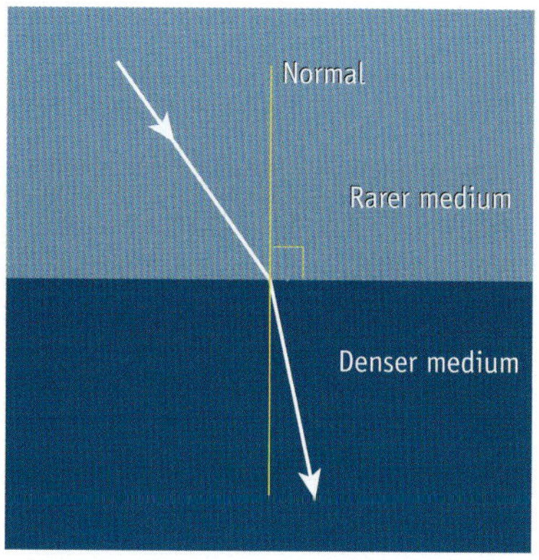

Fig. 4.5

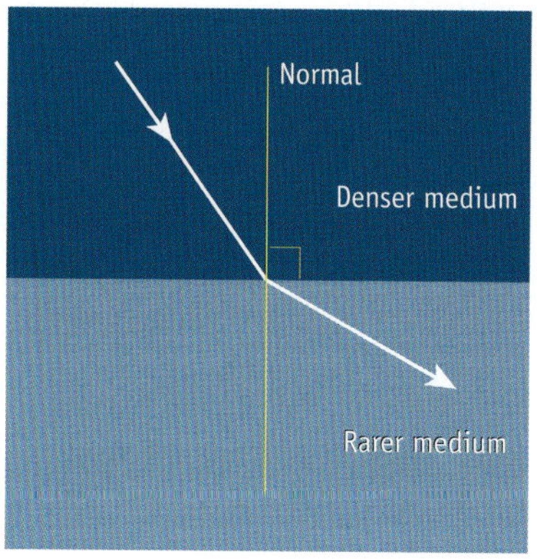

Fig. 4.6

When light travels from a ***rarer to a denser*** medium it is refracted ***towards the normal*** (FIG. 4.5).

When light travels from a ***denser to a rarer*** medium it is refracted ***away from the normal*** (FIG. 4.6).

SOME EFFECTS OF REFRACTION

A spoon dipping into a dish of water appears to bend where it enters the water. This is because light from the parts of the spoon under the water is refracted when it leaves the water. To an eye above the water (FIG. 4.7) this light appears to come from a different point. This is where the eye sees the spoon – thus causing it to appear bent.

Fig. 4.7

Water in a swimming pool or elsewhere always appears to be less deep than it actually is. A fish in water appears to be at a lesser depth than it actually is. The same effect is seen if you look at writing through a thick glass block (FIG. 4.8). The effect is better if you look at the glass at a sharp angle. Through the glass the writing appears to be much nearer to you than it actually is (see also page 33).

Fig. 4.8

THE LAWS OF REFRACTION OF LIGHT

The Laws of Refraction tell us the exact relationship between the angle of incidence and the angle of refraction. These laws can be verified experimentally.

THE LAWS OF REFRACTION OF LIGHT

1st Law The incident ray, the normal at the point of incidence and the refracted ray all lie in the **same plane** (FIG. 4.9).

2nd Law The ratio of the sine of the angle of incidence to the sine of the angle of refraction is a constant

i.e. $\frac{\text{Sin } i}{\text{Sin } r} = n$ where n is a constant

The value of n depends on the two media in question and it is called the **refractive index between the two media**.

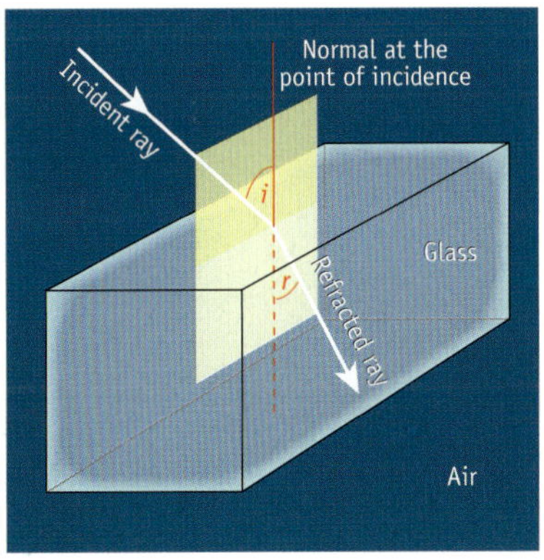

Fig. 4.9
The incident ray, the normal at the point of incidence and the refracted ray all lie in the same plane.

SNELL'S LAW

The Second Law of Refraction is sometimes called **Snell's Law** in honour of the Dutch mathematician Willebrord Snell (1591–1626) who discovered it in 1621.

REFRACTIVE INDEX OF A MEDIUM

The ***Refractive Index of a Medium*** is the ratio of the sine of the angle of incidence to the sine of the angle of refraction when light travels ***from a vacuum*** into that medium.

Substance	Refractive Index
Vacuum	1
Air	1.0003 ≈ 1
Water	1.33
Glass (Crown)	1.5
Glass (Flint)	1.6
Diamond	2.4

Fig. 4.10

The difference between when light travels from air or from a vacuum into a medium is very small and for our purposes can be neglected. Thus **the refractive index of medium is also equal to the ratio of the sine of the angle of incidence to the sine of the angle of refraction when light travels from AIR into that medium.** FIG. 4.10 gives the refractive index of some common media.

NOTE

- The bigger the refractive index of a medium the greater the amount by which that medium bends the light for a given angle of incidence
- n (the refractive index of a medium) is always 1 or bigger. This is so because when light goes from a vacuum into another medium it is always bent towards the normal

$$\Rightarrow r < i \Rightarrow \text{Sin } r < \text{Sin } i \Rightarrow \frac{\text{Sin } i}{\text{Sin } r} > 1$$

THE REFRACTIVE INDEX BETWEEN TWO MEDIA

The ***refractive index between two media*** is the ratio of the sine of the angle of incidence to the sine of the angle of refraction when light travels from one of those media into the other.

For any two media x and y:

$$_{x}n_{y} = \frac{1}{_{y}n_{x}}$$

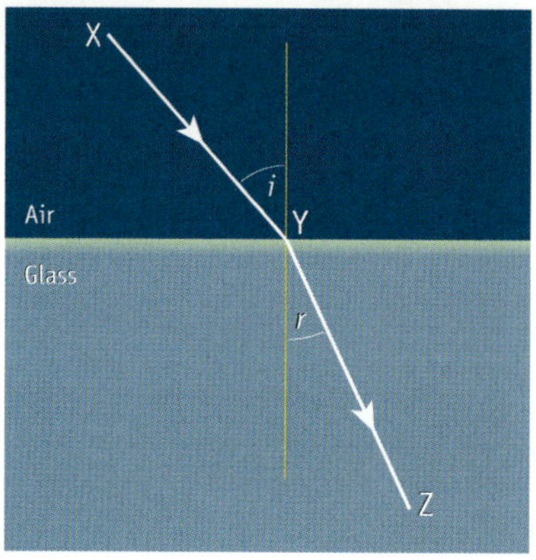

Fig. 4.11

The order in which the light passes through the media is important. For example, when light travels from glass into water the refractive index is 0.91 whereas when it travels from water into glass the refractive index is 1.1. We write this as: $_{g}n_{w} = 0.91$ and $_{w}n_{g} = 1.1$

The relationship between these two numbers is a simple one, namely:

$$0.91 = \frac{1}{1.1} \quad \text{or} \quad 1.1 = \frac{1}{0.91}$$

This result is generally true, e.g. $_{a}n_{g} = 3/2$ and $_{g}n_{a} = 2/3$

It is an experimental fact that a ray of light will retrace its path. Thus in Fig. 4.11 if a ray is shone along ZY it will emerge in air along YX. From the diagram:

$$_{a}n_{g} = \frac{\text{Sin } i}{\text{Sin } r} \quad \text{and} \quad _{g}n_{a} = \frac{\text{Sin } r}{\text{Sin } i}$$

$$\text{But} \quad \frac{\text{Sin } i}{\text{Sin } r} = \frac{1}{\text{Sin } r\,/\,\text{Sin } i} \qquad \text{Thus } _{a}n_{g} = \frac{1}{_{g}n_{a}}$$

Problem 1: A ray of light enters glass from air. The angle of incidence is 30° and the angle of refraction is 19°. What is the refractive index of the glass?

Solution: Refractive index of glass $= {_{a}n_{g}} = \frac{\text{Sin } i}{\text{Sin } r} = \frac{\text{Sin } 30°}{\text{Sin } 19°} = \frac{0.5}{0.326} = 1.53$

Problem 2: A ray of light enters water from air (Fig. 4.12). If the angle of incidence is 40°, find the angle of refraction. The refractive index of water is 1.33.

Solution: In Fig. 4.12: $_{a}n_{w} = \frac{\text{Sin } 40°}{\text{Sin } r} = 1.33$

$\Rightarrow \text{Sin } r \Rightarrow \frac{\text{Sin } 40°}{1.33} = 0.4833$

$\Rightarrow$ angle of refraction $r = \text{Sin}^{-1}(0.4833) = 28.9°$

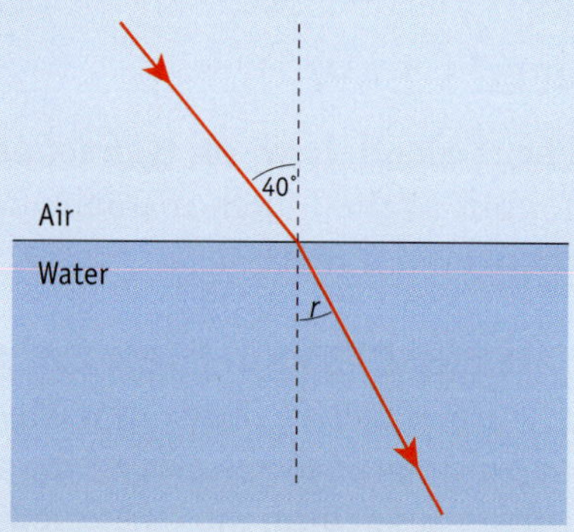

Fig. 4.12

Problem 3: The refractive index between glass and water is 0.91. What is the refractive index between water and glass?

Solution: $_{w}n_{g} = \frac{1}{_{g}n_{w}} = \frac{1}{0.91} = 1.1$

Problem 4: $_{w}n_{g} = 9/8$. Light enters water from glass. If the angle of incidence is 40°, find the angle of refraction.

Solution: Draw a diagram to be clear on exactly what is being asked (Fig. 4.13).

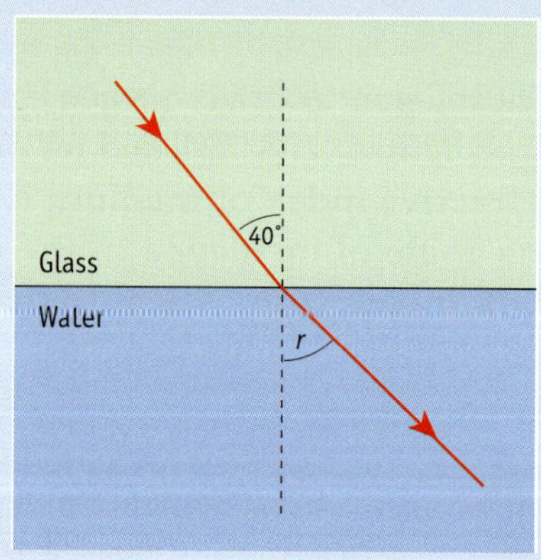

Fig. 4.13

$_{w}n_{g} = 9/8 \Rightarrow {_{g}n_{w}} = 8/9$

In going from glass to water: $_{g}n_{w} = \frac{\text{Sin } 40°}{\text{Sin } r} = \frac{8}{9} \qquad \text{Sin } r = \frac{9 \text{ Sin } 40°}{8}$

$\Rightarrow$ angle of refraction $r = \text{Sin}^{-1}\left(\frac{(9)(0.6428)}{(8)}\right) = 46.3°$

EXERCISE 4.1

1. A ray of light enters glass from air. The angle of incidence is 25° and the angle of refraction is 16.4°. Calculate the refractive index of glass.
2. A ray of light enters diamond from air. The angle of incidence is 60° and the angle of refraction is 21°. Find the refractive index of diamond.
3. The refractive index of water is 1.33. A ray of light entering water from air makes an angle of incidence of 30°. Find the angle of refraction.
4. The refractive index of glass is 1.5. What is the refractive index between glass and air?
5. The refractive index between water and glass is 1.13. What is the refractive index between glass and water?
6. A ray of light enters air from glass. The angle of incidence is 35° and the angle of refraction is 69°. Find the refractive index of the glass.
7. A ray of light leaves a rectangular diamond and enters air. It makes an angle of refraction of 15°. If the refractive index of diamond is 2.42, find the angle of incidence.
8. The refractive index of diamond is 2.42. A ray of light passing from the inside of a diamond into air makes an angle of incidence of 20°. Find the angle of refraction.

EXPERIMENT LIGHT 2

To Verify Snell's Law and Hence to Measure the Refractive Index of Glass.

Summary of Method

In this experiment you will use two pins to mark the direction of a ray of light which strikes a glass block. You will use two other pins to mark the same ray as it emerges from the block (FIG. 4.14). The angle of incidence *i* of the ray on the block and the **angle of refraction *r* within the block** can then be measured. You will repeat this for different values of *i*, and verify that the ratio Sin *i* / Sin *r* is a constant. Thus Snell's Law is verified. The constant value is the refractive of the glass. You will plot a graph of Sin *i* against Sin *r*. A straight line through the origin will result, again verifying Snell's Law. The slope of the graph is the refractive index.

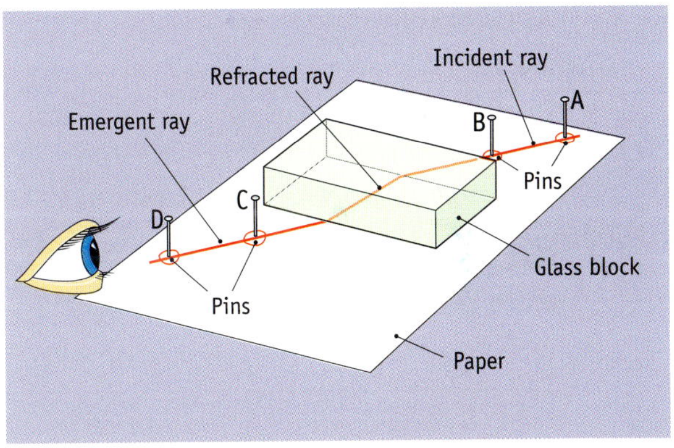

Fig. 4.14

Equipment Needed

- A rectangular glass block
- A sheet of paper and drawing board
- Four pins, a pencil, a ruler and a protractor

Method

1. Place the sheet of paper on the drawing board and place the glass block on the paper.
2. Draw the outline of the glass block on the paper with a finely-pared pencil. **Make sure that you do not accidentally move the block from the marked position.**
3. Stick two pins *A* and *B* into the paper and board (FIG. 4.14) so that the line *AB* makes an acute angle with one face of the block.
4. Looking in through the opposite face of the block, move your eye so that pin *A* cannot be seen because of pin *B* being in the way. Mark this direction with two more pins *C* and *D*.

5. Mark the position of each pin on the paper by drawing a small circle in pencil around it. Remove the pins and the block from the board.
6. Draw in the incident ray and the emergent ray. **Draw a straight line joining the point of incidence to the point of emergence – this is the refracted ray within the block.** Draw in the normal at the point of incidence using a protractor (FIG. 4.15). With the protractor measure the angle of incidence *i* and the angle of refraction *r*. Record these values.
7. Place the block on a different part of the paper or use a different sheet and **repeat** Steps 2 to 6 at least five times for different values of the angle of incidence. This will give a series of angles of incidence *i* and a series of corresponding angles of refraction *r*.

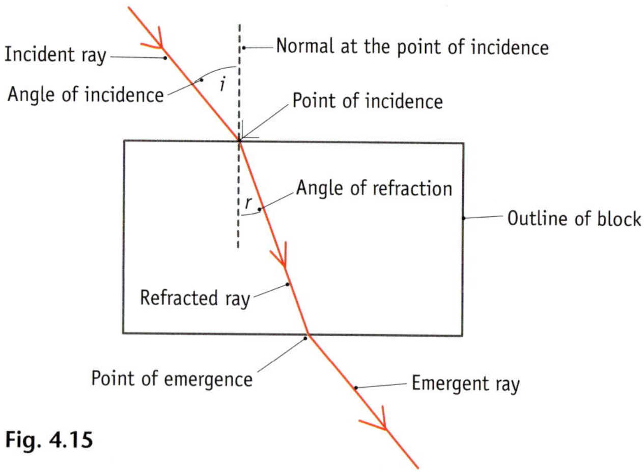

Fig. 4.15

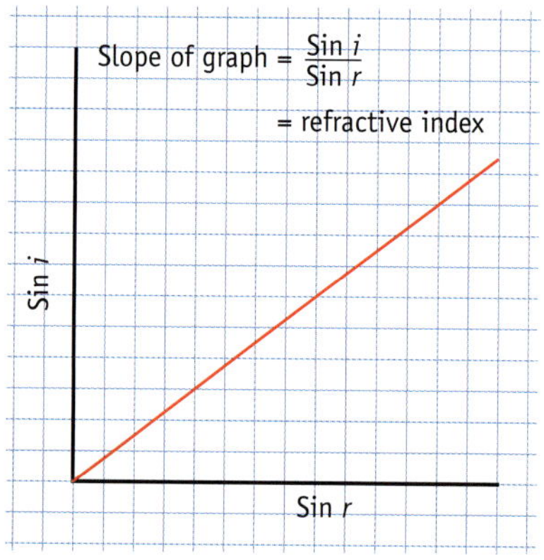

Fig. 4.16

Handling the Data

Method 1

- Complete the Table using your calculator.
- To a high degree of accuracy the values in the last column of the Table will all be the same – thus verifying Snell's Law.
- Calculate the average value of $\frac{\text{Sin } i}{\text{Sin } r}$ (last column in table). It is the refractive index of the glass in the block.

i	*r*	Sin *i*	Sin *r*	Sin *i*/ Sin *r*

Refractive index of material in block = Average value of $\frac{\text{Sin } i}{\text{Sin } r}$ =

Method 2

- On graph paper plot a graph of Sin *i* (on the *y*-axis) against Sin *r* (on the *x*-axis). Within the limits of experimental error all the points will lie on a straight line passing through the origin (FIG. 4.16). **The fact that the graph is a straight line through the origin verifies that Sin *i* $\propto$ Sin *r* (i.e. it verifies Snell's Law).**
- Measure the slope of the graph. It is the refractive index of the glass in the block.

Sources of Error

- Make sure pins *A* and *B* are not too near each other to reduce errors in marking the incident ray. The same applies to pins *C* and *D*.
- Use a finely-pared pencil for all drawing.
- Do not use a value of *i* that is very small, as both *i* and *r* will then be very difficult to measure accurately.

EXPERIMENT LIGHT 2 – ALTERNATIVE METHOD

To Verify Snell's Law and Hence to Measure the Refractive Index of Glass.

Summary of Method

In this experiment you will use a ray box to produce a ray of light which strikes a glass block (Fig. 4.17). You will draw the outline of the block and the rays on a sheet of paper. The angle of incidence *i* of the ray on the block and the **angle of refraction r within the block** can then be measured. You will repeat this for different values of *i*, and verify that the ratio Sin *i* / Sin *r* is always the same. Thus Snell's Law is verified. The constant value of the ratio is the refractive index of the glass. You will plot a graph of Sin *i* against Sin *r*. A straight line through the origin will result again verifying Snell's Law. The slope of the graph is the refractive index.

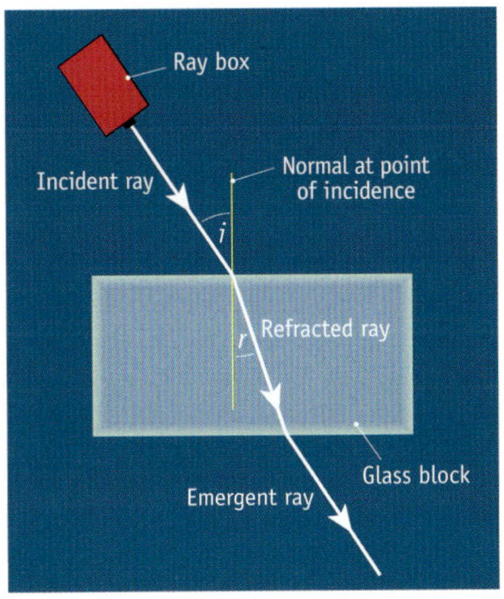

Fig. 4.17

Equipment Needed

- A ray box and power supply
- A rectangular glass block
- A sheet of drawing paper
- A pencil, a ruler and a protractor

Method

1. Place the drawing paper on the bench and place the glass block on the paper.
2. Trace the outline of the glass block onto the paper with a finely-pared pencil. During the rest of the experiment take care that you **do not accidentally move the block from the marked position**.
3. Set up the ray box as in Fig. 4.17 and shine a ray of light onto the block. With a pencil, mark the incident ray and the ray emerging from the other side of the block. Switch off the ray box and remove the block from the paper.
4. Draw in the incident ray and the emergent ray.
5. With your pencil, join the point where the incident ray meets the glass block with the point where the emergent ray meets the glass block. **Thus you have drawn the refracted ray within the block.** Draw in the normal at the point of incidence using a protractor (Fig. 4.15). The angle of incidence and the corresponding angle of refraction are now drawn on the paper.
6. With the protractor measure these angles. Record their values.
7. By placing the block on different parts of the paper or by using different sheets of paper, repeat steps 2 to 6 for at least five different angles of incidence – thus giving a series of angles of incidence *i* and the corresponding angles of refraction *r*.

Handling the Data

The data is handled by the same methods used in the previous experiment.

Real Depth and Apparent Depth

You saw on page 29 that an object viewed through a glass block appears to be nearer to you than it really is. It is also true that water in a swimming pool appears to be less deep than it actually is. These effects are due to refraction and are explained as follows. Fig. 4.18 (page 34) shows a small object at the bottom of a trough of water. A number of rays of light from the object are shown striking the surface of the water. At the surface they undergo refraction. To an observer in the air, these rays appear to be diverging from the point Q and thus P appears to be at Q, i.e. nearer to the surface than it actually is. Q is a virtual image of P. |SP| is called the **real depth** of the object and |SQ| is called the **apparent depth** of the object in the water.

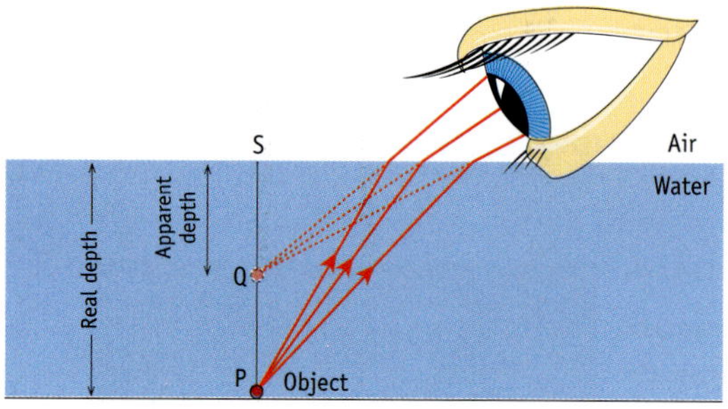

Fig. 4.18

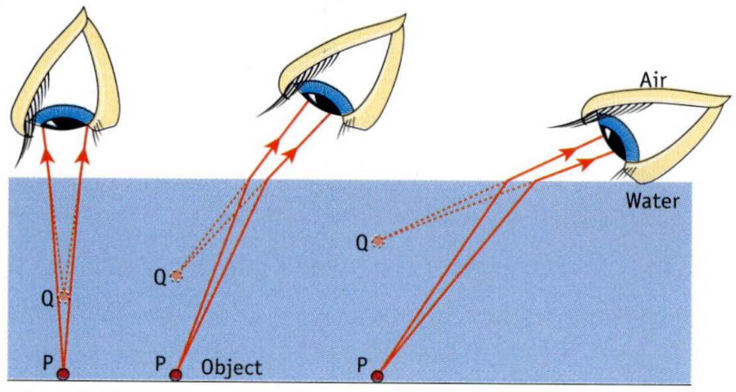

Fig. 4.19

From FIG. 4.19 you can see that the apparent depth decreases as you view the object more from the side and that the apparent depth is greatest when the object is viewed from vertically above it. Try this in the laboratory or at home.

REFRACTIVE INDEX IN TERMS OF REAL DEPTH AND APPARENT DEPTH

It can be proved (you do not need to know the proof) that refractive index n = real depth / apparent depth

If an object is viewed from air through another medium and the line of view is perpendicular to the surface of the medium then:

$$\textbf{Refractive Index of Medium} = \frac{\textbf{Real depth of object}}{\textbf{Apparent depth}} \quad \textbf{i.e.} \quad n = \frac{\textbf{Real depth}}{\textbf{Apparent depth}}$$

Problem 5: A block of glass of thickness 4 cm is placed on top of a mark on the bench. When the mark is viewed perpendicularly through the glass a virtual image of it appears 2.67 cm from the top of the block. Find the refractive index of the glass.

Solution: Refractive Index $n = \frac{\text{Real depth}}{\text{Apparent depth}} = \frac{4}{2.67} = 1.5$

Problem 6: A pool of water is 12 m deep. If the bottom of the pool is viewed perpendicularly from air, how deep does it appear? Refractive index of water = 1.33.

Solution: ${}_an_w = \frac{\text{Real depth}}{\text{Apparent depth}} \Rightarrow 1.33 = \frac{12}{\text{Apparent depth}} \Rightarrow \text{Apparent depth} = \frac{12}{1.33} = 9.02\text{m}$

EXERCISE 4.2

1. A block of glass of thickness 5 cm is placed on top of a mark on the bench. When the mark is viewed perpendicularly through the glass a virtual image of it appears 3.33 cm from the top of the block. Find the refractive index of the glass.
2. A pool of water is 10 m deep. If the bottom of the pool is viewed perpendicularly from air how deep does it appear (refractive index of water = 1.33)?
3. A pool of water appears to be 0.8 m deep when viewed perpendicularly from above. What is its actual depth? Refractive index of water = 4⁄3.
4. A rectangular glass block of dimensions 5 cm × 5 cm × 6 cm is viewed perpendicularly from one side and it appears to be in the shape of a cube. What is the refractive index of the material in the glass?
5. A block of glass of refractive index 1.5 is placed on top of a mark on a sheet of paper. When the mark is viewed perpendicularly through the glass, a virtual image of it appears 3.33 cm from the **bottom** of the block. What is the thickness of the glass block?

EXPERIMENT LIGHT 3

To Find the Refractive Index of a Liquid by Measuring the Real Depth and Apparent Depth of an Object in the Liquid.

Summary of Method

In this experiment you will view a pin at the bottom of a beaker of water, through the water (FIG. 4.20). You will use the image of a second pin in a plane mirror as the search pin. You will adjust the height of the second pin until there is no parallax between the image in the water and the image in the plane mirror. From FIG. 4.20 it is obvious that distance X = the real depth and distance Y = the apparent depth. The refractive index of the water is the ratio of these two measurements,

i.e. Refractive index of water = $_{air}n_{water} = \dfrac{\text{Real depth}}{\text{Apparent depth}}$

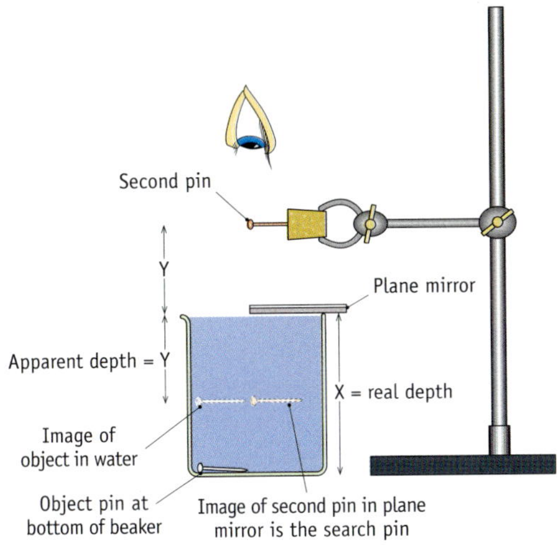

Fig. 4.20

Equipment Needed

- A number of beakers of various depths. Make sure the beakers can be filled right to the top with water. Some cannot because of the shape of their spouts.
- A retort stand, a cork and two pins
- A plane mirror and a metre stick

Method

1. Fill one of the beakers with water and place a pin at the centre of the base of the beaker (FIG. 4.20). This pin is the object.
2. Place the plane mirror across the top of the beaker. Make sure the water is touching the back of the mirror.
3. Stick the second pin in the cork on the retort stand. Adjust the position of the second pin until its image in the plane mirror is in line with the image of the first pin in the water (FIG. 4.21).
4. Adjust the height of the second pin until there is no parallax between the two images.
5. With the metre stick measure:
   - the distance from the object pin to the top of the beaker (X),
   - the distance from the second pin to the top of the water (Y).
6. Record these values.
7. Repeat the experiment at least four times using beakers of different depths.
8. Complete the Table and calculate the average value of Real depth/Apparent depth. This is the refractive index of the water.

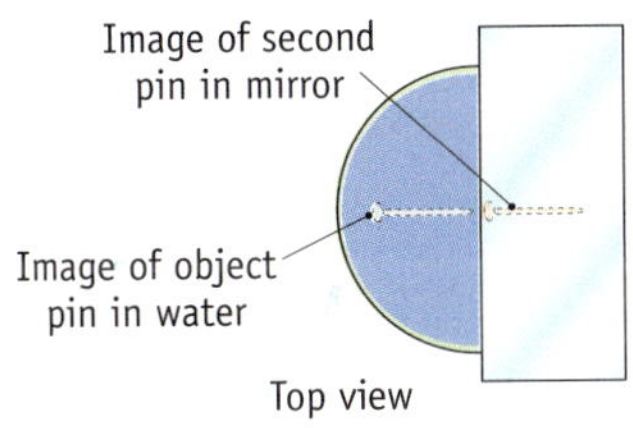

Fig. 4.21

View of beaker and mirror from above.

Real depth X/cm	Apparent depth Y/cm	$\dfrac{\text{Real depth}}{\text{Apparent depth}}$

Average value of $\dfrac{\text{Real depth}}{\text{Apparent depth}}$ = Refractive index of liquid =

Sources of Error

Errors arise both in locating the position of no parallax and in measuring the distances with the metre stick.

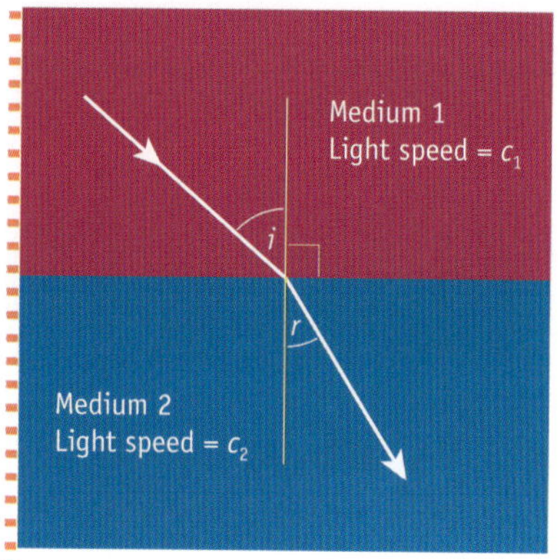

Fig. 4.22

Refractive Index in Terms of Relative Speeds

When light, which is a wave, travels from one medium to another, its speed usually changes (it is fastest in a vacuum at 3×10^8 m s^{-1}). It is because of this that it changes direction on entering the second medium. This is explained more fully on page 180. It can be proved that if the speed in medium 1 is c_1 and the speed in medium 2 is c_2 then:

$${}_1n_2 = \frac{c_1}{c_2} \quad \text{i.e.} \quad \frac{\textbf{Sin } i}{\textbf{Sin } r} = \frac{c_1}{c_2}$$ (see Fig. 4.22)

It follows that for any medium x, ${}_{\text{air}}n_x = \frac{c_{\text{air}}}{c_x}$

i.e. Refractive index of medium $= \dfrac{\text{Speed of light in air}}{\text{Speed of light in medium}}$

It follows that speed of light in a medium $= \dfrac{\text{Speed in air}}{\text{Refractive index of medium}}$

⇒ The greater the refractive index the slower the speed of the light.

Problem 7: What is the refractive index of a medium in which light travels at a speed of 2×10^8 m s^{-1}? Speed of light in air $= 3 \times 10^8$ m s^{-1}.

Solution: $${}_{\text{air}}n_x = \frac{c_{\text{air}}}{c_x} = \frac{\text{Speed in air}}{\text{Speed in medium}} = \frac{3 \times 10^8}{2 \times 10^8} = 1.5$$

Problem 8: The refractive index of water $= 1.33$ and the speed of light in air $= 3 \times 10^8$ m s^{-1}. Calculate the speed of light in water.

Solution: $${}_{a}n_w = \frac{c_{\text{air}}}{c_{\text{water}}} \Rightarrow \frac{\text{Speed in air}}{\text{Speed in water}} = 1.33$$

$$\Rightarrow \text{Speed in water} = \frac{\text{Speed in air}}{1.33} = \frac{3 \times 10^8}{1.33} = 2.26 \times 10^8 \text{ m s}^{-1}$$

EXERCISE 4.3

Take speed of light in air to be 3×10^8 m s^{-1}

1. What is the speed of light in glass whose refractive index is 1.5?
2. What is the speed of light in diamond whose refractive index is 2.42?
3. What is the refractive index of a medium in which the speed of light is 2×10^8 m s^{-1}?
4. What is the refractive index of a medium in which light travels at a speed of 1.25×10^8 m s^{-1}?

Total Internal Reflection

CRITICAL ANGLE
When light travels from a ***denser to a rarer*** medium the angle of incidence whose corresponding angle of refraction is 90° is called the ***critical angle (C)*** for those two media.

When a light ray travels from a denser to a rarer medium it is refracted away from the normal (page 28). As the angle of incidence is increased so is the angle of refraction (Fig. 4.23). Eventually a stage is reached when the angle of refraction is 90°. The angle of incidence for which this occurs is called the **critical angle** for those two media. Its usual symbol is C.

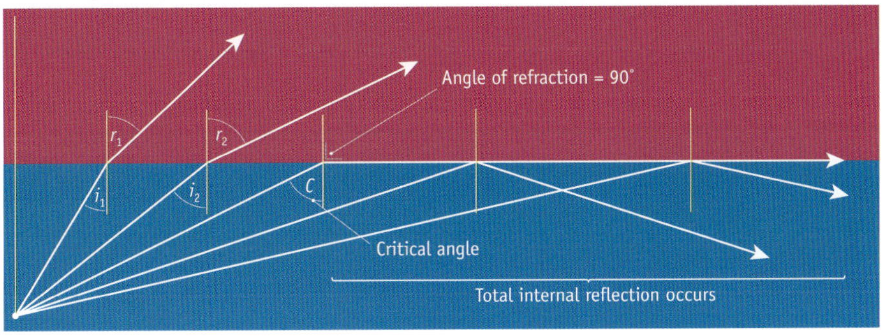

Fig. 4.23

TOTAL INTERNAL REFLECTION

When light going from a ***denser to a rarer*** medium strikes the second medium with an angle of incidence greater than the critical angle, it does not enter the second medium. It is all reflected back in the denser medium. This is called ***total internal reflection***.

If the angle of incidence is increased beyond the critical angle, the ray of light does not enter the second medium at all. Instead it is all reflected back into the first medium obeying the laws of reflection as it does. This phenomenon is called **total internal reflection***.

EXPERIMENT

To Demonstrate Total Internal Reflection

- Set up a ray box and a semicircular slab of glass as in Fig. 4.24(A).
- Starting with a small angle of incidence, slowly increase this angle.
- Soon the critical angle will be reached and the refracted ray skims along the flat face of the glass.
- If the angle of incidence is then increased any further, the refracted ray suddenly jumps from that shown in Fig. 4.24(B) to that in Fig. 4.24(C) to become the totally internally reflected ray.

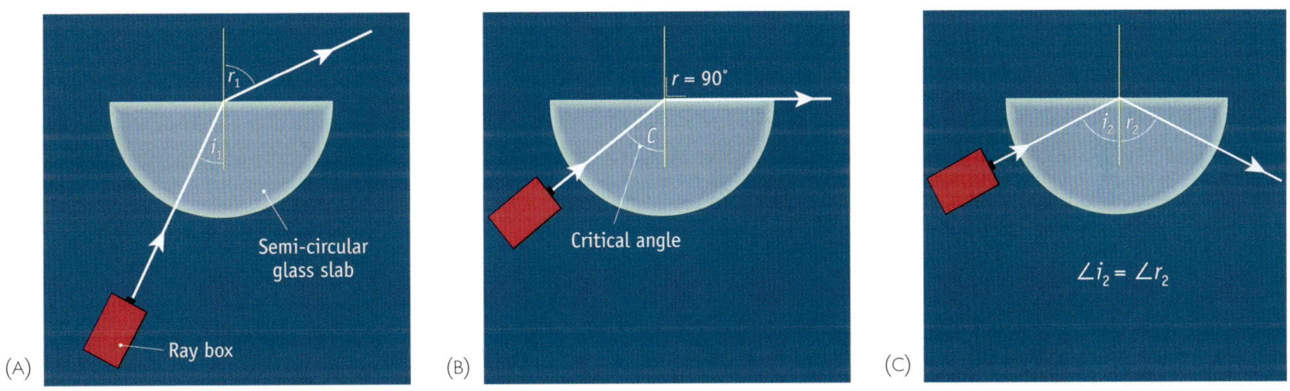

Fig. 4.24

Relationship between Refractive Index and Critical Angle

Fig. 4.25 shows a ray of light going from a denser to a rarer medium and striking the rarer medium at the critical angle. Applying Snell's Law:

$${}_{x}n_{y} = \frac{\text{Sin } C}{\text{Sin } 90^\circ} = \frac{\text{Sin } C}{1} = \text{Sin } C$$

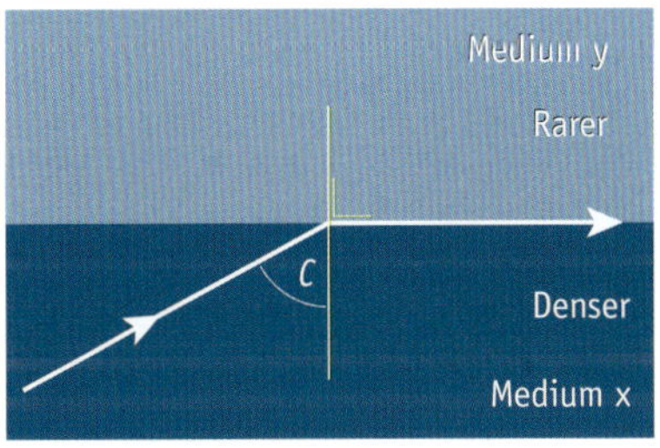

Fig. 4.25

* The word total means exactly what it says here. **All** the light is reflected, none is absorbed. In the best ordinary mirror available, about 4% of the light incident on it is absorbed, the rest is reflected.

If the second medium (the rarer one) is air (or a vacuum) this equation becomes:

$$_{x}n_{a} = \text{Sin } C \quad \Rightarrow \quad _{a}n_{x} = \frac{1}{\text{Sin } C}$$

i.e. Refractive index of a medium $\boldsymbol{n} = \dfrac{\mathbf{1}}{\mathbf{Sin\ C}}$ where C is the critical angle.

Problem 9: The critical angle for a certain medium is 50°. Find its refractive index.

Solution: Refractive index $n = \dfrac{1}{\text{Sin } C} = \dfrac{1}{\text{Sin } 50°} = 1.3$

Problem 10: The refractive index of glass is 1.5. What is the critical angle for glass?

Solution: $_{a}n_{g} = \dfrac{1}{\text{Sin } C} \Rightarrow \text{Sin } C = \dfrac{1}{_{a}n_{g}} = \dfrac{1}{1.5} = 0.6667$

i.e. $\text{Sin } C = 0.6667 \Rightarrow C = \text{Sin}^{-1}\, 0.6667 \Rightarrow$ critical angle $C = 41.8°$

Using a 45°–90°–45° Prism to turn a Ray of Light through (i) 90° (ii) 180°

If a ray of light enters a glass prism as shown in Fig. 4.26(a), it strikes the right-hand side of the prism at an angle of incidence of 45°. This angle of incidence is greater than the critical angle, therefore total internal reflection occurs at 0 and the ray is reflected as shown. Since the angle of reflection is also 45°, the angle between the incident ray and the reflected ray is 90°, thus the prism turns the ray through 90°. Similarly, the same prism can turn a ray through 180° (Fig. 4.26(b)).

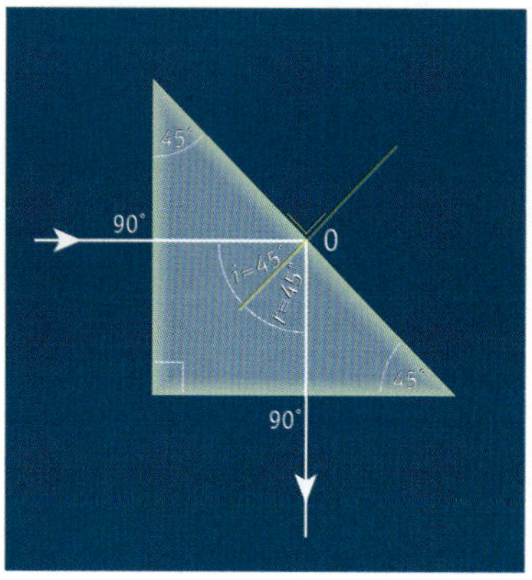

(A) Turning a ray of light through 90° with a prism.

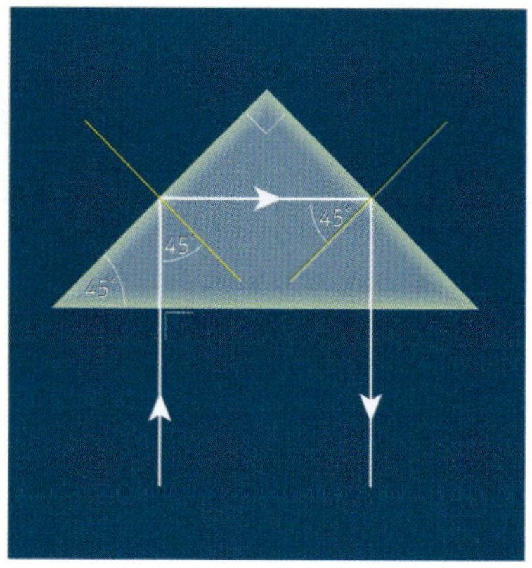

(B) Turning a ray of light through 180° with a prism.

Fig. 4.26

Problem 11: If a prism is to turn a ray of light through 90° as in Fig. 4.26(a), what is the smallest value the refractive index of the material in the prism could be?

Solution: The angle of incidence of the ray on the glass air surface is 45°. In going from glass to air the angle of incidence must be greater than the critical angle if total internal reflection is to occur.

i.e. for total internal reflection $45° > C$ i.e. $C < 45°$

$\Rightarrow \text{Sin } C < \text{Sin } 45° \Rightarrow \text{Sin } C < 0.707 \Rightarrow \dfrac{1}{\text{Sin C}} > \dfrac{1}{0.707}$

$\Rightarrow$ But $1/\text{Sin } C$ = refractive index of the glass and $1/0.707 = 1.414$

$\therefore$ Refractive index must be greater than 1.414

Problem 12: A point source of light P is 2 m below the surface of the water in a still pond. Find the radius of the circle at the surface through which light can pass into air given that the refractive index of water = 4⁄3

Solution: Fig. 4.27 shows the situation. Light from the source will not leave the water if it strikes the water at an angle of incidence greater than the critical angle.

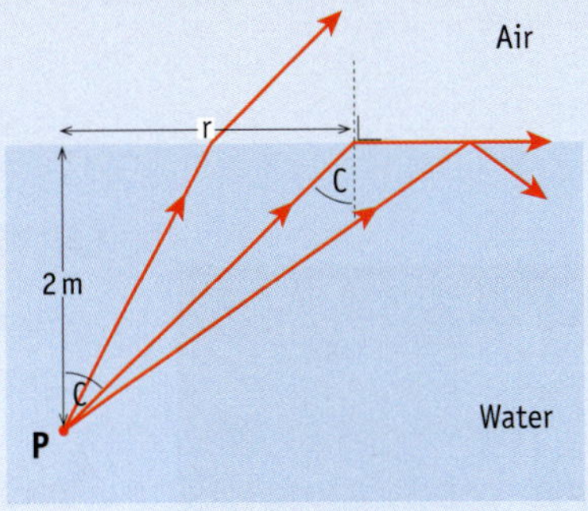

Fig. 4.27

Refractive index of water $= 4/3$

$$\Rightarrow \text{Sin } C = \frac{1}{(4/3)} = 3/4$$

$$\Rightarrow \text{Critical angle } C = 48.59^\circ$$

From the diagram:

$$\text{Tan } C = \frac{r}{2} \Rightarrow r = 2\text{Tan } C \Rightarrow r = 2 \text{ Tan } 48.59^\circ = 2.268 \text{ m}$$

Since a light ray can retrace its path it is found that if light is to enter the water from air and arrive at the point P below the surface, only light which strikes the water within a circle of radius r found above will reach P. Thus if you are underwater and looking upwards, light from outside can only enter the water through the circle of radius r. Fig. 4.28 shows this quite well. Divers refer to this circle as **Snell's Window**.

Fig. 4.28

Suppose that three plane mirrors are held at right angles to each other. If a ray of light is shone onto one of the mirrors so that it is reflected off each in turn, the ray emerging from the mirrors will be travelling back in the direction from which the original ray came. The safety reflector on a bicycle or on a car makes use of this fact (Fig. 4.29). Instead of mirrors, the plastic or other material from which the reflector is made is shaped into many small right-angled prisms. The light is reflected by **total internal reflection** in each prism and hence back in the direction from which it came. Some reflective road signs are based on the same principle.

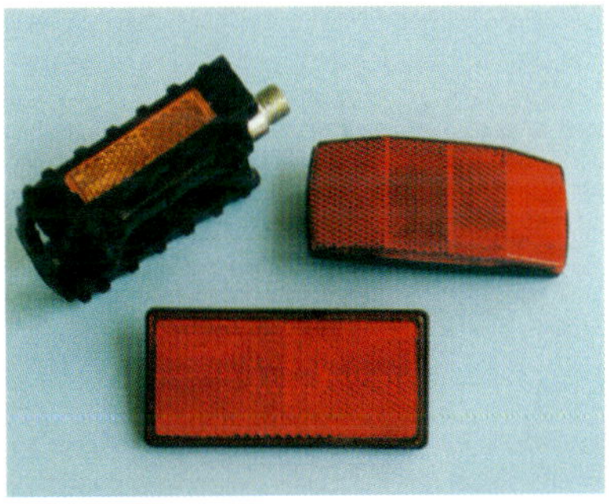

Fig. 4.29

EXERCISE 4.4

1. The critical angle for a certain medium is 40°. What is its refractive index?
2. The critical angle for diamond is 24.6°. What is its refractive index?
3. The refractive index of a certain type of glass is 1.2. What is the critical angle between the glass and air?
4. The refractive index of water is 1.33. What is the critical angle between water and air?
5. Find the critical angle for a medium whose refractive index is 1.66.
6. What is the apparent depth of an object in a block of a material whose critical angle is 40°? The object is viewed perpendicularly from air and its real depth is 12 cm.
7. A source of light at the bottom of a pond 4 m deep emits light in all upward directions. The rays that leave the water and enter the air pass through a disc of radius r at the surface of the water. Find the radius of this disc. Refractive index of water = 1.33.

8. A point source of light is 2 m below the surface of a pool of water. The radius of the circle through which light from the source can enter air is 2.3 m. Calculate the refractive index of water.

9. If a 45° –90° –45° prism is to turn a ray of light through 90° as in FIG. 4.26(A) (page 38), what is the smallest value the refractive index of the material in the prism could be?

OPTICAL FIBRES

An interesting and important application of total internal reflection is the optical fibre.

OPTICAL FIBRE

A very thin transparent rod (usually of glass) through which light can travel by total internal reflection is called an ***optical fibre***.

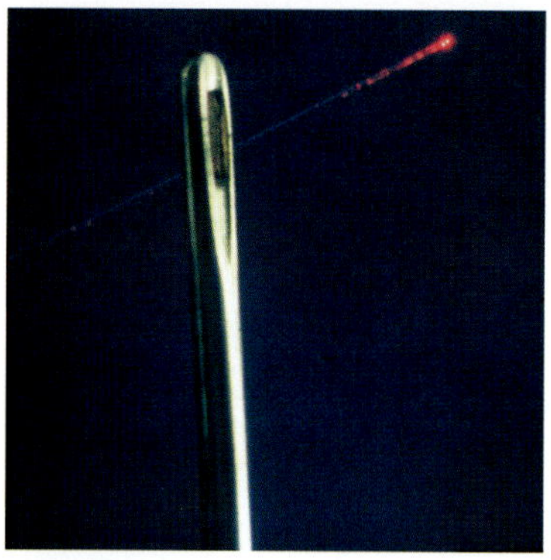

Fig. 4.30
An optic fibre passing through the eye of a needle.

Optical fibres are sometimes called light pipes. This is a bad choice of word since **optical fibres are solid, not hollow**. Since the fibres are very thin they can bend easily and thus can carry light around corners.

TRANSMISSION OF LIGHT THROUGH AN OPTICAL FIBRE

- Light enters the fibre and strikes the inside of the fibre at an angle **greater than the critical angle**. Total internal reflection occurs.
- The ray is reflected to the opposite side and total internal reflection occurs again (FIG. 4.31).
- This process continues and the light travels through the fibre.

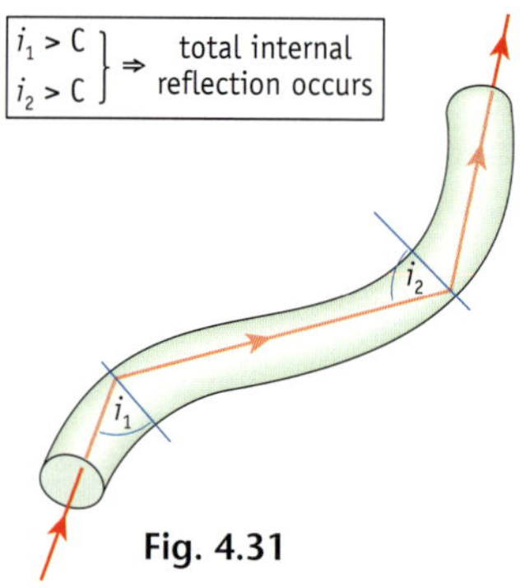

Fig. 4.31

Light can escape from optical fibres in two ways:

- If an optical fibre is bent through too large an angle, the ray travelling inside it may strike it at an angle less than the critical angle. If this happens the light will not be totally internally reflected and will leave the fibre.
- If an optical fibre comes into contact with another fibre (or another part of itself) a ray of light inside it may no longer be travelling from glass to air but instead from glass to glass. Total internal reflection will not then occur and the ray will travel from one fibre into the other (FIG. 4.32). One method to prevent this occurring is to coat the optical fibre with a layer of glass of lower refractive index. Even if two fibres then come in contact, total internal reflection still occurs as the light is going from a denser to a rarer glass at an angle of incidence greater than the critical angle (FIG. 4.33).

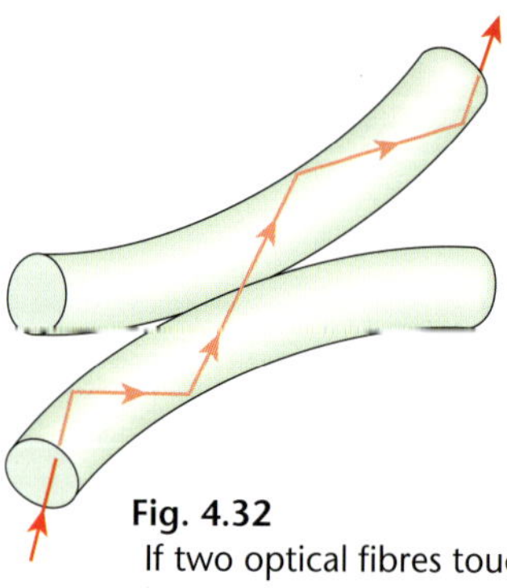

Fig. 4.32
If two optical fibres touch light may escape from one and pass into the other.

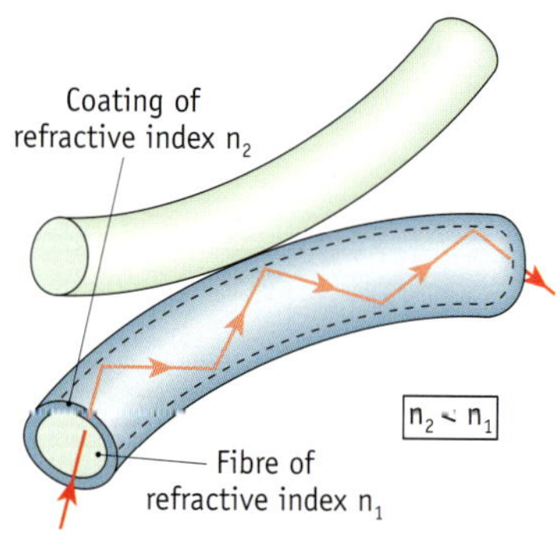

Fig. 4.33
By coating the optical fibre with a material of lower refractive index, light will not escape from it if it touches another fibre.

USES OF OPTICAL FIBRES

Optical fibres are used in telecommunications to transmit telephone signals in the form of pulses of light. The electrical telephone signals are used to create corresponding light signals which can travel through optical fibres. At the other end of the fibre the light signals are used to create electrical signals which operate telephones.

The advantage of doing this are as follows:

(i) Energy losses in optical fibres are much smaller than losses in electrical cables.

(ii) Optical fibres are much smaller than the electric wires needed to carry the same amount of signals.

(iii) Interference in optical fibres is much less.

In medicine, optical fibres are used to bring light to and from inaccessible parts of the body – e.g. the stomach – to examine them. An **endoscope** is a group of fibres used for this purpose (FIG. 4.34).

A dentist's drill uses optical fibres to carry light to very near the drill bit. Using this, the dentist can see inside the patient's mouth while drilling.

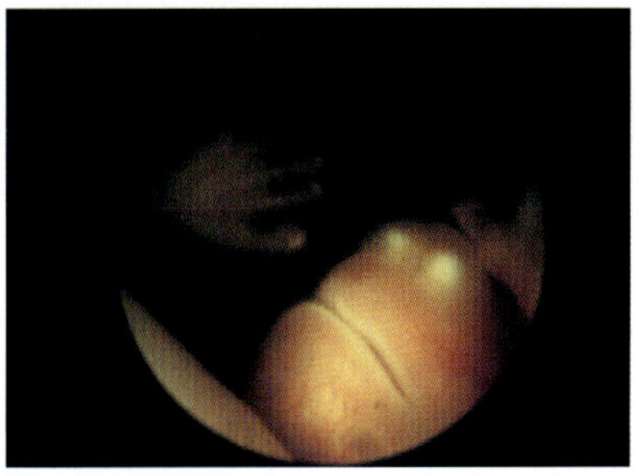

Fig. 4.34
Endoscopic image of the human foetus.

MIRAGES

The refractive index of air changes slightly with temperature. This change is responsible for the mirage which can be seen on a road on a hot Summer's day. The mirage appears as a shiny puddle on the road in which images of nearby objects can be seen (FIG. 4.35).

FIG. 4.36 shows the sun shining on a road. The road absorbs heat and strongly heats the air just above it. As the height above the road increases the temperature decreases. Hot air is less dense than cool air and thus has a slightly lower refractive index. Consider a ray of light from the sky heading towards the road. It is travelling from a denser to a rarer medium and will be refracted away from the normal and will bend as shown. If, at some point, the angle of incidence is greater than the critical angle, the ray will undergo **total internal reflection** and be reflected upwards. An observer will see an image of the sky in the ground. This is the shiny puddle, i.e. the mirage. A nearby object may be seen upside down in the puddle due to the same effect.

Fig. 4.35

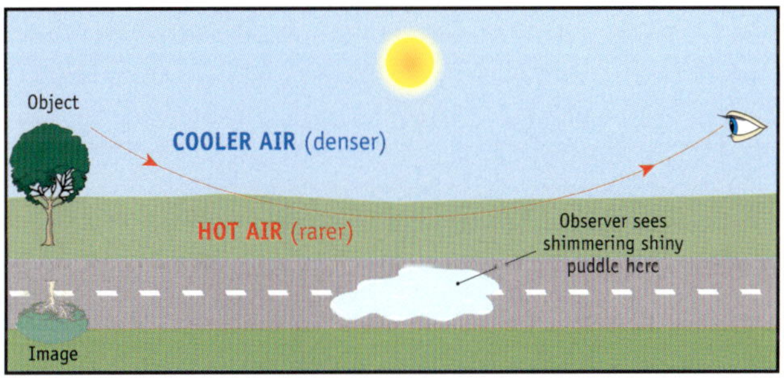

Fig. 4.36

CHAPTER CHECKLIST

- **Define** each of the following: Refraction; Refractive Index; Critical Angle; Total Internal Reflection.
- **State** the Laws of Refraction of Light and Snell's Law.
- **Explain** what is meant by: Real depth; Apparent depth; Optical fibre.
- **Describe** and carry out an experiment to:
  Demonstrate refraction; Verify Snell's Law; Measure refractive index; Demonstrate total internal reflection.
- **Recall** and use the formulae:

$$n = \frac{\text{Sin } i}{\text{Sin } r};\quad {}_xn_y = \frac{1}{{}_yn_x};\quad n = \frac{\text{Real depth}}{\text{Apparent depth}};\quad n = \frac{1}{\text{Sin}\,C}$$

$${}_1n_2 = \frac{c_1}{c_2}$$

- **List** some practical effects due to refraction.
- **Show** how a prism can be used to turn a ray of light through 90° and 180°.
- **Recall** that reflective road signs, mirages, prism reflectors and optical fibres are based on total internal reflection.

Lenses

CHAPTER 5

Lenses

In this chapter we look at refraction of light by **lenses**. We shall study two types of lens – a **convex lens** and a **concave lens** (Fig. 5.1). The convex lens is thicker at the centre than at the sides and the concave lens is thicker at the sides than at the centre.

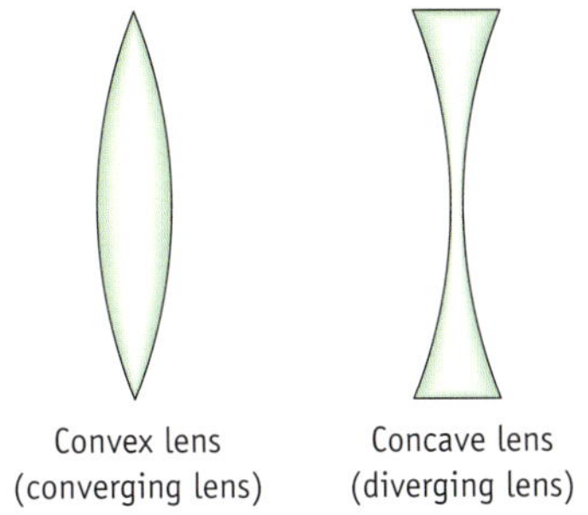

Fig. 5.1

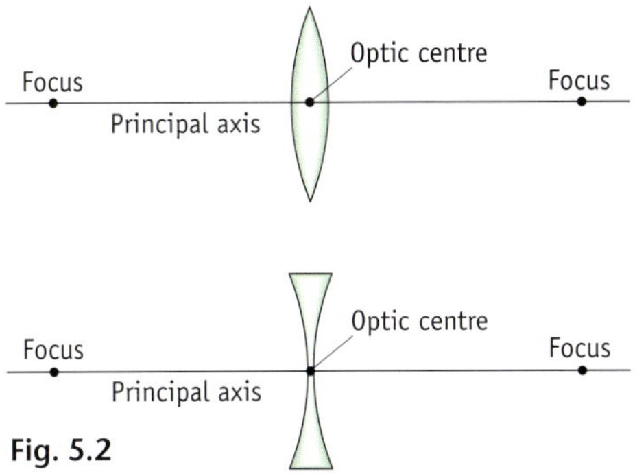

Fig. 5.2

Lenses are found in such everyday objects as spectacles, contact lenses, cameras, projectors, telescopes, binoculars and microscopes. A magnifying glass is simply a convex lens.

- The centre of the lens is called the **optic centre** (Fig. 5.2).
- The straight line passing through the optic centre which is at right angles to the face of the lens is called the **principal axis** or simply the **axis** of the lens (Fig. 5.2).

Focusing Property of Lenses

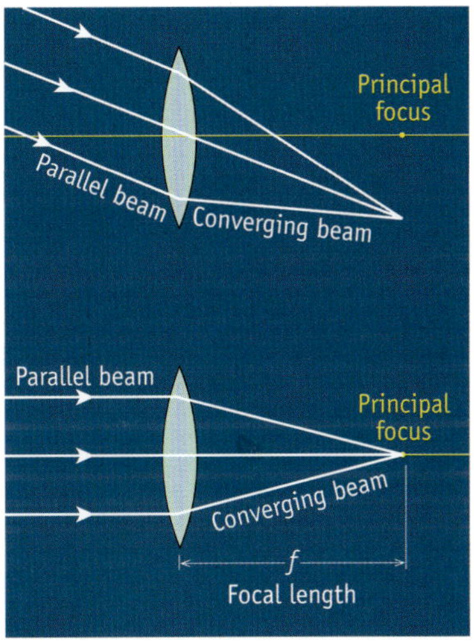

Fig. 5.3

If a convex lens is thin, it is found that a parallel beam of light falling on the lens is brought to a focus at a particular point (Fig. 5.3). If the beam is parallel to the axis of the lens, the point to which it converges is called the **principal focus** or simply the **focus** of the lens. The distance from the optic centre to the principal focus is called the **focal length** f of the lens. Note that:

- There is a focus at each side of the lens.
- Each focus is the same distance from the optic centre.
- Since a convex lens changes a parallel beam into a converging beam, it is also called a **converging lens**.

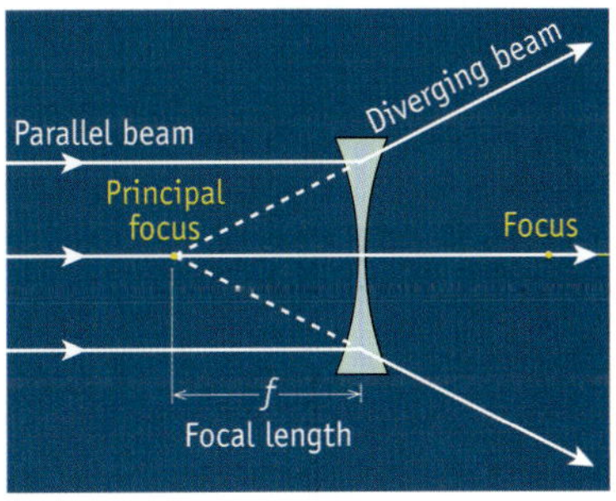

Fig. 5.4

For a thin concave lens, a parallel beam of light falling on the lens spreads out as if it were coming from a particular point (Fig. 5.4). If the beam is parallel to the axis this point is called the **principal focus** or simply the **focus** of the lens. There is a focus at each side of the lens and each focus is the same distance from the optic centre. Since a concave lens changes a parallel beam into a diverging beam, it is also called a **diverging lens**.

Refraction of Light by a Thin Convex Lens

When locating an image in a convex lens we shall use the following results (Fig. 5.5), which you do not need to be able to prove. Simply accept and **remember** them.

A ray which strikes the optic centre passes straight through the lens.

A ray which strikes the lens travelling parallel to the axis passes through the focus at the other side of the lens.

A ray which passes through the focus and then strikes the lens emerges parallel to the principal axis.

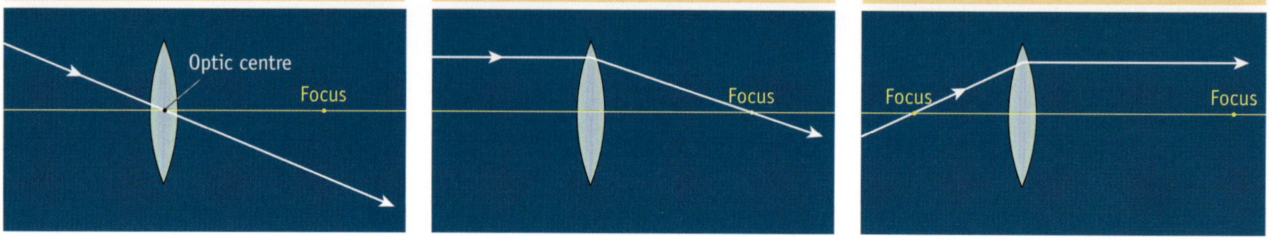

Fig. 5.5

How an Image is Formed in a Convex Lens

Fig. 5.6 shows an object in front of a convex lens. Whether it is self-luminous or diffusely reflecting, every point on the object is giving out light in all directions. Light rays from a point P at the top of the object are shown striking the lens and being refracted.

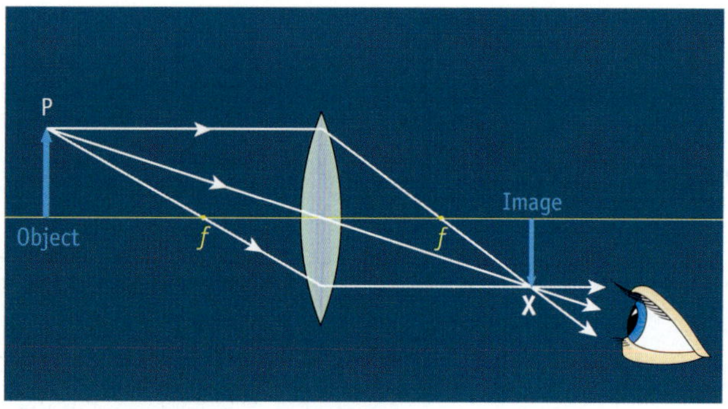

Fig. 5.6

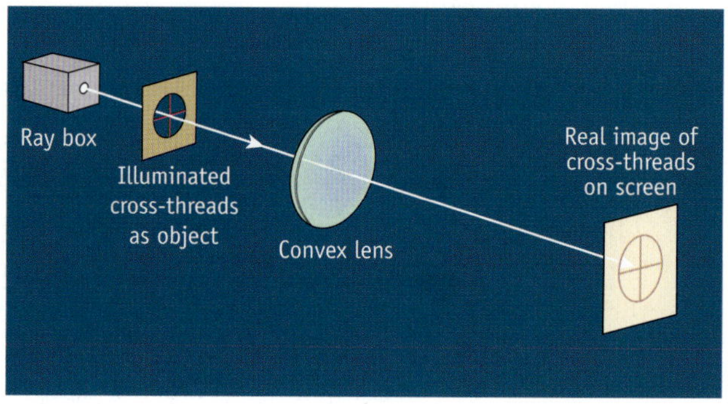

Fig. 5.7

Recall from pages 16 and 17 the following definitions:

REAL IMAGE

A ***real image*** is an image formed by the actual intersection of rays. It can be located on a screen.

VIRTUAL IMAGE

A ***virtual image*** is an image formed by the apparent intersection of rays. It cannot be located on a screen.

What do we notice? After refraction at the lens each of the rays passes through the same point X. It can also be shown that **any other ray** from P would also pass through X after refraction. It follows that an eye (i.e. an observer) as shown in Fig. 5.6 will see an **image** of the top of the object at X, since rays which originally were diverging from P are now diverging from X.

By repeating this process for other points on the object we would find that an image of the whole object is formed as in Fig. 5.6. This image is a **real image** since it is caused by the actual intersection of rays. A real image produced by a convex lens can easily be projected onto a screen or a wall as shown in Fig. 5.7.

How the Image Changes as the Position of the Object Changes

To find where the image is located, we need not draw all three rays as in Fig. 5.6. Any two will do. We will now see where the image is when the object is: outside $2f$; at $2f$; between f and $2f$; at the focus; inside the focus.

Try to draw the relevant ray diagrams yourself before looking at FIG. 5.8. From FIG. 5.8 we see the following:

FOR A CONVEX LENS:
If the object is ***outside the focus*** the image is ***real*** and located at the opposite side of the lens to the object. The image is ***inverted***.

If the object is ***inside the focus*** the image is ***virtual*** and is located at the same side of the lens as the object. The image is ***upright (erect)***.

Image of a Distant Object in a Convex Lens

If the object is a large distance from the lens (we say at infinity) the image formed is real and at the focus (to be precise, at the focal plane). Recall that light from a point on a distant object arrives as a parallel beam and that the beams from two different points on the distant object will not be parallel to each other. In FIG. 5.9 light from the top (P) and the bottom (Q) of a distant object falls on a convex lens. The beam parallel to the axis is brought to a focus at the principal focus of the lens. The other beam is brought to a focus as shown – on what is called the focal plane of the lens. Thus a real image of the distant object is formed at the focal plane of the lens. This can easily be shown in the laboratory as in FIG. 5.10.

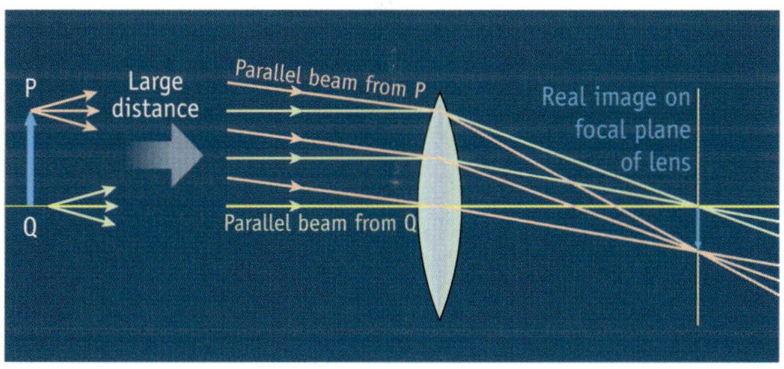

Fig. 5.9

Fig. 5.10
A convex lens forming a real image of a distant object.

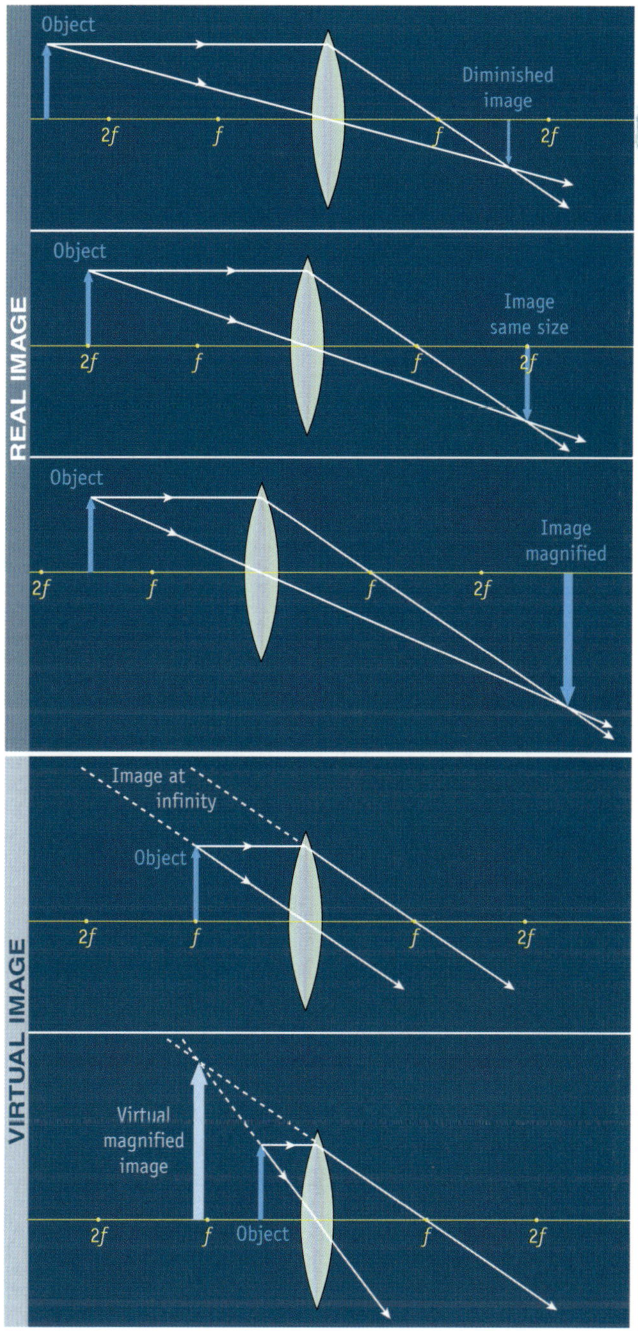

Fig. 5.8

Formula for a Convex Lens

A simple formula, similar to that for spherical mirrors applies to lenses. You need not prove the formula; simply remember it. It applies to both convex and concave lenses.

Let u be the distance from the **object** to the optic centre, let v be the distance from the **image** to the optic centre and let f be the **focal length**. Then the formula is:

$$\frac{1}{u} + \frac{1}{v} = \frac{1}{f}$$

where: u is always +

v is + for a real image v is – for a virtual image

f is + for a convex lens f is – for a concave lens

Magnification

In FIG. 5.8 you saw that the size of the image is usually different to the size of the object. The ratio of the height of image to the height of the object is called the **magnification**, i.e:

$$\text{Magnification } (m) = \frac{\text{Height of image}}{\text{Height of object}}$$

For example, if the image is twice the height of the object the magnification is 2 since: height of image/height of object = 2. It can be proved that the magnification produced by a convex lens = image distance/object distance. You need not worry about the proof.

$$\text{Magnification} = \frac{\text{Image distance}}{\text{Object distance}} \qquad \text{i.e.} \quad m = \frac{v}{u}$$

Problem 1: An object is placed 40 cm from a convex lens of focal length 30 cm. Find the position and nature of the image.

Solution: Here $u = 40$, $f = 30$ and v is to be found.

Since the object is outside the focus the image is real and the formula to be used is: $\frac{1}{u} + \frac{1}{v} = \frac{1}{f}$

Even if we were not sure whether the image was real or virtual we can still use this formula. If v then turns out to be positive, the image is real; if v turns out to be negative the image is virtual.

$$\frac{1}{u} + \frac{1}{v} = \frac{1}{f} \Rightarrow \frac{1}{40} + \frac{1}{v} = \frac{1}{30} \Rightarrow \frac{1}{v} = \frac{1}{30} - \frac{1}{40} \Rightarrow \frac{1}{v} = \frac{4-3}{120} = \frac{1}{120} \Rightarrow v = 120$$

i.e. Image is real and 120 cm at the other side of the lens to the object.

Problem 2: An object is placed 10 cm in front of a convex lens of focal length 20 cm. Find the position, nature and magnification of the image. If the object is 3 cm high, what is the height of the image?

Solution: The object is inside the focus therefore the image is virtual and the formula to use is $\frac{1}{u} - \frac{1}{v} = \frac{1}{f}$. However, let us assume we do not know whether the image is real or virtual and use the formula $\frac{1}{u} + \frac{1}{v} = \frac{1}{f}$

$$\frac{1}{u} + \frac{1}{v} = \frac{1}{f} \Rightarrow \frac{1}{10} + \frac{1}{v} = \frac{1}{20} \Rightarrow \frac{1}{v} = \frac{1}{20} - \frac{1}{10} \Rightarrow \frac{1}{v} = \frac{1-2}{20} = -\frac{1}{20} \Rightarrow v = -20$$

A negative value for v tells us the image is virtual. It is 20 cm from the lens at the same side of the lens as the object.

Magnification $m = \frac{v}{u} = \frac{20}{10} = 2$

$\Rightarrow$ Image is two times as high as the object, therefore height of image = (2)(3) = 6 cm high

Problem 3: An image which is four times the size of the object is formed in a convex lens of focal length 30 cm. Where must an object be placed if:

(i) the image is real,

(ii) the image is virtual?

Solution: Magnification $m = \frac{v}{u} = 4 \Rightarrow v = 4u$

Real image

$$\frac{1}{u} + \frac{1}{v} = \frac{1}{f} \Rightarrow \frac{1}{u} + \frac{1}{4u} = \frac{1}{f} \Rightarrow \frac{5}{4u} = \frac{1}{30} \Rightarrow u = 37.5 \text{ cm}$$

Virtual image

$$\frac{1}{u} - \frac{1}{v} = \frac{1}{f} \Rightarrow \frac{1}{u} - \frac{1}{4u} = \frac{1}{f} \Rightarrow \frac{3}{4u} = \frac{1}{30} \Rightarrow u = 22.5 \text{ cm}$$

For the required real image the object must be placed 37.5 cm from the lens and for the virtual image the object must be 22.5 cm from the lens.

EXERCISE 5.1

1. An object is 40 cm from a convex lens and the image formed is 25 cm from the lens. What is the focal length of the lens. Is the image real or virtual?
2. An object is placed 100 cm from a convex lens of focal length 30 cm. If the object is 4 cm high, find the position, nature and height of the image.
3. An object is placed 20 cm from a convex lens of focal length 30 cm. If the object is 2 cm high, find the position, nature and height of the image.
4. An image twice the height of the object is formed in a convex lens of focal length 50 cm. Find the position of the object if:
   (i) the image is real,
   (ii) the image is virtual.
5. A convex lens has a focal length of 20 cm. Find two positions at which the object may be placed so that the image formed is twice the size of the object.
6. A square picture of side 8 cm is projected on a square screen of side 3 m by means of a lens, so that a well defined image of the picture just fills the whole screen. If the screen is at a distance of 10 m from the lens find the focal length of the lens.
7. It is desired to use a lens to form a real image, the length of which will be 15 times the length of the object. If the distance from the lens to the image is 4 m, what type of lens must be used and what is its focal length?
8. An object and a screen are fixed at a distance of 80 cm apart and a convex lens forms a real image of the object on the screen. When the lens is moved along its axis a distance of 16 cm, a real image of the object is again formed on the screen. Find the focal length of the lens and the magnification produced in each case.
9. FIG. 5.11 shows a ray box R placed at a fixed distance from a screen S. A convex lens is placed between R and S. Explain why there will in general be two positions of the lens for which a sharp image of the opening of the ray box will be formed on the screen. One of these positions is shown in FIG. 5.11. Using the values given on the diagram find:
   (i) the distance between the two positions of the lens for which a sharp image is formed on the screen,
   (ii) the focal length of the lens.

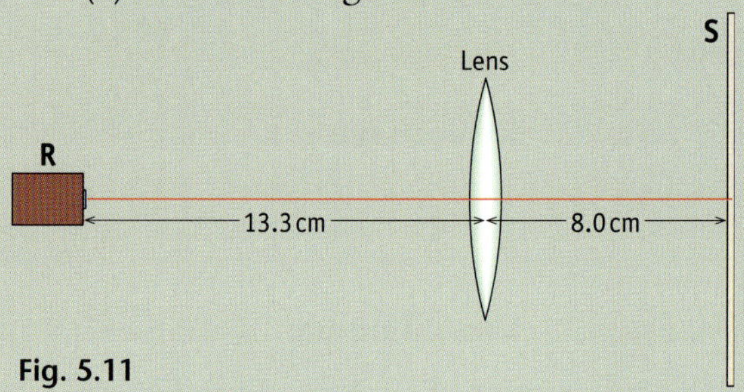

Fig. 5.11

EXPERIMENT LIGHT 4

To Measure the Focal Length of a Converging (Convex) Lens.

Summary of Method

In this experiment you will place an illuminated cross-threads as object in front of a convex lens and locate the image on a screen. You will measure the distance u from the cross-threads to the centre of the lens and the distance v from the image on the screen to the centre of the lens.

Using the formula: $\frac{1}{u} + \frac{1}{v} = \frac{1}{f}$ the focal length f of the lens can be found.

Equipment Needed

- A convex lens with stand
- A retort stand and clamp
- A ray box
- A sheet of white cardboard and a metre stick
- An illuminated object. This can be made by cutting a small hole (about 1 cm in diameter) in a sheet of cardboard and sticking two pieces of thread across it. It is illuminated by the ray-box.

Method

1. Find an approximate value for the focal length by focusing an image of a distant object (such as a window several metres away) on a sheet of paper. The distance from the image to the lens is its focal length approximately. Measure and record this distance.
2. Set up the equipment as shown in FIG. 5.12. The illuminated object should be a distance from the lens greater than the rough focal length found in Step 1.
3. Adjust the position of the cardboard screen until the image of the cross-threads is in sharpest focus and most clearly visible on it.
4. With a metre stick, measure the distance *u* from the cross-threads from the centre of the lens and then measure the distance *v* from the real image on the cardboard screen to the centre of the lens. Record these values.
5. Change the value of *u* and repeat steps 3 and 4. Do this at least four more times. Some values of *u* should be chosen to give a diminished image and some to give a magnified one.

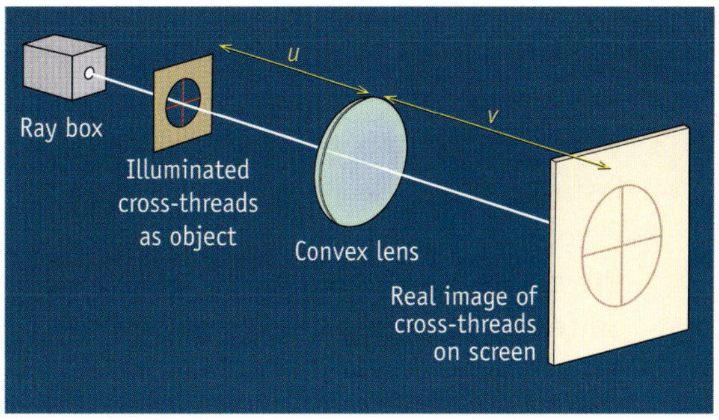

Fig. 5.12

Experimental Data and Results:

Approximate focal length found by focusing image of distant object on screen =

u/cm	v/cm	$\frac{1}{u}$	$\frac{1}{v}$	$\frac{1}{f} = \frac{1}{u} + \frac{1}{v}$	f/cm

Average value of focal length f =

Handling the Data

Method 1

Complete the Table and calculate the average value of *f*. This is the focal length of the lens.

Method 2

1. On graph paper, plot a graph of values of $1/u$ against corresponding values of $1/v$. Draw the straight line that best fits the points and produce it to intersect both axes.
2. Read from the graph the value of $1/u$ where the graph cuts the $1/u$ axis and the value of $1/v$ where it cuts the $1/v$ axis. Get the average of these two values ($= 1/f$). Find its reciprocal. It should agree closely with the value obtained by method 1.

Sources of Error

Errors arise when:

(i) measuring the distance between the optic centre of the lens and the cross-threads,

(ii) determining when the image of the cross-threads is in clearest focus.

Obviously there is a larger error in *v* than in *u* since *u* involves only the error in using the metre stick.

Questions

1. If you could not find any clear image of the cross-threads on the screen, what is likely to be the problem? How would you rectify it?
2. Why is the value for *u* obtained more accurately than the value for *v*?
3. What would happen if the object were placed inside the focus?
4. List two precautions that should be taken in measuring *u*.
5. Errors occur both in locating the position of the image and measuring its distance from the optic centre.
   State one precaution in each case you could take to reduce these errors.

---

1. A ray which strikes the optic centre passes straight through the lens.
2. A ray which is heading for the focus and then strikes the lens emerges parallel to the principal axis.
3. A ray which strikes the lens travelling parallel to the axis is refracted as if it came from the focus.

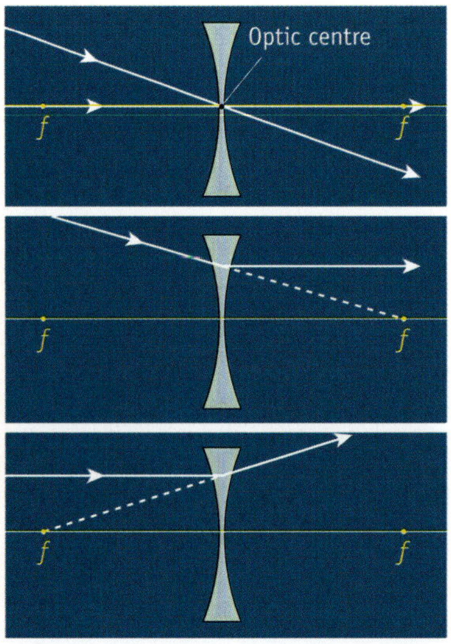

Fig. 5.13

FOR A CONCAVE LENS

1. The image is always **virtual and upright** (i.e. erect) and located at the same side of the lens as the object.
2. The image is always **diminished**. The nearer the object is to the lens the bigger the image is.

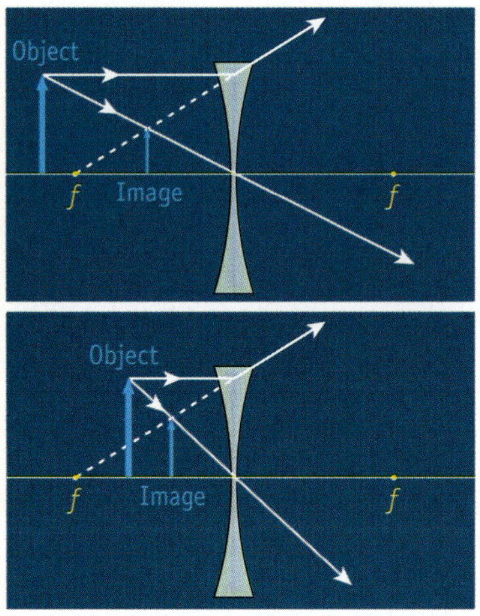

Fig. 5.14

How an Image is Formed by a Thin Concave Lens

When locating an image in a concave lens we shall use the three results given in Fig. 5.13. We do not need to prove these, but simply accept and **remember** them! Fig. 5.14 shows how an image is formed in a concave lens. This image is a **virtual image** since it is caused by the apparent intersection of rays. Fig. 5.14 also shows how the position and size of the image change when the position of the object changes.

Formula for a Concave Lens

Recall from page 46 that:

For a concave lens:

$$\frac{1}{u} - \frac{1}{v} = -\frac{1}{f}$$

As with the convex lens we also have:

$$\text{Magnification} = \frac{\text{Image distance}}{\text{Object distance}}$$

$$\text{i.e. } m = \frac{v}{u}$$

Problem 4: An object is placed 40 cm from a diverging lens of focal length 50 cm. Find the position and nature of the image.

Solution: Here $u = 40$ cm $\quad f = 50$ cm and v is unknown

$$\frac{1}{u} - \frac{1}{v} = -\frac{1}{f} \Rightarrow \frac{1}{40} - \frac{1}{v} = -\frac{1}{50} \Rightarrow -\frac{1}{v} = -\frac{1}{50} - \frac{1}{40}$$

$$\Rightarrow \frac{1}{v} = \frac{1}{50} + \frac{1}{40} \Rightarrow \frac{1}{v} = 0.045 \Rightarrow v = \frac{1}{0.045} = 22.2 \text{ cm}$$

The image is virtual and 22.2 cm from the lens.

Problem 5: A concave lens of focal length 10 cm produces an image which is half the size of the object. How far is the object from the lens? Find also the position and nature of the image.

Solution: Magnification $= \frac{1}{2} \Rightarrow \frac{v}{u} = \frac{1}{2} \Rightarrow v = \frac{u}{2}$

$$\frac{1}{u} - \frac{1}{v} = -\frac{1}{f} \Rightarrow \frac{1}{u} - \frac{1}{u/2} = -\frac{1}{10} \Rightarrow \frac{1}{u} - \frac{2}{u} = -\frac{1}{10}$$

$$\Rightarrow \frac{1-2}{u} = \frac{-1}{10} \Rightarrow \frac{-1}{u} = \frac{-1}{10} \Rightarrow u = 10$$

$\Rightarrow$ The object is 10 cm from the lens.

$v = \frac{u}{2} = \frac{10}{2} = 5$ cm. $\therefore$ the image is virtual and 5 cm from the lens.

EXERCISE 5.2

1. An object is placed 30 cm from a concave lens of focal length 20 cm. Find the position and nature of the image.
2. An object is placed 20 cm from a diverging lens of focal length 20 cm. If the object is 5 cm high, find the position, nature and height of the image.
3. An image one third the height of the object is formed in a concave lens of focal length 60 cm. Find the distance of the object and image from the lens.
4. A concave lens is used to form an image which is half the height of the object. Prove that whatever the focal length may be, the object distance will be equal in length to it.

Power of a Lens

A convex lens of short focal length has a greater converging effect on a parallel beam of light than a lens of long focal length (Fig. 5.15). We say that the **power of the lens** with the short focal length is larger. The same effect is seen for concave lenses. The shorter the focal length, the greater the diverging power. The power P of a lens is defined as follows:

$$\text{The Power of a lens} = \frac{1}{\text{Focal length}} \quad \text{i.e.} \quad \boldsymbol{P} = \frac{\mathbf{1}}{\boldsymbol{f}}$$

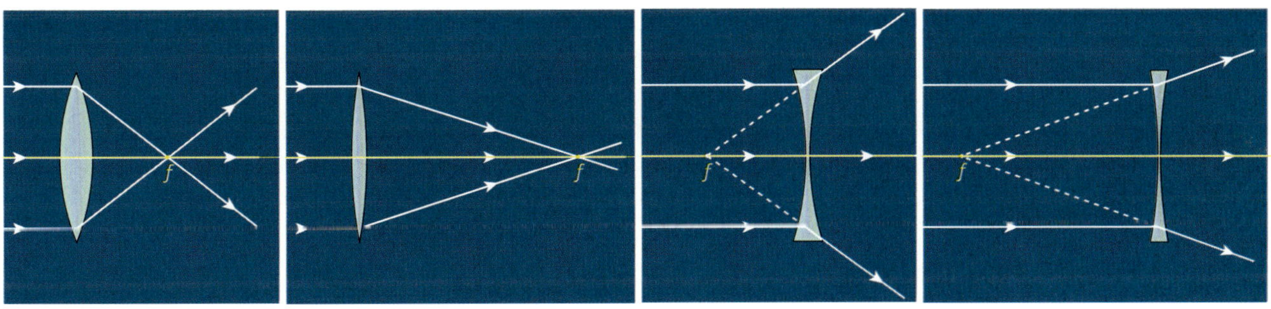

Fig. 5.15
The shorter the focal length the greater the power of the lens.

The power of a converging (convex) lens is taken as positive (+).
The power of a diverging lens (concave) is taken as negative (–).

UNIT OF POWER OF A LENS
The unit of power of a lens is the ***per metre***, i.e. the (metre)$^{-1}$ or m^{-1}

Problem 6: Find the power of:
(i) a convex lens of focal length 30 cm,
(ii) a concave lens of focal length 20 cm.

Solution:
(i) 30 cm = 0.3 m

$$P = \frac{1}{f} = \frac{1}{0.3} = 3.33 \text{ m}^{-1}$$

(ii) 20 cm = 0.2 m

$$P = \frac{1}{f} = \frac{-1}{0.2} = -5 \text{ m}^{-1}$$

Two Lenses in Contact

By experiment it is found that if two lenses are placed in contact, the power of the combination is the algebraic sum of the powers of the individual lenses i.e.:

If two lenses of power P_1 and P_2 are placed in contact, the power P of the combination is given by: $\boldsymbol{P = P_1 + P_2}$

It follows from this that:

If two lenses of focal length f_1 and f_2 are placed in contact the focal length f of the combination is given by:

$$\frac{1}{f} = \frac{1}{f_1} + \frac{1}{f_2}$$

Where f is given a '**+**' sign for a **convex** lens and a '**–**' sign for a **concave** lens.

Problem 7: Two convex lenses of power 5 m^{-1} and 8 m^{-1} are placed in contact. Find:
(i) the power of the combination,
(ii) the focal length of the combination.

Solution:
(i) $P = P_1 + P_2 = 5 + 8 = 13 \text{ m}^{-1}$

(ii) $P = \frac{1}{f} \Rightarrow f = \frac{1}{P} = \frac{1}{13} = 0.0769 \text{ m} = 7.69 \text{ cm}$

Problem 8: A concave lens of power – 0.06 m^{-1} and a convex lens of power 0.04 m^{-1} are placed in contact. Find the power of the combination. Does the combination behave as a concave or a convex lens?

Solution: $P = P_1 + P_2 = -0.06 + 0.04 = -0.02 \text{ m}^{-1}$

Since the power of the combination is negative, the combination behaves as a concave (diverging) lens.

EXERCISE 5.3

1. Find the power of:
   (i) a convex lens of focal length 40 cm,
   (ii) a concave lens of focal length 60 cm.
2. A lens has a power of 12 m^{-1}. Find its focal length.
3. A diverging lens has a power of –0.025 m^{-1}. Find its focal length.
4. Two convex lenses of power 6 m^{-1} and 10 m^{-1} are placed in contact. Find:
   (i) the power of the combination,
   (ii) the focal length of the combination.
5. Two concave lenses of power 4 m^{-1} and 8 m^{-1} are placed in contact. Find:
   (i) the power of the combination,
   (ii) the focal length of the combination.
6. A concave lens of power –0.02 m^{-1} and a convex lens of power 0.05 m^{-1} are placed in contact. Find the power of the combination. Does the combination behave as a concave or a convex lens?
7. An object of height 2 cm is placed 20 cm from a lens combination consisting of a convex lens of focal length 10 cm and a concave lens of focal length 15 cm. Find the position, nature and height of the image.
8. A combination of two convex lenses has a focal length of 20 cm. If the focal length of one of the lenses is 60 cm, what is the focal length of the other?
9. A combination of convex and a concave lenses behaves as a convex lens of focal length 40 cm. If the focal length of the concave lens is 40 cm, what is the focal length of the convex lens?
10. A combination of a convex and a concave lens behaves as a concave lens of focal length 40 cm. If the focal length of the convex lens is 20 cm, what is the focal length of the concave lens?

THE EYE

FIG. 5.16 shows a simplified diagram of the eye. When light from an object enters your eye it is brought to a focus on or near the back of the eye. The result is that you see the object.

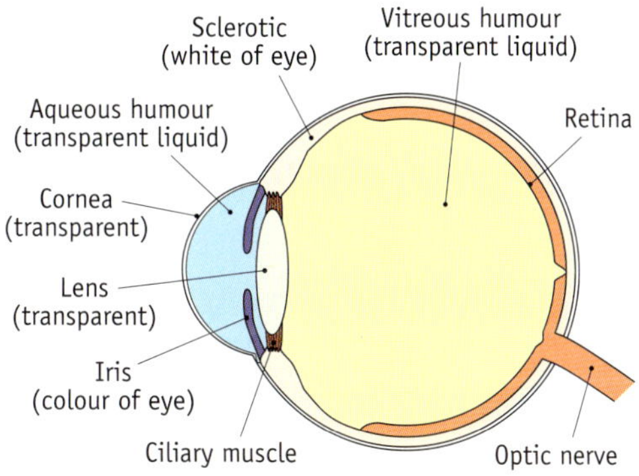

Fig. 5.16

CONTROLLING THE LIGHT ENTERING THE EYE

The **iris** controls the amount of light entering the eye by making the size of the hole in its centre – called the **pupil** – larger or smaller. In bright light the pupil is small, not allowing too much light into the eye. In dim light it is large to allow as much light as possible into the eye. The colour of the iris is the colour of your eye.

THE FOCUSING SYSTEM OF THE EYE

The **cornea** and the **crystalline lens** together form a converging system. The jelly-like liquids, the **aqueous humour** and the **vitreous humour**, also form part of this focusing system. Light from the object being viewed is focused by this system onto the back of the eye.

THE IMAGE FORMED ON THE RETINA

The **retina** is a light-sensitive screen at the back of the eye. It consists of many nerve endings which are sensitive to light. When light strikes the retina, electrical messages are sent to the brain through the optic nerve. The result of this is sight. When a real image is brought into focus on the retina that object is seen clearly. The image on the retina is always upside down. However the brain 'sees' it the right way up. If the image is brought to focus in front of or behind the retina, the object is seen blurred.

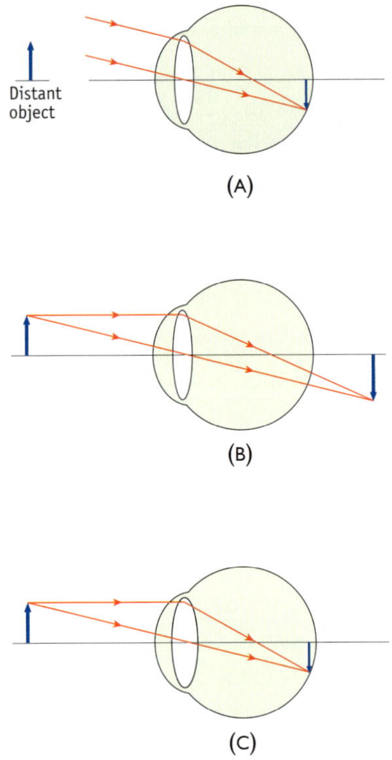

Fig. 5.17

POWER OF ACCOMMODATION

The **power of accommodation** of the eye is its ability to focus a real image of an object on the retina, whether the object is far from or near to the eye. It does this by changing the shape and hence the focal length of the lens. The **ciliary muscles** change the shape of the lens.

You can easily see this as follows: Place your hand about 30 cm in front of your eye and look carefully at one particular line on your hand. Without taking your gaze from that line notice that the room in the background is visible but out of focus – you cannot see it clearly. If, however, you decide to look carefully at the wall of the room it comes into focus almost immediately. As it does the hand in front of you becomes blurred. You can cause your eye to focus on your hand or on the wall but not on both at the same time.

If the object being viewed is a large distance from a normal eye, the image is brought to focus on the retina with the ciliary muscles relaxed. The lens is then at its thinnest and its focal length is greatest (Fig. 5.17(a)). If the object is then brought nearer the eye, the image would be brought into focus behind the eye and would be seen blurred (Fig. 5.17(b)).

However, the ciliary muscles contract fattening the lens and thus shortening its focal length, bringing the image back into focus on the retina (Fig. 5.17(c)).

THE LEAST DISTANCE OF DISTINCT VISION

The least distance of distinct vision is the smallest distance between an object and the eye for which that object can be seen clearly without eye strain. When something is this distance from the eye it is said to be at the near point. If an object is brought inside the near point the lens cannot shorten its focal length enough to bring its image onto the retina and the object appears blurred.

Simple Defects of Vision

SHORT SIGHT

A ***short-sighted*** person can see nearby objects clearly but cannot bring distant objects into focus.

In a short-sighted eye, the image of a distant object is formed in front of the retina with the ciliary muscles relaxed and the lens at its thinnest. **Short sight can be corrected with a concave lens** Fig. 5.18((a)).

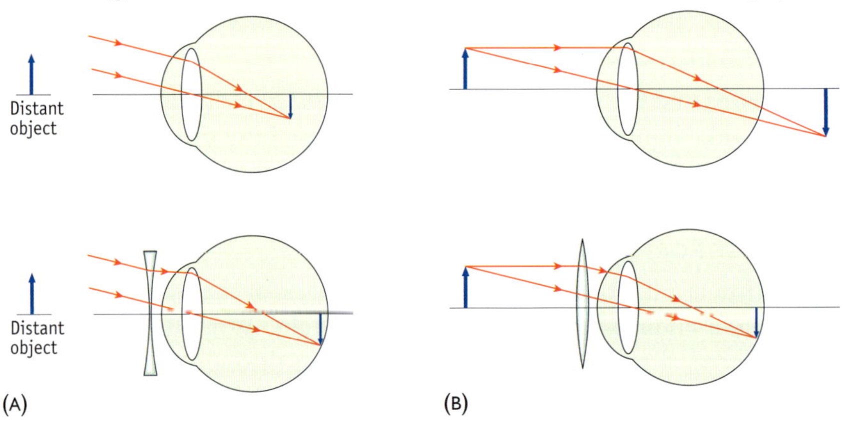

Short-sighted eye viewing a distant object without and with a correcting lens.

Long-sighted eye viewing a nearby object without and with a correcting lens.

Fig. 5.18

LONG SIGHT

A ***long-sighted*** person can see distant objects clearly but cannot bring nearby objects into focus.

In a long-sighted eye, the image of a nearby object is brought to a focus behind the retina even with the lens at its fattest. **Long sight can be corrected with a convex lens** (FIG. 5.18(B)).

CHAPTER CHECKLIST

- **Define** each of the following in the context of lenses: Converging lens; Diverging lens; Optic centre; Axis; Focus; Focal length; Real image; Virtual image; Magnification; Power of a lens.
- **Answer** these questions:

  What is short sight? What type of lens can be used to correct it?

  What is long sight? What type of lens can be used to correct it?
- **Draw** a ray diagram to locate the image formed in a convex or concave lens, no matter what the position of the object.
- **Recall** and use the formulae:

  $$\frac{1}{u} + \frac{1}{v} = \frac{1}{f}; \qquad m = \frac{v}{u}; \qquad P = \frac{1}{f};$$

  $$P = P_1 + P_2; \qquad \frac{1}{f} = \frac{1}{f_1} + \frac{1}{f_2}$$
- **Describe** and carry out an experiment to measure the focal length of a converging lens.
- **List** practical uses of lenses.
- **Describe** the optical structure of the eye.

CHAPTER 6

Speed, Displacement and Velocity

> **UNIT OF TIME**
> The unit of time is the ***second* (s)**.

TIME

The unit of time is the **second (s)**. Time is a **scalar** quantity*. Its symbol is ***t***. Small amounts of time are measured in milliseconds or microseconds. Larger time intervals are usually measured in hours, days or years.

When doing numerical problems in physics, time should always be expressed in seconds. In the laboratory, we measure time with a stop-watch, stop-clock or an electronic timer (FIG. 6.1).

> **NOTE**
> 1 millisecond (ms) = 10^{-3} s
> 1 microsecond (μs) = 10^{-6} s
> (Maths tables p. 5)

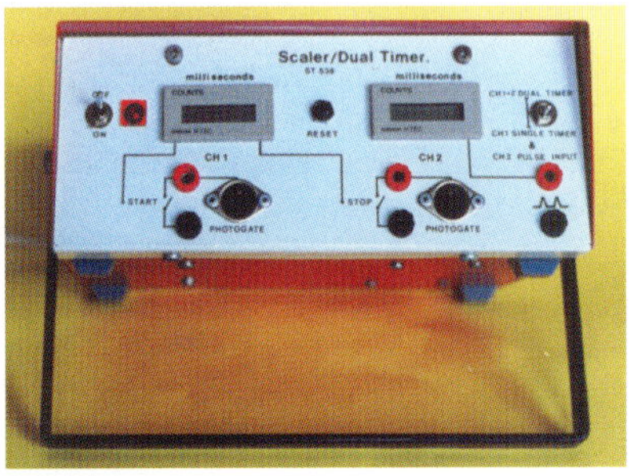

Fig. 6.1
With the scaler timer time can be measured to the nearest millisecond.

DISTANCE

In the laboratory, we use a metre stick to measure large distances. We use a vernier callipers or a micrometer (FIG. 6.2) to measure smaller ones.

Distance is a **scalar** quantity. Its symbol is ***s*** or ***d***. The unit of distance is the **metre (m)**. When doing numerical problems in physics, distances should always be expressed in metres.

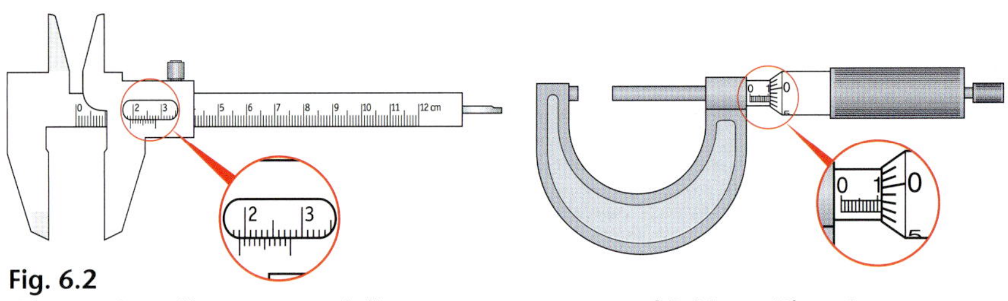

Fig. 6.2
The vernier callipers can read distance to an accuracy of 0.01 cm. The micrometer can read to an accuracy of 0.01 mm.

> **UNIT OF DISTANCE**
> The unit of distance is the ***metre* (m)**.

SPEED

Speed is the rate of change of distance with respect to time. The meaning of this definition will become clear as you read on.

> **SPEED**
> ***Speed*** is the rate of change of distance with respect to time.

AVERAGE SPEED

If an object travels a distance s in a time t, its average speed during that time is, by definition, equal to s/t.

$$\text{Average speed} = \frac{\text{Distance travelled}}{\text{Time taken}}$$

The symbol for speed is ***v*** or ***u***. Speed is a **scalar** quantity. Its unit is the **metre per second (m s⁻¹)**. This can be seen from the above as follows.

> **UNIT OF SPEED**
> The unit of speed is the ***metre per second*** **(m s^{-1} or m/s)**.

$$\text{Average speed} = \frac{\text{Distance}}{\text{Time}} \Rightarrow \text{Unit of speed} = \frac{\text{Unit of distance}}{\text{Unit of time}} = \frac{\text{metre}}{\text{second}} = \text{metre per second}$$

* A quantity that has a direction in space associated with it is called a vector quantity. A quantity that does not have a direction in space associated with it is called a scalar quantity. This is explained fully in Chapter 8.

Problem 1: A sprinter runs 200 metres in 22 seconds. Calculate her average speed over the 200 m.

Solution: $\text{Average speed} = \dfrac{\text{Distance}}{\text{Time}} = \dfrac{200}{22} = 9.09 \text{ m s}^{-1}$

Problem 2: A man walks with an average speed of 2 m s^{-1}. How long will it take him to walk 1 km?

Solution: $\text{Time taken} = \dfrac{\text{Distance travelled}}{\text{Average speed}} = \dfrac{1000}{2} = 500 \text{ seconds}$

CONSTANT SPEED

An object has a constant speed if it neither speeds up nor slows down. A more precise definition is: An object moves with **constant speed** if its average speed is always the same, no matter over which part of the journey it is measured. The words **steady** or **uniform** are sometimes used instead of constant.

Fig. 6.3
The fastest land speed over one mile is 1227.985 km/h (763.035 mph). This was achieved by Andy Green in Thrust SSC in October 1997. It is the first car to have travelled faster than sound.

VARYING SPEED

Most moving objects that we see around us have varying speed. Sometimes they are speeding up, at other times they are slowing down. For example, if you drop something, it moves downwards with its speed increasing.

SPEED AT AN INSTANT (i.e. INSTANTANEOUS SPEED)

Suppose you are travelling along in a car and at a certain instant the reading on the speedometer is 10 m s^{-1}. It does not mean that the car travelled 10 m in the last second. Neither does it mean that the car will travel 10 m in the next second. It does mean that the car would travel 10 m in the next second if the car's speed stopped changing at that instant and remained constant from then on. 10 m s^{-1} is called the car's **instantaneous speed** (Fig. 6.4). If the average speed is measured over a very small time interval or a very small distance, its value is the instantaneous speed to a high degree of accuracy.

Fig. 6.4
The speedometer on a car shows the car's speed at any instant.

Problem 3: A dog runs along a road at a constant speed of 3 m s^{-1}.
(i) How far will it travel in 10 s? (ii) How far will it travel in ¼ hour?

Solution: (i) Distance = Speed × Time = (3)(10) = 30 m

(ii) First convert hours to seconds:

¼ hour = 15 minutes = (15)(60) seconds = 900 s

Distance = Speed × Time = (3)(900) = 2700 m

Problem 4: A car has a steady speed of 63 km h^{-1}.
(i) How far does it travel in 12 seconds? (ii) How long does it take it to travel 400 m?

Solution: Convert km h^{-1} to m s^{-1}:

$63 \text{ km per hour} = 63\,000 \text{ m per hr} = \dfrac{63\,000}{(60)(60)} \text{ m s}^{-1} = 17.5 \text{ m s}^{-1}$

(i) Distance = Speed × Time = (17.5)(12) = 210 m

(ii) $\text{Time} = \dfrac{\text{Distance}}{\text{Speed}} = \dfrac{400}{17.5} = 22.9 \text{ seconds}$

EXERCISE 6.1

1. Express each of the following in seconds:
   (i) 20 ms (ii) 500 ms
   (iii) 4000 ms (iv) 1 μs
   (v) 50 μs (vi) ½ hour
   (vii) 2 days (viii) 1 year
2. How many days in 1 million seconds?
3. How many microseconds in a millisecond?
4. Convert each of the following to metres:
   (i) 6.25 km (ii) 1 cm
   (iii) 20 mm (iv) 4 nm
5. Find the average speed in metres per second of a car that travels:
   (i) 2000 m in 4 minutes,
   (ii) 100 m in 10 seconds.
6. A woman travels a distance of 200 km with an average speed of 25 m s^{-1}. How long does the journey take?
7. If the average speed of a plane over a 2000 km journey is 200 m s^{-1}, how long does the journey take?
8. A boy walks once around a circle of radius 30 m in 1 minute 30 seconds. What is the average speed for the journey?
9. A girl walks along a road at a constant speed of 2 m s^{-1}. How far will she travel in:
   (i) 25 s, (ii) ½ hour,
   (iii) 1 ms, (iv) *t* seconds?

DISPLACEMENT

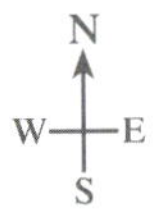

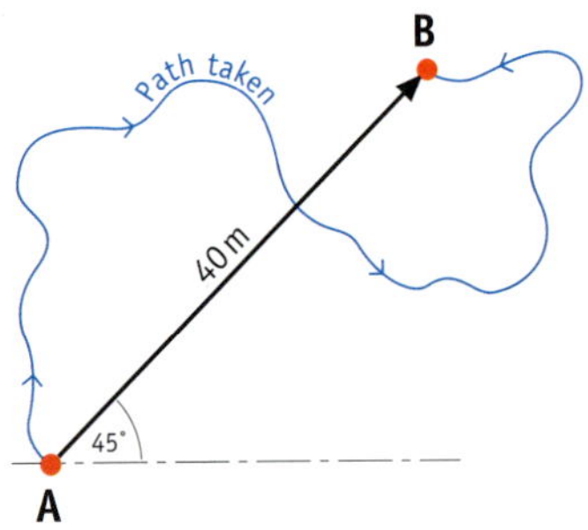

Fig. 6.5
The displacement of B from A is 40 m North East

Suppose a man walks from A to B by the path shown in FIG. 6.5. When he is at B, he is 40 metres North East of A. We say that his displacement from A is now 40 metres North East. We also say that he underwent a displacement of 40 metres North East in going from A to B. Note that the displacement undergone by the man here is different from the distance travelled by him since the distance is the length of the curved path he took. Displacement is defined as follows:

Displacement **is distance in a given direction.**

- The symbol for displacement is ***s***.
- The unit of displacement is the **metre (m)**.
- Displacement is a **vector** quantity since it has a direction associated with it (see Chapter 8).

VELOCITY

A quantity very like speed is velocity. Very simply, velocity is speed in a given direction. For example, if a car is moving Northwards at a speed of 20 m s^{-1} we say it has a velocity of 20 m s^{-1} North. The precise definition of velocity is as follows:

Velocity **is the rate of change of displacement with respect to time.**

UNIT OF VELOCITY
The unit of velocity is the ***metre per second*** **(m s^{-1} or m/s).**

- The usual symbol for velocity is ***v*** or ***u***.
- Velocity is a **vector** quantity.

AVERAGE VELOCITY

The average velocity of a moving object is defined as follows:

$$\text{Average velocity} = \frac{\text{Distance travelled in a given direction}}{\text{Time taken}}$$

or

$$\text{Average velocity} = \frac{\text{Displacement undergone}}{\text{Time taken}}$$

These definitions are clearly the same since distance travelled in a given direction is displacement.

Problem 5: In 12 seconds a girl travels from A to B by the path shown in Fig. 6.5 (page 58). The total distance travelled is 68 m. The overall displacement she undergoes is 40 m North East. Calculate: (i) her average speed (ii) her average velocity for the journey.

Solution:

(i) $$\text{Average speed} = \frac{\text{Distance travelled}}{\text{Time taken}} = \frac{68}{12} = 5.67\ \text{m s}^{-1}$$

(ii) $$\text{Average velocity} = \frac{\text{Displacement undergone}}{\text{Time taken}}$$

$$= \frac{\text{40 metres North East}}{\text{12 seconds}} = 3.33\ \text{m s}^{-1}\ \text{North East}$$

CONSTANT VELOCITY

If a body has constant velocity it does not speed up, slow down or change direction. Its average velocity is the same no matter over what time interval during the motion it is measured.

VARYING VELOCITY

If a moving object does not have constant velocity, it is said to have **varying velocity**. Fig. 6.6 shows a freely falling object near the Earth's surface. It moves in a straight line downwards with its speed increasing. Its velocity is thus varying. Fig. 6.7 shows a stone moving in a circular path with a constant speed. Its speed does not change but its velocity is changing because the direction in which it moves is always changing.

CONSTANT VELOCITY
If an object moves in a straight line and does not speed up or slow down it has a ***constant velocity***.

Just Like For Speed:
Average Velocity = ***Instantaneous Velocity*** (to a high degree of accuracy) provided the average velocity is measured over a very small time interval or a very small distance.

Fig. 6.6
As the bone falls its speed increases, thus its velocity is varying even though it moves in a straight line.

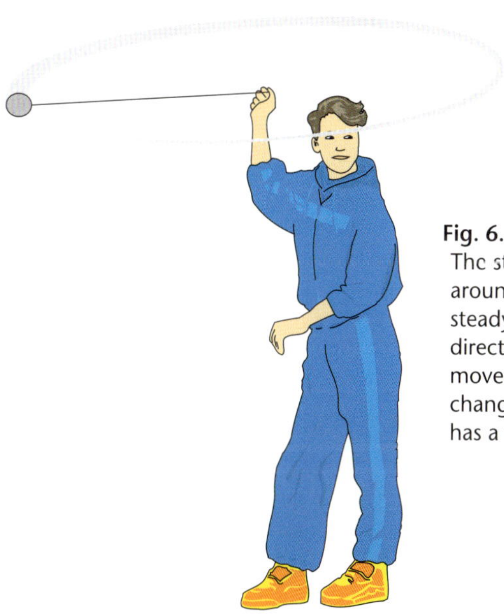

Fig. 6.7
The stone is moving around in a circle at steady speed. The direction in which it moves is continually changing. It therefore has a varying velocity.

EXERCISE 6.2

1. A particle undergoes a displacement of 2 km South East in 5 minutes. Calculate the average velocity of the particle in metres per second.
2. A car moves South along a straight road with a constant velocity of 30 m s^{-1}. Find the displacement it undergoes in 10 s.
3. A car has a constant velocity of 10 m s^{-1} East. How far East does it travel in:
   (i) 1 s, (ii) 10 s,
   (iii) t seconds, (iv) 2 μs?
4. In FIG. 6.8 what is the displacement of
   (i) A from B, (ii) B from A,
   (iii) B from C, (iv) C from B,
   (v) A from C, (vi) C from A?

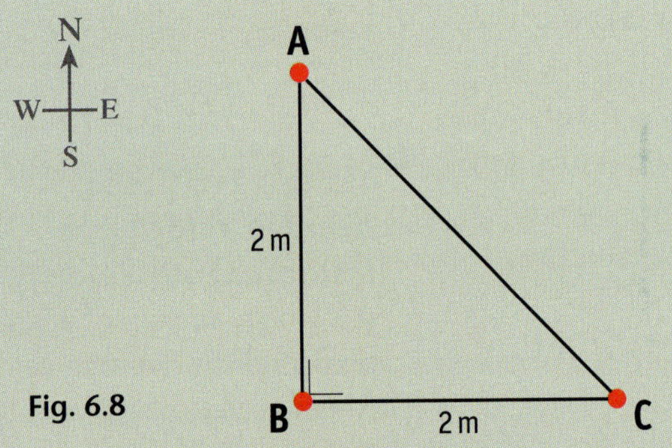

Fig. 6.8

5. A man walks from A to B by the circular path shown in FIG. 6.9. What distance did he travel? What is his displacement from A when the journey is completed?

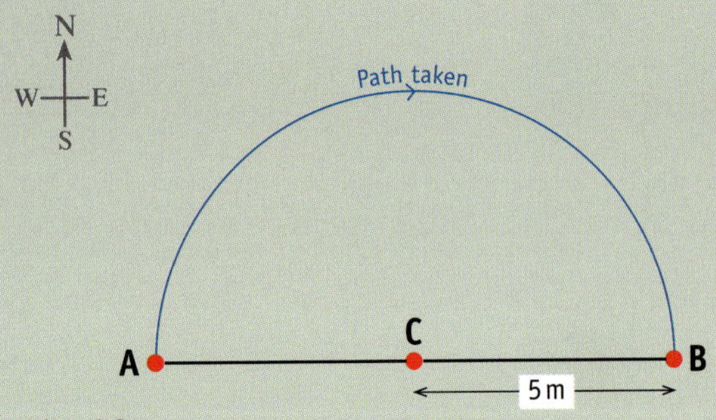

Fig. 6.9

If the time for the journey was 10 seconds find:
   (i) his average speed for the journey,
   (ii) his average velocity for the journey.

6. 'The speed of a car at a given instant is 30 m s^{-1}'. Explain exactly what this statement means.
7. Can a particle have a constant velocity but a varying speed? Can a particle have a constant speed but a varying velocity? If the answer to either is yes, give an example.
8. Could a particle have an Eastward displacement from a point and a Westward velocity?
9. A particle undergoes a displacement of 3 m East followed by a displacement of 4 m North in a total time of 5 seconds. Calculate the overall displacement it undergoes. Calculate the average velocity of the particle during the 5 seconds.

MEASURING VELOCITY

To measure the velocity of an object moving in a straight line, measure the time taken for the object to move a known distance. Divide the distance by the time and the result is its average velocity.

MEASURING VELOCITY WITH A TICKER TIMER AND TICKER TAPE

A piece of ticker tape is a thin strip of paper on which a black dot appears if the tape is struck by a sharp object. The tape is attached to the moving object (e.g. a trolley) with a thumbtack. As the trolley moves it pulls the tape through the ticker timer (FIG. 6.11). The ticker timer operates on a 50 Hz a.c. source and a striker in it strikes the ticker tape 50 times per second. It produces a dot on the tape every one fiftieth of a second. From the tape, the distance travelled and the time taken can be measured. Thus the speed of the trolley can be calculated.

Problem 6: Find the velocity of the trolley that produced the ticker tape shown in Fig. 6.10. The trolley moved in a straight line.

Fig. 6.10
A piece of ticker tape from an object moving with constant velocity.

Solution: On the tape the dots are evenly spaced. This means the velocity is constant.

Using a metre stick, the distance from A to B = 15 cm = 0.15 m

Number of spaces between A and B = 10. Each space represents 1/50th of a second

$\Rightarrow$ Time taken to travel from A to B = 10 × (1/50) s = 0.2 s

$$\text{Velocity} = \frac{\text{Distance}}{\text{Time}} = \frac{0.15}{0.2} = 0.75 \text{ m s}^{-1}$$

EXPERIMENT — MECHANICS 1 PART 1

To Measure the Constant Velocity of a Trolley using a Ticker Timer and Ticker Tape.

Summary of Method

In this experiment you will cause a trolley to move along a runway at a constant velocity. From the evenly spaced dots produced on the ticker tape attached to it you will calculate its velocity.

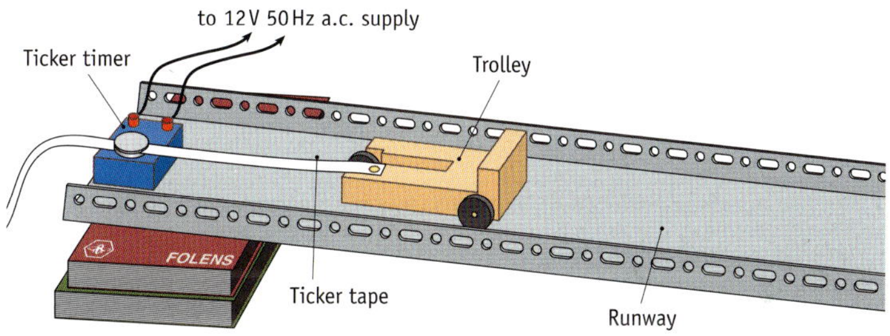

Fig. 6.11
Using a ticker timer and ticker tape to measure the constant velocity of a trolley.

Equipment Needed

- A trolley and a runway
- A ticker timer and ticker tape
- 12 volt, 50 Hz a.c. power supply (the voltage depends on the timer used)
- 2 connecting leads, a metre stick and a thumbtack or Sellotape.

Method

1. Dust the runway and trolley wheels to remove any grit or dirt as this can cause the velocity to vary.
2. Compensate the runway for friction, i.e. raise one end of the runway by resting it on a wooden block or a book (Fig. 6.11). Adjust the height until the trolley, if given a gentle push, will be seen to move down the runway at constant velocity. (The component of the trolley's weight down the runway now cancels the force of friction acting on the trolley.)
3. Place the timer and the trolley on the raised end of the runway. Pass a suitable length of tape through the timer and attach it to the trolley with Sellotape or a thumbtack. Ensure that the tape can run freely through the timer.
4. Switch on the timer and give the trolley a gentle push to get it moving. It should then move with constant velocity.
5. Stop the trolley, switch off the timer and remove the ticker tape. It should have a series of equally spaced dots printed on it like that shown in Fig. 6.10.

6. Ignoring the unevenly spaced dots at the start of the tape (which were produced when you were pushing the trolley), measure and record the distance (s) between two dots (A and B) with a metre stick. Be careful to avoid parallax error.
7. Count the number (*n*) of spaces between A and B (this is the number of fiftieths of a second that it took the trolley to move through distance (*s*). Record this.
8. Complete the Table calculating the time *t* and the velocity *v* from the formula: $v = s/t$.
9. Repeat the experiment with the trolley travelling at different velocities.

Table of Results

Distance	No. of spaces	Time taken	Velocity
s / m	n	t / s $(= n \times 1/50)$	v / m s^{-1} $(v = s / t)$

EXPERIMENTAL NOTES

- You could measure the distance between two consecutive dots on the tape and divide it by 1/50th of a second to calculate the velocity. However, this result would tend to be inaccurate since errors in measuring a small length will be a significant percentage of this length.
- You should also calculate the average velocity over the first 5 spaces from A and the average velocity over the last 5 spaces ending at B. Within experimental error these should be the same, verifying that the trolley was moving at constant velocity.

Questions

1. Why should you dust the runway and trolley wheels?
2. Why is the runway raised at one end?
3. How do you know when it is raised to the correct height?
4. If the dots on the tape are not evenly spaced what do we know about the velocity of the trolley. What should you do to remedy this?
5. List two precautions you should take in this experiment to ensure an accurate result.

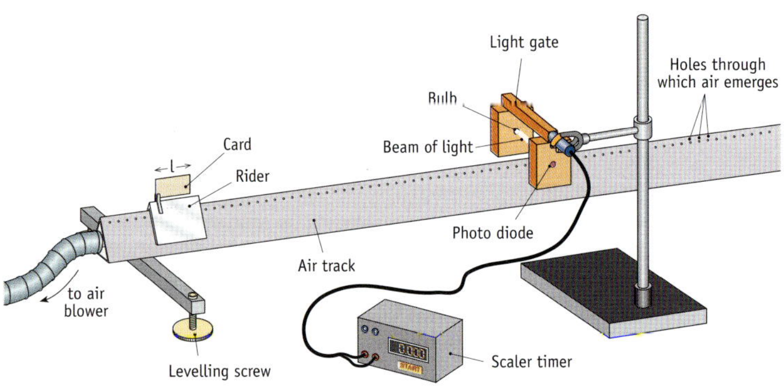

Fig. 6.12
A linear air track, a light gate and a scaler timer.

The Linear Air Track

A linear air track is a long rigid tube with a series of holes along both of its upper faces (Fig. 6.12). Air from a blower – like a vacuum cleaner in reverse – comes out of the holes and lifts a v-shaped metal rider just up off the track. The rider thus floats on a cushion of air like a Hovercraft. There is then almost no friction between the rider and the track. Give the rider a gentle push and it moves off readily. If the track is level it keeps going with constant velocity until it reaches the end of the track. The track may be levelled by means of levelling screws on the legs supporting it.

Measuring Velocity with a Light Gate and a Scaler Timer

A light gate and a scaler timer can also be used to measure velocity. The light gate consists of a bulb (or a light emitting diode) which shines a beam of light onto a photo diode (Fig. 6.12). When the beam of light is broken, i.e. when something stops light getting from the bulb to the photo diode, an electrical signal is sent to the scaler timer. The signal turns on the timer. When the interruption of the beam of light stops, the timer is turned off. Thus the time for which the beam was broken is known.

A piece of card of known length (say 4 cm) is mounted on the rider, and it is the card that interrupts the light beam. Thus the time for the rider to travel 4 cm is known and its velocity can be found. Since the time interval over which the velocity is measured is small, the measured velocity will be very near to the instantaneous velocity of the rider at any instant during the time interval.

Fig. 6.13
The time taken by each sprinter to finish the race can be measured with light beams and electronic timers.

EXPERIMENT **MECHANICS 1 PART 1**

To Measure the Constant Velocity of a Rider on a Linear Air Track Using a Light Gate and a Scaler Timer.

Summary of Method

In this experiment you will cause a metal rider to move along a linear air track at a constant velocity. By measuring directly the time taken for the rider to travel a known distance you will calculate its velocity.

Equipment Needed

- A linear air track, an air blower and a rider
- A scaler timer, a light gate and a metre stick
- A retort stand and a clamp (depending on the light gate used)

Method

1. Set up the air track on the bench and adjust the levelling screws until the track looks to be reasonably level. Connect on the air blower and place the rider on the track.
2. Turn on the blower and make further adjustments to the levelling screws until the rider will stay stopped at any place on the track. The track is now level. If the rider is now given a gentle push and released, it will move along the track at constant velocity.
3. Mount the light gate on a retort stand and connect it to the scaler timer (Fig. 6.12). Check with the manual or with your teacher if necessary. Turn on the timer and check that the light gate and timer are operating properly. The numbers on the clock of the timer should be changing when the light beam is broken. Place the timing gate about half way down the track at such a height that the card on the rider will interrupt the light beam. Reset the timer to zero.
4. Connect the piece of card onto the rider and make sure it is parallel to the direction in which the rider will move. Turn on the blower and give the rider a push. It will move along the track at a constant velocity. Record the value on the timer (t) in the Table.
5. With a metre stick, measure the length (l) of the card if it is not already known. Record this. Make sure to avoid the error of parallax when using the metre stick.
6. Complete the Table calculating the value of the velocity.
7. If a second timing gate is available you may find the velocity of the rider at another point on the track. See if the rider is in fact moving at constant velocity.
8. Repeat the experiment a number of times with the rider travelling at a different velocity each time.

Table of Results

Length of card	Transit time	Velocity
l / m	t / s	v / m s^{-1} ($v = l / t$)

Questions

1. Why should the track be level?
2. How would you be sure that the track was level?
3. Other than by watching it move, how would you know if the rider was not moving at constant velocity?
4. List three precautions you would take in this experiment to ensure an accurate result.

DISTANCE-TIME GRAPHS

A car moves along a road. Its distance from a point P on the road is plotted against time as in the distance-time graph in FIG. 6.14. From the graph you can see that as time increases, the distance of the car from P also increases. Since the graph passes through the origin, the car was at P when $t = 0$. Since the graph is a straight line, the speed of the car is constant (if you find the average speed from the graph over any time interval it will always be the same).

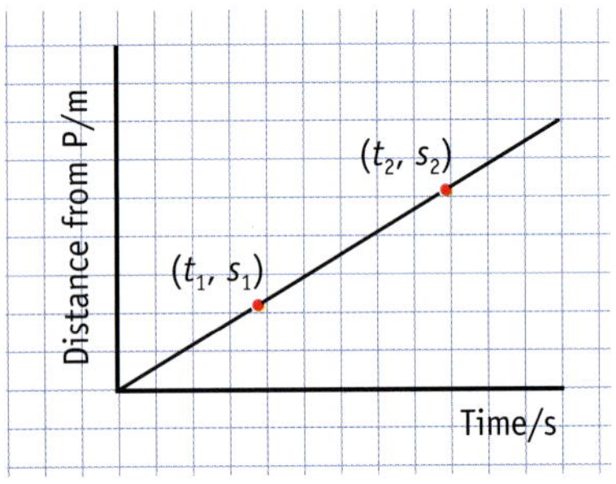

Fig. 6.14

A **distance-time graph** for an object moving with constant velocity is a straight line.

The ***slope*** of the graph is the *speed* of the object.

THE SLOPE OF A DISTANCE-TIME GRAPH IS THE SPEED

From FIG. 6.14: Slope of graph $= \dfrac{s_2 - s_1}{t_2 - t_1}$ (Slope $= \dfrac{y_2 - y_1}{x_2 - x_1}$)

$$= \frac{\text{Distance travelled in time } t_2 - t_1}{\text{Time taken}} = \text{Speed}$$

EXERCISE 6.3

1. A dog runs in a straight line down a road. Its distance from a signpost on the road is measured.
   FIG. 6.15 is a distance-time graph of the motion.
   (i) Describe the motion of the dog.
   (ii) When was the dog 20 m from the post?
   (iii) What was the distance of the dog from the post after 4.5 s?
   (iv) Find the slope of the graph (show your work).
   (v) What is the speed of the dog?

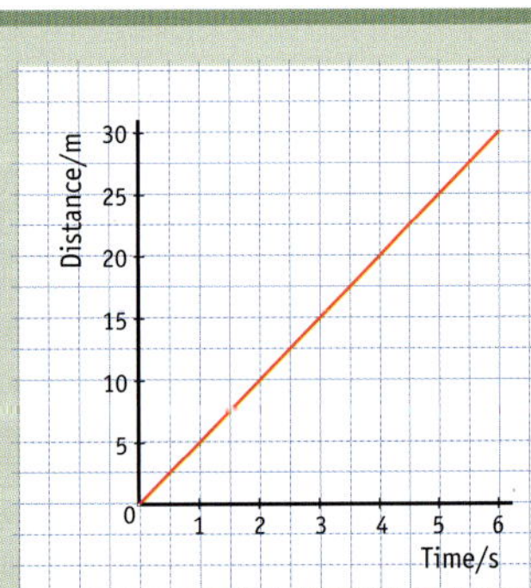

Fig. 6.15

2. A car is on a straight road. Its distance from a pothole on the road is measured. Each graph in FIG. 6.16 represents a different motion of the car. Describe the motion of the car in each case.

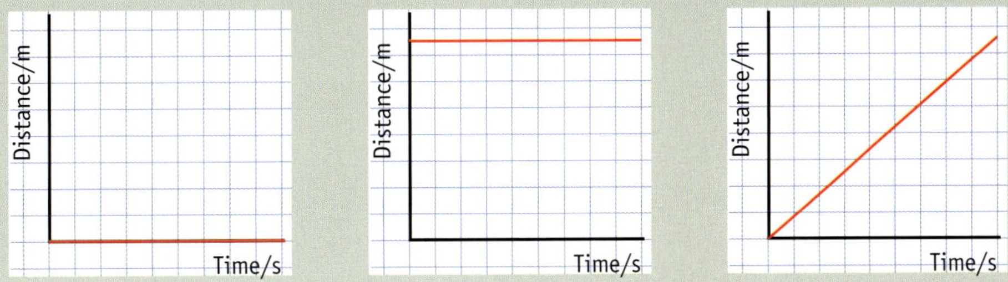

Fig. 6.16 (i) (ii) (iii)

3. FIG. 6.17 shows distance-time graphs for two cars moving along a straight road. The distance is from a fixed point P on the road. From the graphs determine:
   (i) the speed of each car,
   (ii) when the distance of each car from P is the same.

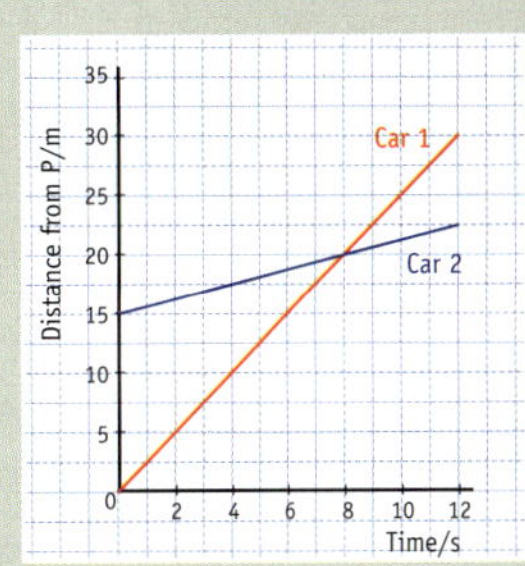

Fig. 6.17

CHAPTER CHECKLIST

- **Define** each of the following: Speed; Average speed; Constant speed; Displacement; Velocity; Average velocity; Constant velocity.
- **State** the unit of each of the following: Time; Distance; Speed; Displacement; Velocity.
- **Describe** and be able to carry out an experiment to measure velocity.
- **Recall** that: A metre stick, a vernier callipers and a micrometer can be used to measure distance; Distance and speed are scalar quantities whereas displacement and velocity are vector quantities; A distance-time graph for an object moving with constant velocity is a straight line and its slope is the speed.
- **Recall** and use the formulae:

$$\text{Average speed} = \frac{\text{Distance travelled}}{\text{Time taken}}$$

$$\text{Average velocity} = \frac{\text{Displacement undergone}}{\text{Time taken}}$$

CHAPTER 7

Acceleration

Fig. 7.1
This McLaren can go from 0 to 100 km h^{-1} in 4.8 seconds.

ACCELERATION

An object is said to be **accelerating** if its velocity is changing in any way. Thus an object is accelerating if it is:

- speeding up,
- slowing down,
- changing direction,
- speeding up as it changes direction,
- slowing down as it changes direction.

This use of the word 'acceleration' in physics is different from its use in everyday English. For example, to a physicist, a car slowing down or going around a corner at a steady speed is accelerating, but in everyday English we would say that a car speeding up is accelerating. Acceleration is defined as follows:

ACCELERATION
Acceleration is the rate of change of velocity with respect to time.

Acceleration is a **vector** quantity since it has a direction (see Chapter 8). The symbol for acceleration is $\boldsymbol{a}$. Average acceleration is defined as follows:

$$\text{Average acceleration} = \frac{\text{Change in velocity}}{\text{Time taken}} = \frac{\text{Final velocity} - \text{Initial velocity}}{\text{Time taken}}$$

If a body moves in a straight line and its velocity changes from u to v in t seconds its average acceleration is given by:

$$\text{Average acceleration} \quad a = \frac{v-u}{t}$$

UNIT OF ACCELERATION
The unit of acceleration is the ***metre per second squared*** **(m s^{-2}).**

UNIT OF ACCELERATION

$$\text{Since: Acceleration} = \frac{\text{Change in velocity}}{\text{Time taken}}$$

$$\text{Unit of acceleration} = \frac{\text{Unit of velocity}}{\text{Unit of time}}$$

$$= \frac{\text{metre per second}}{\text{second}} = \text{metre per second squared } (\mathbf{m/s^2 \text{ or } m\ s^{-2}})$$

Problem 1: The velocity of a car changes from 10 m s^{-1} East to 40 m s^{-1} East in 5 seconds. Calculate its average acceleration.

Solution: Average acceleration =

$$\frac{\text{Final velocity} - \text{Initial velocity}}{\text{Time taken}} = \frac{40 \text{ m s}^{-1} \text{ East} - 10 \text{ m s}^{-1} \text{ East}}{5 \text{ s}} = 6 \text{ m s}^{-2} \text{ East}$$

Thus this car is gaining speed at the rate of 6 m s^{-1} every second. Put another way, the car gains 6 m s^{-1} of extra speed every second.

Problem 2: The speed of a bicycle moving in a straight line decreases from 15 m s^{-1} to 3 m s^{-1} in 6 seconds. Find its average acceleration.

Solution: $$\text{Average acceleration} = \frac{\text{Final velocity} - \text{Initial velocity}}{\text{Time taken}} = \frac{3 - 15}{6} = -2 \text{ m s}^{-2}$$

Here the speed of the bike is decreasing. The acceleration has a negative sign indicating this. A slowing down or a negative acceleration is sometimes called a **deceleration**. While the bicycle is slowing down it loses on average 2 m s^{-1} of speed every second. It is losing speed at the rate of 2 m s^{-1} per second.

EXERCISE 7.1

1. The velocity of a car changes from 10 m s^{-1} North to 30 m s^{-1} North in 3 seconds. Find the average acceleration of the car.
2. The velocity of a car changes from 0 m s^{-1} to 10 m s^{-1} West in 2 seconds. Find the average acceleration of the car.
3. The velocity of a plane changes from 40 m s^{-1} to 25 m s^{-1} in the same direction in 10 seconds. Find the average acceleration of the plane.
4. A motorbike has a constant acceleration of 3 m s^{-2} in a certain direction. If the initial velocity of the bike is 0 m s^{-1}, find the velocity of the bike after:
   (i) 1 s, (ii) 4 s,
   (iii) 13.5 s, (iv) t seconds.
5. A runner starting from rest reaches a speed of 10 m s^{-1} in a time of 6 seconds. Calculate the average acceleration of the runner.
6. The velocity of a car changes from 5 m s^{-1} East to 9 m s^{-1} West in 2.5 seconds. Find the average acceleration of the car.

CONSTANT (UNIFORM) ACCELERATION

If an object is picking up speed or losing speed at the same steady rate, its acceleration is constant (also called uniform). Put another way: If the average acceleration of a body is always the same that body has **constant acceleration**. In this chapter we shall only deal with constant acceleration.

EQUATIONS OF MOTION FOR AN OBJECT MOVING IN A STRAIGHT LINE WITH CONSTANT ACCELERATION

When an object (with **initial velocity** u) moves in a straight line with constant **acceleration** a, its **displacement** s from its starting point, and its **velocity** v, change with **time** t. The following equations tell us how these quantities are related:

$$v = u + a t \qquad s = u t + \tfrac{1}{2} a t^2 \qquad v^2 = u^2 + 2 a s$$

These equations, which can be found on page 40 of your Maths tables, are used in problems 3–6 below. They are derived (proved) on page 69.

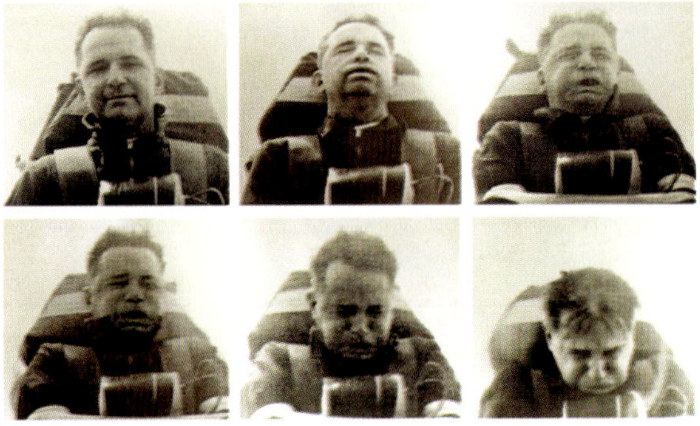

Fig. 7.2
Time-lapse images of John Stapp strapped in a sled seat during high g-force acceleration and deceleration testing. Stapp reached 632 mph in 5 seconds before being brought back to rest in just over a second. Despite initial blindness and collapsed lungs, he quickly recovered proving that it was possible to survive an ejection from a supersonic aircraft.

NOTE

When solving problems with these equations it is useful to:

- list the quantities you know and the ones which you need to calculate,
- write down the three equations,
- decide which of the three equations has only one unknown in it,
- substitute in the known values in this equation and solve for the unknown.

Problem 3: A car starting from rest has an acceleration of 4 m s^{-2}. Find:
(i) its velocity after 8 seconds, (ii) the distance it travels in 8 seconds.
After how many seconds is its velocity 100 m s^{-1}?

Solution: Here $u = 0$ (it starts from rest), $a = 4$, $t = 8$, $v = ?$, $s = ?$

(i) To find v:

Use $v = u + at$

$v = 0 + (4)(8)$

i.e. $v = 32 \text{ m s}^{-1}$

(ii) To find s:

Use $s = ut + \frac{1}{2}at^2$

$s = (0)(8) + \frac{1}{2}(4)(8)^2$

$\Rightarrow s = 0 + (2)(64)$

i.e. $s = 128$ m

For the last part of the question we have: $u = 0$, $a = 4$, $v = 100$ and t is unknown.
Use $v = u + at$:

$$100 = 0 + 4t \Rightarrow t = \frac{100}{4} = 25 \text{ s}$$

Problem 4: A car with constant acceleration travels a distance of 100 m while its velocity changes from 10 m s^{-1} to 25 m s^{-1}. Calculate its acceleration.

Solution: Here $u = 10$, $v = 25$, $s = 100$, $a = ?$ To find a, use $v^2 = u^2 + 2as$

$$v^2 = u^2 + 2as \Rightarrow (25)^2 = (10)^2 + 2a(100) \Rightarrow a = 2.625 \text{ m s}^{-2}$$

Problem 5: A car travelling with a speed of 10 m s^{-1} passes a pole on the roadside at a certain instant. It immediately accelerates with an acceleration of 2 m s^{-2}. How far is it from the pole when its speed is 30 m s^{-1}?

Solution: Here $u = 10$, $a = 2$, $v = 30$, $s = ?$
To find s use $v^2 = u^2 + 2as$

$30^2 = 10^2 + (2)(2)(s)$

$900 = 100 + 4s$

$800 = 4s$

$s = \frac{800}{4} = 200$ m

Problem 6: A train which has a constant acceleration travels a distance of 2 km in 50 s. If its initial velocity is 20 m s^{-1}, find its acceleration.

Solution: Here $u = 20$, $s = 2000$, $t = 50$, $a = ?$
To find a use $s = ut + (\frac{1}{2})at^2$

$2000 = (20)(50) + (\frac{1}{2})(a)(50)^2$

$2000 = 1000 + 1250a$

$1250a = 1000$

$\Rightarrow a = \frac{1000}{1250} = 0.8 \text{ m s}^{-2}$

DERIVING THE EQUATIONS OF MOTION

To prove: $v = u + at$

From the definition of acceleration we have:

$$a = \frac{v - u}{t} \Rightarrow at = v - u \Rightarrow v = u + at$$

To prove: $s = ut + \frac{1}{2}at^2$

It can be proved (and you may assume it to be true) that when the acceleration is constant, the average velocity is given by:

$$\frac{\text{Initial velocity} + \text{Final velocity}}{2} \quad \text{i.e. average velocity} = \frac{u + v}{2}$$

On page 58 you saw that: $\text{average velocity} = \dfrac{\text{displacement}}{\text{time}}$

$$\Rightarrow \text{displacement} = \text{average velocity} \times \text{time}$$

i.e. $s = \dfrac{(u + v)}{2} \times t$ Substituting $u + at$ for v in this equation gives:

$$s = \frac{(u + u + at)t}{2} = \frac{(2u + at)t}{2} \quad \text{i.e. } s = ut + (\tfrac{1}{2})at^2$$

To prove: $v^2 = u^2 + 2as$

Squaring both sides of $v = u + at$ gives: $v^2 = (u + at)^2$

$$\begin{aligned} \Rightarrow v^2 &= u^2 + 2uat + (at)^2 \\ &= u^2 + 2uat + a^2t^2 \\ &= u^2 + 2a\{ut + (\tfrac{1}{2})at^2\} \end{aligned}$$

But $ut + (\frac{1}{2})at^2 = s \quad \therefore \quad v^2 = u^2 + 2as$

EXERCISE 7.2

1. A car with an initial speed of 10 m s^{-1} is given an acceleration of 2 m s^{-2}. What is the speed of the car after 12 seconds? How far does it travel in the 12 seconds?
2. The speed of a lorry increases from 14 m s^{-1} to 30 m s^{-1} in 20 seconds. Find:
   (i) the acceleration,
   (ii) the distance travelled by the lorry in this time.
3. The speed of a bicycle increases uniformly from 2 m s^{-1} to 12 m s^{-1} as it travels a distance of 50 m. Find its acceleration and the time taken to travel this distance.
4. The speed of a racing car decreases uniformly from 60 m s^{-1} to 20 m s^{-1} in 4 seconds. Find:
   (i) the deceleration,
   (ii) the distance travelled while slowing down.
5. A car travelling with constant acceleration has a velocity of 6 m s^{-1} at a given instant. In the next minute it travels 3 km. Find its acceleration and the velocity after the minute.
6. A car travels a distance of 200 m as its speed changes from 30 m s^{-1} to 10 m s^{-1}. Find its deceleration.
7. A lorry travelling at 20 m s^{-1} is given a deceleration of 3 m s^{-2}. How long does it take the lorry to come to rest?

8. A plane with a velocity of 120 m s^{-1} in a certain direction is given a constant acceleration of 10 m s^{-2} in:

(i) the direction of the velocity,
(ii) the opposite direction to the velocity.

In each case find the velocity of the plane in magnitude and direction after 12 seconds.

9. A car with a velocity of 5 m s^{-1} in a certain direction is given a constant acceleration of 4 m s^{-2} in:

(i) the direction of the velocity,
(ii) the opposite direction to the velocity.

In each case find the velocity of the car in magnitude and direction after 5 seconds.

10. The rear of a bicycle passes a point P on a road and travels at a steady speed of 12 m s^{-1} down the road. At the same instant the front of a car starts from rest at P and moves in the same direction as the bicycle with an acceleration of 2 m s^{-2}. When and how far from P does the front of the car catch up with the rear of the bicycle?

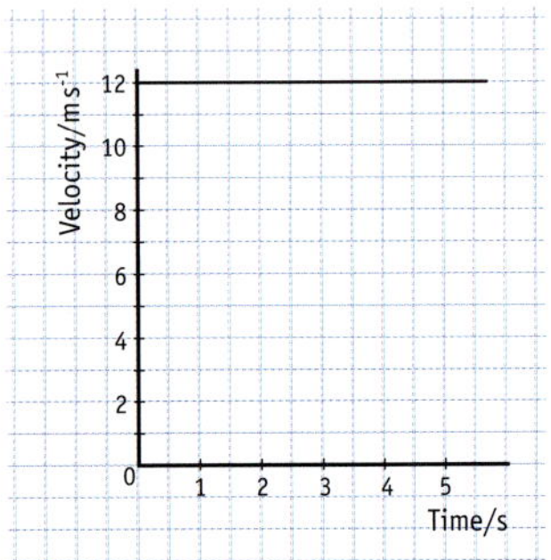

Fig. 7.3

Velocity-Time Graphs

A graph showing the magnitude of the velocity of an object plotted against time is called a **velocity-time graph**. FIG. 7.3 shows the velocity-time graph of a body moving with a constant velocity of 12 m s^{-1}.

FIG. 7.4 shows a velocity-time graph of a body which starts from rest and moves with a constant acceleration of 2 m s^{-2}. The graph is a straight line since the velocity increases by 2 m s^{-1} every second.

THE SLOPE OF A VELOCITY-TIME GRAPH IS THE ACCELERATION

In FIG. 7.5 slope of graph $= \dfrac{v_2 - v_1}{t_2 - t_1}$ (Slope $= \dfrac{y_2 - y_1}{x_2 - x_1}$)

$= \dfrac{\text{Change in velocity in time } t_2 - t_1}{\text{time taken}} =$ Acceleration of body

In FIG. 7.4 the slope of the graph is 2. This is also the acceleration of the body.

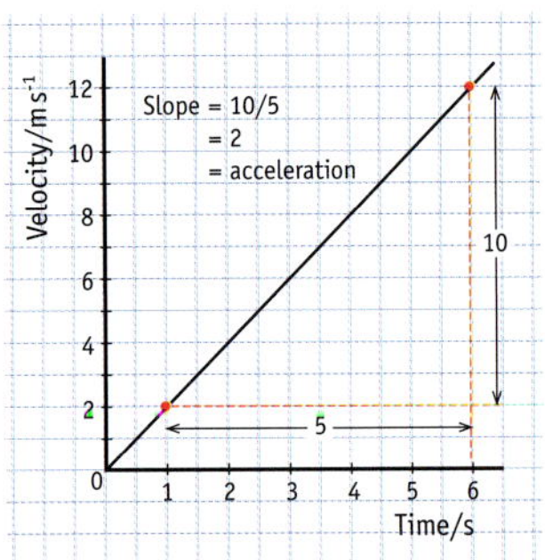

Fig. 7.4

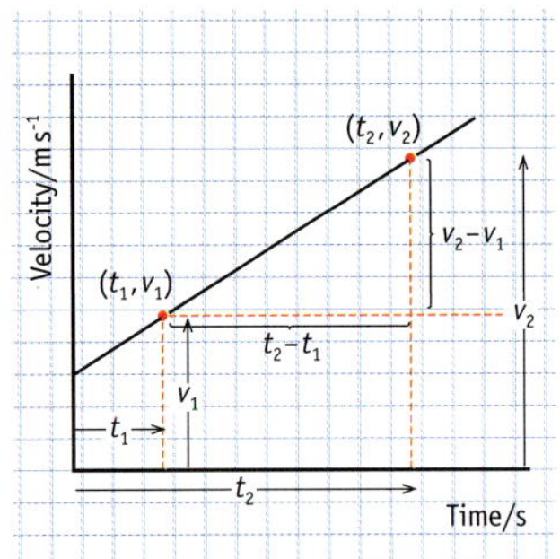

Fig. 7.5

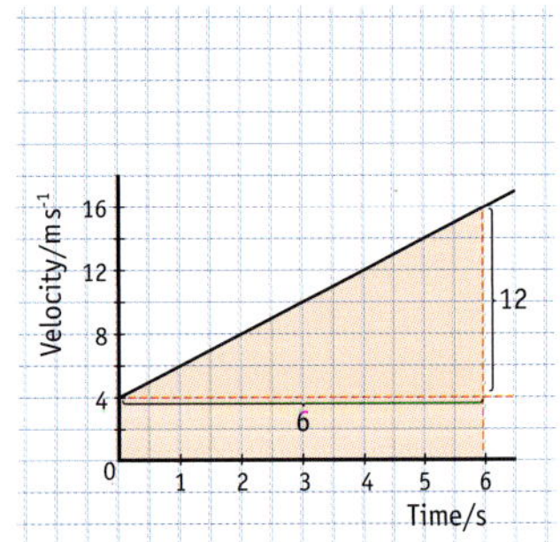

Fig. 7.6

A ***velocity-time graph*** of a body moving with constant acceleration is a straight line.

The ***slope*** is the ***acceleration***.

The ***area*** under the graph is the ***distance travelled***.

THE AREA UNDER A VELOCITY-TIME GRAPH IS THE DISTANCE TRAVELLED

No matter how complicated the motion, the area under a velocity-time graph is the distance travelled. In the case of motion in a straight line with constant acceleration, it is easy to see why this is so. FIG. 7.6 shows the graph of a body with initial velocity 4 m s^{-1} and an acceleration of 2 m s^{-2}.

Area under graph from $t = 0$ to $t = 6$	Distance travelled in 6 seconds
= Area of ▭ + Area of ◺	$s = ut + \tfrac{1}{2}at^2$
$= (4)(6) + \tfrac{1}{2}(6)(16 - 4)$	$= (4)(6) + \tfrac{1}{2}(2)(6)^2$
$= 24 + 36 = 60$	$= 24 + 36 = 60$ m

This result is always true and you can assume it to be so without proof.

EXERCISE 7.3

1. FIG. 7.7 shows a number of velocity-time graphs. In each case describe the motion of the moving object.

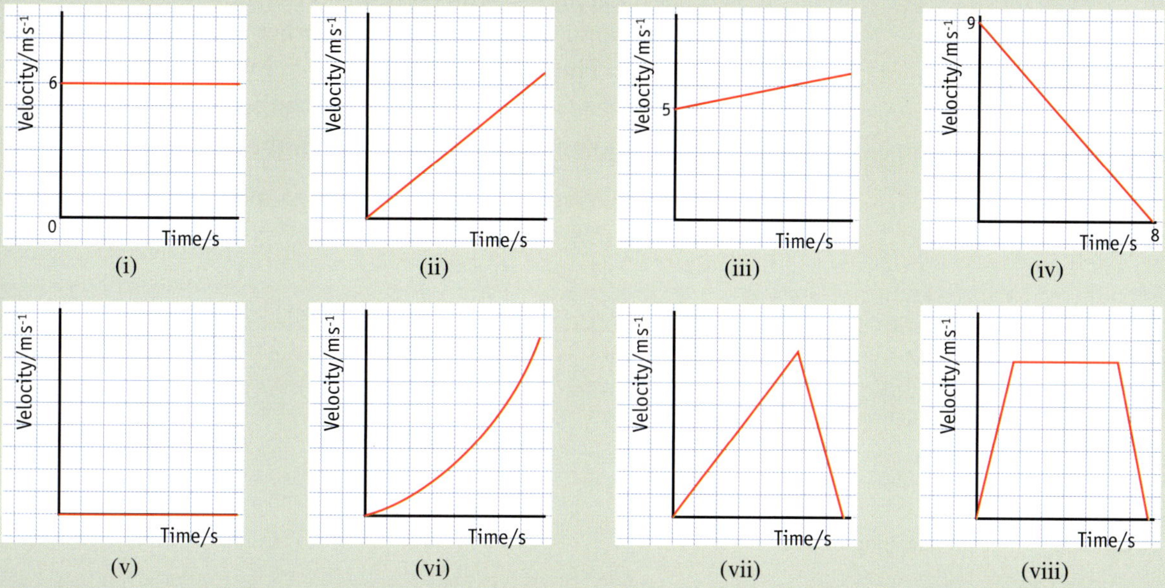

Fig. 7.7

2. A lorry is travelling in a particular direction along a road. FIG. 7.8 is a graph of its velocity plotted against time. From the graph find the:
   (i) velocity of the lorry after 4.5 s,
   (ii) time taken for the lorry to reach a velocity of 7.0 m s^{-1},
   (iii) acceleration of the lorry,
   (iv) distance travelled by the lorry in 5 seconds.

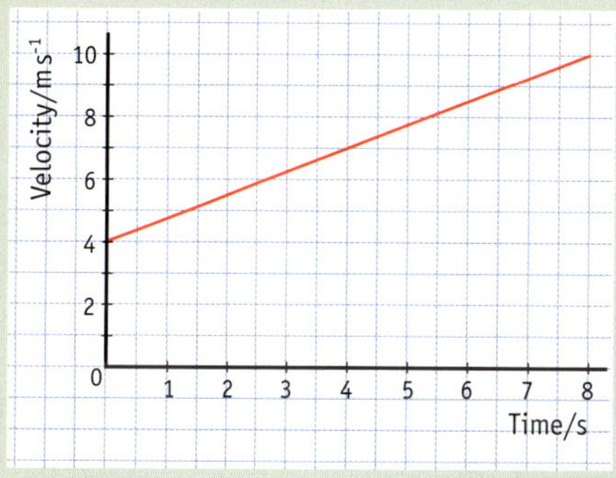

Fig. 7.8

3. In a laboratory, the velocity of an object was measured at various times and the following table was drawn up:

Velocity / m s^{-1}	0.84	1.58	2.32	3.06	3.80	4.54	5.28
Time / s	0	1.5	3.0	4.5	6.0	7.5	9.0

   (i) Draw a velocity-time graph for the motion.
   (ii) From the graph, calculate the acceleration of the object.
   (iii) From the graph, find the distance travelled by the object in the first 5.25 s.
   (iv) From the graph, find the distance travelled in the first 4 s of motion.

Measuring Constant Acceleration using a Ticker timer and Ticker tape

A piece of ticker tape is passed through a ticker timer and attached to a trolley. If the trolley then moves with constant acceleration, the dots produced on the tape will be found to be progressively further apart as one moves along the tape (FIG. 7.9). This is because the trolley moves a different distance during every fiftieth of a second as the velocity changes.

The acceleration may be calculated from the tape by using $a = \frac{v-u}{t}$

Measure the average velocity over say two spaces near the start of the tape to find u (FIG. 7.9).

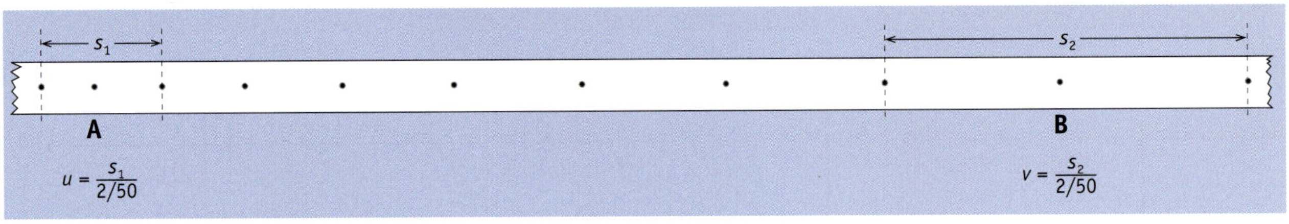

Fig. 7.9
Ticker tape from an accelerating object.

Measure the average velocity over say two spaces near the end of the tape to find v.

With negligible error, t is the time taken to go from A to B in FIG. 7.9. This can be read from the tape. Thus the acceleration can be calculated using: $a = (v - u)/t$

Problem 7: Find the acceleration of the ticker tape in FIG. 7.9.

Solution: $s_1 = 1.6$ cm $= 0.016$ m; $s_2 = 4.8$ cm $= 0.048$ m. Each distance is travelled in $\frac{2}{50}$ s $= 0.04$ s.

Time (t) for speed to change from u to v = Time taken to go from A to B = $8(\frac{1}{50}) = 0.16$ s.

$$u = \frac{s_1}{0.04} = \frac{0.016}{0.04} = 0.4 \text{ m s}^{-1}$$

$$v = \frac{s_2}{0.04} = \frac{0.048}{0.04} = 1.2 \text{ m s}^{-1}$$

$$a = \frac{v-u}{t} = \frac{(1.2-0.4)}{0.16}$$

$\Rightarrow$ Acceleration $a = 5 \text{ m s}^{-2}$

Measuring Acceleration Using Two Timing Gates and a Scaler Timer

Consider FIG. 7.11 on the next page. Using the formula:

$$a = \frac{v^2 - u^2}{2s} \quad (\text{From } v^2 = u^2 + 2as)$$

u is measured with the first timing gate. v is measured with the second timing gate. s is the distance between the two light beams and is measured with a metre stick. Putting these measured values in the formula the acceleration a can be calculated.

EXERCISE 7.4

1. Find the acceleration recorded on the ticker tape in FIG. 7.10.

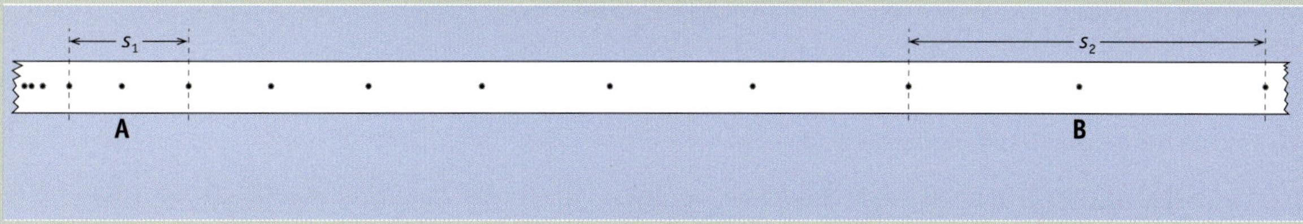

Fig. 7.10

2. A card of length 4 cm was attached to a rider on a linear air track as in FIG. 7.11 and the rider was allowed to accelerate along the track. The time for which the first beam was interrupted by the card was 0.1333 s and the time for which the second beam was interrupted was 0.0167 s. The distance between the light beams was 1.4 m. Calculate the acceleration of the rider.

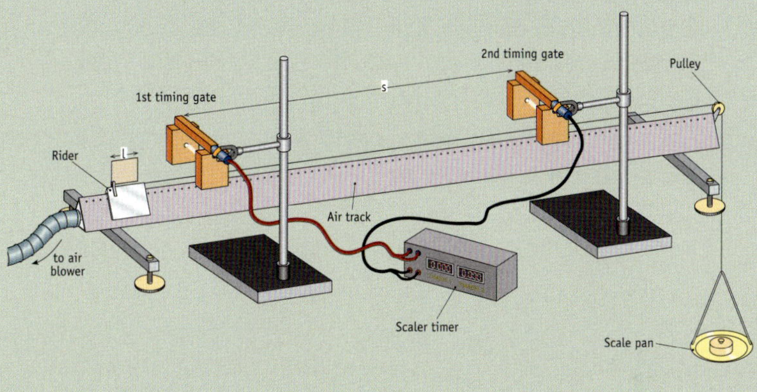

Fig. 7.11

EXPERIMENT — MECHANICS 1 PART 2

To Measure the Constant Acceleration of a Trolley Using a Ticker Timer and Ticker Tape.

Summary of Method

In this experiment you will accelerate a trolley along a runway by raising one end of the runway. From the unevenly spaced dots produced on the ticker tape attached to it, you will calculate its acceleration.

Equipment Needed

- A trolley and a runway
- A ticker timer and ticker tape
- 12V, 50 Hz a.c. power supply (depending on the ticker timer)
- Connecting leads, metre stick, thumbtack or sticky tape

Method

1. Dust the runway and trolley wheels to remove any grit or dirt as this can cause the acceleration to vary.
2. Raise one end of the runway and rest it on blocks or books. Adjust the height of the runway so that the trolley will accelerate down the runway if released.
3. Set up the ticker timer and trolley at the high end of the runway. Pass the tape through the timer and attach it to the trolley.
4. Switch on the timer and allow the trolley to run down the runway, pulling the ticker tape with it.
5. Stop the trolley at the end of the runway, switch off the timer and remove the tape. The tape should look something like that in FIG. 7.12 on the next page.

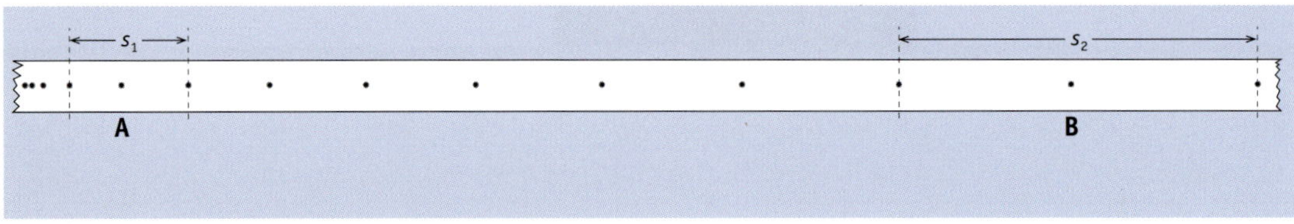

Fig. 7.12
Finding the acceleration from the ticker tape.

6. Ignoring the first few blurred dots on the tape (FIG. 7.12), measure the length of two adjacent spaces near the start of the tape (s_1) and measure the length of two adjacent spaces near the end of the tape (s_2). Record these values.
7. Count the number of spaces (n) between the middle dot of s_1 and the middle dot of s_2 (i.e. between A and B in FIG. 7.12). Record this in the Table.
8. Complete the Table calculating the acceleration a.
9. The experiment may be repeated with the runway at different slopes, thus giving different values for the acceleration.

s_1 / cm	s_2 / cm	n	Initial velocity u / cm s^{-1} [$u = s_1 / (2/50)$]	Final velocity v / cm s^{-1} [$v = s_2 / (2/50)$]	Time t / s [$t = n(1/50)$]	Acceleration a / cm s^{-1} [$a = \frac{v-u}{t}$]

Questions

1. Why should you dust the runway and trolley wheels?
2. Other than by looking at it, how would you know that the trolley was accelerating?
3. If you used the formula $s = \frac{1}{2} at^2$ to find the acceleration, why must s be measured from the first dot produced on the tape? What is the disadvantage of using this method?
4. List two precautions you would take in this experiment to ensure an accurate result.

EXPERIMENT — MECHANICS 1 PART 2

TO MEASURE THE CONSTANT ACCELERATION OF A RIDER ON AN AIR TRACK USING TWO TIMING GATES AND A SCALER TIMER.

Summary of Method

In this experiment you will accelerate a rider along the air track by exerting a constant force on it. The force is caused by weights hanging on a scale pan at the end of the track (FIG. 7.13).

You will calculate the acceleration from the formula:

$$a = \frac{v^2 - u^2}{2s}$$

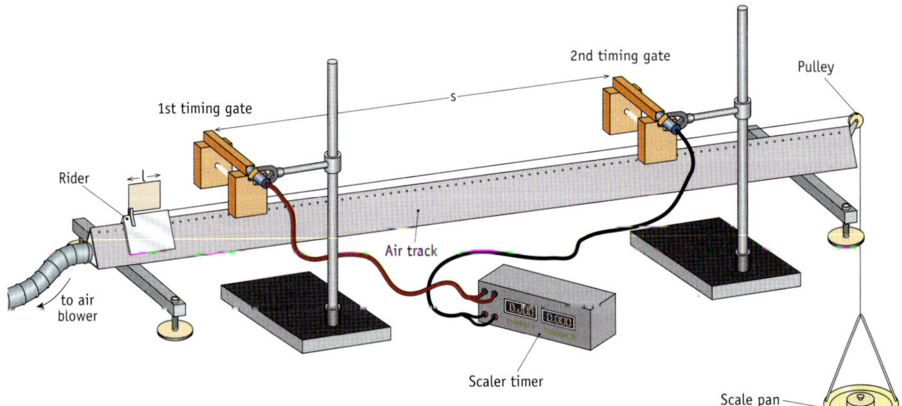

Fig. 7.13
Measuring acceleration with two light gates and a scaler timer.

Equipment Needed

- A linear air track, an air blower and a rider
- A scaler timer and 2 timing gates
- 2 retort stands (depending on timing gates used) and a metre stick
- A scale pan, thread and some weights

Method

1. Set up the air track on the bench and adjust the levelling screws until it looks to be reasonably level. Connect the air blower and place the rider on the track. Further adjust the levelling screws until the rider will remain at rest at any position on the track.
2. Mount each timing gate on a retort stand and connect them to the scaler timer (check with the manual or with your teacher if necessary). Turn on the timer and check that each light gate is operating properly – i.e. the numbers on the timer change when the light beam is broken and remain the same when the beam is complete. Place one timing gate near each end of the track. Adjust the heights of the gates so that the card on the rider will interrupt the light beam.
3. Connect the rider to the scale pan with the thread and place some small weights on the scale pan. Place the rider at the opposite end of the track to the pulley. Place the thread over the pulley and hold the pan in place with your hand.
4. Connect the piece of card to the rider and turn on the blower. Release the pan, the trolley should accelerate along the track, interrupting each beam as it does so.
5. Catch the rider at the end of the track. Switch off the air blower.
6. Record the readings on the scaler timer (t_1 and t_2). Switch off the scaler and measure with a metre stick the distance (s) between the two light beams. Record this value also. If the length (l) of card on the rider is not known, measure and record its value too.
7. Complete the Table and calculate the acceleration of the rider.
8. If time permits, repeat the experiment with different weights on the pan, giving different values for the acceleration.

Transit time 1	Transit time 2	Distance between gates	Length of card	Initial velocity	Final velocity	Acceleration
t_1 / s	t_2 / s	s / cm	l / cm	u / cm s^{-1} ($u = l/t_1$)	v / cm s^{-1} ($v = l/t_2$)	a / cm s^{-2} $\left(a = \dfrac{v^2 - u^2}{2s}\right)$

Acceleration Due To Gravity '*g*'

If you drop an object, the force of gravity causes it to accelerate downwards. This force is called the **weight of the object**. If you drop a coin and a feather from the same height, the coin reaches the ground first. The feather takes longer to fall because the air stops it from picking up as much speed as the coin. The effect of the air on the feather is called **air resistance** or **air friction**.

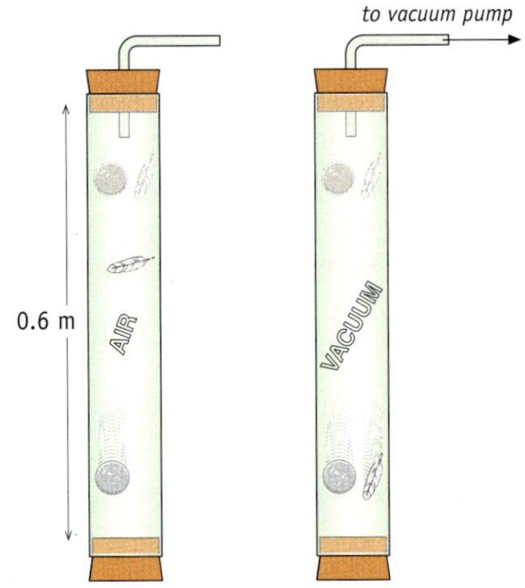

Fig. 7.14
When the air is removed from the tube the coin and the feather fall with the same acceleration.

If the coin and the feather are placed in a container (Fig. 7.14) and the air removed from the container with a vacuum pump, both the coin and feather will be seen to fall together and reach the bottom together. The acceleration of the coin and the feather are the same. It is an experimental fact that:

> In the absence of **air resistance**, all objects near the Earth's surface, if released, will fall downwards with the same acceleration. This acceleration is called **acceleration due to gravity**. Its symbol is '*g*'.

The exact value of '*g*' varies slightly from place to place on the Earth's surface. Correct to one place of decimals $\mathbf{g = 9.8\ m\ s^{-2}}$ anywhere near the surface of the Earth.

Location	Equator	North Pole	London	Top of Mt Everest
Acceleration due to gravity '*g*' (in m s^{-2})	9.78	9.83	9.81	9.77

- The value of '*g*' decreases as you move away from the centre of the Earth. The decrease is only very small unless the distance is very large (e.g. 33 km above the surface the value of *g* is about 0.1 m s^{-2} less than its value on the surface).
- On different planets or other heavenly bodies, the value of acceleration due to gravity is different. For example, on the Moon it is about 1.6 m s^{-2} whereas on planet Jupiter it is about 25 m s^{-2}. You will see on page 115 that the value depends on the mass and radius of the planet. Since acceleration due to gravity is constant, the three **Equations of Motion** for an object moving with constant acceleration apply to an object falling or to an object thrown vertically upwards near the Earth's surface.

Problem 8: An object is dropped from the top of a building which is 30 m high. With what speed does the object hit the ground? How long does it take to reach the ground?

Solution: For the downward motion of the object we have: $u = 0$ (it is dropped from rest), $s = 30$ m, $a = 9.8$ ms^{-2}, $v = ?$, $t = ?$

To find v use: $v^2 = u^2 + 2as$

$$v^2 = 0^2 + 2(9.8)(30)$$
$$\Rightarrow v^2 = 588$$
$$\Rightarrow v = \sqrt{588} = 24.2 \text{ m s}^{-1}$$

To find t use: $v = u + at$

$$24.2 = 0 + 9.8t$$
$$t = \frac{24.2}{9.8} = 2.47 \text{ s}$$

Problem 9: An object falls from the top of a building and strikes the ground 5 seconds later. Find the height of the building.

Solution: $u = 0$, $t = 5$, $a = 9.8$. Let s = height of building

Use $s = ut + ½ at^2$

$$s = (0)(5) + ½(9.8)(5)^2$$
$$\Rightarrow s = 0 + 122.5$$
$$\Rightarrow s = 122.5 \text{ m}$$

Problem 10: A stone is thrown upwards with an initial velocity of 20 m s^{-1}. Find the greatest height reached by the stone.

Solution: Since the stone is thrown upwards and acceleration due to gravity acts downwards, the acceleration must be given a negative sign.

At the highest point the stone is instantaneously stopped. Thus for the upward journey $v = 0$.

$u = 20$, $a = -9.8$, $v = 0$, $s = ?$

Use $v^2 = u^2 + 2as$

$$0^2 = 20^2 - 2(9.8)s$$
$$\Rightarrow s = \frac{20^2}{(2)(9.8)}$$
$$\Rightarrow s = 20.41 \text{ m} = \text{Greatest height}$$

Note that even though the stone is instantaneously stopped at the greatest height, it still has an acceleration of 9.8 m s^{-2} vertically downwards. This is an example of **an object with zero velocity but non-zero acceleration**.

Problem 11: A body is thrown upwards with an initial speed u from a point P which is 20 m above the ground. After 4 seconds it is at a point Q and has a velocity of 6 m s^{-1} downwards (FIG. 7.15). Find the value of u and the height of Q above the ground.

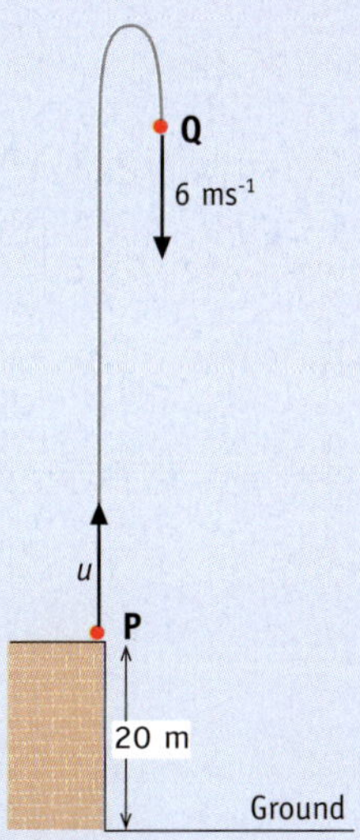

Fig. 7.15

Solution: $u = ?$, $t = 4$, $a = -9.8$, $v = -6$, $s = ?$

Use $v = u + at$ to find u

$$\Rightarrow -6 = u + (-9.8)(4)$$
$$\Rightarrow u = 33.2 \text{ m s}^{-1}$$

Use $s = ut + ½ at^2$ to find height

$$= (33.2)(4) + ½(-9.8)(4^2)$$
$$= 54.4 \text{ m}$$

i.e. Q is 54.4 m above P

$\Rightarrow$ Q is 54.4 + 20 = 74.4 metres above ground

EXERCISE 7.5

In the following, assume all falling objects have an acceleration of 9.8 m s^{-2} *vertically downwards.*

1. A stone is dropped from the top of a cliff 60 m high. How long does it take to reach the ground? With what speed does it hit the ground?
2. A body is projected vertically upward from the ground with an initial velocity of 200 m s^{-1}. Find the greatest height reached and the time taken to reach that height.
3. A stone is dropped from the top of a vertical cliff and reaches the water at the foot of the cliff in 3 seconds. Find the height of the cliff.
4. A cheetah starting from rest reaches a speed of 22 m s^{-1} while it travels a distance of 30 m. Assuming the acceleration is constant, find its value and the time taken to travel this distance.
5. A stone is projected vertically upwards from the ground with an initial velocity of 80 m s^{-1}. Find the greatest height reached and the time taken. When will the stone be 96 m above the ground?
6. A stone is projected vertically upwards from the ground and reaches a height of 100 m in 2 seconds. Find:
   (i) its initial velocity,
   (ii) the greatest height reached,
   (iii) how long after being at a height of 100 m above the ground it is again 100 m above the ground.
7. From a point 16 m above the ground a ball is thrown vertically upwards with an initial velocity of 24 m s^{-1}. Find:
   (i) the greatest height which it reaches,
   (ii) the total time it takes to reach the ground,
   (iii) the velocity with which it strikes the ground.
8. A person can throw a ball 2 m vertically upwards on Earth. How far could an astronaut throw the ball vertically upwards on the Moon, giving the ball the same initial vertical velocity. Take the acceleration due to gravity on the Moon to be one sixth that on Earth.
9. A body leaves a point A and moves in a straight line with a constant velocity of 40 m s^{-1}. Ten seconds later another body which is at rest at A is given an acceleration of 2 m s^{-2} and moves in the same direction as the first body. How long does it take the second body to catch up with the first? How far from A does this occur?
10. A body is thrown upwards with an initial speed u from a point P which is 40 m above the ground. After 6 seconds it is at a point Q and its velocity 8 m s^{-1} downwards. Find the value of u and the height of Q above the ground.

Measuring Acceleration Due To Gravity 'g' Using a Free Fall Apparatus and a Scaler Timer

The value of 'g' can be found by measuring the time t it takes a steel ball, dropped from rest, to fall through a known distance s. As it is dropped from rest, $u = 0$; thus the formula $s = ut + ½ at^2$ becomes $s = ½ gt^2$. Since s and t are known, 'g' can be calculated. Fig. 7.16 shows one type of equipment specially made for this purpose. When the switch is at A, the electromagnet is energised and it holds the ball in place. When the switch is quickly turned to position B, the solenoid ceases to be an electromagnet and the ball immediately drops. At the same instant, the timer is switched on. When the ball strikes the trapdoor, the timer is switched off. Thus the time taken to fall the distance s is known. The distance is measured with a metre stick.

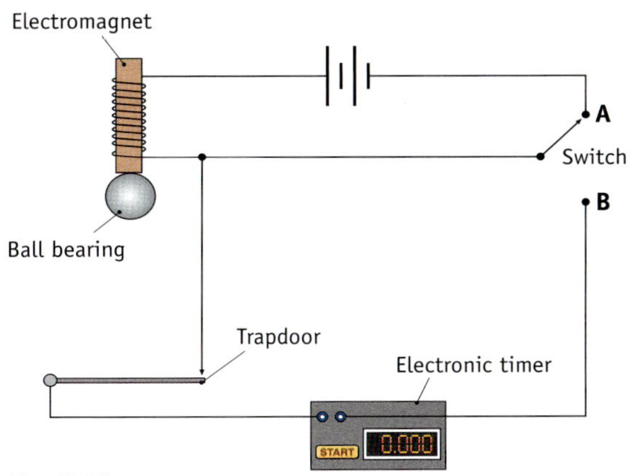

Fig. 7.16

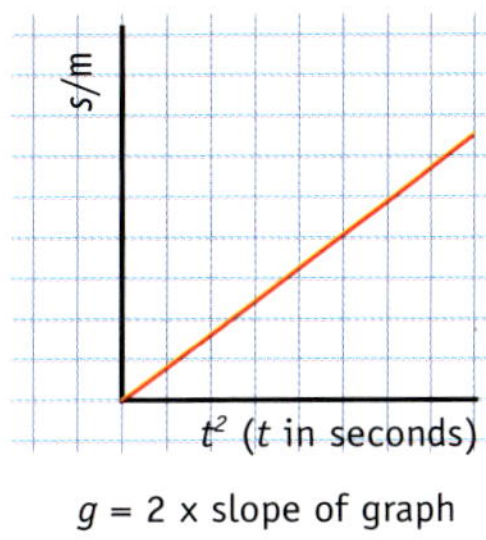

Fig. 7.17

The whole procedure is then repeated four or five times more, each time for a different value of s. The average value of 'g' is then calculated.

HANDLING THE DATA GRAPHICALLY

If a graph of s against t^2 is drawn, it will be found to be a straight line through the origin (FIG. 7.17). Acceleration due to gravity is equal to the slope of the graph multiplied by 2. This is so because the equation of a line through the origin of slope m is: $y = mx$. Comparing this with $s = ½ gt^2$, we see that the slope of the graph of s against t^2 is equal to $½ g$. Therefore $g = 2 \times$ slope of graph.

EXPERIMENT MECHANICS 4

TO MEASURE ACCELERATION DUE TO GRAVITY 'g' USING A FREE FALL APPARATUS AND A SCALER TIMER.

Summary of Method

Using a 'g' by free fall apparatus and a scaler timer you will allow a steel ball to fall through a known distance s. You will measure the time taken t for the fall. Using the formula $s = (½)gt^2$, acceleration due to gravity 'g' can be calculated.

Equipment Needed

- A 'g' by free fall apparatus
- A scaler timer and connecting leads
- A retort stand, a clamp and a metre stick

Method

1. Set up the equipment shown in FIG. 7.16. The switch is in position A and the electromagnet holds the ball in place.
2. Make sure the timer is connected correctly (check with the manual or with your teacher). Check that the timer comes on and goes off when it should.
3. With the metre stick, measure the distance s from the bottom of the ball (when in position in the electromagnet) to the top of the trapdoor. Record this value.
4. Set the timer to zero. Flick the switch to B, releasing the ball and starting the timer. When the ball strikes the trapdoor the timer stops. Record the reading on the timer.
5. Reset the timer and repeat step 4 at least three times. The value of t to be used is the smallest one recorded.
6. Repeat the whole procedure four or five times, each time with a different value for s.

Handling the Data

7. Complete the Table and calculate the average value of $\frac{2s}{t^2}$ (= g)

8. On graph paper, plot a graph of s (on the y–axis) against t^2 (on the x–axis). Draw the best straight line that fits the points and measure its slope. Acceleration due to gravity g = 2 × slope of graph. Compare this value with that obtained in step 7.

Distance fallen s / m	Times for fall in seconds t_1 t_2 t_3 t_4	Smallest time t / s	Acceleration due to gravity g / m s^{-2} (= $2s/t^2$)

Average value of 'g' = m s^{-2}

Questions

1. Why should the smallest time for a given distance fallen be used in the calculations to find 'g'?
2. What would be the effect on the value of the final result if the time as read on the timer was bigger than the actual time taken for the fall?
3. Prove that the slope of the graph multiplied by two is 'g'.
4. Why is the formula $s = (\frac{1}{2})gt^2$ rather than $s = ut + (\frac{1}{2})gt^2$ used in this experiment?
5. List two disadvantages in allowing the metal ball to only fall through a small distance (say 40 cm) in this experiment.
6. State three precautions that should be taken to ensure an accurate result.

CHAPTER CHECKLIST

- **Define:** Acceleration; Average acceleration.
- **State** the unit of acceleration.
- **List** the different ways in which a body could be accelerating.
- **Recall** that: The velocity-time graph of a body moving with constant acceleration is a straight line; The slope of the graph is the acceleration and the area under the graph is the distance travelled; In the absence of air resistance all objects fall with the same acceleration (g) near the Earth's surface.
- **Describe** the motion of a body if given its velocity-time graph.
- **Describe** and carry out an experiment to: Measure constant acceleration; Measure acceleration due to gravity (g) by free fall.
- **Recall** and use the formulae:
  $v = u + at$; $s = ut + \frac{1}{2}at^2$; $v^2 = u^2 + 2as$
- **Derive** the formulae:
  $v = u + at$; $s = ut + \frac{1}{2}at^2$; $v^2 = u^2 + 2as$

Vectors and Scalars

CHAPTER 8

Physical Quantities

You saw in Chapter 1 that a **quantity** is any physical property that can be measured. You also saw that when you measure something, the size of the measurement is called its **magnitude**. In physics, the quantities we meet can be divided into two classes. Those in one class are called **scalar quantities** and those in the other are called **vector quantities**.

SCALAR QUANTITY
A quantity that has magnitude only and has no direction in space is called a ***scalar quantity***.

Scalar Quantities

A scalar quantity is one that has **magnitude only**. It has no direction in space associated with it.

Examples of Scalar Quantities

- length
- area
- volume
- time
- frequency
- mass
- density
- pressure
- energy
- work
- power
- temperature
- electric charge
- electric current
- resistance
- potential difference

The combined result – the resultant – of two scalar quantities is found by normal addition, e.g. 1000 kg + 500 kg = 1500 kg.

VECTOR QUANTITY
A quantity that has both magnitude and a direction in space is called a ***vector quantity***.

Vector Quantities

A quantity that has both **magnitude and a direction** in space is called a vector quantity. To fully specify a vector quantity we must give both its magnitude and its direction. For every quantity on your course you need to know whether it is a vector or a scalar quantity. An easy way to do this is to memorise the examples of vector quantities below. All the other quantities that you meet are scalar quantities.

Examples of Vector Quantities

- displacement
- velocity
- acceleration
- momentum
- force
- electric field strength
- magnetic flux density

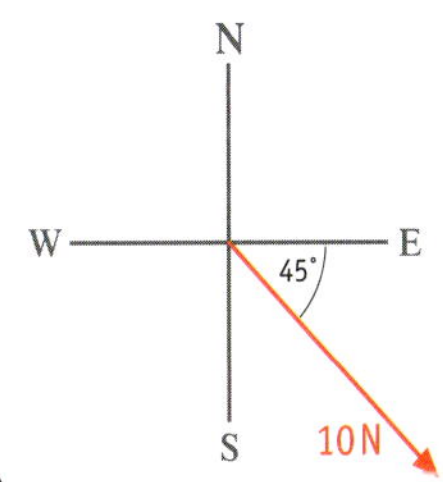

(A)

A force of 10 N acting in a South-Easterly direction.

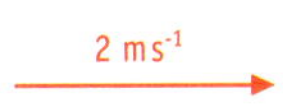

(B)

A velocity of 2 m s^{-1} to the right.

Fig. 8.1

Representing Vector Quantities

A vector quantity is represented on a diagram by an arrow. The length of the arrow represents its magnitude and the direction of the arrow shows its direction, e.g. a force of 10 newtons acting in a South-Easterly direction is represented as in Fig. 8.1(A). A velocity of 2 m s^{-1} to the right is represented as in Fig. 8.1(B).

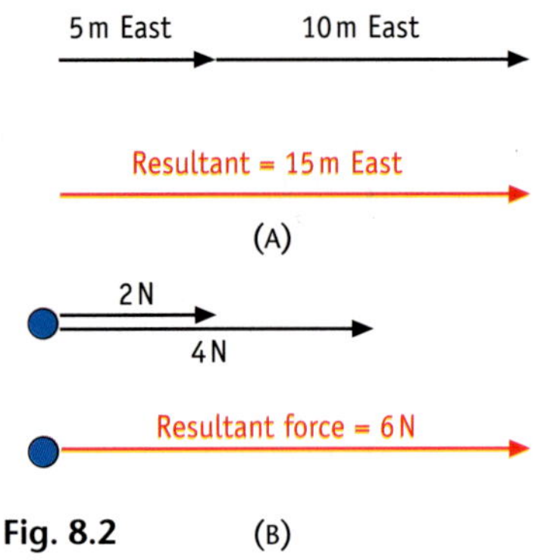

Fig. 8.2

RESULTANT (COMBINED EFFECT) OF TWO VECTOR QUANTITIES

- The combined effect of a displacement of 5 m East followed by a displacement of 10 m East is a displacement of 15 m East (FIG. 8.2(A))
- The combined effect of a force of 2 newtons and a force of 4 newtons in the same direction, is a force of 6 newtons in that direction (FIG. 8.2(B)).
- If a train moves forward at 5 m s^{-1} and a girl walks from the front to the back of the train at 3 m s^{-1}, the velocity of the girl over the ground is 2 m s^{-1} (FIG. 8.3(A)).
- The resultant of a 10 N force and a 6 N force acting in the opposite direction is a force of 4 N in the direction of the 10 N force (FIG. 8.3(B)).

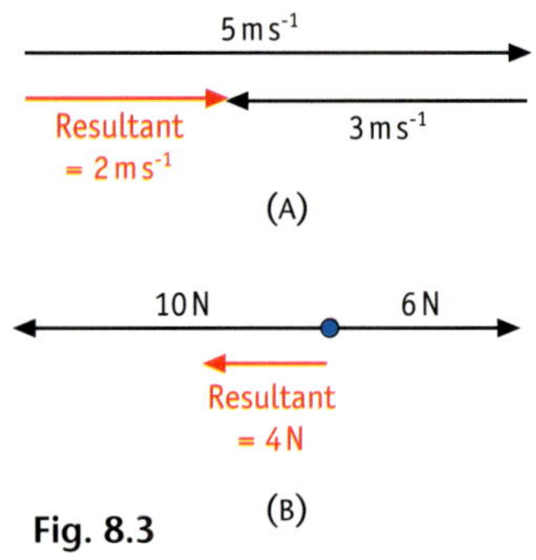

Fig. 8.3

VECTORS

The arrows used to represent amounts of a vector quantity are called **vectors** (you may have met this in your maths course). The resultant of two vectors is found from the **Parallelogram Law** or the **Triangle Law**.

THE PARALLELOGRAM LAW FOR FINDING THE RESULTANT OF TWO VECTORS

If two vectors, drawn tail to tail, are the adjacent sides *ab* and *ad* of the parallelogram *abcd* (FIG. 8.4), the diagonal from *a* to *c* of this parallelogram is their resultant.

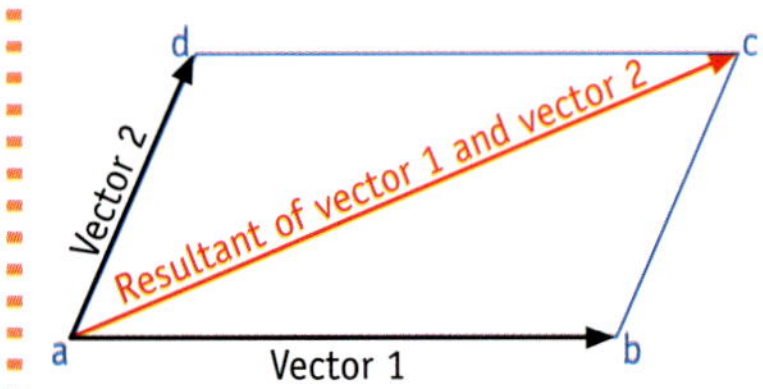

Fig. 8.4

TRIANGLE LAW FOR FINDING THE RESULTANT OF TWO VECTORS

If two vectors are drawn head to tail (FIG. 8.5) the vector from the tail of the first to the head of the second is their resultant. It makes no difference whether we use the Parallelogram Law or the Triangle Law to find the resultant of two vectors. From FIG. 8.4 and FIG. 8.5 it is clear that each law gives the same vector as the resultant.

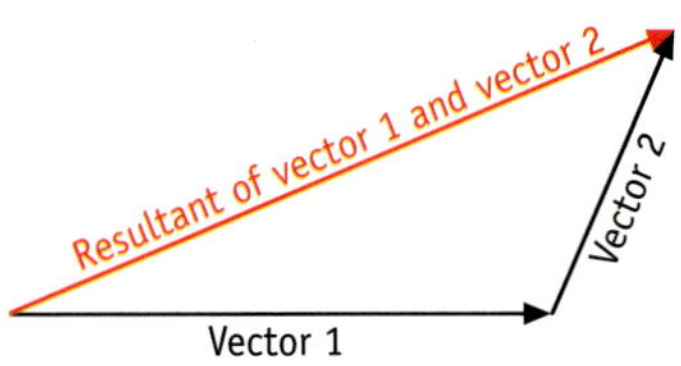

Fig. 8.5

VECTOR NATURE OF VECTOR QUANTITIES

The combined result of two displacements is found by the Parallelogram Law or Triangle Law. The combined result of two velocities or two forces is found by the Parallelogram Law or Triangle Law. In fact: The combined result of two amounts of any vector quantity is found by the **Parallelogram Law** or **Triangle Law**.

Problem 1: A horse undergoes a displacement of 3 km East followed by a displacement of 5 km North. What is its overall displacement from its starting point, i.e. find the resultant of the two displacements?

Solution: FIG. 8.6 shows the 3 km displacement represented by the vector arrow *ab* and the 5 km displacement by the vector *ad*. By the Parallelogram Law, the vector *ac* represents the resultant displacement.

The magnitude of the resultant is found by using Pythagoras' Theorem on the triangle *abc*.

Magnitude of resultant = length of arrow $ac = \sqrt{3^2 + 5^2}$

$= 5.83$ km

Direction of arrow *ac* : $\tan\theta = 5/3 \Rightarrow \theta = \tan^{-1} 5/3 = 59°$

∴ Resultant displacemant is 5.83 km East 59° North

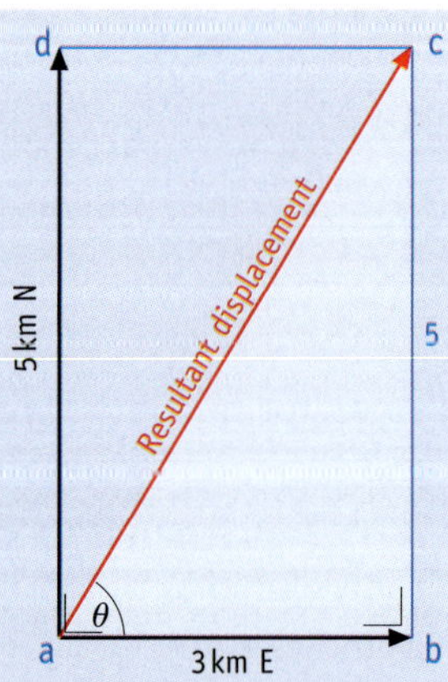

Fig. 8.6

Problem 2: A ship moves parallel to a straight river bank at 4 m s^{-1} (FIG. 8.7). A man walks across the ship at right angles to the direction of forward motion of the ship at 3 m s^{-1}. Find the overall velocity of the man in magnitude and direction.

Solution: The overall velocity of the man is the resultant of the two given velocities. FIG. 8.7 shows these represented by vectors and the diagonal of the parallelogram represents the resultant.

Magnitude of resultant = length of arrow

$ac = \sqrt{4^2 + 3^2} = 5 \text{ m s}^{-1}$

Direction of resultant = $\theta°$

with forward direction of motion of ship

Where $\tan\theta = ¾ \Rightarrow \theta = 36.87°$

Resultant velocity of man = 5 m s^{-1} at 36.87° with forward direction of motion of ship.

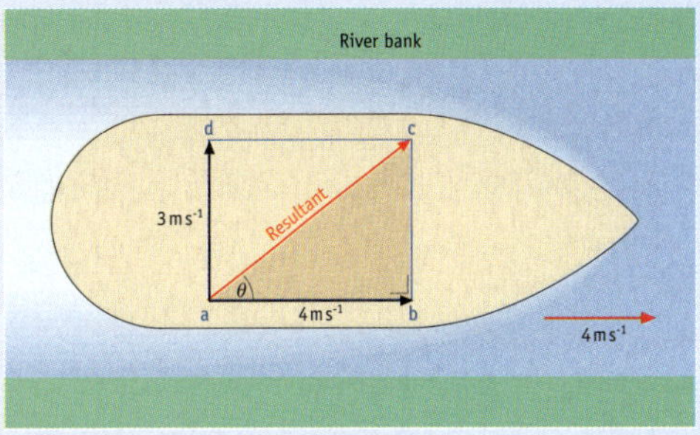

Fig. 8.7

EXERCISE 8.1

1. Copy each of the diagrams in FIG. 8.8 onto your paper and draw in the resultant. In each case state its magnitude and direction.

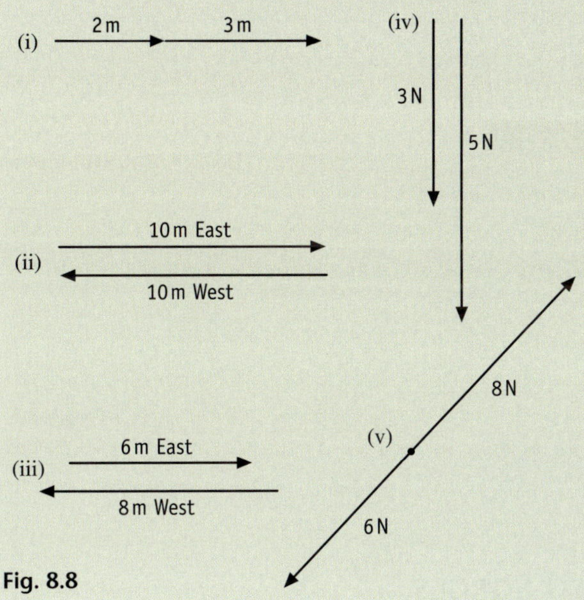

Fig. 8.8

2. Copy each of the diagrams in FIG. 8.9 onto your paper and draw in the resultant for each pair of vectors.

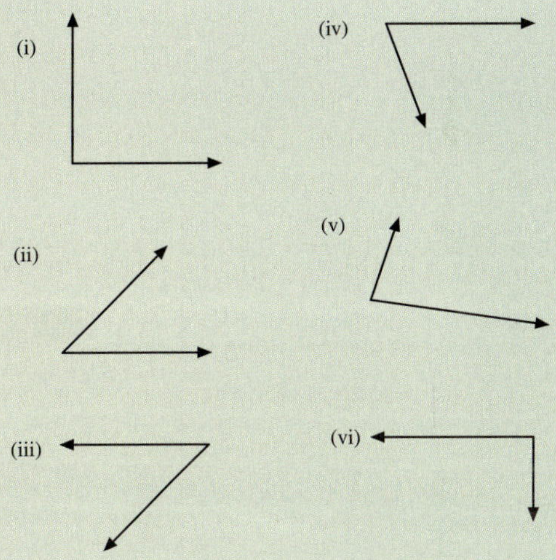

Fig. 8.9

3. Find the magnitude and direction of the resultant of each pair of forces in FIG. 8.10:

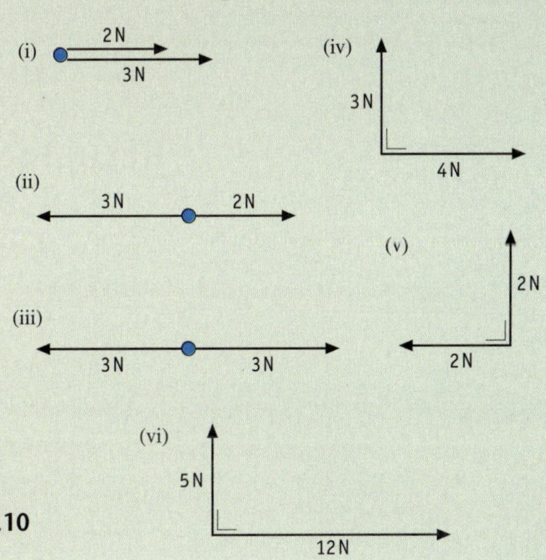

Fig. 8.10

FINDING THE RESULTANT OF THREE (OR MORE) VECTORS

To do this, first find the resultant of any two of them using the Parallelogram Law. Call this R_1. Then, using the Parallelogram Law again, find the resultant of R_1 and the third vector. This is the resultant of the three.

Problem 3: Find the magnitude and direction of the resultant of the three forces shown in FIG. 8.11(A).

Solution: The easiest way to do this is to find the resultant of the two 4 N forces first, since these are at right angles to each other. In FIG. 8.11(B) we have:

$R_1^2 = 4^2 + 4^2 \Rightarrow R_1 = \sqrt{32} \Rightarrow R_1 = 5.66\text{ N} \qquad \tan\theta = 4/4 = 1 \Rightarrow \theta = 45°$

We see that R_1 is now in the same straight line as the 8 N force but in the opposite direction. Thus the overall resultant is $8 - 5.66 = 2.34$ N (FIG. 8.11(C)).

Its direction in that of the 8 N force.

In a question like this, make sure that you find the resultant of the two vectors that are at right angles first!

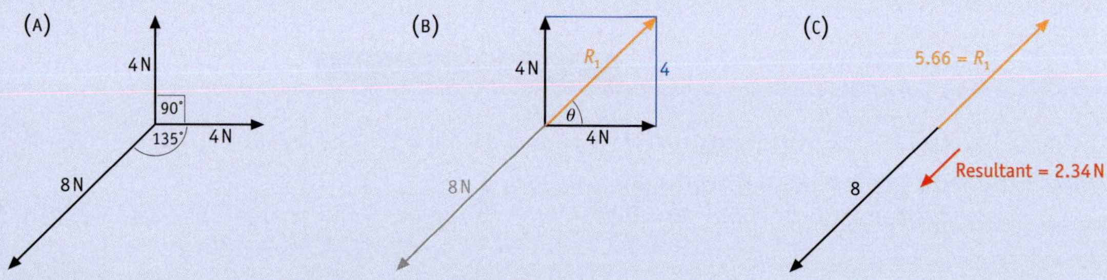

Fig. 8.11

EXERCISE 8.2

1. Find the resultant of each 3 forces in FIG. 8.12 in magnitude and direction.

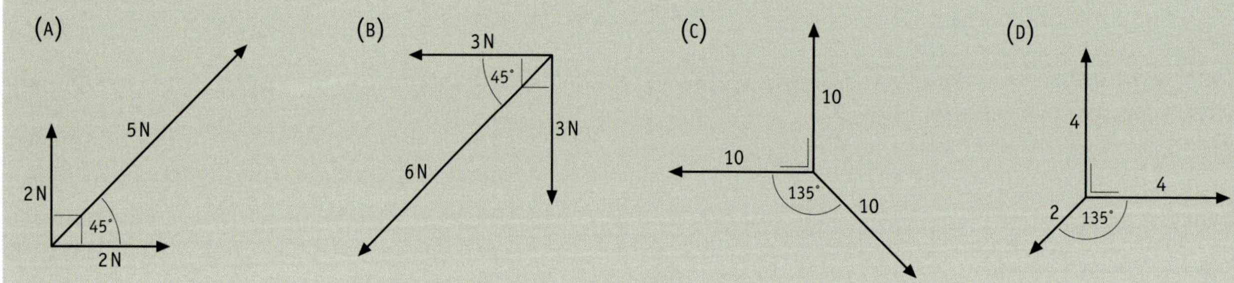

Fig. 8.12

RESULTANT OF TWO FORCES

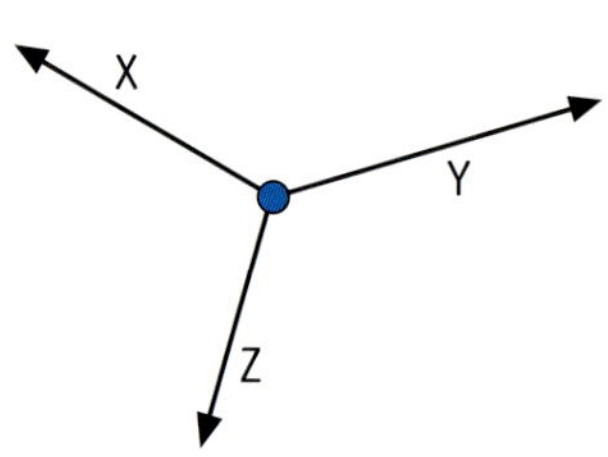

Fig. 8.13

Force is a vector quantity. The unit of force is the **newton**. The resultant of two forces is that single force which, acting alone, has the same effect as the two forces acting together. How can we find the resultant of two forces? Consider the following: Suppose three forces X, Y and Z act on a small object and keep it at rest (FIG. 8.13). Suppose the resultant of X and Y is a force R. Then it is obvious that R must be the same size as Z but in the opposite direction, since if it was not the object would not remain at rest. Thus **if we have three forces keeping a particle at rest we know that the resultant of any two of them is equal in magnitude but opposite in direction to the third force**. We use this fact to find the resultant of two forces as in the next experiment.

EXPERIMENT

TO FIND THE RESULTANT OF TWO FORCES.

- Set up the equipment as in FIG. 8.14 using newton balances (spring balances graduated in newtons).
- Adjust the size and direction of the three forces until the knot in the thread remains at rest.
- If we want the resultant of the two forces F_1 and F_2, its magnitude is the reading on the third balance (F_3). The direction of the resultant is in the opposite direction to F_3.

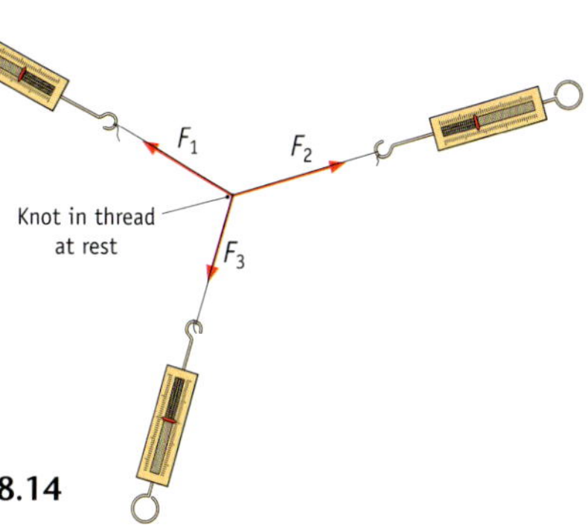

Fig. 8.14

RESOLVING A VECTOR INTO COMPONENTS

Given two vectors, you have seen how to find their resultant. We now look at the reverse process. Given a vector, we want to express it in terms of two other vectors so that it is the resultant of these two.

Resolving a vector into **components** is expressing it in terms of two other vectors so that it is the resultant of these two. These two are called components of the given vector. For example, in FIG. 8.15 *a* and *b* represent two components of *x*, since by the Parallelogram Law *x* is the resultant of *a* and *b*. Likewise *c* and *d* are two other components of *x*.

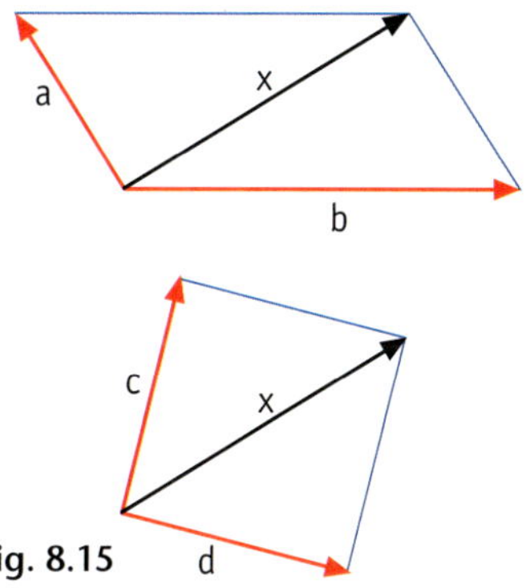

Fig. 8.15

PERPENDICULAR COMPONENTS

If we resolve a vector into components which are at right angles to each other, these are called **perpendicular components**. In Leaving Certificate Physics we will deal only with perpendicular components. FIG. 8.16 shows three different pairs of perpendicular components of the same vector *x*.

Suppose we represent a vector quantity by a vector, we then resolve this vector into perpendicular components. For example, in FIG. 8.17 a 100 N force acts on the cart in the direction shown. This force can be resolved into a horizontal component *x* and a vertical component *y*. (Check using the Parallelogram Law that 100 N at 30° to the horizontal is the resultant of 86.6 N acting horizontally and 50 N acting vertically). Each component represents the complete effect of the given vector quantity in that direction.

The cart now behaves exactly the same as if there was a 50 N force trying to lift it vertically upwards and a force of 86.6 N trying to pull it along the ground. To prevent the cart from moving horizontally along the ground, a force of 86.6 N acting to the left is required. If the weight of the cart is less than 50 N it would be lifted off the ground.

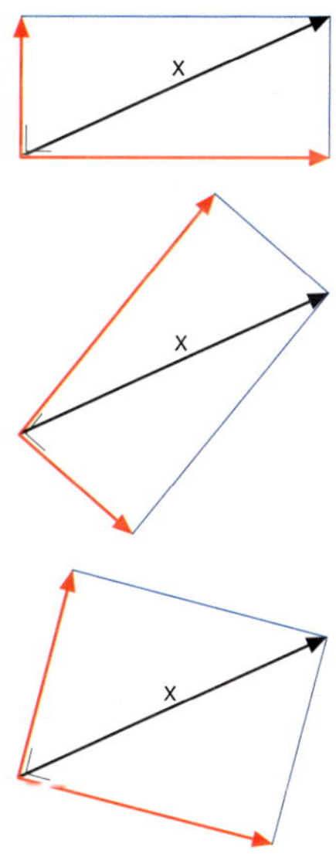

Fig. 8.16

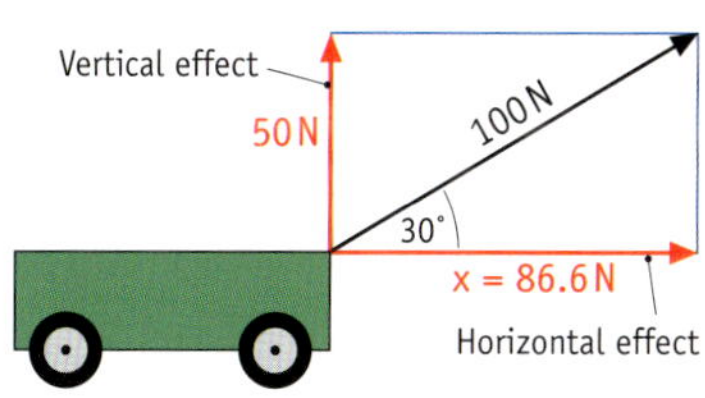

Fig. 8.17

CALCULATING THE MAGNITUDE OF THE PERPENDICULAR COMPONENTS

If a vector of magnitude v has two perpendicular components x and y, and v makes an angle θ with the component x then the magnitudes of the components are:

$x = v \text{ Cos } \theta \qquad y = v \text{ Sin } \theta$ (See FIG. 8.18)

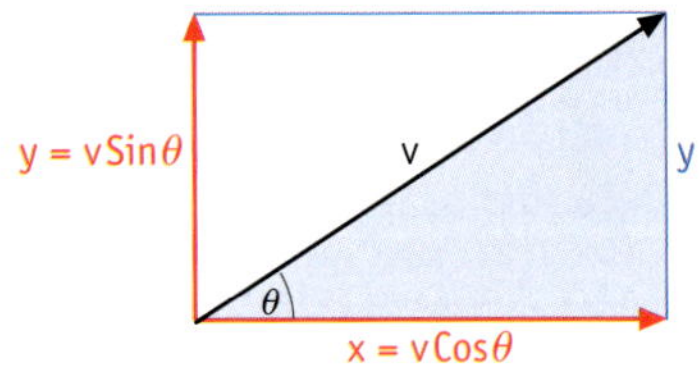

Fig. 8.18

Proof: In the shaded triangle in FIG. 8.18 we have:

$\text{Cos } \theta = \dfrac{\text{adjacent}}{\text{hypotenuse}}$	Also $\text{Sin } \theta = \dfrac{\text{opposite}}{\text{hypotenuse}}$
$\Rightarrow \text{Cos } \theta = \dfrac{x}{v}$	$\Rightarrow \text{Sin } \theta = \dfrac{y}{v}$
$\Rightarrow x = v \text{ Cos } \theta$	$\Rightarrow y = v \text{ Sin } \theta$

Problem 4: Find the vertical and horizontal components of a vector of magnitude 20 N acting at 60° to the horizontal.

Solution: FIG. 8.19 shows the components.

Horizontal component $= x = 20 \text{ Cos } 60° = 10$ N

Vertical component $= y = 20 \text{ Sin } 60° = 17.32$ N

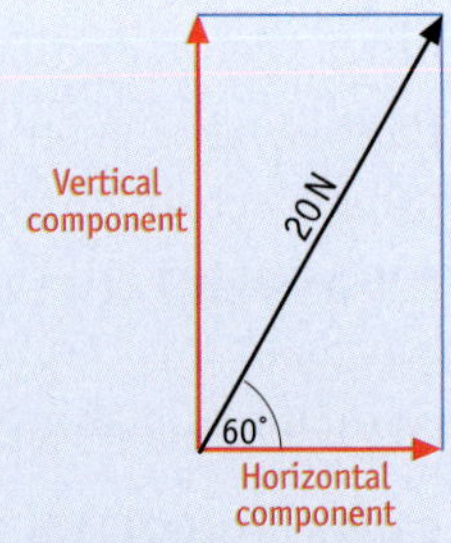

Fig. 8.19

Problem 5: A man pulls a rope which is tied to a cart with a force of 300 N. The rope makes an angle of 20° with the horizontal. Find the effective vertical force on the cart due to the rope and the effective horizontal force on the cart due to the rope.

Solution: FIG. 8.20 shows the situation. The required forces are the vertical component of the 300 N force and the horizontal component of it.

Effective vertical force $= 300 \text{ Sin } 20°$
$= (300)(0.342) = 102.6$ N

Effective horizontal force $= 300 \text{ Cos } 20°$
$= (300)(0.940) = 282$ N

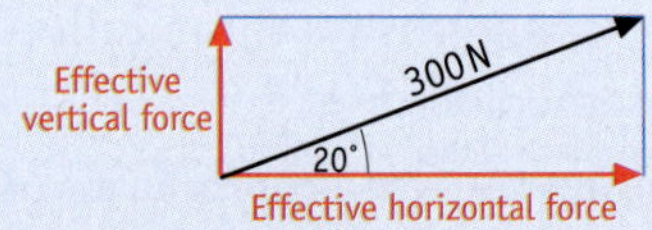

Fig. 8.20

NOTE A downward vertical force of at least 102.6 N is needed to keep the cart on the ground. A horizontal force of 282 N is needed to prevent the cart from moving along the ground.

When resolving a vector into two perpendicular components proceed as follows:

- Through the tail of the vector draw in the two required perpendicular directions.
- Complete the rectangle (parallelogram) with the given vector as the diagonal.
- The components are the adjacent sides of this rectangle.
- Use information given in the question to work out the angle between one of the components and the given vector.
- Calculate the magnitude of the components using:

$x = v \text{ Cos } \theta \qquad y = v \text{ Sin } \theta$

Problem 6: A stone of weight 50 N rests on a sloped roof. The roof is inclined at 20° to the horizontal. Resolve the weight of the stone into components parallel and perpendicular to the roof.

Solution: FIG. 8.21 shows the weight vector and its two components. **Take care to draw these correctly.**

Component of weight perpendicular to roof

= 50 Cos 20° = 46.98 N

Component of weight parallel to roof

= 50 Sin 20° = 17.10 N

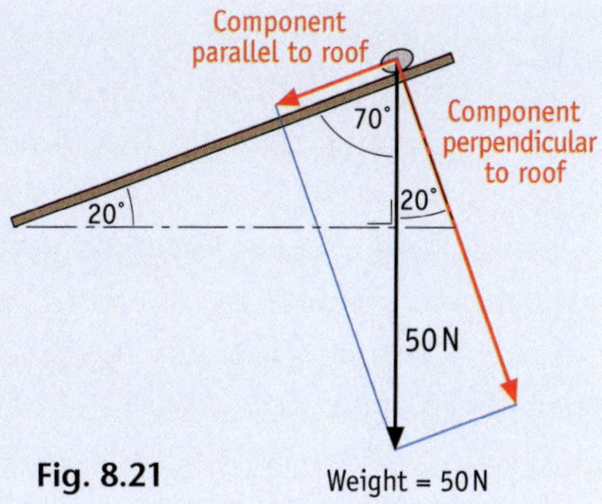

Fig. 8.21

Problem 7: At a certain instant, a pendulum is in the position shown in FIG. 8.22 inclined at 50° to the vertical. The weight of the pendulum bob is 200 N. Resolve this weight into components parallel and perpendicular to the pendulum.

Solution: Again taking care to draw in the components correctly, we have:

Parallel component = 200 Cos 50° = 128.6 N

Perpendicular component = 200 Sin 50° = 153.2 N

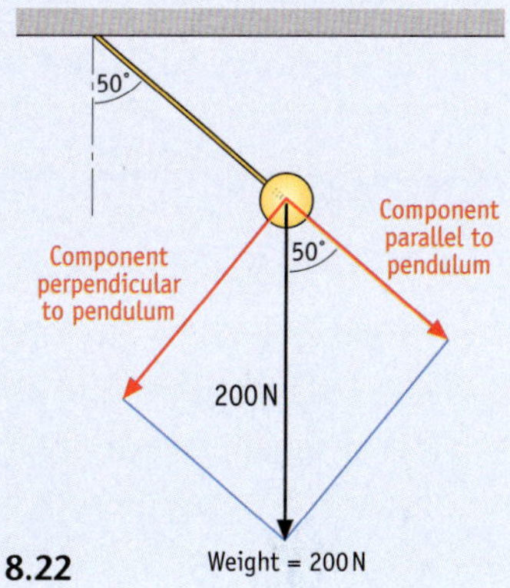

Fig. 8.22

EXERCISE 8.3

1. A force of 200 N is inclined at an angle of 70° to the horizontal. Resolve the force into its horizontal and vertical components.
2. A rope pulls a cart with a force of 3500 N. If the rope is inclined at 25° to the horizontal find:
   (i) the horizontal force on the cart due to the rope,
   (ii) the vertical force on the cart due to the rope.
3. Find the horizontal and vertical components of a vector of magnitude 100 acting at 60° to the horizontal.
4. Find the horizontal and vertical components of a vector of magnitude 200 acting at 40° to the vertical.
5. A boat heads across a still lake as shown in FIG. 8.23 with a constant velocity of 6 m s^{-1} at 40° to one side of the lake. If the lake has parallel sides 2 km apart, how long does it take the boat to reach the opposite side. How far does the boat travel parallel to the lakeside in this time? How far is the boat from its starting point when it reaches the other side?

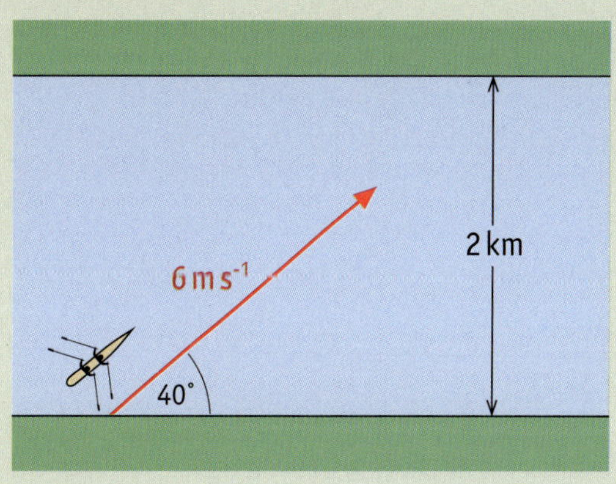

Fig. 8.23

6. A particle rests on the sloped roof of a house which is inclined at an angle of 30° to the horizontal (FIG. 8.24). A downward vertical force of 40 N acts on the particle. Resolve this force into perpendicular components – one parallel to the roof and the other perpendicular to the roof.

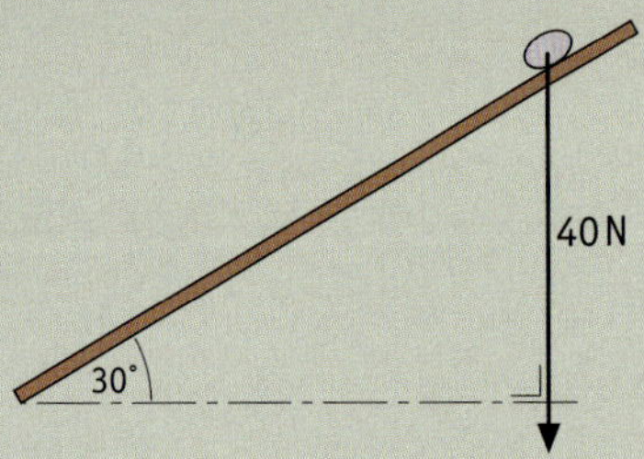

Fig. 8.24

7. A bob-sleigh accelerates down a hill inclined at 30° to the horizontal with an acceleration of 4 m s^{-2} (FIG. 8.25). Resolve this acceleration into its horizontal and vertical components.

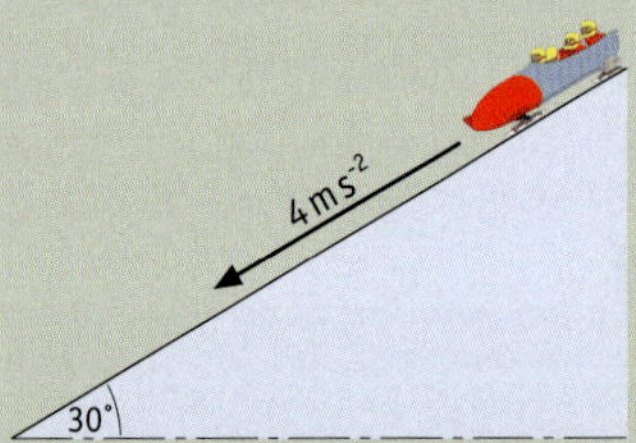

Fig. 8.25

8. At a certain instant a pendulum makes an angle of 20° with the vertical as in FIG. 8.26. Resolve its weight (which acts vertically downwards) into components parallel and perpendicular to the string.

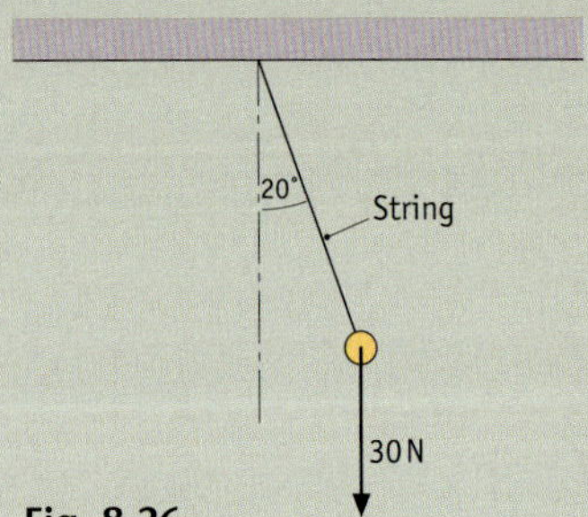

Fig. 8.26

9. At a certain instant, a pendulum makes an angle θ with the vertical as in FIG. 8.27. Resolve its weight W into components parallel and perpendicular to the string.

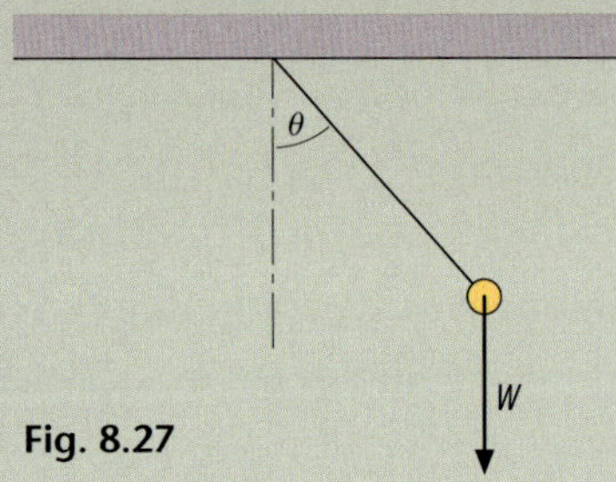

Fig. 8.27

10. A particle P is moving in a circle as shown in FIG. 8.28. At any instant the direction in which the particle moves is perpendicular to the radius from the centre of the circle to the particle. If the radius of circle is 2 m, the speed of particle is 5 m s^{-1} and the magnitude of acceleration is 3 m s^{-2}.
   (i) Resolve its displacement from O into components parallel to the x-axis and to the y-axis.
   (ii) Resolve its velocity into components parallel to the x-axis and to the y-axis.
   (iii) If its acceleration is directed towards the centre of the circle resolve it into components parallel to the x-axis and the y-axis.

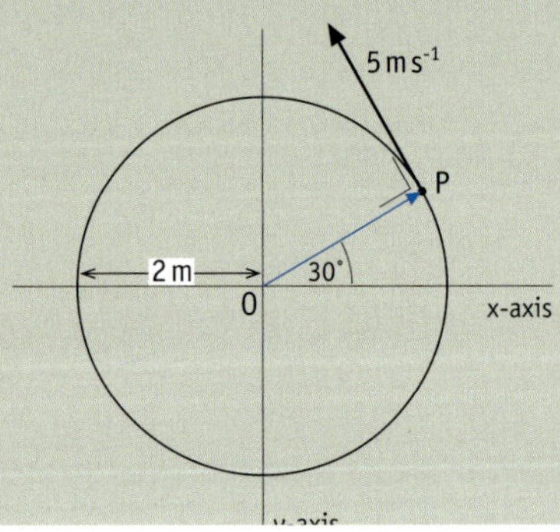

CHAPTER CHECKLIST

- **Define**: Scalar quantity; Vector quantity; Resultant.
- **State:** Whether any given quantity is a scalar or a vector;
  The Parallelogram Law; The Triangle Law.
- **Calculate:** The resultant in magnitude and direction of any two perpendicular vectors;
  The components of a vector in two given perpendicular directions.
- **Describe** and carry out an experiment to find the resultant of two forces.

Force, Mass and Momentum

CHAPTER 9

FORCE

FIG. 9.1 shows a block on a bench. If the block is given a push and let go, it moves along the bench. Thus, **a force can cause an object to start moving**.

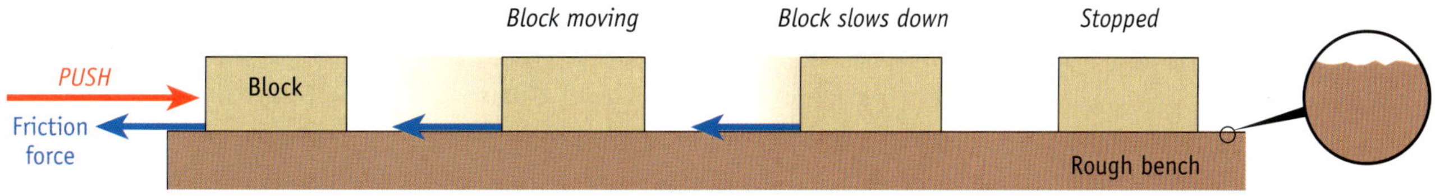

Fig. 9.1

The moving block soon comes to a stop because the bench exerts a force on the block and slows it down (called the **force of friction**.) Thus **force can cause a moving object to slow down** and stop. To keep the block moving at a steady speed you must keep pulling (or pushing) it with a force to overcome the force of friction (FIG. 9.2).

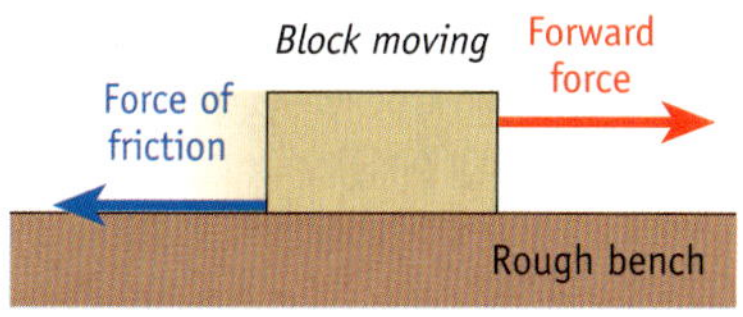

Fig. 9.2
If the block is moving and if the forward force equals the force of friction, then the block will move at a constant speed.

If we polish the bench and the block and put a **lubricant** such as oil between them and give the block the same strength push as before, what will happen? The block will move much, much farther before stopping. This is because the friction force slowing it down is now much less.

(A)

The force the cable exerts on the person comes from outside the person. It is called an external force and lifts the person.

Fig. 9.3

(B)

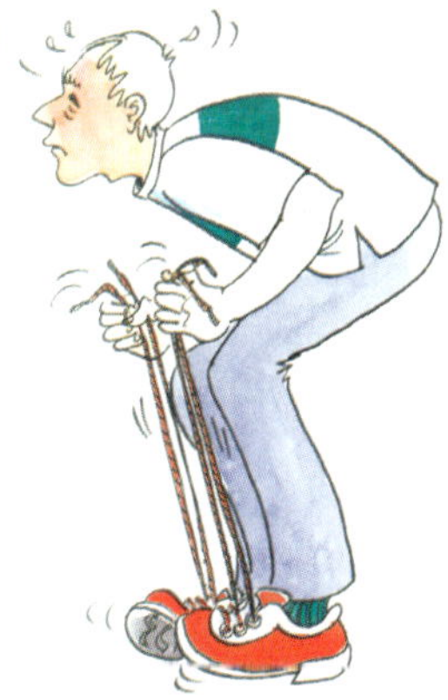

The man will never be able to lift himself by pulling up on his shoe laces because the force he is exerting on the shoe laces does not come from outside himself.

In the 16th century the Italian scientist Galileo Galilei realised that if you could remove all the forces of friction acting on a body and give that body a push, it would continue moving at a steady speed in a straight line forever. He concluded that **what a force does to a body is to cause its velocity to change** i.e. **force causes acceleration.** A linear air track (page 62) shows clearly the way an object behaves when the forces of friction are almost zero.

FORCE – Anything that causes the velocity of an object to change (i.e. to speed up, slow down or change direction) is called a ***force***.

Fig. 9.4
One newton is about the weight of a 100 gram object.

FORCE IS A VECTOR QUANTITY

Since a force has a direction – namely the direction in which it acts – it is a **vector quantity**. The symbol for force is ***F***. The unit of force is the **newton (N)**. One newton is quite a small force (FIG. 9.4). One newton is defined below.

> **UNIT OF FORCE**
> The unit of force is the ***newton*** **(N)**.

FORCE AND THE ACCELERATION IT CAUSES

The bigger the force you exert on an object, the more speed it picks up in a given time, i.e. the bigger the force the bigger the acceleration. Accurate experiments show that for a given body, **the acceleration *a* is directly proportional to the force *F* causing it.** (We write this as: $a \propto F$ (see Appendix 1).)

This means that two times the force causes two times the acceleration, three times the force causes three times the acceleration etc. (FIG. 9.5).

> Acceleration of a body is directly proportional to the force causing it, i.e. for a given body:
>
> $$\boldsymbol{a \propto F} \qquad (1)$$

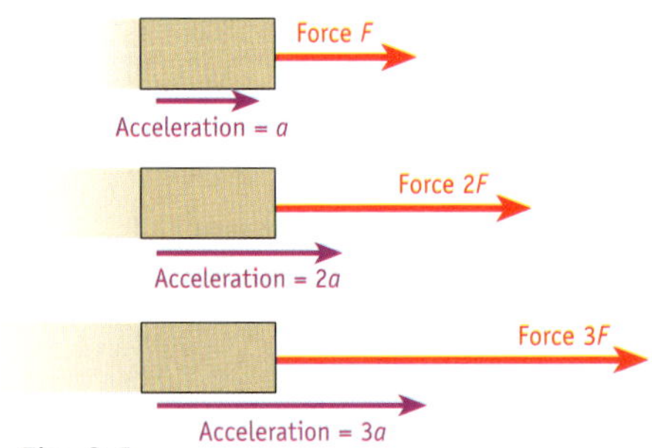

Fig. 9.5

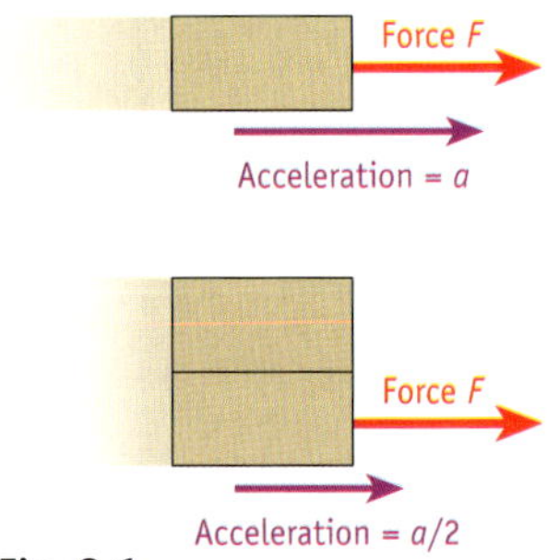

Fig. 9.6

MASS

Different bodies do not get the same acceleration from the same force. If a force gives a body an acceleration a, then the same force will give two of those bodies together half that acceleration (FIG. 9.6). The greater the amount of matter being accelerated by a given force, the less acceleration it gets. The **mass** of a body is a measure of how difficult it is to accelerate that body.

> **MASS**
> The ***mass (m)*** of a body is a measure of how difficult it is to accelerate that body.

Mass is a **scalar quantity**. Its unit is the **kilogram (kg)**. For the present we shall take it that the mass of an object does not change.

> **UNIT OF MASS**
> The unit of mass is the ***kilogram*** **(kg)**.

It follows from the above that **for a given body the acceleration produced by a given force is inversely proportional to the mass of the body**.

> Acceleration of a body $\propto \dfrac{1}{\text{Mass}}$
> i.e. for a given force:
>
> $$\boldsymbol{a} \propto \frac{\mathbf{1}}{\boldsymbol{m}} \qquad (2)$$

THE NEWTON

The unit of force, the newton, is defined as follows:

> **THE NEWTON**
> 1 ***newton*** is the force that gives a mass of 1 kilogram an acceleration of 1 m s^{-2}

From $\boldsymbol{a} \propto \boldsymbol{F}$ and $\boldsymbol{a} \propto \dfrac{\mathbf{1}}{\boldsymbol{m}}$ it follows that:

$$a \propto \frac{F}{m} \Rightarrow F = kma \quad \text{where } k \text{ is a constant.}$$

On page 95 you will see that k = 1; thus:

> $\boldsymbol{F = ma}$ i.e. Force = Mass × Acceleration

> The newton in terms of basic units:
> From $F = ma$ it follows that:
> Unit of Force =
> Unit of mass x Unit of acceleration
> **i.e. 1 N = 1 kg m s^{-2}.**

Weight and Mass

Without air friction all objects, if released, fall with an acceleration due to gravity (g) of 9.8 m s^{-2}. The force causing this acceleration is the **force of gravity** and is called the **weight** of the object. Since weight is a force, **weight is measured in newtons**.

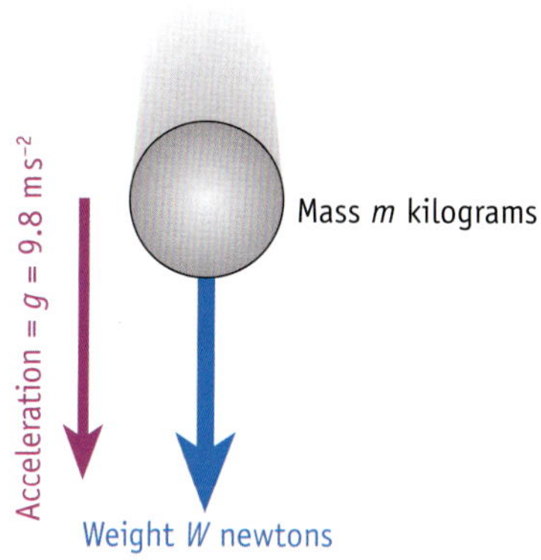

Fig. 9.7

WEIGHT – The ***weight*** of an object is the force of the Earth's gravity acting on it.

Fig. 9.7 shows an object of mass m kilograms falling with an acceleration g. The force acting on it is its weight W.

Applying $F = ma$ gives:

$W = mg$ i.e. Weight = Mass × Acceleration due to gravity

Problem 1: Find the weight of an object of mass 24 kg.
Solution: $W = mg = (24)(9.8) = 235.2$ newtons.

Experiment — Mechanics 2

To Show that the Acceleration of a Body is Directly Proportional to the Force Acting on it, i.e. $a \propto F$.

Summary of Method

In this experiment you will exert a force F on a trolley by means of a scale pan of weights hanging vertically (Fig. 9.8). You will measure the acceleration a produced with a ticker timer and ticker tape. By transferring weights from the trolley to the pan you will increase the accelerating force and again measure the acceleration. You will repeat this a number of times. By plotting a graph of a against F, a straight line through the origin will result, verifying that $a \propto F$.

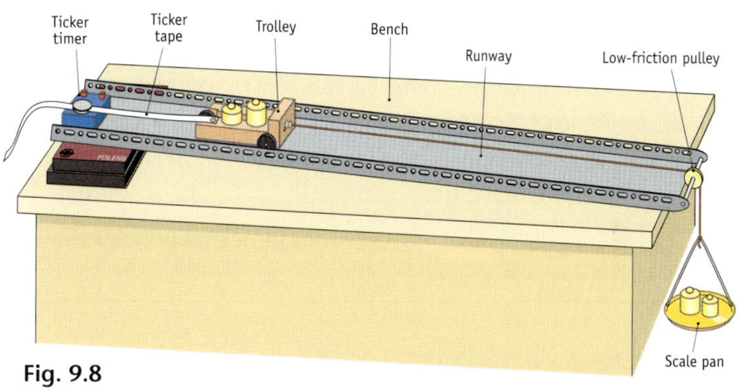

Fig. 9.8

Equipment Needed

- A trolley, a runway, a low-friction pulley and a set of masses (e.g. 0–800 grams)
- A ticker timer, ticker tape, power supply and connecting leads

Method

1. Place all except one of the masses on the trolley and connect the ticker tape to it.
2. Raise one end of the runway so that the trolley with the tape attached will move along the runway at constant speed when given a push.
3. Connect the scale pan to the trolley and place the other mass on it (100 grams). Hold the trolley in place.
4. Turn on the ticker timer, release the trolley and it will accelerate down the runway.
5. Find and record the total mass (m) of the scale pan and contents. The weight of this mass is the accelerating force.
6. Label the ticker tape with this value of the mass.
7. Transfer a mass from the trolley to the scale pan (say 100 grams) and repeat steps 4, 5 and 6.

8. Repeat step 7 a number of times with a larger accelerating force each time. Increase the mass on the pan by about 100 grams each time. Any weights put on the pan must come from the trolley and if a weight is removed from the pan it must be put back on the trolley. In this way the total mass being accelerated remains constant.
9. Complete the table calculating the accelerating **force** and the **acceleration** produced each time. The acceleration may be calculated from each tape by the method used on page 72.
10. On graph paper plot a graph of acceleration a on the y-axis against Force F on the x-axis.

Table

s_1/cm	s_2/cm	Initial velocity u / cm s^{-1} ($u = s_1/(2/50)$)	Final velocity v / cm s^{-1} ($v = s_2/(2/50)$)	Time t / s $\left(t = n \times \frac{1}{50}\right)$	Acceleration a / cm s^{-2} $\left(a = \frac{v-u}{t}\right)$	Mass m/grams	Force F/N $\left(\frac{m}{1000} \times 9.8\right)$

Result

A graph similar to FIG. 9.9 should result. This is a straight line through the origin, verifying that $a \propto F$.

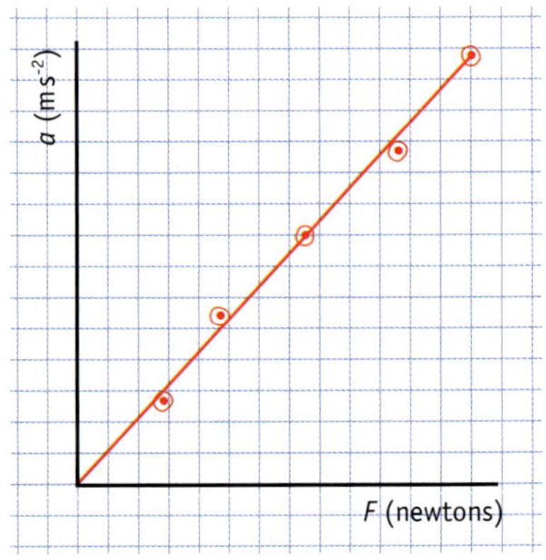

Fig. 9.9

Questions

1. Why must the masses be transferred from the trolley to the scale pan and vice versa in this experiment?
2. Why must the runway be compensated for friction by raising one end?
3. Draw a diagram showing the forces on the trolley when it is being accelerated.
4. If the graph does not pass through the origin what precaution is likely to have been omitted?
5. Explain how, from the graph, you could find the mass being accelerated.

Problem 2: What force gives a 10 kg mass an acceleration of 4 m s^{-2}?

Solution: $F = ma = (10)(4) = 40$ N

Problem 3: Find the acceleration of a car of mass 1000 kg and its two occupants of masses 60 kg and 110 kg under the action of a force of magnitude 2000 N.

Solution: $F = ma \Rightarrow a = F/m$

$$\therefore a = \frac{(2000)}{(1000 + 60 + 110)} = \frac{2000}{1170} = 1.71 \text{ m s}^{-2}$$

Problem 4: A force of 20 N acts on a body of mass 5 kg which is at rest. What is the velocity of the body after 8 seconds?

Solution: The body moves with constant acceleration $a = \frac{F}{m} = \frac{20}{5} = 4 \text{ m s}^{-2}$

$u = 0,\ v = ?\ t = 8$ and $a = 4$ $\qquad v = u + at = 0 + (4)(8) = 32 \text{ m s}^{-1}$

Problem 5: A body of mass 105 kg has a velocity of 10 m s^{-1}. What force is required to stop it in 0.5 s?

Solution: Here $u = 10,\ v = 0,\ t = 0.5,\ a = ?$ and $F = ?$

Find a: $v = u + at \Rightarrow a = \frac{v-u}{t} = \frac{0-10}{0.5} = -20 \text{ m s}^{-2}$

Find F: $F = ma = 105 \times 20 = 2100 \text{ N}$

Problem 6: Find the acceleration of each mass in FIG. 9.10 under the action of the forces shown.

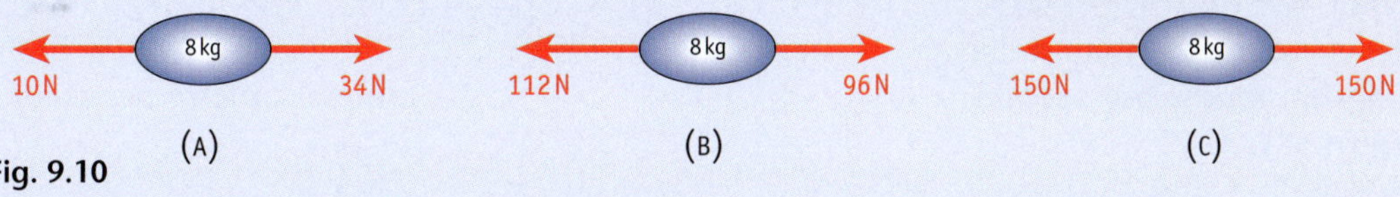

Fig. 9.10

Solution: Apply $F = ma$. F is the **net** or **resultant** force on the mass. a is the acceleration produced.

(i) $F = 34 - 10 = 24$ N to the right $\Rightarrow a = F/m = 24/8 = 3 \text{ m s}^{-2}$ to the right

(ii) $F = 112 - 96 = 16$ N to the left $\Rightarrow a = F/m = 16/8 = 2 \text{ m s}^{-2}$ to the left.

(iii) Net force $= 150 - 150 = 0$ N $\Rightarrow$ Acceleration $= 0 \text{ m s}^{-2}$

EXERCISE 9.1

1. What force would give a body of mass 20 kg an acceleration of 5 m s^{-2}?
2. A body of mass 10 kg is acted on by a force of 4 N. Find the acceleration of the body.
3. What force will give a body of 100 kg an acceleration of 2 m s^{-2}?
4. A force of 4 kN gives a body an acceleration of 3 m s^{-2}. What is the mass of the body?
5. A body of mass 10 kg has an initial velocity of 6 m s^{-1} in a certain direction. A constant force of 40 N is then applied to the body in the same direction for 12 seconds. Find:
   (i) its acceleration,
   (ii) its velocity after 12 s,
   (iii) the distance travelled in 12 s.
6. A force changes the velocity of an object from 2 m s^{-1} to 10 m s^{-1} in 4 seconds. If the mass of the object is 20 kg find the force.
7. Find the weight of a body of mass:
   (i) 1 kg (iii) 105 kg
   (ii) 1 gram (iv) m kg
8. A force of 2000 N gives an acceleration of 4 m s^{-2} to a stone. What is the mass of the stone? If the stone is initially at rest and the force acts for 20 seconds find:
   (i) the velocity acquired by the stone,
   (ii) the distance travelled while the force acts.

   What other force acting on its own would then stop the stone in 0.1 seconds?
9. Find the acceleration of each of the blocks in FIG. 9.11:

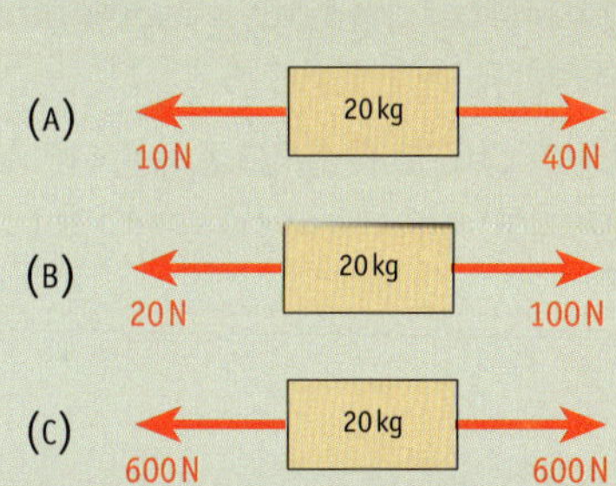

Fig. 9.11

If at a given instant each block has a velocity of 20 m s^{-1} to the right, find the velocity of each 2 seconds later.

10. A wooden block of mass 6 kg is being pulled along a rough horizontal table by a horizontal force of 500 N. If the frictional forces opposing motion amount to 400 N find the acceleration of the block.

11. A car of mass 1200 kg which is travelling at 100 km h^{-1} wants to stop in a distance of 100 m to avoid a collision. If the maximum retarding force due to the brakes is 2000 N, will the car avoid a collision? What is the minimum force needed to avoid collision?

MOMENTUM
Momentum =
Mass × Velocity $p = mv$

UNIT OF MOMENTUM
The unit of momentum is the ***kilogram metre per second*** **(kg m s^{-1}).**

Momentum

The momentum of a body is, by definition, equal to its mass multiplied by its velocity. The symbol for momentum is $\boldsymbol{p}$. Momentum is a **vector quantity.** Its direction is that of the velocity.

$$\text{Since } p = mv$$
$$\text{unit of momentum} = (\text{unit of mass}) \times (\text{unit of velocity})$$
$$= \text{kilogram metre per second (kg m s}^{-1})$$

Problem 7: A bus of mass 5000 kg moves East at 15 m s^{-1}. Find its momentum.
Solution: $p = mv = (5000)(15) = 75\ 000$ kg m s^{-1} East

The table below shows the new quantities and units in this chapter.

Quantity	Symbol	Unit	Symbol	Unit in terms of basic units
Mass	m	kilogram	kg	kg
Force	F	newton	N	kg m s^{-2}
Momentum	p	kilogram metre per second	kg m s^{-1}	kg m s^{-1}

Newton's Laws of Motion

The way in which force affects the motion of a body was first stated completely by Isaac Newton in 1687 in his three laws of motion:

NEWTON'S FIRST LAW OF MOTION states that every body will remain in a state of rest or travelling with a constant velocity unless an unbalanced external force acts on it.

NEWTON'S SECOND LAW OF MOTION states that when an unbalanced force acts on a body the rate of change of the body's momentum is directly proportional to the force and takes place in the direction of the force.

NEWTON'S THIRD LAW OF MOTION states that if a body A exerts a force on a body B, then body B exerts an equal but opposite force on body A, i.e. action and reaction are equal and opposite.

NEWTONS FIRST LAW

This law means that if there is an overall (resultant) force on a body, that body will accelerate. If there is no force or no resultant force acting on the body, it will remain travelling with a constant velocity or it will remain at rest. Recall that if a body has a constant velocity it travels in a fixed direction with a steady speed. The truth of Law 1 can easily be demonstrated with a linear air track. If the track is level and a rider is placed on it with zero velocity, it remains at rest on the track. If the rider is given a push and let go, it is seen to move along the track with a steady speed until it comes to the end of the track. If any horizontal force acts on the rider, its speed either increases or decreases or it tries to turn off the track.

Suppose you are travelling along in a car at 30 m s^{-1} and the car stops suddenly (e.g. it crashes into solid wall). By Newton's 1st Law you were moving at 30 m s^{-1} and will continue to do so (perhaps through the windscreen of the car!) until some force causes your velocity to change (FIG. 9.12). The purpose of a seatbelt and/or an air bag in a car is to provide this force in a fashion that will do you least harm.

Fig. 9.12

$F = ma$ IS A SPECIAL CASE OF NEWTON'S SECOND LAW

Newton's 2nd Law states: Force $\propto$ Rate of change of momentum

$$\text{i.e.} \quad F \propto \frac{\text{Change in momentum}}{\text{Time taken for change}}$$

$$\text{i.e.} \quad F \propto \frac{\text{Final momentum} - \text{Initial momentum}}{\text{Time taken}}$$

$$\Rightarrow \quad F \propto \frac{mv - mu}{t} \quad \Rightarrow \quad F = \frac{km(v-u)}{t} \qquad (3)$$

But acceleration $a = \dfrac{v-u}{t}$ $\quad \therefore \quad F = kma$ where k is a constant (4)

Remembering that 1 newton gives 1 kg an acceleration of 1 m s^{-2} and substituting these values into (4) we get:

$1 = k(1)(1) \Rightarrow k = 1$. Thus we have: $F = ma$

It follows from this that the experiment to show that $a \propto F$ (page 91) demonstrates that Law 2 is indeed correct. Also since $k = 1$, equation (3) becomes:

$$\boldsymbol{F = \frac{mv - mu}{t}}$$ i.e. Force is equal to the rate of change of momentum

NEWTONS THIRD LAW

This law tells us that forces always occur in pairs. If we see a body accelerating under the action of a force, there must be another body which experiences a force of the same size but acting in the opposite direction to the first force. Note that **the two forces act on different bodies**. Some examples are as follows:

- When a car crashes into a wall, the wall exerts a force on the car that slows the car down abruptly and damages the car. The car exerts an equal but opposite force on the wall that pushes the wall forward, thus damaging the wall.
- When a rocket (or a jet) wants to accelerate forward, it expels a mass of hot gas at high speed backwards from it (FIG. 9.13). The rocket exerts a force on the gas pushing the gas backwards. The gas exerts an equal but opposite force on the rocket (by Law 3) and thus the rocket accelerates forwards.

Fig. 9.13
The rocket exerts a force on the gas pushing the gas backward. The gas exerts an equal but opposite force on the rocket and the rocket accelerates forwards.

THE FORCE OF FRICTION

When you slide or try to slide one body along another, a force appears opposing this sliding. This force is called the **force of friction**. Friction occurs with solids, liquids and gases. Without friction you would not be able to walk, neither would car tyres be able to grip the ground. When objects – such as cars – are moving, friction forces always tend to slow them down. Energy must be supplied to overcome friction and keep the object moving. Air also exerts friction – called air resistance – on any object moving through it.

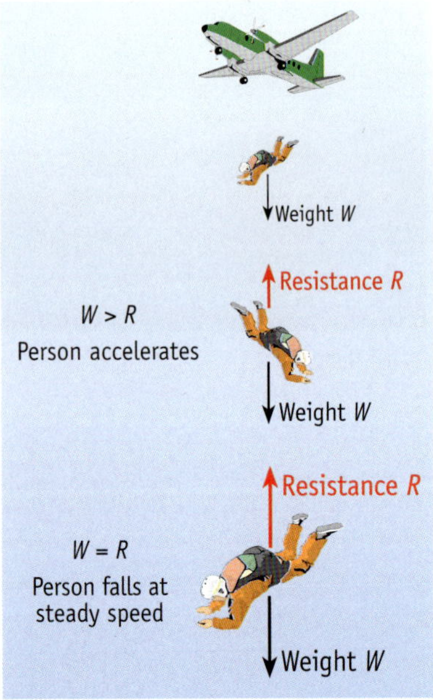

Fig. 9.14

If the friction between the moving parts in machines were large, the machines would rapidly overheat and wear out. By putting a slippery substance called a **lubricant** between moving parts, the friction can be reduced to small levels. Common lubricants are oils, greases and graphite. Air is used as a lubricant in the linear air track and in a hovercraft.

Fig. 9.15
Falling at a steady speed.

FIG. 9.14 shows a person falling from an aeroplane. Initially the only force acting on him is his weight and he accelerates down. As he speeds up, air resistance (air friction) begins to act. Air resistance opposes motion and thus acts upwards. As the speed increases, the size of the air resistance increases. Eventually a speed is reached where the size of the air resistance is equal to the weight of the man. He now no longer accelerates since the resultant force on him is zero. Since he is moving, by Newton's 1st Law, he continues to move with constant velocity FIG. 9.15. This velocity is called **terminal velocity**.

Problem 8: A crane lifts a block of mass 20 kg from the ground to the top of a building by means of a cable.

(i) What force does the cable exert on the block if the block is accelerating upwards at 1.5 m s^{-2}?

(ii) What force does the cable exert on the block if it is moving upwards at constant speed?

Solution: FIG. 9.16 shows the forces acting on the block. They are its weight ($W = 20 \times g$) and the tension in the string (T).

(i) If the block accelerates up, T must be bigger than W and the net upward force is $T - W$. Applying $F = ma$ gives:

$T - W = ma$ i.e. $T - (20)(9.8) = (20)(1.5)$

$\Rightarrow T = 226$ N

(ii) Constant velocity $\Rightarrow$ acceleration = 0 m s^{-2}. Applying $F = ma$ gives:

$T - W = ma$ i.e. $T - W = (20)(0)$

$\Rightarrow T = W = 196$ N (the weight of the block)

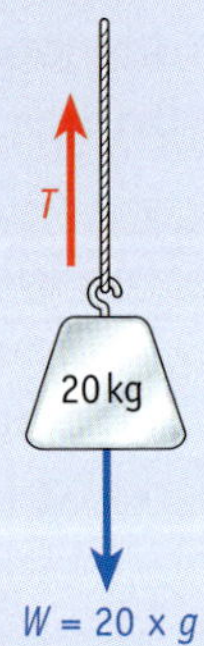

Fig. 9.16

Problem 9: A man of mass 80 kg stands on a weighing scales in a lift. If the scales reads newtons, find the reading on it when the lift:

(i) is at rest,

(ii) moves upwards with a steady speed of 3 m s^{-1},

(iii) moves downwards with a steady speed of 4 m s^{-1},

(iv) accelerates upward at 2 m s^{-2},

(v) accelerates downwards at 2 m s^{-2},

(vi) accelerates downwards at 9.8 m s^{-2}.

Solution:

FIG. 9.17 shows the forces acting on the man. The upward force on the man due to the floor of the lift is called the **normal reaction**. This is the reading on the scales. In each part of the Problem we apply $F = ma$ in the direction of the acceleration.

In each of (i), (ii) and (iii), the acceleration is zero.

Therefore: $80g - N = (80)(0)$

$\Rightarrow \quad N = 80g = 784$ newtons = weight of the man

(iv) $N - 80g = (80)(2) \Rightarrow N = 160 + 784 = 944$ newtons

(v) $80g - N = (80)(2) \Rightarrow N = 784 - 160 = 624$ newtons

(vi) $80g - N = (80)(9.8) \Rightarrow N = 0$

The reading on the scales is zero and relative to the lift the man is **weightless**. Astronauts make use of this fact in training. A plane flies to certain height and then falls freely. As the plane falls the astronauts experience weightlessness (FIG. 9.18).

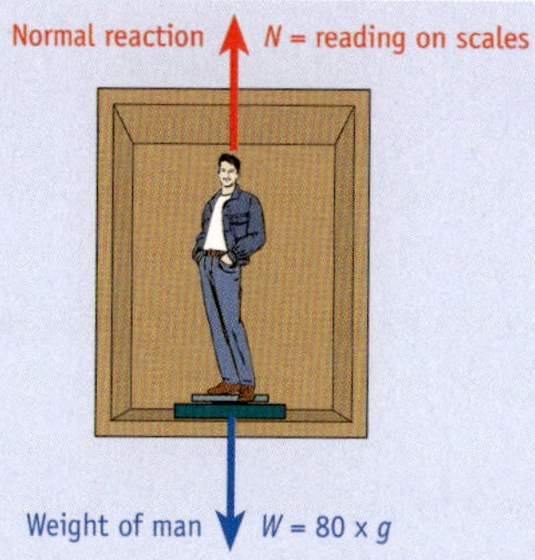

Fig. 9.17

Fig. 9.18

EXERCISE 9.2

1. A car of mass 725 kg is travelling along a straight level road at a uniform speed of 30 m s^{-1}. If the total resistance to its motion is 600 N, what is the driving force due to the engine? Give a reason for your answer.

   If the engine is capable of exerting a force of 1000 N, what is the least time it would take the car to reach a velocity of 100 km h^{-1} if it starts from rest and the forces opposing motion remain the same?

2. A bullet travelling at 200 m s^{-1} enters a block of wood and emerges from the other side travelling at 50 m s^{-1}, 0.005 seconds later. If the mass of the bullet is 0.002 kg find the average force exerted on the bullet by the block.

3. An object of mass 2000 kg is being lowered from the top of a building by means of a metal cable. Find the tension in the cable if:
   (i) the object is being lowered at constant velocity,
   (ii) the object is being lowered with a downward acceleration of 2 m s^{-2}.

4. A mass of 10 kg hangs from a spring balance which is fixed to the ceiling of a lift. If the balance reads newtons, find the reading on it when the lift:
   (i) ascends at a constant speed of 3 m s^{-1},
   (ii) ascends with an acceleration of 2 m s^{-2},
   (iii) descends with an acceleration of 2 m s^{-2},
   (iv) descends at 3 m s^{-1},
   (v) descends with an acceleration of 9.8 m s^{-2}.

5. A man of mass 100 kg stands on a weighing scales in a lift. If the scales reads newtons, find the reading on it when the lift:
   (i) is at rest,
   (ii) moves upwards with a steady speed of 4 m s^{-1},
   (iii) moves downwards with a steady speed of 4 m s^{-1},
   (iv) accelerates upward at 3 m s^{-2},
   (v) accelerates downwards at 3 m s^{-2},
   (vi) accelerates downwards at 9.8 m s^{-2}.

6. When a tennis ball is struck by a racquet its momentum changes by 2.25 kg m s^{-1} in a certain direction. If the ball is in contact with the racquet for 0.05 s what is the average force exerted by the racquet on the ball?

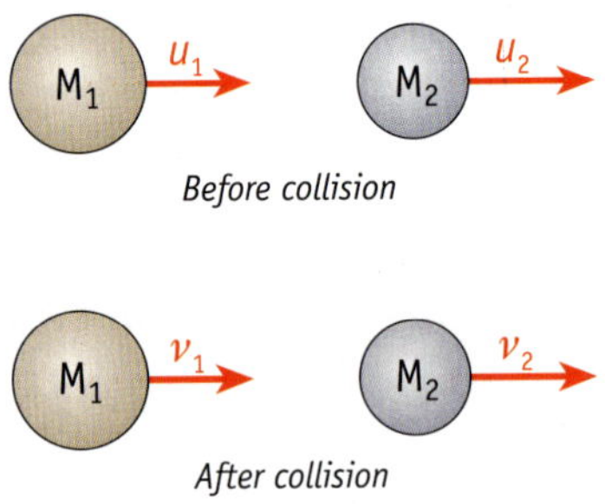

Fig. 9.19

THE PRINCIPLE OF CONSERVATION OF MOMENTUM

Experimentally it is found that when bodies collide, the total momentum of the bodies before and after the collision is the same. This is an example of the principle of conservation of momentum:

THE PRINCIPLE OF CONSERVATION OF MOMENTUM states that in any interaction between two or more bodies the total momentum of the bodies before the interaction is equal to the total momentum after the interaction ***provided*** no external forces act on the system of bodies.

In FIG. 9.19 by the Principle of Conservation of Momentum we have:

$$m_1u_1 + m_2u_2 = m_1v_1 + m_2v_2$$

Problem 10: A body of mass 20 kg moving at a speed of 4 m s^{-1} collides with a body of mass 14 kg which is at rest. If the two bodies stick together on colliding, find the velocity with which the combined mass moves off.

Solution: FIG. 9.20 shows the situation before and after the collision.
The mass of the combined body after the collision is 34 kg.
Let its unknown velocity be v.
'Conservation of Momentum' ⇒
Momentum before collision = Momentum after collision
i.e. $(20)(4) + (14)(0) = 34 \times v$ i.e. $80 + 0 = 34v$
⇒ $v = 80/34 = 2.35$ m s^{-1} in the direction that the 20 kg mass was originally moving.

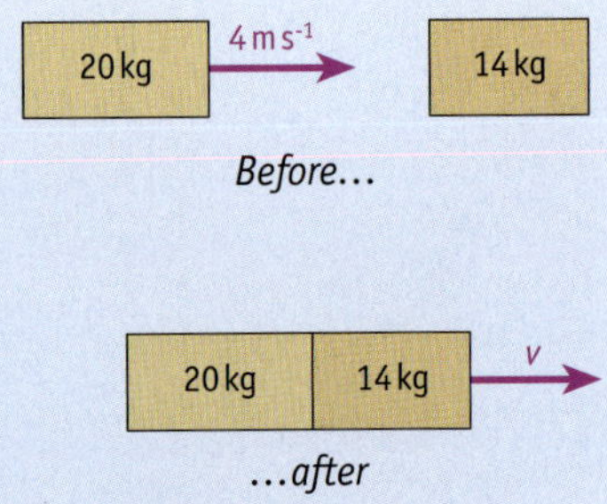

Fig. 9.20

To find the total momentum of a number of bodies we must **take account of the fact that momentum is a vector quantity**. For bodies moving in the same straight line, momentum in one direction is given a positive sign and momentum in the opposite direction is given a negative sign. The total momentum is then found by algebraic addition.

Problem 11: A 10 kg mass is moving at a speed of 5 m s^{-1}. A mass of 4 kg is moving at a speed of 20 m s^{-1} in the opposite direction. Find their total momentum.

Solution: Total momentum $= (10)(5) + (4)(-20) = -30$
∴ Total momentum = 30 kg m s^{-1} in the original direction of motion of the 4 kg mass.

Problem 12: A body of mass 20 kg with a velocity of 3 m s^{-1} collides with another body of mass 15 kg which is moving at 6 m s^{-1} in the opposite direction.
The two bodies stick together on colliding. Find the velocity with which the combined mass moves off.

Solution: FIG. 9.21 shows the situation before and after the collision:
Momentum before = Momentum after ⇒
$(20)(3) + (15)(-6) = 35v$
⇒ $60 - 90 = 35v$ ⇒ $v = -30/35$
i.e. $v = 0.86$ m s^{-1} in direction of motion of 15 kg mass before the collision.

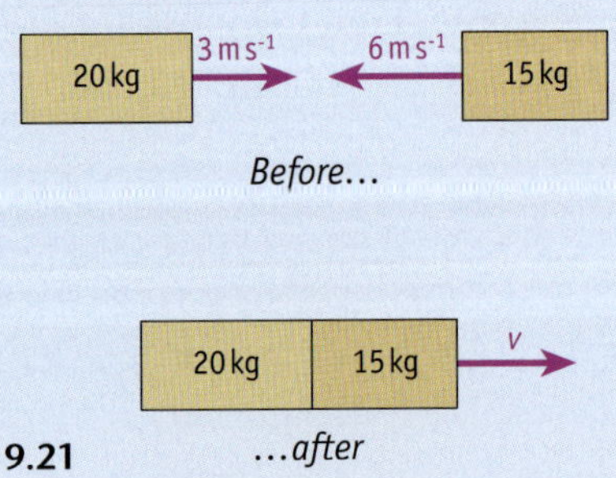

Fig. 9.21

Problem 13: A gun of mass 3 kg fires a bullet of mass 10 grams with a velocity 500 m s^{-1}. Calculate the recoil velocity of the gun.

Solution: Fig. 9.22 shows the situation before and after the explosion. Before the explosion the system had no momentum. After the explosion both the gun and the bullet have momentum – but in the opposite direction to each other. The sum of these momenta must be zero.

Let v metres per second be the recoil velocity of the gun. Then by the Principle of Conservation of Momentum we have:

Momentum before = Momentum after

$\Rightarrow \quad (3)(0) + (0.01)(0) = 3v + (0.01)(500)$

$\Rightarrow \quad 0 = 3v + 5 \Rightarrow v = -1.67$ m s^{-1}

i.e. the gun recoils **backwards** with a speed of 1.67 m s^{-1}.

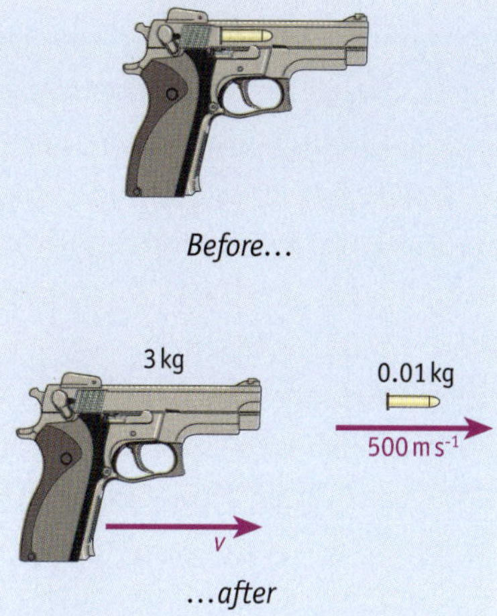

Fig. 9.22

ACCELERATION OF SPACECRAFT

Conservation of momentum can also be used to explain the acceleration of a rocket or jet. The rocket 'fires' hot gases backwards with a certain momentum. The rocket gets the same momentum in the opposite direction, i.e. the change in the rocket's momentum is equal to the momentum given to the gases. The rocket therefore accelerates forward. The same holds for a jet aircraft. Since the Principle of Conservation of Momentum follows from Newton's three laws, this explanation of how a spacecraft accelerates and that on page 95 are equivalent.

Problem 14: A spacecraft of mass 400 kg moving at 1000 m s^{-1} ejects an object of mass 20 kg at a speed of 2000 m s^{-1} at right angles to the direction in which the craft is moving. Calculate the resultant velocity of the craft in magnitude and direction.

Solution: Since no forces act in the original direction of motion the velocity of the craft in this direction remains 1000 m s^{-1}. Let x be the recoil velocity acquired by the craft at right angles to the 1000 m s^{-1} (Fig. 9.23).

Apply conservation of momentum perpendicular to the original direction of motion:

$(400)(0) = (380)(x) + (20)(-2000) \Rightarrow$

$0 = 380x - 40\,000 \Rightarrow x = 105.26$ m s^{-1}

Resultant velocity v, of spacecraft (Fig. 9.24) is found as follows:

$v^2 = (105.26)^2 + (1000)^2 \Rightarrow$

$v = 1005.5$ m s^{-1}

$$\text{Tan } \theta = \frac{105.26}{1000} \Rightarrow \theta = 6°$$

i.e. $v = 1005.5$ m s^{-1} at 6° to original direction of motion.

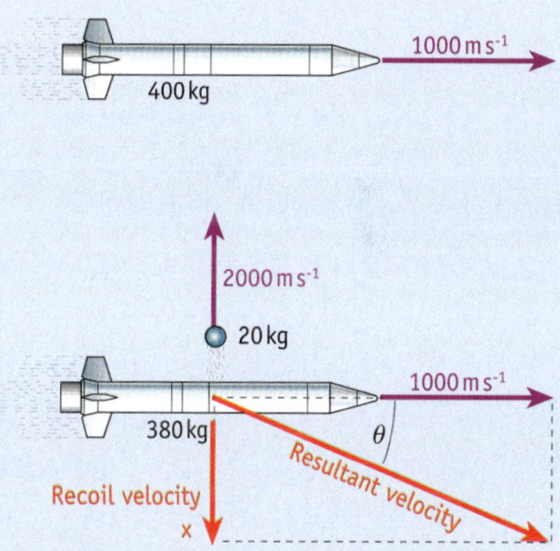

Fig. 9.23

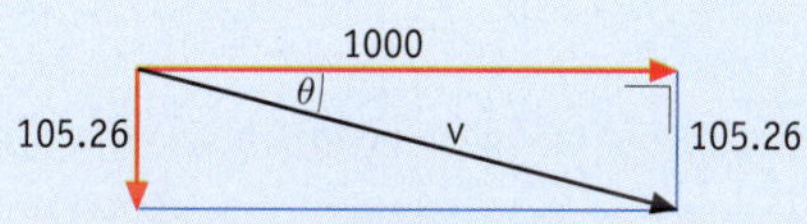

Fig. 9.24

EXERCISE 9.3

1. What is the momentum of a car of mass 800 kg moving at a speed of 20 m s^{-1}?
2. What is the momentum of a car of mass 1200 kg moving with a velocity of:
   (a) 30 m s^{-1}, (b) 0 m s^{-1},
   (c) 100 km/h, (d) 500 cm s^{-1}?
3. An object has a momentum of 40 000 kg m s^{-1}. What is the speed of the object if its mass is 1200 kg? What would its speed be if its mass was 2 kg?
4. What is the total momentum of a mass of 20 kg moving with a velocity of 40 m s^{-1} and a mass of 50 kg moving with a velocity of 20 m s^{-1} in the opposite direction?
5. A car of mass 800 kg moving at 20 m s^{-1} collides with another car of mass 1500 kg which is at rest. If both cars stick together, find their initial velocity after the collision.
6. A railway carriage of mass 6000 kg moving at 10 m s^{-1} catches up with and becomes coupled to another carriage of mass 2000 kg moving in the same direction at 2 m s^{-1}. Find their initial velocity after the collision.
7. A railway carriage of mass 6000 kg moving at a speed of 10 m s^{-1} collides with and becomes coupled to another carriage of mass 2000 kg moving in the opposite direction at 2 m s^{-1}. Find their initial velocity after the collision.
8. A gun of mass 2 kg fires a bullet of mass 10 grams with a velocity of 400 m s^{-1}. Find the initial recoil velocity of the gun.
9. A 100 kg block moving at 10 m s^{-1} collides with a 60 kg block moving in the opposite direction at 15 m s^{-1}. After the collision the 60 kg block initially moves with a speed of 8 m s^{-1} in the opposite direction to its original velocity. If no external forces acted on the system, find the velocity of the 100 kg mass just after the collision.
10. A sphere A of mass m moving with a speed of 0.6 m s^{-1}, collides with a sphere B of mass 3 m which is at rest. After the collision, A and B move in the same direction and A's initial velocity is 0.2 m s^{-1}. Find the initial speed of B.
11. A gun of mass 500 kg fires a shell of mass 2 kg with a muzzle velocity of 500 m s^{-1}. Calculate:
    (i) the velocity of recoil of the gun,
    (ii) the force required to stop the gun in a distance of 0.25 m.
12. A bullet of mass 12 grams travelling at 200 m s^{-1} enters a block of wood which is at rest. It emerges from the other side at a speed of 50 m s^{-1}, 0.002 seconds later. If the block has a mass of 10 kg, find the velocity it acquires. Assume the mass of the block remains unchanged. Verify that the force exerted by the block on the bullet and the force exerted on the block by the bullet have the same magnitude.
13. A car and a lorry are travelling along roads that intersect at right angles. The car of mass 1000 kg is travelling at 50 m s^{-1} and crashes into the lorry which has a mass of 4000 kg and is travelling at 20 m s^{-1} when they reach the intersection. If the vehicles coalesce on impact, find the magnitude and direction of the velocity of the wreckage immediately after the collision.
14. A rocket of mass 5000 kg is travelling at 40 m s^{-1}. It emits a mass of 10 kg at 2000 m s^{-1} at right angles to its direction of motion. Assuming you can neglect the loss in mass of the rocket, find the velocity of the rocket in magnitude and direction after the expulsion.
15. A body of mass 80 g travelling with a speed of 5 m s^{-1} collides with another body of mass 200 g which is at rest. After the collision both move off together. Calculate:
    (i) the change in momentum of each body,
    (ii) the average magnitude of the force exerted by each body on the other if the change in momentum occurs in 0.1 s.

EXPERIMENT MECHANICS 3

To Verify the Principle of Conservation of Momentum

Summary of Method

In this experiment you will measure the velocity of a trolley before and after it collides with another trolley to which it sticks. You will measure the mass of each trolley. You will calculate the momentum of the trolley before the collision and the momentum of the combined trolleys after the collision and see that they are the same.

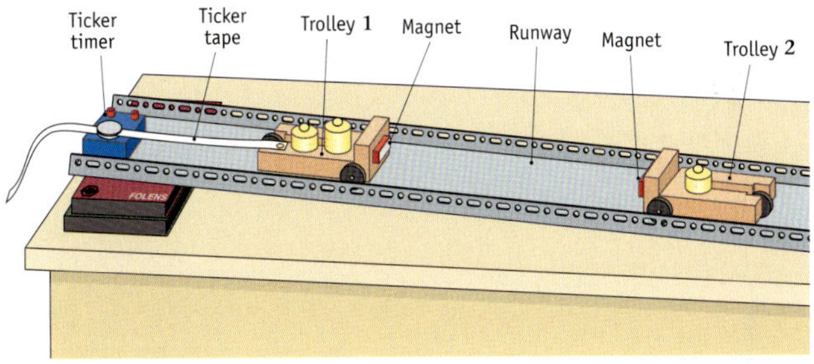

Fig. 9.25

Equipment Needed

- A runway, two trolleys and some known masses (e.g., 250, 500 and 1000 grams)
- A ticker timer, ticker tape and power supply
- Two magnets that can be mounted on the fronts of the trolleys (or some other means of making the trolleys stick when they collide)

Method

1. Use the equipment as in Fig. 9.25.
2. Raise one end of the runway so that trolley 1 with the ticker tape attached will move down the runway at constant speed if given a push.
3. Place trolley 1 at the upper end of the runway and trolley 2 about half way down. Turn on the ticker timer. Give trolley 1 a push, so that it moves down the runway and collides with and sticks to trolley 2.
4. Stop the trolleys at the end of the runway and remove the ticker tape. It should look like that in Fig. 9.26.

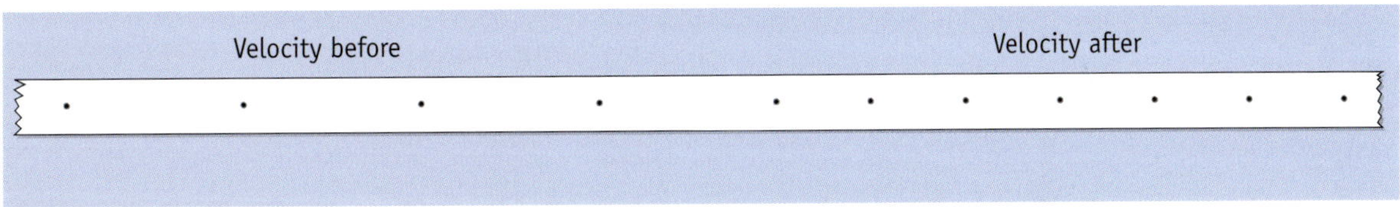

Fig. 9.26

5. Calculate the velocity of trolley 1 before (u) and after (v) the collision from the ticker tape. (Use the method shown on page 61).
6. Using a balance, find the mass of trolley 1 (m_1) and trolley 2 (m_2).
7. Find the momentum before the collision (m_1u) and the momentum after ($(m_1 + m_2)v$).

Result

8. The momentum before the collision will be equal to the momentum after the collision, i.e. $m_1u = (m_1+m_2)v$
9. Repeat the experiment with different masses on each trolley and with the trolleys travelling at different speeds. All results can be recorded in the Table.

Before collision					After collision				
Velocity from Ticker tape			Mass	**Momentum**	Velocity from Ticker tape			Mass	**Momentum**
s_1	number of spaces (n) in length s_1	$u = s_1/(n/50)$	m_1	m_1u	s_2	number of spaces (n) in length s_2	$v = s_2/(n/50)$	m_2	$(m_1 + m_2)v$

Questions

1. If magnets are used in this experiment, why is it that the forces they exert do not change the total momentum of the system?
2. List three precautions that should be taken in this experiment to ensure an accurate result.

CHAPTER CHECKLIST

- **Define:** Force; The newton; Mass; Weight; Momentum.
- **State** the unit of: Force; Mass; Weight; Momentum.
- **State:** Newtons three laws of motion; The principle of conservation of momentum.
- **Show** that: $F = ma$ is a special case of Newton's second law.
- **Recall** that: Force is a vector; Momentum is a vector; The acceleration of a body is directly proportional to the force causing it; The acceleration a body gets from a given force is inversely proportional to its mass; Force is equal to the rate of change of momentum; Friction opposes motion and can be reduced by using lubricants.
- **Describe** and carry out an experiment to: Show that $a \propto F$; Verify the principle of conservation of momentum.
- **Explain**: The accelertion of spacecraft and jet aircraft using Newton's third law; The acceleration of spacecraft and jet aircraft using the principle of conservation of momentum.
- **Express** the newton in terms of basic units.
- **List** examples of the importance of friction in everyday life.
- **Recall** and use the formulae:

$$a \propto F; \quad a \propto \frac{1}{m}; \quad F = ma; \quad W = mg; \quad p = mv;$$

$$F = \frac{mv - mu}{t}; \quad m_1u_1 + m_2u_2 = m_1v_1 + m_2v_2$$

Pressure, Gravity and Moments

CHAPTER 10

Density

To compare the masses of the same volumes of different substances we use the quantity **density**. By definition, the density of a substance is its **mass per unit volume**. Thus the density of a substance is the mass of 1 m^3 of it. The symbol for density is ρ (the Greek letter rho). Density is a **scalar quantity**. For any object it follows that:

DENSITY
The ***density*** of a substance is its mass per unit volume.

UNIT OF DENSITY
The unit of density is the ***kilogram per cubic metre (kg m^{-3})***.

$$\textbf{Density} = \frac{\textbf{Mass}}{\textbf{Volume}} \qquad \textbf{i.e.}\ \rho = \frac{m}{V}$$

Since $\rho = \frac{m}{V}$ Unit of density $= \frac{\text{Unit of mass}}{\text{Unit of volume}} = \frac{\text{kg}}{\text{m}^3}$

= the **kilogram per cubic metre (kg m^{-3})**

NOTE
Note that:
1 kg = 10^3 grams ⇒ 1 gram = 10^{-3} kg and
1 m^3 = 10^6 cm^3 ⇒ 1 cm^3 = 10^{-6} m^3

Problem 1: A piece of wood has a volume of 0.02 m^3 and a mass of 10 kg. Find its density.

Solution: $\rho = \frac{m}{V} = \frac{10}{0.02} = 500\ \text{kg m}^{-3}$

Problem 2: A piece of copper of mass 420 grams has a volume of 47.19 cm^3. What is its density?

Solution: 420 grams = 420 × 10^{-3} kg = 0.42 kg. 47.19 cm^3 = 47.19 × 10^{-6} m^3

$$\rho = \frac{m}{V} = \frac{0.42}{47.19 \times 10^{-6}} = 8.9 \times 10^3\ \text{kg m}^{-3}$$

Problem 3: The density of aluminium is 2.7 × 10^3 kg m^{-3}. What is the volume of 120 grams of aluminium?

Solution: 120 grams = 120 × 10^{-3} kg

$$\rho = \frac{m}{V} \Rightarrow V = \frac{m}{\rho} = \frac{120 \times 10^{-3}}{2.7 \times 10^3} = 4.44 \times 10^{-5}\ \text{m}^3$$

EXERCISE 10.1

1. Convert each of the following to kg:
   (i) 200 g (ii) 4 g
   (iii) 2×10^5 g (iv) 24 mg
2. Convert each of the following to m^3:
   (i) 1 cm^3 (ii) 120 cm^3
   (iii) 4 litres (iv) 2×10^6 cm^3
3. Convert each of the following to m^2:
   (i) 1 cm^2 (ii) 220 cm^2
   (iii) 4 mm^2 (iv) 3×10^4 cm^2
4. Find the density of a piece of wood of mass 4 kg and volume 0.012 m^3.
5. A piece of metal has a volume of 1.61 m^3 and a mass of 1.8×10^4 kg. Find its density.
6. The density of silver is 1.05×10^4 kg m^{-3}. What is the volume of 4 kg of silver? What is the mass of 3 cm^3 of silver?
7. What is the mass of 1 cm^3 of Mercury? The density of Mercury is 1.36×10^4 kg m^{-3}.
8. A sheet of steel has a layer of tin electroplated onto one side of it. The thickness of the layer is 2×10^{-6} m and 20 grams of tin is used in all. What is the area of one side of the sheet of steel? (Density of tin = 7.3×10^3 kg m^{-3})

PRESSURE

PRESSURE
Pressure is ***force per unit area***.

UNIT OF PRESSURE
The unit of pressure is the ***pascal* (Pa)**.

1 pascal = 1 newton per square metre
i.e. 1 Pa = 1 N m^{-2}

The pascal can be expressed in basic units as:
1 Pa = 1 N m^{-2}
= 1 (kg m s^{-2}) m^{-2}
= 1 kg m^{-1} s^{-2}

If you walk on soft snow, your weight will cause you to sink into it. If, however, you wear snow shoes that spread your weight over a much larger area of snow you will not sink. In a similar way, it is easier to push a sharp thumbtack into a piece of wood than to push a blunt nail into the same wood. In situations like these it is the **pressure** that determines what happens. Pressure is defined as **force per unit area**.

Thus if a force F acts evenly over an area A, the pressure P at any point on that area is given by:

$$\textbf{Pressure} = \frac{\textbf{Force}}{\textbf{Area}} \qquad \textbf{i.e.} \quad P = \frac{F}{A}$$

Pressure is a **scalar quantity**, its symbol is P or p.

Since $P = \frac{F}{A}$, Unit of pressure = $\frac{\text{Unit of force}}{\text{Unit of area}} = \frac{\text{Newton}}{\text{Square metre}}$

= **newton per square metre (N m^{-2} or N/m^2)**

The newton per square metre is called the **pascal (Pa)**.

Problem 4: A force of 600 N acts evenly on a surface of area 5 m^2. What pressure does it exert?

Solution: $P = \frac{F}{A} = \frac{600}{5} = 120$ Pa

Problem 5: A force of 40 N acts on an area of 25 cm^2. What pressure does it exert?

Solution: $1\ m^2 = 10\ 000\ cm^2 \Rightarrow 1\ cm^2 = \frac{1}{10\ 000}\ m^2 = 1 \times 10^{-4}\ m^2$

$\Rightarrow 25\ cm^2 = 25 \times 10^{-4}\ m^2$

$P = \frac{F}{A} = \frac{40}{25 \times 10^{-4}} = 16\ 000$ Pa

Problem 6: FIG. 10.1 shows a rectangular block of mass 5 kg resting on a horizontal table. Find the value of the pressure on the table if (i) side A and (ii) side B is on the table.

Solution: Force acting on table = weight of block = (5)(9.8) = 49 N

(i) Side A: $P = \frac{F}{A} = \frac{49}{(0.04)(0.12)} = 1.02 \times 10^4$ Pa

(ii) Side B: $P = \frac{F}{A} = \frac{49}{(0.1)(0.12)} = 4.08 \times 10^3$ Pa

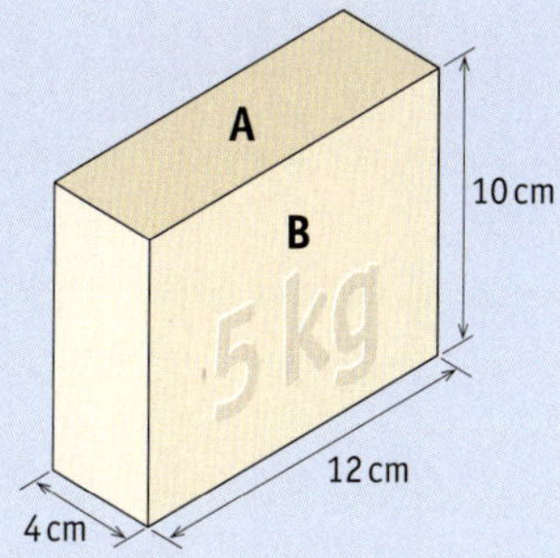

Fig. 10.1

EXERCISE 10.2

1. A force of 100 N acts uniformly over a surface of area 5 m^2. Calculate the pressure at any point on the surface.
2. A force of 60 N acts uniformly over an area of 25 cm^2. What is the pressure at any point on that area?
3. Calculate the pressure on the table caused by the block shown in FIG. 10.2 if:

   (i) side A, (ii) side B and (iii) side C

   is on the table. The mass of the block is 4 kg and acceleration due to gravity is 9.8 m s^{-2}.

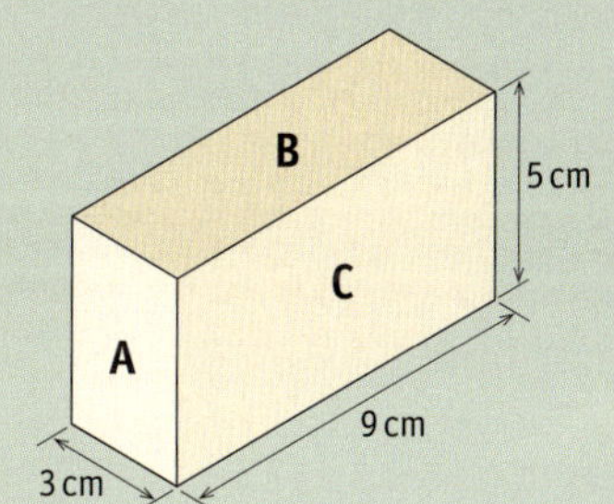

Fig. 10.2

4. The pressure on a metal sheet is 400 Pa. If the area of the sheet is 0.06 m^2, calculate the force on the sheet.
5. The pressure on the cover of a book due to the Earth's atmosphere is 1×10^5 Pa. If the area of the cover of the book is 621 cm^2, calculate the force on the cover of the book due to the atmosphere.
6. The pressure at any point on one side of a disc of radius 10 cm is 500 Pa. What is the total force on this side of the disc?
7. A tank of oil, in the shape of a cube of side 2.2 m, sits on a horizontal surface. The pressure at the bottom of the tank due to the oil is 556 Pa. What is the weight of the oil? What is the density of the oil?
8. A cylindrical metal block of mass 32 kg stands with its base on a horizontal table. The radius of the base of the cylinder is 4 cm. Calculate the value of the pressure on the table.

PRESSURE DUE TO A LIQUID

In FIG. 10.3 the weight of the liquid is acting on the base of the beaker and thus exerts pressure on the base. It can be proved that:

At depth h in a liquid of density ρ where acceleration due to gravity is g, the pressure P due to the liquid is given by:

$$P = \rho g h$$

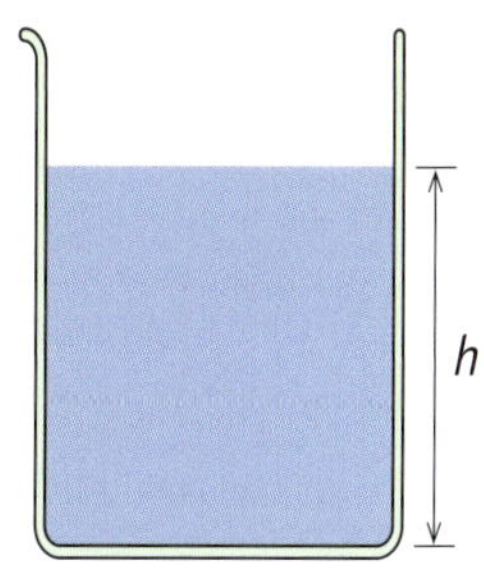

Fig. 10.3 The weight of the liquid causes pressure in the liquid.

Problem 7: Find the pressure due to the Mercury at a depth of 0.76 m in a container of Mercury. (Density of Mercury = 1.36×10^4 kg m^{-3})

Solution: $P = \rho g h = (1.36 \times 10^4)(9.8)(0.76) = 1.01 \times 10^5$ Pa

If an object is placed in a liquid it is found that the liquid exerts a pressure on that object too. It also exerts a pressure on the walls of the container.

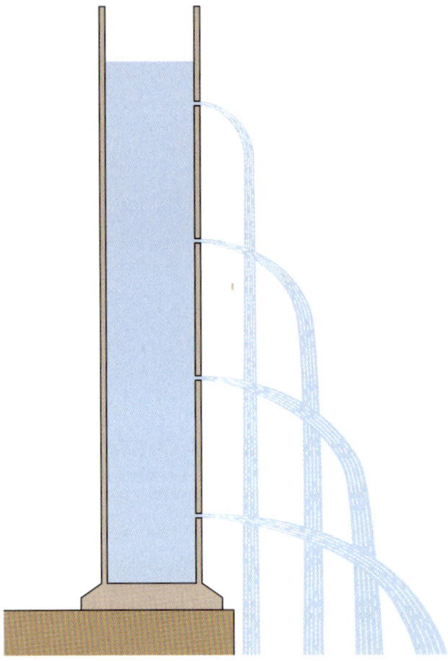

Fig. 10.4
Pressure increases with depth.

Pressure in a liquid obeys the following laws:

- **Pressure increases with depth**
  This follows from the above since $P = \rho gh$. It may be demonstrated as in FIG. 10.4, where the water from the deepest hole squirts out the farthest.
- **Pressure acts perpendicular to any surface put in the liquid.**
  This may be demonstrated as shown in FIG. 10.5. The water emerges from each hole perpendicular to the surface.
- **At a given depth the value of the pressure is the same in all directions.**

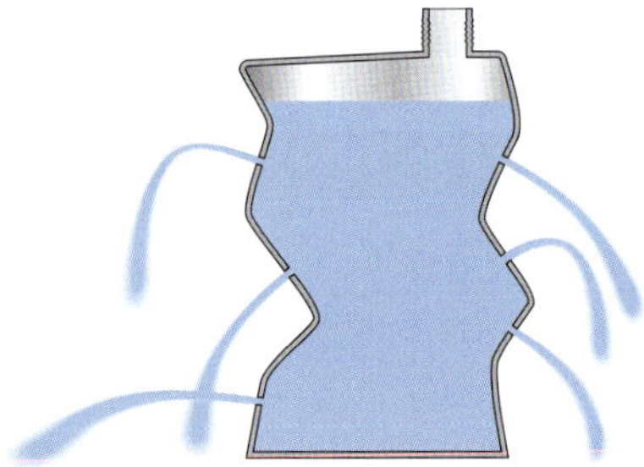

Fig. 10.5
Pressure acts at right angles to any surface in the liquid.

Problem 8: A rectangular wooden block has dimensions 14 cm × 12 cm × 20 cm. It is placed with its upper surface at a depth of 30 cm in a basin of water as in FIG. 10.6. Find:

(i) the pressure at the upper surface of the block due to the water,

(ii) the pressure at the lower surface of the block due to the water,

(iii) the force on the upper surface due to the water,

(iv) the force on the lower surface due to the water.

If the weight of the block is 26 N, determine whether the block will float or sink.
(density of water = 1000 kg m^{-3})

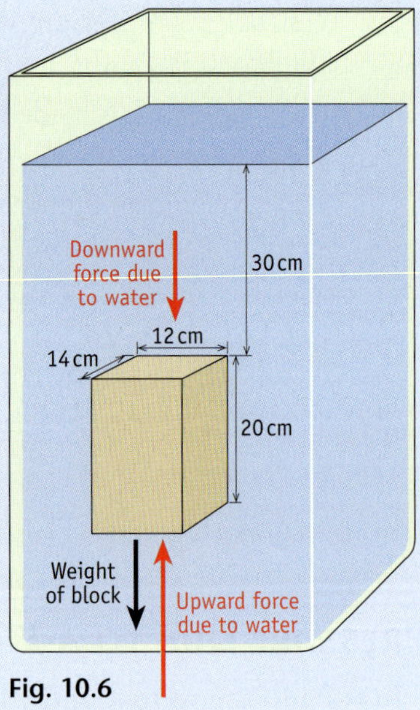

Fig. 10.6

Solution:

(i) Pressure at top = ρgh = (1000)(9.8)(0.3) = 2940 Pa

(ii) Pressure at bottom = ρgh = (1000)(9.8)(0.5) = 4900 Pa

(iii) $F = PA \Rightarrow$ Downward force on top = (2940)(0.12 × 0.14) = 49.4 N

(iv) Upward force on bottom = PA = (4900)(0.12 × 0.14) = 82.3 N

Resultant force on block (FIG. 10.6) is upward and of magnitude
= 82.3 – (49.4 + 26.0) = 6.9 N

Therefore, if the block is released it will float.

EXERCISE 10.3

(g = 9.8 m s^{-2})

1. A tank contains 100 kg of water. If the area of the base of the tank is 0.5 m^2, calculate the pressure of the water at the bottom of the tank.
2. Find the pressure, due to the water, at a depth of 33 m in water. (Density of water = 1000 kg m^{-3})
3. Find the pressure at a depth of 20 cm in a graduated cylinder full of (i) water (ii) Mercury. Assume the pressure at the top of each liquid is zero. (Density of water = 10^3 kg m^{-3}, Density of Mercury = 13.6 × 10^3 kg m^{-3})
4. A rectangular block has dimensions 10 cm × 20 cm × 30 cm. It is placed with its upper surface at a depth of 1m in water as in FIG. 10.7. Find:
   (i) the pressure at the upper surface of the block due to the water,
   (ii) the pressure at the lower surface of the block due to the water.
   (iii) the force on the upper surface due to the water,
   (iv) the force on the lower surface due to the water.
   If the weight of the block is 70 N determine whether the block will float or sink. (Density of water = 1000 kg m^{-3})

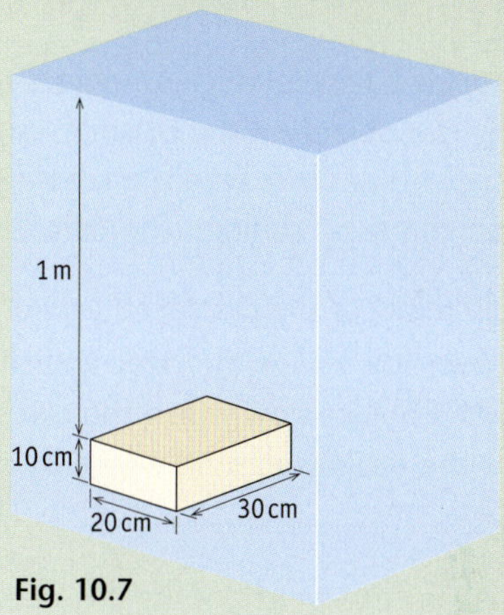

Fig. 10.7

UPTHRUST ON AN OBJECT IMMERSED IN A FLUID

FIG. 10.8(A) shows an object immersed in a liquid. The pressure of the liquid acts inward on the object as shown. Because pressure increases with depth, the pressure underneath is greater than the pressure on top. The liquid therefore exerts an overall upward force on the body. The last numerical problem showed an example of this. This upward force due to the liquid is called the **buoyancy force** or the **upthrust**.

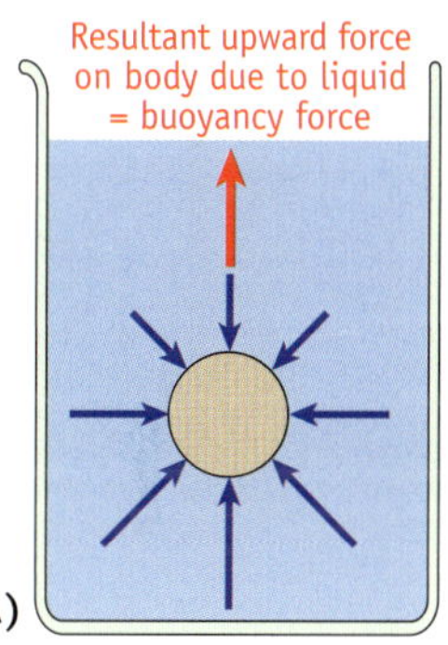

Fig. 10.8 (A)

The ancient Greek, Archimedes (287 BC – 212 BC) realised that the size of the upthrust is the same as the weight of the liquid displaced by the object. This fact is now known as **Archimedes' Principle**:

ARCHIMEDES' PRINCIPLE states that when an object is partially or completely immersed in a fluid it experiences an upthrust equal in magnitude to the weight of the fluid displaced.

Fig. 10.8 (B)
The upthrust on the submerged part of the ship keeps it afloat.

EXPERIMENT

TO DEMONSTRATE THE PRINCIPLE OF ARCHIMEDES

In this experiment you will measure:

(i) the upthrust on a submerged object and

(ii) the weight of liquid it displaces.

The upthrust and the weight will be the same, thus verifying Archimedes' Principle.

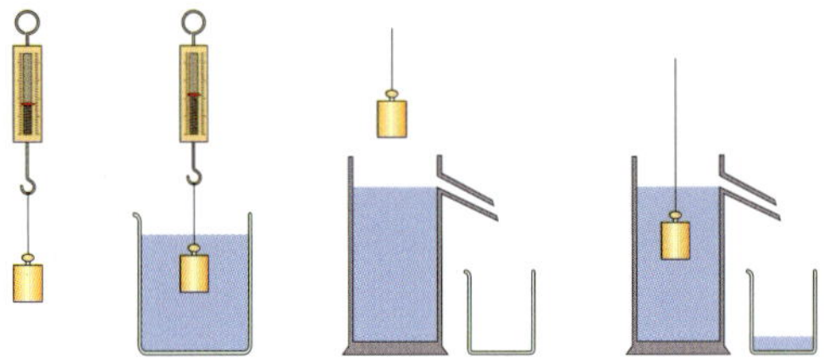

Fig. 10.9

To Find the Upthrust

- On a spring balance, weigh an object that sinks in water (FIG. 10.9).
- Keeping the object on the balance, lower the object into a beaker of water until it is completely submerged (FIG. 10.9). Take the reading on the balance.
- Subtract the two readings. The difference between the two readings is the upthrust.

To Find the Weight of the Displaced Liquid

- Fill an overflow can with water and allow it to settle.
- Lower the object slowly into the can. Collect the water displaced in a previously weighed beaker.
- Weight the beaker and the water. By subtraction find the weight of the displaced water.

Result

The upthrust and the weight of the displaced water will be the same, thus verifying Archimedes' Principle.

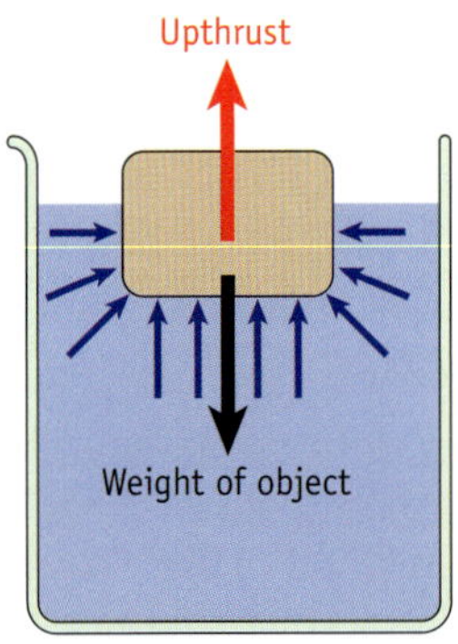

Fig. 10.10
The upthrust on the submerged part of a floating object is equal to the weight of the object.

THE LAW OF FLOTATION

When an object is floating in a liquid (FIG. 10.10) there is still an upthrust on the submerged part of the object. By Archimedes' Principle the upthrust equals the weight of the liquid displaced. Since the object is not accelerating vertically, this upthrust must be equal in magnitude to the weight of the object. Thus a floating object displaces its own weight of liquid. This fact is called the Law of Flotation.

THE LAW OF FLOTATION states that the weight of a floating body is equal to the weight of the fluid it displaces.

The truth of the Law of Flotation can easily be demonstrated in the laboratory as follows:

- On a spring balance weigh an object that floats in water.
- Fill an overflow can with water and allow the water to settle.
- Lower the object very slowly into the can until it floats. Collect the water displaced in a previously weighed beaker.
- Weigh the beaker with the displaced water in it. By subtraction find the weight of the displaced water.

Result: The weight of the object will be equal to the weight of the water displaced, thus verifying the Law.

Hydrometers

It follows from the Law of Flotation that a floating object will sink less in a denser liquid. The hydrometer (Fig. 10.11) is based on this fact. The denser the liquid, the higher the hydrometer floats. A scale calibrated on its side reads the density of the liquid. Hydrometers are used to find:

- the percentage of alcohol in beers, wines and spirits (alcohol is less dense than water),
- the density of sulphuric acid in a lead acid battery and hence determine the state of charge of the battery,
- the percentage of fat in milk and to check that water has not been added to the milk.

Fig. 10.11
A hydrometer.

Pressure in Gases and Atmospheric Pressure

The Earth is covered with a layer of air called the **atmosphere**. As you move farther from the surface of the Earth the air becomes less dense. Although there is no actual upper limit to the atmosphere, most of it lies below 200 km. The gases of the Earth's atmosphere have weight, and due to this weight there is pressure – called **atmospheric pressure** – acting on any object near the Earth. A gas exerts a pressure on a surface in the same way as a liquid exerts a pressure. The average value of atmospheric pressure at sea level is taken to be 1×10^5 Pa. Its actual value varies depending on the type of weather system present. Its value decreases as you move away from the surface of the Earth.

Atmospheric Pressure and Weather

In Ireland, the type of weather we get depends largely on the prevailing atmospheric pressure. When the pressure is low, the weather is usually cloudy, wet and windy. When the pressure is high the weather tends to be dry, with clear skies and little wind. In Summer, high pressure results in settled fine weather. In Winter, high pressure brings bright sunny days and cold frosty nights with clear skies.

Diving and Pressure – The Bends

As you dive deeper under water the pressure increases. It increases by about 1×10^5 Pa (the value of standard atmospheric pressure) every 10 m. The human body is designed to operate at normal atmospheric pressure. If divers breathe normal air (79% Nitrogen) while diving deeply, the pressure causes too much Nitrogen to become dissolved in their blood. If they returned to the surface too quickly, this nitrogen could form bubbles in their blood and they would experience 'the bends'. This is very painful and can easily kill divers. The cure for this is to place deep sea divers in a decompression chamber and allow the pressure to revert back to normal atmospheric pressure very slowly.

Demonstration of Atmospheric Pressure

The Collapsing Can

Get a clean empty can and place a small amount of water in it. Place it over a Bunsen and bring the water to the boil. The can becomes full of steam which pushes all the air out of it. Turn off the heat source quickly and carefully place the cap on the can taking care not to scald yourself.

> **WARNING!** Under no circumstances must you continue to heat the can with the cap on it – it will explode causing serious injury.

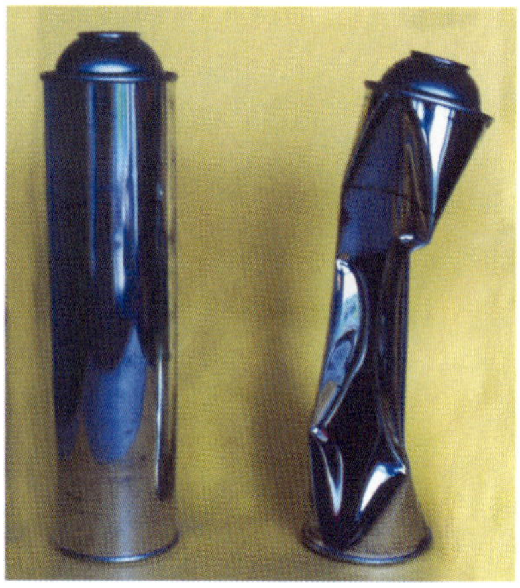

Fig. 10.12

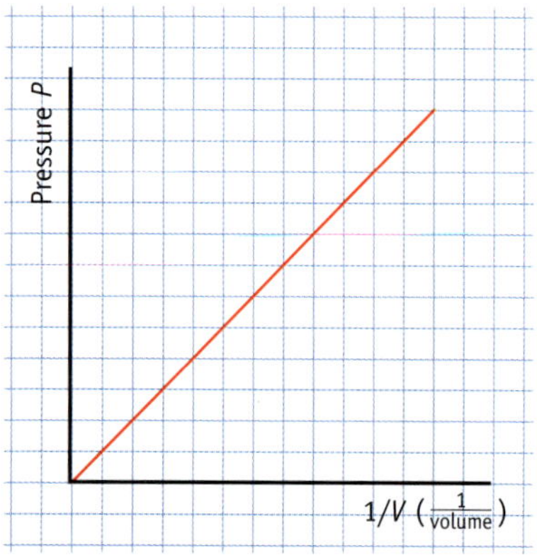

Fig. 10.13

As the can cools down, the steam inside turns back into water, creating a partial vacuum (i.e. very low pressure) in the can. Atmospheric pressure acts on the can from outside and is much bigger than the pressure inside. Thus the can collapses dramatically (FIG. 10.12).

GASES ARE COMPRESSIBLE

It is easy to change the volume of an amount of gas by changing its pressure. We say gases are highly compressible. This can be shown with a bicycle pump (). If you block the opening at the front of the pump, a fixed mass of air is trapped in the pump. If you push in the handle, you increase the pressure of the gas in the pump and at the same time decrease its volume. The greater the pressure the smaller the volume. Likewise, as you release the handle the volume increases and the pressure decreases.

BOYLE'S LAW

The relationship between the pressure p and the volume V of a fixed mass of gas is called **Boyle's Law**:

BOYLE'S LAW states that at constant temperature the volume of a fixed mass of gas is inversely proportional to its pressure.

This means that:

If the pressure is doubled the volume is halved.

If the pressure is trebled the volume is decreased to a third.

If the pressure is quartered the volume is increased four times etc...

Mathematically:

Boyle's Law $\Rightarrow \quad p \propto \frac{1}{V} \quad \Rightarrow \quad p = k\left(\frac{1}{V}\right) \quad \Rightarrow \quad pV = k$

For a fixed mass of gas at constant temperature
$\boldsymbol{pV = k}$ where k is a constant.

It follows that a graph of p against $\frac{1}{V}$ is a straight line through the origin (FIG. 10.13).

Problem 9: The volume of a certain mass of gas is 600 cm^3 at a pressure of 1×10^5 Pa. Find its volume if the pressure changes to 3.2×10^5 Pa, the temperature remaining constant.

Solution: $pV = \text{constant} \Rightarrow p_1V_1 = p_2V_2$, where V_2 is the required volume

Thus: $(1 \times 10^5)(600) = (3.2 \times 10^5)(V_2) \Rightarrow V_2 = 187.5$ cm^3

EXERCISE 10.4

1. A fixed mass of Oxygen has a volume of 3 m^3 at a pressure of 1×10^5 Pa. What is its volume if the pressure is increased to 3×10^5 Pa, the temperature remaining the same?
2. A certain mass of methane has a volume of 40 cm^3 at a pressure of 1×10^5 Pa. At what pressure would its volume be:
   (i) 160 cm^3, (ii) 80 cm^3,
   (iii) 1 cm^3?

   Assume the temperature remains the same.

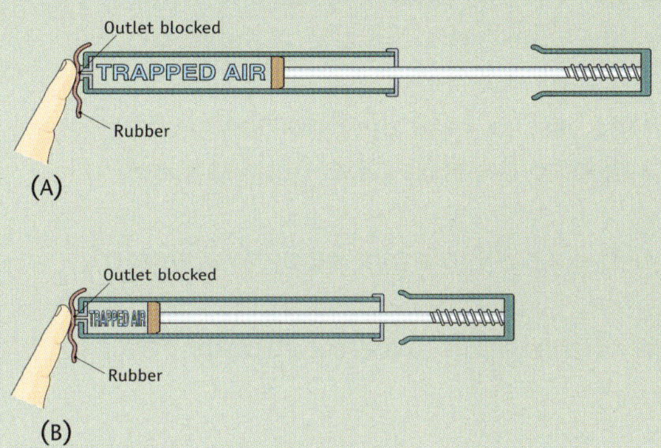

Fig. 10.14

3. FIG. 10.14(A) shows some air trapped in a bicycle pump. No air escapes and the temperature remains constant when it is compressed (FIG. 10.14(B)). Describe what changes, if any, take place in:
   (i) the mass of the air,
   (ii) the volume of the air,
   (iii) the pressure of the air,
   (iv) the density of the air.
4. The volume of a certain mass of gas is 700 cm^3 at a pressure of 1×10^5 Pa. Find its volume if the pressure changes to:
   (i) 2×10^5 Pa, (ii) 7×10^5 Pa and
   (iii) 5×10^4 Pa

   with the temperature remaining constant throughout.
5. The value of pressure × volume (i.e. pV) of a fixed mass of gas is 20 Pa m^3 at a temperature of 20°C. If the pressure is trebled and the temperature remains the same, find the new value of pV for the mass of gas.

VERIFICATION OF BOYLE'S LAW IN THE LABORATORY

FIG. 10.15 and FIG. 10.16 show a piece of equipment used to verify Boyle's Law. It is called a Boyle's Law apparatus. It consists of a thick-walled glass tube which contains the fixed mass of gas – air. The air is trapped in the tube by oil which also fills most of the reservoir. Air is pumped into the reservoir with a pump. This increases the air pressure above the oil in the reservoir and hence the pressure of the air trapped in the glass tube. The pressure of the air can be read from the Bourdon gauge (which may be graduated in atmospheres.)* The volume of the trapped air can be read from the scale next to the glass tube. An outer perspex shield is fitted in case the glass tube might shatter under pressure.

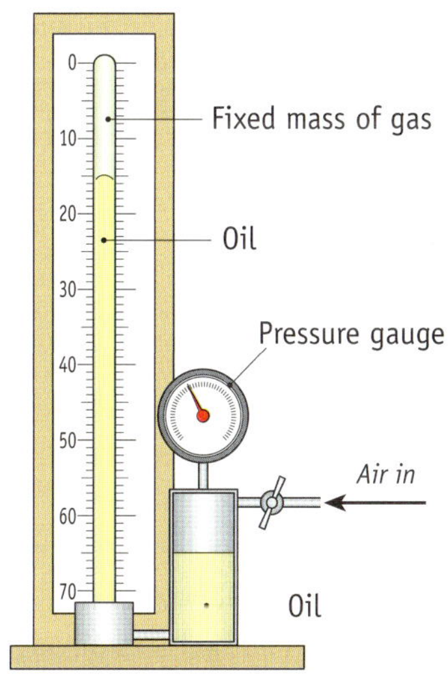

Fig. 10.15

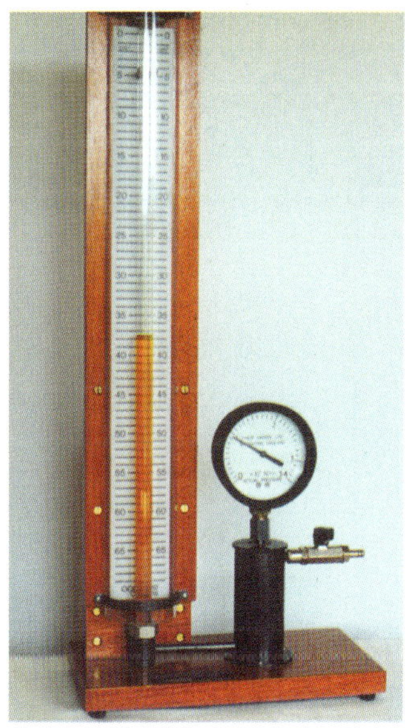

Fig. 10.16

After changing the pressure of the trapped air, always wait a minute or so before reading the pressure or the volume. This allows the air to reach room temperature. This is necessary because when the air is compressed or expanded, there may be slight changes in temperature.

---

* Sometimes it is convenient to express the pressure of a gas in terms of standard atmospheric pressure. Standard atmospheric pressure (1 atmosphere) is taken to be 1×10^5 Pa. Thus if the pressure of a gas is 2.2 atmospheres its pressure in pascals is: $= (2.2)(1 \times 10^5) = 2.2 \times 10^5$ Pa

EXPERIMENT **MECHANICS 5**

To Verify Boyle's Law.

Summary of Method

In this experiment, using a Boyle's Law apparatus, you will measure the pressure and volume of a fixed mass of a gas for a series of different values of the pressure. You will plot a graph of p against $1/V$. This will be a straight line through the origin (Fig. 10.13, page 110), thus verifying Boyle's Law.

Equipment Needed

- A Boyle's Law apparatus
- An air pump
- Optional – a hand vacuum pump

Method

1. Connect the air pump to the inlet of the oil reservoir (Fig. 10.15). Open the tap and pump air in until the pressure gauge reads its maximum value. Close the tap and remove the pump.
2. Wait a minute or two, then measure the pressure of the gas, i.e. read the Bourdon gauge. Measure the volume of the gas from the vertical scale next to the glass tube. Record these values.
3. Gently open and then close the tap to release some of the air. Wait a minute or two, then read and record the pressure and volume again.
4. Repeat step 3 at least six times until the pressure of the gas is back to atmospheric pressure.
5. Complete the remaining columns of the table.
6. On graph paper and using suitable scales, plot a graph of p against $\frac{1}{V}$.

Result

Within the limits of experimental error you will find that:

- the values of pV in the last column of the table will all be the same.
- the graph of p against $1/V$ will be a straight line passing through the origin.

This verifies that p is inversely proportional to V, i.e. it verify's Boyle's Law.

Pressure p/atms	Volume V/cm^3	$\frac{1}{\text{Volume}}$ $1/V$	Pressure × Volume pV

Experimental Notes

1. The oil in the Boyle's apparatus smells. When releasing pressure from the apparatus a certain amount of oil leaks out sometimes as a fine mist. Do not get this on your clothes. Cover the outlet with a cloth to prevent this.
2. Be careful when reading the volume of the enclosed gas. Make sure your eye is level with the horizontal through the top of the Mercury meniscus.
3. If a hand suction pump is available you will be able to reduce the pressure of the gas below atmospheric pressure. You can then take a further series of values of p and V and include them in the table and graph.

Questions

1. When reading the volume of the gas, your eye should be on the same horizontal level as the top of the meniscus of the mercury. Why is this?
2. How would you expect the measured values of pV to change if the temperature of the laboratory rose by 10°C half way through the experiment?

Gravity – The Cement of the Universe

UNIVERSAL GRAVITATION

In 1666 Sir Isaac Newton first stated his theory of gravitation. He realised that the force that causes an object, when released, to fall downwards towards the Earth was the same kind of force that keeps the planets of the Solar System orbiting the Sun. This force he called the **Force of Gravity**. In the theory he announced the following law, which is now called **Newton's Law of Universal Gravitation**:

NEWTON'S LAW OF UNIVERSAL GRAVITATION states that any two point masses in the universe attract each other with a force that is directly proportional to the product of their masses and inversely proportional to the square of the distance between them.

This means that if two point masses of mass m_1 and m_2 kilograms are a distance d apart (Fig. 10.17), each one attracts the other with a gravitational force F, where:

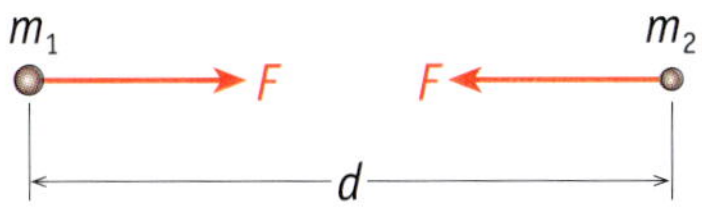

Fig. 10.17

$F \propto m_1 m_2$ and $F \propto \frac{1}{d^2}$ It follows that: $F \propto \frac{m_1 m_2}{d^2} \Rightarrow$

$$F = \frac{Gm_1 m_2}{d^2} \qquad (1)$$

where G is a constant, G is called the **Universal Gravitation Constant**

- The gravitational force is **always attractive**. The force on m_1 pulls it towards m_2 and the force on m_2 pulls it towards m_1.
- The **size of the force on each mass is the same**, even if one mass is much larger than the other.
- Newton showed that the gravitational force between two spherical bodies is the same as if each had all its mass at its centre and the distance between the bodies is the **distance between their centres**. In the problems below, we shall assume that the Earth, Moon and planets are spheres and use this fact.
- The value of the universal gravitation constant G is the same anywhere in the universe. Its value is found by experiment. This was first done by Henry Cavendish in 1731 in Cambridge. The value of G is $G = 6.7 \times 10^{-11}$ N m^2 kg^{-2}. This is an **extremely small number**.
- Since G is such a very small number (0.000 000 000 067), **the gravitational force between two bodies is negligible unless at least one of them has a very large mass.** For objects on Earth, the gravitational forces between them are so small that they are normally completely hidden by the much larger forces of friction and electrical forces.
- $F \propto \frac{1}{d^2}$ means that the size of the force is inversely proportional to the square of the distance between the two bodies. This means that:
  (i) If the distance between the bodies is **doubled,** the force becomes **four times smaller**.
  (ii) If the distance between them is trebled, the force becomes **nine times smaller** etc.

- The unit of G can be found by solving (1) for G, giving $G = \frac{Fd^2}{m_1 m_2}$

Thus, the unit of G is $= \frac{\text{N m}^2}{\text{kg}^2} = \text{N m}^2\ \text{kg}^{-2}$ called the **newton metre squared per kilogram squared**.

Problem 10: Find the gravitational force of attraction between two steel spheres, each of mass 80 kg if the distance between their centres is 0.5 m. ($G = 6.7 \times 10^{-11}$ N m^2 kg^{-2})

Solution: $F = \frac{Gm_1 m_2}{d^2} = \frac{(6.7 \times 10^{-11})(80)(80)}{(0.5)^2} = 1.72 \times 10^{-6}$ N

$= 0.000\ 0017$ N. Note that this is a very small force.

Problem 11: Find the gravitational attraction between a man of mass 90 kg and the Earth when the man is standing on the surface of the Earth.
(Mass of Earth $= 6 \times 10^{24}$ kg; Radius of Earth $= 6.4 \times 10^6$ m; $G = 6.7 \times 10^{-11}$ N m^2 kg^{-2})

Solution: $F = \frac{Gm_1 m_2}{d^2} = \frac{(6.7 \times 10^{-11})(6 \times 10^{24})(90)}{(6.4 \times 10^6)^2} = 883.3$ N

This force is equal to the weight of the man: $W = mg = (90)(9.81) = 833$N

Gravity and the Solar System

The force of gravity is the force that keeps the planets of the solar system orbiting the Sun (Fig. 10.18). It keeps the Moon and man-made satellites orbiting the Earth (page 142). The gravitational forces involved are large since the masses of the bodies are large.

The force of gravity keeps the Earth's **atmosphere** attached to the Earth. On some planets and moons the gravity is too weak to keep an atmosphere present and any gas escapes into space. This occurs on our Moon and explains why the Moon has no atmosphere (Fig. 10.19).

Fig. 10.18
The force of gravity keeps the planets orbiting the Sun.

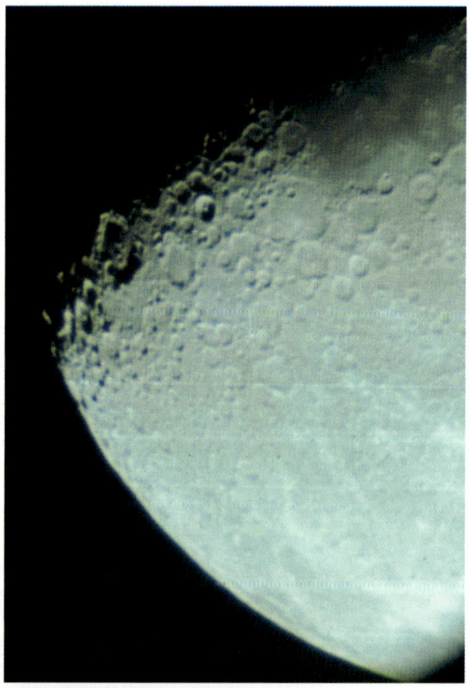

Fig. 10.19
The Moon's gravity is too weak to maintain an atmosphere. Any gas molecules placed there would escape into space.

EXERCISE 10.5

(*Take*: G = 6.7 × 10^{-11} N m^2 kg^{-2}, g = 9.8 m s^{-2}, *Radius of the Earth* = 6.4 × 10^6 m, *Mass of Earth* = 6 × 10^{24} kg, *Mass of Moon* = 7 × 10^{22} kg, *Radius of Moon* = 1.7 × 10^6 m.)

1. Find the gravitational attraction between two steel spheres each of mass 1 kg, when the distance between their centres is 1 m.
2. Find the gravitational attraction between a woman of mass 76 kg and the Earth when she is standing on the surface of the Earth.
3. Find the gravitational force of attraction between a sphere of mass 90 kg and another sphere of mass 1000 kg when the distance between their centres is 2 m.
4. Find the gravitational attraction between a man of mass 76 kg and the Moon when he is standing on the surface of the Moon.
5. Find the gravitational attraction between the Earth and the Moon. (Distance from centre of Earth to centre of Moon = 3.8 × 10^8 m)
6. Find the gravitational attraction between the Earth and the Sun given that the distance between their centres is 1.5 × 10^{11} m. Compare with the answer to Q5. (Mass of Sun = 1.9 × 10^{30} kg)

GRAVITY AND WEIGHT

Recall that weight is a force (measured in newtons) and that the weight W of an object of mass m on the surface of the Earth is given by:

$W = mg$ (2) (where g is acceleration due to gravity on the surface.)

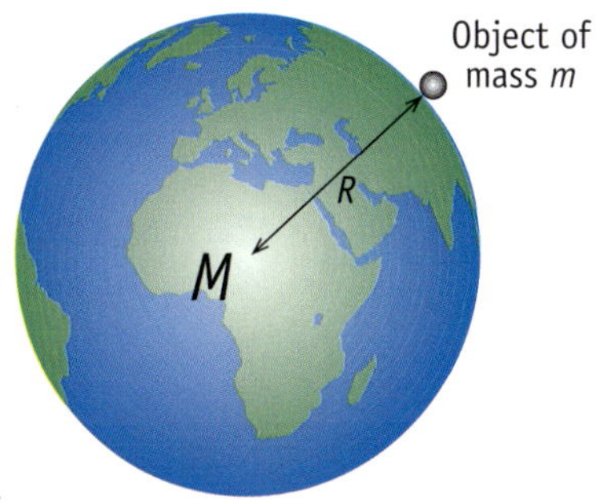

Fig. 10.20

FIG. 10.20 shows an object of mass m on the surface of the Earth. Its weight is also the force of gravitational attraction between it and the Earth, which from Newton's Law of Gravitation is given by:

$W = \frac{GMm}{R^2}$ (3) (where M is the mass of the Earth and R its radius.)

Equating (2) and (3) gives: $mg = \frac{GMm}{R^2} \Rightarrow g = \frac{GM}{R^2}$

$$g = \frac{GM}{R^2}$$

From this equation you can see that the value of acceleration due to gravity, and hence the weight of any object, depends on the mass of the Earth and its distance from the centre of the Earth, i.e. the Earth's radius. The earth is not a perfect sphere. Its polar radius is less than the radius at its equator. Thus acceleration due to gravity and the weight of any object is slightly greater at the poles than at the equator (page 76). Correct to one place of decimals g = 9.8 m s^{-2} anywhere near the surface of the Earth.

This formula obviously applies on other planets also, using the correct values for M and R.

A formula almost identical to the above applies to any object above the surface of the Earth (FIG. 10.21). If d is the distance from the centre of the Earth to the object and g_d is acceleration due to gravity at that point, then the formula is:

$$g_d = \frac{GM}{d^2}$$

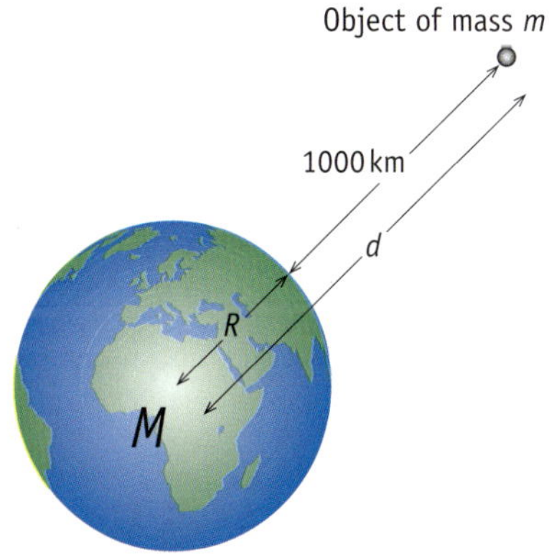

Fig. 10.21

It follows that the value of acceleration due to gravity, and hence weight, decreases as you move away from the Earth (d increases).

In the following problems take: $G = 6.7 \times 10^{-11}$ N m^2 kg^{-2}, $g = 9.8$ m s^{-2},
Radius of the Earth $= 6.4 \times 10^6$ m, *Mass of Earth* $= 6 \times 10^{24}$ kg.

Problem 12: Find the value of acceleration due to gravity on Mars, given that the radius of Mars is 3.4×10^6 m and the mass of Mars is 6.6×10^{23} kg. Hence find the weight of a 90 kg man on Mars.

Solution:

$$g = \frac{GM}{R^2} = \frac{(6.7 \times 10^{-11})(6.6 \times 10^{23})}{(3.4 \times 10^6)^2} = 3.8 \text{ m s}^{-2}$$

Weight of man $= W = mg = (90)(3.8) = 342$ N

Problem 13: Find acceleration due to gravity 1000 km above the Earth's surface. What is the weight of a woman of mass 60 kg at this height?

Solution:

$$g_d = \frac{GM}{d^2}$$ From FIG. 10.23 $d = R + 1000 \times 10^3 = 6.4 \times 10^6 + 1 \times 10^6$

$$g_d = \frac{(6.7 \times 10^{-11})(6.6 \times 10^{24})}{(6.4 \times 10^6 + 1 \times 10^6)^2} = 7.3 \text{ m s}^{-2}$$

Weight $= mg_d = (60)(7.3) = 438$ N

Problem 14: Given that the value of g on the surface of the earth is 9.8 m s^{-2}, at what height above the Earth's surface is acceleration due to gravity equal to half its value on the surface of the Earth?

Solution: Let d be the distance from the centre of the Earth to the point where acceleration is half the value at the surface.

$$g_d = \frac{GM}{d^2}$$ Half the value of acceleration on the surface $= 9.8 / 2 = 4.9$ m s^{-2}

$$\therefore\ 4.9 = \frac{(6.7 \times 10^{-11})(6 \times 10^{24})}{d^2} \Rightarrow d = \sqrt{\frac{(6.7 \times 10^{-11})(6 \times 10^{24})}{4.9}} = 9.1 \times 10^6 \text{ m}$$

Radius of Earth $= 6.4 \times 10^6$ m

$\therefore$ Height above surface $= 9.1 \times 10^6 - 6.4 \times 10^6 = 2.7 \times 10^6$ m above surface.

Problem 15: Find the value of acceleration due to gravity on the surface of Jupiter, given that the mass of Jupiter is 318 times the mass of the Earth and that the radius of Jupiter is 11 times the radius of the Earth. Take $g = 9.8$ m s^{-2}.

Solution: Mass of Jupiter M_j = (318)(Earth's mass). Radius of Jupiter R_j = (11)(Earth's radius).
Let g_j be acceleration on the surface of Jupiter.

$$g_j = \frac{GM_j}{R_j^2} = \frac{(6.7 \times 10^{-11})(318)(6 \times 10^{24})}{(11 \times 6.4 \times 10^6)^2} = 25.79 \text{ m s}^{-2}$$

Note that even though Jupiter is 318 times the mass of the Earth, the value of acceleration due to gravity on the surface of Jupiter is only 2.6 times that on Earth. This is because Jupiter's radius is much bigger than Earth's.

EXERCISE 10.6

($G = 6.7 \times 10^{-11}$ N m^2 kg^{-2}, g = 9.8 m s^{-2}, *Radius of Earth* = 6.4×10^6 m, *Mass of Earth* = 6×10^{24} kg)

1. Find the value of acceleration due to gravity on Jupiter, given that the radius of Jupiter is 7×10^7 m and the mass of Jupiter is 1.9×10^{27} kg. Hence find the weight of a 90 kg man on Jupiter.
2. Find the value of acceleration due to gravity on the Moon, given that the radius of the Moon is 1.7×10^6 m and the mass of the Moon is 7×10^{22} kg. Hence find the weight of a 60 kg woman on the Moon.
3. Calculate the value of acceleration due to gravity on the surface of the Sun, given that mass of the Sun = 1.9×10^{30} kg and the radius of the Sun = 7×10^8 m.
4. Prove the formula $g = GM / R^2$
5. Prove that acceleration due to gravity g_d at a distance d from the centre of a planet of mass M and radius R is given by: $g_d = GM / d^2$. Hence derive a formula for the acceleration due to gravity at height h above the surface of a planet of radius R in terms of G, R, M and h.
6. Find acceleration due to gravity 100 km above the Earth's surface.
7. At what height above the surface of the Earth is acceleration due to gravity equal to:
   (i) half its value and (ii) one tenth its value
   on the surface of the Earth? Take g = 9.8 m s^{-2}
8. Find the mass of the Earth given that $G = 6.7 \times 10^{-11}$ N m^2 kg^{-2}, g = 9.8 m s^{-2}, and the radius of the Earth = 6.4×10^6 m.
9. Given that the value of g at the surface is 9.8 m s^{-2}, calculate a value for acceleration due to gravity at a height above the Earth equal to twice the radius of the Earth.
10. Is the value of acceleration due to gravity a constant on the surface of the Earth? Explain your answer.
11. The mass of Mercury is 0.04 times the mass of the Earth. The radius of Mercury is 0.37 times the radius of the Earth. If acceleration due to gravity on the surface of the Earth is 9.8 m s^{-2}, find the value of acceleration due to gravity on the surface of Mercury.
12. Describe how the weight of an object changes as it moves from the surface of the Earth to the surface of the Moon. At what point between them is the resultant gravitational force on an object zero? Where is its acceleration zero? (Distance between Earth and Moon = 3.8×10^8 m and Mass of Earth = 81 × mass of Moon.)

THE TURNING EFFECT OF A FORCE

FORCE MAY CAUSE ROTATION

As well as causing a body to accelerate a force may cause a body to turn or **rotate**, i.e. a force can have a **turning effect**.

When a force causes something to rotate about an axis, the size of the turning effect depends on the size of the force and the distance from the force to the axis, e.g. it is easier to undo a nut with a spanner with a long handle than with one with a short handle. It is much easier to open or close a door from a point on the door far away from the hinges than it is from a point near the hinges (FIG. 10.22). This leads us to a new quantity – called **moment of a force**.

Fig. 10.22

MOMENT OF A FORCE

The ***moment*** of a ***force*** about an axis is equal to the magnitude of the force multiplied by the perpendicular distance from the axis to the line of action of the force (FIG. 10.23).

i.e. Moment of force = Force × Perpendicular distance

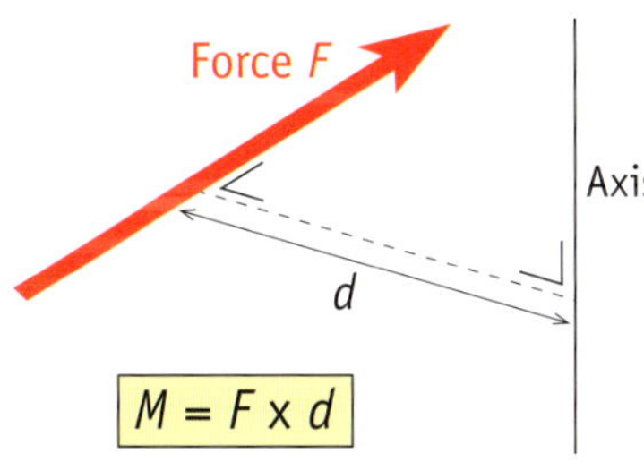

Fig. 10.23

UNIT OF MOMENT
The unit of moment is the ***newton metre* (N m).**

Moment is a **scalar quantity**. Its symbol is M. Since $M = F \times d$

Unit of moment = Unit of force × Unit of distance

i.e. Unit of moment = newton meter (N m)

Problem 16: Find the moment of a 10 N force about an axis that is 3 m from it (Fig. 10.24).

Solution: Moment = Force × Perpendicular distance

= 10 × 3 = 30 N m

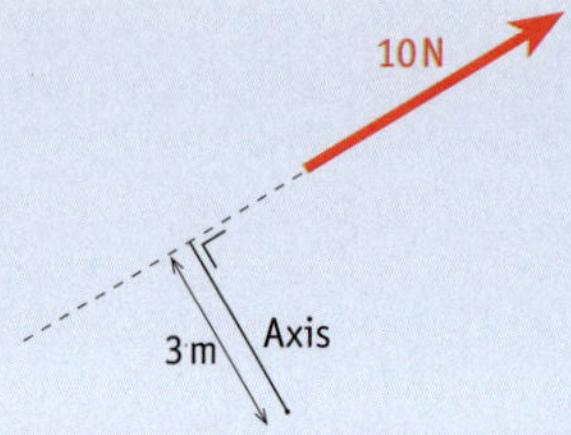

Fig. 10.24

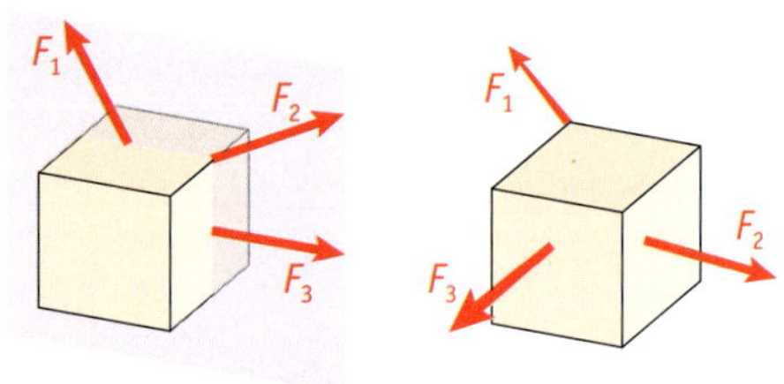

(A) Coplanar forces (B) Non-coplanar forces

Fig. 10.25
Forces that all act in the same plane are called **Coplanar Forces.**

The moment of a force about an axis determines the turning effect of that force about that axis. Forces that have the same moment about an axis have the same turning effect. Forces that have the same moment but tend to produce rotation in opposite directions cancel each other out.

Clockwise and Anticlockwise Moments

When a number of coplanar forces (Fig. 10.25) act on a body, those tending to rotate it in a clockwise direction are given a different sign to those that tend to rotate it in an anticlockwise direction. The **sum of the moments** is then **their algebraic sum**.

Problem 17: Fig. 10.26 shows a number of forces acting on a metre stick. Find the sum of the moments of these forces about the centre of the metre stick.

Solution: Clockwise moments = (50)(30) + (60)(50) = 4500 N m

Anticlockwise moments = (20)(40) + (30)(10) + (80)(40)

= 4300 N m

Taking clockwise as positive:

Sum of moments = + 4500 - 4300 = + 200 N m

The metre stick has an overall clockwise moment hence it tends to turn in a clockwise direction.

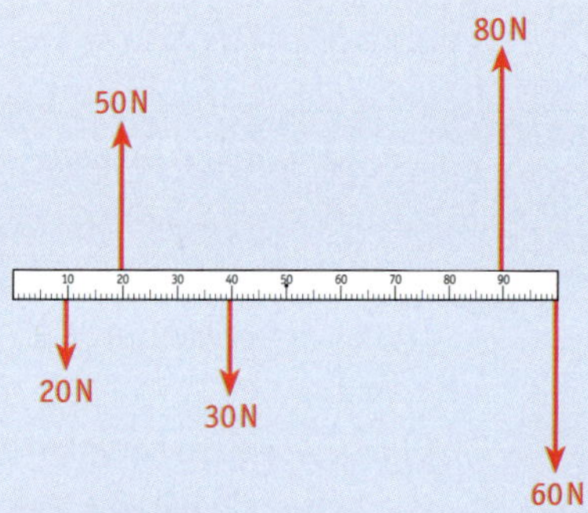

Fig. 10.26

Conditions for Equilibrium of a Body Under Coplanar Parallel Forces

A body that is at rest is said to be in equilibrium. If the resultant of the forces acting on a body is zero, that body may not be in equilibrium. Fig. 10.27(A) shows this. Whether or not the body is in equilibrium depends on the positions of the lines of action of the forces. In Fig. 10.27(B), the sum of the forces is zero but the forces still cause rotation. If the forces do not cause any rotation, it is found that the algebraic sum of the moments of the forces about any point in their plane is zero. We shall verify this fact experimentally.

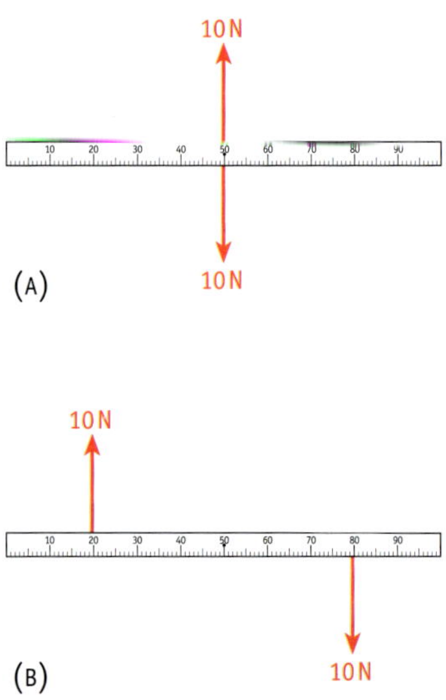

Fig. 10.27

If a body is in ***equilibrium*** then:

- The vector sum of the forces in any direction is zero.
- The sum of the moments about any point is zero.

Problem 18: A metre stick is in equilibrium under the action of the forces shown in FIG. 10.28. Find the value of the force labelled X.

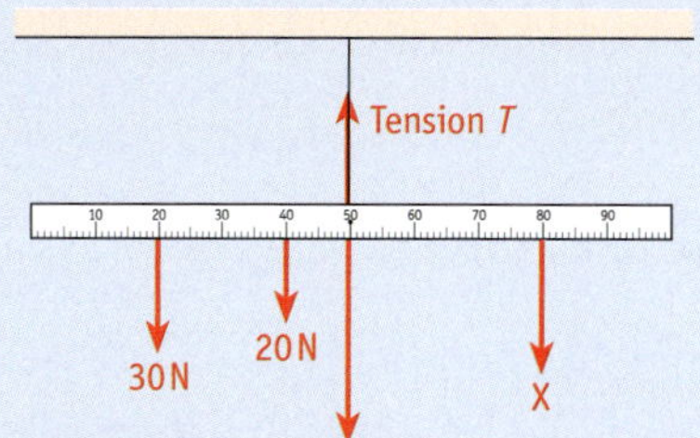

Fig. 10.28

Solution: Acting down through the centre of gravity (i.e. the geometrical centre) of the metre stick is its weight W. The value of W is unknown. Also acting up through this point is the tension T in the string. This is also unknown. If we take moments about the centre of the metre stick these forces have no moment and will not effect the resulting equation.

Anticlockwise moments $= (30)(30) + (20)(10)$ Clockwise moments $= (X)(30)$

Since the stick is in equilibrium these are equal, thus: $30X = 900 + 200 \Rightarrow X = 36.67$ N

Problem 19: A uniform beam ab of length 4 m and weight 200 N rests on two supports. One support is at the end a of the beam and the other is 1 m from the other end. Find the values of the upward force each support exerts on the beam.

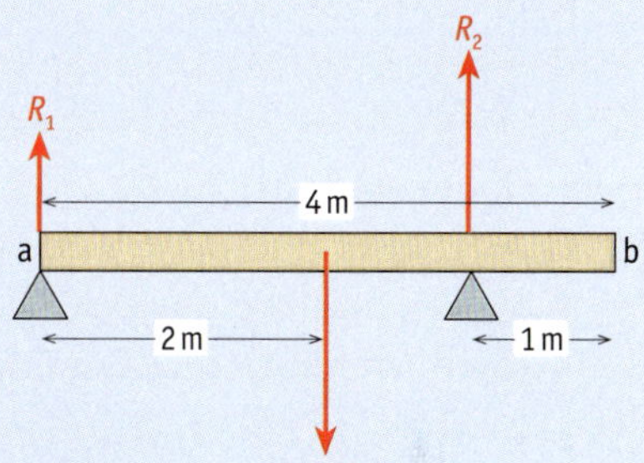

Fig. 10.29

Solution: FIG. 10.29 shows the beam and the three forces acting on it.

Sum of upward forces $=$ Sum of downward forces

$\Rightarrow R_1 + R_2 = 200$

Take moments about a:

Clockwise moments $=$ Anticlockwise moments

$\Rightarrow (200)(2) = (R_2)(3) \Rightarrow R_2 = 400/3 = 133.33$ N

$R_1 + R_2 = 200 \Rightarrow R_1 = 200 - R_2 = 200 - 133.33 \Rightarrow R_1 = 66.67$ N

Problem 20: A man and a boy hold up opposite ends of a rod 2 m long. The rod is horizontal and has a weight of 1000 N hanging from it. The weight of the rod is negligible. If the man is to exert three times the upward force as the boy, where on the rod must the weight be hung?

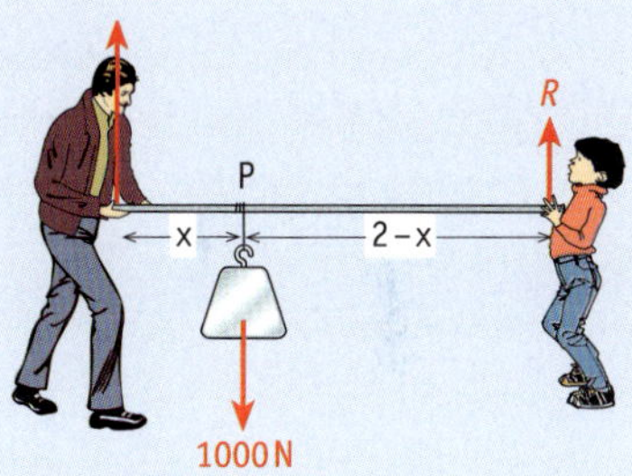

Fig. 10.30

Solution: Suppose the weight is a distance x from the man (FIG. 10.30). If the boy exerts an upward force of R then the man exerts an upward force of $3R$. The rod is in equilibrium, therefore the sum of the moments about any point is zero. Take moments about the point p:

$(3R)(x) - (R)(2 - x) = 0 \Rightarrow 3x - 2 + x = 0 \Rightarrow 2x = 2 \Rightarrow x = 1$ m

EXERCISE 10.7

1. What is the moment of the force shown in FIG. 10.31 about:
   (i) the point A,
   (ii) the point B,
   (iii) the point C?

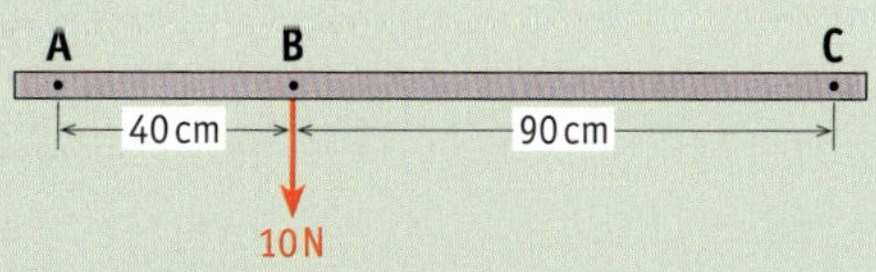

Fig. 10.31

2. Find the sum of the moments of the forces shown in FIG. 10.32 about the point:
   (i) a (ii) b (iii) c?

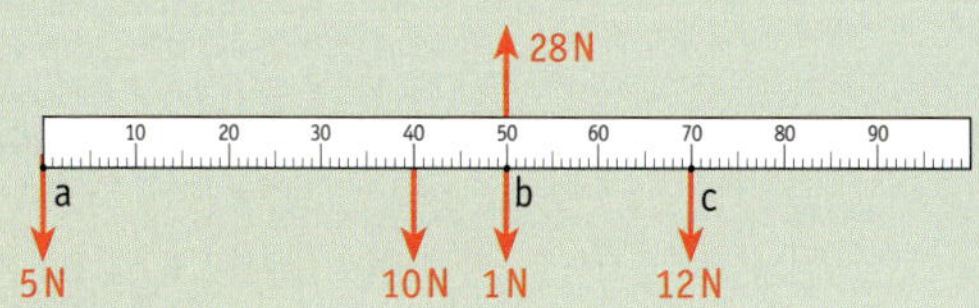

Fig. 10.32

3. In each part of FIG. 10.33 the metre stick is in equilibrium. Find the value of X in each case:

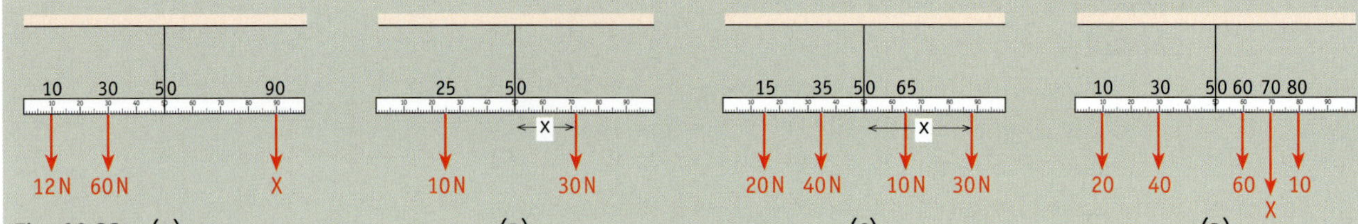

Fig. 10.33 (A) (B) (C) (D)

4. Find the weight of the uniform metre stick shown in FIG. 10.34. It is in equilibrium.

Fig. 10.34

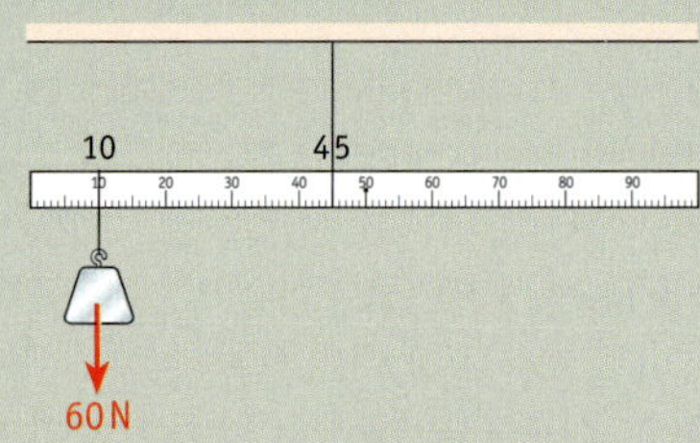

5. If the uniform metre stick in FIG. 10.35 is in equilibrium, find the weight of the metre stick. Verify that the sum of the moments of the forces about each of the points A, B, C and D is zero.

Fig. 10.35

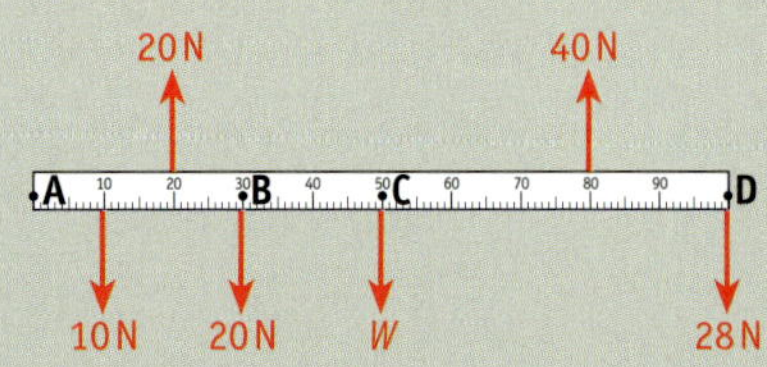

6. A uniform wooden beam of length 1 m is resting on supports as shown in FIG. 10.36. Find the force that each support exerts on the beam.

Fig. 10.36

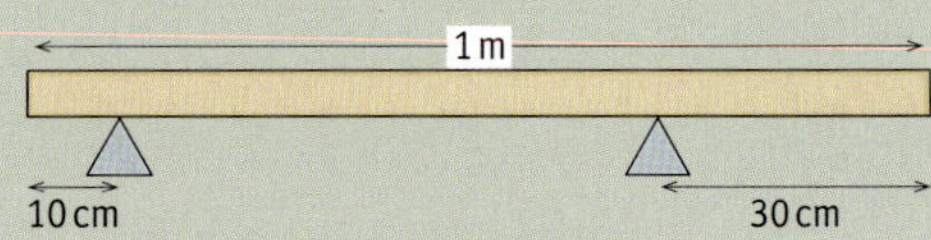

7. A uniform mast 10 m long and of mass 100 kg is being carried horizontally by a man and a boy. If the man is 3 m from one end of the mast and the boy is at the far end, find the upward force exerted by the boy and the man on the mast. Where should the man support the mast if he is to carry three times the weight of the boy, the boy remaining at one end?

8. A man and a boy hold opposite ends of a pole 8 m long which is horizontal. A mass of 200 kg is tied to the pole. If the man is to support four times as much weight as the boy, where along the pole must the weight be tied? Neglect the weight of the pole.

The strict definition of equilibrium tells us that a body is in equilibrium if:

1. The body as a whole is not accelerating*.
2. The body is not rotating with angular acceleration.

Condition 1 means the body has constant velocity or is stopped.

Condition 2 means the body is either rotating at a constant rate or is not rotating. Obviously, therefore, a spinning top that is rotating at a constant rate and that is moving at a constant velocity is in equilibrium. This body is in **dynamic equilibrium**. We shall not be concerned with cases like this. A body in equilibrium that is completely at rest, is said to be in **static equilibrium**.

* To be precise we should say the centre of mass (the centre of gravity) is not accelerating and the angular acceleration about the centre of mass is zero.

EXPERIMENT **MECHANICS 6**

To Investigate the Laws of Equilibrium for a Set of Coplanar Forces.

Summary of Method

In this experiment you will weigh a metre stick. You will then place it in equilibrium under the action of a number of coplanar parallel forces. You will calculate the sum of the forces acting on the stick. It will turn out to be zero. By calculating the sum of the moments of these forces about any chosen point on the metre stick you will verify that the sum of the moments of the forces is zero.

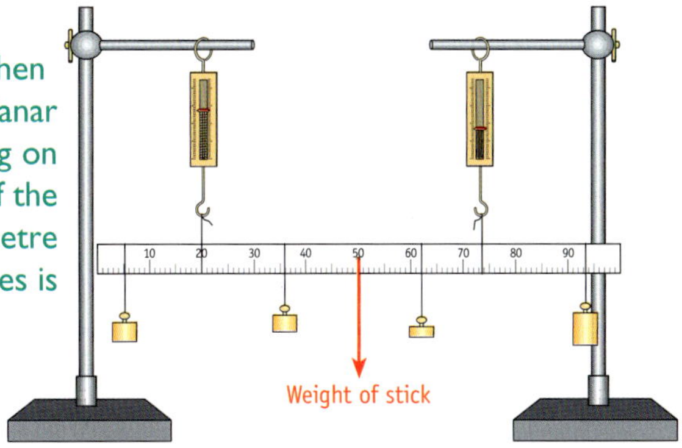

Fig. 10.37

Equipment Needed

- A metre stick, a set of weights and some thread
- Two (or more) spring balances (preferably graduated in newtons) and a retort stand for each balance

Method

1. Find the weight of the metre stick using one of the spring balances.
2. Mount the spring balances on the retort stands and suspend the metre stick from them with thread (Fig. 10.37).
3. Hang (with loops of thread) known weights from the metre stick at various positions along its length and adjust their positions and the heights of the spring balances until the metre stick remains in horizontal equilibrium.
4. Choose any point on the lower edge of the metre stick and measure the distance of each force from this point (the weight of the metre stick is one of these forces, it acts vertically down through its centre of gravity, i.e. the 50 cm mark).
5. Record all weights and distances in the Table.
6. Calculate the resultant upward force on the stick (i.e. find sum of upward forces – sum of downward forces).
7. Complete the Table and calculate the sum of the clockwise moments and the sum of the anticlockwise moments.
8. Change the positions of the spring balances and the positions and values of the weights and repeat the experiment.

Result

- The resultant force on the stick will be zero.
- The sum of the clockwise moments will be equal to the sum of the anticlockwise moments.

Table

Force tending to turn stick clockwise	Distance of force from point	Clockwise moment	Force tending to turn stick anti-clockwise	Distance of force from point	Anti-clockwise moment

Sum of clockwise moments = Sum of anti-clockwise moments =

Questions

1. Why must the metre stick be balanced horizontally?
2. What is meant by the centre of gravity of an object?
3. If the metre stick were not horizontal when balanced, how would you use the information obtained to verify that the sum of the moments was zero?

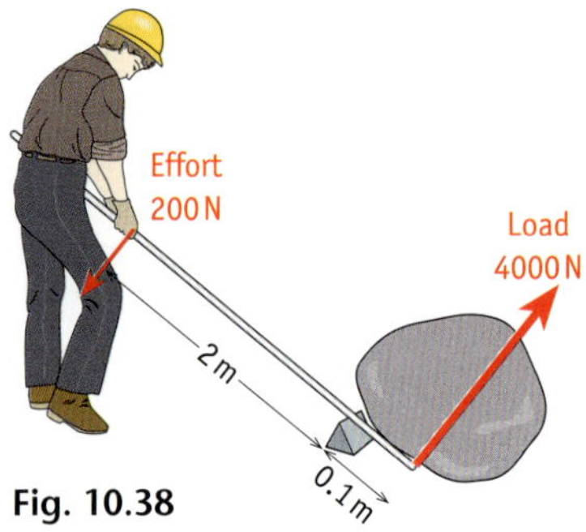

Fig. 10.38

Levers

A rigid body which is free to rotate about a fixed point is called a lever. The point about which it rotates is called the **fulcrum**. Very often a lever is used to exert a force called the **load**. The force exerted on the lever is called the **effort**. A lever can be used to magnify the size of a force. In Fig. 10.38 the man exerts a force of 200 N on the crowbar. This force is 2 m from the fulcrum. The moment of this force about the fulcrum is 400 N m. The boulder is 10 cm (0.1 m) from the fulcrum. If the man just moves the boulder, the moment of the load must also be 400 N m. Thus if L is the load $L \times 0.1 = 400 \Rightarrow L = 4000$ N. Thus by using the lever the man can cause a force of 4000 N to be exerted on the boulder.

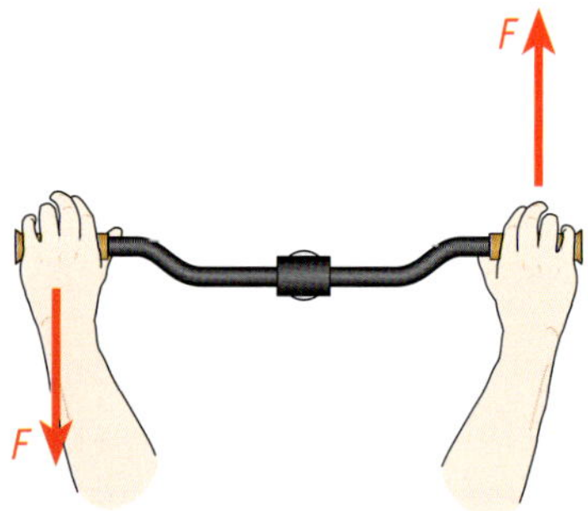

Fig. 10.39

Couples

In Fig. 10.39 two forces of the same magnitude acting in opposite directions are applied to the handlebars of the bicycle. The resultant force is zero, therefore the handlebars as a whole do not accelerate. The forces do, however, cause the handlebars to turn. Very often when we want to turn or cause an object to rotate we exert a pair of equal but opposite forces on it. For example; turning on a tap, turning a knob on a door, opening a pot of jam. Two parallel forces with the same magnitude acting in opposite directions are called a **couple** (Fig. 10.39). The resultant of the two forces is zero. A couple has only a turning effect.

COUPLE
Two parallel forces with the same magnitude acting in opposite directions is called a ***couple***.

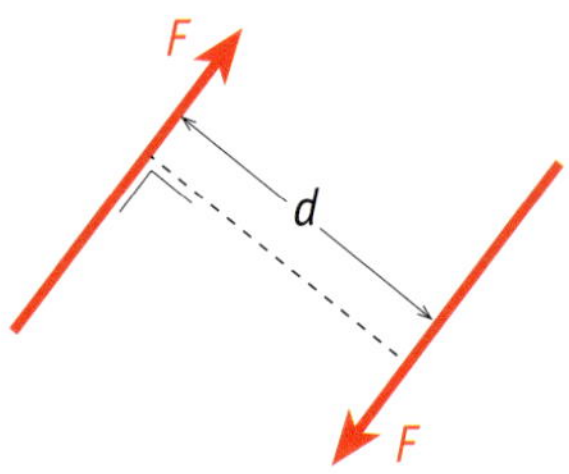

Fig. 10.40

Since a couple has a turning effect, it has a moment about any point. The moment of a couple is sometimes called the **torque** of the couple. Its symbol is T. Its unit is the **newton metre (N m)**.

It can be proved that the moment of a couple about any point in its plane is the same. The moment is equal to the product of one of the forces and the perpendicular distance between them. Thus in Fig. 10.40 the moment of couple $T = Fd$.

Moment of Couple (Torque) $T = Fd$

Problem 21: Find the moment of the couple in Fig. 10.41.

Solution: $T = Fd = (20)(0.3) = 6$ N m

Problem 22: A force of magnitude 6 N and 4 N act on a stick as shown in Fig. 10.42. Describe the resulting motion of the stick.

Solution: The resultant force on the stick is 2 N (= 6 – 4). Thus the stick accelerates. The stick also starts to rotate about its centre of gravity because the 4 N force and 4 N of the 6 N force form a couple which acts on the stick.

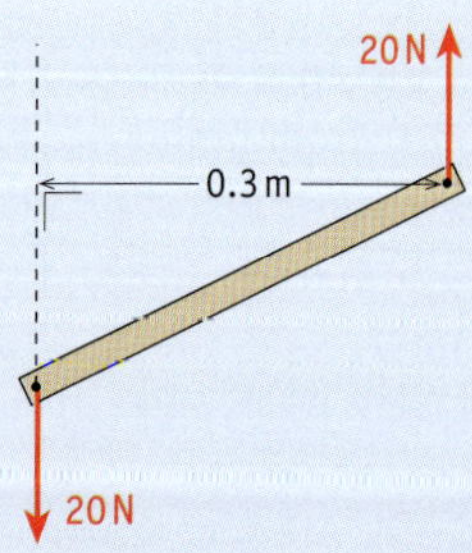

Fig. 10.41

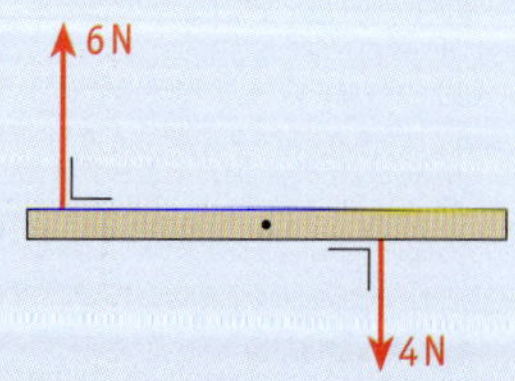

Fig. 10.42

In a **simple d.c. electric motor** and in a **moving coil galvanometer**, a pair of parallel forces on a coil of wire exert a couple on the coil. The coil then rotates (see Chapter 33).

EXERCISE 10.8

1. In FIG. 10.43 the man exerts a force of 300 N on the crowbar. What is the force exerted on the stone by the crowbar?

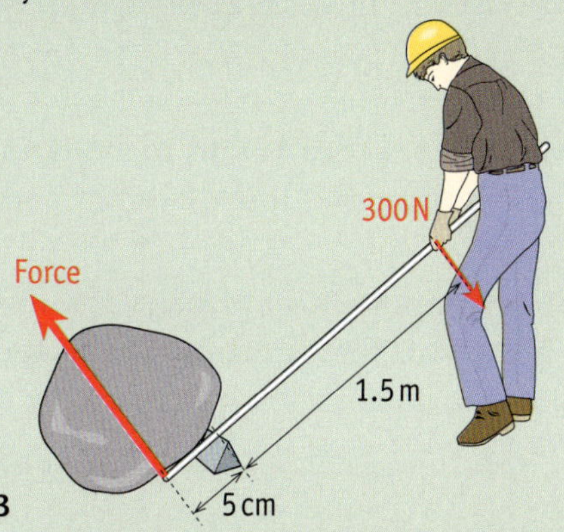

Fig. 10.43

2. Two forces of magnitude 40 N acting in opposite directions act on a steering wheel. If the perpendicular distance between the forces is 0.5 m, find the torque on the wheel.

3. One of the nuts holding a wheel on a car is to be tightened to a torque of 85 N m. A woman uses a wrench of length 40 cm to tighten the nut (FIG. 10.44). If she exerts two equal but opposite forces on the wrench find the size of one of the forces.

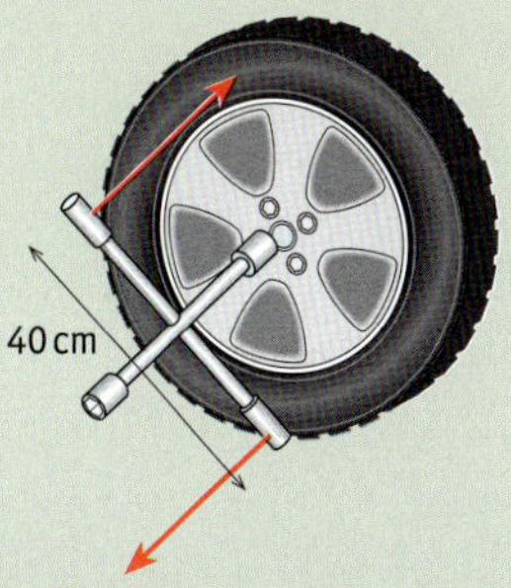

Fig. 10.44

4. The torque of the couple acting on an electric motor is 600 N m. If the perpendicular distance between the forces is 10 cm, find the magnitude of one of the forces.

CHAPTER CHECKLIST

- **Define:** Density; Pressure; The pascal; Upthrust; Buoyancy force; Moment of a force; Lever; Couple; Torque; Universal gravitation constant.
- **State:** The unit of density; The unit of pressure; Archimedes' Principle; The Law of Flotation; Boyle's Law; Newton's Law of Universal Gravitation; The unit of moment; The conditions for equilibrium of a body; The unit of torque.
- **Recall:** 1 kg = 1000 grams; 1 $m^3 = 10^6$ cm^3; That a force can have a turning effect; That the torque of the couple on a coil turns a simple electric motor and moving-coil meter.
- **Recall** that pressure in a fluid: Increases with depth; Acts perpendicular to any surface; At a given depth is the same in all directions.
- **Express** the unit of pressure in basic units.
- **Describe** and carry out an experiment to: Demonstrate Archimedes' Principle; Demonstrate the Law of Flotation; Demonstrate atmospheric pressure; Verify Boyle's Law; Investigate the Laws of Equilibrium for a set of coplanar forces.
- **Recall** and use the formulae: $\rho = \frac{m}{V}$ $P = \frac{F}{A}$ $P = \rho gh$

  $pV = k$ $F = \frac{Gm_1m_2}{d^2}$ $W = mg$ $g = \frac{GM}{R^2}$ $M = Fd$
- **Recall:** What a hydrometer is used for; The general relationship between atmospheric pressure and weather; What the 'bends' in diving is; That the value of acceleration on the surface of a planet or moon determines whether it can maintain an atmosphere or not.
- **Derive** the formula: $g = \frac{GM}{R^2}$

CHAPTER 11

Work, Energy and Power

UNIT OF WORK
The unit of work is the ***joule* (J)**

NOTE

Note that:
1 millijoule (mJ) = 10^{-3} J
1 kilojoule (kJ) = 10^{3} J
1 megajoule (MJ) = 10^{6} J

The joule in **basic units:**
1 J = 1 N m = 1 (kg m s^{-2}) m
(1 N = 1 kg m s^{-2})
$\therefore$ **1 J = 1 kg m^2 s^{-2}**

Work

A man pushing a car along a level road is said to be doing work. The amount of work done depends on the force exerted and the distance through which the force acts.

This concept of work is very important in physics. If a force F moves a body through a displacement s in the direction of the force, the work done W is by definition equal to the force multiplied by the displacement.

WORK
When a ***force F*** moves a body through a ***displacement s*** in the direction of the force, the ***work W*** done is equal to the force multiplied by the ***displacement***.

Work = Force × Displacement $\qquad$ ***W = Fs***

Work is a **scalar quantity**. Its usual symbol is W.

UNIT OF WORK

Since $W = Fs$, Unit of work = (unit of F) × (unit of s)
= newton × metre (N m)

This unit is called the **joule** in honour of James Joule (1818–89) who did much early research in the area of work and the related quantity energy.

THE JOULE
One joule is the work done when a force of 1 newton acts for a distance of 1 metre in the direction of the force.

1 J = 1 N m

Fig. 11.1
Considerable work is done in bringing the lift and the people upwards.

Problem 1: Find the work done when a horse exerts a horizontal force of 400 N on a cart and pulls it 20 m in the direction of the force.

Solution: $W = Fs = (400)(20) = 8000\text{ J} = 8\text{ kJ}$

Problem 2: Find the work done in raising an object of mass 5 kg through a vertical distance of 10 m at a steady speed. If this is repeated 50 times how much work is done?

Solution: Force needed = weight of object = (5)(9.8) = 49 N

Work done in one lift $W = Fs = (49)(10) = 490\text{ J}$

Work done in 50 lifts = (50)(490) = 24 500 J

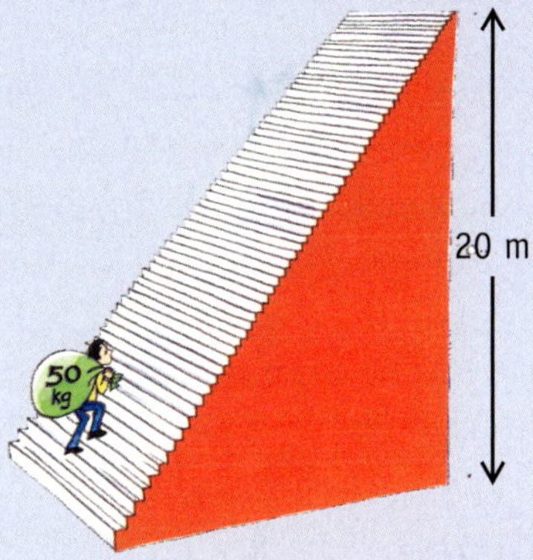

Fig. 11.2

Problem 3: A man carries a 50 kg bag of cement up a stairs at a steady speed. The vertical height through which he rises is 20 m (FIG. 11.2). He exerts an upward force on the bag equal to its weight. Find the work done by the man on the bag of cement.

Solution: Force exerted by man = Weight of bag = (50)(9.8) = 490 N

Displacement in direction of force = vertical height through which bag is raised = 20 m

Work done = Fs = (490)(20) = 9800 J

The man actually does more work than 9800 J in bringing the bag of cement up the stairs since he must also exert forces to bring himself up. The work done in bringing an object through a vertical height h can be considerable. Runners, cyclists and hill climbers all must work much harder to maintain the same speed on a hill that they had when moving along the horizontal. Similarly, **lifts** and **escalators** do a lot of work in bringing people upwards.

Problem 4: Find the work done in bringing a 1000 kg car from rest to 30 m s^{-1} in 10 seconds.

Solution: First find the acceleration and the distances travelled.

$u = 0,\ v = 30,\ t = 10,\ s = ?\ \ a = ?$

To find a use:

$$v = u + at$$
$$30 = 0 + 10a$$
$$\Rightarrow a = 3\text{ m s}^{-2}$$

To find s use:

$$v^2 = u^2 + 2as$$
$$30^2 = 0^2 + 2(3)s$$
$$\Rightarrow s = \frac{30^2}{6} = 150\text{ m}$$

$F = ma \Rightarrow$ force exerted on car = (1000)(3) = 3000 N

Work Done = Force × Distance = (3000)(150) = 450 000 = 450 kJ

EXERCISE 11.1

1. Find the work done when a 10 N force undergoes a displacement of 30 m in the direction of the force.
2. A woman exerts a horizontal force of 400 N on a cart and pushes it horizontally through a distance of 30 m. Calculate the work done by the force.
3. A husky dog does 20 kJ of work as he pulls a sleigh with a force of 340 N. Through what distance does he pull it?
4. What is the weight of a mass of 100 kg? Find the work done in raising a mass of 100 kg through a vertical height of 60 m at a constant velocity.

5. A body of mass 20 kg is raised through a vertical distance of 1 m. Find the work done.
6. A man carries a 60 kg bag of coal up a staircase at a steady speed. The vertical height through which he rises is 8 m. The man exerts an upward force on the coal equal to its weight throughout the journey. Calculate the work done by the man on the bag of coal.
7. Find the work done in bringing a car of mass 800 kg from rest to a speed of 30 m s^{-1} in 10 seconds.
8. A car of mass 1000 kg accelerates from 0 to 80 km h^{-1} in 20 seconds. Find the work done.
9. A man of mass 105 kg climbs up a ladder of length 10 m which is inclined at an angle of 40° to the horizontal. Find the work done by the man in going from the bottom to the top of the ladder at constant speed. (Neglect the effects of friction). How much work would the man do if the ladder was vertical?

ENERGY
Energy (E) is the ability to do work. The amount of energy something has is the amount of work it can do.

UNIT OF ENERGY
The unit of energy is the ***joule* (J)**.

Energy

Energy is the ability to do work. Anything that can do work has energy, i.e. anything that can exert a force through a distance has energy. Energy is a **scalar quantity**. Since the energy of a body is the amount of work it can do, the unit of energy is the same as the unit of work, i.e. the joule.

Different Forms of Energy

Since a body can do work for different reasons it can have energy for different reasons. It is useful to classify energy according to the reasons that bodies have it.

KINETIC ENERGY

All moving objects have energy due to their motion. **The energy a body has due to its motion is called kinetic energy** (see page 128).

POTENTIAL ENERGY

Potential energy is the energy a body has due to its position in a force field (see page 129).

ELECTROMAGNETIC ENERGY (ALSO CALLED RADIANT ENERGY)

This is the energy transmitted from one point to another in the form of electromagnetic waves that travel at the speed of light (see page 217). You have probably heard of the following kinds of electromagnetic energy:

Radio waves	Microwaves	Radiated heat (infra red)	Visible light waves	Ultra violet waves	X-rays	Gamma rays

INTERNAL ENERGY (HEAT ENERGY)

The molecules of every substance, whether solid, liquid or gas are in perpetual motion. If heat is supplied to a substance, the motion (the vibration) of its molecules increases as its temperature rises. The kinetic energy of the molecules therefore increases. Thus heat is a form of energy.

SOUND ENERGY

Any source of sound is always some object vibrating. As sound travels through a substance, the vibration from the source is passed on from molecule to molecule in that substance. Each molecule acquires energy in the form of kinetic and potential as the sound wave passes (page 192).

Fig. 11.3
Mobile phones transmit signals in the form of microwaves, which are a form of electromagnetic energy.

CHEMICAL ENERGY

The energy given out (or sometimes taken in) in the form of heat, light or kinetic energy during a chemical reaction is called **chemical energy**. For example, when something is burned, some of the chemical energy stored in it is changed into heat and light energy (FIG. 11.4).

Fig. 11.4
As these people warm themselves by the fire, chemical energy is being converted to heat energy.

ELECTRICAL ENERGY

When an electric current flows in a metal wire subatomic particles called electrons move in the wire. As they move, they lose energy called **electric potential energy**. This potential energy appears as heat and other forms of energy in the wire.

NUCLEAR ENERGY

The energy given out from the nuclei of certain atoms during nuclear reactions (fission or fusion for example) is called **nuclear energy**. This energy is used practically in the nuclear reactor (Chapter 31).

PRINCIPLE OF CONSERVATION OF ENERGY

Energy cannot be created or destroyed but can only be transferred from one body to another or changed from one form to another. This fact is one of the most important discoveries ever made in physics. In every chemical or physical reaction, there is no net gain or loss of energy; any loss is always accompanied by an equal gain. This experimental fact is called the **principle of conservation of energy***.

PRINCIPLE OF CONSERVATION OF ENERGY states that energy cannot be created or destroyed but can only be converted from one form to another.

EXAMPLES OF ENERGY CONVERSIONS

Energy can be changed from one form to another. Sometimes these changes occur by themselves; sometimes special equipment or conditions are needed to bring about a particular conversion. The following list shows some of the forms of energy. Also given are some examples of particular energy conversions. You should complete this list of examples.

1. Kinetic
2. Potential
3. Internal (Heat)
4. Sound
5. Light
6. Chemical
7. Electrical

Examples of Energy Conversion:

Conversion	Example	Conversion	Example
1 to 2	A stone thrown upwards	2 to 1	A stone falling
1 to 3	Rubbing your hands together	3 to 1	A hot object expanding
1 to 4	A vibratinq tuning fork	4 to 1	Resonance between tuning forks
1 to 5 etc...		5 to 1	

* In this context mass must be considered as a form of energy. See Chapter 31 for further details.

KINETIC ENERGY

KINETIC ENERGY
The ***kinetic energy*** **(E_k)** of a body is the energy that body has due to its motion.

The kinetic energy of a body of mass m moving with a speed v is given by:

$$E_k = \frac{1}{2}mv^2$$

You need not worry about the proof of this formula.

Problem 5: A body of mass 1000 kg is moving with a velocity of 40 m s^{-1}. Calculate its kinetic energy.

Solution: $E_k = \frac{1}{2}mv^2 = \frac{1}{2}(1000)(40)^2 = 800\ 000$ J

Problem 6: A mass of 30 kg has a kinetic energy of 1000 J. Find its speed.

Solution: $E_k = \frac{1}{2}mv^2 \Rightarrow 1000 = \frac{1}{2}(30)v^2$

$$\Rightarrow v = \sqrt{\frac{2000}{30}} = 8.16 \text{ m s}^{-1}$$

WORK AS THE TRANSFER OF ENERGY

When a body does work, it loses energy. For example, if you carry a block upstairs, you do work and the chemical energy in your muscles decreases. If work is done on a body it gains energy. The block gains energy as it moves up. Any time work is done energy is always transferred.

Problem 7: A bullet of mass 2 grams travelling with a speed of 150 m s^{-1} penetrates 0.6 m into a block of wood before coming to rest. Find the force that the wood exerts on the bullet.

Solution: Kinetic energy of bullet $E_k = \frac{1}{2}mv^2 = (\frac{1}{2})(2 \times 10^{-3})(150^2) = 22.5$ J

This kinetic energy is lost by the bullet and must equal the work done by the bullet in coming to rest. If F is the force that the bullet exerts on the wood then the wood exerts an equal but opposite force on the bullet.

Work $= F \times d \Rightarrow F \times 0.6 = 22.5 \Rightarrow$ Force $F = 37.5$ J

EXERCISE 11.2

1. A body of mass 20 kg has a velocity of 12 m s^{-1}. Find its kinetic energy.
2. A car of mass 1000 kg has a speed of 28 m s^{-1}. Calculate its kinetic energy.
3. The picture on a television screen is produced by a beam of electrons striking the screen. Calculate the kinetic energy of one of the electrons if its mass is 9.1×10^{-31} kg and its speed is 1.5×10^7 m s^{-1}.
4. A car travelling at a speed of 20 m s^{-1} has a kinetic energy of 160 000 J. Calculate its mass.
5. A 200 kg mass has a kinetic energy of 4000 J. What is its speed?
6. Find the kinetic energy of:
   (i) a bullet of mass 4 grams moving at 400 m s^{-1},
   (ii) a car of mass 800 kg moving at 28 m s^{-1},
   (iii) a lorry of mass 10 tonnes moving at 100 km per hour,
   (iv) a tennis ball of mass 30 grams moving at 150 km per hour.

7. A particle of mass 20 kg moving at a speed of 3 m s^{-1} accelerates to a speed of 30 m s^{-1} under the action of a force of magnitude 12 N in the direction of the motion. Calculate:
   (i) the initial kinetic energy of the particle,
   (ii) the final kinetic energy of the particle,
   (iii) the change in kinetic energy of the particle,
   (iv) the distance travelled,
   (v) the work done by the force.

8. A bullet of mass 4 grams with a velocity of 200 m s^{-1} penetrates 0.5 m into a block of wood. Find the force that the wood exerts on the bullet. Calculate the velocity with which the bullet emerges if the block is only 0.25 m thick, assuming the same opposing force.

9. Show that if the speed of an object doubles its kinetic energy quadruples.

10. Two moving bodies have the same kinetic energies. The speed of the first is four times the speed of the second. Find the ratio of the mass of the first to the mass of the second.

POTENTIAL ENERGY

The **potential energy (E_P)** of a body is the energy it has due to its position in a force field.

The force field may be gravitational or electric or due to the object being in a state of strain.

There are thus different forms of potential energy.

POTENTIAL ENERGY
The ***potential energy (E_P)*** of a body is the energy it has due to its position in a force field.

STRAIN POTENTIAL ENERGY

If a force is applied to an elastic body it may change shape, i.e. it may become distorted. Some of the work done by the force becomes stored in the body as **strain potential energy**. If the force is removed, the strained body does work in coming back to its original shape. FIG. 11.5 and FIG. 11.6 show two examples.

Fig. 11.5
The strained bow and string have potential energy stored in them. If they are allowed to come back to their original shape, this energy appears as kinetic energy of the arrow.

Fig. 11.6
At the instant a ball is hit with a tennis racket, both the ball and the racket start to change shape. As they do the racket loses kinetic energy. This energy is stored as strain potential energy in the racket, strings and ball for a very short amount of time. The diagram shows a ball as it is just about to rebound from the racket. As it rebounds the strain potential energy becomes the kinetic energy of the ball.

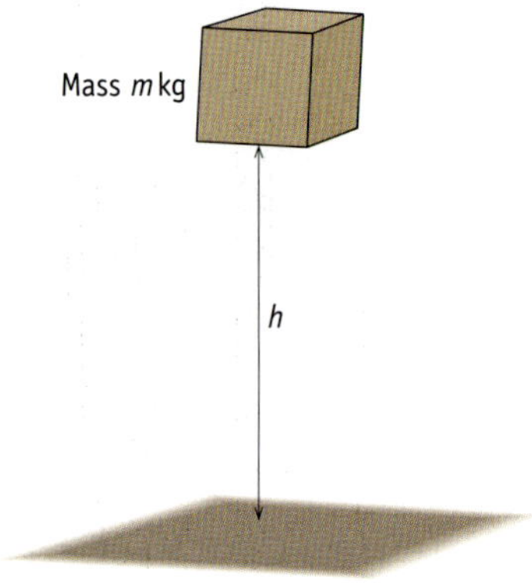

Fig. 11.7

ELECTRICAL POTENTIAL ENERGY

This will be discussed more fully in Chapter 20.

GRAVITATIONAL POTENTIAL ENERGY

Gravitational potential energy is the energy a body has because it can fall to a lower level due to its weight. For example, a concrete block held 3 m above the ground has more energy than an identical block on the ground. This is due to the fact that if the first block is released, it will fall with its weight doing work. The lower block cannot do this. If two bodies are a vertical distance *h* apart, the higher one has more potential energy than the lower one. The larger *h* is, the greater the potential energy difference.

FORMULA FOR GRAVITATIONAL POTENTIAL ENERGY

The difference in potential energy E_p between two points that are a vertical distance h apart in a gravitational field is given by:

$$E_p = mgh$$

If the mass in Fig. 11.7 is released, it falls through the vertical height h under the action of its weight.

Work done by weight = Force × distance = $mg \times h$ = mgh joules.

This must be the potential energy lost by the mass.

Problem 8: A particle of mass 35 kg is projected upwards through a vertical height of 1200 metres. Find the increase in its potential energy.

Solution: $E_p = mgh = (35)(9.8)(1200) = 411\,600\text{ J}$

Problem 9: How far above the surface of the Earth must 1 kg be raised so that its potential energy is increased by 1 MJ.

Solution: $E_p = mgh \Rightarrow 1 \times 10^6 = (1)(9.8)h \Rightarrow h = \dfrac{1 \times 10^6}{9.8} = 102\,040.8\text{ m}$

LOSS IN POTENTIAL ENERGY = GAIN IN KINETIC ENERGY FOR A FREELY-FALLING OBJECT

If friction is absent, or if it is so small that it can be neglected, the kinetic energy that a falling object gains as it speeds up is equal to the potential energy that it loses (due to decreasing height). Likewise, if an object is thrown vertically upwards, it gains potential energy as it gets higher; however, this gain is accounted for by an equal loss in kinetic energy as it slows down.

Problem 10: A stone of mass 4 kg is dropped from a height of 200 m. Find its kinetic energy and speed just before it hits the ground.

Solution: We could use the equations of motion here. However, it is easier to use energy methods.

At the top: Kinetic energy $E_k = 0$ and

Potential energy $E_p = mgh = (4)(9.8)(200) = 7840\text{ J}$

At the bottom, the potential energy has changed to kinetic energy, $\Rightarrow E_k = 7840\text{ J}$

i.e. $\frac{1}{2}mv^2 = 7840 \Rightarrow \frac{1}{2}(4)v^2 = 7840 \Rightarrow v = \sqrt{\dfrac{(7840)(2)}{4}} = 62.61\text{ m s}^{-1}$

Problem 11: A stone is projected vertically upwards with an initial speed of 40 m s^{-1}. Find the greatest height reached.

Solution: Let m be mass of stone. At greatest height speed is zero,

$\Rightarrow E_k = 0$

On upward journey: Gain in E_p = Loss in E_k

i.e. $mgh = (½)mv^2 \Rightarrow m(9.8)h = (½)m(40^2)$

$\therefore$ Greatest height reached $h = \dfrac{40^2}{(2)(9.8)} = 81.63$ m

Problem 12: A pendulum bob is released at position A in FIG. 11.8. Find the speed of the bob as it passes through the vertical position B.

Solution: In going from A to B: Gain in E_k = Loss in E_p

i.e. $(½)mv^2 = mgh \Rightarrow v^2 = 2gh$

$\Rightarrow v = \sqrt{(2)(9.8)(0.4)} = \sqrt{7.84} = 2.8$ m s^{-1}

Note that the loss in potential energy depends only on the **vertical height** through which the object falls – in this case 40 cm – and it does not depend on the actual path taken. This result is generally true.

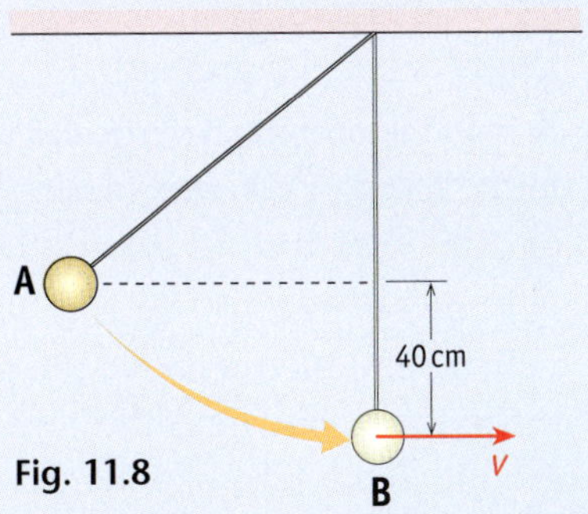

Fig. 11.8

LOSS OF KINETIC ENERGY IN COLLISIONS

When two objects collide, **momentum is always conserved** provided no external forces act. **Usually kinetic energy is not conserved**. The total kinetic energy of the objects after the collision will usually be less than the total before. The lost kinetic energy is converted into other forms such as **heat energy and sound energy**. How much is lost depends on the elastic properties of the colliding bodies. In pool, when one ball hits another ball almost no kinetic energy is lost and almost all the first ball's energy is shared between the two balls as kinetic energy (FIG. 11.9). In a collision between a lump of plasticine and a tennis ball, a considerable amount of kinetic energy is lost.

Fig. 11.9

Problem 13: A mass of 5 kg travelling at 20 m s^{-1} collides with and sticks to a mass of 2 kg which is at rest. Find the velocity of the combined mass after the collision. Find also the loss in kinetic energy.

Solution: Let v be the speed of the combined mass after the collision (FIG. 11.10).

Momentum before = Momentum after collision:

$\Rightarrow (5)(20) + (2)(0) = 7v \Rightarrow 100 + 0 = 7v$

$\Rightarrow v = 14.29$ m s^{-1}

Total E_k before collision $= (½)(5)(20^2) + (½)(2)(0^2)$

$= 1000 + 0 = 1000$ J

Total E_k after collision $= (½)(7)(14.29^2) = 714.7$ J

$\therefore$ Loss in Kinetic energy

$= 1000 - 714.7 = 285.3$ J

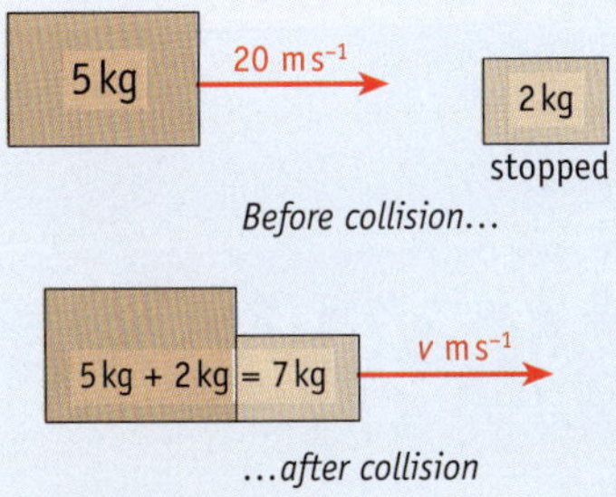

Fig. 11.10

EXERCISE 11.3

Take $g = 9.8$ m s^{-2}

1. A particle of mass 20 kg is thrown upwards through a vertical height of 600 m. Find the increase in its potential energy.
2. How far above the surface of the Earth must a 4 kg mass be raised so that its potential energy is increased by 4000 J?
3. A stone of mass 3 kg is dropped from a height of 60 m. Find its kinetic energy and speed just before it hits the ground.
4. A stone is projected vertically upwards with an initial speed of 100 m s^{-1}. Using energy methods or otherwise find the greatest height reached.
5. A body of mass 100 kg is at rest at a point A, which is 50 metres above the ground. What is the potential energy of the body at A, taking the ground as the zero of potential energy? What is the kinetic energy of the body when it has fallen a distance of 20 m?
6. A pendulum bob of mass m is released at position A in FIG. 11.11. Find the speed of the bob as it passes through the vertical position B.

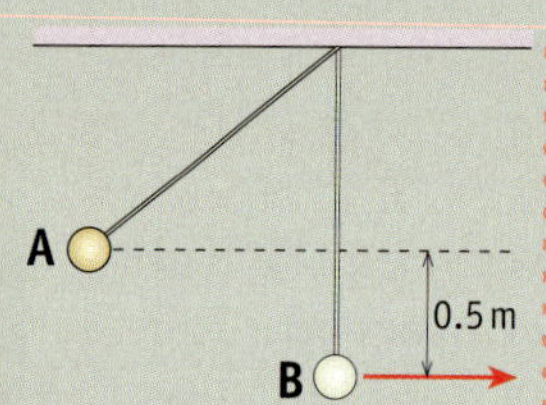

Fig. 11.11

7. A pendulum bob of mass m is released at position X in FIG. 11.12. Find the speed of the bob as it passes through the vertical position Y.

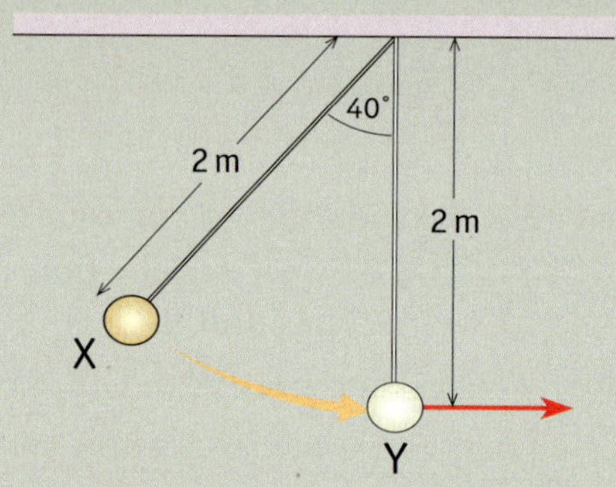

Fig. 11.12

8. A mass of 10 kg travelling at 30 m s^{-1} collides with and sticks to a mass of 15 kg which is at rest. Find the velocity of the combined mass after the collision. Find also the loss in kinetic energy.
9. A small mass of 5 kg is suspended from a fixed point by a light string 2 m long. Another mass of 6 grams moving horizontally at 100 m s^{-1} strikes the 5 kg mass and sticks to it. The combined mass then swings in a vertical plane. Find the greatest height it reaches.

RENEWABLE AND NON-RENEWABLE SOURCES OF ENERGY

Everyday we use energy for many purposes, e.g. transport and heating. The sources of this energy are many, e.g. oil, coal and hydroelectricity. Sources of energy are classified as renewable or non-renewable.

RENEWABLE SOURCE OF ENERGY

A source of energy that does not get used up is called a ***renewable source of energy.***

Examples of renewable energy sources are wind energy (FIG. 11.13), solar energy (FIG. 11.14), hydroelectric energy, wave energy and biomass. Non-renewable sources include oil, coal, peat and natural gas.

Fig. 11.13
Wind Turbines.

Fig. 11.14
Solar Panels.

EFFICIENT USE OF ENERGY IN THE HOME

There are good reasons for using energy efficiently in the home. It reduces the amount of money spent on energy. It reduces the rate at which valuable non-renewable sources of energy are used. This can be done, for example, by insulating walls, attics and floors. Double glazing in windows and draught-proofing in doors and windows also helps. Energy efficient miniature fluorescent bulbs use considerably less energy than a conventional filament bulb to produce the same amount of light.

POWER

A bulb in a desk lamp may convert electrical energy into heat and light at the rate of 40 joules per second; a bulb in a floodlight may convert electrical energy to heat and light at the rate of 500 joules per second. We say that the second bulb is more powerful than the first.

A builder brings 1000 blocks each of mass 20 kg from the ground, up a ladder, to the top of a house. It takes him 5 hours to do so. A crane lifts all the blocks together from the ground to the top in a time of 10 seconds. Even though the man and the crane do the same work on the blocks the crane does it in a much shorter time. The crane is much more powerful than the man.

In physics, **power** is defined as **the rate at which work is done or the rate at which energy is converted from one form to another.**

These two definitions are really the same thing. Whenever energy is converted, work is always done. The symbol for power is ***P***. Power is a **scalar quantity**.

POWER
Power is the rate at which work is done.
or
Power is the rate at which energy is converted from one form to another.

AVERAGE POWER

The average power developed in a time t is given by:

$$\text{Average Power} = \frac{\text{Work done}}{\text{Time taken}} \qquad P = \frac{W}{t}$$

or

$$\text{Average Power} = \frac{\text{Energy converted}}{\text{Time taken}} \qquad P = \frac{E}{t}$$

If the power is being developed at a constant rate each of these formula represent the power at any instant.

UNIT OF POWER

Since $P = \frac{W}{t}$, Unit of power $= \frac{\text{Unit of work}}{\text{Unit of time}} = \frac{\text{joule}}{\text{second}}$

$=$ **joule per second (J s^{-1})**

This unit is called the **watt *(W)*** in honour of the English scientist James Watt (1736–1819).

UNIT OF POWER
The unit of power is the **watt** (*W*).
1 watt = 1 joule per second

The watt **in terms of basic units:**

$1\ W = 1\ \text{J s}^{-1}$
$= 1(\text{kg m}^2\ \text{s}^{-2})\ \text{s}^{-1}$
$= \mathbf{1\ kg\ m^2\ s^{-3}}$

(From page 124: $1\ \text{J} = 1\ \text{kg m}^2\ \text{s}^{-2}$)

Fig. 11.15
The outboard engine on this boat has a maximum power of 110 kW when operating at full throttle.

Problem 14: A man raises an object of mass 40 kg through a vertical height of 20 m in 4 seconds. Find his average power.

Solution: $P = \frac{W}{t} = \frac{\text{Energy gained}}{\text{Time taken}} = \frac{mgh}{t} = \frac{(40)(9.8)(20)}{4} = 1960\text{ W}$

Problem 15: A weight lifter raises a mass of 30 kg through a height of 0.6 m. He does this 50 times in half a minute. Find the average power he develops.

Solution: Work done in one lift $= mgh = (30)(9.8)(0.6) = 176.4\text{ J}$

$\therefore$ Total work done $= (50)(176.4) = 8820\text{ J}$

$\text{Power} = \frac{\text{Work}}{\text{Time}} = \frac{8820}{30} = 294\ W$

Problem 16: A woman of mass 80 kg runs up a flight of 120 stairs in 30 seconds. If the height of each stair is 15 cm, find her average power.

Solution: Vertical height through which woman runs = (0.15)(120) = 18 m

$P = \frac{\text{Work}}{\text{Time}} = \frac{mgh}{t} = \frac{(80)(9.8)(18)}{30} = 470.4\ W$

Problem 17: A 100 watt electric heater is used to heat water and bring it to the boil. If 5000 J are needed to bring the water to the boil, for how long is the heater operating?

Solution: $\text{Power} = \frac{\text{Energy transferred}}{\text{Time taken}} \Rightarrow \text{Time} = \frac{\text{Energy}}{\text{Power}} = \frac{5000}{100} = 50\text{ s}$

EXERCISE 11.4

Take $g = 9.8\text{ ms}^{-2}$

1. A car engine produces 600 000 joules of energy in 12 seconds. At what power is it operating?
2. A light bulb gives out 18 000 J of heat and light in 30 minutes. Calculate the power rating of the bulb.
3. A bar on an electric fire gives out 60 000 J of energy in one minute. Calculate its power.
4. A lawnmower does 2×10^7 J of work in 2 hours. Calculate the power of the lawnmower.
5. How much energy is given out by a 60 W bulb in 5 hours?
6. A car has a maximum power of 130 kW. While operating at this power how much energy does it expend in 5 minutes?
7. A man raises an object of mass 50 kg through a vertical height of 50 m in 12 seconds. Find his average power.
8. A weight lifter raises a mass of 50 kg through a height of 2 m. He does this 25 times in one minute. Find his average power.
9. A boy of mass 40 kg sprints up 50 stairs each of height 16 cm in 40 seconds. Find his average power.
10. How long does it take a 77 kW engine to use 1 MJ of energy?
11. The unit of electrical energy that the ESB use for domestic power supplies is called the kilowatt-hour. One kilowatt-hour is the amount of energy given out by a 1000 W appliance in one hour. Calculate the number of joules in a kilowatt-hour.

Efficiency

In machines and motors of all types, energy is input and output. For example, in an electric motor, electrical energy is supplied and kinetic energy is produced. In all such devices there are always some energy losses. Thus the useful energy output is always less than the energy input. The difference in energy is converted to some other unwanted forms, usually heat. Such devices are not therefore 100% efficient. In terms of power, the useful output power is less than the input power.

Efficiency is defined as follows:

$$\text{Percentage efficiency} = \frac{\text{Power Output}}{\text{Power Input}} \times 100$$

Problem 18: An engine has an input power of 4500 W. The useful output power is 3000 W. Calculate its percentage efficiency.

Solution: $\text{Efficiency \%} = \frac{\text{Power output}}{\text{Power input}} \times 100 = \frac{(3000)(100)}{4500} = 66.67\%$

Problem 19: A 60 kW motor in a crane lifts a mass of 4000 kg through a height of 10 m in 20 s. Calculate the efficiency % of the motor if it is working at full power.

Solution: $\text{Power output} = \frac{\text{Work}}{\text{Time}} = \frac{mgh}{t} = \frac{(4000)(9.8)(10)}{20} = 19\,600 \text{ W}$

$\text{Efficiency \%} = \frac{\text{Power output}}{\text{Power input}} \times 100 = \frac{(19\,600)(100)}{60\,000} = 32.67\%$

Problem 20: A petrol car engine is 30 % efficient. Kinetic energy is produced at the rate of 105 kW. The rest of the energy input appears as heat. Calculate the rate at which heat is produced.

Solution: 30% of the input power is 105 kW $\therefore$ 70% of the power is converted to heat.

$\text{Amount converted to heat} = \frac{(105)(70)}{30} = 245 \text{ kW}$

EXERCISE 11.5

1. An engine has an input power of 5000 W. The useful output power is 4000 W. Calculate its percentage efficiency.
2. A 60 kW motor in a crane lifts a mass of 2000 kg through a height of 20 m in 10 s. Calculate the percentage efficiency of the motor if it is working at full power.
3. A petrol car engine is 25% efficient. Kinetic energy is produced at the rate of 130 kW. The rest of the energy input appears as heat. Calculate the rate at which heat is produced.

CHAPTER CHECKLIST

- **Define:** Work; Energy; Power; Percentage efficiency; The joule; The watt; Kinetic energy; Potential energy; Renewable source of energy; Non-renewable source of energy.
- **State:** The unit of work; The unit of energy; The Principle of Conservation of Energy; The unit of power.
- **Express:** The joule in basic units; The watt in basic units.
- **Recall** that: Whenever work is done energy is transferred; In the absence of air resistance for a freely falling body the loss in potential energy equals the gain in kinetic energy and vice versa; In collisions, momentum is always conserved provided no exteral forces act, but kinetic energy is usually not conserved.
- **Be able to:** Give an example of any named energy conversion; State what energy conversion or conversions are taking place in a given situation.
- **Recall** and use the formulae: $W = Fs$; $E_k = ½mv^2$;

  $E_p = mgh$ $\quad P = \frac{W}{t}$ $\quad \text{Percentage efficiency} = \frac{\text{Power Output}}{\text{Power Input}} \times 100$
- **List:** Six forms of energy; Three different forms of potential energy; Three examples of renewable and three examples of non-renewable sources of energy; Five ways of making energy use in the home more efficient.

CHAPTER 12

Circular Motion

Measuring Angles in Radians

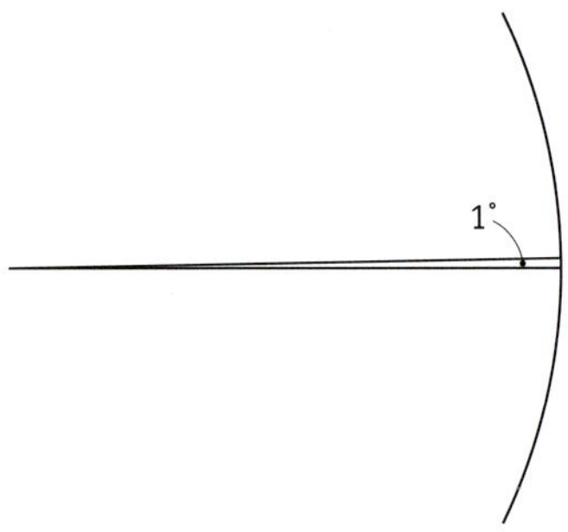

Fig. 12.1
1 degree (1°) is $^1/_{360}$ of a full circle.

You are, no doubt, familiar with measuring angles in degrees, where $1° = \frac{1}{360}$ of a full circle (Fig. 12.1). **Angles are also measured in a unit called the radian.** What is the size of one radian? In Fig. 12.2 three different arcs of length s_1, s_2 and s_3 are drawn on a circle of radius r. From the diagram it is obvious that arc length s_1 is less than the radius r and arc length s_2 is greater than the radius r. For one particular angle (about 57.3°) it is found that the arc length and the radius are the same length. The angle for which this occurs is called **one radian**. In Fig. 12.2(c) arc length s_3 = radius r and therefore the angle θ is one radian.

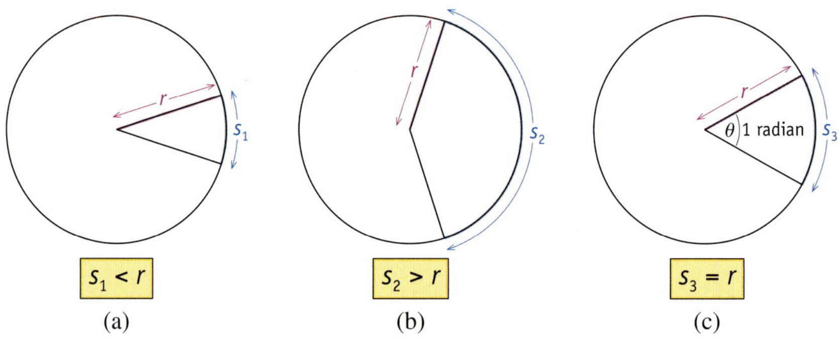

Fig. 12.2
The angle for which the arc length is equal to the radius is called one radian. One radian is about 57.3°.

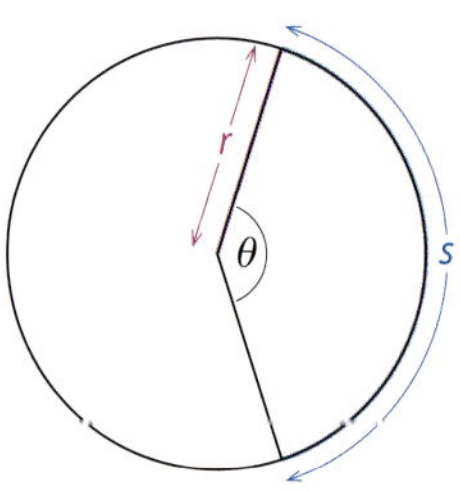

Fig. 12.3

Suppose that in Fig. 12.3 the arc length is 2.5 times the length of the radius. This would mean that there are 2.5 radians in the angle θ. In general:

Number of radians in an angle = Number of times radius fits into the arc.

$$\text{i.e. Angle in radians} = \frac{\text{Arc length}}{\text{Radius}} \qquad \theta = \frac{s}{r}$$

A version of this formula appears on page 6 of your maths tables. When using this formula in problems to calculate s or r, make sure that θ is always in radians, not in degrees.

Problem 1: In Fig. 12.4 arc length = 200 m and r = 75 m. What is the angle θ in radians?

Solution: $\theta = \frac{s}{r} = \frac{200}{75} = 2.67$ rad

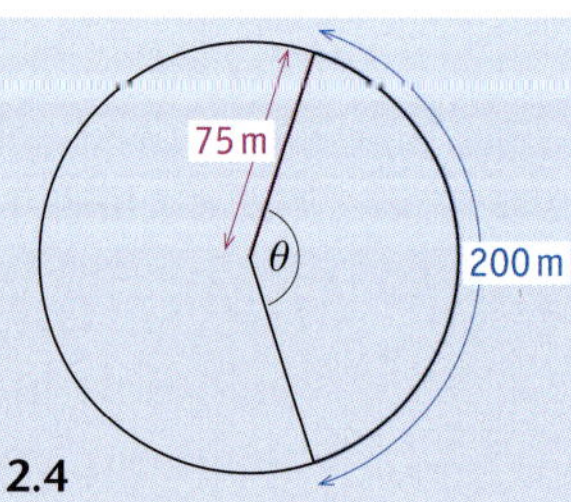

Fig. 12.4

Problem 2: How many radians in 180°?

Solution: FIG. 12.5 shows an arc of length s with radius r about the point o.

Length of arc s = half of circumference of circle

$$= (½)(2\pi r) = \pi r$$

$$\theta = \frac{\text{Arc length}}{\text{Radius}} = \frac{s}{r} = \frac{\pi r}{r} = \pi \text{ radians}$$

Thus: 180° = π radians, i.e. 180° = 3.14159 rad

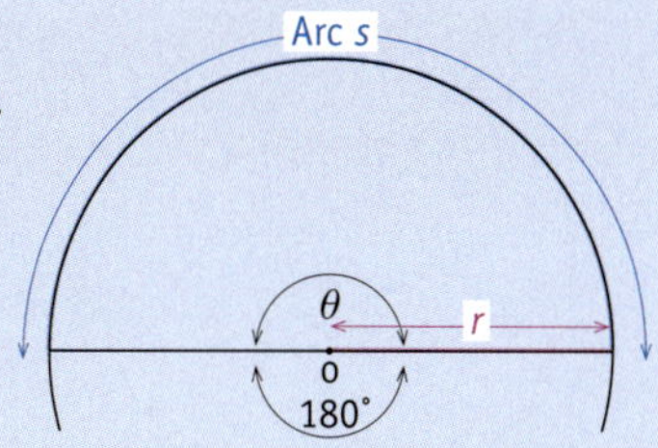

Fig. 12.5

180° = π radians i.e. 180° = 3.14159 rad

Using this formula you can convert degrees to radians or radians to degrees.

Problem 3: (i) Find the number of radians in 17° (ii) Find the number of degrees in 1 radian.

Solution: (i) $180° = \pi \text{ rad} \Rightarrow 1° = \frac{\pi}{180} \text{ rad} \Rightarrow 17° = \frac{17\pi}{180} = 0.2967 \text{ rad}$

(ii) $\pi \text{ rad} = 180° \Rightarrow 1 \text{ rad} = \frac{180}{\pi} \text{ degrees} = \frac{180}{3.14159} = 57.296°$

Problem 4: Find the length of the arc s in FIG. 12.6.

Solution: We shall use the formula $\theta = \frac{s}{r}$ to find s. First **49° must be converted to radians**:

$$180° = \pi \text{ rad} \Rightarrow 1° = \frac{\pi}{180} \text{ rad}$$

$$\Rightarrow 49° = \frac{49\pi}{180} = 0.8552 \text{ rad}$$

Find s: $\theta = \frac{s}{r} \Rightarrow s = r\theta = (12)(0.8552) = 10.26 \text{ m}$

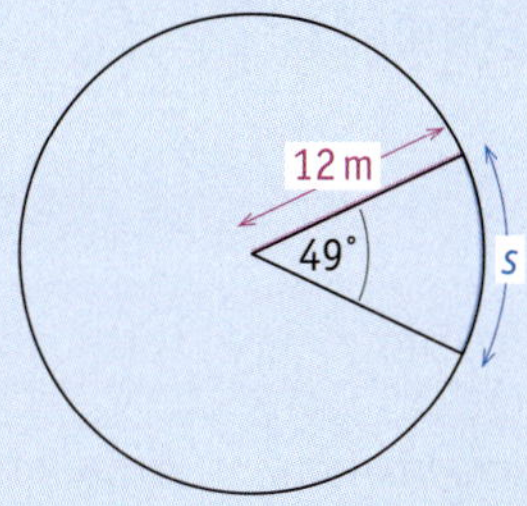

Fig. 12.6

EXERCISE 12.1

1. Convert to radians each of the following: 180°, 90°, 45°, 30°, 60°, 360°, 10°, 48°.
2. Convert to degrees each of the following: $\pi/5$ rad, $\pi/7$ rad, 1 radian, 4.2 rad, $\pi/3$ rad, $\pi/6$ rad, $\pi/4$ rad.
3. The radius of a circle is 3 m. What is the length of the arc of the circle subtending an angle of (i) 1 rad (ii) π rad (iii) 45° and (iv) 123° at the centre of the circle?
4. An arc of length 3 cm of a circle subtends an angle of 3.5 radians at the centre of the circle. What is the radius of the circle?
5. The Sun subtends an angle of 0.00932 radians to an observer on the Earth. If the sun is 93 million miles away, find the approximate diameter of the Sun.
6. A pendulum 2.2 m long oscillates through an angle of 10°. What is the length of the arc traced out?

PARTICLE MOVING IN A CIRCLE

A small object P (called a particle) is moving at a constant speed of v metres per second in an anticlockwise direction around the circle in FIG. 12.7 (page 138). At any instant P is moving along a tangent to the circle, i.e. when it is at A it is moving along the tangent at A; when it is at B it is moving along the tangent at B. v is called the **tangential speed** or the **linear speed** of the small object.

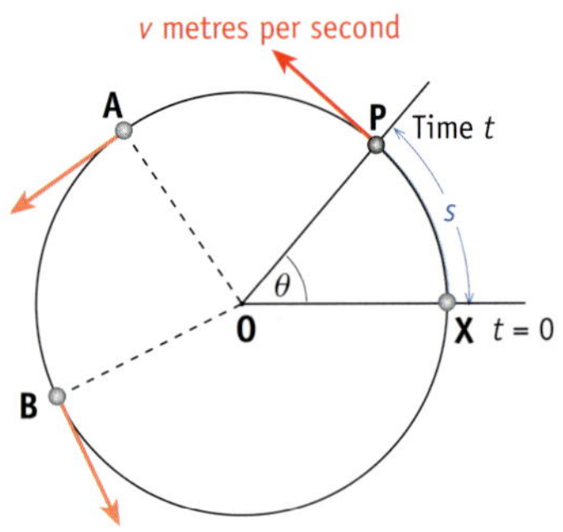

Fig. 12.7

ANGULAR VELOCITY
Angular velocity is the rate of change of angle with respect to time.

UNIT OF ANGULAR VELOCITY
The unit of angular velocity is the ***radian per second*** **(rad s^{-1}).**

Since the direction of the velocity of P is continually changing, ***P* is accelerating**. The magnitude of P's velocity, i.e. its speed, is constant.

Suppose that P moves from X to the position shown in t seconds. Let the distance travelled along the circle in this time be s metres.

Then since P has a constant speed v we have: $v = \dfrac{s}{t}$

The radius OP traces out an angle θ with the line OX in t seconds. Thus as t increases so does the angle θ.

Angular Velocity

The rate at which the angle θ is changing with respect to time is called the **angular velocity** of P about O.

- Angular velocity is a **scalar quantity**.
- The symbol for angular velocity is ω (the Greek letter omega).
- Angular velocity ω is measured in **radians per second (rad s^{-1})**. It could also be measured in degrees per second but we shall never do this.
- Since P is moving at a steady speed, the angle θ is also changing at a steady rate. Thus P has a constant angular velocity.
- Just like constant speed $= \dfrac{\text{Distance travelled}}{\text{Time taken}}$ $\left(v = \dfrac{s}{t}\right)$

Constant angular velocity $= \dfrac{\textbf{Angle traced out}}{\textbf{Time taken}}$

i.e. $\omega = \dfrac{\theta}{t}$ ***or*** $\theta = \omega t$

Problem 5: If the angle traced out by P (Fig. 12.7) in 4 seconds is 10 radians, find its angular velocity.

Solution: Angular velocity $\omega = \dfrac{\theta}{t} = \dfrac{10}{4} = 2.5$ rad s^{-1}

Problem 6: A small object moving in a circle with a steady speed does 3000 revolutions of the circle per minute. Find its angular velocity (in rad s^{-1}).

Solution: 1 revolution $= 2\pi$ radians $\Rightarrow$ 3000 rpm $= 3000 \times 2\pi$ radians per minute

$\therefore$ angular velocity $\omega = \dfrac{3000 \times 2\pi}{60}$ rad s^{-1} $= 314.16$ rad s^{-1}

Problem 7: A particle moving in a circle has a constant angular velocity of 2.2 rad s^{-1}. What angle does it trace out in 6 seconds?

Solution: $\omega = \dfrac{\theta}{t} \Rightarrow \theta = \omega t = (2.2)(6) = 13.2$ rad

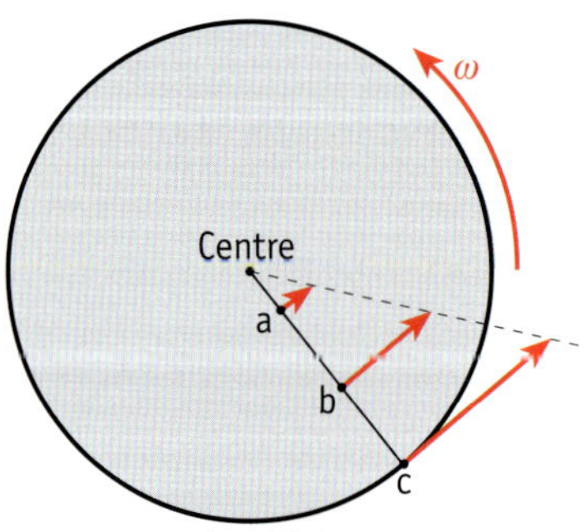

Fig. 12.8

Relationship Between Linear Speed and Angular Velocity $v = r\omega$

In Fig. 12.8 a wheel is rotating about its centre with constant angular velocity ω. Since the points a, b and c take the same time to undergo a full circle, **they all have the same angular velocity**. Since the point c has to travel a longer path than a or b to trace out a full circle, it has the greatest linear speed v. We shall now prove the following equation relating v, r and ω:

If an object is moving in a circular path of radius r with constant angular velocity ω rad s^{-1} and constant linear speed v m s^{-1} then:

$$v = r\omega$$

Proof: Suppose the object traces out an angle θ in time t seconds as it moves a distance s (FIG. 12.9). Then from the definition of the radian we have: $\theta = \frac{s}{r}$

Dividing both sides of this equation by t gives:

$$\frac{\theta}{t} = \frac{s}{rt} \Rightarrow \frac{\theta}{t} = \frac{s}{t} \times \frac{1}{r}$$

But $\frac{\theta}{t} = \omega$ and $\frac{s}{t} = v$

$\therefore \quad \omega = \frac{v}{r}$ It follows that $v = r\omega$

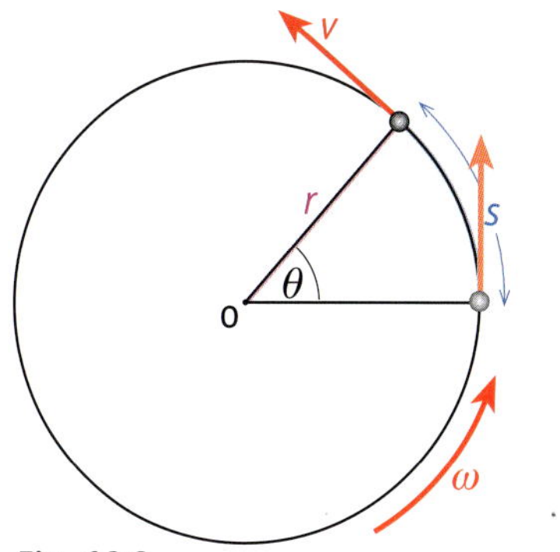

Fig. 12.9

Problem 8: A wheel of radius 50 cm rotates with a constant angular velocity. If a point on the rim of the wheel has a speed of 10 m s^{-1}, what is the angular velocity of the wheel?

Solution: $\omega = \frac{v}{r} = \frac{10}{(0.5)} = 20 \text{ rad s}^{-1}$

Problem 9: A stone moving at a steady speed in a circle of radius 2 m performs 10 complete revolutions of the circle per second. Find:

(i) its angular velocity, (ii) its tangential or linear speed.

Solution: (i) 10 revolutions per second $= 10 \times 2\pi$ radians per sec i.e. $\omega = 20\pi \text{ rad s}^{-1}$

(ii) Tangential or linear speed $v = r\omega = (2)(20\pi) = 40\pi = 125.66 \text{ m s}^{-1}$

Problem 10: The radius of the Earth at the Equator is 6.4×10^6 m. Assuming that the Earth rotates once about its polar axis every 24 hours find:

(i) the angular velocity of the Earth.

(ii) the linear speed of a point on the Equator.

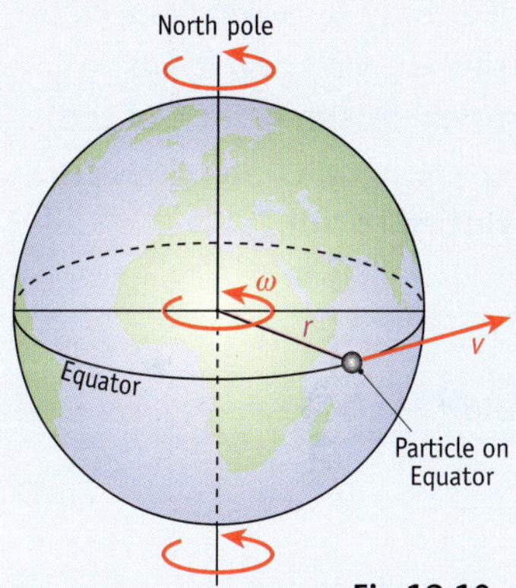

Fig.12.10

Solution: (i) $\omega = \frac{\text{Angle traced out}}{\text{Time taken}} = \frac{2\pi \text{ rad}}{(24)(60)(60) \text{ seconds}}$

$= 7.27 \times 10^{-5} \text{ rad s}^{-1}$

(ii) A point on the Equator is moving in a circle whose radius is the radius of the Earth with an angular velocity equal to that of the Earth (FIG. 12.10). Therefore:

$v = r\omega = (6.4 \times 10^6)(7.27 \times 10^{-5}) = 465.28 \text{ m s}^{-1}$

EXERCISE 12.2

1. A small object moving with a constant speed in a circular path of radius 3 m, travels a distance of 40 m in 2 minutes. Find:
   (i) its linear speed,
   (ii) its angular velocity.

   How long does it take for the radius from the centre of the circle to the small object to trace out an angle of:
   (i) $\pi/2$ rad, (ii) 60°?

2. A small object moving in a circle of radius 10 m has a constant angular velocity of 4 rad s^{-1}. How long does it take to travel 1 km?

3. A wheel of radius 40 cm rotates with constant angular velocity. If a point on the rim of the wheel is travelling at 3 m s^{-1}, find the angular velocity.

4. A flywheel rotates with a constant angular velocity of 6 rad s^{-1}. Find the linear speed of a point at:
   (i) its centre,
   (ii) 30 cm from its centre.
5. A flywheel revolves at a constant rate of 400 rpm. Find the linear speed of a point on the wheel 2 m from its centre.
6. The radius of the Earth at the Equator is 6.4×10^6 m. Find:
   (i) the angular velocity of the Earth about its polar axis,
   (ii) the linear speed of a small object at the equator,
   (iii) the linear speed of a small object at latitude 53° North.

CENTRIPETAL FORCE
If a body is moving in a circle the force towards the centre needed to keep it moving in that circle is called ***centripetal force***.

CENTRIPETAL ACCELERATION
If a body is moving in a circle the acceleration it has towards the centre of the circle is called ***centripetal acceleration***.

Centripetal Force

We saw that when a particle moves at constant speed around a circle, the direction of its velocity is changing and hence it is **accelerating**. By Newton's 2nd Law the particle must therefore have a resultant force acting on it. The force must be at right angles to the direction of motion at any instant because if the force had a component parallel to the direction of the motion, the particle would speed up or slow down. The force must be acting towards the centre of the circle rather than in the opposite direction. This force is called **centripetal force**.

This can easily be demonstrated in the laboratory. A light rubber bung is tied to the end of about 0.5 m of thread and is then whirled in a circle. The centripetal force acting on the bung is the tension in the thread* which is obviously acting towards the centre of the circle. If the string is released from your hand, the tension immediately disappears and the bung moves off along the tangent. With a bit of practice you can get the bung to move in the direction of whatever tangent you want (be careful!).

Since centripetal force acts towards the centre of the circle the acceleration is towards the centre. This acceleration is called **centripetal acceleration**.

FACTORS ON WHICH CENTRIPETAL FORCE DEPENDS

To cause an object moving in a straight line to turn a circular corner, common experience tells us that:

- the greater the mass of the object the greater the force needed,
- the smaller the radius of the turn the greater the force needed,
- the faster the object is moving the greater the force needed.

It can be proved that:

To keep an object of mass m moving at a constant speed v in a circle of radius r, a force F acting towards the centre of the circle is needed, where F is given by:

$$F = \frac{mv^2}{r} \quad \text{or} \quad F = mr\omega^2 \quad (\omega \text{ is the angular velocity})$$

Since acceleration = Force/mass it follows that:

Centripetal acceleration $a = \frac{v^2}{r}$ or $a = r\omega^2$ (See Maths tables p40)

Note that centripetal force is only a name for some real force (or the resultant of a number of forces) such as, for example, the tension in a string, the force of gravity or the force of friction. In a given problem, you must be able to spot what real force is supplying the centripetal force needed to keep the object moving in a circle.

* Gravity also effects the centripetal force but for the moment we shall ignore this fact.

Problem 11: A 2 kg stone moves in a circle of radius 3 m with a constant speed of 4 m s^{-1}. Find: (i) its acceleration, (ii) the resultant force acting on it.

Solution: (i) Centripetal acceleration $= \frac{v^2}{r} = \frac{4^2}{3} = \frac{16}{3} = 5.333$ m s^{-1}

(ii) Resultant force acts towards the centre $F = \frac{mv^2}{r} = \frac{(2)(4^2)}{3}$ 10.667 N

Alternative solution to (ii): $F = ma = (2)(5.333) = 10.667$ N

Problem 12: A particle of mass 10 kg travelling at 8 m s^{-1} is moving in a circular path with a constant angular velocity of 2 rad s^{-1}.

(i) What is the radius of the circle?

(ii) What is the acceleration of the particle in magnitude and direction?

Solution: (i) $v = r\omega \Rightarrow r = \frac{v}{\omega} = \frac{8}{2} = 4$ m

(ii) $a = r\omega^2 = (4)(2)^2 = 16$ m s^{-2} towards the centre of the circle.

Problem 13: An electron in a magnetic field moves in a circular path of radius 1 cm. If its speed is 2×10^6 m s^{-1} find the centripetal force on the electron. (Mass of electron $= 9.1 \times 10^{-31}$ kg.)

Solution: $F = \frac{mv^2}{r} = \frac{(9.1 \times 10^{-31})(2 \times 10^6)^2}{1 \times 10^{-2}} = 3.64 \times 10^{-16}$ N

EXERCISE 12.3

Take $g = 9.8$ m s^{-2}

1. A car of mass 1200 kg travels around a circular track at 25 m s^{-1}. If the radius of the track is 200 m find the centripetal force.
2. Find the centripetal force on a mass of 4 kg which is moving in a circle of radius 20 cm with constant angular velocity 4.5 rad s^{-1}.
3. Find the centripetal acceleration of a particle moving in a circle with a speed of 12 m s^{-1} if the radius of the path is:
   (i) 1 m
   (ii) 30 cm
   If the mass of the particle is 5 kg find the centripetal force acting on it in each case.
4. A mass of 10 kg moves in a circular path of radius 40 cm with a constant linear speed of 4 m s^{-1}. Find:
   (i) its angular velocity,
   (ii) its centripetal acceleration,
   (iii) the force acting on it.
5. A car of mass 950 kg travels around a circular bend of radius 100 m at 100 km hr^{-1}. What is the force on the car towards the centre of the circle?
6. A bicycle wheel turns with an angular velocity of 20 rad s^{-1}. If the radius of the wheel is 0.5 m, find the linear speed of a point on the rim of the wheel.
7. A human body can safely withstand a sustained acceleration of $9g$ m s^{-2}. What is the minimum radius within which a pilot can turn his plane if its speed is 500 m s^{-1}? Neglect gravity.
8. Fig. 12.11 shows a body of mass 0.24 kg attached to a fixed point P by a light string of length 0.80 m. When the body is at A, vertically below P, it is given an initial horizontal velocity of 5 m s^{-1} as shown. It then follows a circular path to the position B. When it is at B calculate:
   (i) the velocity of the body,
   (ii) the centripetal acceleration of the body,
   (iii) the force exerted by the string on the body.

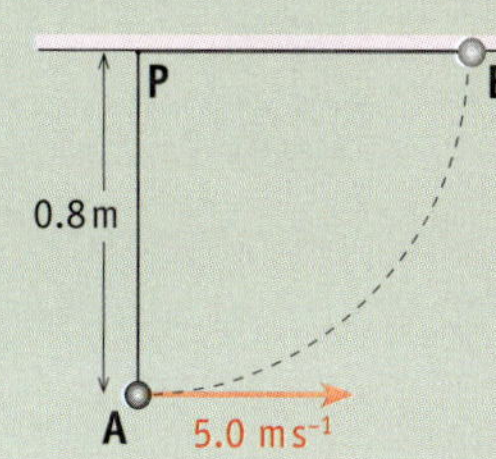

Fig. 12.11

Fig. 12.12

CIRCULAR SATELLITE ORBITS

The Earth and the other planets of the solar system orbit the Sun; some of them with orbits that are approximately circular. Many artificial satellites orbit the Earth (FIG. 12.12). The Moon orbits the Earth and is its only natural satellite. **What supplies the necessary centripetal force to keep these satellites in their orbits?**

Answer: **The force of gravity**.

Newton's Law of Universal Gravitation (page 113) tells us that the gravitational force F of attraction between two spherical bodies is given by

$$F = \frac{Gm_1m_2}{d^2}$$

Where G is the universal gravitation constant, m_1 and m_2 are the masses of the bodies and d is the distance between their centres. Thus in FIG. 12.13 if we have a satellite of mass m moving with speed v in an orbit of radius R about a central body of mass M then:

Centripetal force = Force of Gravity i.e. $\frac{mv^2}{R} = \frac{GMm}{R^2}$

$$\Rightarrow v^2 = \frac{GM}{R}$$

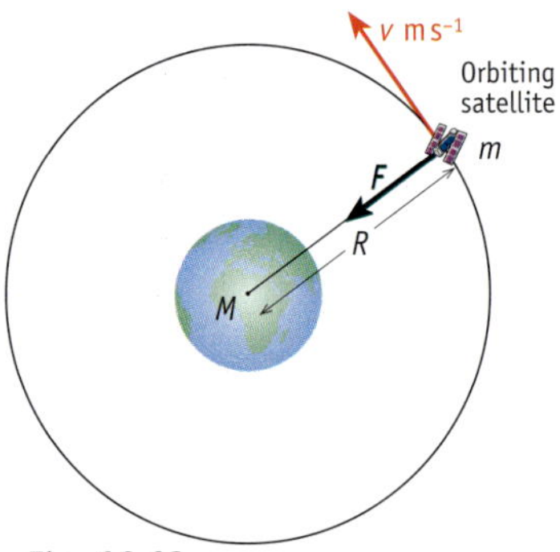

Fig. 12.13

From this equation we see that:

- If you pick the height of the orbit, this equation tells you the speed of the satellite. If you pick the speed, this equation gives you the height.
- The speed of the satellite is independent of the mass of the satellite.
- $v = \sqrt{\frac{GM}{R}} \Rightarrow v = \sqrt{G}\frac{\sqrt{M}}{\sqrt{R}}$

  Thus, since G is constant, the speed of a satellite is directly proportional to the square root of the mass of the body it orbits and is inversely proportional to the square root of the radius of the orbit.

Problem 14: A military spy satellite orbits the Earth at a height of 40 000 km above the Earth's surface. At what speed is it travelling?

$G = 6.7 \times 10^{-11}\ \text{N m}^2\text{kg}^{-2}$, Radius of Earth $= 6.4 \times 10^6$ m, Mass of Earth $= 6 \times 10^{24}$ kg

Solution: Radius of orbit = radius of Earth + height of satellite above Earth

$= (6.4 \times 10^6 + 40 \times 10^6)$ metres

$$\frac{mv^2}{R} = \frac{GMm}{R^2} \Rightarrow v^2 = \frac{GM}{R} \Rightarrow v = \sqrt{\frac{(6.7\times10^{-11})(6\times10^{24})}{(6.4\times10^6 + 40\times10^6)}} = 2943.4\ \text{m s}^{-1}$$

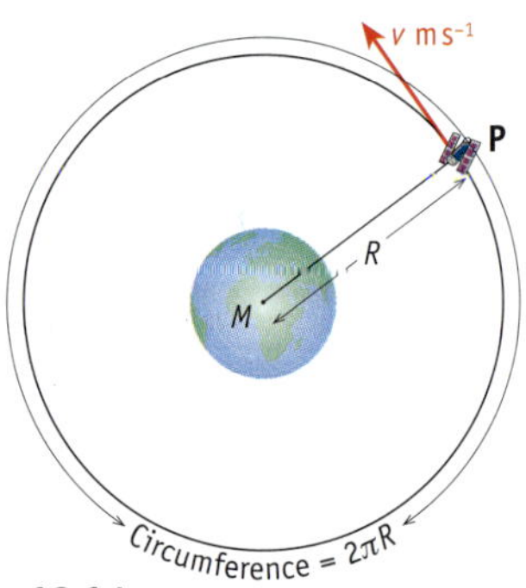

Fig. 12.14

PERIOD OF AN ORBIT

The time taken for a satellite to go once around the central body is called the periodic time or simply the ***period*** **(*T*)** of the orbit.

In FIG. 12.14 the time taken to go from P right around the orbit and back again to P is the period.

T = Time for one revolution = $\frac{\text{Length of orbit}}{\text{Speed}}$

$$\text{i.e. } T = \frac{2\pi R}{v}$$

Problem 15: What is the period of the satellite in the last problem?

Solution: Period $T = \dfrac{2\pi R}{v} = \dfrac{2\pi\,(6.4 \times 10^6 + 40 \times 10^6)}{2943.4} = 99048.65\text{ s} = 27.51\text{ hours}$

Relationship Between the Period of an Orbit and its Radius

Before Newton discovered his Law of Gravity, it was known that: for the planets and the Moon, the square of the period of the orbit was directly proportional to the cube of the radius of the orbit. (This was known as Kepler's 3rd Law.) Newton was able to show that the Law followed mathematically from his Law of Gravity – thus showing that his Law was most likely correct. This was a startling prediction at the time, especially since direct verification of the Law of Gravitation was not yet possible.

Let M be the mass of the body being orbited (the central body). Let m be the mass of the satellite and v its speed. Let R be the radius of the orbit (Fig. 12.15).

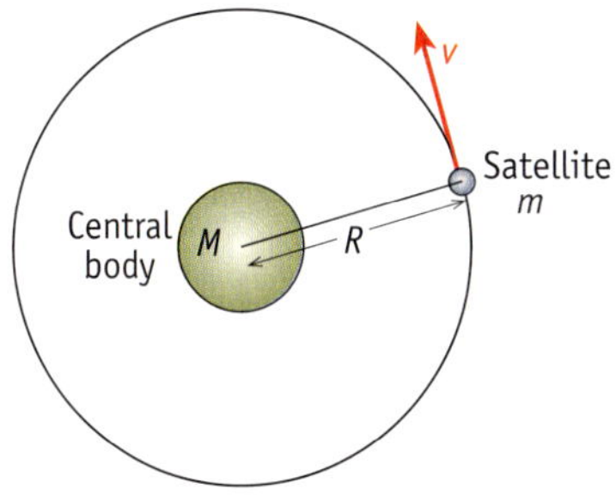

Fig. 12.15

$$\frac{mv^2}{R} = \frac{GMm}{R^2} \Rightarrow v^2 = \frac{GM}{R} \quad \text{(as above)}$$

Period of orbit T = Time for one revolution = $\dfrac{\text{Length of orbit}}{\text{Speed}}$

i.e. $T = \dfrac{2\pi R}{v} \Rightarrow T^2 = \dfrac{4\pi^2 R^2}{v^2}$

Substituting the expression for v^2 into this gives: $T^2 = \dfrac{4\pi^2 R^2}{(GM/R)}$

i.e. $T^2 = \dfrac{4\pi^2 R^3}{GM}$ Thus: $T^2 \propto R^3$

i.e. The square of the period is directly proportional to the cube of the radius of the orbit.*

It follows that: $T \propto \sqrt{R^3}$ and that $T \propto \dfrac{1}{\sqrt{M}}$

$$\mathbf{T^2 = \frac{4\pi^2 R^3}{GM}}$$

Problem 16: The radius of the orbit of the planet Neptune about the Sun is 30 times longer than the Earth's radius of orbit around the Sun. Calculate the time it takes Neptune to complete one orbit of the Sun.

Solution: $T^2 = \dfrac{4\pi^2 R^3}{GM} \Rightarrow \dfrac{T_N^2}{T_E^2} = \dfrac{\frac{4\pi R_N^3}{GM}}{\frac{4\pi R_E^3}{GM}} = \dfrac{R_N^3}{R_E^3}$ where M is the mass of the Sun.

$$\therefore \frac{T_N^2}{T_E^2} = \frac{R_N^3}{R_E^3} = \left(\frac{R_N}{R_E}\right)^3 = 30^3 \Rightarrow \frac{T_N}{T_E} = \sqrt{30^3}$$

$\Rightarrow T_N = T_E \times 164.32 = 164.32$ years

(since period of Earth's orbit = 1 year)

* The orbits of some of the planets are not exactly circular, instead they are elliptical. It may still be proved using Newton's Law of Gravity that $T^2 \propto R^3$. In any questions we deal with we shall assume the orbits are circular.

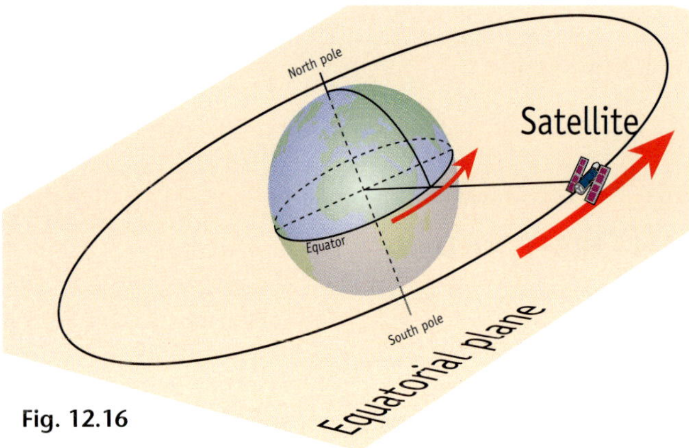

Fig. 12.16

The orbit is geostationary if:
- the satellite orbits in the equatorial plane,
- its period is 24 hours,
- the satellite moves in the same direction as the Earth rotates.

SATELLITE STOPPED ABOVE THE EQUATOR

In FIG. 12.16 a satellite is orbiting the Earth above the Equator and is moving in the equatorial plane. The Earth rotates once about its polar axis every 24 hours. Suppose the period of the orbit of the satellite is also 24 hours and suppose it moves in the same direction as the Earth. The satellite will therefore remain stationary above the same point on the equator. Such a satellite is said to be in a **geostationary orbit** or a **parking orbit**. Most communication satellites are geostationary, thus allowing uninterrupted TV and radio communication around the Earth.

- From the equation $T^2 = \frac{4\pi^2 R^3}{GM}$ we can see that there is only one height above the Earth for a stationary orbit.

Problem 17: At what height above the Equator will a satellite in a parking orbit be found?

Solution: Solving $T^2 = \frac{4\pi^2 R^3}{GM}$ for R gives: $R = \sqrt[3]{\frac{T^2 GM}{4\pi^2}}$

The period of a stationary orbit is 24 hours = 86 400 seconds. The values of all the other quantities in the equation are known. Thus R can be found.

$$= \sqrt[3]{\frac{(86\,400)^2(6.7 \times 10^{-11})(6 \times 10^{24})}{4\pi^2}} = 4.24 \times 10^7 \text{ m}$$

Height of satellite above Earth = R – radius of Earth

$4.24 \times 10^7 - 6.4 \times 10^6 = 3.6 \times 10^7$ m = 36 000 km

EXERCISE 12.4

In the following questions where necessary take $G = 6.7 \times 10^{-11}$ N m^2 kg^{-2}, *Radius of Earth* $= 6.4 \times 10^6$ m, *Mass of Earth* $= 6 \times 10^{24}$ kg

1. A satellite orbits the Earth in a circular orbit at a height of 50 000 km. With what speed does it travel? How long does it take to make one revolution of the Earth?
2. A satellite orbiting the Earth in a circular orbit takes 30 minutes to go around once. What is the radius of the orbit? With what speed does the satellite travel?
3. The planet Jupiter orbits the Sun in an orbit of radius 7.8×10^{11} m. Given that the mass of the Sun is 2.0×10^{30} kg, calculate the time for Jupiter to complete one orbit of the sun.
4. The radius of the orbit of the planet Saturn is 9.5 times the radius of the Earth's orbit. Find the time taken for Saturn to make one complete orbit of the Sun.
5. One of the moons of Saturn is in an orbit which has approximately the same radius as that of the Earth's moon. Given that the speed of the Saturn moon is 10 times the speed of the Earth's moon, calculate a value for the mass of Saturn (mass of Earth = 6×10^{24} kg).
6. Find the radius of orbit and speed of a satellite stationary above the Equator.
7. The radius of the Earth at the equator is 6.4×10^6 m. Find:
   (i) the linear speed of a particle at the Equator,
   (ii) the acceleration of a particle on the Earth's Equator.

CHAPTER CHECKLIST

- **Define:** The radian; Angular velocity; Centrepetal force; Centripetal acceleration; Period of an orbit.
- **State:** The unit of angular velocity; The factors on which centripetal force depends.
- **Explain** what is meant by: Linear speed or tangential speed; Geostationary orbit or parking orbit.
- **Recall** that gravity is the centripetal force needed to keep satellites in orbit.
- **Recall** and use the formulae:

$$\theta = \frac{s}{r} \qquad 180° = \pi \text{ radians} \qquad \omega = \frac{\theta}{t}$$

$$v = r\omega \qquad F = \frac{mv^2}{r} \qquad F = mr\omega^2$$

$$a = \frac{v^2}{r} \qquad a = r\omega^2 \qquad v^2 = \frac{GM}{R}$$

$$T = \frac{2\pi R}{v} \qquad T^2 = \frac{4\pi^2 R^3}{GM}$$

- **Derive** the formulae:

$$v = r\omega \qquad T^2 = \frac{4\pi^2 R^3}{GM}$$

CHAPTER 13

Simple Harmonic Motion and Hooke's Law

ELASTICITY

Many objects change shape when a force is applied to them, e.g. an elastic stretches, a rod bends and a spring stretches or gets compressed. When the force is removed, the object may return to its original shape. If this happens, the object is said to be **elastic**. If the applied force is too great, the object may not return to its original shape when the force is removed (FIG. 13.1). It becomes permanently strained. If this happens, the object has exceeded its **elastic limit**.

Fig. 13.1
If the spring is stretched too far, it will not return to its original length when released.

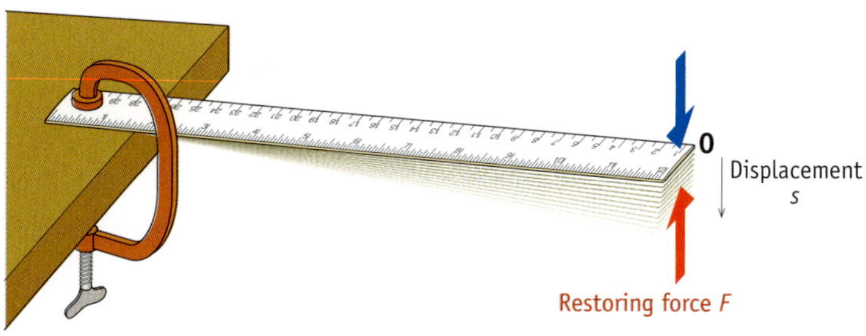

Fig. 13.2
The restoring force is proportional to the displacement.

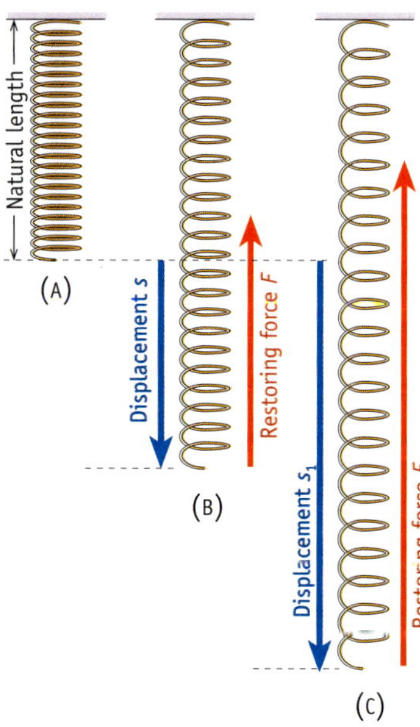

Fig. 13.3
The restoring force is proportional to the displacement.

FIG. 13.2 shows a ruler clamped to a bench. The end 0 is pulled downwards a small distance s (i.e. it undergoes a displacement s). When it is pulled down the ruler exerts an upward force trying to pull it back to its original position. This force is called the **restoring force**. Provided the displacement is small, **the restoring force is directly proportional to the displacement** and acts in the opposite direction to the displacement.

FIG. 13.3(A) shows a spring hanging vertically. The length of the spring is called its **natural length**. In FIG. 13.3(B) the spring is stretched beyond its natural length. The end of the spring is stretched by a displacement s. The spring then exerts a force trying to restore it to its original length. This force is the **restoring force**. When stretched by a greater displacement s_1 (FIG. 13.3(C)) the restoring force is greater. The restoring force is proportional to the displacement undergone by the end of the spring. If the spring is compressed smaller than its natural length it also exerts a force which tries to restore it to its natural length. This force is also proportional to the displacement.

The two examples above are particular cases of **Hooke's Law**, which in a simple form is as follows:

HOOKE'S LAW states that when an object is bent, stretched or compressed by a displacement s, the restoring force F is directly proportional to the displacement – provided the elastic limit is not exceeded.

Mathematically, Hooke's law states that: $F \propto -s \Rightarrow F = -ks$ where k is a constant, called the **elastic constant**. The negative sign indicates that the displacement and the restoring force are always in the opposite direction.

HOOKE'S LAW
Restoring force $\propto$ Displacement
$\Rightarrow \ F = -ks$

Problem 1: When a spring is stretched 10 cm beyond its natural length, the restoring force is 50 N. Assuming that the spring obeys Hooke's Law, find the restoring force when the extension is 12 cm.

Solution: In problems like this we can ignore the minus sign.

$F = ks \Rightarrow 50 = k(0.1) \Rightarrow k = 500$

thus: $F = 500\,s$

When $s = 12$ cm $= 0.12$ m Restoring force $F = (500)(0.12) = 60$ N

EXERCISE 13.1

1. A system obeying Hooke's Law has elastic constant $k = 4000$ N m^{-1}. Find:
   (i) the force when the displacement is 2 cm,
   (ii) the displacement when the force is 1000 N.

2. When a spring is stretched 6 cm beyond its natural length the restoring force is 8 N. Find:
   (i) the restoring force when the extension is 2 cm,
   (ii) the extension when the restoring force is 15 N.

SIMPLE HARMONIC MOTION (SHM)

FIG. 13.4 shows a mass m hanging on a spring. There is one position O where the upward force on the mass due to the spring is equal to its weight. If the mass is placed at O at rest it will remain there. O is called the **equilibrium position**.

If the mass is pulled down beyond O, say to B, and released, it vibrates up and down between A and B.

A particle vibrating up and down like this is said to be moving with **simple harmonic motion** (SHM). Simple harmonic motion is defined as follows:

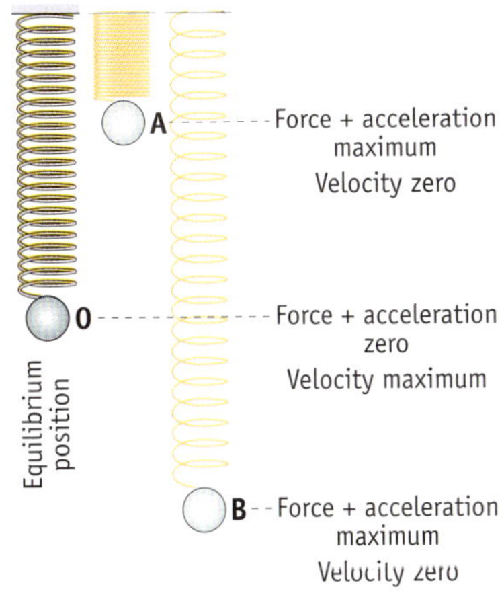

Fig. 13.4

A body is said to be moving with ***simple harmonic motion*** if:
(i) its acceleration is directly proportional to its distance from a fixed point on its path and
(ii) its acceleration is always directed towards that point.

Let a be the acceleration of the particle and let O be the fixed point on its path.

Let s be the displacement of the particle from O.

Then since a is always in the opposite direction to s, the equation defining Simple Harmonic Motion is:

$$a = -\omega^2 s$$

where ω^2 is a constant. The negative sign shows that a and s are always in opposite directions.

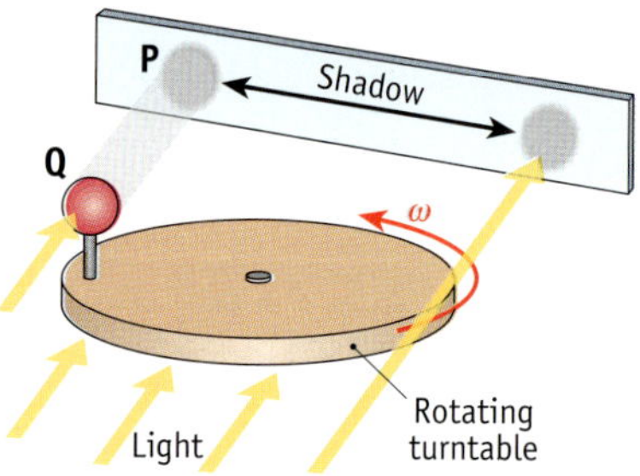

Fig. 13.5
Q is a small sphere moving in a circle at constant speed. The shadow of Q, caused by a distant light source, is P.
As Q moves around the circle at steady speed, P moves with SHM on the screen.

1 hertz
= 1 cycle per second

ANY SYSTEM THAT OBEYS HOOKE'S LAW WILL EXECUTE SHM

If a system obeys Hooke's Law then:

$$F = -ks$$

$$\Rightarrow \quad ma = -ks \quad (\text{Since } F = ma)$$

$$\Rightarrow \quad a = -\frac{k}{m}s \quad \text{i.e.} \quad a = -\omega^2 s \text{ where } \omega^2 = k/m$$

$\Rightarrow$ The system moves with simple harmonic motion.

EXAMPLES OF BODIES MOVING WITH SIMPLE HARMONIC MOTION

- A mass vibrating up and down at the end of a spring moves with SHM.
- Each prong on a vibrating tuning fork moves with SHM.
- The projection of uniform circular motion on a diameter is SHM (FIG. 13.5).
- For a small angle of swing a **pendulum** moves with SHM.
- The tides coming in and out every 6 hours move with SHM.
- A magnet suspended horizontally from a piece of thread moves with SHM if it is displaced slightly from being aligned North-South.

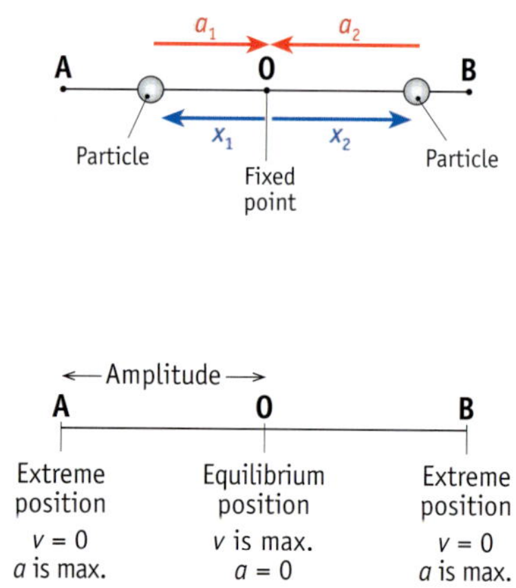

Fig. 13.6

TERMS USED TO DESCRIBE SIMPLE HARMONIC MOTION

If a body is moving with simple harmonic motion (FIG. 13.6) then:

- A **cycle** or an **oscillation** is the movement from A to B and back again to A.
- The **periodic time** or the **period** T of a particle executing SHM is the time for one complete oscillation. T is measured in seconds.
- The **frequency** f is the number of cycles occurring per second. Frequency is measured in cycles per second. One cycle per second is called one **hertz (Hz)**.

Since T is the time for one oscillation and f is the number of oscillations occurring per second, it is obvious that:

$$T = \frac{1}{f} \quad \text{and} \quad f = \frac{1}{T}$$

- The **amplitude** is the greatest displacement that the particle has from the equilibrium position. In FIG. 13.6 the length $|OA|$ is the amplitude.

Problem 2: A particle moving with SHM makes 10 full oscillations in 4 seconds. Find:
(i) the period and (ii) the frequency of the motion.

Solution: (i) Period $= \text{Time for one oscillation} = \dfrac{\text{Time for 10 oscillations}}{10} = \dfrac{4}{10}$

$= 0.4 \text{ s}$

(ii) Frequency $f = \dfrac{1}{T} = \dfrac{1}{0.4} = 2.5 \text{ Hz}$

Problem 3: The acceleration of a particle executing SHM is 4 m s^{-2} when it is 20 cm from the equilibrium position.
(i) What is its acceleration when it is 12 cm from the equilibrium position?
(ii) What is its distance from the equilibrium position when its acceleration is 1 m s^{-2} ?

Solution: First find ω^2: $a = \omega^2 s$ (We can ignore the minus sign in these calculations)

$\Rightarrow 4 = \omega^2(0.2) \Rightarrow \omega^2 = \dfrac{4}{0.2} = 20$ $\therefore$ the equation of the motion is: $a = 20s$

(i) When s = 12 cm = 0.12 m find a: $a = 20s = (20)(0.12) = 2.4 \text{ m s}^{-2}$

(ii) When $a = 1$ find s: $a = 20s \Rightarrow s = \dfrac{a}{20} = \dfrac{1}{20} = 0.05 \text{ m}$

Problem 4: The maximum force acting on a particle moving with SHM is 40 N. The mass of the particle is 8 kg. If the acceleration of the particle is 2 m s^{-2} when its displacement from the equilibrium position is 12 cm, find the amplitude of the motion.

Solution: First find ω^2

$a = \omega^2 s$

$\Rightarrow 2 = \omega^2(0.12)$

$\Rightarrow \omega^2 = 16.667$

$\therefore a = (16.667)s$

$F = ma$ and $F_{max} = 40 \text{ N}$

$\Rightarrow a_{max} = \dfrac{F_{max}}{m} = \dfrac{40}{8}$

$= 5 \text{ m s}^{-2}$

The amplitude is the displacement when the acceleration is greatest,

$\therefore 5 = (16.667)(\text{amplitude}) \Rightarrow \text{amplitude} = \dfrac{5}{16.667} = 0.3 \text{ m}$

Energy of a Body Moving with SHM

As a body moves with SHM its energy changes continually from potential to kinetic and back again to potential. Fig. 13.7 shows how the energy changes as the body moves.

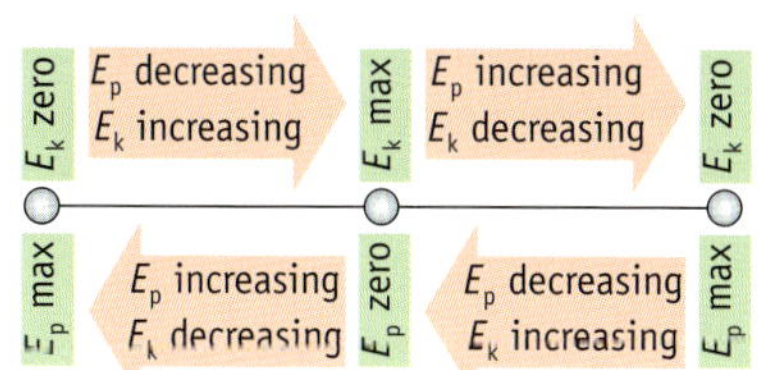

Fig. 13.7

Formula for the Period of SHM

The following result is important and must be remembered. You do not need to be able to prove it:

If a particle moves with SHM whose equation is: $a = -\omega^2 s$ then the period T of the motion is given by:

$$T = \frac{2\pi}{\omega}$$

Problem 5: The period of a particle executing SHM is 2 seconds. What is its acceleration when it is 15 cm from the equilibrium position?

Solution:

$$T = \frac{2\pi}{\omega} \Rightarrow \omega = \frac{2\pi}{T} = \frac{2\pi}{2} = 3.142$$

$$a = \omega^2 s = (3.142)^2(0.15) = 1.481 \text{ m s}^{-2}$$

Problem 6: When the displacement of a particle executing SHM is 12 cm, its acceleration is 2 m s^{-2}. What is the period of the motion?

Solution:

$$a = \omega^2 s$$

$$\Rightarrow 2 = \omega^2(0.12)$$

$$\Rightarrow \omega = \sqrt{16.66666} = 4.08$$

$$T = \frac{2\pi}{\omega} = \frac{2\pi}{4.08} = 1.54 \text{ s}$$

Problem 7: A particle executing SHM has a maximum acceleration of 3 m s^{-2}. If the total distance travelled in one oscillation is 0.5 m, find the period of the motion.

Solution: Total distance travelled in one oscillation = 0.5 m $\Rightarrow$ Amplitude $= \frac{0.5}{4} = 0.125$ m

$$a = \omega^2 s \Rightarrow a_{max} = \omega^2 s_{max} = \omega^2 \times \text{(amplitude)}$$

$$\therefore 3 = \omega^2 (0.125) \Rightarrow \omega = \sqrt{\frac{3}{0.125}} = 4.90$$

$$T = \frac{2\pi}{\omega} = \frac{2\pi}{4.90} = 1.28 \text{ s}$$

EXERCISE 13.2

1. A particle executing SHM makes 20 full oscillations in 4 seconds. Find:
   (i) the period, and
   (ii) the frequency of the motion.
2. A pendulum in a school laboratory executing SHM makes 50 oscillations in 20 seconds. What is the period of the pendulum?
3. The acceleration of a particle executing SHM is 2 m s^{-2} when it is 50 cm from the equilibrium position. What is its acceleration when it is 10 cm from the equilibrium position? What is its displacement when its acceleration is 0.5 m s^{-2}?
4. The maximum force acting on a particle which moves with SHM is 60 N. The mass of the particle is 12 kg. If the acceleration of the particle is 1 m s^{-2} when its displacement from the equilibrium position is 10 cm, find the amplitude of the motion.
5. A mass of 4 kg moves with SHM of period 0.5 s. Find the force acting on the mass when it is 4 cm from the equilibrium.
6. A particle executes SHM. What is the period if the acceleration is 3 m s^{-2} when the displacement from the equilibrium position is 0.5 m?
7. When the displacement of a particle executing SHM is 14 cm, the acceleration is 2.5 m s^{-2}. What is the frequency of the motion?
8. The period of a particle executing SHM is 1.5 s. What is its acceleration when its displacement from the mean position is 25 cm?

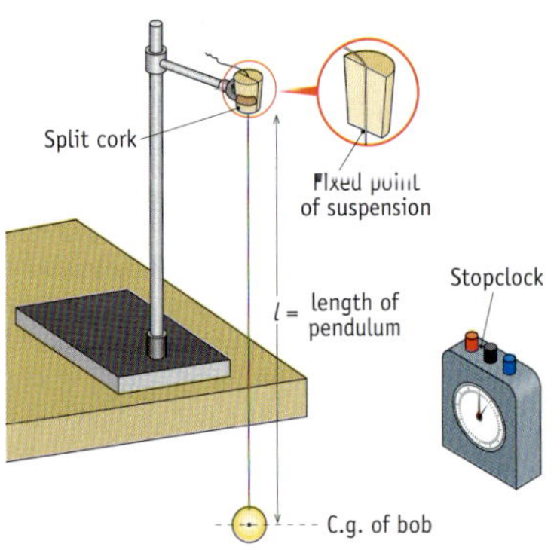

Fig. 13.8 A simple pendulum.

The Simple Pendulum

FIG. 13.8 shows a simple pendulum. It consists of a small mass called a **bob** attached to a fixed point by a light string. The string cannot stretch, i.e. its length does not change. The point where the thread enters the cork is the fixed point of suspension. The **length of the pendulum** l is the distance from the fixed point of suspension to the centre of the bob.

If the bob is moved to the right or the left of the position shown and then released, it will swing back and forth between the points A and B as in FIG. 13.9. The time taken for the bob to go from A to B and back again to A is called the **period** T of the pendulum. The period is the time taken for one oscillation.

Due to air friction, the amplitude of the oscillations decrease as time goes on. However, as the amplitude decreases, the period of the pendulum remains the same. This fact was discovered by Galileo Galilei.

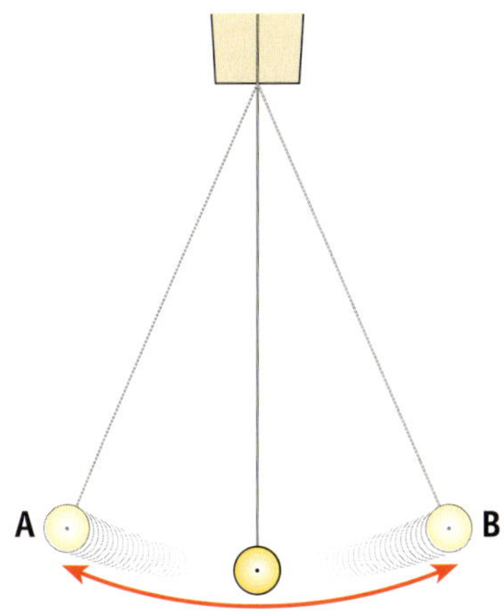

Fig. 13.9
The time taken for the bob to go from A to B and back again to A is called the period of the pendulum.

A Simple Pendulum Can Move with Simple Harmonic Motion

The following is an important fact which you do not need to be able to prove:

> **NOTE** For a small angle of swing (i.e. not more than 5° from the vertical at either side) a simple pendulum moves with simple harmonic motion.

It can also be proved (you don't need to know the proof) that when a pendulum moves with SHM:

> The period T of a pendulum of length l is given by: $\boldsymbol{T = 2\pi\sqrt{\frac{l}{g}}}$ where g is acceleration due to gravity.

Problem 8: What is the period of a pendulum of length 60 cm?

Solution: $T = 2\pi\sqrt{\frac{l}{g}} = 2\pi\sqrt{\frac{0.6}{9.8}} = 1.55$ s

Problem 9: What is the length of a pendulum whose period is 2 s?

Solution: $T = 2\pi\sqrt{\frac{l}{g}}$ and $T = 2$ seconds $\Rightarrow 2 = 2\pi\sqrt{\frac{l}{g}} \Rightarrow \frac{1}{\pi} = \sqrt{\frac{l}{g}}$

$\Rightarrow \frac{l}{g} = \left(\frac{1}{\pi}\right)^2 \Rightarrow l = \frac{9.8}{\pi^2} = 0.9929$ m

EXERCISE 13.3

1. What is the period of a pendulum of length 2 m at a place on Earth where acceleration due to gravity is 9.81 m s^{-2}?
2. A simple pendulum of length 0.8 m makes 50 oscillations in 90 seconds. What is the value of acceleration due to gravity at this location?
3. What is the length of the pendulum whose period is 2 s? Take $g = 9.8$ m s^{-2}.
4. The period of a simple pendulum on the surface of the Earth is 0.4 s. What would be its period at a height above the Earth equal to three times the radius of the Earth?

Measuring Acceleration due to Gravity Using a Simple Pendulum

Squaring both sides of $T = 2\pi\sqrt{\frac{l}{g}}$ gives $T^2 = 4\pi^2\frac{l}{g}$

Solving for g gives: $g = 4\pi^2\left(\frac{l}{T^2}\right)$

Therefore if the period T and the length l of a pendulum are known or can be measured, acceleration due to gravity g can be found.

Experiment — Mechanics 7

To Investigate the Relationship between the Period and the Length of a Simple Pendulum and from this to Calculate Acceleration due to Gravity g.

Summary of Method

In this experiment you will measure the period T of a simple pendulum a number of times. Each time, the pendulum will have a different length l which you will also measure. You will plot a graph of l against T^2. A straight line through the origin will result showing that $l \propto T^2$. You will measure the slope of the graph. You will calculate a value for g using $g = 4\pi^2 \times$ slope.

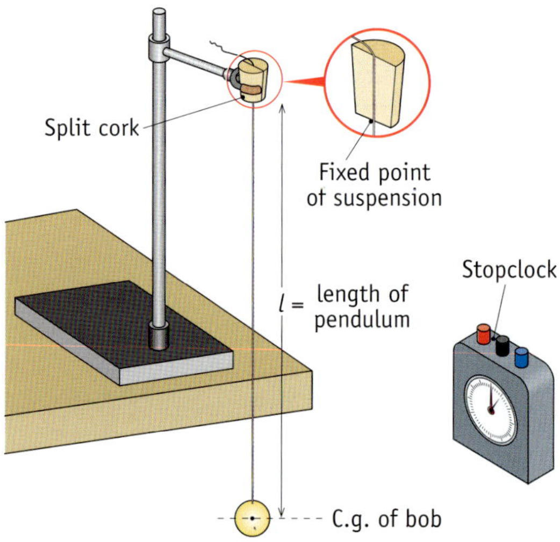

Fig. 13.10

Equipment Needed

- A small spherical mass to which thread can be tied (i.e. a pendulum bob)
- A split cork, some thread, a retort stand and a clamp
- A timer or a stopwatch and a metre stick

Method

1. Tie the bob to one end of the thread and clamp the other end between the flat faces of the split cork.
2. Set up the pendulum as in Fig. 13.10.
3. Make the pendulum as long as conveniently possible by hanging it over the edge of the bench.
4. Carefully measure and record the distance from the fixed point of suspension to the **centre** of the bob with the metre stick. This is the length l of the pendulum.
5. Set the bob swinging, but make sure that:
   - the greatest angle that the string makes with the downward vertical is less than 5° and
   - the pendulum oscillates in one plane only (i.e. ensure that the bob moves along the same path in going from A to B as from B to A).
6. With the stopwatch or timer, measure and record the time for 50 oscillations ($50T$).
7. Decrease the length of the pendulum by about 10 cm and repeat steps 4, 5 and 6 recording all values in the table.
8. Repeat step 7 at least six times. Do not make the pendulum much less than 40 cm as percentage errors in measuring its length then become significant.

Handling the Data

1. Complete the columns for T, T^2 and l/T^2 in the table. Within the limits of experimental error all the values of l/T^2 should be the same, showing that: $l \propto T^2$.
2. On graph paper plot a graph of l against T^2 (l along the y-axis). The points will lie on a straight line that passes through the origin, again showing that: $l \propto T^2$. Draw the straight line that best fits the points (Fig. 13.11). Find the slope of the line. It is the average value of l/T^2.
3. Calculate g by substituting the slope of your graph for l/T^2 in the formula: $g = 4\pi^2\left(\frac{l}{T^2}\right)$

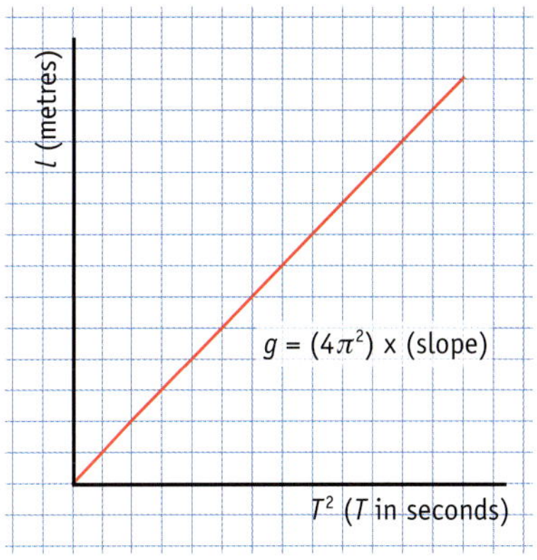

Fig. 13.11

Length of pendulum l/m	Time for 50 oscillations $50T$/s	Time for one oscillation T/s	T^2	l/T^2

Questions

1. What is meant by the length of a pendulum?
2. Why should the length of the pendulum not be less than 40 cm?
3. Why must the angle that the pendulum makes with the vertical be small ($< 5°$)?
4. What is meant by one oscillation of the pendulum?
5. What should you do if the pendulum is not moving in one plane?
6. What is meant by the period of the pendulum?
7. Why is the time for one oscillation not measured directly?
8. Why could the number of oscillations be reduced when the length of the pendulum was very long?
9. While counting the number of oscillations for each value of the length, the amplitude may decrease slightly. Does this affect the accuracy of the result? Explain your answer.
10. How does the number of oscillations timed affect the accuracy of the experiment?

CHAPTER CHECKLIST

- **Define:** Simple harmonic motion; Oscillation; Period; Frequency; Amplitude.
- **State:** Hooke's Law; The unit of frequency; The relationship between the period and length of a pendulum.
- **Explain** what is meant by: Elastic limit; Restoring force; Natural length; Elastic constant; Equilibrium position; The length of a pendulum.
- **Show** that a system that obeys Hooke's Law executes simple harmonic motion.
- **Describe** and carry out an experiment to investigate the relationship between the period and length of a pendulum and hence measure g.
- **List** five examples of a body moving with SHM.
- **Recall** and use the formulae: $F = -ks$ $\quad a = -\omega^2 s$

$$T = \frac{1}{f} \qquad T = \frac{2\pi}{\omega} \qquad T = 2\pi\sqrt{\frac{l}{g}}$$

CHAPTER 14

Temperature and Thermometers

TEMPERATURE
Temperature is the measure of the hotness or coldness of a body.

UNIT OF TEMPERATURE
The SI unit of temperature is the ***kelvin*** (**K**).

CONCEPT OF TEMPERATURE

We all know the difference between hot and cold. We can tell if one object is hotter or colder than another, i.e. we can recognise different degrees of hotness. A scientist would say we are able to recognise differences in temperature. **Temperature is the measure of the hotness or coldness of a body.**

UNITS OF TEMPERATURE

The SI unit of temperature is the **kelvin** (**K**). Temperatures expressed in kelvins are symbolised by T. We need not worry about the exact definition of the kelvin.

In practice, we normally measure temperature in **degrees Celsius**. The Celsius scale is the practical scale of temperature. Temperatures in degrees Celsius are symbolised by t or by θ.

RELATIONSHIP BETWEEN DEGREES CELSIUS AND KELVINS

The kelvin scale of temperature is related to the Celsius scale as follows:

0 °C = 273.15 K and 100 °C = 373.15 K (FIG. 14.1).

Thus there are 100 kelvins between 0 °C and 100 °C, i.e. **the size of 1 kelvin is the same as 1 °C**. Thus:

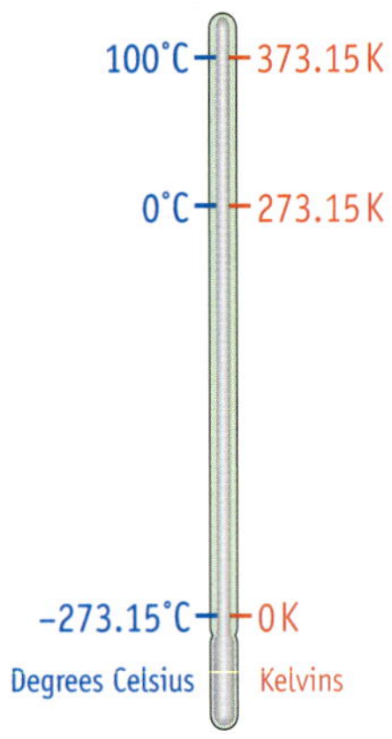

Fig. 14.1 The relationship between the kelvin scale and the celsius scale.

Temperature in degrees Celsius = Temperature in kelvins − 273.15

$$t\,/\,°\text{C} = T\,/\,\text{K} - 273.15$$

Problem 1: (i) Convert 423 K to degrees Celsius (ii) Convert 32 °C to kelvins

Solution: (i) 423 K = (423 – 273.15) °C = 149.85 °C
(ii) 32 °C – (32 + 273.15) K = 305.15 K

EXERCISE 14.1

1. Convert the following temperatures to kelvins:
   (i) 0 °C (ii) –100 °C
   (iii) 20 °C (iv) 100 °C
2. Convert the following temperatures to degrees Celsius:
   (i) 100 K (ii) 273 K (iii) 373 K (iv) 500 K

Thermometric Properties

For the accurate measurement of temperature, scientists choose some physical property that changes measurably as temperature changes. They use the changing value of this property to indicate the temperature. Such a property is called a **Thermometric Property**. An instrument based on a particular thermometric property and used to measure temperature is called a **Thermometer**.

THERMOMETRIC PROPERTY
Any physical property that changes measurably with temperature is called a ***thermometric property***.

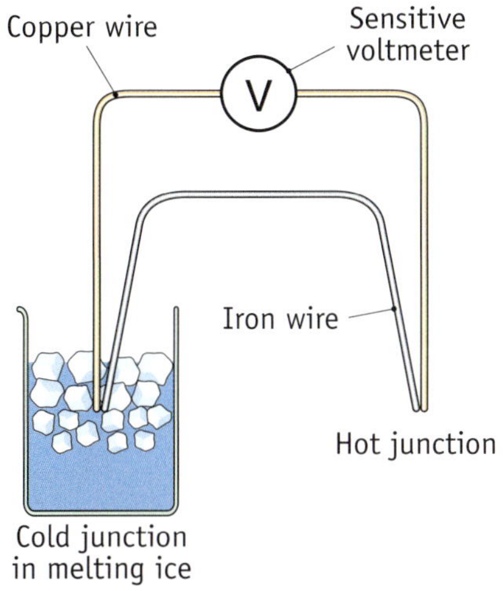

Fig. 14.3
A thermocouple. The cold junction is kept at a fixed temperature and the emf produced is measured as the temperature of the other junction is varied.

Fig. 14.2
The thermometric property on which a mercury in glass thermometer is based is the length of the column of mercury.

DEMONSTRATION OF SOME THERMOMETRIC PROPERTIES

The variation of the following thermometric properties with temperature can easily be demonstrated in the laboratory:

- **Length of a column of liquid**. When a liquid is heated it expands (i.e. its volume increases). If the liquid is in a thin capillary tube, the length of the column of liquid increases as the liquid expands. Thus as the temperature increases, the length of the column of liquid increases. This is the thermometric property on which a **Mercury-in-glass thermometer** is based (Fig. 14.2).
- **Electrical resistance**. The electrical resistance of a conductor changes with temperature (page 262). For a metal, the resistance increases with increasing temperature whereas for a semiconductor or carbon, the resistance decreases with increasing temperature. Resistance is the thermometric property on which the **resistance thermometer** is based. A thermistor (page 263) is a semiconductor whose electrical resistance decreases rapidly with a small rise in temperature. These facts can easily be demonstrated in the lab. The resistance is measured with an ohmmeter (page 256)
- **Emf of a thermocouple**. If two different metals are joined together to form a complete circuit (Fig. 14.3) and the two junctions maintained at different temperatures, a small emf (a few millivolts) appears in the circuit which causes a very small electric current to flow. The emf can be measured with a very sensitive voltmeter. The greater the temperature difference between the junctions, the greater the emf. This device is called a **thermocouple** and can be used to measure temperature.
- **Colour**. The colour of certain crystals changes with temperature. This is the basis of one form of thermometer used to measure the temperature of your body (Fig. 14.4).

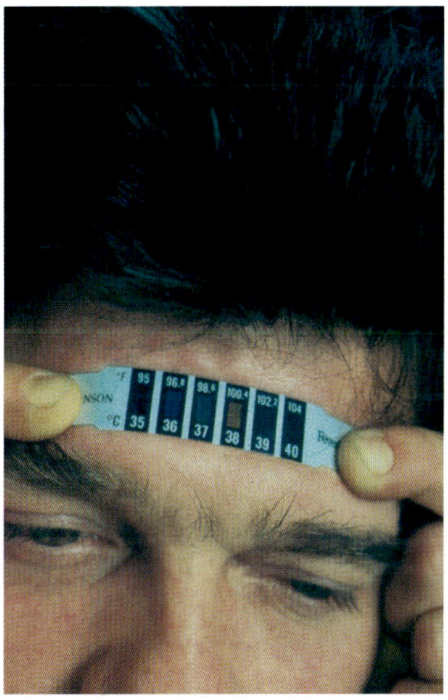

Fig. 14.4

The colour of a very hot object changes as its temperature rises; e.g. if a piece of iron is heated, its colour changes from a dull brown to red, to bright red, then yellow, then white. Similarly, increasing the current through the filament of a bulb causes its colour to change. The colour can be used to indicate temperature inside furnaces or volcanoes, from a safe distance.

- **Volume of a gas at constant pressure.** FIG. 14.5 shows a gas syringe. It contains a fixed mass of gas and a rubber cap seals one end. Atmospheric pressure acts on the piston at the other end; thus the gas is at constant pressure. If the temperature of the gas is increased by heating it, the volume of the gas will increase and push the piston out. If the gas is cooled, its volume decreases and the piston moves in.

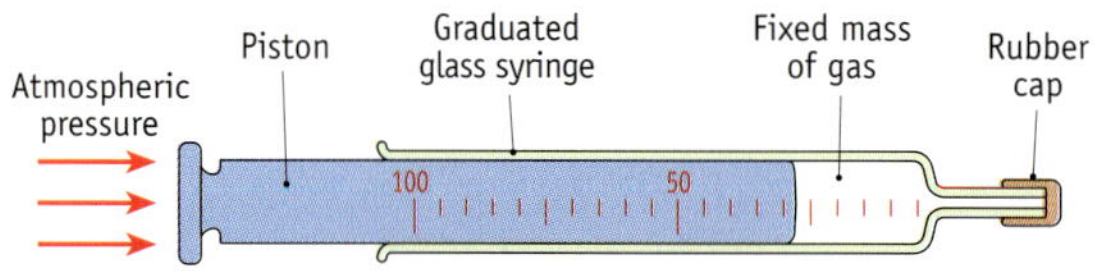

Fig. 14.5
At constant pressure the volume of the gas changes with temperature.

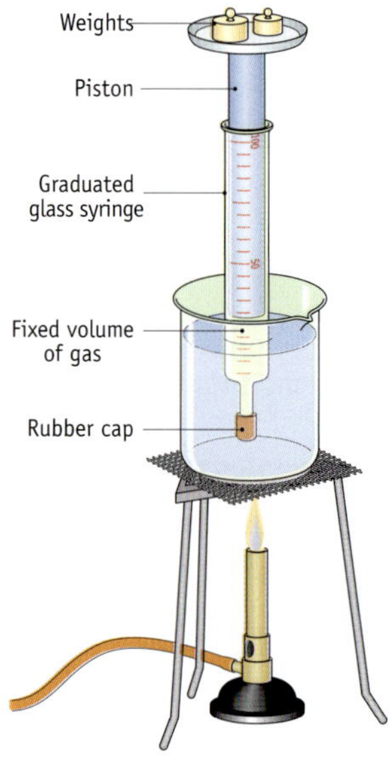

Fig. 14.6
At constant volume the pressure of a gas changes with temperature.

- **Pressure of a gas at constant volume.** A gas syringe can also be used to show how the pressure of a fixed volume of gas varies with temperature. The syringe is set up as in FIG. 14.6 and the volume of the gas noted. The gas is then heated. To keep the volume at the same value, weights will have to be placed on the end of the syringe. The larger the rise in temperature, the greater the weights that must be added and thus the greater the pressure.

USING A THERMOMETRIC PROPERTY TO MEASURE TEMPERATURE

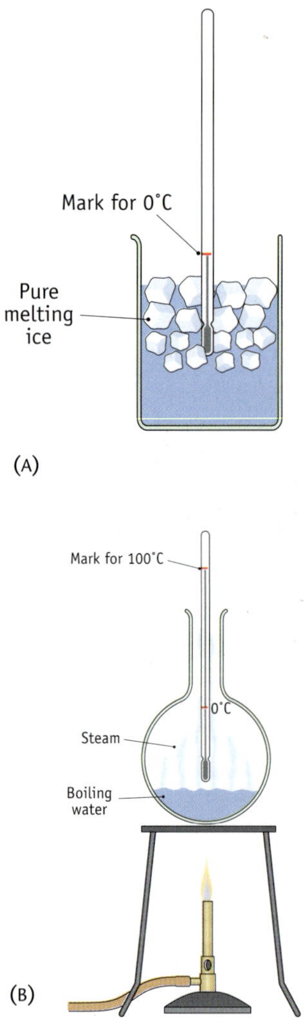

Fig. 14.7

To see how to set up a Celsius scale for a given thermometric property let us use an ungraduated mercury-in-glass thermometer. The length of the column of mercury is the thermometric property.

- Place the thermometer in pure melting ice and mark the position of the top of the column of mercury (FIG. 14.7(A)). On the Celsius scale the temperature at which pure ice melts is by definition zero degrees Celsius (0 °C).
- Place the thermometer in the steam above pure boiling water (FIG. 14.7(B)) and mark the position of the top of the column of Mercury. On the Celsius scale the temperature of steam above pure boiling water at standard atmospheric pressure is by definition one hundred degrees Celsius (100 °C).
- Remove the thermometer and measure the lengths of the columns of mercury; call these lengths L_{ice} and L_{steam}.

- On graph paper plot the points (L_{ice},0) and (L_{steam},100) as in FIG. 14.8. and draw a straight line through these points.
- The temperature corresponding to any length L_θ can then be found from the graph as shown in the diagram.

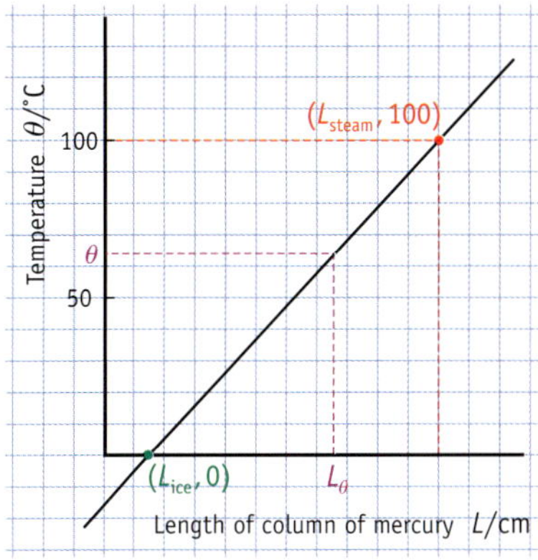

Fig. 14.8

DISAGREEMENT BETWEEN THERMOMETERS

If two different kinds of thermometer (e.g. a mercury-in-glass and a resistance thermometer) are created in this way, they will both give the temperature as 0 °C in melting ice and 100 °C in steam.

They will however, in general, **disagree** at other temperatures.

This is because different thermometric properties do not change proportionally with the same change in the degree of hotness (i.e. different thermometric properties vary differently with temperature).

Each thermometer is correct on its own scale. No one reads the 'true value' for the temperature.

To get agreement on the quantity **temperature** we therefore need a standard thermometer, i.e. we must pick one thermometer and decide to take temperatures given by this one as the 'true value' of the temperature. In the school laboratory we can use any commercial laboratory thermometer as a standard. We usually use a commercial laboratory mercury-in-glass thermometer as a school standard.

EXERCISE 14.2

1. The length of the mercury column in a capillary tube is 3.2 cm when the tube is placed in melting ice. In the steam above boiling water its length is 22.3 cm. If the length of the column when placed in a beaker of water is 10 cm, by using a suitable graph, calculate the temperature of the water in °C, according to this thermometer.

2. Two methods were used to find the temperature of a liquid. The following table shows the readings obtained when a length of wire and a tube containing a column of mercury were inserted in

(a) melting ice,
(b) the steam above boiling water,
(c) a liquid at unknown temperature.

Thermometric property	ice	steam	liquid
Resistance/Ω	11.0	14.3	12.0
Length of mercury column/cm	40	240	100

By drawing suitable graphs find the value of the unknown temperature on:

(i) the resistance scale,
(ii) the mercury scale.

Why is the temperature measured on each scale different?

PRACTICAL THERMOMETERS

MEASURING BODY TEMPERATURE

FIG. 14.9 shows one type of thermometer used to measure body temperature. It is often called 'the' clinical thermometer, though other types are used more often in hospitals and by doctors. The clinical thermometer has a narrow constriction in, so that when removed from the patient the mercury does not fall and it can be read accurately. In another type, called an infra-red radiation thermometer, a probe is inserted into the ear. Infra-red radiation emitted from the ear drum is detected and a reading of the patient's temperature is obtained. The plastic strip thermometer was shown in FIG. 14.4.

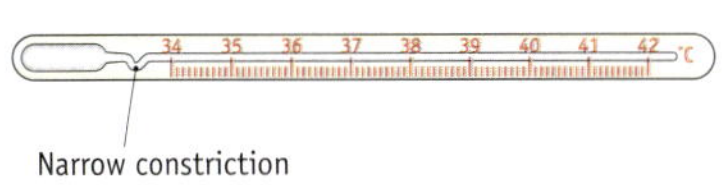

Fig. 14.9
Clinical thermometer.

We meet different types of thermometers in many everyday appliances; e.g. cookers, ovens, boilers and car engines all have thermometers. They can be based on different thermometric properties.

EXPERIMENT HEAT 1

TO PLOT THE CALIBRATION CURVE OF A THERMOMETER USING THE LABORATORY MERCURY THERMOMETER AS A STANDARD.

Summary of Method

In this experiment, you will place an ungraduated alcohol-in-glass thermometer in melting ice and measure the length of the alcohol column. You will have a laboratory mercury standard thermometer in the ice also. You will read the temperature from the mercury thermometer. By gently heating the water you will repeat the above at a number of different temperatures. You will plot a graph of 'length of alcohol' column against 'temperature'.

Equipment Needed

- An ungraduated alcohol-in-glass thermometer
- A mercury-in-glass thermometer
- Some ice
- A beaker (0–250 ml)
- A retort stand and clamp
- A Bunsen burner, tripod and gauze

Method

1. Place both thermometers in the beaker of melting ice. Leave them there until both columns stop moving. Read and record the temperature on the mercury thermometer. Mark the position of the alcohol on the glass. Remove the alcohol thermometer from the beaker and measure the length of the column of alcohol. Record its value.
2. Set up the Bunsen, tripod, gauze and beaker of ice/water as in FIG. 14.10.
3. Heat the water by about 10 °C.
3. Read and record the temperature on the mercury thermometer. Mark the position of the alcohol on the glass. Remove the alcohol thermometer from the water and measure the length of the column of alcohol. Record its value.
4. Repeat steps 3 and 4 at least six times, raising the temperature by about 10 °C each time.
5. On graph paper, plot a graph of length of alcohol column against temperature. This is the calibration curve of the alcohol thermometer.
6. The temperature of any environment can be read with the alcohol thermometer, by measuring its length at the unknown temperature and using the calibration curve to determine the temperature.

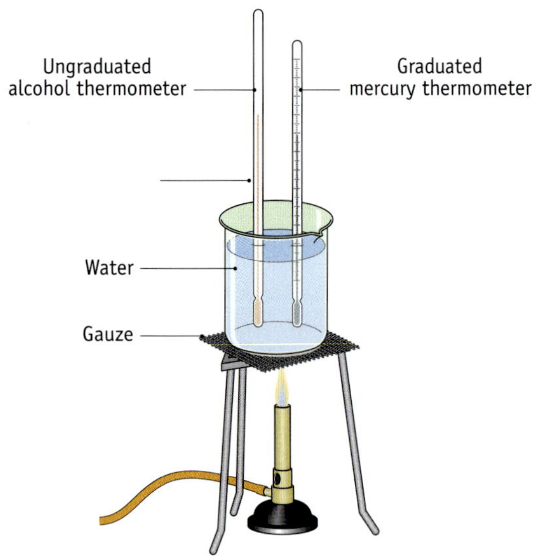

Fig. 14.10

EXPERIMENTAL NOTES

A calibration curve for any other type of thermometer, such as a thermocouple or a resistance thermometer can be made by the above method. The graph plotted is the value of the thermometric property of the thermometer against the reading of temperature on the mercury thermometer.

CHAPTER CHECKLIST

- **Define**: Temperature; Thermometric property.
- **State:** The SI unit of temperature; The practical unit of temperature.
- **Recall** and use the formula: $t\,/\,°\mathrm{C} = T\,/\,\mathrm{K} - 273.15$.
- **Describe** and carry out an experiment to: Demonstrate thermometric properties; Graduate two thermometers at ice and steam points and compare their readings at other temperatures; Plot the calibration curve of a thermometer using the laboratory mercury thermometer as standard.
- **List** six thermometric properties.
- **List** five everyday practical thermometers.

CHAPTER 15

Quantity of Heat and Heat Transfer

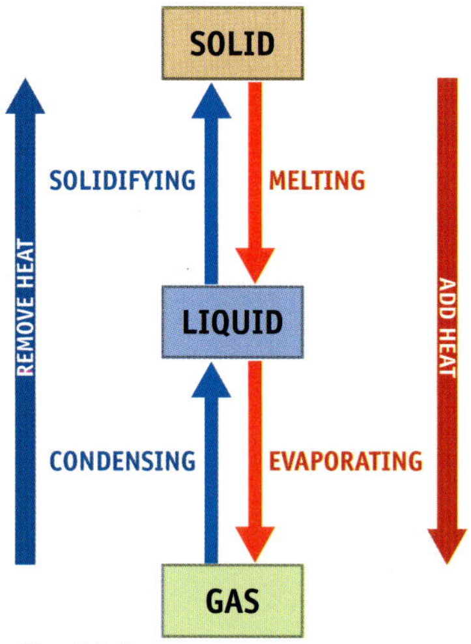

Fig. 15.1

The States of Matter

Matter can exist in three different states, **solid**, **liquid** or **gas**. You should be familiar with the terms used in Fig. 15.1.

Heat Capacity

When you add heat energy to a substance its temperature usually rises* and when you take heat energy away, its temperature usually falls. Experimentally it is found that the amount of heat needed to raise the temperature of an object by 1 °C (1 K) is the same as the amount of heat given out if its temperature falls by 1 °C. This amount of heat is called the **heat capacity** of the object. Its value is different for different objects.

HEAT CAPACITY
The ***heat capacity*** of an object is the heat energy needed to change its temperature by 1 K (1 °C).

The symbol for **heat capacity** is ***C*** (capital '*C*'). Its unit is the **joule per kelvin (J K^{-1})**.

If an object has heat capacity *C*, the heat energy *Q* needed to change its temperature by $\Delta\theta$ degrees Celsius is given by:

Heat = Heat capacity × Change in temperature

i.e. $Q = C\,\Delta\theta$

Problem 1: A beaker of water has a heat capacity of 2000 J K^{-1}. How much heat energy must be added to raise its temperature from 4 °C to 96 °C?

Solution: $Q = C\Delta\theta = (2000)(96 - 4) = 184\,000$ J

Specific Heat Capacity

To compare the heat needed to produce a given temperature change in different substances, scientists use the amount of heat necessary to change the temperature of **1 kilogram** of that substance by 1 °C (1 K). This amount is called the **specific heat capacity** ***c*** (lower case '*c*') of that substance (Fig. 15.2).

* If the substance is changing state as you add (or remove) heat there may not be any change in temperature (page 164). For the present, we will assume no change of state takes place when heat is added to or taken from a substance.

SPECIFIC HEAT CAPACITY

The ***specific heat capacity (c)*** of a substance is the heat energy needed to change the temperature of one kilogram of that substance by one kelvin.

Substance	Specific heat capacity (J kg^{-1} K^{-1})
Water	4180
Copper	390
Iron	451
Glass	674
Aluminium	910
Paraffin Oil	2100
Alcohol	2500
Wood	1700
Ice	2100
Air	1000

Fig. 15.2
Specific heat capacities of some common substances.

- *c* is the symbol for specific heat capacity.
- The unit of specific heat capacity is the **joule per kilogram per kelvin**. Its symbol is **J $kg^{-1}K^{-1}$**.

SPECIFIC HEAT CAPACITY FORMULA

Experimentally it is found that:

- the heat energy Q needed to produce a given rise in temperature is directly proportional to the rise in temperature,
- the heat energy needed to produce a given rise in temperature in a body is directly proportional to the mass of that body.

UNIT OF SPECIFIC HEAT CAPACITY

The unit of specific heat capacity is the ***joule per kilogram per kelvin*** **(J kg^{-1} K^{-1}).**

It follows that, provided a substance does not change state:

Heat energy added = Mass × Specific heat capacity × Rise in temperature

Heat energy lost = Mass × Specific heat capacity × Fall in temperature

In either case: $\mathbf{Q = m\,c\,\Delta\theta}$

In the problems: specific heat capacity of water = 4180 J kg^{-1} K^{-1} *and specific heat capacity of copper* = 390 J kg^{-1} K^{-1}

Problem 2: How much heat energy is needed to raise the temperature of 2 kg of water from 5 °C to 100 °C?

Solution: $Q = m\,c\,\Delta\theta = (2)(4180)(100-5) = 794\,200$ J

Problem 3: 5544 J of heat energy are given out when the temperature of 800 grams of copper cools from 25 °C to 7 °C. Calculate the specific heat capacity of copper.

Solution: $Q = m\,c\,\Delta\theta \Rightarrow 5544 = (0.8)(c)(25-7) \Rightarrow c = \dfrac{5544}{(0.8)(18)} = 385$ J kg^{-1} K^{-1}

Note: Mass must be in kilograms when using $Q = m\,c\,\Delta\theta$. 800 grams = 0.8 kg.

Problem 4: A 2 kW electric heater raises the temperature of 10 kg of water from 15 °C to 80 °C. If there is no heat loss to the surroundings, how long does it take?

Solution: Heat supplied $Q = m\,c\,\Delta\theta = (10)(4180)(80-15) = 2\,717\,000$ J

The heat is supplied at the rate of 2 kW = 2000 joules per second

$$\text{Time taken} = \frac{\text{Energy supplied}}{\text{Energy supplied per second}} = \frac{2\,717\,000}{2000} = 1358.5 \text{ s}$$

Problem 5: 20 grams of copper at 100 °C is added to 150 grams of water at 15 °C. There is no loss of heat to the surroundings. Find the final temperature of the mixture.

Solution: Let θ = final temperature of mixture. Then:

Rise in temperature of water $\Delta\theta\uparrow = \theta - 15$

Fall in temperature of the copper $\Delta\theta\downarrow = 100 - \theta$

Heat lost by copper = Heat gained by water, i.e. $m_c c_c \Delta\theta\downarrow = m_w c_w \Delta\theta\uparrow$

$\Rightarrow (20 \times 10^{-3})(390)(100 - \theta) = (150 \times 10^{-3})(4180)(\theta - 15)$

$\Rightarrow 7.8(100 - \theta) = 627(\theta - 15) \Rightarrow 780 - 7.8\theta = 627\theta - 9405$

$\Rightarrow 780 + 9405 = 627\theta + 7.8\theta \Rightarrow \theta = \dfrac{10\,185}{634.8} = 16.04\ °C$

EXERCISE 15.1

Where needed, specific heat capacities: Copper 390 J kg^{-1} K^{-1}; Water 4180 J kg^{-1} K^{-1}

1. A calorimeter of oil has a heat capacity of 1500 J K^{-1}. How much heat energy is needed to raise its temperature from 10 °C to 80 °C?
2. A metal block of heat capacity 600 J K^{-1} loses 1 MJ of heat energy to its surroundings. Find the decrease in its temperature.
3. 4000 J of heat energy raises the temperature of a calorimeter of water by 10 °C. Find its heat capacity.
4. How much heat energy is needed to raise the temperature of 2 kg of water from 12 °C to 80 °C?
5. How much heat energy is needed to raise the temperature of 12 kg of copper from 10 °C to 120 °C?
6. 11 088 J of heat energy are needed to raise the temperature of 1.6 kg of copper from 7 °C to 25 °C. Calculate the specific heat capacity of copper.
7. One megajoule of heat energy is supplied to a piece of iron and raises its temperature from 20 °C to 200 °C. Find the mass of the piece of iron. (Specific heat capacity of iron = 451 J kg^{-1} K^{-1})
8. 4000 J of heat energy are added to 400 grams of water at 20 °C. What is the resulting temperature of the water?
9. If 40 000 joules of heat energy is supplied separately to both 2 kg of copper at 0 °C and 2 kg of aluminium at 20 °C, which reaches the higher temperature? (Specific heat capacity of aluminium = 910 J kg^{-1} K^{-1})
10. A copper calorimeter of mass 80 grams contains 120 cm^3 of water. Assuming no heat loss to the surroundings, how much heat energy is needed to raise its temperature from 18 °C to 60 °C? (Density of water = 1 g cm^{-3})
11. A 1 kilowatt electric heater supplies heat to 5 kg of water and raises its temperature from 10 °C to 100 °C. Assuming there is no heat loss to the surroundings, how long does it take?
12. 30 grams of water at 100 °C are added to 100 grams of water at 20 °C. If there is no heat loss to the surroundings, what is the resulting temperature of the mixture?
13. 120 grams of copper at 100 °C is added to 80 grams of water in a 60 gram copper calorimeter at 20 °C. If there is no heat loss to the surroundings, find the final temperature of the system.

Fig. 15.3 A modern domestic storage heater.

STORAGE HEATERS

A storage heater (FIG. 15.3) usually consists of an electric heater surrounded by bricks of high specific heat capacity. The bricks are usually heated by night when electricity is cheap. During the day the bricks slowly give out their heat, thus heating the room.

EXPERIMENT HEAT 2

To Measure the Specific Heat Capacity of Water by an Electrical Method.

Summary of Method

In this experiment you will send an electric current through a heating coil which is immersed in a copper calorimeter of water. With a joulemeter, you will measure the amount of energy supplied. The rise in temperature is measured with a thermometer. Knowing the mass of the water, the mass of the calorimeter and the specific heat capacity of copper, you will calculate the specific heat capacity of water.

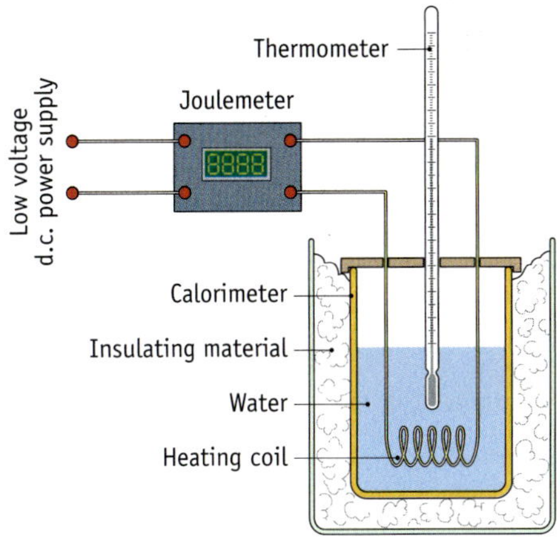

Fig. 15.4

Equipment

- A copper calorimeter, a stirrer and an insulating lid
- Insulating material (e.g. cotton wool or polystyrene beads)
- A container (e.g. a large beaker) to hold the calorimeter and insulating material
- A heating coil and a thermometer (0–50 °C in steps of 0.1 °C)
- A d.c. power supply unit and connecting leads
- A joulemeter and a balance

Method

1. Find the mass of the empty calorimeter m_c. Place sufficient water in it to completely cover the heating coil, then find the mass of the calorimeter and water. By subtraction find the mass of the water.
2. Place the calorimeter of water in the insulating material and set up the equipment as in Fig. 15.4.
3. Wait a minute or two then measure the temperature θ_1 of the cold water and the calorimeter.
4. Switch on the joulemeter and the electric current. Allow the current to flow until a temperature rise of about 15 °C has occurred. Stir the water continuously throughout the experiment.
5. Switch off the current and joulemeter. Take the reading on the joulemeter Q.
6. Stir the water and wait until the temperature stops rising. Then read the final temperature θ_2 of the water and the calorimeter.

Table

Mass of empty calorimeter m_c	=	
Mass of calorimeter and water	=	
Mass of water m_w	=	
Temperature of cold water and calorimeter θ_1	=	
Reading on joulemeter Q	=	
Final temperature of water and calorimeter θ_2	=	

Calculations

On the assumption that no heat is given to or taken from the surroundings we have:

Electrical energy supplied = Heat gained by water + Heat gained by calorimeter

$$\text{i.e.}\quad Q = m_w c_w(\theta_2 - \theta_1) + m_c c_c(\theta_2 - \theta_1)$$

Since all these quantities are known except the specific heat capacity of water (c_w), it can easily be calculated.

EXPERIMENTAL NOTES

1. A rise in temperature of about 15 °C in 10 minutes or so is suitable. Check with the maker's instructions for the maximum current suitable for the coil. Too high a current can cause the water next to the coil to boil, producing steam which will affect the result significantly.
2. Ensure that the coil is completely immersed in the water or it may burn out when the current is switched on.
3. Errors in measuring the masses of the water and calorimeter are negligible compared with those that arise in measuring the temperature. To keep these to a minimum, use a sensitive thermometer. One that reads in steps of 0.1 °C is suitable.
4. To minimise errors due to heat loss to the surroundings you may pre-cool the water to about 7 °C below room temperature (e.g. by allowing some ice to melt in it). Allow the current to raise the temperature to about 7 °C above room temperature. The heat flowing into the calorimeter and water during the first part of the experiment almost cancels out the heat leaving during the remainder of the experiment.

Questions

1. Why must the heating coil be completely covered with water?
2. How would the accuracy of the experiment be affected if the calorimeter was not well insulated?
3. Why is the water stirred throughout the experiment?
4. How would the accuracy of the experiment be affected if the rise in temperature produced was very large – say 60 °C?
5. How would the accuracy of the experiment be effected if the current used was very small, thus causing the experiment to take a very long time to produce a 15 °C rise in temperature?
6. What is the advantage in pre-cooling the water in the calorimeter to about 7 °C below room temperature?
7. Why is it more important in this experiment to use a thermometer graduated in steps of 0.1 °C than to use a very high precision balance to measure the various masses?
8. List three precautions you should take in this experiment to minimise heat loss to the surroundings.
9. What in this experiment – other than the water and the calorimeter – receives heat? Why can these heat losses be neglected?
10. Why should the calorimeter be polished?
11. Why would using more water and supplying the same amount of energy be less accurate?

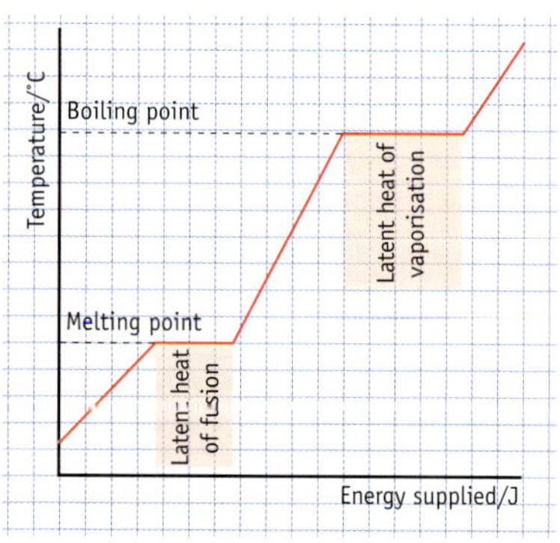

Fig. 15.5
How temperature changes as heat energy is added.

HEAT CHANGE WITHOUT A CHANGE IN TEMPERATURE

When a substance is changing state, it can take in or give out heat energy without its temperature changing. For example, to keep a beaker of boiling water boiling, heat must continually be added to it. A thermometer will show that the water remains at 100 °C while it is boiling. Likewise, melting ice will stay at 0 °C as it melts, even though heat energy is being added to it. FIG. 15.5 shows how the temperature of a pure substance changes as heat energy is added to that substance.

LATENT HEAT

LATENT HEAT
The ***latent heat (L)*** of a substance is the heat energy needed to change its state without a change in temperature.

The latent heat needed to change from a solid to a liquid is called the **latent heat of fusion**.

The latent heat needed to change from a liquid to a gas is called the **latent heat of vaporisation**.

The symbol for latent heat is ***L***. Its unit is the **joule (J)**.

The Human Body is Cooled by Perspiring

When you are hot, your body produces perspiration on your skin (Fig. 15.6). Perspiration is mainly water. As this water evaporates, it takes its latent heat from you and you cool down.

Fig. 15.6

Specific Latent Heat

To compare the amounts of heat needed to change the state of various substances, scientists use the heat needed to change the state of 1 kg of the substance without a change in temperature. This is called the **specific latent heat (*l*)**. Thus we have the following definitions:

SPECIFIC LATENT HEAT
The ***specific latent heat (l)*** of a substance is the amount of heat energy needed to change the state of 1 kg of that substance without a change in temperature.

SPECIFIC LATENT HEAT OF FUSION
The ***specific latent heat of fusion*** of a substance is the amount of heat energy needed to change 1 kg of that substance from a solid to a liquid without a change in temperature (i.e. at its melting point).

The same amount of heat energy is given out when the substance changes from a liquid to a solid at its melting point.

SPECIFIC LATENT HEAT OF VAPORISATION
The ***specific latent heat of vaporisation*** of a substance is the amount of heat energy needed to change 1 kg of that substance from a liquid to a gas without a change in temperature (i.e. at its boiling point).

The same amount of heat energy is given out when the substance changes from a gas to a liquid at its boiling point.

UNIT OF SPECIFIC LATENT HEAT
The unit of specific latent heat of vaporisation and of fusion is the ***joule per kilogram*** **($J\ kg^{-1}$)**

Formula for Latent Heat

Since the heat energy Q needed to change the state of a substance is directly proportional to the mass m of that substance it follows that:

Heat needed to change from solid to liquid or vice versa = Mass × Specific latent heat of fusion

Heat needed to change from liquid to gas or vice versa = Mass × Specific latent heat of vaporisation

In either case: $Q = ml$

In the following problems and exercises take:

Specific heat capacity of water = 4180 J kg^{-1} K^{-1}

Specific heat capacity of copper = 390 J kg^{-1} K^{-1}

Specific heat capacity of aluminium = 910 J kg^{-1} K^{-1}

Specific latent heat of vaporisation of water = 2.3 × 10^{6} J kg^{-1}

Specific latent heat of fusion of ice = 3.3 × 10^{5} J kg^{-1}

Problem 6: How much heat energy is needed to convert 6 kg of ice at 0 °C to water at 0 °C?

Solution: $Q = ml = (6)(3.3 \times 10^5) = 1.98 \times 10^6$ J

Problem 7: How much heat energy is needed to completely convert 50 grams of water at 20 °C to 50 grams of steam at 100 °C?

Solution:

$$\begin{pmatrix}\text{Heat}\\\text{needed}\end{pmatrix} = \begin{pmatrix}\text{Heat needed to bring 50 g of}\\\text{water from 20 °C to 100 °C}\end{pmatrix} + \begin{pmatrix}\text{Heat needed to convert 50 g of water}\\\text{at 100 °C to 50 g of steam at 100 °C}\end{pmatrix}$$

$$= (m_w)(c_w)(\Delta\theta) + (m_w)(l) = (0.05)(4180)(80) + (0.05)(2\,300\,000) = 131\,720\text{ J}$$

Problem 8: 20 grams of ice at 0 °C is added to 80 grams of water in a copper calorimeter of mass 50 grams. The temperature of the calorimeter and water is 25 °C. Find the final temperature of the calorimeter and water assuming that no heat is transferred to or from the surroundings and that all the ice melts.

Solution: Let the final temperature of the water and calorimeter be θ °C. Since no heat is transferred to or from the surroundings we have:

$$\begin{pmatrix}\text{Heat lost}\\\text{by water}\end{pmatrix} + \begin{pmatrix}\text{Heat lost by}\\\text{calorimeter}\end{pmatrix} = \begin{pmatrix}\text{Heat needed to melt ice at}\\\text{0 °C to water at 0 °C}\end{pmatrix} + \begin{pmatrix}\text{Heat needed to raise temp}\\\text{of melted ice from 0 °C}\\\text{to final temp } \theta\text{ °C}\end{pmatrix}$$

i.e. $m_w c_w(\text{fall in temp}) + m_c c_c(\text{fall in temp}) = m_{ice}\, l + m_{ice}\, c_w(\text{rise in temp})$

$\Rightarrow (0.08)(4180)(25 - \theta) + (0.05)(390)(25 - \theta) = (0.02)(330\,000) + (0.02)(4180)(\theta - 0)$

$\Rightarrow 8360 - 334.4\theta + 487.5 - 19.5\theta = 6600 + 83.6\theta$

$\Rightarrow 437.5\theta = 2247.5 \Rightarrow$ Final temperature $\theta = 5.1$ °C

EXERCISE 15.2

Values for c and l are given before the problems above.

1. How much heat energy is needed to completely convert 10 kg of ice at 0 °C to water at 0 °C?
2. How much heat energy is needed to completely melt 500 grams of ice at 0 °C?
3. What is the maximum amount of ice at 0 °C that can be melted by 1 MJ of heat energy?
4. How much heat energy is needed to completely convert 0.4 kg of water at 100 °C to steam at 100 °C?
5. How much heat energy is needed to completely convert 80 grams of water at 100 °C to steam at 100 °C?
6. How much heat energy is needed to bring 3 kg of ice at 0 °C to 99 °C?
7. How much heat energy is needed to convert 1 kg of ice at 0 °C to steam at 100 °C?
8. How much heat energy is needed to completely convert 60 grams of water at 15 °C to steam at 100 °C?
9. How much steam at 100 °C must be added to 60 grams of ice at 0 °C to just bring the resulting mixture to 100 °C?
10. 200 grams of ice at 0 °C is mixed with 500 grams of water at 80 °C. Find the resulting temperature of the mixture assuming no heat loss to the surroundings.

11. 2 grams of steam at 100 °C are added to an insulated calorimeter containing 80 grams of water at 10 °C. Find the resulting temperature of the mixture assuming:

(i) the calorimeter has negligible heat capacity,

(ii) the calorimeter is made from copper and has a mass of 70 grams.

12. 20 grams of ice at 0 °C was added to 80 grams of water which is in a copper calorimeter of mass 50 grams. The temperature of the calorimeter and water was 25 °C. When the ice was melted the temperature of the calorimeter and water was 5.1 °C. Assuming that no heat is transferred to or from the surroundings, calculate the specific latent heat of fusion of ice.

13. An aluminium kettle of mass 0.5 kg is fitted with a 2.5 kW heater. It contains 1.7 litres of water at 10 °C. How long does it take to bring the water to the boil? How much longer does it take for half the water to boil off? (Specific heat capacity of aluminium = 910 J kg^{-1} K^{-1})

14. A hailstone of mass 3.5 grams is at a height of 3 km. What is its potential energy at this height? If all this potential energy is converted into latent heat during its fall to the ground, calculate the mass of the hailstone on reaching the ground.

The Heat Pump

A heat pump transfers energy from a cooler region to a warmer one. Work must be done to bring this about. It is used in refrigerators and in air conditioning systems in buildings and cars. Fig. 15.7 shows a simple diagram of a heat pump in a refrigerator. The circulating liquid has a high specific latent heat of vaporisation and a low boiling point. The liquid is pumped around a closed circuit. On reaching the expansion valve, pressure drops, the liquid vaporises and takes in its latent heat from the inside of the fridge, thus cooling the fridge. When the vapour reaches the compressor, its pressure is increased and it turns back into a liquid, giving out its latent heat as it does so. This heat is given off to the surroundings from the black pipes and cooling fins at the back of the fridge.

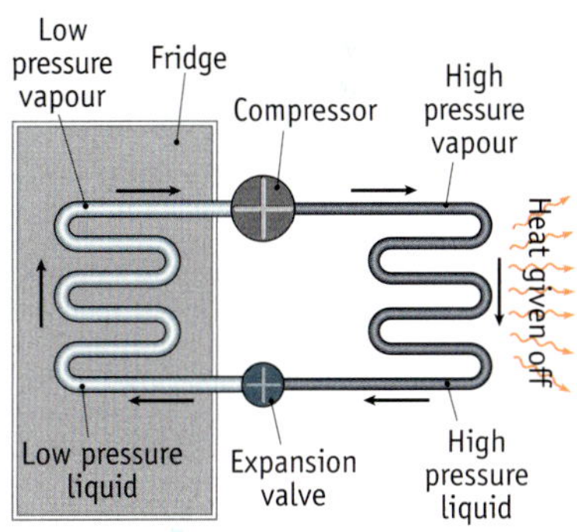

Fig. 15.7

Experiment — Heat 3

To Measure the Specific Latent Heat of Fusion of Ice.

Summary of Method

In this experiment you will add ice at 0 °C to a calorimeter of warm water and allow the ice to melt fully. The water in the calorimeter will be about 5 °C above room temperature and you will add sufficient ice to bring the final temperature to about 5 °C below room temperature. Knowing the mass of the calorimeter, water and ice, as well as the temperature of the calorimeter before and after adding the ice, the specific latent heat of fusion can be found.

Equipment Needed

- Some ice cubes and a clean cloth
- A thermometer (0–50 °C in steps of 0.1 °C)
- A Bunsen burner, tripod and gauze
- Some insulating material (e.g. cotton wool or polystyrene beads)
- A copper calorimeter
- A beaker (250 cm^3)
- A heavy object with which to crush the ice
- A container to hold the calorimeter (e.g. a large beaker)

Method

1. Crush the ice, place it in a beaker and leave it there for a few minutes until it is noticeably melting. The thermometer should then read 0 °C, i.e. initial temperature of ice = 0 °C.
2. Find the mass of the empty calorimeter m_c.
3. Heat some water to about 30 °C in a beaker using the Bunsen. Fill the calorimeter about O full with this water. Find the combined mass of the calorimeter and the warm water m_2.
4. Note the value of room temperature, then place the thermometer in the calorimeter of hot water, stirring it occasionally until its temperature is about 5 °C above room temperature.
5. Dry some of the crushed ice thoroughly with the cloth.
6. Place the calorimeter into the container of insulating material and measure the temperature θ_1 of the water in the calorimeter.
7. Add some of the dried ice to the water and stir it with the thermometer until it melts. Keep adding further pieces of ice and allow them to melt until the temperature of the water is about 5 °C below room temperature.
8. Record this temperature θ_2, i.e. the lowest temperature reached.
9. Find the combined mass m_3 of the calorimeter, water and melted ice. Complete the other rows in the table.

Table

Mass of empty calorimeter m_c	=
Mass of calorimeter and warm water m_2	=
Mass of calorimeter, water and melted ice m_3	=
Mass of water $(m_w = m_2 - m_c)$	=
Mass of melted ice $(m_{ice} = m_3 - m_2)$	=
Temp. of water and calorimeter before ice is added θ_1	=
Final temp. of calorimeter, water and melted ice θ_2	=
Fall in temperature $(\theta_1 - \theta_2)$	=

Calculations

On the assumption that no heat is given to or taken from the surroundings we have:

$$\begin{pmatrix}\text{Heat lost}\\ \text{by water}\end{pmatrix} + \begin{pmatrix}\text{Heat lost by}\\ \text{calorimeter}\end{pmatrix} = \begin{pmatrix}\text{Heat needed to melt ice at}\\ 0\text{ °C to water at }0\text{ °C}\end{pmatrix} + \begin{pmatrix}\text{Heat needed to raise temperature of}\\ \text{melted ice from }0\text{ °C to final}\\ \text{temperature }\theta_2\end{pmatrix}$$

$$\text{i.e.}\quad m_w c_w(\theta_1 - \theta_2) + m_c c_c(\theta_1 - \theta_2) = m_{ice}\, l + m_{ice} c_w(\theta_2 - 0)$$

$$\Rightarrow l = \frac{(m_w c_w + m_c c_c)(\theta_1 - \theta_2) - m_{ice} c_w \theta_2}{m_{ice}}$$

By substituting the values for the variables on the right-hand side of this formula, the latent heat of fusion of ice l can be found.

EXPERIMENTAL NOTES

1. When you get ice from the freezer, it will be at a temperature below 0 °C. You must allow it to reach 0 °C before adding it to the water.
2. The ice must be **crushed** so that it is at the same temperature throughout. A large lump could have its surface at 0 °C and its centre much cooler.
3. The ice must be **dried** before adding it to the calorimeter because our calculations assume only ice and not water are added to the water already in the calorimeter.
4. The calorimeter must not be allowed to reach more than 5 or 6 °C below room temperature. If it does, water vapour in the air may condense on it giving errors in the final weighing. Neither should the final temperature get too near 0 °C as the ice will only melt slowly and a significant amount of heat may enter from the surroundings.
5. The calorimeter is insulated to reduce heat flow to or from the surroundings.
6. Warm water is used in the calorimeter so that heat loss from the calorimeter when it is above room temperature is more or less compensated by heat entering the calorimeter when it is cooler than the surroundings. It also allows the ice to melt quickly and a larger mass of ice to be used.

Questions

1. Why is the ice crushed?
2. Why is the ice allowed to start to melt?
3. Why is the ice dried before adding it to the water?
4. List three other precautions you would take to ensure an accurate result.

EXPERIMENT HEAT 4

TO MEASURE THE SPECIFIC LATENT HEAT OF VAPORISATION OF WATER.

Summary of Method

In this experiment you will pass steam into cold water in a copper calorimeter. There the steam will condense back into water. By measuring the rise in temperature produced and the mass of steam added, the specific latent heat of vaporisation of water can be calculated.

Equipment Needed

- A Bunsen burner, a tripod and gauze
- A round-bottomed flask, a stopper and glass tubing
- 2 retort stands and 2 clamps
- A steam trap (if available)
- A copper calorimeter, a lid and insulating material
- A container to hold the calorimeter and insulating material (e.g. a large beaker)
- A thermometer (0-50 °C in steps of 0.1

Method

1. Find the mass of the empty calorimeter m_c. Find the mass of the same calorimeter I full of cold water m_1. Hence by subtraction find the mass of the cold water m_w. Place the calorimeter in the container of insulating material.
2. Set up the round-bottomed flask as in FIG. 15.8 on the next page. Connect the steam trap to it. Fill the flask about N with water and bring it to the boil. Adjust the Bunsen so that steam is emerging from the trap at a steady rate. If no steam trap is available, insulate the delivery tube by surrounding it with cotton wool.

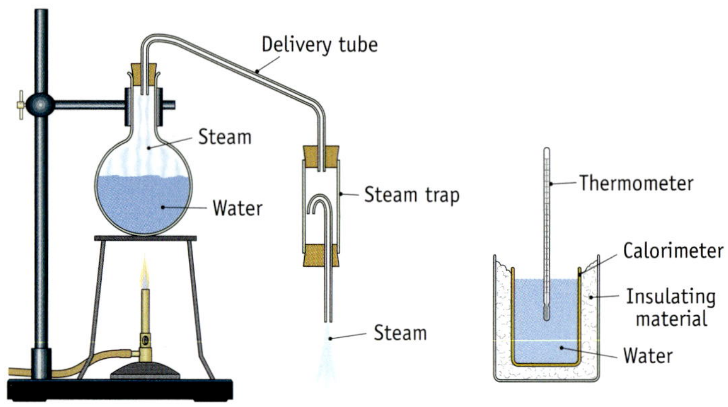

Fig. 15.8

3. Note the temperature θ_1 of the cold water in the calorimeter. Dry the exit tube from the steam trap with a cloth and place it deep into the cold water.
4. Leave it there until there is a rise in temperature of about 12 °C as shown on the thermometer. Turn off the Bunsen and remove the steam delivery tube from the calorimeter.
5. Stir the water in the calorimeter with the thermometer and note and record the highest temperature reached θ_2.
6. Find the combined mass m_2 of the calorimeter, water and condensed steam. Hence find the mass of the condensed steam m_s.

Calculations

On the assumption that no heat is given to or taken from the surroundings we have:

$$\begin{pmatrix}\text{Heat gained}\\ \text{by water}\end{pmatrix} + \begin{pmatrix}\text{Heat gained by}\\ \text{calorimeter}\end{pmatrix} = \begin{pmatrix}\text{Heat given out by steam in}\\ \text{turning to water at 100 °C}\end{pmatrix} + \begin{pmatrix}\text{Heat lost by condensed}\\ \text{steam in cooling from}\\ \text{100 °C to final temperature}\end{pmatrix}$$

i.e. $m_w c_w (\theta_2 - \theta_1) + m_c c_c (\theta_2 - \theta_1) = m_s l + m_s c_w (100 - \theta_2)$

The values of all the variables in this formula, except l (the latent heat of vaporisation of water) are known, hence l can be calculated.

Table

Mass of empty calorimeter m_c	=	
Mass of calorimeter and cold water m_1	=	
Mass of cold water m_w	=	
Temperature of cold water θ_1	=	
Final temperature of water and calorimeter θ_2	=	
Mass of calorimeter, water and steam m_2	=	
Mass of condensed steam $m_s = m_2 - m_1$	=	

EXPERIMENTAL NOTES

1. The experiment will be more accurate if you pre-cool the water in the calorimeter with some ice to about 5 °C below room temperature before putting the steam into it. Allow the steam to raise the temperature to about 5 °C above room temperature. By doing this:
   - errors due to heat loss are minimised because heat flowing into the system at the start of the experiment will almost cancel heat leaving at the end of the experiment,
   - a greater mass of steam can be used, which reduces % errors in measuring its mass,
   - the steam condenses faster, again reducing heat loss.
2. The steam trap ensures that only steam and not water enters the calorimeter. If a trap is not available, the delivery tube must be insulated to prevent steam condensing in it, and sloped towards the round-bottomed flask.
3. To reduce errors in measuring temperature, use a thermometer graduated in steps of 0.1 °C.

WARNING!

The steam emerging from the tube is very hot and will scald you. Be careful not to knock the flask of boiling water or sit anywhere near it.

Questions

1. Why should the end of the delivery tube be dried before putting it into the calorimeter?
2. How would the accuracy of the experiment be affected if the rise in temperature were very large (e.g. 70 °C)?
3. What would be the affect on the value of the final result if steam was seen to be leaving the calorimeter of water?
4. What is the function of the steam trap?
5. Why must the delivery tube be insulated if no steam trap is available?
6. How would the accuracy of the experiment be affected if only a very small rise in temperature were produced (e.g. 2 °C)?
7. What is the advantage in cooling the water before adding the steam?

HEAT TRANSFER

Heat can be transferred from one place to another by **conduction**, **convection** and **radiation**.

CONDUCTION

CONDUCTION

Conduction is the movement of heat energy through a substance by the passing on of molecular vibration from molecule to molecule. There is no overall motion of the substance.

If one end of an object is at a higher temperature than the other, the molecular vibration is passed on from the hotter end to the cooler end. Thus heat energy flows through the material. Substances in which this occurs easily are called good **thermal conductors**, e.g. metals. Substances in which it only occurs a little are called **thermal insulators**.

EXPERIMENT

TO COMPARE THE RATES OF CONDUCTION THROUGH VARIOUS SOLIDS.

1. Use the equipment in FIG. 15.9.
2. Melt some candle wax and stick the coins onto the end of each rod.
3. Place boiling water in the trough and wait until the wax melts and the coins fall off.
4. The better the conductor, the less time it takes for the coin to fall off.

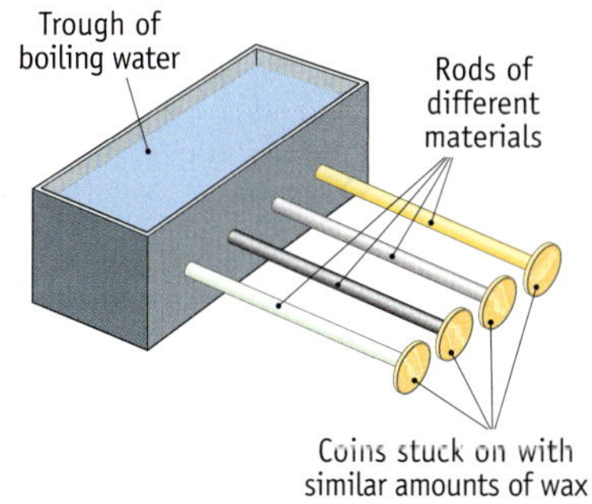

Fig. 15.9

FIG. 15.10 shows a simple experiment to show that water is not a very good conductor. The water at the top of the tube boils but very little heat is conducted to the ice at the bottom and it does not melt.

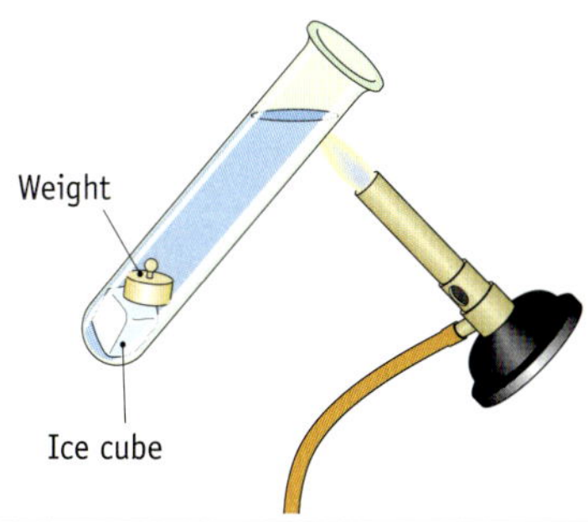

Fig. 15.10

U-VALUE

How well a part of a building, e.g. a roof or a wall, conducts heat is given in terms of what is called its **U-value**.

> **U-VALUE**
> The ***U-value*** of a structure is the amount of heat energy conducted per second through 1 m^2 of that structure when a temperature difference of 1°C (i.e.1 K) is maintained between its ends.

The unit of U-value is the **W m^{-2} K^{-1}**.

A structure that is a bad insulator (i.e. **a good conductor**) **has a high U-value**.

A structure that is **a good insulator has a low U-value**.

Increasing the insulation of a structure reduces its U-value.

RADIATION

> **RADIATION**
> ***Radiation*** is the transfer of heat energy from one place to another in the form of electromagnetic waves.

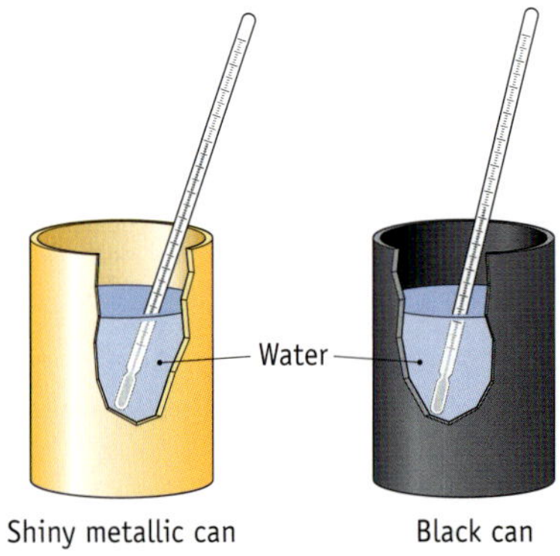

Fig. 15.11

The electrons at the surface of a substance vibrating with thermal energies emit electromagnetic waves. The higher the temperature the shorter the wavelengths emitted. The waves travel at the speed of light, i.e. 3×10^8 m s^{-1}. Heat from the Sun is radiated heat. The darker the colour of an object the better it is at radiating heat. FIG. 15.11 shows two cans, one which is black and the other is shiny metallic. Hot water at the same temperature is placed in each. The black can shows a quicker fall in temperature because heat radiates from it better. The darker can will also absorb heat radiation falling on it better than the shiny one. If both cans containing cold water are placed in direct sunlight (or radiated heat from a heater), the dark can will show a greater rise in temperature.

> **SOLAR CONSTANT (Solar Irradiance)**
> The average amount of the Sun's energy falling **per second** perpendicularly on **1 metre squared** of the Earth's atmosphere is the ***solar constant.*** Its value is about 1.35 kW m^{-2}

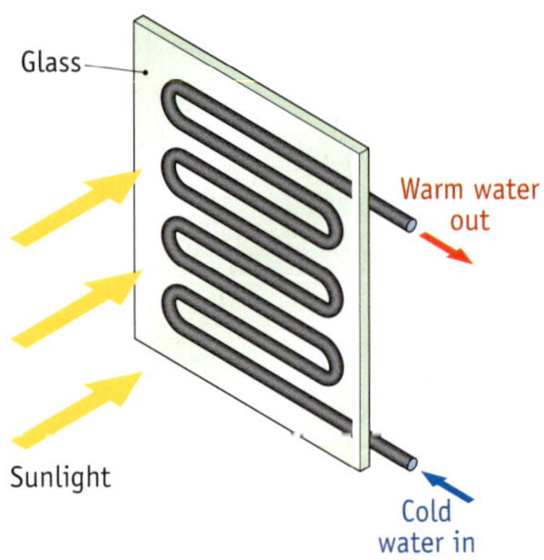

Fig. 15.12

SOLAR HEATING

Solar heating is using the sun's energy to heat something. FIG. 15.12 shows one method of solar heating. Here water flowing in black pipes under glass absorbs heat and heats the water. In another method, a solar panel consisting of photocells converts the Sun's energy into electrical energy which can be used for heating or other purposes.

Problem 9: The solar constant is 1.35 kW m^{-2}. What is the average amount of energy falling on each square metre of the Earth's atmosphere per year?

Solution:

1 year = (365)(24)(60)(60) s = 3.1536×10^7 s

Energy per m^2 per year = Solar constant × number of seconds in a year

= $(1.35)(3.1536 \times 10^7)$ = 4.2573×10^7 J

Convection

CONVECTION

Convection **is the transfer of heat through a fluid by means of circulating currents of fluid caused by the heat.**

When the lower part of the fluid is heated, it expands and becomes less dense. It rises above the cooler fluid. This sets up a movement of the fluid called a **convection current** (FIG. 15.13). Note that this only occurs if there is gravity and would not happen in satellites orbiting the Earth. Hot water in a domestic hot water tank rises by convection to the top of the tank. The hot water outlet is at the top of the tank. Water in domestic heating systems may circulate by convection. FIG. 15.14 shows a simple arrangement. A pump is usually used to speed up the flow of water through the system.

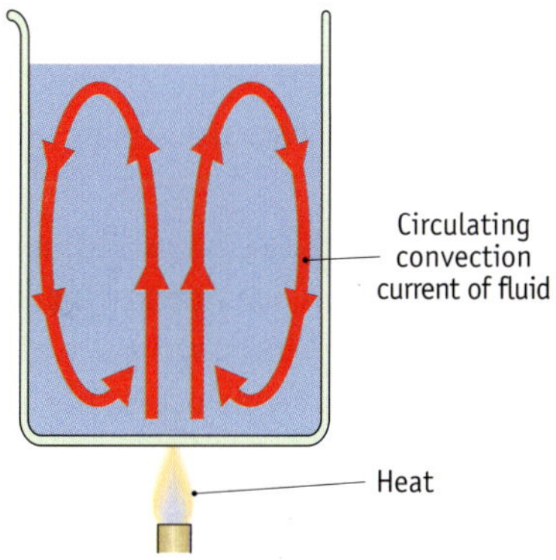

Fig. 15.13

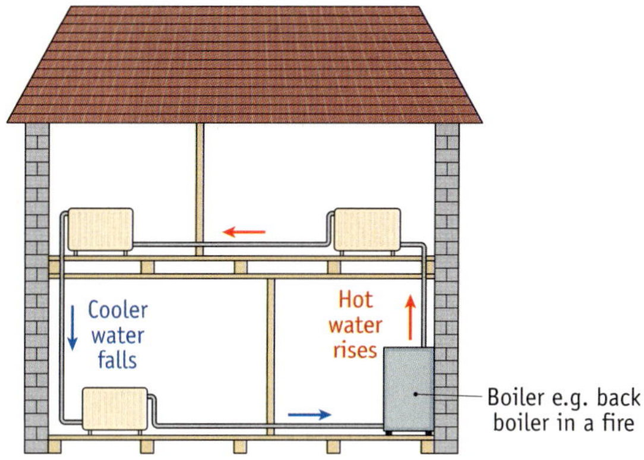

Fig. 15.14

CHAPTER CHECKLIST

- **Define** each of the following and state its unit; Heat capacity; Specific heat capacity; Latent heat; Specific latent heat; Specific latent heat of fusion; Specific latent heat of vaporisation; Solar constant (solar irradiance); U-value.
- **State:** The three states of matter; The relationship between the U-valve of a structure and how good an insulator it is.
- **Explain** what is meant by: Conduction; Convection; Radiation; U-valve.
- **Recall** and use the formulae:
  $Q = C\Delta\theta \qquad Q = mc\Delta\theta \qquad Q = ml$
- **Describe** and carry out an experiment to measure: The specific heat capacity of water; The specific latent heat of fusion of ice; The specific latent heat of vaporisation of water.
- **Describe** simple experiments to: Compare the rates of conduction through solids; Show convection; Show that black is a good radiator.
- **Explain** the physical principles behind the operation of: Storage heaters; Heat pumps; Perspiration; Domestic hot water and heating systems; Solar heating.

CHAPTER 16

Waves and Wave Motion

Fig. 16.1

Waves

You are familiar with different kinds of waves, such as water waves (FIG. 16.1), radio waves or waves on a rope. You probably know that light and sound travel as waves. After an earthquake, shock waves – called seismic waves – travel through the Earth and can be detected thousands of miles away. Waves are very important in physics. Waves can be classified as either **mechanical** or **electromagnetic**.

MECHANICAL WAVES

Examples of these are: water waves, waves on a rope, waves on a spring, sound waves and ultrasonic waves.

Mechanical waves must have a substance (i.e. a solid, a liquid or a gas) to travel through. They cannot travel in a vacuum. The substance through which they travel is called a **medium**. A mechanical wave passing through a medium is vibrations being passed on from molecule to molecule.

ELECTROMAGNETIC WAVES

Examples of these are: radio waves, microwaves, infra-red waves, 'visible' light waves, ultraviolet waves, X-rays and gamma-rays.

Electromagnetic waves can travel through a vacuum. They do not need a medium to travel through, though they can travel through various media. In a vacuum, all electromagnetic waves travel at the incredibly fast speed of 3×10^8 metres per second. This speed is sometimes called the **speed of light**. Electromagnetic waves travel fastest in a vacuum. In other media their speed is always less.

Travelling Waves

In the following examples, the waves move from one place to another. They are thus called travelling waves.

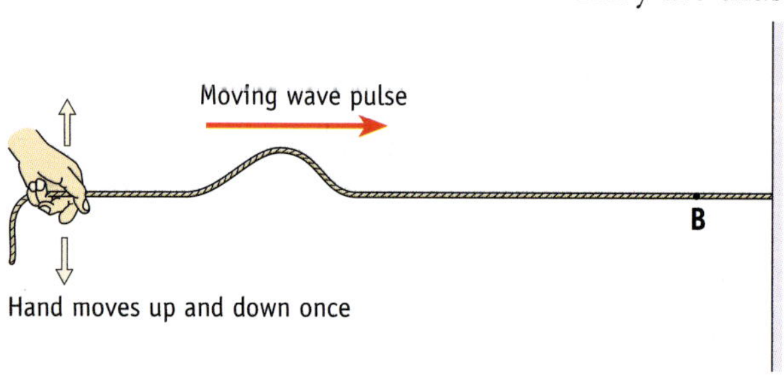

Fig. 16.2
A wave pulse travelling on a rope.

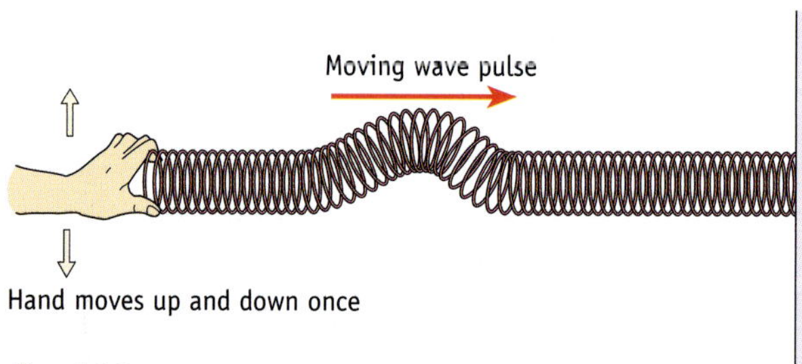

Fig. 16.3
A wave pulse travelling on a slinky.

WAVES ON A ROPE

In FIG. 16.2 the hand moves one end of the rope up and down once. This causes a disturbance on the rope which travels along the rope. Try this at home or in the laboratory. The disturbance is called a wave (or more correctly a wave pulse). The same type of wave can be formed on a spring (called a slinky) as in FIG. 16.3.

WAVES ON WATER

If you drop something into a calm pool of water, a disturbance is produced. The disturbance moves away from the point where the object entered the water. This moving disturbance is a water wave. Water waves are also caused by the action of the wind on water.

COMPRESSION WAVES ON A SPRING

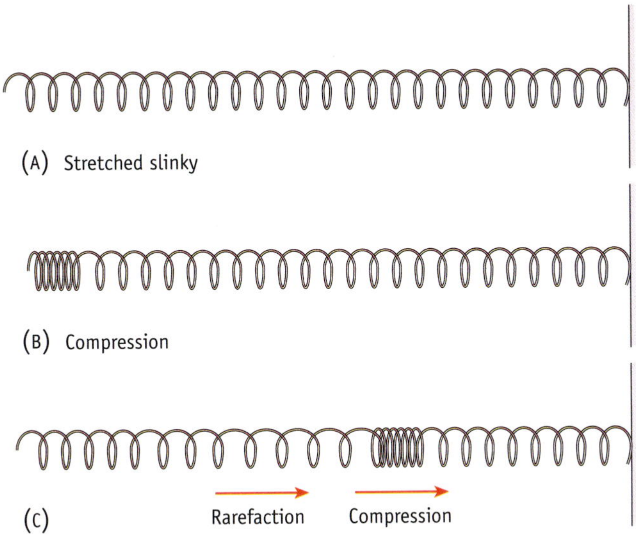

Fig. 16.4
A compression wave on a spring.

FIG. 16.4(A) shows a stretched slinky. If you hold a number of the coils together – called a **compression** – and let them go, the compression moves along the spring (FIG. 16.4(B), (C)). Try this in the laboratory. After the compression passes a point on the spring, the coils in that part become stretched apart more than normal. This region is called a **rarefaction**. The rarefaction follows the compression along the spring. This disturbance travelling along the spring is another type of travelling wave.

WAVES ARE A MEANS OF TRANSFERRING ENERGY FROM ONE PLACE TO ANOTHER

When waves move along water, a rope or a slinky, there is no overall motion medium as the wave passes. As the wave pulse passes a point, the medium around that point is disturbed, but when the pulse has passed, the medium at that point is no longer moving. The rope itself does not move away from the hand; the water does not move away from the point where the object falls into it.

If a small object is attached to the rope at some point (B in FIG. 16.2) this object is at rest until the wave pulse reaches it. Thus it has no kinetic energy. When the wave pulse passes B, the object does move, i.e. it gets kinetic energy. A similar situation occurs for the water wave or the wave on the spring. Thus:

TRAVELLING MECHANICAL WAVE

A ***travelling mechanical wave*** is a disturbance carrying energy through a medium without any overall motion of that medium.

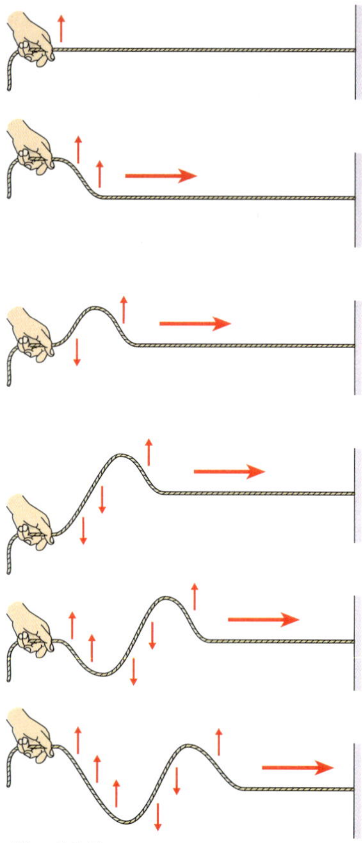

Fig. 16.5

ELECTROMAGNETIC WAVES

It is known that when an electromagnetic wave passes through a region of space, there is a rapidly changing electric and magnetic field in that region. By this means, energy gets transferred from one place to another by the wave. Heat energy from the Sun travels to earth by this means. The following is thus true:

> **TRAVELLING WAVE**
> A ***travelling wave***, either mechanical or electromagnetic, is a disturbance that travels out from the source producing it, transferring energy from the source to other places through which it passes.

Periodic Travelling Waves

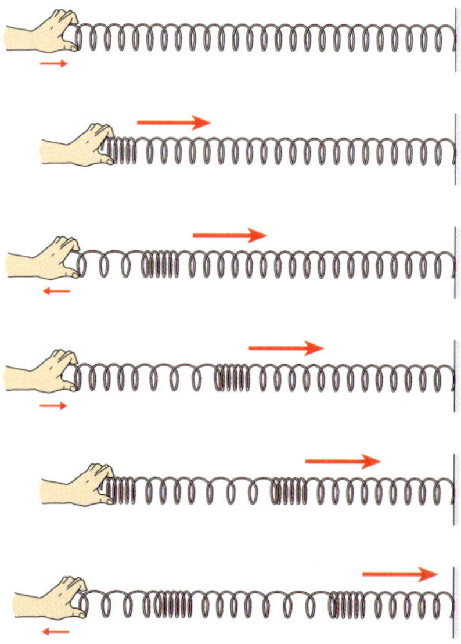

Fig. 16.6

If the hand in Fig. 16.5 repeatedly and regularly moves the end of the rope up and down, a continuous series of identical wave pulses moves along the rope. Such a disturbance is called a **periodic travelling wave**.

In Fig. 16.6 as the hand moves to the right it forms a compression. This compression immediately moves out along the spring. As the hand moves to the left it forms a rarefaction. This also moves out along the spring. As the hand continues to move back and fourth, a series of evenly spaced compressions and rarefactions move out along the spring. This is also a periodic travelling wave.

Transverse and Longitudinal Waves

> **TRANSVERSE WAVE**
> A ***transverse wave*** is a wave where the direction of vibration is perpendicular to the direction in which the wave travels.

As a wave on a rope passes, the particles of the rope vibrate (or oscillate) up and down, i.e. they vibrate at right angles to the direction in which the wave is travelling. Such a wave is called a **transverse wave**.

Examples of transverse waves

- Waves on a rope.
- Waves on a spring as in Fig. 16.3.
- Water waves.

It can be shown that **all electromagnetic waves are transverse waves**.

> **LONGITUDINAL WAVE**
> A ***longitudinal wave*** is a wave where the direction of vibration is parallel to the direction in which the wave travels.

In the compression-rarefaction wave on the slinky, you see that as the waves pass, the particles of the spring vibrate (oscillate) parallel to the direction in which the wave is travelling. Such a wave is called a **longitudinal wave**.

Examples of longitudinal waves

- Compression waves on a spring.
- Sound waves in a solid, liquid or a gas.
- Ultrasonic waves.

TERMS USED TO DESCRIBE A TRANSVERSE PERIODIC TRAVELLING WAVE

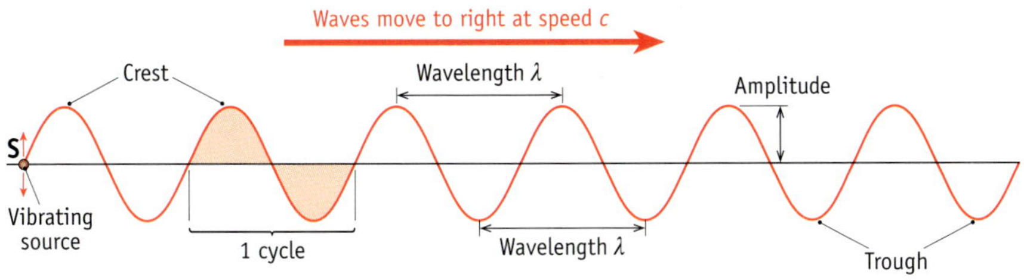

Fig. 16.7
A periodic transverse wave.

In FIG. 16.7:

- The top of a wave is called the **crest** of the wave. The bottom of a wave is called the **trough**.
- The maximum distance of any particle from its undisturbed position is called the **amplitude** (A). Thus the distance from the crest (or trough) to the undisturbed position is the amplitude.
- The disturbance produced by one complete vibration of the source (i.e. one crest and one trough) is called an **oscillation** or a **cycle**.
- The distance from any point on one cycle to the corresponding point on the next cycle is called the **wavelength** (λ) of the wave. Obviously the distance from one crest to the next is the wavelength, as is the distance from one trough to the next. Wavelength is measured in metres.
- The number of cycles passing any point per second is called the **frequency** (f). Frequency is measured in number of cycles per second. 1 cycle per second is called **1 hertz (Hz)**.
- The distance travelled by one cycle in one second is called the **velocity (c)** of the wave. Obviously it is also the distance travelled in one second by a crest, a trough or any other point on a wave.

UNIT OF FREQUENCY
The unit of frequency is the ***hertz* (Hz)**, where:

1 hertz = 1 cycle per second
1 Hz = 1 s^{-1}

NOTE

1 kHz = 1000 Hz
1 MHz = 1 000 000 Hz

Quantity		Unit	
Wavelength	λ	metre	m
Frequency	f	hertz	Hz
		or cycle per sec.	s^{-1}
Velocity	c	metre per sec.	m s^{-1}

TERMS USED TO DESCRIBE A LONGITUDINAL PERIODIC TRAVELLING WAVE

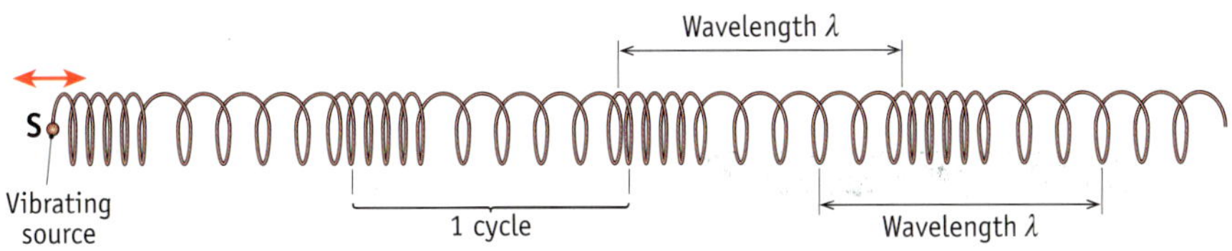

Fig. 16.8
A periodic longitudinal wave on a spring.

In FIG. 16.8:

- The disturbance produced by one complete vibration of the source is called an **oscillation** or a **cycle**. One cycle consists of one compression and one rarefaction.
- The maximum distance of any particle from its undisturbed position is called the **amplitude**.
- The distance from any point on one cycle to the corresponding point on the next cycle is called the wavelength. For example, the distance from the middle of one compression to the middle of the next compression is the **wavelength**.
- The number of cycles passing any point per second is called the **frequency**.

THE RELATIONSHIP $c = f\lambda$

Suppose a wave of wavelength λ and frequency f moves with a speed c. FIG. 16.7 shows such a wave at a particular instant. Consider the complete cycle that is shaded. Start a clock as this cycle begins to pass. One second later f cycles will have passed. Each cycle has a length λ, therefore the first moves a distance $f\lambda$ in this second. Thus the speed of the waves is $f\lambda$, i.e. $c = f\lambda$. The same argument applies to longitudinal waves.

If a periodic wave has frequency f, wavelength λ and velocity c then:

$$c = f\lambda$$

Problem 1: The frequency of a wave is 20 Hz. How many full cycles pass per second? How long does it take one cycle to pass?

Solution: 20 Hz = 20 cycles passing per second. Time for one cycle to pass = $\frac{1}{20}$ s = 0.05 s

Problem 2: If the wavelength of a sound wave emitted from a whistle is 4 m and the speed of sound in air is 340 m s^{-1} find the frequency of the wave.

Solution: Here $\lambda = 4$ m and $c = 340$ m s^{-1}

$$c = f\lambda \Rightarrow f = \frac{c}{\lambda} = \frac{340}{4} = 85 \text{ Hz}$$

Problem 3: If the frequency of the radio station 2 FM is 92.2 MHz and its wavelength is 3.254 metres, find the velocity of radio waves.

Solution: Here $\lambda = 3.254$ m and $f = 92.2$ MHz $= 92.2 \times 10^6$ Hz

$$c = f\lambda = (3.254)(92.2 \times 10^6) = 300 \times 10^6 = 3 \times 10^8 \text{ m s}^{-1}$$

Problem 4: The wavelength of 'visible light' varies from 3.7×10^{-7} m to 7.0×10^{-7} m. Find the lowest frequency that an electromagnetic wave can have and still be in the 'visible' range.

Solution: Since $c = f\lambda$ it follows that f is smallest when λ is biggest. Therefore the required frequency is the one corresponding to a wavelength of 7.0×10^{-7} m.

$$f_{min} = \frac{c}{\lambda_{max}} = \frac{(3 \times 10^8)}{(7.0 \times 10^{-7})} = 4.286 \times 10^{14} = 4.3 \times 10^{14} \text{ Hz.}$$

EXERCISE 16.1

1. The frequency of a water wave is 10 Hz. How many full cycles pass per second? How long does it take one cycle to pass?
2. The distance between a crest and an adjacent trough in a periodic travelling wave is 2 m. What is the wavelength of the wave?
3. When a water wave passes a small boat, the boat moves up and down. If the boat moves through a vertical distance of 2 m, what is the amplitude of the wave?
4. A source emits waves of wavelength 1.5 m with a frequency of 5 Hz. What is the speed of the waves?
5. The frequency of a radio station is 96 MHz and its wavelength is 3.125 m. What is the velocity of radio waves?
6. One radio station operates at a frequency of 50 MHz and a wavelength of 6 m. Another operates at a frequency of 25 MHz. What is the wavelength of the second station?
7. X-rays are electromagnetic waves with wavelengths between about 5×10^{-9} m and 1×10^{-11} m. What is the frequency range of X-rays? Take the speed of light to be 3×10^8 m s^{-1}.

8. A longitudinal wave of wavelength 12 m has a frequency of 40 Hz. What is its speed? On entering a different medium its speed doubles. What happens to:

(i) its frequency,

(ii) its wavelength?

9. The time interval between the arrival of a trough of a tsunami (tide goes very far out) and the crest (a huge wave) is 15 minutes. If the tsunami is moving at 400 km h^{-1}, what is its wavelength?

10. When sound waves travel from air into another medium their speed changes from 340 m s^{-1} to 500 m s^{-1} and their frequency remains the same. If the wavelength of the sound waves in air is 40 cm, what is its wavelength in the other medium?

Wave Phenomena

The different types of waves we meet in physics have certain properties in common. These properties are: **reflection**, **refraction**, **interference**, **diffraction** and **polarisation** (which occurs for transverse waves only).

Reflection

When a wave meets an obstacle in its path, it bounces off that obstacle.

REFLECTION
The bouncing of waves off of an obstacle in their path is called ***reflection*** of waves

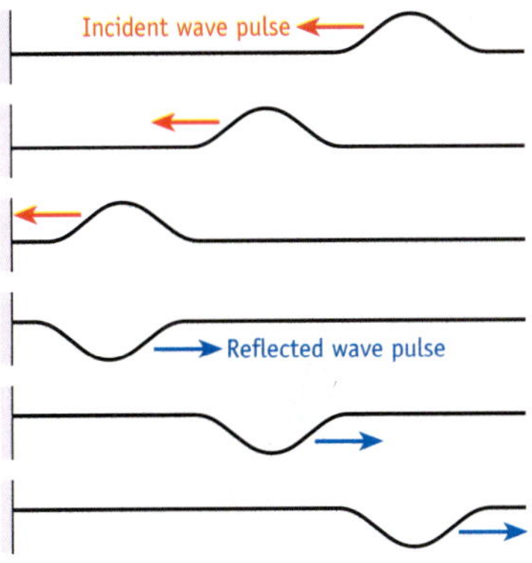

Fig. 16.9
A wave pulse on a rope being reflected from a fixed end.

Fig. 16.9 shows a wave pulse on a rope being reflected from a wall to which it is tied. Try this with a rope of slinky. You may have seen the reflection of water waves in the sea, a lake or in a sink. Reflection of water waves can easily be demonstrated with a ripple tank. The reflection of light waves from a mirror is discussed in Chapter 2.

Longitudinal waves also undergo reflection. A compression sent along a slinky which is fixed at one end will be reflected back.

If you shout or clap your hands some distance away from a large building, an **echo** is heard. The echo is a sound wave being reflected from the building.

When a series of parallel waves – called plane waves – strike a flat obstacle, they are reflected back at the same angle at which they strike the obstacle, i.e. in Fig. 16.10 angles A and B are the same. In Fig. 16.10 the parallel lines represent the crests of the waves and are known as wave fronts*.

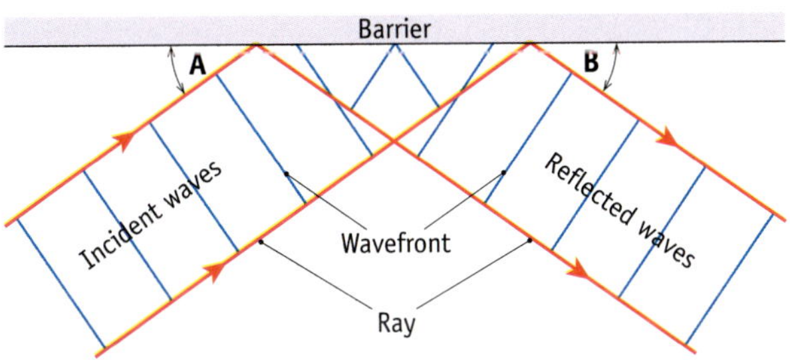

Fig. 16.10

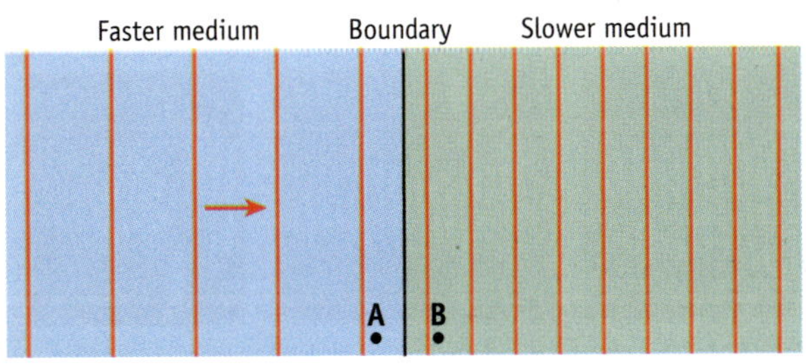

Fig. 16.11
When the wave slows down its wavelength decreases.

* The parallel lines in Fig. 16.10 are called **wave fronts**. Here they represent the crests of the waves. They could just as easily represent the troughs, or indeed they could represent the corresponding parts of each cycle. For simplicity we can assume they are crests.

WAVES CHANGING SPEED

Fig. 16.13
Refraction in a ripple tank.

When waves go from one medium to another, they generally change speed on entering the second medium. When waves move from one medium to another or through regions of the same medium where their speeds differ, their **frequency remains the same**.

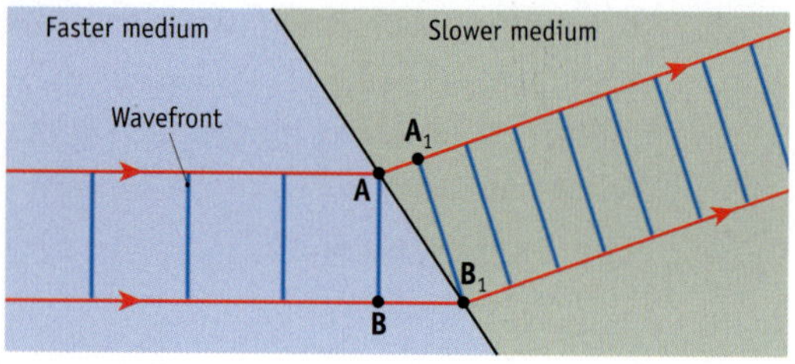

Fig. 16.12
The change in speed causes the waves to change direction.

Since $c = f\lambda$ and f remains the same, it follows that **the wavelength increases if the wave speeds up and the wavelength decreases if the waves slow down** (FIG. 16.11, FIG. 16.12 and FIG. 16.13).

REFRACTION OF WAVES

REFRACTION OF WAVES The changing of direction of a wave when it enters a region where its speed changes is called ***refraction***.

In FIG. 16.12, the waves enter the second medium at an angle other than 90°. The waves change direction on entering the second medium as the speed changes. This phenomenon is called **refraction**. FIG. 16.13 shows the refraction of water waves in a ripple tank. The refraction of light waves is discussed in Chapter 4.

EXPLANATION OF REFRACTION

In FIG. 16.12 A enters the slower medium first and thus slows down, while B continues at the same speed. When B reaches B_1, A will only have reached A_1. Thus the crest in the second medium is along A_1B_1. It is then found that the wave continues to move perpendicular to A_1B_1. The wave changes direction on entering the second medium due to the speed changing.

DIFFRACTION

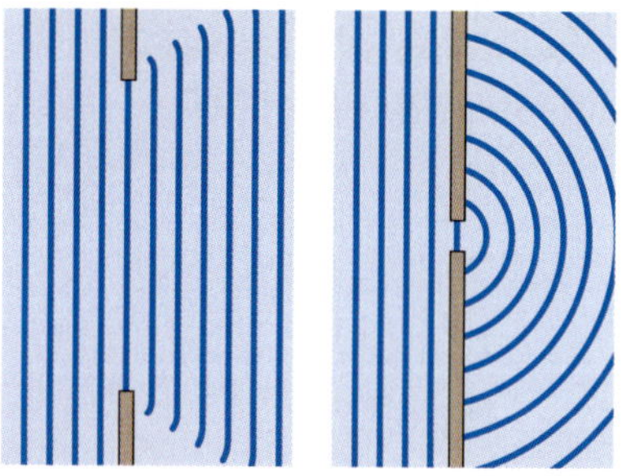

(A) (B)

Fig. 16.14

FIG. 16.14(A) shows parallel water waves meeting a flat obstacle with a gap in it. The waves pass through the gap. At the other side of the gap the waves spread out sideways slightly, but most of the waves pass straight through. However, if the size of the gap is near in size to the wavelength λ of the waves, the waves spread into the whole region beyond the gap (FIG. 16.14(B)). This phenomenon is called **diffraction**. It can readily be seen for water waves in a ripple tank (FIG. 16.15). Waves can also spread around an obstacle in their path. Again, the effect is only significantly noticeable if the object is small relative to the wavelength of the wave involved.

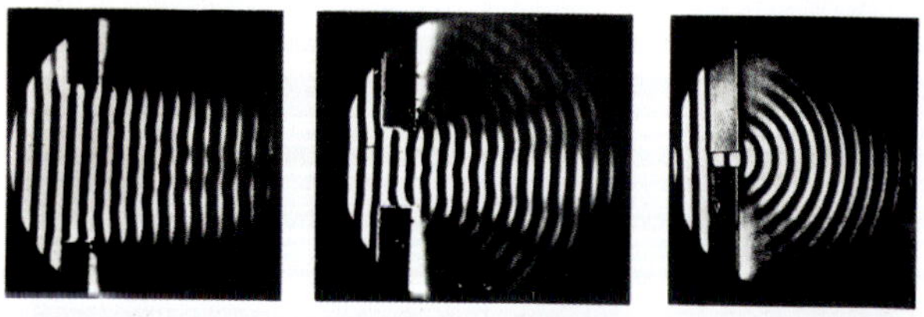

Fig. 16.15
Diffraction of water waves in a ripple tank.

Diffraction occurs for sound waves and accounts for the fact that we can hear around corners and obstacles (the size of many everyday gaps, such as doorways, are near in size to the wavelengths of audible sounds). Likewise, it occurs for electromagnetic waves of all forms. It is only noticeable for light when the size of the gap is very small, since light has a very small wavelength (page 208).

DIFFRACTION
The sideways spreading of waves into the region beyond a gap or around an obstacle is called ***diffraction***.

INTERFERENCE OF WAVES

Consider two sources of water waves S_1 and S_2 (FIG. 16.16) that have the same frequency and amplitude. Suppose that crests from S_1 and crests from S_2 arrive together at X. Troughs from each source will also arrive together. The waves arrive perfectly in step. The two crests combine to give a wave with a crest twice as high as each of the individual waves. The troughs combine to produce a trough which is twice as deep. The waves moving from X to Y have twice the amplitude of the waves from either source. The frequency of the new wave is the same as either of the original two. When waves arrive crest with crest and trough with trough, they are said to be **in phase** with each other.

When waves from two different sources overlap, they produce a new wave. In this new wave the displacement of any particle of the medium at a given instant is the algebraic sum of the displacements that each wave on its own would cause. This phenomenon is called the **interference of waves**.

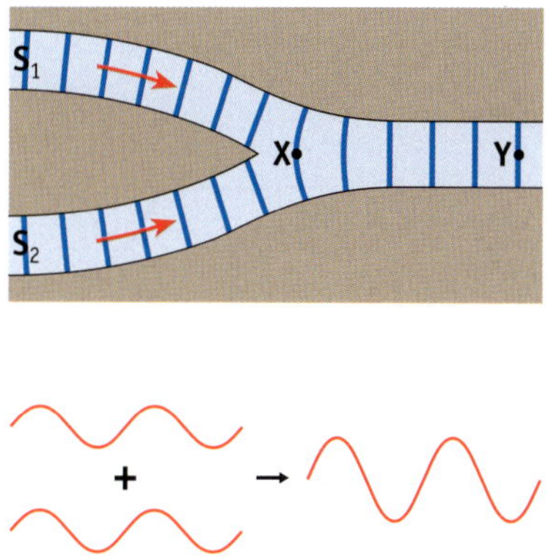

Fig. 16.16
Constructive interference.

INTERFERENCE
When waves from two sources meet, a new wave is produced. The displacement produced at any point by this wave is the algebraic sum of the displacements that each wave would produce on its own. This is called ***interference*** of waves.

In FIG. 16.16 the amplitude of the resulting waves is bigger than the amplitude of each of the individual waves. The waves are said to undergo **constructive interference**.

CONSTRUCTIVE INTERFERENCE
When waves from two sources meet and the amplitude of the resulting wave is ***greater than*** the amplitudes of each of the individual waves, the waves are said to undergo ***constructive interference***.

If crests from one source arrive at X together with troughs from the other source, it is found that the waves completely **cancel** each other (FIG. 16.17). Thus the water between X and Y remains undisturbed. This phenomenon is called **destructive interference**. When waves arrive out of step like this they are said to be **completely out of phase** with each other. The waves are completely out of phase when one set is one half wavelength ($\lambda/2$) ahead of the other.

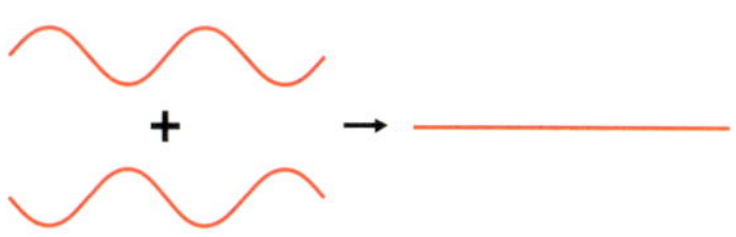

Fig. 16.17
Destructive interference.

DESTRUCTIVE INTERFERENCE
When waves from two sources meet and the amplitude of the resulting wave is less than the amplitude of each of the individual waves, the waves are undergoing ***destructive interference***.

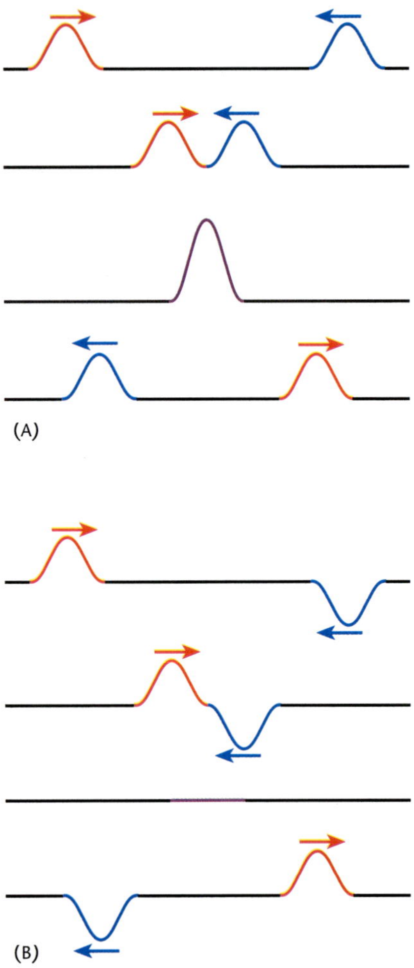

Fig. 16.18

Interference can also be demonstrated with a slinky. Fig. 16.18(A) shows pulses heading towards each other. At the instant the pulses are in the same place, a pulse with twice the height is produced. The pulses then move on unaffected by each other. Similarly, destructive interference can be shown (Fig. 16.18(B)).

NOTE Waves of all types will undergo interference.

If two sources are emitting waves that are a definite fixed amount out of step with each other, there is said to be a **constant phase difference** between them.

COHERENT SOURCES
Two sources of periodic waves are said to be ***coherent*** if they are in phase or if there is a constant phase difference between waves from each of the sources. If this is so the sources must also have the same frequency.

INTERFERENCE PATTERN
When waves from two (or more) coherent sources meet, the resulting wave pattern formed is called an ***interference pattern***.

Interference Pattern from Two Point Sources

A very important interference pattern occurs when waves of equal amplitude from two coherent point sources meet. This can be shown in the laboratory using the ripple tank and two vibrating dippers (Fig. 16.19 and Fig. 16.20). Here two point sources S_1 and S_2 are producing circular water waves. Suppose that the circles drawn represent crests from each source. Then, obviously, half way between two consecutive crests from the same source is a trough.

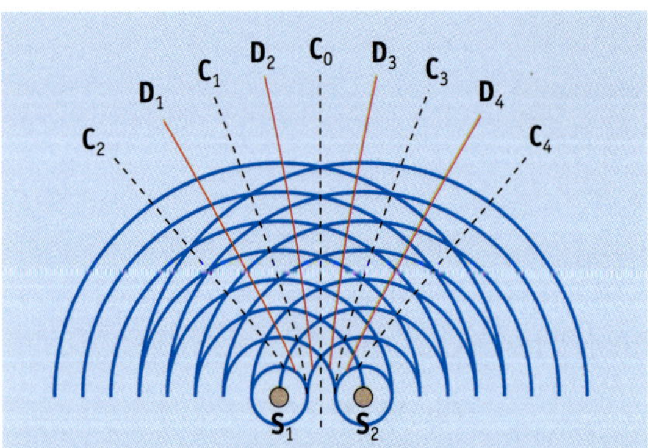

Fig. 16.19

Fig. 16.20 Interference from two point sources in a ripple tank. The regions of maximum disturbance and zero disturbance can be clearly seen.

CONSTRUCTIVE INTERFERENCE

Consider the situation along a line such as C_0 or C_1. Along each of these lines we see that crests from one source always meet crests from the other.

Likewise troughs from one source meet troughs from the other. Thus we get constructive interference along these lines. Such lines are called **antinodal lines**.

DESTRUCTIVE INTERFERENCE

Along the lines D_1, D_2 etc. ... you can see that crests from one of the sources arrive together with troughs from the other. Thus destructive interference occurs. The water remains calm and undisturbed along these lines. Such lines are called **nodal lines**. Between the nodal and antinodal lines the waves still interfere. Near the nodal lines it is almost completely destructive. As you move towards the antinodal lines the amplitude of the wave increases.

This kind of interference pattern also occurs for sound waves (page 192) and for light waves (page 208). It may seem a bit strange at first to be told that sound from two different sources can produce silence in a region or that light shining on light can produce darkness. However, since both light and sound travel as waves it does happen and can easily be shown in the laboratory.

POLARISATION

FIG. 16.7 (page 177) shows a transverse wave on a rope. The rope is vibrating up and down in a vertical plane. This wave is said to be **vertically polarised**. FIG. 16.21 shows a horizontally polarised wave on a rope. The plane in which a transverse wave is vibrating is called its **plane of polarisation**.

Fig. 16.21
A horizontally polarised wave.

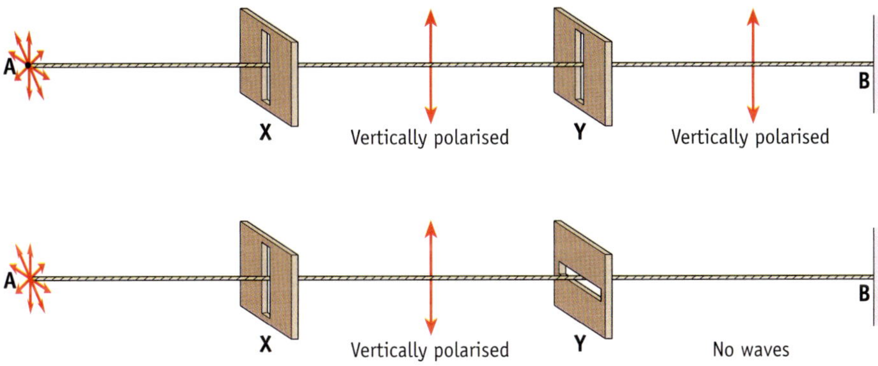

Fig. 16.22

In FIG. 16.22 the end A of the rope is moved to and fro in various planes so as to produce a series of wave pulses on the rope, vibrating in different planes. We say **unpolarised waves** travel between A and the vertical slit at X. At the slit only the waves that are vibrating in a vertical plane pass through. Thus at the other side of the slit, a vertically polarised wave is produced on the rope. These waves are said to be **plane polarised** (in the vertical plane). This can easily be demonstrated in the laboratory. If this vertically polarised wave then meets another slit (Y) it will pass through this slit if the slit is vertical. If the slit is horizontal the vertically polarised waves will not get through. It should be obvious that:

NOTE

Only transverse waves can be polarised.

We shall study the polarisation of light on page 215.

Stationary Waves

If a rope or a slinky is held fixed at one end and the other end is shaken up and down, at certain frequencies of vibration waves will be seen to remain at the same place on the rope. Try this in the laboratory. Some points on the rope will be **at rest**. These points are called **nodes**. Other points will be vibrating with **maximum amplitude**. These points are called **antinodes** (Fig. 16.23).

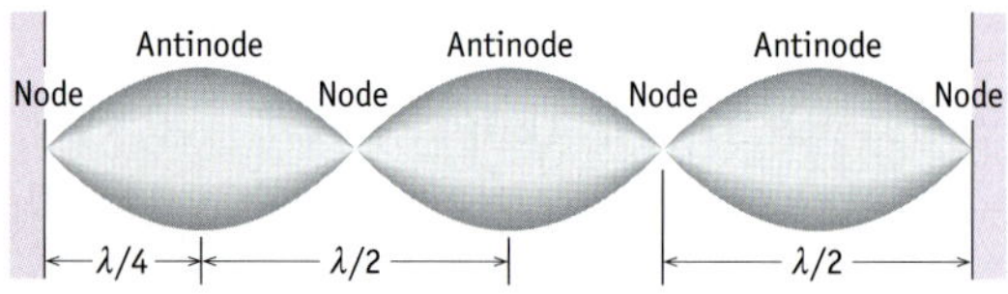

Fig. 16.23

Points between the nodes and the antinodes vibrate up and down with the amplitude decreasing as you move from an antinode to a node. **The frequency of all vibrating particles is the same**. The wave does not travel along the rope but remains at the same position on it. Such waves are called **stationary waves** or **standing waves**. Fig. 16.24 shows a thick rubber band attached to a vibration generator. With the frequency at about 50 Hz, by varying the tension in the band, a number of stationary waves can be set up on it.

Fig. 16.24
Stationary waves on a rubber band.

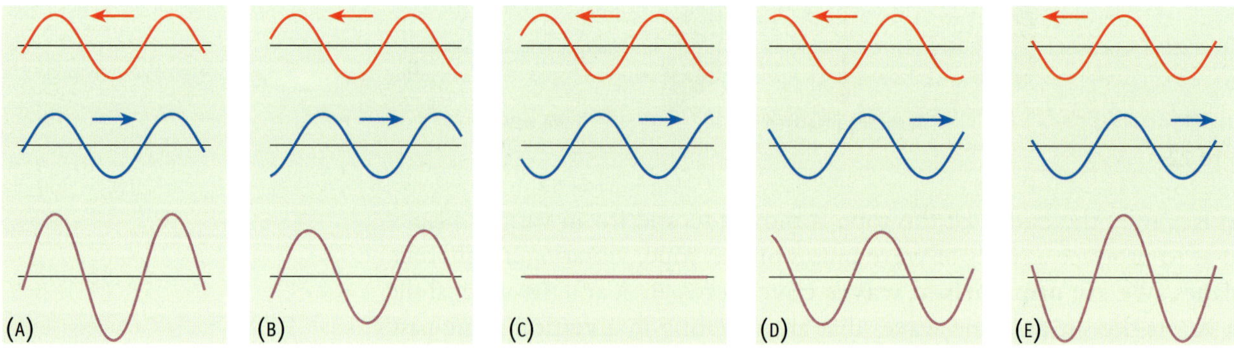

Fig. 16.25
Two travelling waves of the same frequency and amplitude moving in opposite directions produce a stationary wave.

HOW STATIONARY WAVES OCCUR

Stationary waves are formed when two periodic travelling waves of the same frequency and amplitude travelling in opposite directions meet. The two wave trains interfere, producing a stationary wave. Fig. 16.25 shows how this happens. At each instant the resultant of the two waves is found by algebraic addition.

In (A) and (E) crests of each travelling wave coincide giving the resultant wave shown by the purple line.

In (B) and (D) the purple line again shows the resultant of the two waves.

In (C) crests from one wave coincide with troughs from the other. Destructive interference occurs at this instant. By drawing the travelling waves at any instant and finding their resultant, the position of the stationary wave at that instant is found. If you do this you will see that a stationary wave with nodes and antinodes will result.

STATIONARY WAVE

When two periodic travelling waves of the same frequency and amplitude moving in opposite directions meet, they interfere with each other. The resulting wave formed is called a ***stationary wave*** or a ***standing wave***.

In Fig. 16.24 the vibrating source produces a travelling wave which moves along the rubber band and is reflected from the fixed hand. The reflected wave and the original wave then interfere to produce the stationary wave. Stationary waves in water can be produced similarly.

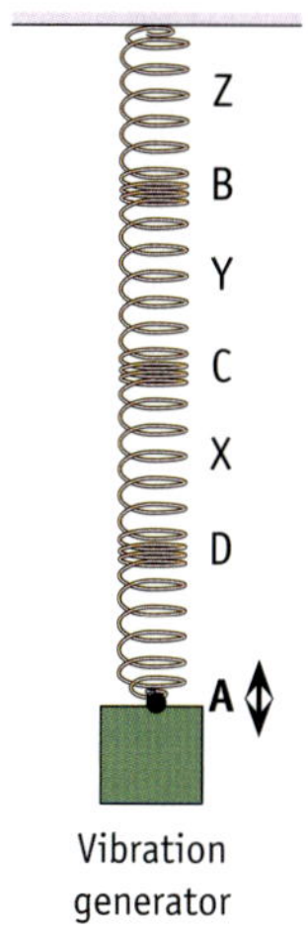

(A) Longitudinal stationary waves on a spring.

Frequency and Wavelength of a Stationary Wave

From the diagrams in Fig. 16.25 we see that:

- The frequency of vibration of every vibrating particle in a stationary wave is the same.
- The frequency of a stationary wave is the same as the frequency of the travelling wave producing it.
- If λ is the wavelength of travelling waves causing the stationary wave then:

The distance between 2 consecutive nodes $= \frac{\lambda}{2}$

The distance between 2 consecutive antinodes $= \frac{\lambda}{2}$

The distance between an antinode and the next node $= \frac{\lambda}{4}$

Stationary Longitudinal Waves

Longitudinal stationary waves can be set up using a vibration generator and a coiled spring as in Fig. 16.26(A). Point A vibrates up and down sending a travelling longitudinal wave up the spring. This wave is reflected from the top which is held fixed. The travelling waves moving in opposite directions produce a stationary wave, with some points on the spring (B, C, D) vibrating vertically with maximum amplitude and others (X, Y, Z) remaining at rest. Longitudinal stationary waves can be set up with sound waves – and play an important part in some methods of measuring the velocity of sound. The relationship between the internode distance and the wavelength is the same as for transverse stationary waves.

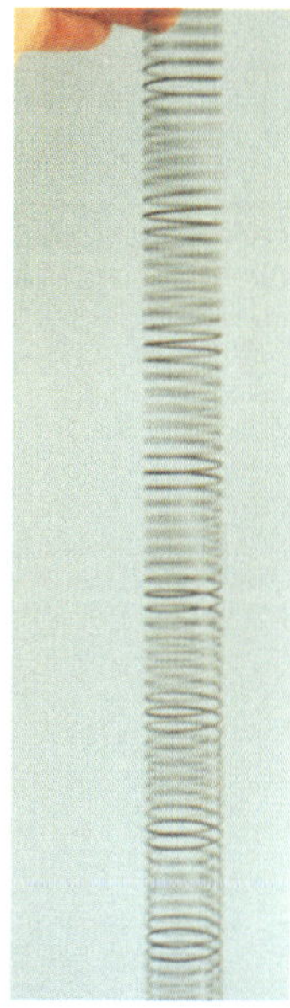

(B) Longitudinal stationary waves on a spring.

Fig. 16.26

Problem 5: Waves on a rope travel at a speed of 3 m s^{-1}. When a stationary wave is set up on the rope the distance between an antinode and the nearest node is 60 cm. What is the frequency of the wave?

Solution: Distance between antinode and adjacent node $= \frac{\lambda}{4}$

$$\Rightarrow 0.6 = \frac{\lambda}{4} \Rightarrow \lambda = 2.4\,\text{m} \qquad c = f\lambda \Rightarrow f = \frac{c}{\lambda} = \frac{3}{2.4} = 1.25\,\text{Hz}$$

Problem 6: A stationary sound wave is set up between a loudspeaker and a wall. The distance between the second and the eighth node is 4 m and the speed of sound in air is 340 m s^{-1}. Find the frequency of the sound wave.

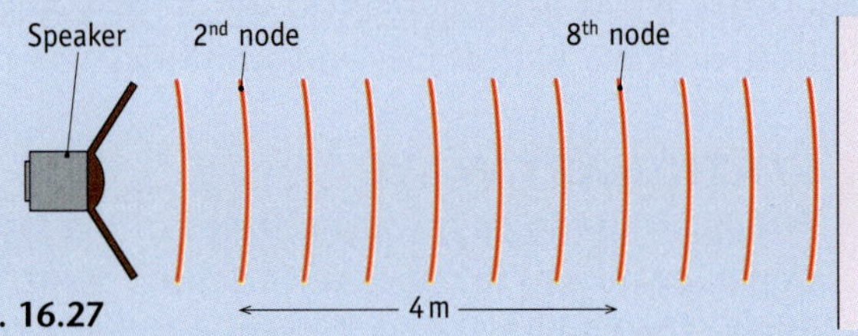

Fig. 16.27

Solution:

Distance between two adjacent nodes $= \frac{\lambda}{2}$

Distance between 2nd and 8th node $= 6\frac{(\lambda)}{2} = 3\lambda$ (FIG. 16.27)

$\therefore \quad 3\lambda = 4\text{ m} \Rightarrow \lambda = \frac{4}{3}\text{ m} \qquad c = f\lambda \Rightarrow f = \frac{c}{\lambda} = \frac{340}{\frac{4}{3}} = 255\text{ Hz}$

EXERCISE 16.2

1. What is the relationship between the frequency of a stationary wave and the frequency of the travelling wave producing it?
2. In a stationary wave, what is the distance between two consecutive nodes equal to? What is the distance between two consecutive antinodes equal to? What is the distance between a node and the next antinode equal to?
3. Waves on a rope travel at a speed of 4 m s^{-1}. When a stationary wave is set up on the rope, the distance between a node and the nearest antinode is 500 cm. What is the frequency of the wave?
4. Stationary waves of frequency 6 Hz are produced on water. If the speed of the waves is 60 cm s^{-1}, what is the distance between adjacent nodes?
5. A stationary sound wave is set up between a loud-speaker and a wall. The distance between the 4th and 12th node is 2 m and the speed of sound in air is 340 m s^{-1}. Find the frequency of the sound wave.

The Doppler Effect

CHANGE IN WAVELENGTH OF WAVES FROM A MOVING SOURCE

FIG. 16.28 shows a source of periodic circular water waves of frequency f. The source is not moving. Let c be the speed of the waves through the water and λ their wavelength. FIG. 16.29 shows the same source moving to the right with a constant speed of u m s^{-1}. Look what happens ahead of the moving source:

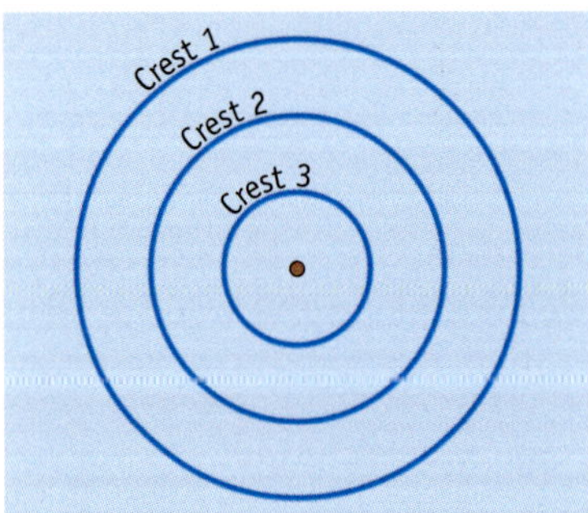

Fig. 16.28
Waves from a source which is not moving.

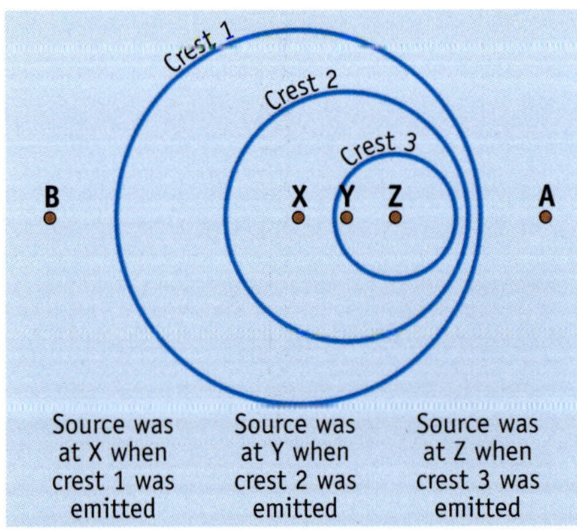

Fig. 16.29
Waves from a source moving to the right at speed u m s^{-1}.

- The **crests are closer together** than crests from the stationary source.
- Thus the **wavelength is less** than the wavelength of waves from the stationary source.
- Since the speed of the waves is the same, it follows that to a stationary observer (e.g. at A), the **frequency of the waves** from the approaching source **must be greater** than the frequency from the stationary source.

Behind the moving source the opposite happens.

- The **crests are further apart** than crests from the stationary source.
- The **wavelength is greater** than the wavelength of the waves from the stationary source.
- Thus the **frequency is less** than the frequency from the stationary source.

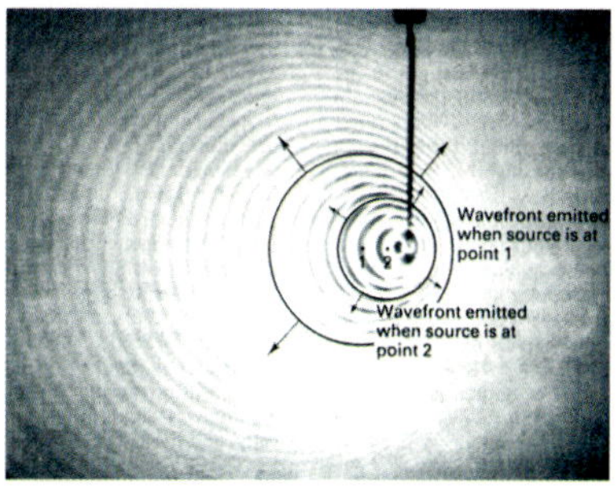

Fig. 16.30
The change in wavelength for water waves in a ripple tank produced by a moving source. The wavefronts represent crests.

A change in observed frequency also occurs if the source is stationary and the observer moves away from or towards the source. In this case, the wavelength does not change, but the speed of the wave relative to the observer and hence the observed frequency does change.

This change in frequency occurs for all kinds of waves and is called the **Doppler Effect**.

> **DOPPLER EFFECT**
> The apparent change in the frequency of waves due to the motion of the source or the observer is called the ***Doppler Effect.***
> If a source emits waves of frequency f, the observed frequency f' will be:
>
> **greater than f** — if the source moves **towards** the observer or the observer moves **towards** the source.
>
> **less than f** — if the source moves **away from** the observer or the observer moves **away from** the source.

Doppler Effect for Sound Waves

The higher the frequency of a sound wave, the higher its pitch, i.e. high notes have high frequency and low notes have low frequency. The Doppler Effect for sound waves is familiar to most people. For example, if an approaching car sounds its horn, its pitch sounds higher than if the car were at rest. When the car passes there is an abrupt decrease in the pitch of the note and as it moves away its pitch is lower than if the car were at rest. Other common examples of this are when a racing car, a train sounding its horn or an ambulance with its siren on. The effect is heard as each passes a stationary observer. The Doppler Effect can easily be demonstrated in the laboratory if an electronic buzzer is whirled in a horizontal circle. The pitch of the note from the buzzer is increased as it moves towards you and is decreased as it moves away.

The Doppler Effect and Electromagnetic Waves

The Doppler Effect also occurs for light waves and other electromagnetic waves. It can be used to determine the speeds at which stars or galaxies are moving towards or away from us. If a star is moving away from us, the wavelength of the light emitted from it will appear longer than light from a similar stationary source. The wavelength is thus shifted towards the red (red light has the longest wavelength). This is known as **red shift**. The wavelength of light from a source moving towards us is shifted towards the violet.

The Doppler Effect for microwaves is used by the Gardaí in speed traps. Microwaves reflected from a moving vehicle will have slightly different wavelengths from those emitted by the radar gun. The difference allows the speed of the vehicle to be determined.

Formula for the Frequency of Waves from a Moving Source

The observed frequency f' of the waves from a moving source of mechanical waves as observed by a stationary observer depends on:

- the actual frequency f of the source,
- the speed of the source u,
- the speed of the waves in the medium c,
- whether the source is approaching or moving away.

If the source is moving towards the observer the observed frequency f' is given by:

$$f' = \frac{fc}{c - u}$$

If the source is moving away from the observer the observed frequency f' is given by:

$$f' = \frac{fc}{c + u}$$

Problem 7: A police car travelling at 30 m s^{-1} passes a stationary observer. Its siren emits a note of frequency 1 kHz. If the velocity of sound is 336 m s^{-1}, what is the frequency heard by the observer when the car is (i) approaching the observer (ii) moving away from the observer?

Solution: Here: $u = 30$ m s^{-1} $\quad c = 336$ m s^{-1} $\quad f = 1000$ Hz $\quad f' = ?$

(i) When the car is approaching: $f' = \frac{fc}{c - u} = \frac{(1000)(336)}{336 - 30} = 1098$ Hz

(ii) When the car is moving away: $f' = \frac{fc}{c + u} = \frac{(1000)(336)}{336 + 30} = 918$ Hz

Note that the increase in frequency (98 Hz) when the source is approaching the observer is different from the decrease in frequency (82 Hz) when it is moving away.

Problem 8: A train's whistle emits a continuous note of frequency 800 Hz as it approaches a person standing near the track. To the person, the frequency of the note appears to be 920 Hz. Find the speed of the train. (Speed of sound in air = 340 m s^{-1})

Solution: $f = 800$ Hz $\quad f' = 920$ Hz $\quad c = 340$ m s^{-1} $\quad u = ?$

$$f' = \frac{fc}{c - u} \Rightarrow 920 = \frac{(800)(340)}{340 - u} \Rightarrow (920)(340 - u) = 272\,000$$

$$\Rightarrow 312\,800 - 920u = 272\,000 \Rightarrow u = 44.35 \text{ m s}^{-1}$$

Problem 9: A train travelling at a constant speed passes through a station. To a person standing on the platform, the note emitted from the horn of the train changes from 1216 Hz to 960 Hz. If the speed of sound in air is 340 m s^{-1}, find the velocity at which the train was travelling and the actual frequency of the note emitted by the train.

Solution: Let f = actual frequency emitted and u the speed of the train, then:

For approaching train:

$$f' = \frac{fc}{c-u}$$

$$\Rightarrow 1216 = \frac{340f}{340-u}$$

$$\Rightarrow (1216)(340-u) = 340f \quad Eqn.1$$

For train moving away:

$$f' = \frac{fc}{c+u}$$

$$\Rightarrow 960 = \frac{340f}{340+u}$$

$$\Rightarrow (960)(340+u) = 340f \quad Eqn.2$$

From these equations we have:

$$(1216)(340-u) = (960)(340+u)$$

$$\Rightarrow 413\,440 - 1216u = 326\,400 + 960u$$

$$\text{i.e. speed of train } u = 40 \text{ m s}^{-1}$$

Substituting this value for u into *Eqn.1* gives: $(1216)(340-40) = 340f$

Solving this for f gives: Actual frequency of horn = 1072.9 Hz

Problem 10: A whistle emitting a note of 2 kHz is whirled in a horizontal circle at an angular velocity of 6 π rad s^{-1}. If the highest note heard by a person a large distance away is 2100 Hz, find:

(i) the radius of the circle,

(ii) the lowest note heard by the observer.

Speed of sound = 340 m s^{-1}.

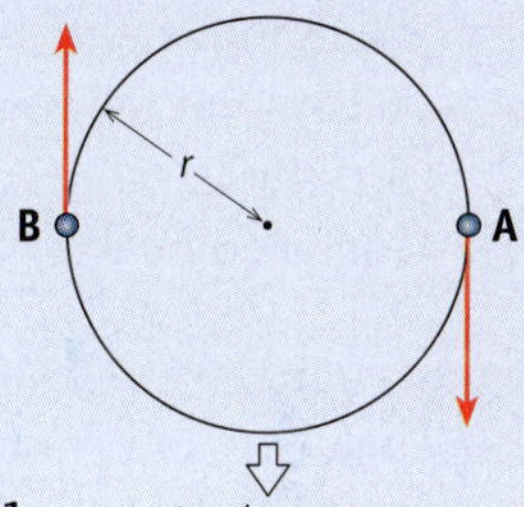

Fig. 16.31

Solution: (i) Highest note occurs when the source is moving towards the observer (i.e. at A in FIG. 16.31).

$$f' = \frac{fc}{c-u} \qquad f = 2000 \text{ Hz}, \quad c = 340 \text{ m s}^{-1}, \quad f' = 2100 \text{ Hz}$$

$$2100 = \frac{(2000)(340)}{(340-u)} \Rightarrow 340 - u = \frac{(340)(2000)}{2100} \Rightarrow u = 16.19 \text{ m s}^{-1}$$

If r = radius of circle then using: $v = r\omega$ where $v = u = 16.19$ we have:

$$r = \frac{v}{\omega} = \frac{16.19}{6\pi} = 0.86 \text{ m}$$

(ii) Lowest frequency occurs when source is moving away from observer at speed 16.19 m s^{-1} (i.e. at B).

$$f' = \frac{fc}{c+u} = \frac{(2000)(340)}{340+16.19} = 1909.1 \text{ Hz}$$

EXERCISE 16.3

1. A source of sound of frequency 2 kHz is approaching an observer at a speed of 50 m s^{-1}. If the velocity of sound is 336 m s^{-1}, what is the frequency heard by the observer?
2. A source of sound of frequency 2 kHz is moving away from an observer at a speed of 50 m s^{-1}. If the velocity of sound is 336 m s^{-1}, what is the frequency heard by the observer?
3. A train's whistle emits a continuous note of frequency 600 Hz as it approaches a person standing near the track. To the person the frequency of the note appears to be 720 Hz. Find the speed of the train. (Speed of sound in air = 340 m s^{-1})
4. A source of sound of frequency 1 kHz approaches an observer at a speed of 40 m s^{-1}, passes the observer and moves away at the same speed. If the velocity of sound is 336 m s^{-1}, what is the change in frequency as the source passes?
5. A police car travelling at 20 m s^{-1} passes a stationary person. Its siren emits a continuous note of frequency 2 kHz. If the velocity of the sound in air is 336 m s^{-1}, find the change in the frequency of the siren as observed by the person as the car passes.
6. A whistle emitting a note of 4 kHz is whirled in a horizontal circle of radius 1 m at a constant speed. If the highest note heard by a person a large distance away is 4200 Hz, find:
   (i) the speed of the whistle,
   (ii) the lowest note heard by the person,
   (iii) the time taken for the whistle to make one complete revolution,
   (iv) the time interval between the person hearing the highest and lowest note.

   Take the speed of sound to be 340 m s^{-1}.
7. A train travelling at constant speed passes through a station. To a person standing on the platform, the note emitted from the horn on the train appears to change from 1000 Hz to 800 Hz. If the speed of sound in air is 340 m s^{-1}, find the speed at which the train is travelling and the actual frequency of the note emitted by the train.
8. A fog horn on a ship emits a note of 200 Hz. To a man on a lighthouse the apparent frequency of the note is 208 Hz. With what speed is the ship moving if it moves either towards or away from the lighthouse? Speed of sound in air 336 m s^{-1}.

CHAPTER CHECKLIST

- **Explain** what is meant by each of the following: Frequency; Wavelength; Amplitude; Transverse wave; Longitudinal wave; Stationary wave; Antinode; Node.
- **Define:** Reflection; Refraction; Diffraction; Interference; Constructive interference, Destructive interference; Polarisation; Doppler Effect.
- **Recall** and use the formula $c = f\lambda$ to solve problems.
- **Recall** that: The distance between adjacent nodes in a stationary wave is $\lambda/2$; The distance between adjacent antinodes in a stationary wave is $\lambda/2$; The distance between a node and the next antinode is $\lambda/4$.
- **Describe** an experiment to demonstrate: Reflection; Refraction; Diffraction; Interference; Polarisation; The Doppler Effect.
- **List** five different everyday examples of waves.
- **Recall** that the Doppler Effect is used in speed traps and explains the red shift of stars.
- **Recall** and **use** the formula: $f' = \frac{fc}{c \pm u}$ to solve problems.

Vibrations and Sound

CHAPTER 17

Vibrating Objects

Fig. 17.1 shows a ruler clamped to a bench. If the free end is pulled down and released it vibrates up and down. The motion from A to B and back again to A is called one cycle. The number of cycles occurring per second is called the **frequency** of the vibration. Frequency is measured in **hertz (Hz)**.

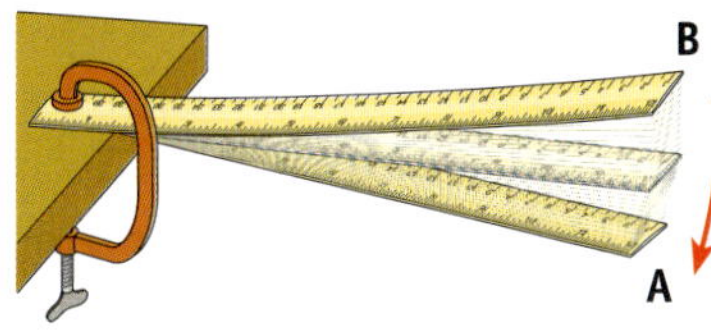

Fig. 17.1

Every Source of Sound is a Vibrating Object

If the length of the ruler in Fig. 17.1 is such that it vibrates with a frequency of 20 Hz or more, the ruler gives off a sound. Every sound, no matter what its source, is produced by something that is vibrating. A string on a guitar or piano, the paper cone in a loud-speaker, a tuning fork or your vocal chords (Fig. 17.2) all produce sound when they are vibrating. For the sound to be audible, the frequency of vibration must be between 20 Hz and 20 000 Hz.

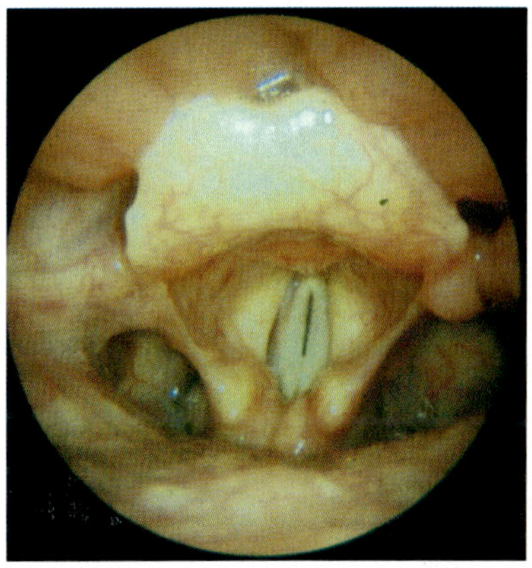

Fig. 17.2
Human vocal chords (also called vocal folds).

Sound Travels as a Wave

The sound produced by a vibrating object travels away from that object as a wave. Evidence for this is that sound shows all the properties of waves; i.e. sound undergoes reflection, refraction, diffraction and interference.

REFLECTION OF SOUND

Stand at least 20 m away from a large high wall or a large cliff face and clap your hands. You will hear an echo. The echo is sound reflected from the wall or cliff. If you are nearer the wall you will not hear the echo separately from the original sound. The further you are from the wall the greater the time between the clap and the echo. In theatres and concert halls the amount of reflection of sound from the surroundings, such as walls, ceilings and seats can either enhance the hearing of music and speech or make it very difficult to hear properly. The science of designing such places with the correct balance of reflection and absorption of sound is known as **acoustics**.

REFRACTION OF SOUND

The refraction of sound waves is responsible for the fact that sounds can be heard more clearly on a cold night or over water (which is cool) than on a hot day or over land. This is explained fully on page 194.

DIFFRACTION OF SOUND

When sound reaches an obstacle with a gap in it, such as a wall with an open door or window, it travels through the gap and then spreads into the region at the other side of it. Because of this, we can hear around corners. The fact that sound travels around corners is evidence that sound undergoes diffraction. The width of doorways and windows is near to that of the wavelength of sound, and thus the diffraction is significant.

INTERFERENCE OF SOUND

Sound can readily be shown to undergo interference. This can be seen in the next experiment. Sound produces both diffraction and interference and since only waves can produce interference and diffraction we can deduce that sound is a wave motion.

EXPERIMENT

To Show the Interference of Sound and Hence that Sound is a Wave Motion.

- Set up the equipment shown in Fig. 17.3 and turn on the signal generator.
- Walk slowly from A to B.
- You will hear the loudness of the sound increase and decrease regularly as you move from A to B.
- This is because the sound from each source produces constructive interference and destructive interference between A and B (producing an interference pattern similar to that on page 182).
- Interference produced is evidence that sound is a wave.

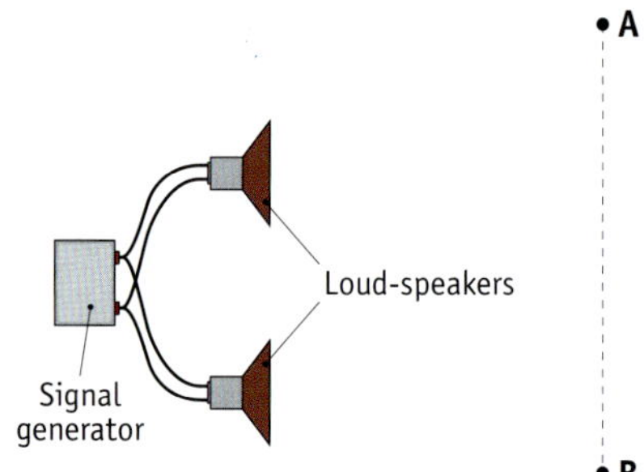

Fig. 17.3

How a Vibrating Object Produces a Sound Wave

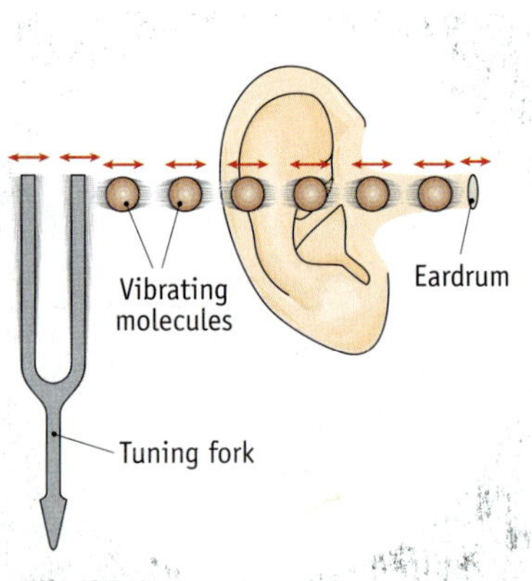

Fig. 17.4

Fig. 17.4 shows a simple picture of a sound wave being emitted from a vibrating prong of a tuning fork. Between the fork and the listener's ear, there are billions of gas molecules which are moving in random directions at high speed. We can ignore this and assume the molecules are at rest. As the prong moves to the right it pushes the molecules which are near it closer together forming a **compression**. This compression then travels out along the rest of the molecules. As the prong moves to the left it pulls the first few molecules further apart from each other, forming a **rarefaction**. This rarefaction also travels out along the rest of the molecules.

As the prong continues to move back and forth, a series of evenly-spaced compressions and rarefactions travels out along the molecules. The moving compressions and rarefactions cause the molecules of the gas to vibrate as they pass. When this vibration reaches the gas molecules next to the eardrum, the eardrum vibrates with the same frequency, with the result that the listener hears a sound. The frequency of the vibrating source and the frequency of the vibrating molecules are the same, thus **the frequency of a sound wave is the same as that of the vibrating source producing it**.

Sound is a Longitudinal Wave

> **NOTE** ***Sound*** travels as a ***longitudinal*** wave.

In Fig. 17.4 we can see that as the wave passes, the molecules vibrate parallel to the direction in which the compressions and rarefactions are travelling. Thus a sound wave is a longitudinal wave. As with other longitudinal waves (page 177) the **wavelength** (λ) of a sound wave is the distance between corresponding points on adjacent cycles. For example, λ = distance between centres of two adjacent compressions.

The **amplitude of a sound wave** is the maximum displacement of any molecule from its rest position.

A Tuning Fork also Shows Interference of Sound

If a vibrating tuning fork is rotated near your ear, the sound heard will be loud in four positions and very weak in four others. As the prongs move apart, a compression leaves the fork and moves along OA and OB, while a rarefaction is produced between the prongs and moves along OX and OY (Fig. 17.5). As the prongs move towards each other the reverse happens. In either case compressions always arrive with rarefactions along each of the directions OZ and **destructive interference** occurs. When your ear is along OZ you hear very little sound. When your ear is along OA, OB, OX or OY you hear the sound quite loudly. Since only waves undergo interference **this also shows that sound travels as a wave motion**.

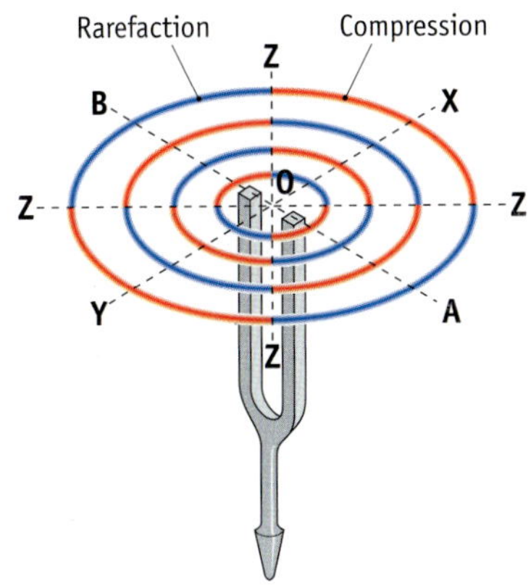

Fig. 17.5

Reduction of Noise by Destructive Interference

Large background noises, such as those from exhaust systems or air conditioning can be significantly reduced by destructive interference. A microphone picks up a sample of the noise and a sound wave of the same frequency and amplitude is created electronically. The crests of this sound wave coincide with the troughs of the noise and the troughs coincide with the crests of the noise. This sound is emitted from a loud-speaker into the region where the noise must be reduced. This sound and the noise interfere destructively producing a region of silence or almost so.

Sound Needs a Medium to Travel Through

As sound is the passing of a vibration from molecule to molecule, it is obvious that sound needs a medium to travel through.

Experiment

To Show That Sound Needs a Medium to Travel Through.

- Set up the equipment in Fig. 17.6 with the bell ringing.
- As the air is removed from the bell jar with a vacuum pump, the loudness of the sound becomes less and less until it is eventually barely audible – some small amount of sound escapes through the supporting wires.
- If the air is let back in, the sound can be heard again.

Conclusion

- Sound needs a medium to travel through.

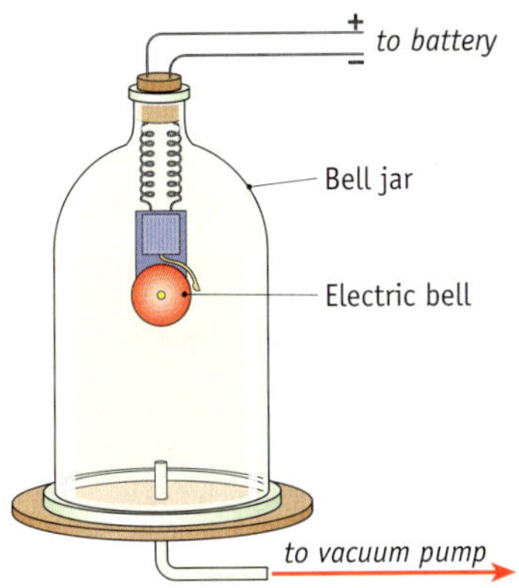

Fig. 17.6

Speed of Sound

The speed at which sound travels through a medium depends on the elastic properties and density of that medium. Fig. 17.7 shows the speed of sound in some common materials. In general, the more dense the medium, the greater the speed. Fig. 17.8 (page 194) shows the speed of sound in air at various temperatures. As the temperature increases so does the speed. The fact that the speed of sound in a gas increases with temperature, sometimes causes refraction, e.g. sounds can be heard more clearly on a cold night than on a warm day. On a warm day the air near the ground is warmer than the air higher up and the sound travels faster near the ground. This causes it to

Material	Approximate Speed (m s^{-1})
Air (0 °C)	331
Water	1500
Copper	3400
Steel	4800

Fig. 17.7
Speed of sound in some materials.

be refracted upwards (FIG. 17.9(A)). At night the air near the ground is cooler and the sound is refracted down (FIG. 17.9(B)) and appears to travel better over the ground.

Temperature (°C)	Approximate Speed (m s^{-1})
0	331
20	344
100	384

Fig. 17.8
Speed of sound in air at various temperatures.

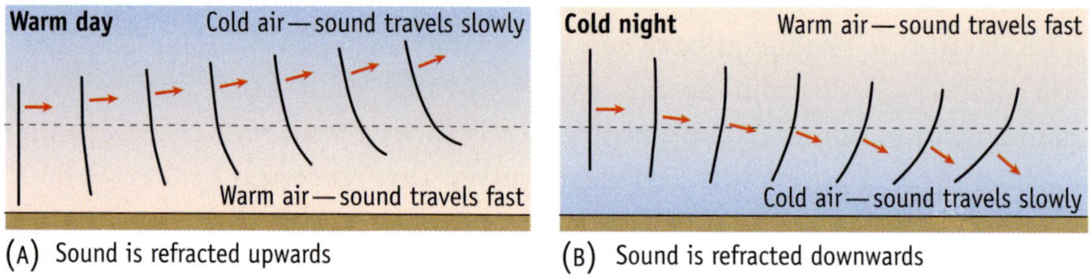

Fig. 17.9

EXERCISE 17.1

Take speed of sound in air to be 340 m s^{-1}

1. A stone is dropped into a well and the splash is heard 2 seconds later. Find the depth of the well.
2. A flash of lightning is seen and 4 seconds later the thunder is heard. How far away was the lightning?

OVERTONES

Sometimes a vibrating object not only gives out sound of one particular frequency, but also emits noises which have frequencies that are multiples of that frequency. These frequencies are called **overtones**.

Suppose f is the lowest frequency emitted.

The frequency $2f$ is called the **1st overtone**.

The frequency $3f$ is called the **2nd overtone** etc...

> **OVERTONES**
> Frequencies which are multiples of a certain frequency are called ***overtones*** of that frequency.
> If f is a given frequency:
> $2f$ is its ***first overtone***,
> $3f$ is its ***second overtone***
> etc...

CHARACTERISTICS OF NOTES

The main characteristics of a note are its **loudness**, its **pitch** and its **quality**. We shall now see how each characteristic depends on a particular wave property of the sound wave.

LOUDNESS

How loud a sound appears to you depends on the amount of energy entering your ear per second. The greater the amplitude of a sound wave the greater the energy it carries. Thus the **loudness** of a sound depends on the **amplitude** of the sound wave.

Your ear is more sensitive to sounds of certain frequencies rather than others. Therefore, two sounds of the same amplitude but of different frequencies may not have the same loudness. Thus the **loudness of a sound also depends on its frequency**.

> **LOUDNESS**
> The ***loudness*** of a sound wave depends on the ***amplitude*** of the sound wave. The greater the amplitude the greater the loudness.

PITCH

If a note is a high note it is said to have high pitch. If it is a low note it is said to have low pitch. The pitch of a note depends on the frequency of the sound wave. The higher the frequency the higher the pitch.

> **PITCH**
> The ***pitch*** of a note depends on the ***frequency*** of the sound wave. The higher the frequency the higher the pitch, the lower the frequency the lower the pitch.

QUALITY

If the same note is played on two different musical instruments it sounds different. The notes are said to have different **quality**. An instrument emits a note of a given frequency. It also emits notes which have frequences that are multiples of that frequency. These higher multiples are called **overtones**. Different instruments emit different numbers of overtones and overtones of different strengths. It is because of this that the same note played on different instruments sounds different. A tuning fork or a signal generator can emit a note of a pure frequency, i.e. with no overtones present.

QUALITY
The ***quality*** of a musical note depends on the ***number of overtones*** present in the note and the ***relative strengths of the different overtones*** present.

FREQUENCY LIMITS OF AUDIBILITY

For a sound wave to be audible its frequency must be between 20 Hz and 20 000 Hz. These values are known as the **frequency limits of audibility**. Most adults, however, cannot hear frequencies of 20 000 Hz. The upper frequency limit decreases with age.

Frequencies above 20 000 Hz are called **ultrasonic** and are inaudible to humans. However, dogs and bats can hear sounds of frequencies up to about 35 000 Hz. Some dog whistles operate at a frequency above 20 000 Hz.

FREQUENCY LIMITS OF AUDIBILITY
The ***frequency limits of audibility*** are the highest and lowest frequencies that can be heard by a normal human ear. The range is **20 Hz – 20 000 Hz**.

NATURAL FREQUENCY

Any object that is free to vibrate, tends to do so at certain frequencies. Usually one of these frequencies predominates, i.e. it is more likely to vibrate at this frequency than the others. This frequency is called its **natural frequency of vibration**.

RESONANCE

FIG. 17.10 shows a person sitting on a swing. If given a push and released, the swing will vibrate back and forth with its own natural frequency. If the person on the swing is pushed every time it is about to move away, the size of the swings will increase. When this happens, the natural frequency of the swing and the frequency of the applied force are the same. This is an example of resonance.

Fig. 17.10

The amount of vibrational kinetic energy picked up by an object is small, unless the frequency of the force acting on it is the same as or near to the natural frequency of vibration of the object. When this is so, the amplitude of the vibration can get very large. This phenomenon is called **resonance**.

RESONANCE
If the frequency of a periodic force applied to a body is the same as or very near to its natural frequency that body will vibrate with very large amplitude. This phenomenon is called ***resonance***.

EXPERIMENT

To Show Resonance.

- Strike a tuning fork to set it vibrating.
- Hold this tuning fork near, but not touching, another tuning fork of the same frequency.
- The second tuning fork will also be found to vibrate and can be heard doing so if you stop the first fork from vibrating.
- Repeat the above with two forks of different frequencies and the effect is much less noticeable.

Conclusion

- The first tuning fork exerts a periodic force on the second tuning fork. The second fork only vibrates with large amplitude if its frequency is the same as or very near to that of the first. This demonstrates **resonance**.

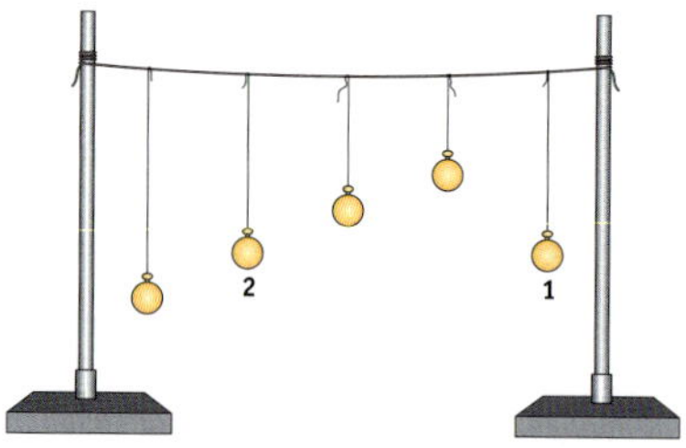

Fig. 17.11
Barton's pendulums.

Fig. 17.12

Fig. 17.13

Other Examples of Resonance:

- If a number of pendulums are set up as in FIG. 17.11 (known as **Barton's pendulums**) and pendulum 1 is set vibrating, it is found that all the other pendulums gain some vibrational energy. However, pendulum 2 (which has the same length and hence the same natural frequency as pendulum 1) starts to swing violently as pendulum 1 slows down. If left swinging, the energy goes from pendulum 1 to 2 and back to 2 again. When swinging, pendulum 1 causes periodic forces in the string which are the same as the natural frequency of pendulum 2 and so resonance occurs.
- Some drinking glasses have natural frequencies which are the same as some high-pitched notes. Opera singers singing high-pitched notes have been known to shatter glasses due to resonance (FIG. 17.12).
- A column of air in a tube can be made to resonate with a tuning fork (page 204).
- The wire on a sonometer can be made to resonate with a tuning fork (page 200).
- Buildings in an earthquake.
- Sound produced by your vocal chords resonates in your larynx, throat, mouth and nose, thus producing a louder sound.

Resonance can cause significant problems in machines and man-made structures such as buildings or bridges. If the frequency of vibration of some part of a machine is the same as some periodic force produced in some other part of it (e.g. from a motor), resonance can occur and cause significant damage. In 1940, wind passing through the Tacoma bridge in the US set up periodic forces on the bridge which resonated with its natural frequency of torsional rotation. The result was that the bridge began twisting with larger and larger amplitude until it broke apart (FIG. 17.13). To avoid resonance, soldiers marching in step usually break out of step when crossing a bridge. In 1850, 500 French soldiers marched in step across the Angers suspension bridge. The bridge collapsed and 226 of them were killed.

Threshold of Hearing

The more sound energy per second entering your ear the louder a sound will appear to you. It does take some minimum amount of sound energy entering your ear per second before your ear (if it is functioning properly) can hear a sound. This minimum amount is called the threshold of hearing. Its precise definition is given below for higher level only.

Intensity of Sound

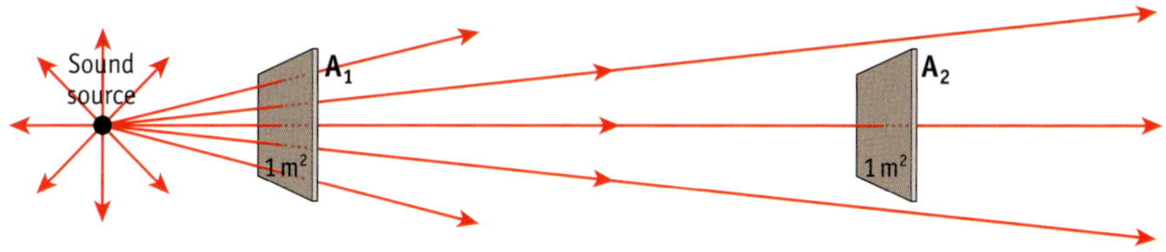

Fig. 17.14

Sound is a mechanical wave transferring energy from one place to another. Fig. 17.14 shows a point source of sound emitting sound in all directions. The further you are from the source, the less sound energy passes per second through an area of 1 m^2 placed at right angles to the direction in which the sound is travelling. In Fig. 17.14, more energy passes through area A_1 per second than through area A_2 per second. By definition, the energy passing per second through an area of 1 m^2 at right angles to the direction in which the sound is travelling is called the **sound intensity (I)** at that point.

SOUND INTENSITY AT A POINT

The ***sound intensity (I)*** at a point is the rate at which sound energy is passing through unit area at right angles to the direction in which the sound is travelling at that point.

$$\text{i.e. Sound Intensity} = \frac{\text{Power}}{\text{Area}} \qquad I = \frac{P}{A}$$

UNIT OF SOUND INTENSITY

The unit of sound intensity is the ***watt per square metre* ($W\ m^{-2}$).**

$$\text{Since } I = \frac{P}{A}, \text{ the unit of } I = \frac{\text{Unit of } P}{\text{Unit of } A} = \frac{\text{watt}}{\text{square metre}} = W\ m^{-2}$$

Threshold of Hearing

The precise definition of threshold of hearing is:

THRESHOLD OF HEARING

The ***threshold of hearing*** is the smallest sound intensity detectable by the average human ear at a frequency of 1 kHz. Its value is $1 \times 10^{-12}\ W\ m^{-2}$.

The Frequency Response of the Ear

Suppose two sounds carrying the same amount of energy per second into your ear have different frequencies. You will not judge them as being of equal loudness because your ear is more sensitive to some frequencies than others. **The ear is most sensitive between 2000 Hz and 4000 Hz.** At frequencies above or below this it is not so sensitive. One reason for this is that sounds of around this frequency can resonate in the ear canal. At frequencies above or below this, there is not as much resonance and the ear is not so sensitive. This means that sounds of these higher or lower frequencies can carry more energy per second into the ear and not appear as loud. In general they will also be less damaging to the ear.

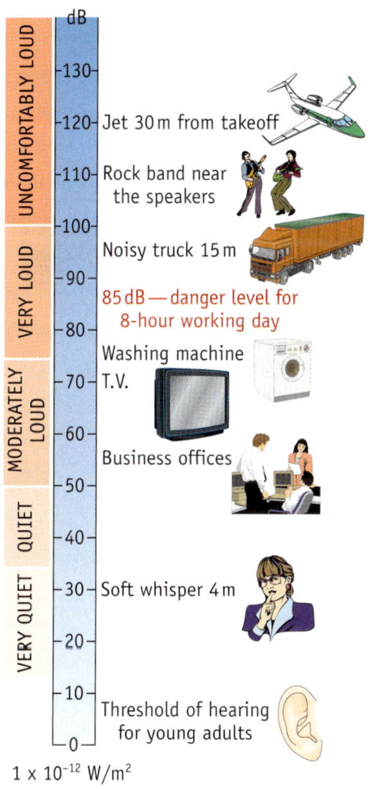

Fig. 17.15

Sound Intensity Level

The amount of energy carried per second into your ear by different everyday sounds varies from the very small to the fairly large. A scale of sound intensity level has been devised to measure this variation. Readings taken on this scale are called **sound intensity level**. They are measured in decibels. The precise difinition of sound intensity level is not on our course. Fig. 17.15 shows some everyday examples of sound intensity level.

SOUND INTENSITY LEVEL

Sound Intensity Level is measured in ***decibels* (dB)**.

Sound intensities measured in W m^{-2} vary from very small (10^{-12} W m^{-2}) to quite large (1 W m^{-2}). This is a very large range (1 = 1 million million times 10^{-12}). A more convenient scale, called the decibel scale, is sometimes used. The range of intensities on this scale is a much handier size, namely 120 decibels. **Thus a large range of intensities can be represented by a small range of numbers.** Readings taken on this scale are called **sound intensity level** as we saw above. The precise definition of intensity level is not on the higher level course either and you need not worry about it. Note, however, that when the sound intensity in W m^{-2} doubles, the sound intensity level increases by 3 dB.

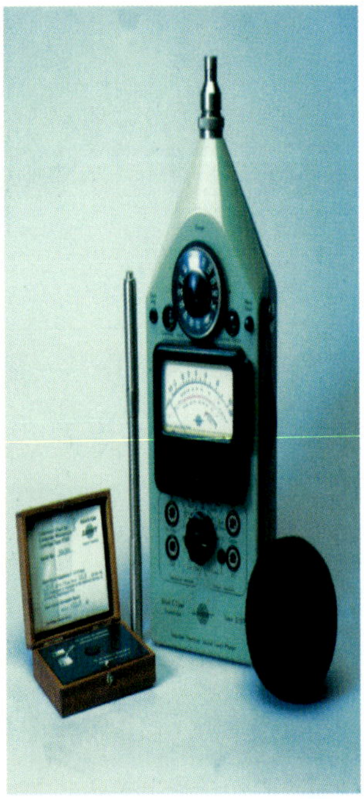

Fig. 17.16
Sound level meter.

The Sound Level Meter

Fig. 17.16 shows a sound level meter. It consists of a microphone, an amplifier and an output meter. It measures sound intensity level in decibels. It uses a scale called the **dBA scale**. This is the **decibel adapted** or frequency weighted scale. It is a variation of the sound intensity level scale. The dBA scale is used to take account of the variation in the human ear's response to sounds of different frequencies. Circuits in the meter suppress or ignore those frequencies that the ear is not sensitive to or does not respond to at all. Such sounds consequently have a less damaging effect on the ear when loud. The meter thus responds more to sounds with frequencies between 2 kHz and 4 kHz, just like the ear.

In the laboratory you should use a sound level meter to measure the sound intensity level of various sounds and see how the level varies with distance from the sound source. You could also see how the meter mirrors the ear's response to sounds of different frequencies from a signal generator. When the frequency is much higher or much lower than 3000 Hz, the sound from a signal generator does not sound so loud to the ear. The low reading on the meter will show this.

SOUND LEVEL dB	MAX. HOURS PER DAY
90	8
93	4
96	2
99	1
102	½
105	¼

Employees must be provided with ear protection if exposed to daily noise levels of 85 dB

Employees must wear ear protection if exposed to daily noise levels of 90 dB

Fig. 17.17

Noise Pollution and Ear Protection

Very loud sounds, e.g. from explosive cartridge operated tools or guns, can immediately permanently damage your hearing. Long-term exposure to not-so-loud sounds will also damage it, e.g. 40% of peoble who have worked most of their lives at a noise level of 90 dBA, will, at 65 years of age, find it difficult to hear other people talking. Some will even be deaf. 90 dBA is roughly equivalent to heavy street traffic. Note also that the type of partial deafness produced from exposure to loud sounds is **incurable**. Some sounds will be heard but they will be muffled and unclear. A hearing aid will not remedy the situation, therefore **prevention** of the damage in the first place is required. In industry and on the farm, ear protection must be worn by people exposed to loud noise from machinery or other sources, e.g. driving some tractors or using chainsaws or lawnmowers. Animals in buildings can be very noisy, e.g. large numbers of pigs in a building can create noise levels above 110 dB, expecially at feeding time. Fig. 17.17 shows the current maximum number of hours for which someone may be exposed to sounds of various intensity without wearing ear protection.

Vibrations on a Stretched String

If a string (or wire) is stretched between two points and plucked or bowed (with a violin bow) a standing wave is set up on the string. The way or **mode** in which the string vibrates depends on where it is plucked and where it is held fixed. Obviously the ends of the string are **nodes** since these cannot undergo any displacement. If the string is plucked at its centre, an **antinode** is formed there and the string vibrates as shown in Fig. 17.18. This string is vibrating at its **fundamental frequency** f. The vibrating string causes a sound wave of the same frequency f in the surrounding air.

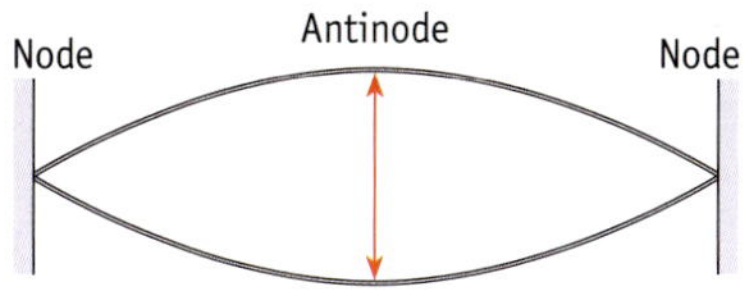

Fig. 17.18
A string vibrating at its fundamental frequency.

FUNDAMENTAL FREQUENCY OF A STRING
A string vibrating with an antinode at its centre and a node at each end (and no other nodes or antinodes) is vibrating at its ***fundamental frequency***.

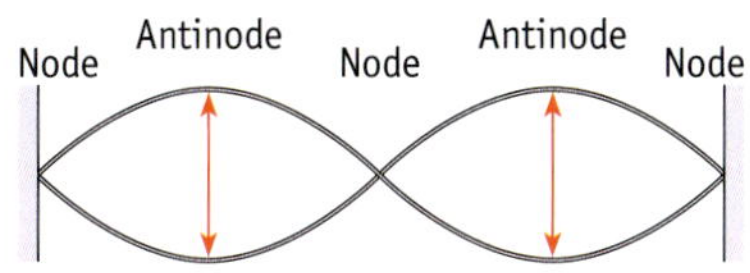

Fig. 17.19

If the string is held lightly at its centre and bowed ¼ way along its length it will vibrate as shown in Fig. 17.19. By suitable bowing or plucking other modes of vibration can also occur (Fig. 17.20). In general, when a string is plucked or bowed it vibrates with some or all of these modes of vibration together.

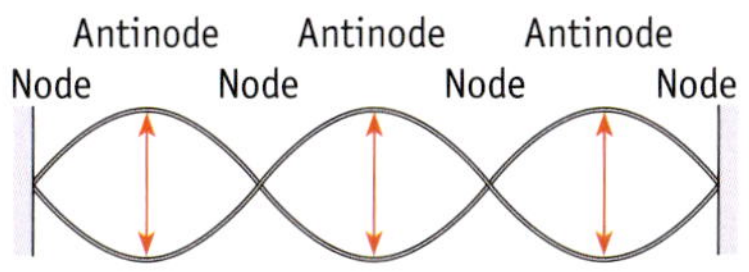

Fig. 17.20

Experimentally it is found that the greater the length of the string, the lower its fundamental frequency. More precisely, it is found that:

Fundamental frequency of string is inversely proportional to its length, i.e. $f \propto \frac{1}{l}$

This means that doubling the length causes its frequency to halve; trebling the length causes its frequency to third; halving the length cause its frequency to double, etc.

In string musical instruments of all kinds, such as the violin or the guitar, the frequency of the note emitted can be varied by changing the length of the vibrating string. Long strings give low notes, short strings give high notes (Fig. 17.21).

Fig. 17.21
String section of an orchestra.

The Sonometer

The fact that $f \propto 1/l$ can be verified in the laboratory using a **sonometer** (Fig. 17.22). This consists of a wooden sounding board with a wire stretched between two movable bridges. The wire is fixed at one end and the tension in the wire is adjusted by means of a winder. The size of the tension is read on a spring balance. In another type, the tension is produced by hanging masses. The tension is equal to the weight of the hanging masses.

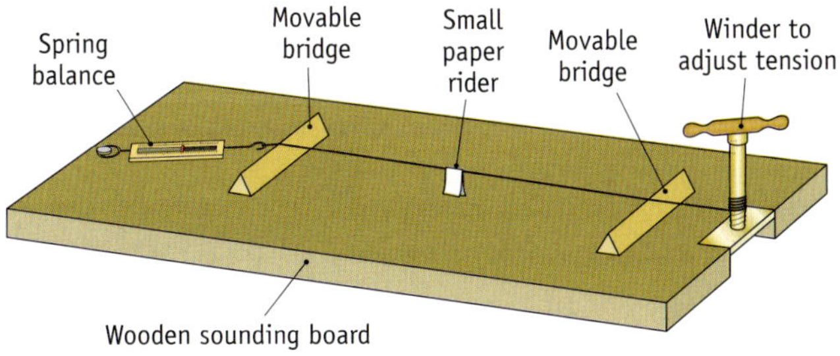

Fig. 17.22
A sonometer.

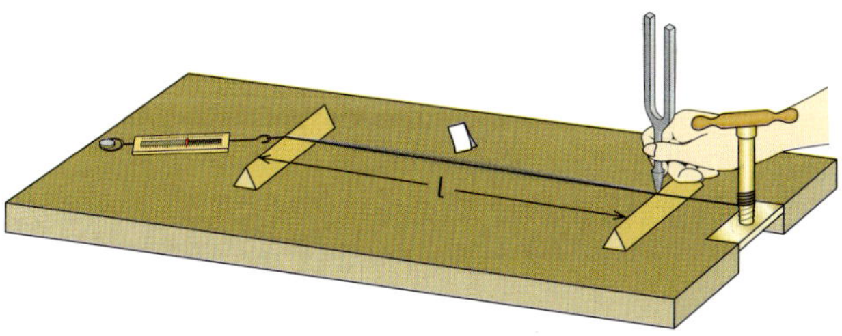

Fig. 17.23
l is the distance from the top of one bridge to the top of the other.

Measuring the Frequency of Vibration of the Wire on a Sonometer

To measure the frequency of vibration of the wire, pluck the wire and strike a tuning fork. Adjust the length or tension of the wire until it sounds to be at the same pitch as the tuning fork. When they are near in pitch, place a small piece of paper on the centre of the wire.

Place the stem of the vibrating tuning fork on one of the bridges. If they are at the same frequency, **resonance** will occur and the wire will start vibrating. The wire will then have an antinode at its centre and the small paper rider will be agitated and jump off the wire (Fig. 17.23 page 199). Vary the length or the tension until this occurs. The **frequency of the wire is then the same as the frequency of the tuning fork**, which is usually written on the fork.

EXPERIMENT SOUND 2

To Investigate the Variation of the Fundamental Frequency of a Stretched String with Length.

Summary of Method

In this experiment you will measure the length l of the wire on a sonometer when it vibrates at the same frequency as a tuning fork of known frequency. You will repeat this for a number of different tuning forks, changing the length of the wire each time. The tension in the wire must be the same each time. You will plot a graph of f against $1/l$. A straight line through the origin will result showing that:

$$f \propto \frac{1}{l}$$

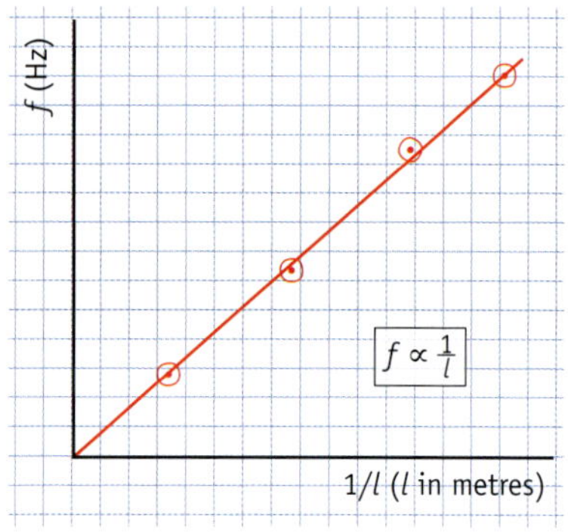

Fig. 17.24

Equipment Needed

- A set of tuning forks of known frequencies
- A sonometer
- A metre stick

Method

1. By moving the bridges on the sonometer, make the wire as long as possible.
2. Strike the tuning fork of lowest frequency on a wooden block and place its stem on one of the bridges. Adjust the tension in the wire until it resonates with the tuning fork.
3. Resonance occurs when a small paper rider placed on the wire is agitated and jumps off the wire when the tuning fork is vibrating. It may take a little practice to find this.
4. The tension **must remain at this value** for the rest of the experiment.
5. Measure the length l of the wire between the tops of the two bridges with the metre stick. Record its value and the frequency of the tuning fork.
6. Take the tuning fork of the next highest frequency. Strike it and place its stem on one of the bridges. Adjust the **length** of the wire (by moving one of the bridges) until the wire again resonates with the fork.
7. Measure the length of the wire between the two bridges with the metre stick. Record its value and the frequency of the tuning fork.
8. Repeat steps 6 and 7 with the remaining tuning forks, ensuring that the value of the tension does not change.
9. Complete the table. Within the limits of experimental error the values in the last column will all be the same, thus verifying that $f \propto 1/l$
10. On graph paper plot a graph of f on the y-axis against $1/l$ on the x-axis.

Result

A straight line through the origin will result (FIG. 17.24) verifying that $f \propto 1/l$

Frequency f/Hz	Length l/m	$\frac{1}{l}$	$f \times l$

Questions

1. Why must the tension be kept constant for each measurement?
2. Why is the piece of paper placed at the centre of the wire and not somewhere else?
3. List two sources of error in this experiment. How may the error be minimised in each case?

HARMONICS ON A STRING

On page 194 you saw that frequencies which are multiples of a certain frequency are called overtones of that frequency. If f is a given frequency, **$2f$ is its first overtone**, **$3f$ is its second overtone**, etc.

The same situation can be described in terms of **harmonics** which are defined as follows:

HARMONICS

Frequencies which are multiples of a certain frequency f are called **harmonics.** f is called the fundamental frequency or the first harmonic. If f is the ***first harmonic***, $2f$ is the ***second harmonic***, $3f$ is ***the third harmonic*** etc ...

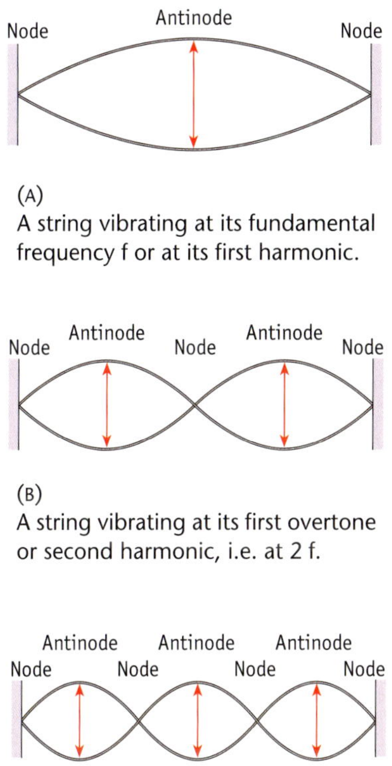

(A) A string vibrating at its fundamental frequency f or at its first harmonic.

(B) A string vibrating at its first overtone or second harmonic, i.e. at 2 f.

(C) A string vibrating at its second overtone or third harmonic, i.e. at 3 f.

Fig. 17.25

Look at FIG. 17.25 which shows stationary waves on a string of fixed length. In FIG. 17.25(B) the wavelength is half what it is in FIG. 17.25(A), thus the frequency is twice as large. Similarly in FIG. 17.25(C), the frequency is three times as large.

FACTORS WHICH DETERMINE THE FUNDAMENTAL FREQUENCY OF A STRING

Experimentally it is found that for a stretched string:

- the greater the length l, the lower the fundamental frequency,
- the higher the tension T, the higher the fundamental frequency,
- the greater its mass per unit length μ, the lower the fundamental frequency.

More precisely:

If T is the tension in the string, μ is its mass per unit length, l its length and f its fundamental frequency of vibration then:

$$f \propto \frac{1}{l} \qquad T \text{ and } \mu \text{ fixed}$$

$$f \propto \sqrt{T} \qquad l \text{ and } \mu \text{ fixed}$$

$$f \propto \frac{1}{\sqrt{\mu}} \qquad T \text{ and } l \text{ fixed}$$

Quantity	Unit
Length l	metre (m)
Tension T	newton (N)
Mass per unit length μ	Kilogram per metre (kg m^{-1})

It follows that:

$$f \propto \frac{1}{l}\sqrt{\frac{T}{\mu}} \Rightarrow f = \frac{k}{l}\sqrt{\frac{T}{\mu}}$$ where k is a constant.

It can be proved that $k = ½$, (you don't need to know the proof) thus:

The fundamental frequency f of a stretched string is given by:

$$f = \frac{1}{2l}\sqrt{\frac{T}{\mu}}$$

Problem 1: A wire of length 3 m and mass 0.6 kg is stretched between two points so that the tension in the wire is 200 N. Calculate its fundamental frequency of vibration.

Solution: The mass per unit length μ of the wire $= \frac{\text{mass}}{\text{length}} = \frac{0.6}{3} = 0.2$ kg m^{-1}

$$f = \frac{1}{2l}\sqrt{\frac{T}{\mu}} = \frac{1}{(2)(3)}\sqrt{\frac{200}{0.2}} = 5.27 \text{ Hz}$$

Problem 2:

(i) If the length of a stretched string is doubled, the tension remaining constant, how does the fundamental frequency of vibration change?

(ii) If the tension T of a stretched string of fundamental frequency f is increases to $9T$, what is the new fundamental frequency?

Solution:

(i) $f \propto \frac{1}{l}$, therefore if the length is doubled the frequency is halved.

(ii) $f \propto \sqrt{T} \Rightarrow f = k\sqrt{T}$, $f_{new} = k\sqrt{T_{new}}$ i.e. $f_{new} = k\sqrt{9T}$

$= 3k\sqrt{T} = 3f$, i.e. the new fundamental frequency is $3f$

EXERCISE 17.2

1. A string of mass per unit length 0.04 kg m^{-1} and length 0.8 m is placed under a tension of 200 N. Calculate its fundamental frequency of vibration.
2. A string on a guitar is vibrating at its fundamental frequency of 500 Hz. Its length is 0.6 m and its mass per unit length is 0.02 kg m^{-1}. Calculate the tension in the string.
3. A wire of length 0.8 m and mass 0.05 kg is stretched between two points, so that the tension in the wire is 100 N. Find its mass per unit length and its fundamental frequency of vibration.
4. A wire of length 4 m and mass 0.04 kg is stretched between two points so that the tension in the wire is 400 N. Calculate its fundamental frequency of vibration.

5. A string on a guitar gives out a note of a certain frequency. By how much must the tension in the string be increased if it is to give a note of twice the frequency, its length remaining the same?
6. When the tension in a stretched string is 40 N, its fundamental frequency is 260 Hz. Find its fundamental frequency if its tension is increased to: (i) 160 N, (ii) 200 N.
7. When the length of a stretched string is 60 cm its fundamental frequency is 460 Hz. Find its fundamental frequency if its length is increased to: (i) 120 cm, (ii) 150 cm.

EXPERIMENT SOUND 3

To Investigate the Variation of the Fundamental Frequency of a Stretched String with Tension.

Summary of Method

In this experiment you will measure the tension in the wire on a sonometer when it vibrates at the same frequency as a tuning fork of known frequency. By changing the tension in the wire, you will repeat this for a number of different tuning forks, the length of the wire being the same each time. You will plot a graph of f against $\sqrt{T}$. A straight line through the origin will result showing that: $f \propto \sqrt{T}$.

Equipment Needed

- A set of tuning forks of known frequencies
- A sonometer
- A metre stick

Method

1. By moving the bridges, make the wire about ⅓ of its maximum length. The **length must not be changed** for the rest of the experiment.
2. Strike the tuning fork of lowest frequency on a wooden block and place its stem on one of the bridges. Adjust the tension in the wire until it resonates with the tuning fork. Resonance occurs when a small paper rider placed on the wire is agitated and jumps off the wire when the tuning fork is vibrating. It may take a little practice to find this.
3. Read the value of the tension from the spring balance. Record the value of the tension in the wire and the frequency of the tuning fork.
4. Repeat steps 2 and 3 with the other tuning forks, the length remaining the same each time.
5. Complete the table. Within the limits of experimental error the values in the last column will all be the same, thus verifying that $f \propto \sqrt{T}$.
6. On graph paper, plot a graph of f on the y-axis against $\sqrt{T}$ on the x-axis.

Result

A straight line through the origin will result (Fig. 17.26) verifying that $f \propto \sqrt{T}$

Frequency f/Hz	Tension T/N	$\sqrt{T}$	$f/\sqrt{T}$

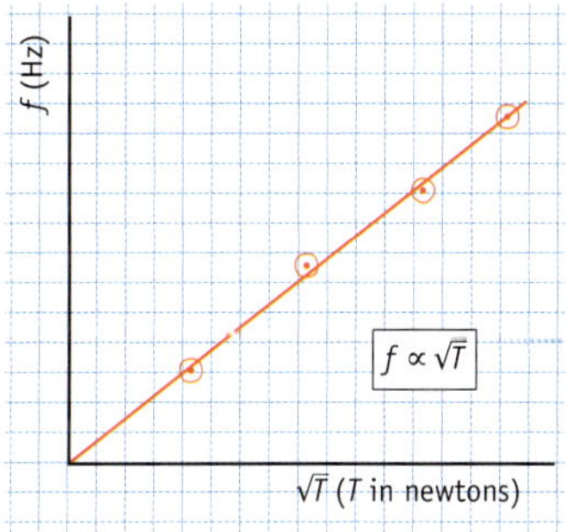

Fig. 17.26

Questions

1. Why must the length be kept constant for each measurement?
2. Why is the piece of paper placed at the centre of the wire and not somewhere else?
3. List two sources of error in this experiment. How may the error be minimised in each case?
4. Given the length of the wire, how could you use the graph to find its mass per unit length?

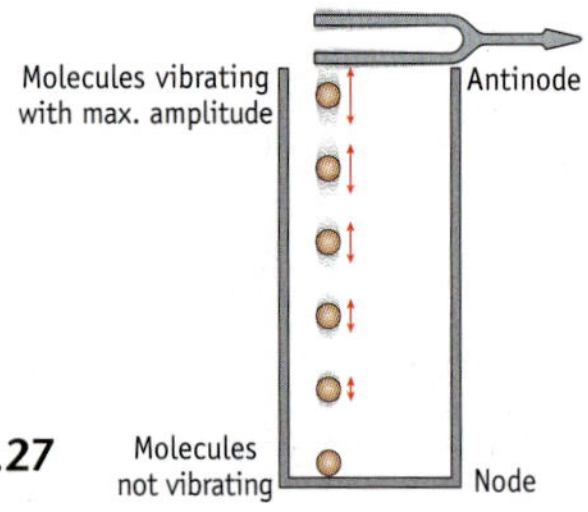

Fig. 17.27

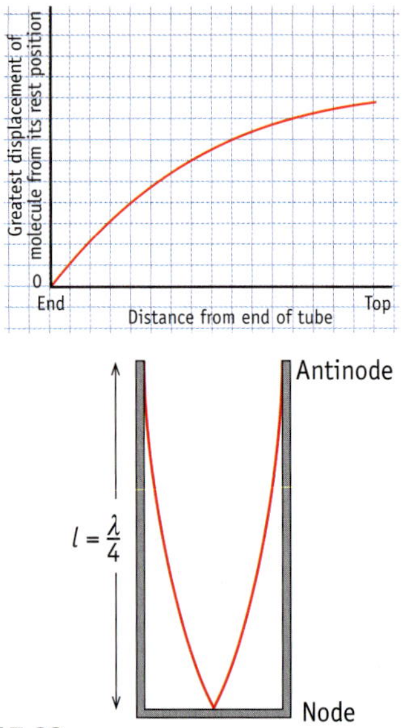

(A)

(B)

Fig. 17.28

STATIONARY SOUND WAVES IN A PIPE CLOSED AT ONE END

A pipe which is closed at one end and open at the other is known as a closed pipe. Suppose a vibrating tuning fork emitting a sound wave of frequency f is placed above its open end. A longitudinal sound wave travels down the pipe, is reflected from the end and travels back up the pipe. The incident and reflected waves interfere with each other. If the length of the pipe is varied, at certain lengths, **resonance** will occur and a stationary longitudinal wave is set up in the pipe. FIG. 17.27 shows the simplest type of stationary wave that can occur. There is a **node** at the bottom of the pipe. As you move up towards the top of the pipe the amplitude of the vibration of the molecules increases, reaching maximum value at the top, with the formation of an **antinode**. The antinode can be heard as a loud sound. If we draw a graph of the maximum displacement of air molecules from their rest positions along the y-axis and the distance of the molecule from the bottom of the tube along the x-axis we get a graph similar to that in FIG. 17.28(A). For convenience, a double version of this graph is drawn in the pipe itself to show the stationary wave (FIG. 17.28(B)). Since the distance between a node and the next antinode in a stationary wave is $\lambda/4$, the length of pipe $= \lambda/4$, i.e. $l = \lambda/4$.

When this simplest type of stationary wave is set up in a pipe, the longer the pipe, the lower frequency of the note emitted.

FIG. 17.29(B) and (C) shows the next simplest stationary waves that can occur in a closed pipe.

HARMONICS IN A PIPE CLOSED AT ONE END

In FIG. 17.29(A) the frequency f of the note emitted is:

$$f_1 = \frac{c}{\lambda} = \frac{c}{4l} \quad \text{i.e.} \quad f_1 = \frac{c}{4l}$$

FIG. 17.29(B) shows the stationary wave of next highest frequency f_2 that can be set up in the pipe.

$$\text{Here } l = \frac{3\lambda}{4} \quad \text{Thus, } f_2 = \frac{c}{\lambda} = \frac{c}{\frac{4l}{3}} = 3\left(\frac{c}{4l}\right) \Rightarrow f_2 = 3f_1$$

Similarly, the stationary wave of next highest frequency f_3 is given by: $f_3 = 5f_1$ Thus we see the following:

NOTE

In a pipe closed at one end only ***odd numbered harmonics*** may be present.

The clarinet, the trombone and the saxophone are examples of **musical instruments** in which a column of air resonates in a pipe closed at one end. As with string instruments, the longer the pipe, the lower the note emitted.

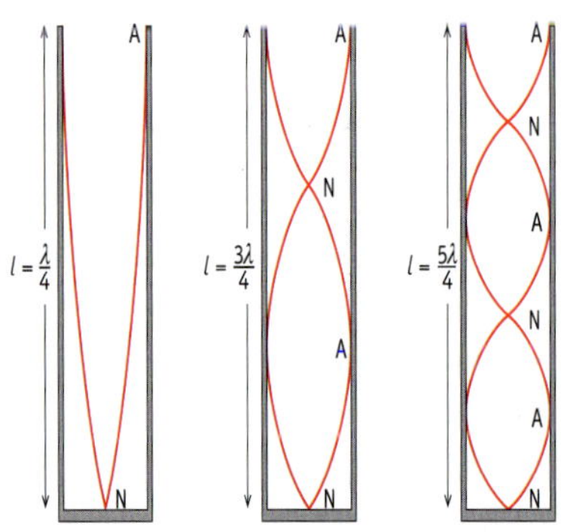

Fig. 17.29
Stationery waves in a pipe closed at one end.

MEASUREMENT OF THE SPEED OF SOUND IN AIR

We can use stationary waves in a pipe to measure the speed of sound in air. However, we must take note of one fact that we so far ignored, namely that the antinode lies a small distance – called the end correction – outside the pipe. It can be shown that the end correction is approximately equal to $0.3d$, where d is the internal diameter of the pipe. Thus we have: $\lambda/4 = l + 0.3d$ instead of $\lambda/4 = l$.

Thus: $\lambda = 4(l + 0.3d)$. l and d can be measured, therefore λ can be found. Since $c = f\lambda$, c can be found. The combined formula is: $c = 4f\,(l + 0.3\,d)$.

EXPERIMENT

TO MEASURE THE SPEED OF SOUND IN AIR USING A RESONANCE TUBE.

Summary of Method

In this experiment you will place a vibrating tuning fork of known frequency f above the column of air in a resonance tube. You will find and measure the length l of the shortest column that resonates with the given tuning fork. You will measure the internal diameter d of the resonance tube.

Using the formula $c = 4f\,(l + 0.3d)$ you will find c, the speed of sound in air.

Equipment Needed

- A large graduated cylinder
- A resonance tube (or tubes)
- A retort stand and clamp
- A set of tuning forks
- A metre stick

Method

1. Set up the resonance tube as in FIG. 17.30 with the air column a few cm long.
2. Strike the tuning fork of the highest frequency and hold it just above the tube.
3. Adjust the length of the tube until the sound becomes loud, i.e. resonance occurs. By making further fine adjustments to the length of the tube above the water, locate the length that gives maximum loudness. Restrike the tuning fork as necessary while doing this.
4. With the metre stick measure the length l of the air column, i.e. the distance from the top of the water to the top of the tube. Record its value and the frequency f of the tuning fork.
5. Repeat the above with tuning forks of different frequencies.
6. With the metre stick measure the internal diameter d of the tube. Record this.
7. Complete the Table and calculate an average value for c, the velocity of sound in air.

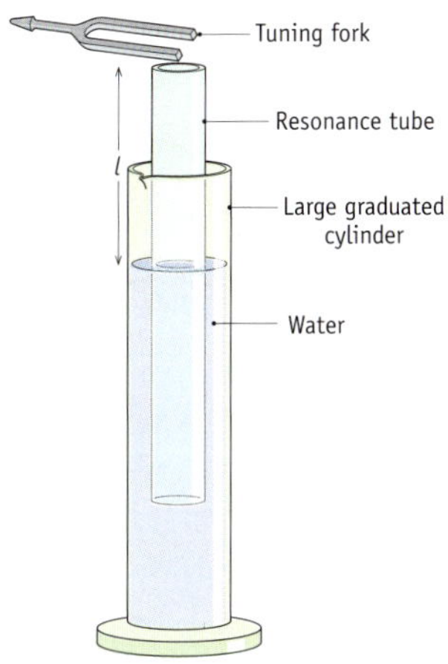

Fig. 17.30

Diameter of tube: d =

Frequency of tuning fork f / Hz	Resonance length l / m	c / m s^{-1} $c = 4f\,(l + 0.3d)$

Handling the Data Graphically

On graph paper plot a graph of l (on the y-axis) against $1/f$ (on the x-axis). A straight line should result (FIG. 17.31). Measure the slope of the graph.

Speed of sound $c = 4 \times$ slope of graph. This is because:

$$c = 4f\,(l + 0.3d) \;\Rightarrow\; \frac{c}{4f} = l + 0.3d \;\Rightarrow\; l = \left(\frac{c}{4}\right)\left(\frac{1}{f}\right) - 0.3d$$

Comparing this with: $y = mx + c$ we see that a graph of l (on y-axis) against $\frac{1}{f}$ (on x-axis) is a straight line of slope $c/4 \;\Rightarrow\; c = 4 \times \text{slope}$

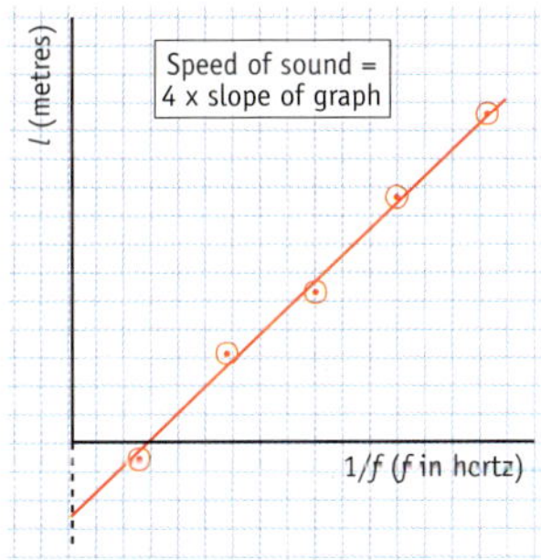

Fig. 17.31

Errors occur in locating the position of loudest sound and in measuring l.

Be careful with parallax error when using the metre stick.

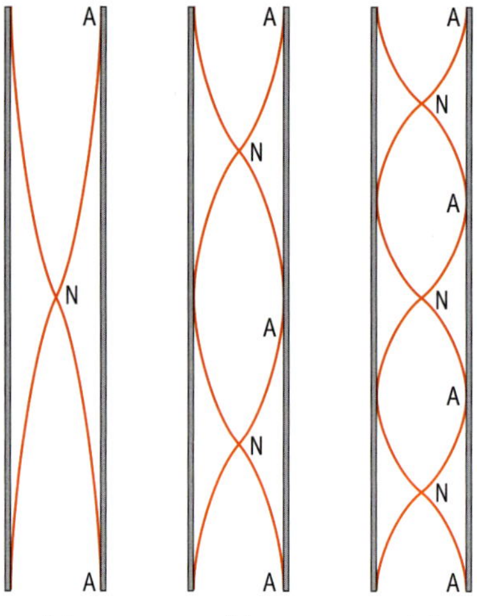

Fig. 17.32
Stationary waves in a pipe open at both ends.

Stationary Waves in a Pipe Open at Both Ends

Stationary waves can also be set up in an open pipe, i.e. in a pipe open at both ends. If a loudspeaker from a signal generator is placed in front of one end of a such a pipe and the frequency of the sound from the generator adjusted, at certain values resonance will be heard. The same thing could be done with a tuning fork except the length of the pipe would have to be changed, e.g. by sliding one pipe inside another. In an open pipe there must be an antinode at each end. Fig. 17.32 shows the three simplest stationary waves that can occur. The wavelength in (B) is half the wavelength in (A). Likewise the wavelength in (C) is one third that in (A). Thus the frequencies are: f, $2f$, $3f$ etc.

From the above we see that:

> **NOTE**
> In an open pipe ***all harmonics*** may be present.

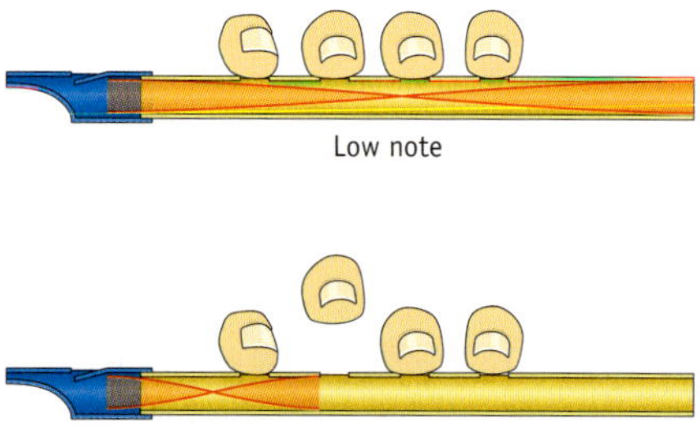

Fig. 17.33
In the tin whistle, which is a pipe open at both ends, the frequency of the standing wave and hence the note produced is varied by blocking or unblocking holes along the pipe.

The flute, the tin whistle (Fig. 17.33) and the recorder are examples of musical instruments in which a column of air resonates in a pipe open at both ends.

CHAPTER CHECKLIST

- **State**: The characteristics of notes; The wave property on which each characteristic depends; The unit of sound intensity; The three factors on which the fundamental frequency f of a string depends; The relationship between f and each factor.
- **Define**: Overtones; Frequency limits of audibility; Resonance; Natural frequency; Threshold of hearing; Fundamental frequency of a string; Sound intensity; Harmonics.
- **Recall** that: Sound is a longitudinal wave; Noise can be reduced by destructive interference; The ear is less responsive at higher and lower frequencies; Sound intensity level is measured in decibels; Frequency of a stretched string is inversely proportional to its length $\left(f \propto \frac{1}{l}\right)$; dBA (decibel adapted) scale is used in a sound level meter; In a closed pipe there is a node at the fixed end an an antinode at the open end; In an open pipe there is an antinode at each end; Only odd-numbered harmonics may be present in a closed pipe; All harmonics may be present in a pipe open at each end; Doubling the sound intensity increases the sound intensity level by 3 dB.
- **Describe** and carry out an experiment to: Show that sound travels as a wave; Show that sound needs a medium; Show resonance; Investigate the variation of fundamental frequency of a stretched string with length; Measure the speed of sound in air; Investigate the variation of fundamental frequency of a stretched string with tension.
- **Explain**: Why sound travels better at night or over water; Why the dBA scale is used in a sound level meter; Why you should protect your ears from loud sounds; The term 'acoustics'.
- **List**: Examples of the reflection, refraction, diffraction and interference of sound; Examples of resonance; Examples of a musical instrument based on a closed pipe; Examples of a musical instrument based on an open pipe; Practical examples of ultrasonic waves.
- **Recall and use the formula** $f = \frac{1}{2l}\sqrt{\frac{T}{\mu}}$ to solve problems.

The Wave Nature of Light

CHAPTER 18

The Spectrometer

The spectrometer (Fig. 18.1) is an optical instrument used in experiments to examine spectra and to measure the wavelength of light. It consists of the following parts:

- A **heavy base** onto which the collimator is fixed. A circular scale graduated in degrees is fixed on the base.
- A **turntable** which is free to rotate about a vertical axis through the centre of the base. The turntable may be levelled by means of three levelling screws attached to it.
- A **collimator** which is made up of two tubes. The larger tube has a convex lens at one end and the narrower tube has a slit at one end. The distance from the slit to the lens can be adjusted by sliding one tube in the other. The width of the slit can be adjusted by means of an adjusting screw. When in use, the distance from the slit to the lens is equal to the focal length of the lens. Thus **light that enters the collimator through the slit comes out through the lens as a parallel beam** (Fig. 18.2). This is the function of the collimator. The word 'collimate' means 'to make parallel'.
- An **astronomical telescope** which is free to rotate about the same axis as the turntable. A vernier scale is attached to the telescope so that the angle through which it rotates relative to the fixed circular scale can be measured accurately. When in use, the slit of the collimator is illuminated by a light source. The telescope is used to view an image of this slit.

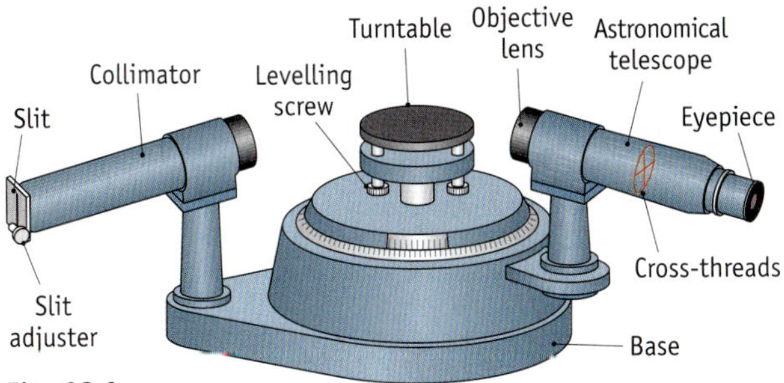

Fig. 18.1

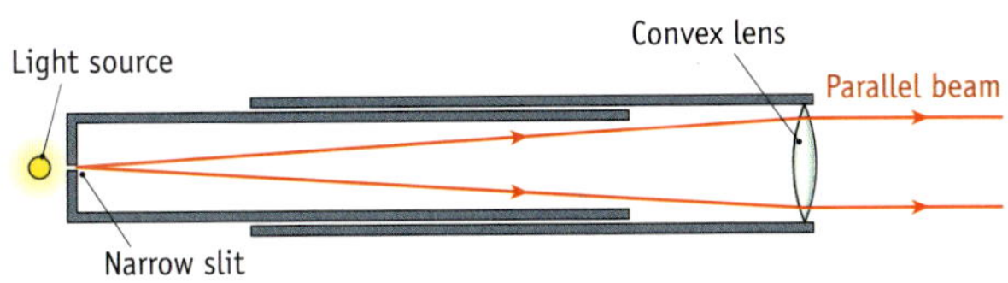

Fig. 18.2
The collimator adjusted for use.

The telescope has two convex lenses. One is called the objective lens and the other the eyepiece. The telescope consists of three tubes. The smallest tube contains the **eyepiece**, the next the **cross-threads** and the largest tube contains the **objective lens**.

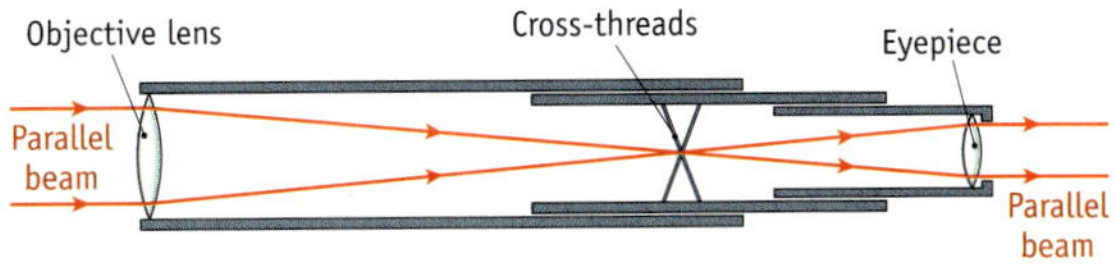

Fig. 18.3
The astronomical telescope adjusted for use.

When in use, the telescope is adjusted so that when you look into the eyepiece, the cross-threads are in focus and a distant object would also be in focus if it were viewed. When this happens, the cross-threads are at the focus of the objective lens and at the focus of the eyepiece (Fig. 18.3)

ADJUSTMENTS TO THE SPECTROMETER BEFORE USE

- Look into the eyepiece. Move the eyepiece relative to the cross-threads **until the cross-threads are in focus**.
- View a distant object with the telescope and move the cross-threads and eyepiece together relative to the objective lens so that **the distant object appears in focus**.
- Illuminate the slit with light and view the slit through the telescope. Move the slit relative to the collimator so that the **slit is in focus**.
- Adjust the **width of the slit**, if it is too wide or too narrow.
- **Level the turntable** by means of the levelling screws.

THE WAVE NATURE OF LIGHT

The Dutch scientist Christiaan Huygens (1629–95) first proposed that light travelled as a wave. At the time others – in particular Isaac Newton – considered light to be a stream of very fast moving particles and this view predominated. Huygens had no way of showing that light was a wave rather than a stream of particles.

If light does travel as a wave then it should show all the characteristic properties of waves, i.e. it should undergo: **reflection**, **refraction**, **interference** and **diffraction**. You saw in Chapters 2, 3 and 4 that light undergoes reflection and refraction. However, a beam of particles could bounce off an object obeying the laws of reflection and could also be made change its direction i.e. undergo refraction, under the right conditions. Thus **to show that light travels as a wave we must show that it undergoes diffraction or interference**. The English scientist Thomas Young was one of the first to demonstrate that light does undergo interference and diffraction, and thus is a wave.

EXPERIMENT

TO DEMONSTRATE THE WAVE NATURE OF LIGHT.

This experiment is similar to that done by Young in 1802. It shows that **light undergoes interference and diffraction**.

Method

- Set up the equipment as in FIG. 18.4. S is a monochromatic light source, i.e. it is light of one particular wavelength only. A sodium vapour lamp is ideal.

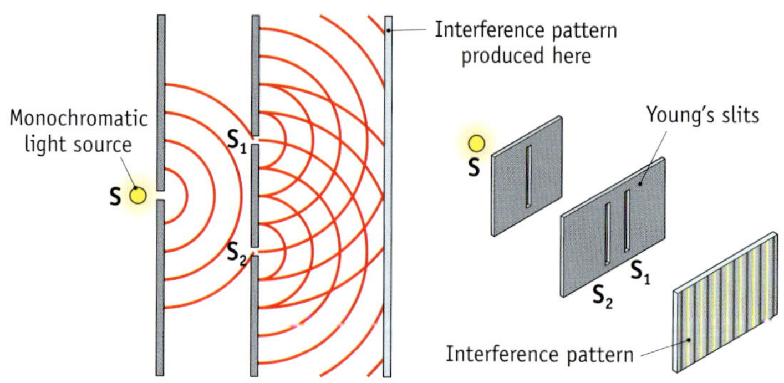

Fig. 18.4
Young's experiment.

- Light from S shines onto the first narrow slit. It undergoes diffraction here and illuminates the pair of slits S_1 and S_2. (These two slits are known as **Young's slits**.)
- **Diffraction occurs** at each of theses slits and in the region to the right of them, where light from each overlaps, interference occurs.
- Since S_1 and S_2 are coherent light sources, the **interference pattern** produced is similar to that on page 182.

- Since the sources are line sources, the interference pattern consists of a series of parallel bright and dark lines, known as interference **fringes**. The fringes are close together and are not very sharp. They may be seen by looking at the slits or can be put on a screen. Viewing the slits with the telescope of a spectrometer is ideal for this (FIG. 18.5). In 1802, from an arrangement like this, Young was able to estimate that the wavelength of light was in the region of 10^{-7} m.

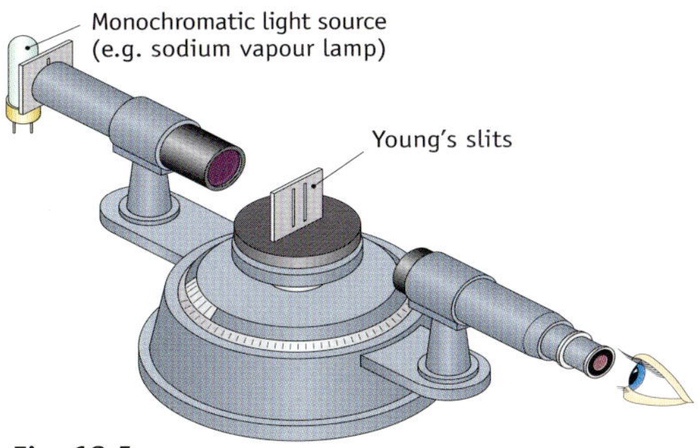

Fig. 18.5

Conclusion

The fact that light shows interference and diffraction tells us that light travels as a wave.

Young's experiment may be demonstrated much more easily and clearly using a **laser** as in FIG. 18.6. The interference pattern produced can be easily seen on a screen or wall. Dust or smoke will show up the beams of light.

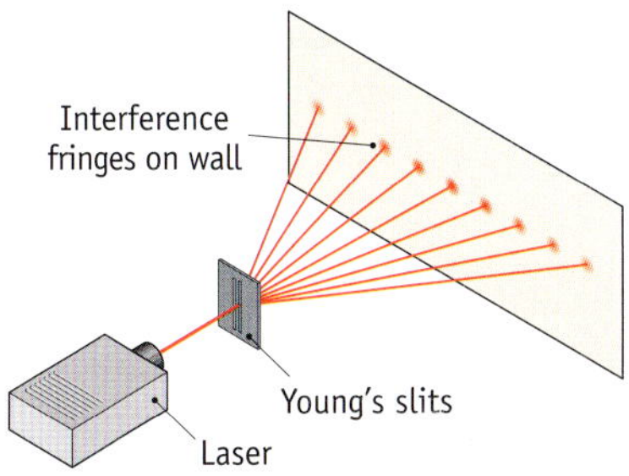

Fig. 18.6

Wavelength and Colour

Light travels as a wave. It is an example of **an electromagnetic wave**. If the wavelength of an electromagnetic wave is between 4×10^{-7} m and 7×10^{-7} m, the wave affects the human eye and is called 'visible light'. A different colour is seen if light of a different wavelength strikes the eye (FIG. 18.7). Normal white light is a mixture of different wavelengths. Light which is not a mixture of different wavelengths is called **monochromatic light** (it cannot be broken down into a number of colours). The light from the sodium vapour lamp used in the above experiment is monochromatic. It has a wavelength of 5.9×10^{-7} m and is orange/yellow in colour.

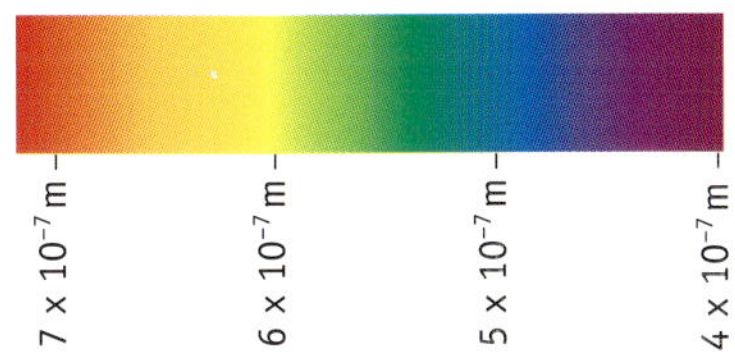

Fig. 18.7

Colours Produced by Interference – Soap Bubbles and Petrol Films

Fig. 18.8
Interference colour on a 35-metre long bubble.

The colours sometimes seen in soap bubbles (FIG. 18.8) or on a film of petrol on water, are due to interference of light waves. FIG. 18.9 is a diagram of a film of petrol on water. When light falls on such a film some is reflected from the top surface of the film and some is reflected from the surface of the water. When light from each surface meets, interference occurs. Different wavelengths are refracted at different angles in the petrol. Depending on the angle at which you view the film or on the thickness of the film, different wavelengths interfere constructively whereas others do not. The colour of the light which is constructively interfering is the colour seen. A similar situation occurs for soap films.

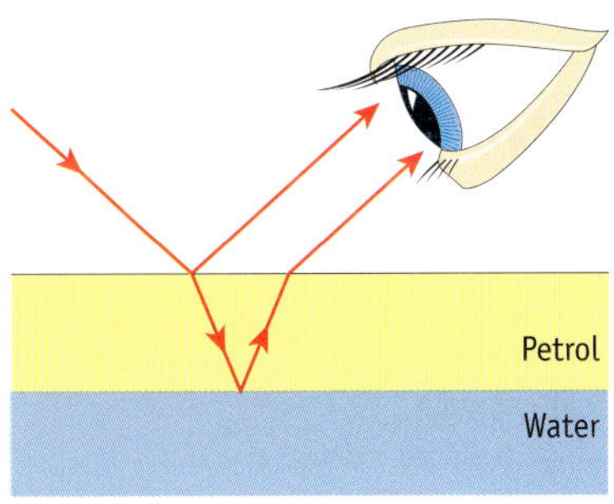

Fig. 18.9

The Diffraction Grating

It is difficult to measure the wavelength of light accurately with Young's slits because the interference fringes are not very clear and are very close together. If, however, the number of slits is increased from two to a few hundred per millimetre, a much clearer pattern is formed. A large number of slits arranged like this is called a **diffraction grating**.

A typical diffraction grating consists of a piece of a transparent medium on which a very large number of parallel lines are engraved. Light cannot pass through these lines. The spaces between the lines behave as slits and allow the light to pass through. Thus the grating has a large number of parallel slits on it. The width of each slit is such that diffraction occurs at it.

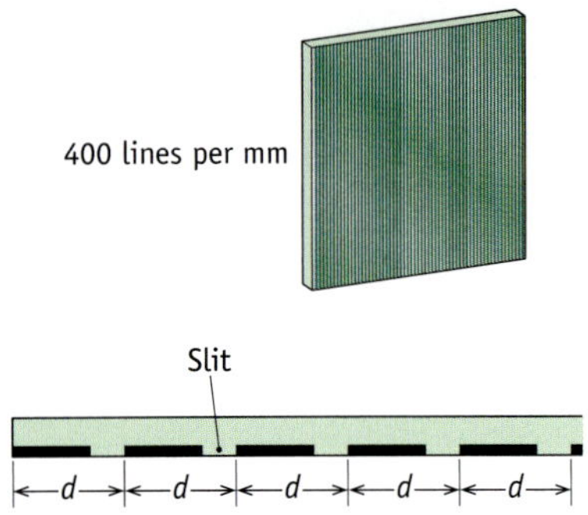

Fig. 18.10

THE GRATING CONSTANT (The Grating spacing)
The distance d between two adjacent slits (i.e. the width of one line and one slit) on the grating is called the ***grating constant*** or the grating spacing (FIG. 18.10).

Problem 1: A fine diffraction grating has 400 lines per mm ruled on it. Find the grating constant d.

Solution: From FIG. 18.10 it is obvious that if the grating has 400 lines per mm it also has 400 slits per mm.

Thus one line and one slit take up $\frac{1}{400}$ mm

$$\Rightarrow \quad d = \frac{1}{400} \text{ mm} = 0.0025 \text{ mm} = 2.5 \times 10^{-6} \text{ m}$$

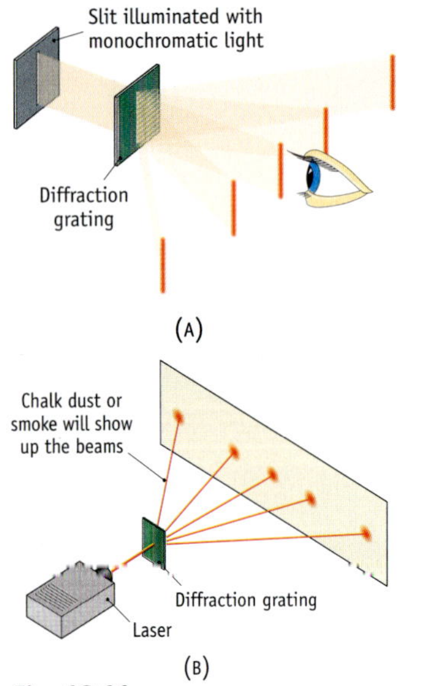

Fig. 18.11

In general, if a grating has n lines per mm then $d = \frac{1}{n}$ millimetres

WHAT HAPPENS WHEN A BEAM OF MONOCHROMATIC LIGHT FALLS NORMALLY ON A DIFFRACTION GRATING?

In FIG. 18.11(A) and (B) a beam of monochromatic light (i.e. of one wavelength) strikes a diffraction grating at right angles (i.e. normally). **Diffraction occurs at each slit** because the width of each slit is small. At the other side of the grating the diffracted **waves overlap and interference occurs**. A number of **bright images are formed due to constructive interference** and can be seen there with the eye. They can also be projected on a screen. The following sections show how these images are formed.

LIGHT THAT PASSES STRAIGHT THROUGH THE GRATING

Consider two corresponding points A and B on adjacent slits (FIG. 18.12). Waves leave these points in phase. Waves from other pairs of corresponding points on these two slits are also in phase. If these waves are brought together – e.g. by means of a convex lens or by the lens in your eye – they will undergo **constructive interference**. If all the light travelling in this direction from every slit is brought together – by means of a convex lens – it will all be in phase and produce a very bright line called the **zero order diffracted image**.

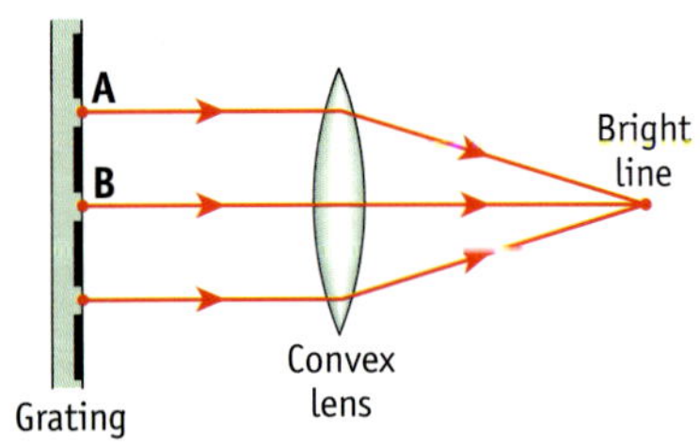

Fig. 18.12
Zero order image; n = 0.

LIGHT THAT PASSES THROUGH THE GRATING MAKING AN ANGLE θ WITH THE NORMAL TO THE GRATING

Suppose the value of θ is chosen so that in FIG. 18.13 $|PQ| = \lambda$, i.e. the path difference between light from corresponding points on adjacent slits is exactly one wavelength. If light from adjacent slits travelling in this direction is brought together, it will undergo constructive interference.

From the diagram it can also be seen that light from corresponding points on one slit and any other slit is also a whole number of wavelengths out of phase. Thus if all the light travelling in this direction from every slit is brought together – by means of a convex lens – it will undergo constructive interference giving another bright line. The same thing happens at the same angle at the other side of the normal to the grating. The images formed are called the **first order diffracted images**.

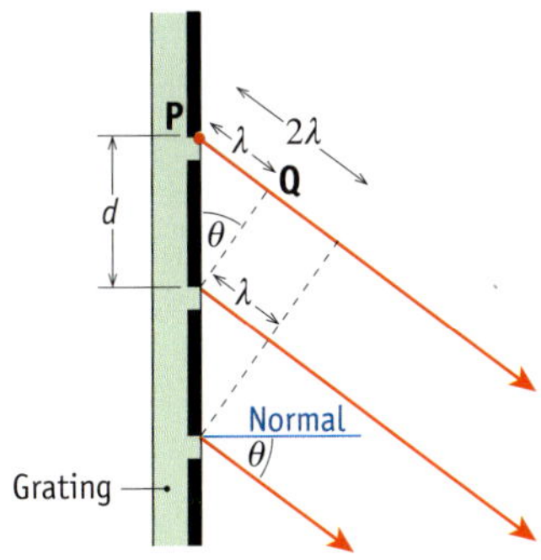

Fig. 18.13
First order image; n = 1.

The same thing happens if the path difference between light from corresponding points on adjacent slits is exactly two wavelengths (2λ). In FIG. 18.14 $|PQ| = 2\lambda$

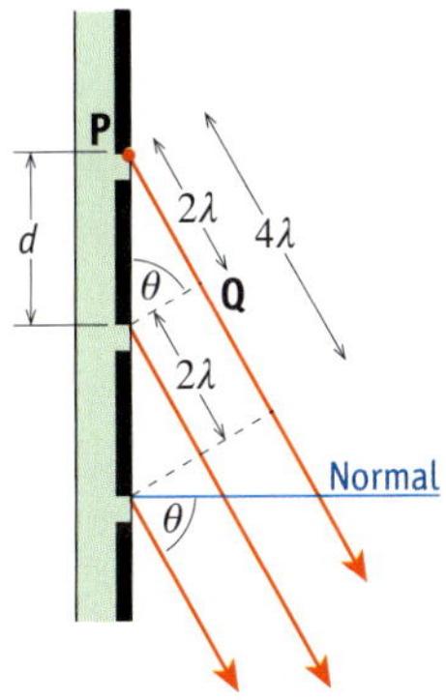

Fig. 18.14
Second order image; n = 2.

In general, constructive interference will occur at an angle θ if light from corresponding points on adjacent slits is a whole number of wavelengths out of phase. Thus for a bright line: $|PQ| = n\lambda$, where n is any whole number.

But $\text{Sin}\ \theta = \dfrac{|PQ|}{d}$ Thus we have:

For the nth bright line: $$n\lambda = d\ \text{Sin}\ \theta$$

When $n = 1$ the image formed is called the **first order diffracted image**, when $n = 2$ the image is called the **second order diffracted image** etc. For other angles some light produces constructive interference and other light produces destructive interference. Thus no bright images are seen.

HOW MANY DIFFRACTED IMAGES ARE FOUND AT EACH SIDE OF THE CENTRAL IMAGE?

Solving the equation $n\lambda = d\ \text{Sin}\ \theta$ for n gives: $n = \dfrac{d\ \text{Sin}\theta}{\lambda}$

The maximum value of Sin θ is 1, therefore the maximum value of n is

$$n_{\text{max}} = \frac{d}{\lambda}$$

Thus, for a given λ, the larger the distance (d) the larger the number of diffracted images (n). Typically there will be two, three or four images at each side depending on the grating being used.

Problem 2: Red light falls normally onto (i.e. at right angles to) a diffraction grating. The grating has 400 lines per mm engraved on it and the second order diffracted image is at 30° from the straight through position. Find the wavelength of the light.

Solution:

400 lines per mm $\Rightarrow d = \dfrac{1}{400}$ mm $= 0.0025$ mm $= 2.5 \times 10^{-6}$ m

$$n\lambda = d\ \text{Sin}\ \theta \Rightarrow \lambda = \frac{d\ \text{Sin}\ \theta}{n} = \frac{(2.5 \times 10^{-6})(\text{Sin}\ 30^\circ)}{2} \Rightarrow \lambda = 6.25 \times 10^{-7}\ \text{m}$$

Problem 3: A diffraction grating has 350 lines per mm ruled on it. Monochromatic light of wavelength 5.2×10^{-7} m is incident normally on it. What is the highest order diffracted image formed?

Solution: $n\lambda = d \text{ Sin } \theta$. Maximum value of n occurs when Sin θ is maximum, i.e. when Sin $\theta = 1$

$$\Rightarrow \quad n_{max} = \frac{d}{\lambda} \qquad d = \frac{1}{350} \text{ mm} = 2.857 \times 10^{-6} \text{ m}$$

$$\Rightarrow \quad n_{max} = \frac{(2.857 \times 10^{-6})}{(5.2 \times 10^{-7})} = 5.49$$

$\Rightarrow$ Max value of n is 5 i.e. the 5th order is the highest.

Problem 4: A diffraction grating having 100 lines per mm is placed between a monochromatic light source which gives out parallel light and a screen as in FIG. 18.15. The distance from the grating to the screen is 2 m. Bright spots are seen on the screen at the points A, B, C, D, E, F, G, H and I. If the distance $|AI|$ = 80 cm, calculate the wavelength of the source.

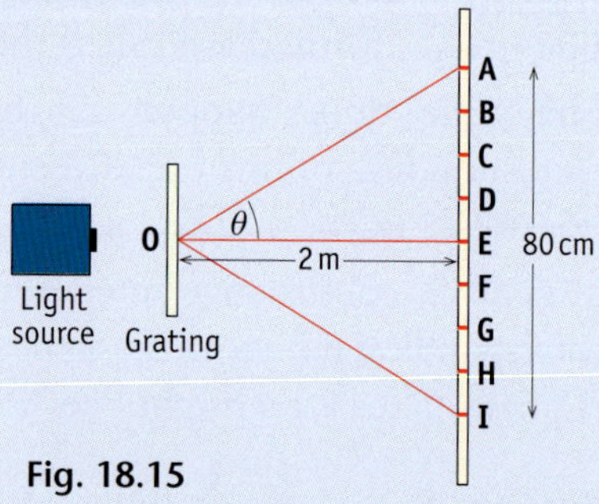

Fig. 18.15

Solution: 100 lines per mm $\Rightarrow$ $d = \frac{1}{100}$ mm $= 1 \times 10^{-5}$ m

$$\text{Sin } \theta = \frac{|AE|}{|OA|} = \frac{0.4}{\sqrt{2^2 + (0.4)^2}} = 0.1961$$

$$n\lambda = d \text{ Sin } \theta \Rightarrow \lambda = \frac{d \text{ Sin } \theta}{n} = \frac{(1 \times 10^{-5})(0.1961)}{4} = 4.90 \times 10^{-7} \text{ m}$$

EXERCISE 18.1

1. Monochromatic light falls onto a diffraction grating. It strikes the grating at right angles. The grating spacing d (the grating constant) is 2.8×10^{-6} m. The second order image is at an angle of 65° with the straight through position. What is the wavelength of the light?
2. Monochromatic light falls onto a diffraction grating. It strikes the grating at right angles. The grating spacing d (the grating constant) is 2.5×10^{-6} m. The first order image is at an angle of 40° with the straight through position. What is the wavelength of the light?
3. A diffraction grating has 200 lines per mm ruled on it. Find d, the grating constant.
4. A fine diffraction grating has 500 lines per mm ruled on it. Find the value of the grating constant.
5. Monochromatic light falls normally on (i.e. at right angles to) a diffraction grating. The grating has 800 lines per mm engraved on it and the first order image is at an angle of 30° from the straight through position. What is the wavelength of the light?
6. When a parallel beam of monochromatic light is incident normally on a diffraction grating having 400 lines per mm on it, the angle between the 2nd order image and the normal to the grating is 31°. What is the wavelength of the light?
7. A diffraction grating having 200 lines per mm is placed between a monochromatic light source (e.g. a laser) and a screen as in FIG. 18.16. The distance from the grating to the screen is 2.5 m. Bright spots are seen on the screen at the points A, B, C, D and E. If the distance $|AE|$ = 60 cm, calculate the wavelength of the source.

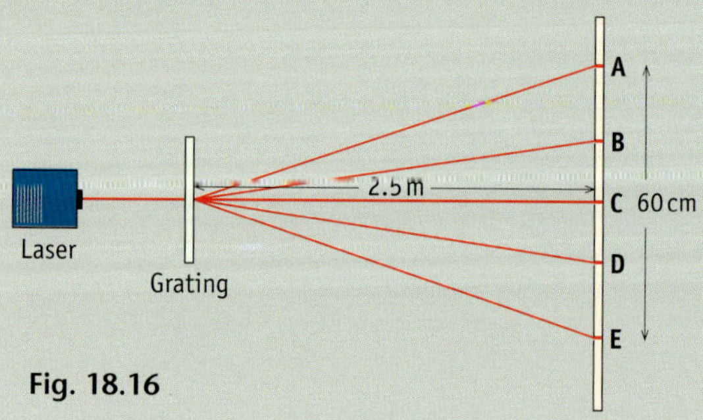

Fig. 18.16

8. A beam of monochromatic light of wavelength 5×10^{-7} m is incident normally on a diffraction grating that has 600 lines per mm ruled on it. A screen is placed 0.4 m from the grating. Find the distance of the second order image from the central image.

9. A diffraction grating has 400 lines per mm ruled on it. Monochromatic light of wavelength 6.2×10^{-7} m is incident normally on it. What is the highest order diffracted image formed?

10. A diffraction grating has 200 lines per mm ruled on it. Monochromatic light of wavelength 6.2×10^{-7} m is incident normally on it. What is the highest order diffracted image formed?

EXPERIMENT LIGHT 5

To Measure the Wavelength of Monochromatic Light.

Summary of Method

In this experiment you will shine light from a sodium vapour lamp onto the slit of the collimator of a spectrometer. Light from the illuminated slit comes out of the collimator as a parallel beam and falls onto a diffraction grating. With the telescope you will locate the diffracted images of the slit. By measuring the angle θ that each image makes with the straight through position, and knowing the grating constant *d*, you will calculate the wavelength of the sodium light using the formula: $n\lambda = d \sin\theta$

Equipment Needed

- A diffraction grating whose grating constant (grating spacing, grating separation) is known
- A sodium vapour lamp
- An HT power supply unit
- A spectrometer

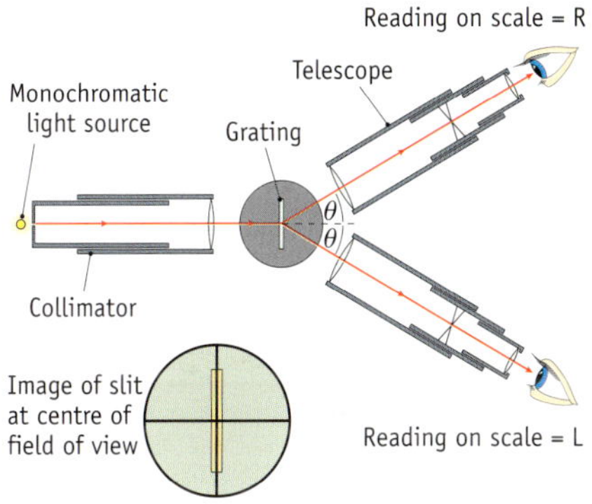

Fig. 18.17

Method

1. Set up and switch on the sodium vapour lamp. Allow it time to warm up until it gives out its characteristic orange/yellow colour.
2. Adjust the spectrometer as follows:
   - Adjust the eyepiece until the cross-threads are clearly in focus.
   - View a distant object with the telescope and adjust its length until the object is in focus.
   - Place the sodium light source behind the slit of the collimator. Ensure the slit is vertical. View the slit with the telescope. Adjust the position of the slit in the collimator until a sharp clear image of the slit can be seen in the telescope.
   - Adjust the width of the slit until it is as narrow as possible, but still bright enough to be seen clearly in the telescope.
   - Level the turntable by means of the levelling screws.
3. Put the grating in its holder and place the holder on the turntable with the lines of the grating vertical and parallel to the slit. Make sure the grating is at right angles to the axis of the collimator.
4. Locate the first order diffracted image at one side. Adjust the levelling screws until the image can be seen in the centre of the field of vision as in Fig. 18.17. Check also the first order image at the other side.
5. Locate the first order diffracted image at one side and place it at the centre of the field of view. Measure and record the reading *L* on the circular scale.
6. Locate the first order image at the other side. Measure and record the reading *R* on the circular scale.
7. Repeat steps 5 and 6 for the second, third etc. ... order diffracted images.

8. Find the grating constant *d*. It should be written somewhere on the grating or its surrounds.
9. Complete the table, calculating the wavelength of the light for each value of *n*.
10. Calculate the average value of the wavelength.

Grating constant d = metres

Order of image n	Reading of angle on left $L/°$	Reading of angle on right $R/°$	$\theta/°$ $\theta = \frac{\lvert L-R \rvert}{2}$	Wavelength λ /m $\lambda = \frac{d \sin\theta}{n}$
1				
2				
3				

Average value of Wavelength λ =

Sources of Error

Errors occur:

1. in adjusting the spectrometer before use,
2. if the slit too wide,
3. if the grating is not at right angles to the light from the collimator or the table is not level,
4. in reading the vernier scale,
5. in judging the cross-threads to be at the centre of the image of the slit.

Questions

1. If the images seen in the telescope of a spectrometer are too faint what adjustment should be made to the spectrometer?
2. If the cross wires are unclear, what adjustment should be made to the spectrometer?
3. If the images seen on one side of the central image are above the centre of the eyepiece and those on the other are below the centre, what adjustment should be made to the spectrometer?

EXERCISE 18.2

1. What is the reading indicated by the spectrometer scales shown in Fig. 18.18?
2. In an experiment to measure the wavelength of light emitted by a sodium lamp, a spectrometer is used in conjunction with a diffraction grating which has 5×10^5 lines per metre. With the telescope in the straight through position, the zero order image is observed at the 200° mark on the spectrometer scale. The first order images are then observed at 182.8° and 217.2°. Calculate the wavelength of sodium light.
3. In using a spectrometer to measure the wavelength of sodium light with a diffraction grating which has 500 lines per mm, on it the following readings were noted for the positions of the images: 243° 30'; 217° 15'; 200°; 182° 45'; 163° 30'. One of the angles was read wrongly. Which one was it? Give a reason for your answer. Calculate the wavelength of the light.

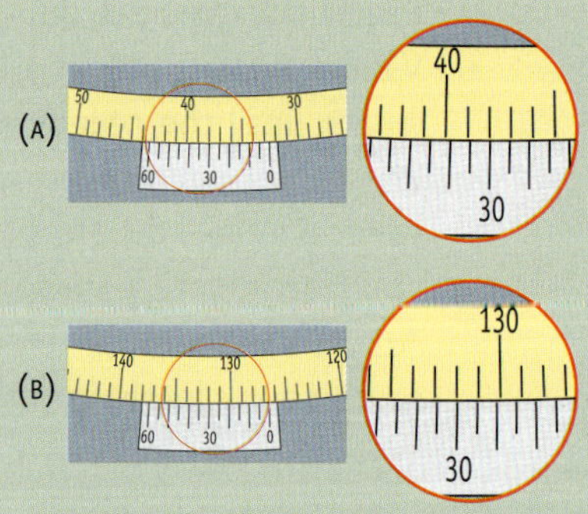

Fig. 18.18

POLARISATION

Light (which is an electromagnetic wave) consists of a vibrating electric and magnetic field. When a substance gives out light, electrons in the substance emit the electromagnetic waves. Light emitted from a particular point on it is emitted as a pulse that lasts for about 10^{-9} seconds. In that pulse the electric field component is vibrating in a particular direction. The direction in which the electric field of the next pulse vibrates is different and changes randomly with each pulse. Other points on the substance are also emitting pulses, which also change their direction of vibration roughly every 10^{-9} s. The pulses from different points are vibrating in different directions. **Thus light from an incandescent source is unpolarised**, i.e. the electric and magnetic fields are vibrating in many different planes. This is represented on a diagram as in FIG. 18.19.

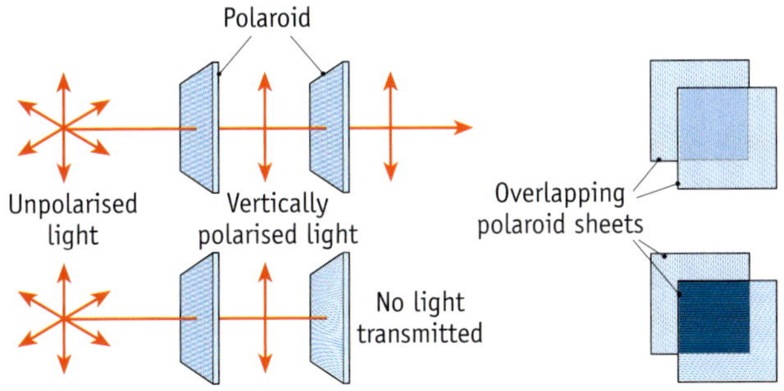

Fig. 18.19

(A)

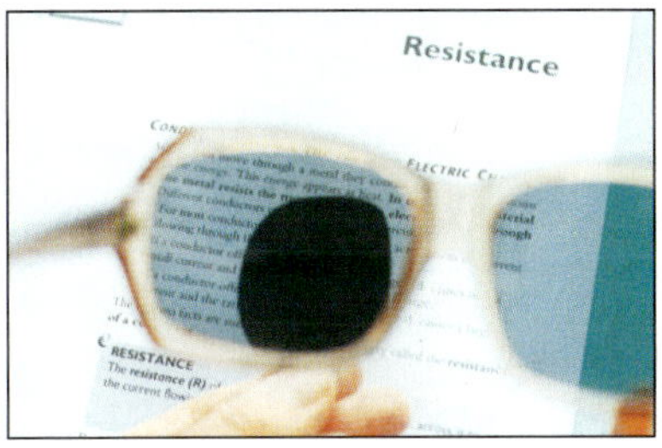

(B)

Fig. 18.20

The fact that light can be polarised shows that ***light is a transverse wave***.

If light from such a source is passed through a substance called a **Polaroid** it becomes **plane polarised**. This means that the light is vibrating in one plane only. Polaroid is a material that does to light waves what the slit did to waves on a rope on page 183. The Polaroid absorbs light vibrating at right angles to its axis and only allows light that is vibrating parallel to its axis to pass through.

If this light is then passed through another Polaroid, it only gets through if the axis of the second Polaroid is parallel to that of the first (FIG. 18.20(A)). If the second Polaroid is then rotated through 90°, virtually no light gets through (FIG. 18.20(B)). Since only transverse waves can be polarised, the fact that light can be polarised shows that light is a **transverse wave**.

POLARISATION BY REFLECTION

Light that is reflected from a glass or a water surface is found to be partially plane polarised. Light reflecting from a horizontal surface will be horizontally polarised. If this is viewed through a piece of polaroid with its axis vertical, this reflected light can be blocked to a large degree. Often this reflected light is a nuisance – reflected sunlight on water prevents you seeing into the water, 'shines' on windows or clocks prevent you from seeing through them and the 'shine on the blackboard' often causes problems in the classroom. By viewing these through a piece of polaroid, the 'shine' can be considerably reduced. Try this yourself with a piece of polaroid. **Polaroid sunglasses** are, as the name suggests, made of polaroid and as well as reducing the brightness of light, they also reduce glare. Polaroid filters can be used on cameras with the same effect (FIG. 18.21)

(A)
Light reflecting sharply off of a window is plane polarised. The reflection makes it almost impossible to see in through the window.

(B)
A piece of polaroid placed in front of the camera blocks out most of the reflection enabling you to see into the room.

Fig. 18.21

(A)

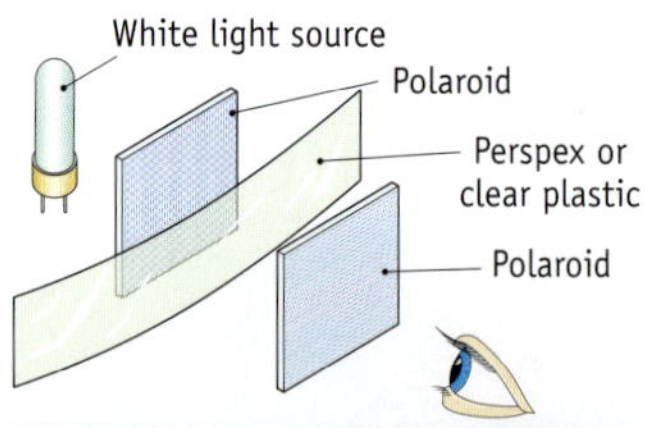

(B)

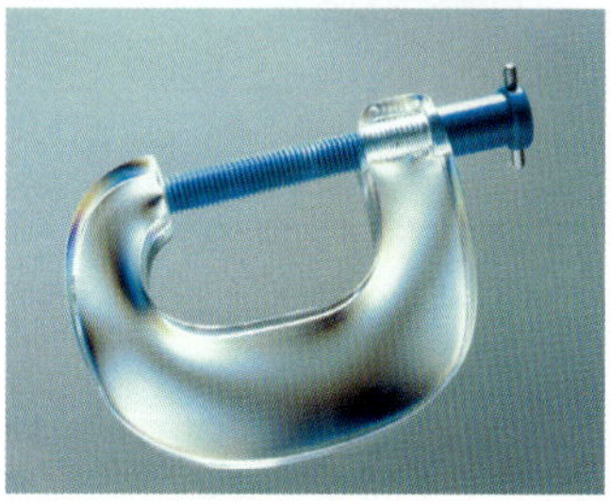

Fig. 18.22

STRESS POLARISATION

In FIG. 18.22(A) white light is passed through a piece of Polaroid and then through a perspex strip. The light is viewed through another piece of Polaroid whose axis is at a right angle to the first piece. If the perspex is put under strain, e.g. by bending it, colours can be seen which show how and where the perspex is under stress. This phenomenon is called photoelasticity and is used by engineers to analyse stresses in components (FIG. 18.22(B)).

DISPERSION

DISPERSION
The separating out of the different wavelengths (colours) present in light is called ***dispersion***.

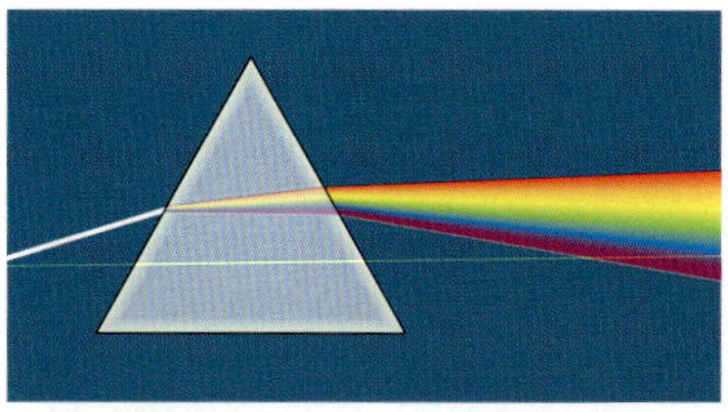

Fig. 18.23
Dispersion in a prism.

Dispersion can be brought about using a prism as in FIG. 18.23. Dispersion with a prism occurs because **the refractive index of a medium is slightly different for different wavelengths**. Thus the different wavelengths (different colours) are turned through different angles by the prism and become separated. When light is dispersed by a prism, red is refracted the least and violet the most.

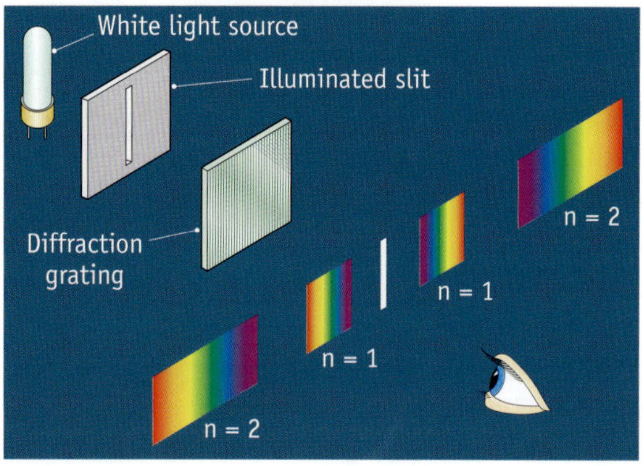

Fig. 18.24
Dispersion in a diffraction grating.

Dispersion can also be brought about using a diffraction grating as in FIG. 18.24. A spectrum is formed for each value of n, other than $n = 0$. This happens because from the formula $n\lambda = d \text{ Sin } \theta$ we see that θ is different if λ is, i.e. the different colours are formed at different angles from the straight through position. When $n = 0$, light of all the wavelengths present in the white light is in phase when brought together (path difference from each slit is the same) and thus a white line is formed.

When white light is dispersed by a prism, blue light is deviated the most and red the least. The opposite is the case with a diffraction grating since from $n\lambda = d \text{ Sin } \theta$, θ is largest for the longest wavelength (λ), namely red.

EXERCISE 18.3

1. What is the angular separation between the red (wavelength 710 nm) and the blue (410 nm) in the second order image when a parallel beam of white light is incident normally on a diffraction grating with 400 lines per mm on it?

2. What is meant by the diffraction of light? White light is passed through a diffraction grating and a number of spectra are formed. Use the diffraction grating formula to explain the following, assuming that the wavelength of visible light lies between 4×10^{-7} m (violet) and 8×10^{-7} m (red).

(i) Why is dispersion greater in the higher orders?

(ii) Why is the red light diffracted more than the violet in any particular order spectrum?

(iii) What is the maximum wavelength that can be obtained with a given grating in (a) the first order, (b) the second order?

(iv) What would be the effect of varying the grating constant?

(v) How does the spectrum obtained from a diffraction grating differ from that obtained from a prism?

Dispersion is responsible for the colour of a **rainbow**. When white sunlight enters a raindrop (or other fine drops of water) it is both refracted and internally reflected. The refraction causes the different wavelengths to be dispersed and the familiar colour is seen (FIG. 18.25). A similar situation occurs when light enters and leaves **polished gemstones,** such as diamonds, and gives them their characteristic coloured sparkle.

The colours seen on a **compact disc** when white light falls on it are due to the disc behaving as a reflection diffraction grating. A compact disc stores information digitally. A laser reads this information from the disc. The disc is made up of a reflective material onto which millions of very small hollows, called pits, are produced. A very fine laser beam falling on the disc and moving across it is scattered from the pits but reflects regularly from the flat parts. Thus pulses of reflected light are produced. These are the information from the disc in digital form. When white light falls on the disc, it is reflected only from the flat parts. It thus behaves as millions of point sources, and diffraction occurs, producing the familiar coloured pattern (FIG. 18.26).

Fig. 18.25

Fig. 18.26

RECOMBINATION

A second prism, as in FIG. 18.27, will recombine the constituents of white light to again form white light.

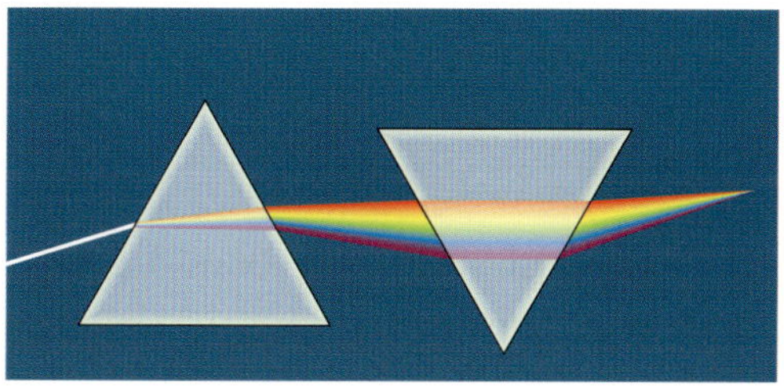

Fig. 18.27

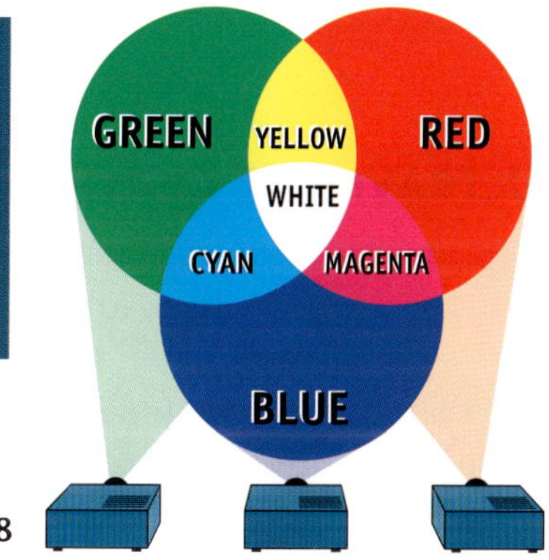

Fig. 18.28

PRIMARY COLOURS

White light can be produced by a combination of the coloured lights red, green and blue of equal intensity. Any other coloured light can be produced by mixing the same three colours in the right proportion. RED, GREEN and BLUE are called the **primary colours**.

When two primary colours are mixed in equal intensity, the resulting colour is called a **secondary colour** (FIG. 18.28). Note that mixing a primary colour and the opposite secondary colour gives white. A primary colour and the secondary colour that together give white light are called **complementary colours**. Creating light of a given colour by mixing the primary colours together is done with stage lighting and colour television.

RED, GREEN and BLUE are the primary colours
YELLOW, CYAN and MAGENTA are the secondary colours

SECONDARY COLOUR
When two primary colours are mixed in equal intensity, the colour formed is a ***secondary colour***.

COMPLEMENTARY COLOUR
A primary colour and the secondary colour that when mixed give white are called ***complementary colours***.

The Electromagnetic Spectrum

Electromagnetic waves were discussed briefly in Chapter 12. There you saw that:

- electromagnetic waves are travelling waves whose speed is 3×10^8 m s^{-1} in a vacuum (the speed of light). In other media their speed is less,
- electromagnetic waves show all typical wave properties, in particular they will undergo interference and diffraction,
- the frequencies and wavelengths of electromagnetic waves are related by the equation $c = f\lambda$, where $c = 3 \times 10^8$ m s^{-1}.

Fig. 18.29 shows the full range of electromagnetic waves in order of increasing wavelength. This is called the electromagnetic spectrum. Electromagnetic waves are also known as **electromagnetic radiation**.

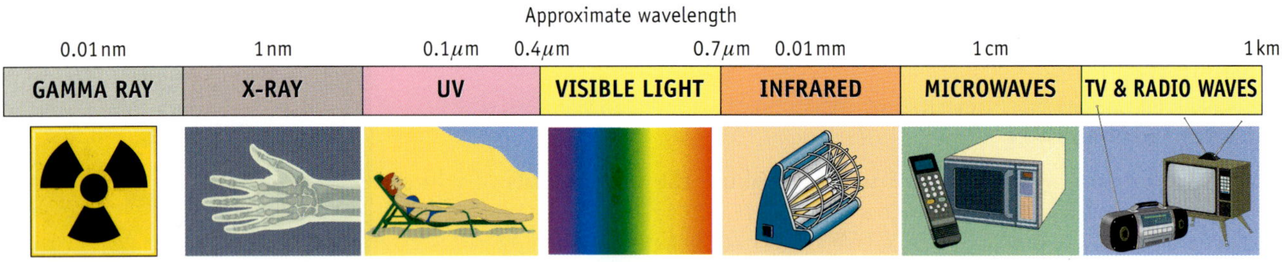

Fig. 18.29
The electromagnetic spectrum.

Ultraviolet Light

Electromagnetic radiation with wavelengths between about 0.4 μm and 5 nm is called **ultraviolet radiation** or **ultraviolet light (UV light)**. UV light is just beyond the violet end of the visible spectrum and thus has shorter wavelengths than visible light. UV light is emitted from the Sun. An ordinary filament lamp emits a small amount of UV light. A UV lamp used in the laboratory is usually a mercury vapour lamp.

Ultraviolet light has the following properties:

- It causes certain substances to fluoresce, i.e. the substance absorbs UV light and then re-emits this light as visible light. Vaseline, dayglow paints and some washing powders emit visible light when UV light falls on them.
- It effects a photographic plate, even more so than visible light.
- It causes sunburn, it is harmful to the eyes and causes skin cancer.
- It produces vitamin D in the skin.
- It does not pass through ordinary glass very well but can pass through quartz glass.
- It can cause photoemission (Chapter 29).

Fluorescence, the effect on a photographic plate or photoemission, can be used to detect the presence of UV radiation. For example, if a UV lamp is turned on in a dark room, Vaseline on a piece of paper will glow in the UV light. A white shirt washed in modern detergents will glow similarly due to fluorescent material in the detergent. In Chapter 29 you will see how UV light can cause electric current to flow in a photocell, indicating the presence of UV light.

Sunlight that reaches Earth contains harmful UV light. **Ozone (O_3)** in the upper atmosphere absorbs much of this UV light, thus protecting us from harmful effects of UV light. In the last number of years there has great concern about 'holes' or gaps in the **ozone layer,** caused probably by a group of chemicals known as CFC's (chlorinated fluorocarbons) which have been used in aerosols.

Infra-Red Light

Electromagnetic radiation with wavelengths between about 0.7×10^{-6} m and 1 mm is called **infra-red radiation (IR radiation)**. Infra-red radiation is just below the red end of the visible spectrum and thus has longer wavelengths than 'visible' red light. Infra-red radiation is the radiated heat discussed in Chapter 11. Infra-red radiation is emitted from the Sun or from an ordinary filament lamp. All objects are emitting some IR radiation. As the temperature of an object increases, it emits more IR radiation and more IR radiation of shorter wavelengths. At about 500 °C most objects start to emit visible red light as well.

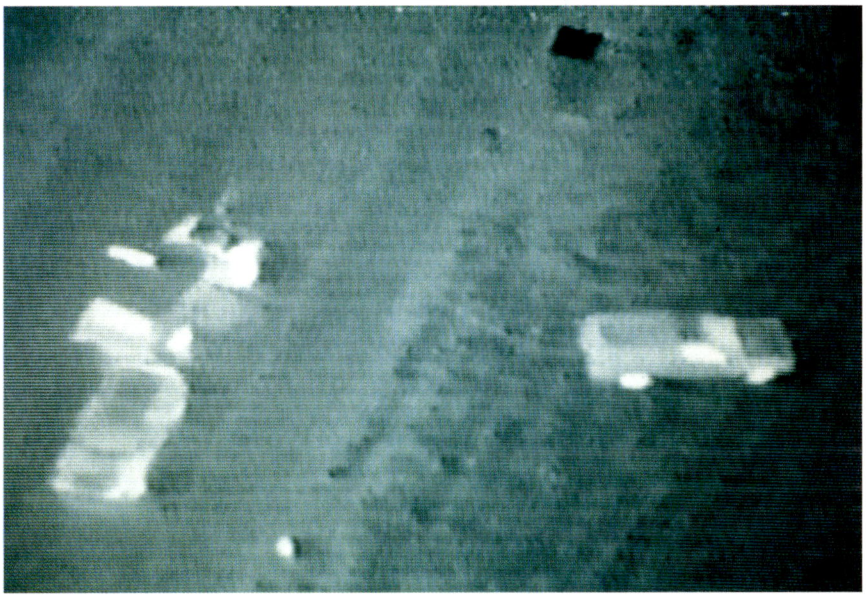

Fig. 18.30

Infra-red radiation can be detected by its heating effect using a blackened thermometer bulb or by the effect it has on a photographic plate. Properties of infra-red light are as follows:

- It effects a photographic plate and can be used to take photographs in darkness.
- It can pass through fog and mist and thus photographs can be taken in such conditions.
- It causes substances on which it falls to heat up.

Fig. 18.30 shows a photograph taken with an infra-red camera. The same photograph taken with an ordinary camera is almost useless due to the lack of visible light. The military and marine search and rescue use infra-red goggles, which enable them to see in the dark.

In medicine, body heat emitted by the skin can be photographed and thermal images of the body produced. These can be used to diagnose abnormalities or other problems under the skin that cannot be detected on X-rays (Fig. 18.31).

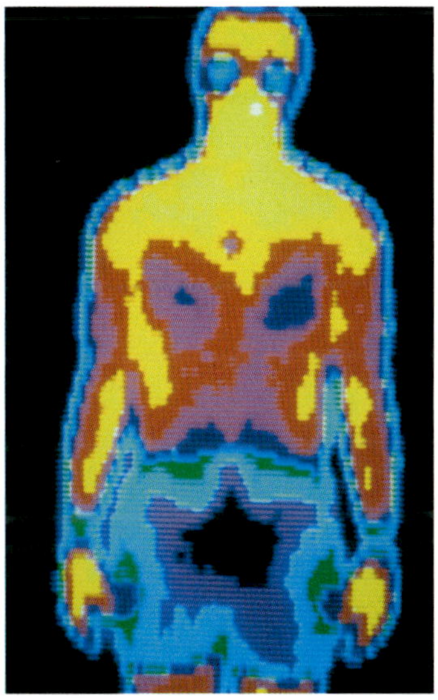

Fig. 18.31
A Thermogram.

THE GREENHOUSE EFFECT

Planet Earth is heated by electromagnetic radiation from the Sun. This radiation passes through the atmosphere and heats the Earth. The earth re-radiates this energy but at a longer wavelength, i.e. in the infra-red range. Gases in the atmosphere, particularly carbon dioxide, do not let this longer wavelength pass through, but reflect it back to Earth, thus the surface of the planet is kept warm. In the last number of years, the amount of carbon dioxide in the atmosphere has increased significantly due to the burning of fossil fuels. This is causing the average temperature of the Earth to increase, possibly with serious environmental consequences.

CHAPTER CHECKLIST

- **Define**: Diffraction; Interference; Grating constant (grating spacing); Polarisation; Dispersion; Primary colour; Secondary colour; Complementary colour; Greenhouse effect.
- **Explain**: The pattern formed when (a) monochromatic light and (b) white light falls on a diffraction grating; Why a prism dispenses white light and how dispersed light may be recombined.
- **State**: The adjustments that should be made to a spectomerer before use; The three primary colours; The three secondary colours; The three pairs of complementary colours.
- **Recall**: That light is an electromagnetic wave whose colour depends on its wavelength; That the colours seen on soap bubbles and petrol films are caused by interference; That the fact that light can be polarised shows light is a transverse wave; The relative positions of the various electromagnetic radiations in terms of wavelength and frequency; The relationship between UV light and the ozone layer.
- **Describe** an experiment to: Show the wave nature of light; Measure the wavelength of monochromatic light; Demonstrate the polarisation of light; Demonstrate the dispersion of light by a prism and a diffraction grating; Demonstrate the mixing of coloured lights; Detect UV and IR radiation.
- **List**: Two practical uses of polaroid and polarised light; Three everyday occurrences of dispersion of light; Two applications of infra-red cameras.
- **Draw** a labelled diagram of a spectrometer and state the function of each part.
- **Recall** and use the formula $n\lambda = d\text{Sin}\,\theta$ to solve problems.
- **Derive** the formula $n\lambda = d\text{Sin}\,\theta$.

Static Electricity

CHAPTER 19

Static Electricity

Some of the effects of static electricity have been known since ancient times. Over 2000 years ago the Greeks knew that a piece of amber rubbed with cloth attracted pieces of dust to itself. One of nature's most dramatic electrical phenomena is lightning, which is caused by a build up of static electricity in clouds. Today static electricity has some important practical applications (e.g. in a photocopier) and some interesting effects (Fig. 19.1). In industry it can be a hazard – sparks due to static electricity can cause fires and explosions.

Fig. 19.1
An interesting effect of static electricity.

Electrification by Contact

When two different materials are placed in close contact, e.g. by rubbing them together*, they may become electrically charged. This may be demonstrated as follows:

Experiment

To Produce Charge by Contact and to Show the Forces Between Charges.

Rub two polythene rods with a woollen cloth. Place one of them in a paper stirrup hanging from a retort stand (Fig. 19.2). Bring the second rod near the first one. The rods will be found to **repel** each other. If two cellulose acetate (or perspex) rods are rubbed with a woollen cloth they are also found to **repel** each other. If a polythene rod and a cellulose acetate rod are rubbed with a woollen cloth they will be found to attract each other.

This and experiments like it are explained as follows:

- When different substances are brought into close contact (by rubbing them together) they become **electrically charged**.
- There are two types of **electric charge**: **positive** and **negative**.
- Positive charge repels positive charge and negative charge repels negative charge. Positive charge is attracted to negative charge, i.e. **like charges repel and unlike charges attract.**

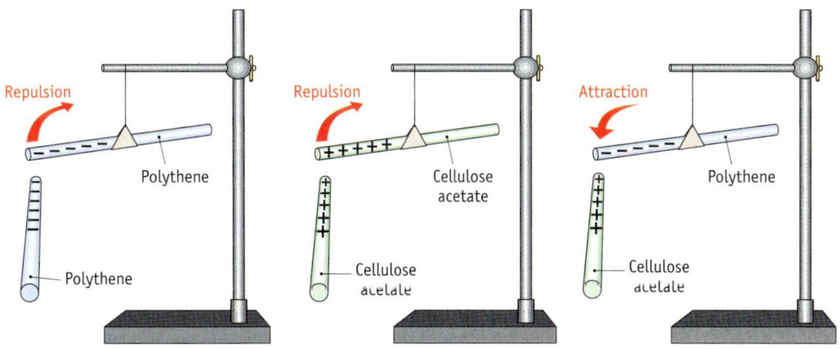

Fig. 19.2 (A) (B) (C)

The above experiment shows the existence of forces between charges. Polythene rubbed with wool becomes negatively charged and cellulose acetate rubbed with wool becomes positively charged.

* The rubbing together of the two substances, enormously increases the contact they make with each other. If they are just pressed together, the amount of charge displaced is small.

Atomic Nature of Electric Charge

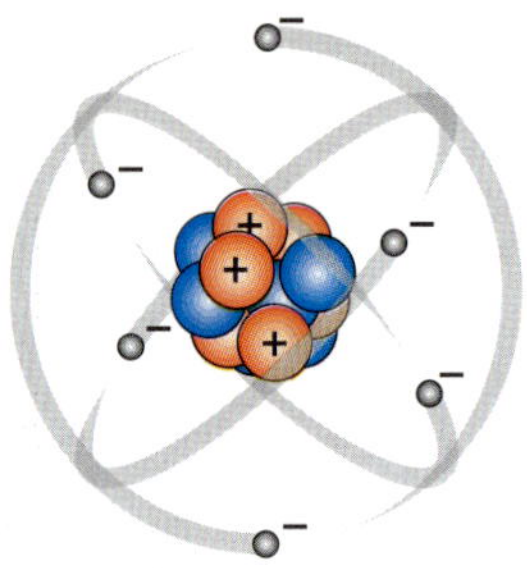

Fig. 19.3
A simple model of an atom.

All matter is made up of atoms. An atom (Fig. 19.3) consists of a nucleus containing **positively charged protons** and uncharged neutrons. Orbiting the nucleus are **negatively charged electrons**. The number of electrons is normally the same as the number of protons. The amount of negative charge on an electron is the same size as the amount of positive charge on a proton. An atom as a whole is therefore electrically neutral, i.e. it has no overall electric charge.

Different substances hold on to their electrons with different degrees of attraction. Suppose two different substances are brought into contact (e.g. by rubbing them together). The substance with the lesser attraction for its electrons loses some of them to the other substance. Thus each becomes electrically charged. The substance that picks up the extra electrons becomes negatively charged. The substance that loses the electrons becomes positively charged. This is because it now has more protons than electrons and thus has an overall positive charge.

The protons in the nucleus of an atom cannot move from one atom to another. **It is only the electrons, i.e. the negative charges, that actually get transferred.** Clearly the amount of positive charge produced by contact is the same as the amount of negative charge produced.

If an object is negatively charged (–) it has **gained** electrons.

If an object is positively charged (+) it has **lost** electrons.

It is only the electrons that actually move when objects become charged.

Unit of Electric Charge

UNIT OF ELECTRIC CHARGE
The unit of electric charge is the ***coulomb (C)***.

Electric charge is measured in a unit called the **coulomb (C)**. 1 coulomb is the amount of charge on about 6.25×10^{18} electrons (6 250 000 000 000 000 000 electrons). You need not remember this huge number.

Conductors and Insulators

INSULATOR
Any substance through which electric charge cannot flow is called an ***insulator***.

CONDUCTOR
Any substance through which electric charge can flow is called a ***conductor***.

Electrons cannot easily move from one atom to another in some substances. If these substances are electrically charged, the charge stays where it is put on them. Electric charge cannot flow through them. Such substances are called **insulators**. Examples are glass, perspex and most plastics. Other substances (e.g. metals) will allow charge to move through them because some electrons can easily move from atom to atom. Such substances are called **conductors**.

Both insulators and conductors can be charged by contact. However, if you want to charge a conductor, it must be held with an insulating handle, otherwise the charge will flow from it through you (the human body is a conductor) to the earth (which is also a conductor). A conductor from which charge cannot escape because it is insulated is called an insulated conductor.

Separation of Charges by Induction

In a metal, usually one electron from each atom is not stuck to that atom but is free to move from atom to atom. Suppose we bring a negatively charged rod, which is an insulator, near but not touching a metal conductor (Fig. 19.4). The electrons in the metal which are free to move are repelled by the negative charge on the rod and move to the opposite side of the metal. Thus a negative charge appears on one side of the metal and a positive charge on the other (from where the electrons have moved). If the charged insulating rod is taken away, the electrons move back again. The charges produced on the metal are called **induced charges**. The size of the induced negative charge is the same size as the induced positive charge.

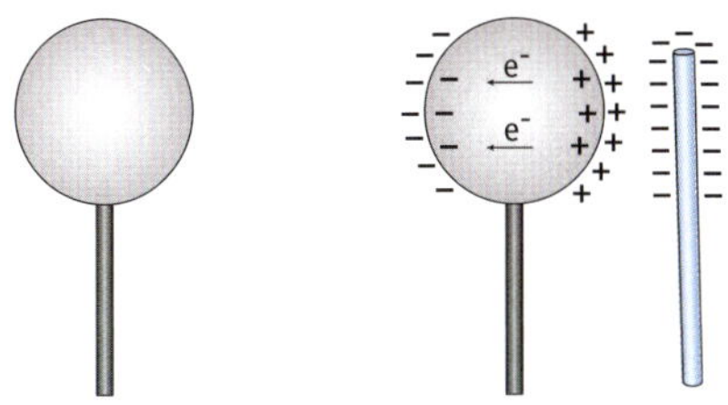

Fig. 19.4
Charging a single object by induction.

Fig. 19.5 shows two conducting spheres which are touching each other. A negatively charged insulating rod is brought near them. The spheres are separated while the negatively charged rod is still in place. When the charged rod is then removed, the electrons cannot get back to the positively charged sphere. Thus each sphere holds onto its charge. The spheres are said to have been **charged by induction**. Draw a diagram yourself showing what would happen if the insulating rod was positively charged.

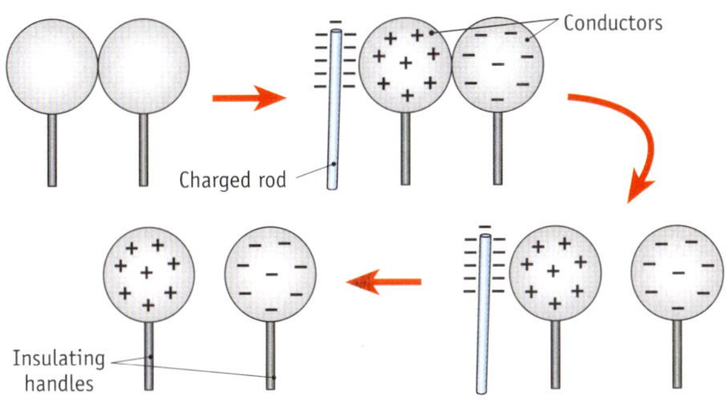

Fig. 19.5
Charging objects by induction.

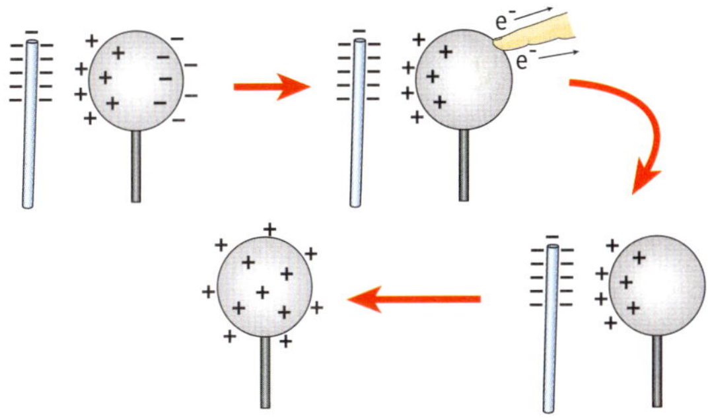

Fig. 19.6
Charging a single object by induction.

A single insulated conductor can also be charged by induction. Bring a negatively charged rod near the conductor (Fig. 19.6). Positive and negative charges are induced on it. Keeping the charged rod in place, touch the metal. The negative charge which was induced on the metal will flow through you to earth. Remove your finger and then remove the rod. The conductor will be positively charged. To charge the conductor negatively, a positive charge must be brought near it. Draw the relevant diagrams yourself.

Experiment

To Show the Separation of Charges by Induction.

- Use the equipment in Fig. 19.5.
- Bring a charged rod near – but not touching – one of the spheres.
- With the charged rod in place, separate the spheres (touch only the insulating handles).
- Test each sphere for electric charge (using an electroscope – see below).
  They will be found to have opposite charges.

The Gold Leaf Electroscope

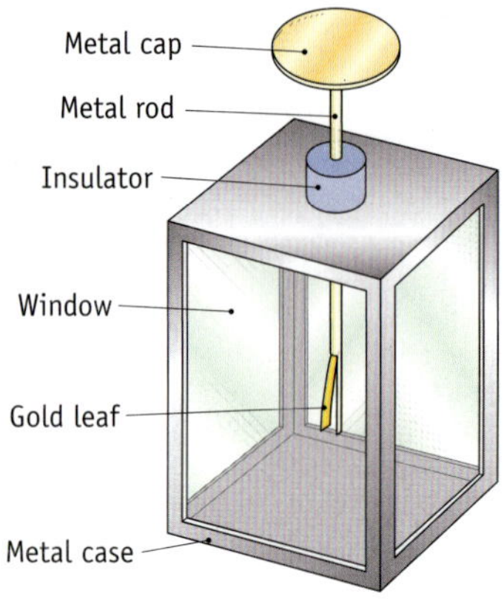

(A)

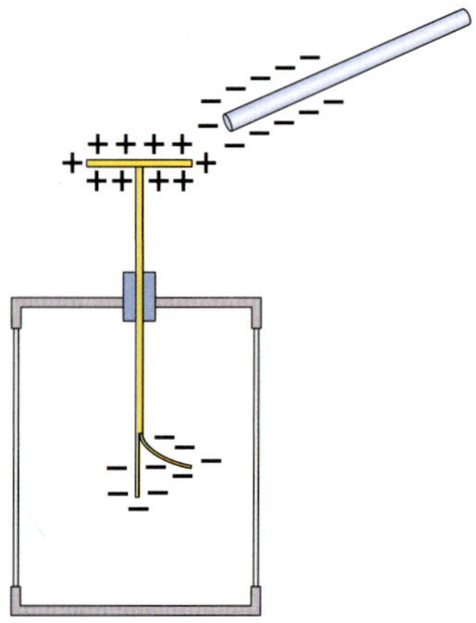

(B)

Fig. 19.7
A gold leaf electroscope.

Fig. 19.7(A) shows a gold leaf electroscope. It consists of a very thin gold leaf attached to one end of a metal rod. The other end of the rod has a metal disc, called the cap, attached to it. The leaf and rod are in a metal case with a window. The case and window stop draughts from causing the leaf to move. The rod is insulated from the case so that any charge put on the rod does not flow away. A gold leaf electroscope can be used to do the following:

Detect charge

If a charged object is brought near an electroscope (Fig. 19.7(B)), induced charges appear on the electroscope. Due to the force of repulsion between like charges, the leaf diverges. Thus the electroscope detects electric charge.

Indicate the approximate size of a charge

Place objects with different charges the same distance from the electroscope. The larger the charge on the object the greater the divergence produced.

Test whether a charge is + or –

Give the electroscope a charge of known sign*. Bring the object with the unknown charge near the cap. If the divergence of the leaf increases, the object and electroscope have charge of the same sign. If the leaf collapses, the electroscope and object have opposite charges (provided that when the charge is removed the leaf diverges again).

Test if an object is an insulator or a conductor

Charge the electroscope, then, holding the object in your hand, touch the cap of the electroscope with the object. If the leaf collapses, the object is a conductor, otherwise it is an insulator.

Indicate the size of a potential difference

See page 235.

Distribution of Charge on an Insulated Conductor

If an insulated conductor is charged, the charges repel each other and move around the conductor to 'get away' as far as possible from each other. When the charges have stopped moving, i.e. **when they are static**, the following is found:

> All **static charge** resides on the **outside** of a conductor.

---

* A gold leaf electroscope can be given a charge of known sign by induction as follows:

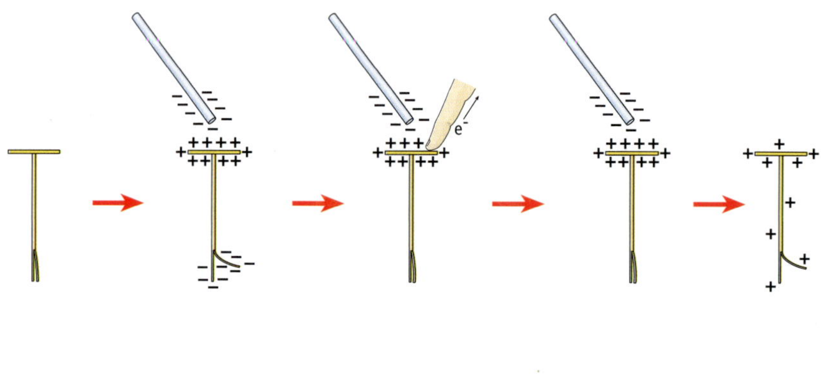

If static charge is placed on a spherical conductor, it is found to be distributed uniformly over the sphere. If it is placed on a pear-shaped conductor, extra charge accumulates at the pointed end (FIG. 19.8). In general the following is true:

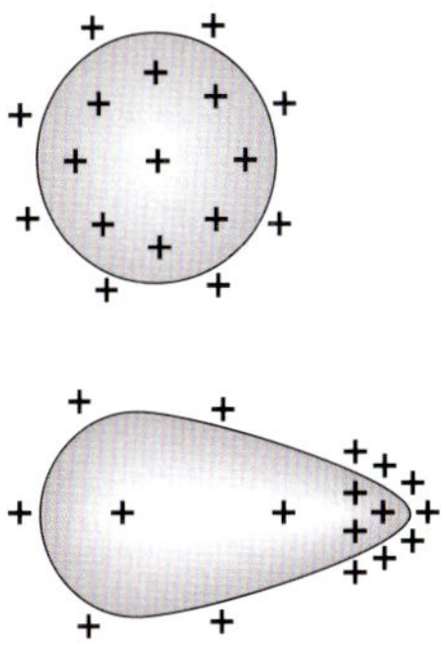

Fig. 19.8

Static *charge* on a conductor tends to *accumulate* where the conductor is most *pointed*.

THE VAN DE GRAAFF GENERATOR

A Van de Graaff generator (FIG. 19.9(A)) is a machine that produces large amounts of static electricity on its dome and is suitable for electrostatic experiments in a school laboratory.

EXPERIMENT

TO SHOW THAT ALL STATIC CHARGE RESIDES ON THE OUTSIDE OF A HOLLOW METAL CONDUCTOR.

- Connect a cylindrical metal can (the hollow conductor) to the dome of a Van de Graaff generator and turn on the generator (FIG. 19.9(B)).
- Touch a proof plane against the **inside** of the can and bring the proof plane very near the cap of an uncharged electroscope.
- The leaf will not diverge.
- Touch a proof plane against the **outside** of the can and bring the proof plane very near the cap of an uncharged electroscope.
- The leaf will diverge – showing that the static charge is on the outside.

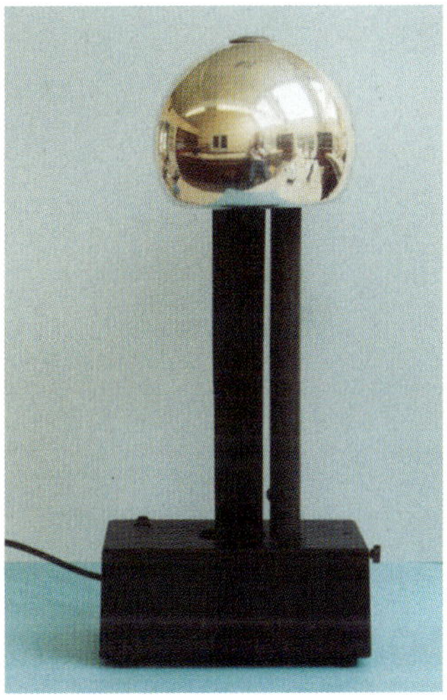

(A)

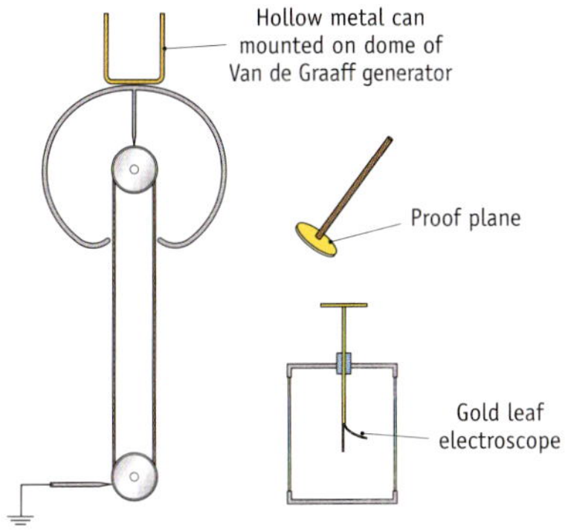

(B)

Fig. 19.9
A Van de Graaff generator is used to produce large amounts of static charge.

EXPERIMENT

TO SHOW THAT STATIC CHARGE ON A CONDUCTOR TENDS TO ACCUMULATE WHERE THE CONDUCTOR IS MOST POINTED.

- Charge an insulated pear-shaped conductor by touching it off the dome of a charged Van de Graaff generator.
- Touch a proof plane against the straight part of the charged pear-shaped conductor and then against the cap of an uncharged electroscope. Note the divergence.
- Earth the proof plane and the electroscope.
- Now touch the proof plane against the pointed end of the pear-shaped conductor and then against the cap of the electroscope.
- The divergence produced will be much greater, showing that the charge tends to accumulate there.

Point Discharge

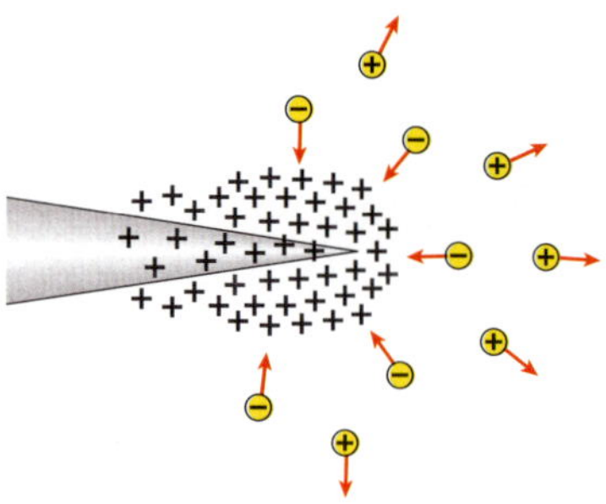

Fig. 19.10

At a very sharp point (Fig. 19.10) there is a large accumulation of charge. This causes a very strong electric force field in the region around the point. Any ions in the air are either strongly attracted to or repelled from the point and move towards or away from the point. The moving ions produce further ions by collision with atoms in the air. Ions with opposite charge to that on the point head towards the point and neutralise the charge on it. Ions with the same charge head away from it creating an 'electric wind'. This may be demonstrated with a candle flame (Fig. 19.11). The ions landing on the point neutralise some of the charge on it. It is as if the charge were removed from the point. The loss of charge from a point by this manner is called **point discharge** or the **point effect**.

A charged object will lose its charge more quickly if there are any points on it. For example, if you place a pin on the cap of a charged electroscope, it will lose its charge quickly. For the same reason the dome of a Van de Graaff generator is smooth so that any charge on it stays there and does not leak off.

Fig. 19.11 An electric wind produced by ions streaming away from the highly-charged point.

Fig. 19.12

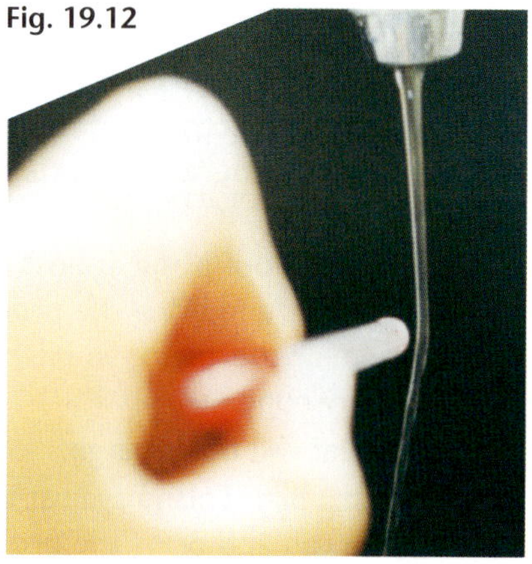

Some Everyday Effects of Static Electricity

- The crackling and clinging of clothes made from synthetic materials is due do static electricity. The crackling heard is due to small sparks which you can see in the dark.
- If you rub a plastic biro on your sleeve or in your hair it becomes charged. If you place it near a small piece of paper, charges are induced on the paper and it is attracted to the biro. Similarly a charged biro causes a fine stream of flowing water from a tap to deflect (Fig. 19.12).
- The picture produced on a television screen is caused by a beam of electrons striking the screen. Electrons are negatively charged. Thus the screen becomes charged and dust is attracted to it in the same manner that the biro attracts paper.
- Sparks produced by static electricity can be dangerous if fine dust, inflammable vapours or gases are present. In oil refineries, flour mills and in the chemical industry, special precautions must be taken to minimise the risk of explosion with these substances due to static electricity.
- An aeroplane (or a helicopter) in flight can build up considerable amounts of static electricity due to rubbing with the air. When the plane lands, this static must be discharged before the plane can be refuelled; otherwise an explosion might occur. Aeroplane tyres are made of conducting rubber to do this.

- Large amounts of static charge can build up in clouds. Lightning is the discharge of this static charge from one cloud to another or from a cloud to the ground (FIG. 19.13(A)). It is a huge flow of electrons (i.e. an electric current). Such a flow can do considerable damage to buildings. To protect them, a metal rod called a lightning conductor is mounted on the top of the building and connected by a thick copper conducting strip to a large copper plate buried in the ground (FIG. 19.13(B)). If a charged thundercloud passes overhead, induced charges on the rod cause point discharge to occur and the voltage (see page 235) between the cloud and the rod is reduced, thus lessening the chance of being struck by lightning. If lightning does strike the building, it can flow harmlessly to the ground through the lightning conductor.

(A)

(B)

Fig. 19.13

The Size of the Force Between Two Static Electric Charges

A static electric charge will exert a force of attraction or repulsion on another static charge in its vicinity. The size of the force that each charge exerts on the other depends on:

- the size of each charge,
- the distance between the charges,
- the material surrounding the charges.

Suppose two charges of Q_1 coulombs and Q_2 coulombs are placed a distance d apart (FIG. 19.14). Let F be the force that one exerts on the other. Then by experiment it is found that:

$$F \propto Q_1 \qquad F \propto Q_2 \quad \text{and} \quad F \propto \frac{1}{d^2}$$

It follows that: $F \propto \frac{Q_1 Q_2}{d^2} \Rightarrow F = \frac{kQ_1Q_2}{d^2}$ where k is a constant.

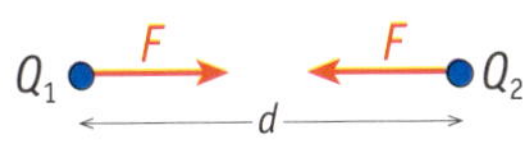

Fig. 19.14

This result is known as **Coulomb's Law**.

The value of k depends on the medium surrounding the charges. For historical and other reasons, about which we need not worry, k is written as:

$k = \frac{1}{4\pi\varepsilon}$ where ε is a constant.

COULOMB'S LAW states that the force of attraction or repulsion between two point charges is directly proportional, the product of the charges and inversely proportional to the square of the distance between them.

If the charges are Q_1 and Q_2 and the distance between them is d, the law states:

$$F = \frac{1}{4\pi\varepsilon}\frac{Q_1Q_2}{d^2}$$

Coulomb's Law is an example of an **inverse square law** since the force is inversely proportional to the square of the distance between the charges. This means that:

If the distance between the charges is **doubled**, the size of the force is **four** times smaller.

If the distance between the charges is made **three** times bigger, the size of the force is **nine** times smaller etc.

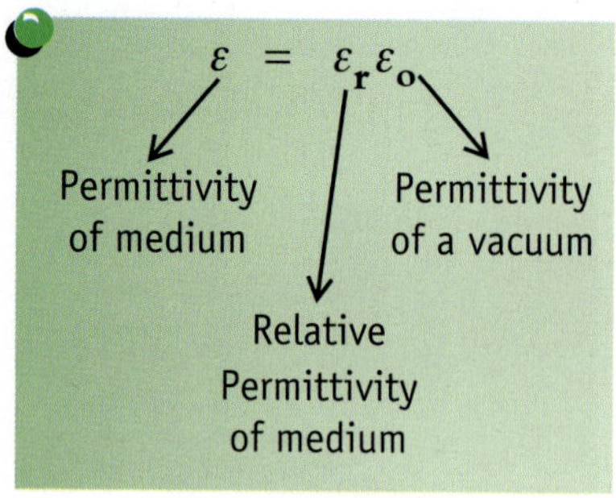

The size of the force depends on the medium surrounding the charges. The force is greatest when the charges are in a vacuum. In any other medium the force is less. Thus the value of ε is different for different media. ε is called the **permittivity** of the medium in question.

If the charges are in a vacuum, the value of the permittivity is written as ε_o. ε_o is called the **permittivity of free space** or the **permittivity of a vacuum**. ε_o = permittivity of air approximately. The **unit of permittivity** is the **Farad per metre (F m^{-1})** – see page 241. The value of ε_o is 8.9×10^{-12} F m^{-1}. You need not remember this value.

The permittivity of any medium can always be written as the permittivity of a vacuum multiplied by some number, i.e. for any medium $\varepsilon = \varepsilon_r \varepsilon_o$. ε_r is called the relative permittivity of that medium.

NOTE

- The **size of the force on each charge is the same** – even if one charge is greater than the other. This follows from Newton's 3rd Law.
- The **direction of the force is always along the line joining the two charges** and is an attractive force if they are unlike charges and repulsive if they are like charges.
- If the charges are not point charges, and if the distance between the charges is large compared to the dimensions of the charges, Coulomb's Law still holds to a high degree of accuracy. If the charges are spherical, it holds if d is the distance between the centres of the spheres.
- Coulomb's Law has the same mathematical form as Newton's Law of Universal Gravitation. Each is an inverse square law. However, gravitational forces are only attractive.
- If a point charge Q is placed in the vicinity of a number of other point charges, the resultant force on Q is the vector sum of the forces that each of the other charges on its own would exert on Q.

In the following problems the charges are in a vacuum and $\varepsilon_o = 8.9 \times 10^{-12}$ F m^{-1} *unless otherwise stated.*

Problem 1: The relative permittivity of water is 81. Find the permittivity of water.

Solution: $\varepsilon = \varepsilon_r\varepsilon_o = (81)(8.9 \times 10^{-12}) = 7.2 \times 10^{-10}$ F m^{-1}

Problem 2: Calculate the force that a charge of +5 *C* exerts on a charge of +6 *C* placed 4 m from it.

Solution: FIG. 19.15 shows the situation.

$$F = \frac{1}{4\pi\varepsilon_o}\frac{Q_1Q_2}{d^2} = \frac{(5)(6)}{4\pi(8.9\times10^{-12})(4)^2} = 1.68 \times 10^{10}\ \text{N}$$

Note the huge size of this force. In practice, static charges of this size are rarely found. In the problems below, the charges are usually in the order of **microcoulombs** (μC), where

1 μC = 1 millionth of a coulomb = 1×10^{-6} C

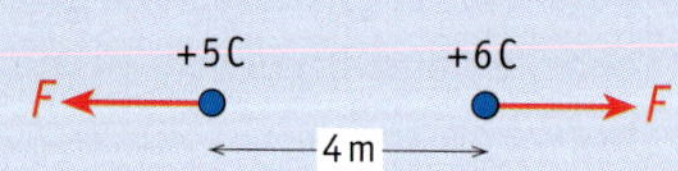

Fig. 19.15

Problem 3: Two point charges of +2 μC and –3 μC are 50 cm apart in air (FIG. 19.16). Find the magnitude and direction of the force on the 2 μC charge.

Fig. 19.16

Solution: The values of the charges must be expressed in coulombs and distances in metres before using Coulomb's Law.

$$F = \frac{1}{4\pi\varepsilon_o}\frac{Q_1Q_2}{d^2} = \frac{(2\times10^{-6})(3\times10^{-6})}{4\pi(8.9\times10^{-12})(50\times10^{-2})^2} = 0.215\ \text{N}$$

Problem 4: Three charges are arranged in air as shown in FIG. 19.17(A).
Find the magnitude and direction of the force on the 2 μC charge.

Solution: FIG. 19.17(B) shows the two forces acting on the +2 μC charge.

They are a force F_1 due to the –1 μC charge and a force F_2 due to the –3 μC charge.

We use Coulomb's Law to find the magnitude of each of these forces.

F_2 = 0.215 N towards the 3 μC charge (from the previous problem).

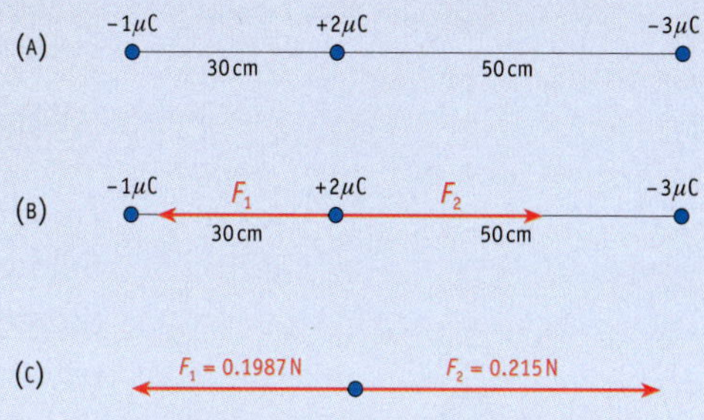

Fig. 19.17

$$F_1 = \frac{1}{4\pi\varepsilon_o}\frac{Q_1Q_2}{d^2} = \frac{(1 \times 10^{-6})(2 \times 10^{-6})}{4\pi(8.9 \times 10^{-12})(30 \times 10^{-2})^2}$$

= 0.1987 N towards the 1 μC charge.

Resultant force on +2 μC charge is 0.215 – 0.1987 = 0.0163 N towards the -3 μC charge.

EXERCISE 19.1

Assume the charges are in a vacuum and $\varepsilon_o = 8.9 \times 10^{-12}$ F m^{-1} *unless otherwise stated.*

1. The relative permittivity of oil is 2.2. What is the permittivity of the oil?
2. The permittivity of a certain insulator is 4×10^{-11} F m^{-1}. What is the relative permittivity of this insulator?
3. Find the force of repulsion between two point charges of +1 C and +3 C if they are placed 1 m apart in a vacuum. On which charge is the greater force?
4. Calculate the force on a charge of +3 μC when placed a distance of 4 m from a negative charge of 6 μC in air. Is the force one of attraction or repulsion?
5. In a vacuum, a charge of +2 μC is 30 cm from a charge of –2 μC. What is the magnitude and direction of the force on the +2 μC charge?
6. In a medium of permittivity 7.2×10^{-10} F m^{-1} a charge of +2 μC is 40 cm from a charge of –4 μC. What is the magnitude and direction of the force on the +2 μC charge?
7. The charge on a proton in the nucleus of an atom is 1.6×10^{-19} C. If the distance between 2 protons in the nucleus of an iron atom is 4×10^{-15} m, find the force of repulsion between them. (Assume the charges are in a vacuum.) If the mass of the proton is 1.67×10^{-27} kg and G = 6.7×10^{-11} N m^2 kg^{-2}, find the gravitational force between the two protons.
8. The force of repulsion between two identically charged particles is 0.2 N. If the charges are 2 cm apart in air, what is the size of the charge on each?
9. The attractive force between two small charges is F when the separation of the charges is x. If the separation is increased to $3x$, what will be the force between the charges?
10. A, B, C are three points on a straight line with |AB| = 10 cm, |BC| = 5 cm, and |AC| = 15 cm. Charges of +8 μC and +10 μC are placed at A and B respectively. What charge must be placed at C so that there is zero resultant force on the charge at B?

Electric Fields

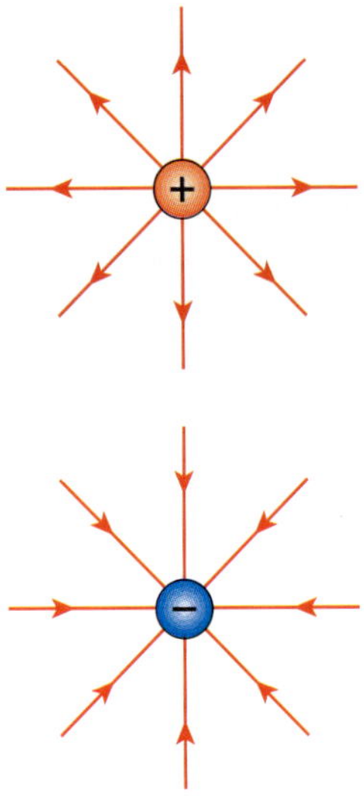

Fig. 19.18
The electric field due to an isolated postive and an isolated negative charge.

ELECTRIC FIELD

An ***electric field*** is any region of space where a static electric charge experiences a force other than the force of gravity.

An ***electric field*** is always caused by other static charges in the vicinity.

An electric field can be represented on a diagram by lines called electric field lines or lines of force. These lines have arrow heads that show the direction of the force that would be exerted on a positive charge if it were placed in the field. Fig. 19.18 shows the shape of the lines of force near an isolated positive charge and an isolated negative charge. Note that **where the electric field is strong, the field lines are close together, where the field is weak the lines are far apart**. Fig. 19.19 shows the shape of other electric fields. If two parallel metal plates are connected to a high voltage source positive charge appears on one plate and an equal amount of negative charge appears on the other. This produces a fairly uniform electric field in the space between the plates (Fig. 19.20).

LINE OF FORCE (ELECTRIC FIELD LINE)

An ***electric field line*** is a line drawn in an electric field showing the direction of the force on a positive charge placed in the field.

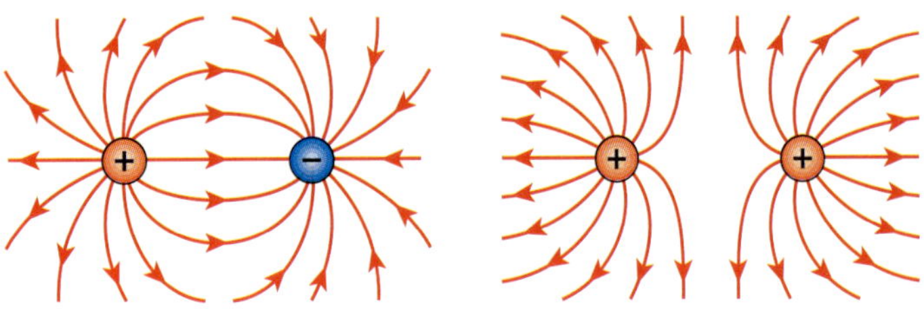

Fig. 19.19

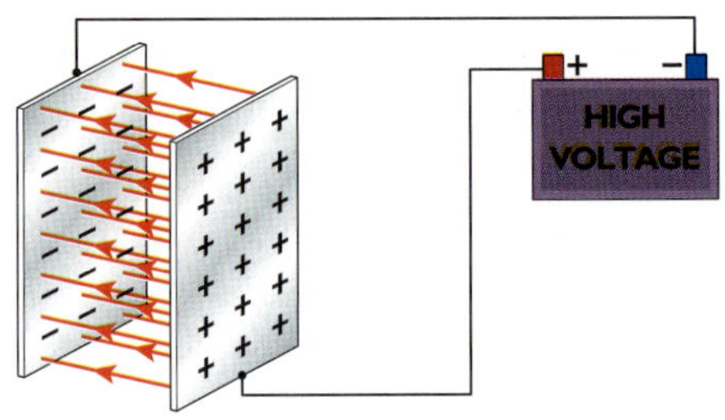

Fig. 19.20
The almost uniform electric field between two oppositely charged parallel plates.

Experiment

To Show Electric Field Patterns.

- Use the equipment in Fig. 19.21.
- Connect a **high voltage source** to the metal plates which are in the oil.
- The semolina lines up in the direction of the field, showing the electric field.

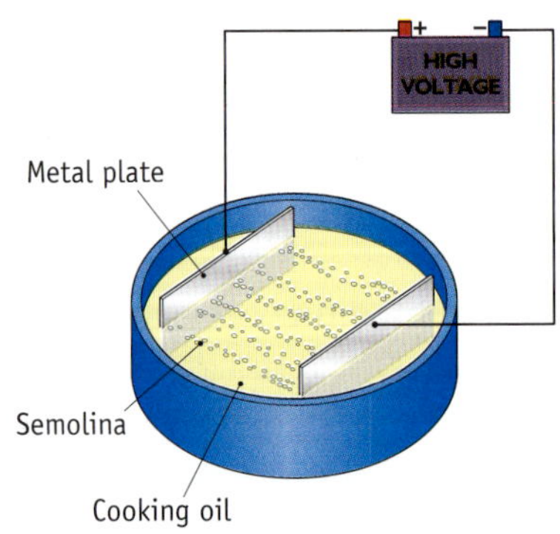

Fig. 19.21
Experiment to show Electric Field Patterns.

Applications of Electric Fields

Electrostatic Precipitators

An electrostatic precipitator is a device that removes dust and other small particles from dirty air. It first transfers electric charge to the dust particles using the point effect and then attracts the charged dust particles to metal plates carrying opposite charge. The particles remain there until a sufficient amount builds up when it is then removed in bulk. Precipitators are often used in industry (Fig. 19.22). Air purifiers and smoke removers used in bars and restaurants often work on the same principle.

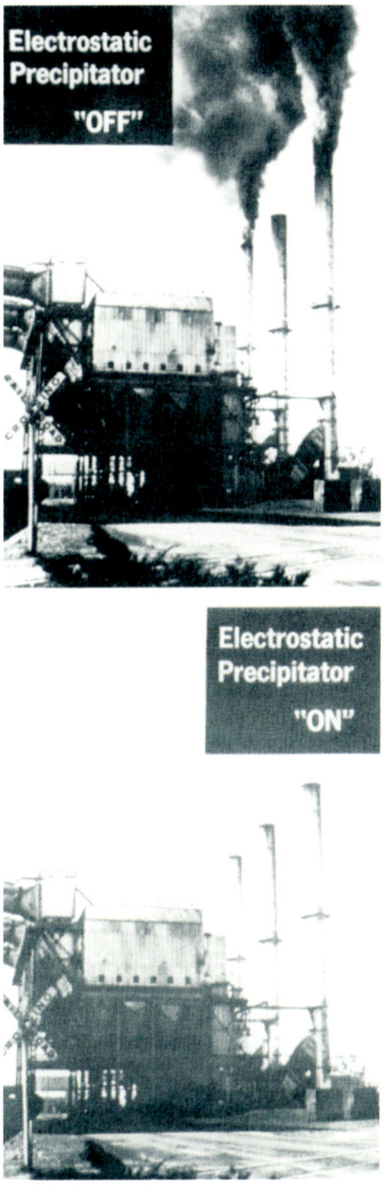

Fig. 19.22
This electrostatic precipitation facility is so effective at removing small particles from smoke that the chimney emissions become invisible when it is on.

Xerography – The Photocopier

In a photocopier based on the xerographic process, a drum (or sometimes a belt) is charged electrostatically. An image of the document to be copied is focused on this. The light causes the electric charge to escape from the regions on the drum where it falls. Toner particles are then attracted to the remaining charged regions. In this way an image of the original document is formed on the drum. This toner image is transferred to a sheet of paper and heat fixes it in place, the result is a 'copy'.

The Effect of an Electric Field on Integrated Circuits

When atmospheric conditions are right, the human body can build up significant amounts of static electric charge when walking around on carpets and other floors. This may be discharged, in the form of a spark, when the person passes very near or touches, for example, a metal radiator or shelf. If this electric charge were discharged through an integrated circuit while working on it, the integrated circuit is very likely to be permanently damaged. To reduce the chances of this happening, personnel working with integrated circuits must make sure they are connected to earth, thereby not building up any static charge. Personnel working with integrated circuits are often connected to earth by a wire attached to their wrist with a wrist band.

Electric Field Strength

The strength or weakness of an electric field can be represented by a vector called the **electric field strength**. This vector is sometimes called **electric field intensity**. It is defined as follows:

ELECTRIC FIELD STRENGTH
The ***electric field strength E*** at a point in an electric field is the force per unit charge at that point, i.e. electric field strength is the force per coulomb.

Suppose a charge of +1 C experiences a force of 6000 N at a particular point in an electric field. We say the electric field strength at this point is 6000 newtons per coulomb.

If a charge of +4 C experiences a force of 80 000 N at a point, the electric field strength at that point is:

$$E = \frac{80\,000}{4} = 20\,000 \text{ NC}^{-1}$$

The unit of electric field strength is obviously the **newton per coulomb (N C^{-1})**.

If a charge Q experiences a force F when placed in an electric field, the field strength E is given by:

$$E = \frac{F}{Q}$$

Electric field strength is a **vector**. The **magnitude of E** at a point in an electric field, is the size of the force that would be exerted on a charge of +1 C placed at that point. The direction of E is the direction of the force on +1 C.

UNIT OF ELECTRIC FIELD STRENGTH

The unit of electric field strength is the ***newton per coulomb*** **($N\ C^{-1}$)**. You will see on page 236 that this is the same as the ***volt per metre*** **($V\ m^{-1}$)**.

In the following take it that $\varepsilon_o = 8.9 \times 10^{-12}$ F m^{-1} *and all charges are in a vacuum unless otherwise stated.*

Problem 5: A charge of 2 μC experiences a force of 34 N when placed at a point in an electric field. Calculate the electric field strength at that point.

Solution:

$$E = \frac{F}{Q} = \frac{34}{2 \times 10^{-6}} = 1.7 \times 10^{7}\ N\ C^{-1}$$

Problem 6: Find the magnitude of the force on a charge of 5 μC when placed in an electric field of strength 2×10^{-3} N C^{-1}.

Solution:

$$E = \frac{F}{Q} \Rightarrow F = EQ$$

$$= (2 \times 10^{-3})\ (5 \times 10^{-6}) = 1 \times 10^{-8} N$$

Problem 7: Find the electric field strength E at a distance d from a point charge +Q.

Solution: Direction of E is radially out from Q (Fig. 19.23)

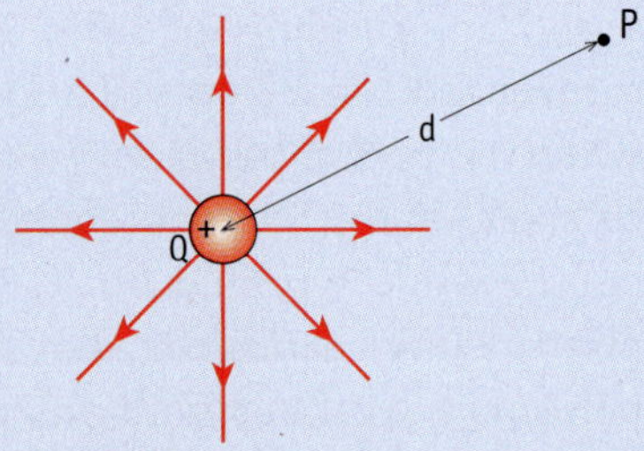

Fig.19.23

Force F on +1 C at point P is:

$$F = \frac{1}{4\pi\varepsilon_o}\frac{(Q)(1)}{d^2} \quad \text{i.e.} \quad E = \frac{Q}{4\pi\varepsilon_o d^2}$$

The direction of E is radially out from +Q. If Q was negatively charged it would be in the opposite direction.

Problem 8: What is the electric field strength half way between a charge of +2 μC and a charge of +9 μC if the distance between the charges is 60 cm?

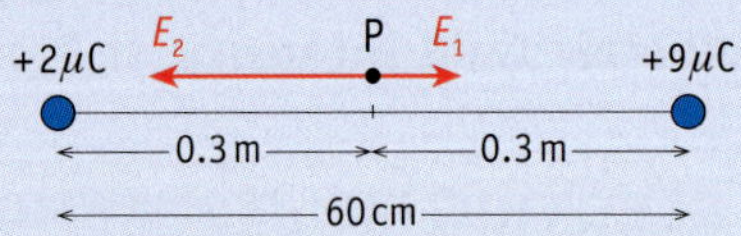

Fig. 19.24

Solution: Fig. 19.24 shows the situation. First calculate the field strength at P due to each charge individually. The resultant field strength is then found.

Field Strength E_1 at P due to +2 μC = Force on +1 C placed at P due to +2 μC

$$= \frac{1}{4\pi\varepsilon_o}\frac{Q_1Q_2}{r^2} = \frac{(2 \times 10^{-6})(1)}{4\pi(8.9 \times 10^{-12})(0.3)^2} = 1.9869 \times 10^{5}\ N\ C^{-1} \text{ to the right.}$$

Field Strength E_2 at P due to +9 μC = Force on +1 C placed at P due to +9 μC

$$= \frac{1}{4\pi\varepsilon_o}\frac{Q_1Q_2}{r^2} = \frac{(9 \times 10^{-6})(1)}{4\pi(8.9 \times 10^{-12})(0.3)^2} = 8.941 \times 10^{5}\ N\ C^{-1} \text{ to the left.}$$

Resultant electric field strength

E at P $= 8.941 \times 10^{5} - 1.9869 \times 10^{5} = 6.954 \times 10^{5}$ N C^{-1} to the left.

EXERCISE 19.2

All charges are in a vacuum unless otherwise stated. Take $\varepsilon_o = 8.9 \times 10^{-12}$ F m^{-1}.

1. A charge of 4 μC experiences a force of 12 N when placed at a point in an electric field. Calculate the electric field strength at that point.
2. What is the force on a charge of 2 μC when placed in an electric field of strength 3×10^3 N C^{-1}?
3. A charged particle experiences a force of 7×10^{-6} N when placed in a field of strength 2×10^3 N C^{-1}. What is the charge on the particle?
4. Find the magnitude and direction of the electric field strength at a distance of 2 m from a charge of +4 μC.
5. Find the magnitude and direction of the electric field strength at a distance of

   (i) 0.1 mm, (ii) 1 mm, (iii) 10 cm.

   from a negative charge of 20 μC.
6. Find the magnitude and direction of the electric field strength at a distance *r* from:

   (i) a positive point charge +Q,
   (ii) a negative point charge −Q.
7. What is the acceleration of an electron if it enters an electric field of strength 3×10^9 V m^{-1}?
   (Charge on electron = 1.6×10^{-19} C and mass of electron = 9.1×10^{-31} kg.)
8. What is the electric field strength half way between a charge of +4 μC and a charge of +2 μC, if the distance between the charges is 20 cm? What is the force on a charge of 5 μC placed at this point?
9. What is the electric field strength half way between a charge of +3 μC and a charge of -7 μC, if the distance between the charges is 40 cm? What is the force on a charge of 2 μC placed at this point?
10. At what point between a charge of +10 μC and a charge of +5 μC is the electric field strength zero if the charges are 1 m apart?
11. A charge of +5 μC and a charge of −12 μC are 30 cm apart. At what point on the line containing the two charges is the electric field strength zero?

CHAPTER CHECKLIST

- **Explain** what is meant by: Electrification (charging) by contact; Conductor; Insulator; Electrification by Induction; Point discharge; An Inverse Square Law; Electric Field; Electric Field Line (Line of Force).
- **Recall** that: Like charges repel and Unlike charges attract; Static charge resides on the outside of a conductor; Charges tend to accumulate where a conductor is most pointed.
- **State**: The unit of charge; Coulomb's Law.
- **Describe** the structure and uses of an electroscope.
- **Describe** and carry out an experiment to: Demonstrate the forces between charges; Charge an insulated conductor by induction: Show that static charge resides on the outside of a hollow conductor; Show that charges tend to accumulate where a conductor is most pointed; Demonstrate electric field patterns.
- **List** everyday occurrences of static electricity and be aware of the hazard that sparks caused by static electricity can be in industry.
- **Recall** that electric fields are used in precipitators and xerography and pose a hazard to integrated circuits.
- **Define** Electric Field Strength and state its unit.
- **Recall** and use the formulae $F = \frac{1}{4\pi\varepsilon}\frac{Q_1Q_2}{d^2}$; $\varepsilon = \varepsilon_r\varepsilon_o$; $E = \frac{F}{Q}$ to solve problems.

CHAPTER 20

Potential Difference and Capacitance

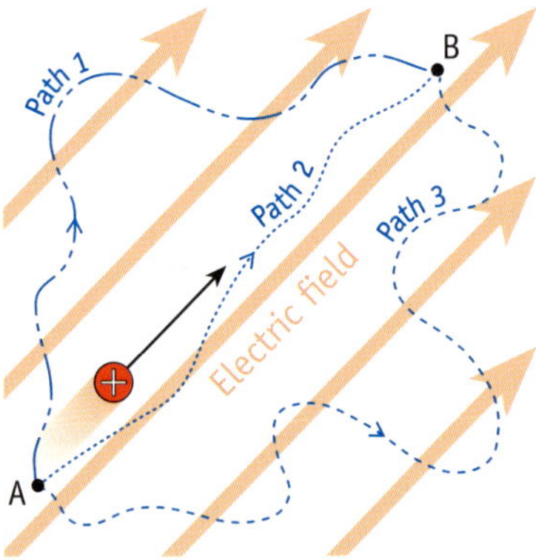

(A)
The electric field does work on the charge as it 'falls' from A to B.

(B)
The man does work in pushing the charge from B to A against the field.

Fig. 20.1

Potential Difference

Imagine an object thrown upwards near the surface of the Earth. As it rises, it slows down, losing kinetic energy. It gains an equal amount of gravitational potential energy. If an object is dropped, as it falls, it loses potential energy and gains an equal amount of kinetic energy. Something similar happens for a charged particle in an electric field. In Fig. 20.1(A) if a positive charge is released at the point A, the field will exert a force on it and it will move in the direction of the electric field. As it moves it loses electrical potential energy and gains an equal amount of kinetic energy. The field does work on the charge.

If we want to move a positive charge from B to A we must exert a force on the charge. We do work on the charge to move it against the force of the electric field (Fig. 20.1(B)). If a positive charge is projected from B to A it loses kinetic energy and gains potential energy as it moves towards A.

Experimentally it is found that the work done (i.e. the amount by which the potential energy changes) in bringing a charge from one point to another in an electric field does not depend on the path taken. For example, the work done in going from A to B by any of the paths shown in Fig. 20.1(A) is the same. The amount of work that must be done to bring a charge of **one coulomb** from A to B is thus a fixed number. It is called the **Potential Difference** between A and B.

POTENTIAL DIFFERENCE
The potential difference (V) between two points in an electric field is the work done in bringing a charge of +1 C from one point to the other.

It should be clear that potential difference is also the amount by which the potential energy of +1 C changes in going from one point to the other.

UNIT OF POTENTIAL DIFFERENCE
The unit of potential difference is the ***joule per coulomb*** **($J\ C^{-1}$)**. This unit is also called the ***volt*** **(V)**.

The volt is defined as follows:

THE VOLT
The potential difference between two points is 1 ***volt*** if 1 joule of work is done when 1 coulomb is brought from one point to the other.

1 VOLT = 1 JOULE per COULOMB $\quad$ **$1\ V = 1\ J\ C^{-1}$**

- The symbol for potential difference is V.
- Potential difference is a **scalar quantity** since work and energy are scalar quantities.
- Potential difference is sometimes called **voltage**.
- Potential difference is the potential energy difference per coulomb.

Experimentally it is found that the work done in bringing a charge Q from one point to another is proportional to the charge Q. It follows that:

$$\begin{pmatrix}\text{Work done in}\\ \text{bringing a charge } Q\\ \text{through voltage } V\end{pmatrix} = \begin{pmatrix}\text{Number of}\\ \text{coulombs}\\ \text{transferred}\end{pmatrix} \times \begin{pmatrix}\text{Work done in}\\ \text{transferring one}\\ \text{coulomb}\end{pmatrix}$$

i.e. Work done = Charge transferred × Voltage

$$W = QV \quad \text{or} \quad V = \frac{W}{Q}$$

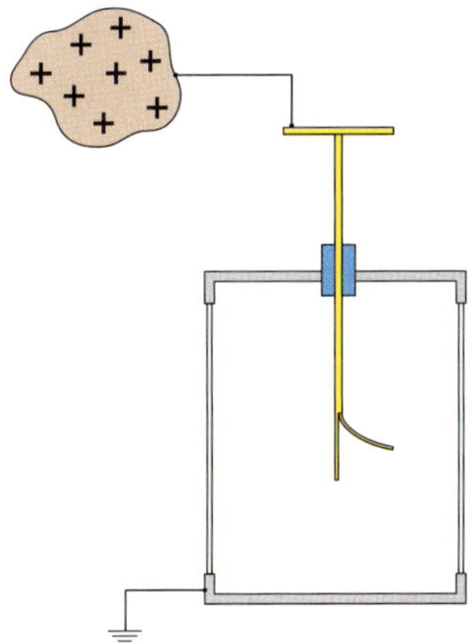

Fig. 20.2
The size of the divergence of the leaf indicates the size of the potential difference between the charged conductor and the ground.

Measuring Potential Difference

Potential difference can be measured with a voltmeter (page 253). Potential difference can also be measured approximately with an electroscope (Fig. 20.2). When the cap is connected to one point and the metal case to another, the size of the divergence of the leaf is a measure of the potential difference between the two points.

Relationship Between Potential Difference And Electric Field Strength

Suppose the points A and B in Fig. 20.1 are a fixed distance apart. Suppose the electric field between A and B is weak; then the work done in bringing +1 C from A to B is small. Thus if the electric field strength is small, the potential difference between the two points is small. Similarly if the field strength is large, then the potential difference between A and B will be large.

Fig. 20.3 shows a battery and its circuit symbol. It has two places on it called its terminals. For the present we can assume that one of the terminals, the negative terminal, has an excess of negative charge and the other, the positive terminal, has an excess of positive charge. There is thus an electric field in the space between the terminals. The higher the voltage between the terminals of the battery the stronger the electric field.

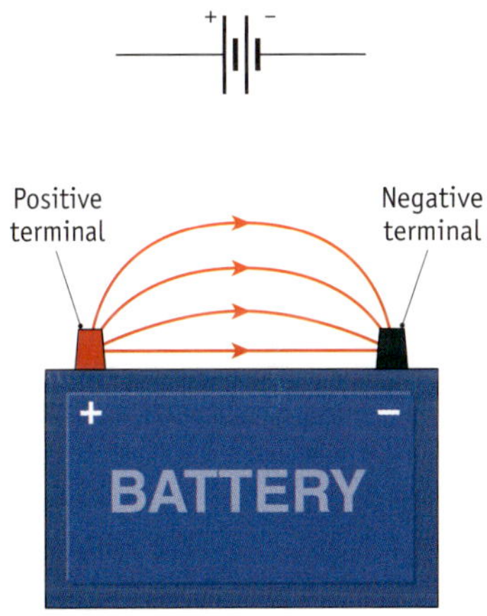

Fig. 20.3
A battery and its circuit symbol.

Electric Current

If the terminals of the battery in Fig. 20.3 are connected with a conductor (e.g. a piece of copper wire), there will be an electric field in the conductor. Free electrons in the conductor thus move under the influence of this field. The electrons move from the negative terminal through the wire back to the positive terminal. The battery continually maintains an excess of electrons on the negative terminal. The flow of electrons – which is also a flow of electric charge – is called an **electric current**. You will study this in more detail in Chapter 21.

ELECTRIC CURRENT
An ***electric current*** is a flow of electric charge.

Problem 1: The potential difference between two points is 12 V. Find the work done in transferring a charge of 8 C between two points.

Solution: $W = QV = (8)(12) = 96\ \text{J}$

Problem 2: The work done in bringing a charge of 4 C from one point to another is 10 J. What is the potential difference between the points?

Solution: $W = QV \Rightarrow V = \frac{W}{Q} = \frac{10}{4} = 2.5\ \text{V}$

Problem 3: The potential difference between two points is 100 kV. Find the work done when a charge of 3 μC is moved from one point to the other.

Solution: $W = QV = (3 \times 10^{-6})(100 \times 10^{3}) = 0.3\ \text{J}$

Problem 4: An electron of charge 1.6×10^{-19} C loses 4×10^{-16} J of energy as it moves from one point to another. What is the potential difference between the two points?

Solution: $W = QV \Rightarrow V = \frac{W}{Q} = \frac{(4 \times 10^{-16})}{(1.6 \times 10^{-19})} = 2500\ \text{V}$

Problem 5: Two oppositely charged plates are 3 cm apart. There is a uniform electric field of strength 2×10^{3} N C^{-1} between them.

(i) What is the force on a charge of +1 C if it is placed between the plates?
(ii) Find the work done in bringing a charge of 1 C from one plate to the other.
(iii) What is the potential difference between the plates?

Solution:

(i) Force on charge of +1 C, $F = EQ = (2 \times 10^{3})(1) = 2 \times 10^{3}\ \text{N}$
(ii) Work = Force × Distance = $(2 \times 10^{3})(3 \times 10^{-2}) = 60\ \text{J}$
(iii) Potential difference = Work done in bringing 1C from one plate to the other = 60 V

Problem 6: The potential difference between two points is 2000 V. An electron (of charge 1.6×10^{-19} C and mass 9×10^{-31} kg) is released at one of the points and moves towards the other under the action of the field. Find its speed when it arrives at the second point.

Solution: Potential energy lost by electron = work done on it by field and is given by: $W = QV$

Potential Energy lost $= W = QV = (1.6 \times 10^{-19})(2000) = 3.2 \times 10^{-16}\ \text{J}$

This lost potential energy becomes the kinetic energy of the electron.

Let v = speed of electron on reaching the second point.

$\frac{1}{2} mv^2 = 3.2 \times 10^{-16} \Rightarrow \frac{1}{2}(9 \times 10^{-31})\, v^2 = 3.2 \times 10^{-16}$

$\Rightarrow v^2 = \frac{(2)(3.2 \times 10^{-16})}{(9 \times 10^{-31})} = 7.111 \times 10^{14} \Rightarrow v = 2.7 \times 10^{7}\ \text{m s}^{-1}$

Problem 7: Prove that the unit of electric field strength, the N C^{-1}, is equivalent to the volt per metre (V m^{-1})

Solution: $1\ \text{V} = 1\ \text{J C}^{-1}$ and recall from Chapter 11 that $1\ \text{J} = 1\ \text{N m}$

$\therefore\ 1\ \text{V m}^{-1} = 1\ \text{J C}^{-1}\ \text{m}^{-1} = 1\ \text{N m C}^{-1}\ \text{m}^{-1} = 1\ \text{N C}^{-1}$

Potential at a Point

So far we have considered potential difference. It is sometimes convenient to talk about the potential at a point. To do this we must choose some zero of potential. Scientists have agreed to use the Earth itself as a convenient reference point and to measure the potential difference between other points and the Earth. The potential difference between a point and the Earth is called the potential of that point.

With this definition **the Earth is at zero potential**. Note also that as you add positive (+) charge to a conductor it becomes more difficult to bring a charge of +1 C from Earth to it, thus its potential increases.

POTENTIAL AT A POINT
The potential difference between a point and the Earth is called the ***potential*** of that point.

All Points on a Conductor Carrying Static Charge are at the Same Potential

If two points on a conductor were not at the same potential there would be a potential difference between them. There would thus be an electric field in the conductor and the charges on the conductor would move under its influence. But the charges are not moving, therefore all points on the conductor must be at the same potential, called the **potential of the conductor**.

EXERCISE 20.1

Take charge on electron = -1.6×10^{-19} C; *Mass of electron* = 9×10^{-31} kg.

1. The work done in bringing a charge of 2 C from one point to another is 6 J. What is the potential difference between the two points?
2. The work done in bringing a charge of 6 μC from one point to another is 6×10^{-5} J. What is the potential difference between the two points?
3. Calculate the work done in transferring a charge of 4 C between two points when the potential difference between the points 20 V.
4. Calculate the work done when a charge of 8 μC moves between two points if the potential difference between the points is 12 V.
5. 4.8×10^{-16} J of work is done in moving an electron between two points. What is the potential difference between the points?
6. Find the work done when an electron passes through:
   (i) 1 volt, (ii) 300 volts.
7. Two oppositely charged parallel plates are a distance of 20 cm apart. There is a uniform electric field of intensity 2×10^4 N C^{-1} between the plates. Find the work done in bringing +1 C from one plate to the other. What is the potential difference between the plates?
8. The potential difference between two oppositely charged parallel plates 2 cm apart is 400 V. Find:
   (i) the force acting on a charge of 1 C when placed between the plates,
   (ii) the electric field intensity between the plates,
   (iii) the force acting on an electron put between the plates,
   (iv) the potential energy lost by an electron if it moves from the negative to the positive plate,
   (v) the kinetic energy of an electron on arriving at the positive plate if its speed when released at the negative plate was zero,
   (vi) the speed of the electron on arriving at the positive plate.

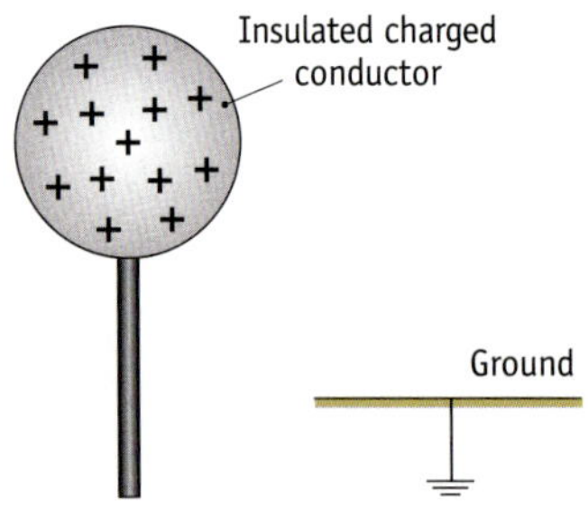

Fig. 20.4
As charge is added to the conductor the potential difference between it and the ground increases. Thus its potential increases.

CAPACITANCE

FIG. 20.4 shows an insulated conductor. As positive charge is added to the conductor, its potential increases. It is found that the potential of the conductor and the charge on it are directly proportional to each other, i.e. $Q \propto V$. It follows that: $Q = CV$, where C is a constant. The value of C depends on the shape and size of the conductor. It is called the **capacitance of the conductor**.

CAPACITANCE
The ***capacitance (C)*** of a conductor is the ratio of the charge on the conductor to its potential, i.e. $C = \frac{Q}{V}$

UNIT OF CAPACITANCE
The unit of capacitance is the ***farad (F)***.

THE FARAD
A conductor has a capacitance of 1 ***farad*** if placing a charge if 1 C on it raises its potential by 1 volt, i.e.

1 farad = 1 coulomb per volt (C V^{-1}).

The farad is a very large unit. In practice we use the microfarad, the nanofarad and the picofarad where:

1 microfarad	=	1 μF	=	10^{-6} F	
1 nanofarad	=	1 nF	=	10^{-9} F	
1 picofarad	=	1 pF	=	10^{-12}F	(see Maths tables p. 5)

Problem 8: A conductor has a potential of 6 V when a charge of 6 μC is placed on it. What is its capacitance?

Solution: $C = \frac{Q}{V} = \frac{(6 \times 10^{-6})}{6} = 1 \times 10^{-6}\ \text{F} = 1\ \mu\text{F}$

Problem 9: The capacitance of a conducting sphere is 20 pF. If its potential is 5000 volts, find the charge on it.

Solution: $C = \frac{Q}{V} \Rightarrow Q = CV = (20 \times 10^{-12})(5000) = 0.1 \times 10^{-6}\text{C} = 0.1\ \mu\text{C}$

THE CAPACITANCE OF A CHARGED CONDUCTOR IS INCREASED BY BRINGING AN OPPOSITELY CHARGED CONDUCTOR OR AN EARTHED CONDUCTOR NEAR IT

FIG. 20.5(A) shows a positively charged conductor. In FIG. 20.5(B) a negatively charged conductor is brought near it. The presence of the negatively charged plate reduces the potential of the positively charged conductor since it is now easier to bring a charge if +1 C from Earth to the conductor. Since $C = Q / V$ and V is decreased then the **capacitance** C of the conductor is increased. A similar argument holds if an earthed conductor (FIG. 20.5(C)) is brought near it. An induced negative charge appears on the earthed conductor, thus we are effectively bringing a negatively charged conductor near it and its capacitance is increased.

Fig. 20.5

Parallel Plate Capacitor

A parallel plate capacitor consists of two parallel plates separated by an insulator. **This insulator is called a dielectric.** A capacitor can store charge. When it is charged, the plates carry equal amounts of charge but of opposite sign. The capacitance of a parallel plate capacitor is defined as follows:

$$\text{Capacitance of parallel plate capacitor} = \frac{\text{Charge on one of the plates}}{\text{Potential difference between the plates}} \quad \text{i.e.} \quad C = \frac{Q}{V}$$

Problem 10: A capacitor has a capacitance of 50 μF. What is the potential difference between its plates if it stores a charge of 1.2 μC?

Solution: $C = \dfrac{Q}{V} \Rightarrow V = \dfrac{Q}{C} = \dfrac{(1.2 \times 10^{-6})}{(50 \times 10^{-6})} = 0.024$ volts

EXERCISE 20.2

Take $\varepsilon_0 = 8.9 \times 10^{-12}$ F m^{-1}

1. Placing a charge of 2 μC on a conductor raises its potential by 10 000 V. Calculate the capacitance of the conductor.
2. What is the capacitance of a conductor if placing a charge of 4 μC on it raises its potential by 12 V?
3. The capacitance of an insulated conducting sphere is 8 pF. If it is raised to a potential of 1 000 000 volts, find the charge on it.
4. The capacitance of the dome of a typical school laboratory Van de Graff generator is 2×10^{-11} F. It typically reaches a potential of 300 kV. How much charge is on the dome?
5. A charge of 4 μ*C* is placed on a conductor of capacitance 3 pF. Find the increase in potential of the conductor.
6. Explain with the aid of a diagram how the capacitance of an insulated charged conductor is increased by bringing an earthed conductor or an oppositely charged conductor near it.
7. What is the capacitance of a capacitor which has a charge of 5 μC when the potential difference between its plates is 12 V?
8. A capacitor has a capacitance of 50 μF. What is the charge on one of its plates if the potential difference between them is 100 V?

Charging a Capacitor

Fig. 20.6(a) shows a circuit containing a battery, a switch and a capacitor. Note the circuit symbol for the capacitor. Closing the switch in the circuit in Fig. 20.6(a) causes the following to happen:

- Electrons flow from the negative terminal of the battery to the right-hand plate of the capacitor. Thus negative charge builds up on this plate.
- Electrons also flow from the left-hand plate of the capacitor to the positive terminal of the battery. Thus positive charge builds up on the left-hand plate of the capacitor (Fig. 20.6(b)).
- As charge builds up on the plates of the capacitor the voltage across the capacitor increases. As this happens the flow of electrons decreases.
- The flow of electrons, i.e. the current, stops when the voltage across the plates of the capacitor is equal to the voltage of the battery.
- If the battery is now removed, the capacitor remains charged, i.e. it has negative charge on one plate and an equal amount of positive charge on the other (Fig. 20.6(c)).
- If the plates are at some later stage connected by a conductor, the capacitor discharges as the electrons flow back from the negative plate to the positive plate.

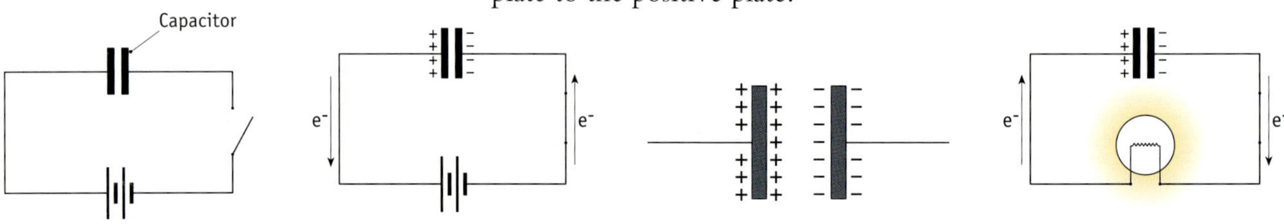

(a) Uncharged capacitor. (b) Capacitor charging. (c) Charged capcitor. (d) Bulb flashes as capacitor discharges.

Fig. 20.6

Experiment

To Show that a Charged Capacitor Stores Energy.

- Set up the equipment shown in Fig. 20.6(a).
- Charge the capacitor by connecting the battery across it.
- Remove the source and connect the plates to a bulb with pieces of wire (Fig. 20.6(d)).
- The bulb will flash as the capacitor discharges, showing that it stores energy.
- Alternatively, connect the charged capacitor to a low-voltage electric motor. As it discharges, the motor will rotate.The energy that was stored in the capacitor is converted to rotational kinetic energy.

Common Uses of Capacitors

- You will see on page 325 that a capacitor allows alternating current (a.c.) to flow through it but blocks direct current (d.c.).
- A variable capacitor is used to tune to a particular station on a radio. As you move from one station to another you are actually adjusting the value of a variable capacitor.
- Capacitors are used to smooth out variations in direct current (see Chapter 33).

- Capacitors are used in a camera flash gun. The capacitor is charged slowly from a battery and discharged quickly through the bulb giving the flash.
- Capacitors can be used to allow alternating signals of certain frequencies past, but block others in a process called filtering.

Formula for Capacitance of Parallel Plate Capacitor

The capacitance of a parallel plate capacitor is given by the formula:

$$C = \frac{\varepsilon A}{d}$$

Where:
- A is the area of overlap of the plates
- d is the distance between the plate
- ε is the permittivity of the dielectric.

You need to know this formula, but you do not need to be able to derive it. Fig. 20.7 shows two parallel plates. The area of overlap is the shaded area, i.e. it is the area of one of the plates that overlaps the other.

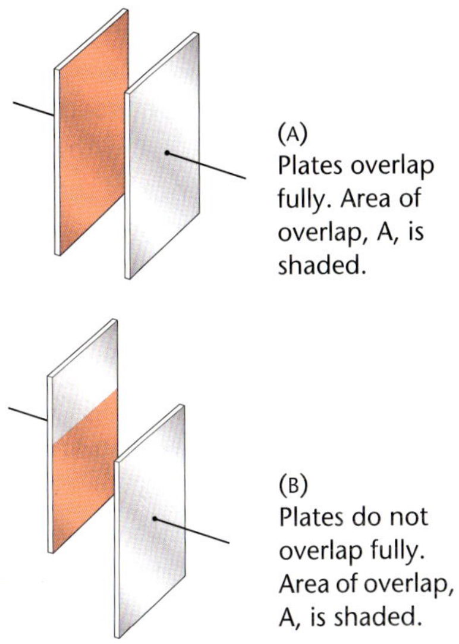

Fig. 20.7

Problem 11: The area of overlap of the plates of an air spaced capacitor is 20 cm^2. The distance between the plates is 1 mm.

(i) Given $\varepsilon_o = 8.9 \times 10^{-12}$ F m^{-1}, find the capacitance of the capacitor.

(ii) If the space between the plates is now filled with mica of relative permittivity 7 calculate the capacitance of the capacitor.

Solution: $A = 20 \text{ cm}^2 = 20 \times 10^{-4} \text{ m}^2$ $\quad d = 1 \text{ mm} = 1 \times 10^{-3} \text{ m}$

(i) $C = \frac{\varepsilon A}{d} = \frac{(8.9 \times 10^{-12})(20 \times 10^{-4})}{(1 \times 10^{-3})} = 1.78 \times 10^{-11} \text{ F} = 0.178 \text{ pF}$

(ii) Permittivity of mica $\varepsilon = \varepsilon_r \varepsilon_o = (7)(8.9 \times 10^{-12}) = 6.23 \times 10^{-11}$ F m^{-1}

$$C = \frac{\varepsilon A}{d} \Rightarrow \frac{(6.23 \times 10^{-11})(20 \times 10^{-4})}{(1 \times 10^{-3})} = 1.25 \times 10^{-10} \text{ F}$$

Problem 12: Find the distance between the plates of an air-spaced capacitor of 2 pF if the area of one side of one of the plates is 100 cm^2 ($\varepsilon_o = 8.9 \times 10^{-12}$ F m^{-1}).

Solution: $100 \text{ cm}^2 = 100 \times 10^{-4} \text{ m}^2 = 0.01 \text{ m}^2$

$$C = \frac{\varepsilon A}{d} \Rightarrow d = \frac{\varepsilon A}{C} = \frac{(8.9 \times 10^{-12})(0.01)}{(2 \times 10^{-12})} = 0.0445 \text{ m} = 4.45 \text{ cm}$$

Problem 13: Prove that the **unit of permittivity is the farad per metre (F m^{-1})**.

Solution:

$$C = \frac{\varepsilon A}{d} \Rightarrow \varepsilon = \frac{Cd}{A} \Rightarrow \text{Unit of } \varepsilon = \frac{(\text{Unit of } C)(\text{Unit of } d)}{(\text{Unit of } A)}$$

$$= \frac{(\text{farad})(\text{metre})}{(\text{metre squared})} = \text{farad per metre} = \text{F m}^{-1}$$

EXPERIMENT

TO SHOW THAT THE CAPACITANCE OF A PARALLEL PLATE CAPACITOR DEPENDS ON:

- the Distance Between the Plates,
- the Common Area of the Plates,
- the Nature of the Dielectric.

Use the equipment in FIG. 20.8. The divergence of the leaf is a measure of the potential difference between the plates.

Since $C = Q / V$ and the amount of charge Q is fixed, it follows that the greater the divergence of the leaf the smaller the capacitance and vice versa.

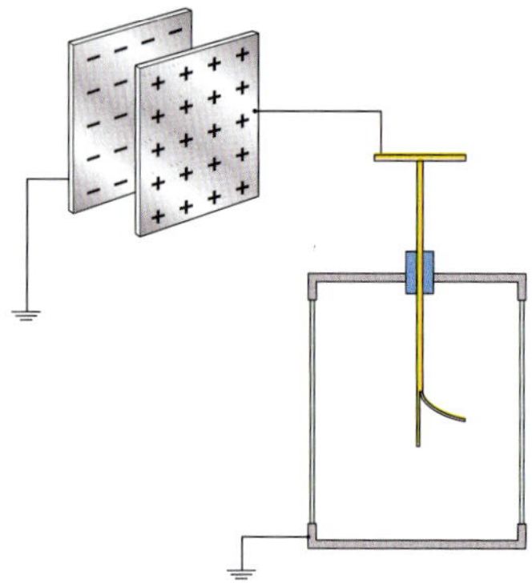

Fig. 20.8

- Charge the plates by connecting them across a high voltage source (say 2000 V).
- Move the plates closer together, i.e. decrease d.
  The divergence of the leaf decreases $\Rightarrow$ C increases.
  If d is increased, the opposite happens.
- Decrease the overlap area, and divergence increases $\Rightarrow$ C decreases.
  If the overlap area is increased the opposite happens.
- Place different slabs of insulating material between the plates. The divergence will be seen to be less than what it is for air.
- Different materials cause the capacitance to increase over its value when the dielectric is air.

FORMULA FOR THE ENERGY STORED IN A CHARGED CAPACITOR

The **energy stored (W)** in a charged capacitor is given by:

$$W = \tfrac{1}{2}\, C\, V^2$$

You need to know this formula, but do not need to be able to derive it.

Problem 14: A capacitor of capacitance 2 μF is charged to a potential difference of 200 V. Find the energy stored in it.

Solution: $E = \tfrac{1}{2} CV^2 = \tfrac{1}{2}(2 \times 10^{-6})(200)^2 = 0.04\ \text{J}$

Problem 15: A capacitor of capacitance 0.47 μF carries a charge of 2.0 μC. Calculate:

(i) the potential difference between the plates,

(ii) the energy stored.

Solution:

(i) $C = \dfrac{Q}{V} \Rightarrow V = \dfrac{Q}{C} = \dfrac{(2 \times 10^{-6})}{(0.47 \times 10^{-6})} = 4.26\ \text{V}$

(ii) $W = \tfrac{1}{2} CV^2 = \tfrac{1}{2}(0.47 \times 10^{-6})(4.26)^2 = 4.26 \times 10^{-6}\ \text{J}$

Problem 16: A capacitor has a capacitance of 6.3 μF. What is the charge on the plates when the energy stored is 0.44 mJ?

Solution: $C = \dfrac{Q}{V} \Rightarrow V = \dfrac{Q}{C}$ Substituting this for V in: $W = \tfrac{1}{2} CV^2$ gives: $W = \tfrac{1}{2}\dfrac{Q^2}{C}$

$\therefore\ 0.44 \times 10^{-3} = \dfrac{Q^2}{2(6.3 \times 10^{-6})} \Rightarrow Q^2 = (2)(0.44 \times 10^{-3})(6.3 \times 10^{-6}) = 5.544 \times 10^{-9}$

$\Rightarrow Q = \sqrt{5.544 \times 10^{-9}} = 7.4458 \times 10^{-5}\ \text{C} = 74\ \mu\text{C}.$

Study the next problem after you have studied current electricity in Chapters 22 and 23.

Problem 17: When the switch is closed, the current flowing at a particular instant in the circuit shown in FIG. 20.9 is 0.5 mA. Find:

(i) the potential difference across the capacitor at this instant,

(ii) the charge on the capacitor at this instant,

(iii) the work done in placing a charge of 1 μC on the capacitor.

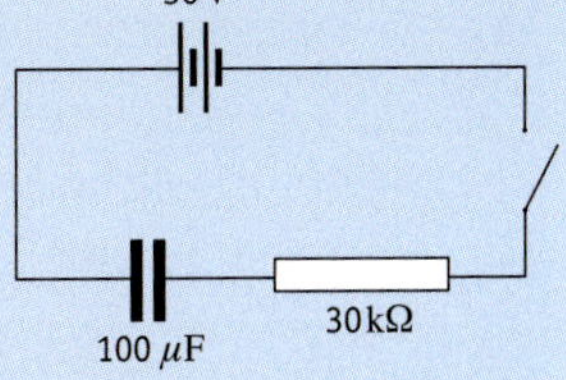

Fig. 20.9

Solution:

(i) Potential difference across resistor

$= IR = (0.5 \times 10^{-3})(30\ 000) = 15$ V

$\therefore$ Potential difference across capacitor $= 30 - 15 = 15$ V

(ii) Charge on capacitor $Q = CV = (100 \times 10^{-6})(15) = 1.5 \times 10^{-3}$ C

(iii) When charge on capacitor is 1 μC the potential difference across it is:

$$V = \frac{Q}{C} = \frac{(1 \times 10^{-6})}{(100 \times 10^{-6})} = 0.01 \text{ V}$$

Work done = Energy stored in capacitor

$= \frac{1}{2} CV^2 = \frac{1}{2}(100 \times 10^{-6})(0.01)^2 = 5 \times 10^{-9}$ J

EXERCISE 20.3

Take $\varepsilon_0 = 8.9 \times 10^{-12}$ F m^{-1}

1. The common area of the plates of an air-spaced parallel plate capacitor is 0.02 m^2. The distance between the plates is 0.001 m. Calculate the capacitance of the capacitor.
2. The common area of the plates of an air-spaced parallel plate capacitor is 150 cm^2. The distance between the plates is 1 mm. Calculate the capacitance of the capacitor.
3. A parallel plate air-spaced capacitor is to have a capacitance of 1 F. If the distance between the plates is 1 mm, find the area of one of the plates.
4. Find the capacitance of a parallel plate air-spaced capacitor if the area of one of the plates is 100 cm^2 and the distance between the plates is 2 mm. Find the capacitance if the space between the plates is filled with perspex which has a relative permittivity of 2.6.
5. A parallel plate capacitor has a distance of 1 mm between the plates, each of which has an area of 25 cm^2. It has a mica dielectric. Find the charge on either of the plates when the potential difference between the plates is 500 V. (Relative permittivity of mica = 7.)
6. Find the energy stored in a capacitor of capacitance 6 mF if the potential difference between its plates is 200 V.
7. Find the energy stored in a capacitor of capacitance 2 μF if each plate has a charge of 4 μC.
8. The capacitance of a capacitor is 4.6 μF. What is the energy stored in it if the potential difference between the plates is 8 V?
9. A capacitor stores a charge of 7 μC and has potential difference of 30 V across it. What energy does it store?
10. A capacitor has a capacitance of 2.4 μF. What is the charge on the plates if the energy stored is 23 m J?

Take $\varepsilon_0 = 8.9 \times 10^{-12}$ F m^{-1}

11. What capacitance is required to store 0.01 J of energy when there is a potential difference of 12 V across the plates?

12. Find the potential difference across and the charge on each capacitor shown in FIG. 20.10.

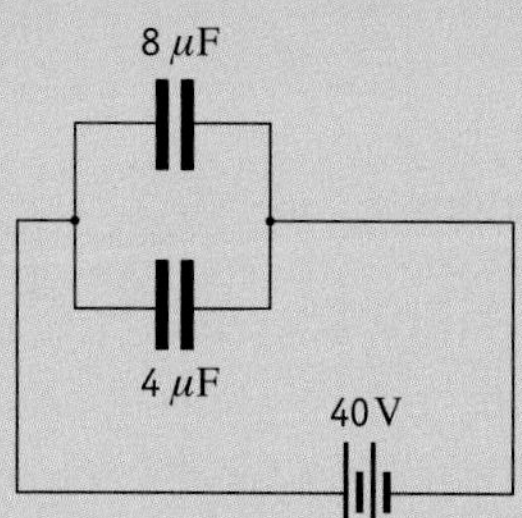

Fig. 20.10

13. In FIG. 20.11 the charge on each capacitor is the same. Find the potential difference across and the charge on each capacitor.

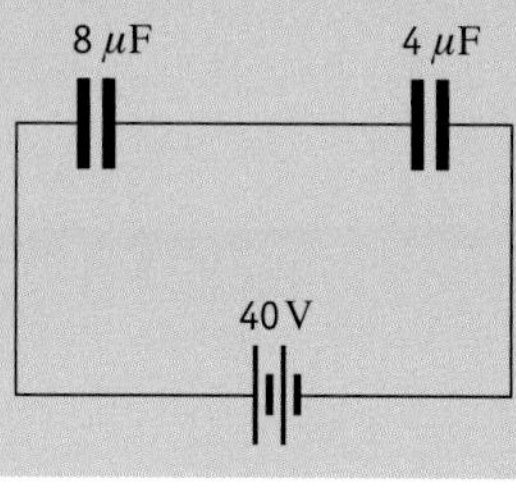

Fig. 20.11

14. When the switch is closed, the current flowing at a particular instant in the circuit shown in FIG. 20.12 is 0.2 mA. Find:

(i) the potential difference across the capacitor at this instant,

(ii) the charge on the capacitor at this instant,

(iii) the work done in placing 2 μC on the capacitor.

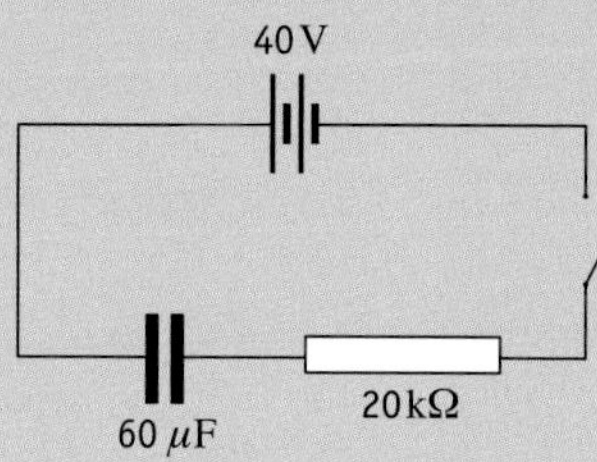

Fig. 20.12

CHAPTER CHECKLIST

- **State** the unit of: Potential difference; Potential; Capacitance.
- **Define**: Potential difference; Potential; Capacitance; The volt; The farad.
- **Recall** and use the formulae $C = \frac{Q}{V}$ and $W = QV$ to solve problems.
- **List** four common uses of a capacitor.
- **Describe** and carry out an experiment to show that a charged capacitor stores energy.
- **Describe** and carry out an experiment to show that the capacitance of a parallel plate capacitor depends on the common area, the distance between the plates and the nature of the dielectric.
- **Recall** and use the formulae: $C = \frac{\varepsilon A}{d}$ and $W = \frac{1}{2}CV^2$ to solve problems.

Current and Charge

21 CHAPTER

Three Effects of an Electric Current

If we connect a bulb, a battery, a switch and a beaker of dilute sulphuric acid with pieces of copper wire (Fig. 21.1) and close the switch, each of the following effects is seen to occur:

A HEATING EFFECT

Heat is given out from the bulb. Some heat is also given out from the wires and from the inside of the battery. However, the heat produced in the wires and the battery is small and difficult to detect.

A MAGNETIC EFFECT

A magnetic compass needle, which normally lines up North-South, deflects from its N-S position when it is brought near the wire, indicating the presence of a magnetic field around the wire.

A CHEMICAL EFFECT

A chemical reaction occurs in the beaker and bubbles of gas are seen coming from the wires dipped in the acid.

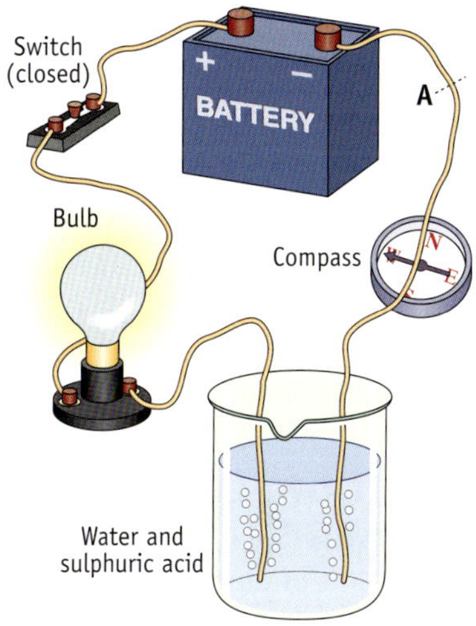

Fig. 21.1

If the wire is broken anywhere, the switch opened or the acid removed from the beaker, the three effects stop. If the break is repaired, they start again. This indicates that something flows through the wire and through the acid; that something is called **electric charge**. A flow of electric charge is called an **electric current**. The path around which the charge flows is called a **circuit**.

ELECTRIC CURRENT
An ***electric current*** is a flow of electric charge.

Conductors and Insulators

Any substance through which electric charge can flow is called **an electrical conductor**. In Fig. 21.1 the wire and the acid are conductors. A substance through which electric charge cannot flow is called **an electrical insulator**. Plastic, glass and rubber are good insulators. If one of the pieces of wire in Fig. 21.1 was replaced with a piece of plastic, the current would not flow and the three effects would stop.

CONDUCTOR
A substance through which electric charge can flow is called a ***conductor***.

INSULATOR
A substance through which electric charge cannot flow is called an ***insulator***.

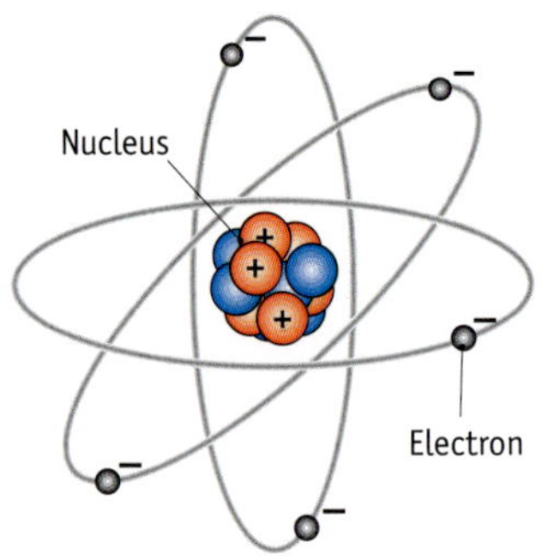

Fig. 21.2
An atom

What is Electric Charge?

Matter is made up of atoms. Every atom has a central part called the **nucleus** which contains particles called protons. Particles called **electrons** orbit in the space surrounding the nucleus (Fig. 21.2). Electrons are attracted to the protons and are repelled by other electrons. Protons repel protons. These forces of attraction and repulsion are not gravitational. They are much stronger and are called **electrostatic forces** (See page 221). Particles that exert electrostatic forces on each other are said to be **electrically charged**. The **electrons are said to be negatively charged** and the **protons positively charged**. The amount of positive charge on a proton is the same size as the amount of negative charge on an electron. **An electric current is charged particles moving.**

Electric Current in a Metal Conductor

In a metal conductor, some electrons are free to wander from atom to atom. If the ends of a piece of metal are connected to a battery, the electrons are found to move through the metal. Thus:

An ***electric current*** in a ***metal*** conductor is a ***flow of electrons***.

UNIT OF ELECTRIC CHARGE
The unit of electric charge is the ***coulomb*** **(C)**.

Unit of Electric Charge

The unit of electric charge is the **coulomb (C)**. The definition of the coulomb is on page 308. One coulomb is the amount of charge on about 6.25×10^{18} electrons. You do not need to remember this number.

UNIT OF ELECTRIC CURRENT
The unit of electric current is the ***ampere*** **(A)**.

1 ampere
= 1 coulomb per second
$1\ A = 1\ C\ s^{-1}$

Size of an Electric Current

SIZE OF AN ELECTRIC CURRENT
The ***size of an electric current*** in a conductor is by definition the amount of charge passing any point of that conductor per second.

The letter I is the symbol for electric current. The unit of current is the **ampere (A)**. If in Fig. 21.1 two coulombs of charge pass any point (such as A) per second, the current is 2 amperes. The exact definition of the ampere is given on page 308.

Since current is the amount of charge passing per second it follows that if a steady current of I amperes flows for a time of t seconds, the amount of charge Q (in coulombs) that passes is given by: $Q = It$.

i.e. Charge gone past = Steady current × Time

$$Q = It$$

Problem 1: If a charge of 10 C passes a point in a circuit at a steady rate in a time of 5 s, what current flows in the wire?

Solution: $Q = It \Rightarrow I = \frac{Q}{t} = \frac{10}{5} = 2\text{ A}$

Problem 2: How much charge passes a point in a circuit in which a steady current of 5 A flows for:
(a) 1 s, (b) 12 s, (c) 2 hours?

Solution:
(a) $Q = It = (5)(1) = 5\text{ C}$
(b) $Q = It = (5)(12) = 60\text{ C}$
(c) Convert 2 hours to seconds: $2\text{ h} = (2)(60)(60) = 7200\text{ s}$
$Q = It = (5)(7200) = 36\,000\text{ C}$

CHARGE ON THE ELECTRON

In a metallic conductor the moving charges are electrons. Each electron has a charge of 1.6×10^{-19} coulombs. This is a very **small** amount of charge.

Charge on one electron $e = 1.6 \times 10^{-19}$ coulombs $= 0.000\,000\,000\,000\,000\,000\,16$ coulombs

Problem 3: The charge on one electron is 1.6×10^{-19} C. How many electrons are needed to give a charge of 1 coulomb? Find the number of electrons passing any point in a circuit, per second, in which a current of 10 A is flowing.

Solution: Let n = the number of electrons needed.
Total charge on n electrons = (n)(charge on 1 electron) = $(n)(1.6 \times 10^{-19})$
If this is one coulomb, $(n)(1.6 \times 10^{-19}) = 1$

$$\Rightarrow n = \frac{1}{1.6 \times 10^{-19}} = 6.25 \times 10^{18} \text{ electrons}$$

$10\text{ A} = 10\text{ C s}^{-1} \Rightarrow (10)(6.25 \times 10^{18})$ electrons passing per second
i.e. In a 10 A current 6.25×10^{19} electrons pass any point per second.

CONVENTIONAL CURRENT

A battery has two terminals onto which we connect the wires. One is called the **positive terminal (+)** and the other the **negative terminal (-)**. FIG. 21.3 shows the circuit symbol for a battery. Note that the longer side is the positive terminal and the shorter side is the negative terminal.

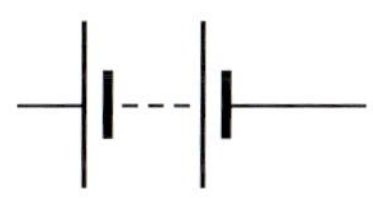

Fig. 21.3
Circuit symbol for a battery. The longer side is the positive terminal.

Scientists knew a lot about electric current even before they knew about the existence of electrons. Over 150 years ago they knew that something flowed in a circuit such as FIG. 21.1. They did not know in which direction an electric current in a metal conductor actually flowed and unfortunately guessed incorrectly. They said an electric current flowed from the positive terminal to the negative terminal. Today we still say that electric current (called **conventional current**) flows from + to – , even though in a metal it is actually a flow of electrons moving in the opposite direction, i.e. from – to + (FIG. 21.4) . Whenever the word current is used it will always be understood to mean conventional current.

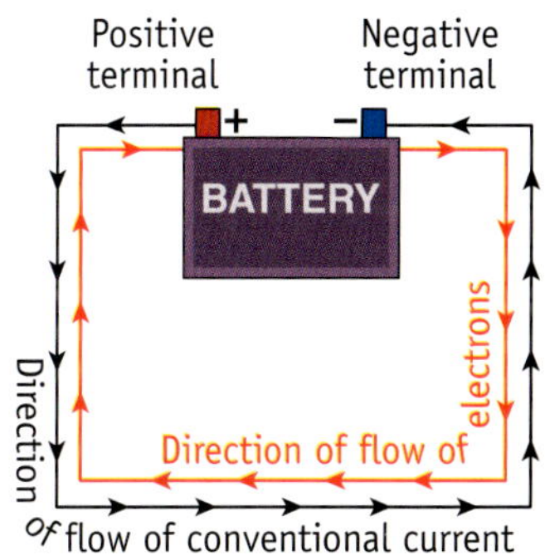

Fig. 21.4

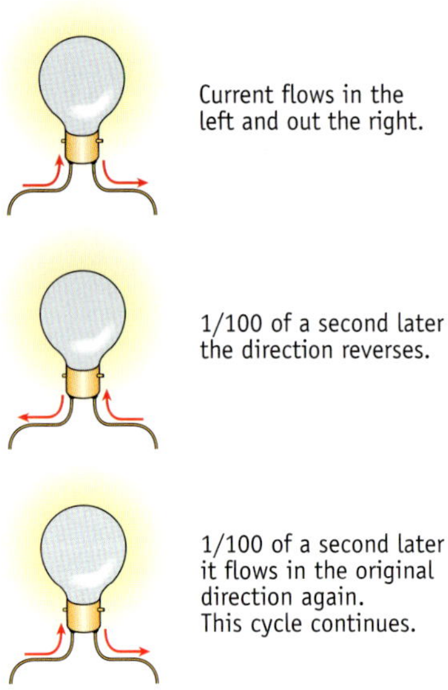

Fig. 21.5
A bulb connected to a.c. mains. The current reverses direction periodically.

Fig. 21.6
An ammeter.

> The ***current is the same*** at every point ***in a series circuit***.

$1\text{ mA} = \frac{1}{1\,000}\text{A} = 10^{-3}\text{A}$

$1\,\mu\text{A} = \frac{1}{1\,000\,000}\text{A} = 10^{-6}\text{A}$

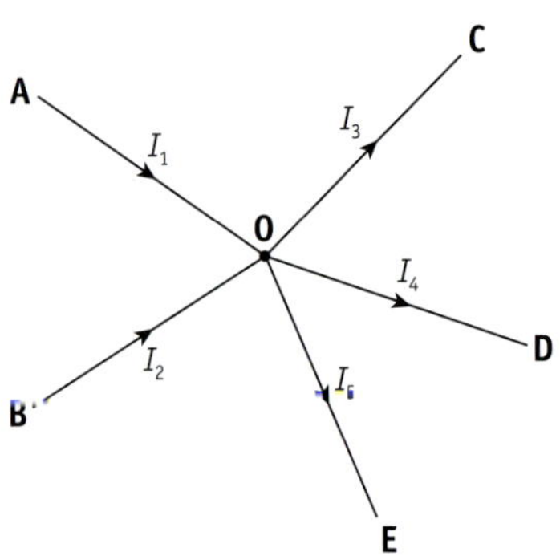

Fig. 21.8

> **NOTE**
> Conventional current is said to flow from + to –
> In a metal, electrons actually flow in the opposite direction.

Direct Current (d.c.) and Alternating Current (a.c.)

In the above circuits, the current always flows in one direction. Such an electric current is called a **direct current (d.c.)**. An electric current caused by a chemical cell or a battery is a direct current.

If an electric current in a circuit reverses direction every so often, it is called **alternating current (a.c.)**. For example, the current that flows through an ordinary domestic light bulb when connected to the mains electricity supply reverses direction 100 times every second and is therefore an alternating current (Fig. 21.5). For the present, we shall deal only with direct current.

Measuring Direct Current

To measure the size of an electric current we use an **ammeter** (Fig. 21.6) which gives the size of the current in amperes. Smaller currents are measured on meters called **milliammeters** or **microammeters**. A **galvanometer** is another word for a milliammeter or microammeter (Fig. 21.7).

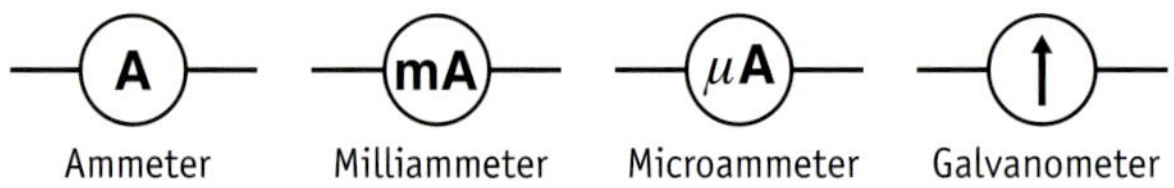

Fig. 21.7

Current in a Series Circuit

The circuit in Fig. 21.1 (page 245) is called a series circuit since there is only one path for the current to follow. Charge does not build up or leak away from the wire at any point, thus the same amount of charge passes any point every second. **The current is the same at every point in the wire in a series circuit.** The size of the current flowing out of the battery is the same size as the current flowing back into the battery.

As the charges move around the circuit they lose potential energy. Their average speed and average kinetic energy remain the same. The charges themselves do not get used up; the current does not get used up.

Current at a Junction of Conductors

In Fig. 21.8 five conductors meet at a point O. O is called a **junction**. If the wires A and B are carrying currents I_1 and I_2 towards O, and if wires C, D and E carry currents I_3, I_4 and I_5 away from O, then no charge is lost or stored up at the junction and it follows that:

$$I_1 + I_2 = I_3 + I_4 + I_5$$

> i.e. (The sum of the currents flowing into a junction) = (The sum of the currents leaving the junction.)

Problem 4: Find the current flowing through wire W in FIG. 21.9.

Solution: Sum of currents entering junction

$$= \text{Sum of currents leaving}$$

$$\Rightarrow \quad 1 + 4 + x = 6 + 3$$

$$\Rightarrow \quad x = 4\text{ A}$$

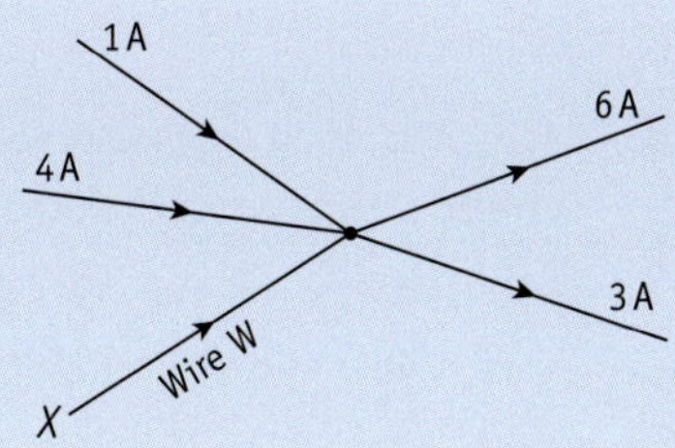

Fig. 21.9

MEASURING CURRENT WITH AN AMMETER

To measure the current flowing in a circuit such as that in FIG. 21.10(A) with an ammeter, the circuit must be broken at some point and the ammeter inserted, i.e. **an ammeter must be connected in series** with the circuit in which the current to be measured is flowing. It doesn't matter where in a series circuit an ammeter is inserted since the current is the same everywhere. In FIG. 21.10(B) each ammeter gives the same reading.

An ***ammeter must be connected in series*** with the part of the circuit in which the current to be measured is flowing.

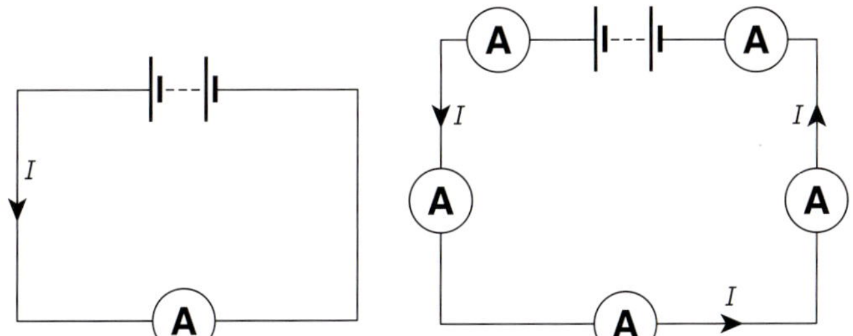

(A) (B)

Fig. 21.10

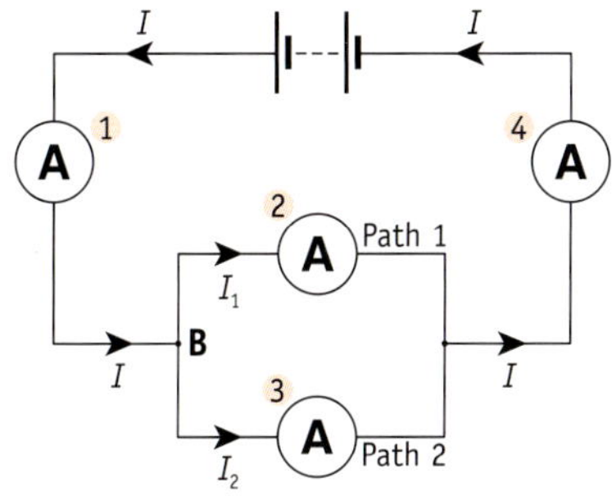

Fig. 21.11 Path 1 and path 2 are said to be in parallel with each other. $I = I_1 + I_2$

CURRENT IN PARALLEL CIRCUITS

At the point B in FIG. 21.11 the current splits up, some going through path 1 and some going through path 2. Ammeter 1 reads the same current as ammeter 4 and the sum of the values on ammeters 2 and 3 is equal to the value read on ammeter 1 (and hence on ammeter 4 also), i.e. $I = I_1 + I_2$. Path 1 and path 2 are said to be **'in parallel'** with each other. The same current does not necessarily pass through each path. The current in each path would be the same if the paths were made from identical pieces of wire.

EXERCISES 21.1

1. Convert each of the following to amperes:
   (i) 1 mA (ii) 0.05 mA
   (iii) 50 μA (iv) 1000 mA (v) 0.2 μA
2. Convert each of the following to milliamperes:
   (i) 1 A (ii) 100 A
   (iii) 0.025 A (iv) 1 μA (v) 0.0006 A
3. Find the value of the current X in each part of FIG. 21.12.
4. A current of 3 A flows through a bulb. How much charge enters the bulb in:
   (i) 1 second, (ii) 1 minute, (iii) 1 hour?
5. Electric charge passes a point in a circuit at the rate of:
   (i) 10 coulombs every 10 seconds, (ii) 1 coulomb per minute, (iii) 10 coulombs per second.
   Find the current flowing in each case.

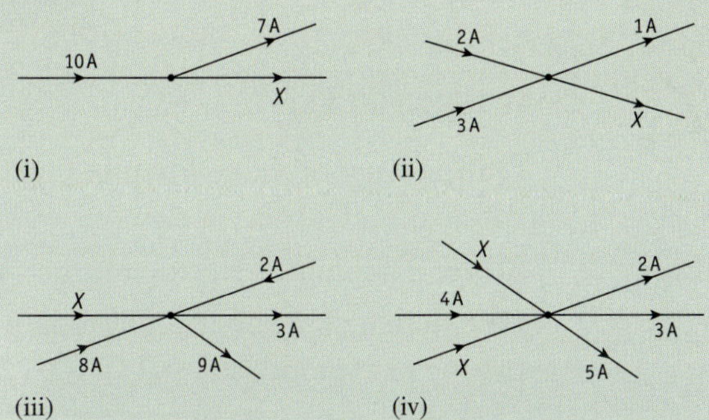

Fig. 21.12

6. How much charge passes a point in a circuit for which a current of 6 A flows for 4 hours?

7. If the charge on one electron is 1.6×10^{-19} C, how many electrons are needed to give a charge of one coulomb?

8. If the charge on one electron is 1.6×10^{-19} C, how many electrons flow past any point in a circuit for which a 20 A current flows for 6 seconds?

9. 2×10^{20} electrons pass a point in a circuit every second. What current flows in the circuit? Assume that the charge on one electron is 1.6×10^{-19} C?

10. How long does it take a current of 5 A to transfer a charge of 36 000 C?

11. In the tube of a television set, a beam of fast moving electrons travelling through a vacuum strikes a screen producing a picture. If the average beam current is 1 mA, how many electrons strike the screen during a 1 hour TV program?
(Charge on one electron = 1.6×10^{-19} C)

12. What is the reading on each of the ammeters in FIG. 21.13?

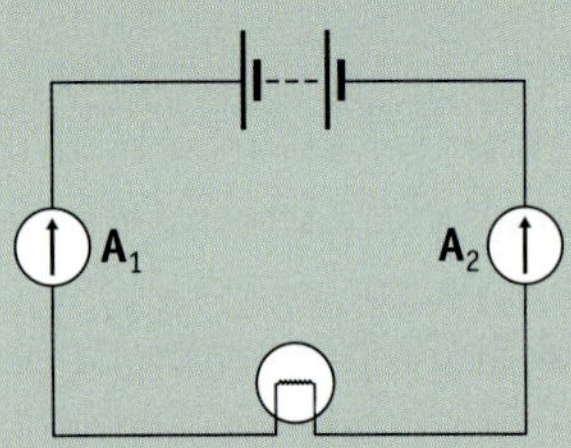

(i) A_1 reads 2 A

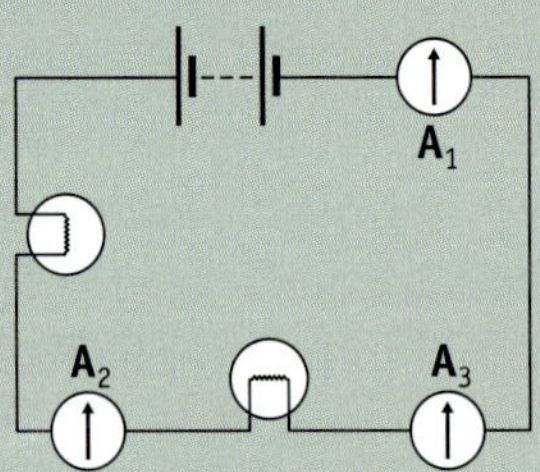

(ii) A_1 reads 2 A

Fig. 21.13

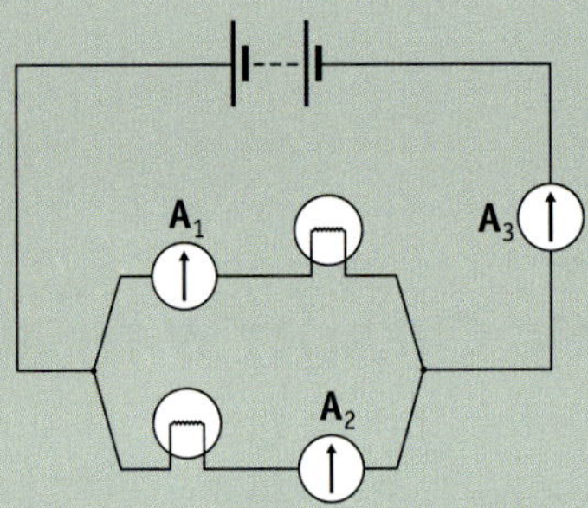

(iii) A_3 reads 4 A; the bulbs and meters are identical

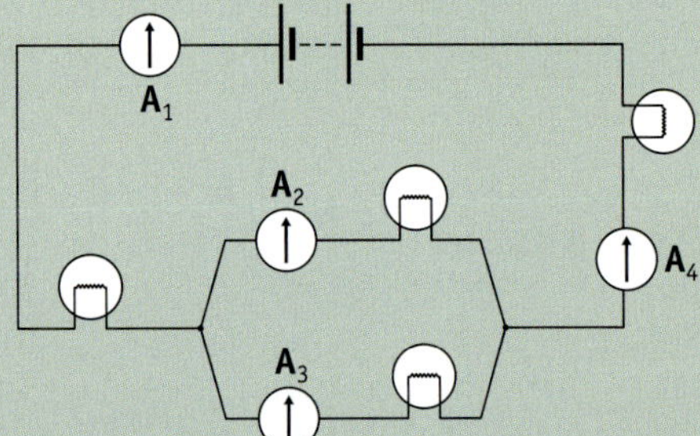

(iv) A_3 reads 4 A; the bulbs and meters are identical

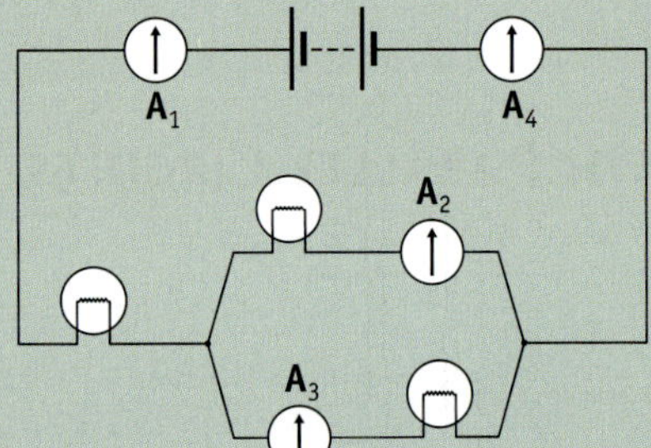

(v) The bulbs are different; A_2 reads 2 A and A_1 reads 10 A

Fig. 21.13 (contd.)

CHAPTER CHECKLIST

- **Define** each of the following: Electric current; Conductor; Insulator; Direct Current (d.c.); Alternating Current (a.c.).
- **State**: The unit of electric charge; The unit of electric current.
- **Recall** that: An electric current in a metal is a flow of electrons; Current is the amount of charge passing per second ($1 \text{ A} = 1 \text{ C s}^{-1}$); Conventional current is said to flow from + to –, but in a metal electrons flow in the opposite direction; The current is the same at every point in a series circuit; An ammeter must be connected in series; The sum of the currents entering a junction equals the sum of the currents leaving the junction.
- **Recall** and use the formula: $Q = It$ to solve problems.
- **List** three effects of an electric current.

Electromotive Force and Potential Difference

CHAPTER 22

Energy Changes in a Simple Electric Circuit

When current flows in a conductor, heat energy is given out at every point in that conductor. This energy comes from the battery (or other source, e.g. a generator) that is driving the current through the conductor. In a battery, chemical energy is converted into electrical energy. The electric charges get electric potential and kinetic energy as they pass through the battery. As they move around the circuit their average speed and average kinetic energy remain constant, but they lose electric potential energy. This lost potential energy appears as heat. FIG. 22.1 shows this in a simplified way.

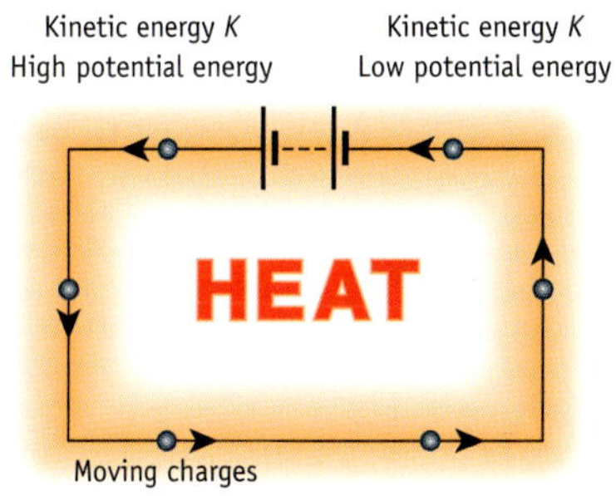

Fig. 22.1
The average kinetic energy of the moving charges remains the same. The lost potential energy is converted into heat as the charges move around the circuit.

Potential Difference Between Two Points in a Circuit

The amount of energy lost by one coulomb in passing between two points of a circuit is called the **potential difference (p.d.)** between those two points.

> **UNIT OF POTENTIAL DIFFERENCE**
> The unit of potential difference is the ***joule per coulomb*** **(J C^{-1})** which is also called the ***volt*** **(V)**.

> **POTENTIAL DIFFERENCE**
> The ***Potential Difference*** **(p.d.)** between two points in a circuit is the energy lost by one coulomb as it moves from one point to the other.

Since potential difference is measured in volts, potential difference is also called **voltage**.

Suppose the potential difference between the points A and B in the circuit in FIG. 22.2 is 4 volts. This means that 4 joules of potential energy are lost by each coulomb as it moves from A to B.

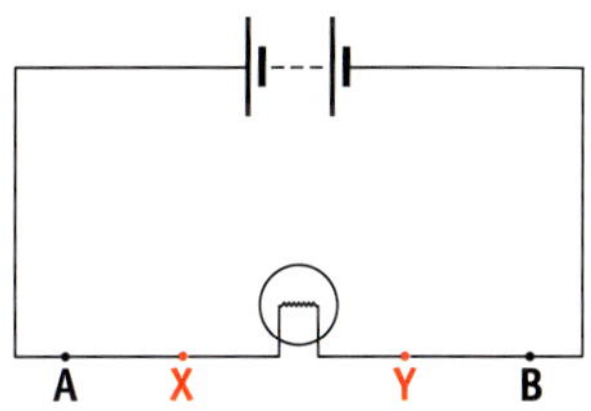

Fig. 22.2

Another Way of Looking at p.d.

Consider the conductor in FIG. 22.3(A), where for simplicity the moving charges have been drawn as coulombs. Suppose we watch the circuit and wait until 1 C of charge passes A (FIG. 22.3(B)). As this happens, all of the other charges between A and B also give out energy. It should be obvious that since each coulomb is losing energy at the same rate we can say:

> The ***potential difference*** between two points in a circuit is the amount of energy converted from electrical to other forms between the two points when 1 coulomb of charge passes any point of that circuit.

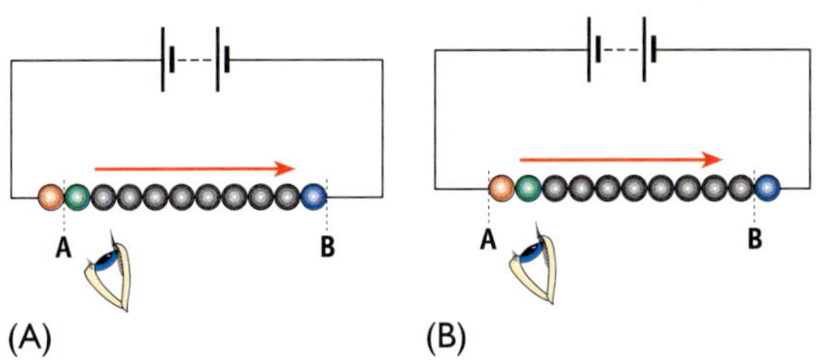

Fig. 22.3
The total energy given out between A and B when one coulomb passes A is the same as the energy given out by one single coulomb as it moves all the way from A to B.

Thus in FIG. 22.2 as 1 C passes any point (such as X or Y) 4 J of energy in total are lost by all the moving charges between A and B.

It follows that if the potential difference between two points in a circuit is V volts and if the energy given out between the two point is W joules when a charge of Q coulombs passes any point in that circuit, then:

$$V = \frac{W}{Q} = \frac{\text{Energy given out}}{\text{Charge gone past}}$$

Problem 1: In FIG. 22.4 the p.d. between A and B is 20 volts. In a certain time 40 C of charge passes A. How much heat energy is produced between A and B?

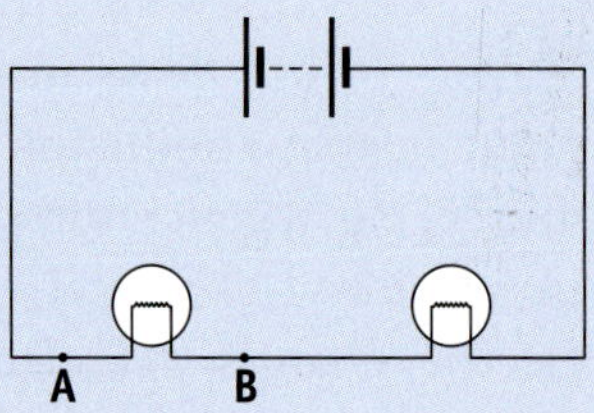

Fig. 22.4

Solution: $V = \frac{W}{Q} \Rightarrow W = QV = (40)(20) = 800\text{ J}$

Problem 2: 300 J of heat are given off from a wire when charge of 60 C passes a point in the wire. What is the p.d. across the wire?

Solution: $V = \frac{W}{Q} = \frac{300}{60} = 5\text{ V}$

If we divide both sides of $W = QV$ by t we get: $\frac{W}{t} = \frac{V \times Q}{t}$

Now $\frac{W}{t}$ is the rate at which energy is converted from electrical to other forms in the circuit between A and B and is called the **power*** (P) dissipated between A and B. $\frac{Q}{t}$ is the current I that flows (from $Q = It$ page 246). We therefore have:

$$P = VI$$

Power dissipated between A and B = (p.d. between A and B)(Current flowing between A and B)

Problem 3: The current in a bulb is 2 A when the p.d. across it is 230 V. What is the power dissipated in the bulb?

Solution: $P = VI = (230)(2) = 460\text{ W}$

Problem 4: What current flows through a 60 W bulb when connected to 230 V mains supply.

Solution: $P = VI \Rightarrow I = \frac{P}{V} = \frac{60}{230} = 0.261\text{ A}$

---

* **POWER** You should already know from Chapter 11 that the rate at which energy is converted from one form to another (or the rate at which work is done) is called **power**, and that if W joules of energy are converted from one form to another at a constant rate in time t seconds then:

Power $P = \frac{\text{Energy converted from one form to another}}{\text{time taken}}$ i.e. $P = \frac{W}{t}$

Recall that the unit of power is the watt (W), where **1 watt = 1 joule per second**.

Voltages in Series and Parallel

VOLTAGES IN SERIES

The potential difference across two (or more) consecutive parts of a series circuit is equal to the sum of the potential differences across each part, i.e. in Fig. 22.5.

$$V = V_1 + V_2 + V_3$$

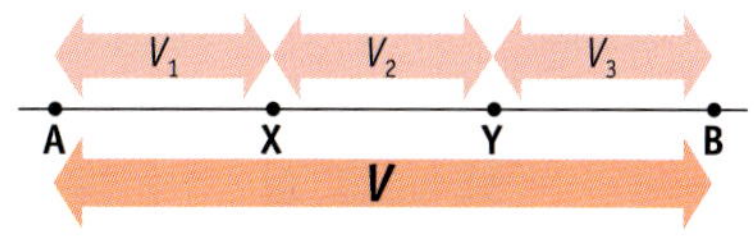

Fig. 22.5

This is because the total energy lost per coulomb between A and B is equal to the energy lost per coulomb between A and X plus the energy lost between X and Y plus the energy lost between Y and B.

VOLTAGES IN PARALLEL

The potential difference across each of a number of conductors connected in parallel with each other is the same. i.e. in Fig. 22.6 p.d. across path 1 = p.d. across path 2 = p.d. across path 3.

$$\text{i.e. } V_1 = V_2 = V_3$$

We do not need to prove this result; simply remember it.

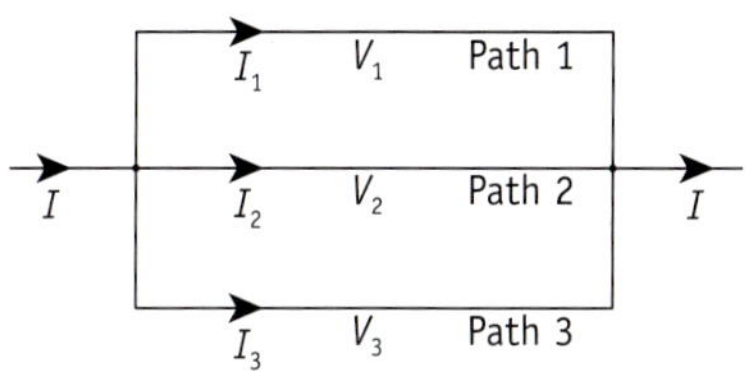

Fig. 22.6

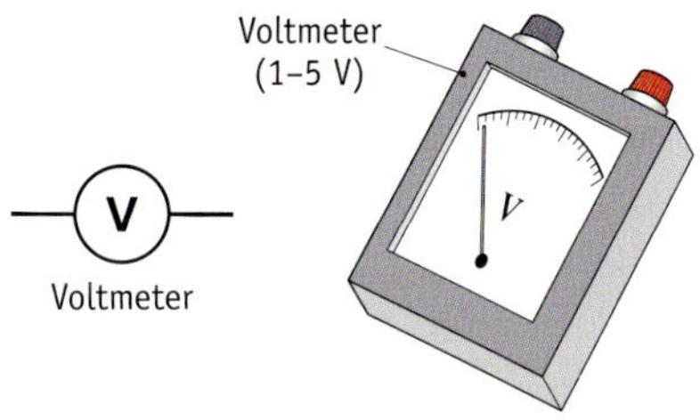

Fig. 22.7
A voltmeter and its circuit symbol.

Measuring Potential Difference

To measure potential difference we use a **voltmeter** (Fig. 22.7).

VOLTMETER

A ***voltmeter*** is always connected in parallel with the part of the circuit across which the p.d. is to be measured.

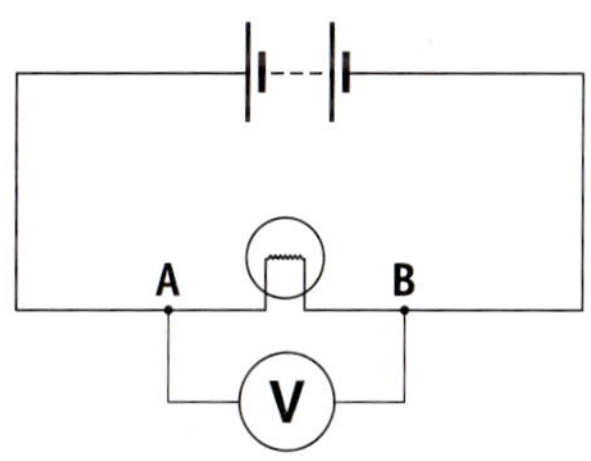

Fig. 22.8

For example, in Fig. 22.8 to measure the potential difference between A and B (i.e. across the bulb) the voltmeter is connected as shown. The meter is connected between two points without breaking the circuit between them. The above results for voltages in series and parallel can be easily verified in the laboratory with voltmeters.

Electromotive Force (emf)

To keep a current flowing in a circuit, an electric field must be maintained in that circuit. This means that there must be a potential difference (a voltage) between the ends of the circuit. This voltage, applied to a circuit, is called an emf. The symbol for emf is E, its unit is obviously the **volt**.

ELECTROMOTIVE FORCE (emf)

A voltage when applied to a circuit is called an ***emf***.

Some Sources of Electromotive Force (emf)

ELECTRIC CELLS

An electric cell is a device that converts chemical energy into electrical energy. It is thus a source of emf. It usually consists of two different metals (or carbon and a metal), called electrodes, immersed in a substance called an electrolyte.

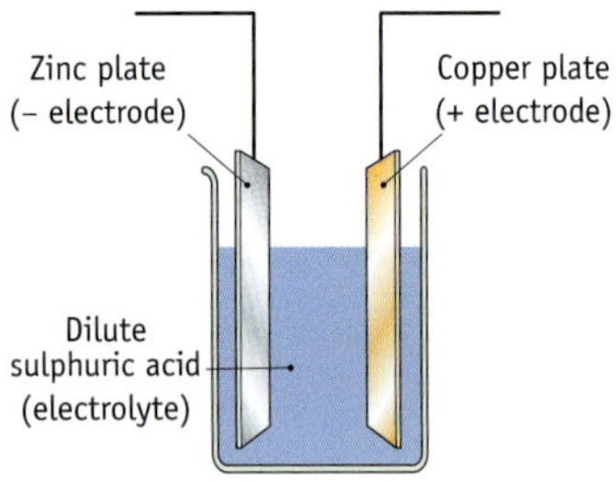

Fig. 22.10
A simple cell

The emf of a cell depends on the materials from which the electrodes and electrolyte are made. FIG. 22.9(A) shows the circuit symbol for a cell. A number of cells connected in series is called a **battery** (battery of cells). Its emf is the sum of the emfs of the cells from which it is made. FIG. 22.9(B) shows the circuit symbol for a battery.

(A) (B)

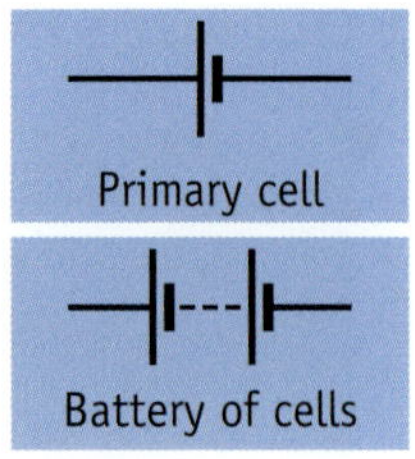

Fig. 22.9

A SIMPLE CELL

A typical simple cell consists of a copper plate and a zinc plate in a beaker of dilute sulphuric acid (FIG. 22.10). The plates react chemically with the acid causing the zinc plate to become negatively charged and the copper plate positively charged. As current is drawn from the cell, the chemicals are used up. When they are fully used, no more current can be got from the cell. Such a cell cannot be recharged. This cell is not very practical and its emf is about 1 V.

PRIMARY AND SECONDARY CELLS

A cell that cannot be recharged is called a **primary cell**. Almost all primary cells used today have electrolytes that are pastes rather than liquids. Such cells are often called **dry batteries**, the zinc-carbon and the alkaline-manganese being the most common.

Some cells can be recharged. A cell that can be recharged is called a **secondary cell** or an **accumulator**. The most common types are the lead acid and the nickel-cadmium cells. These can be recharged hundreds of times before they fail. The lead-acid accumulator consists of two lead plates in a solution of sulphuric acid. It is initially charged by passing direct current from another source through it. In use it becomes discharged. By sending current through it backwards (i.e. in through the + and out the –) it recharges. A common use of a lead-acid accumulator is a car battery which consists of six lead acid cells in a battery (FIG. 22.11).

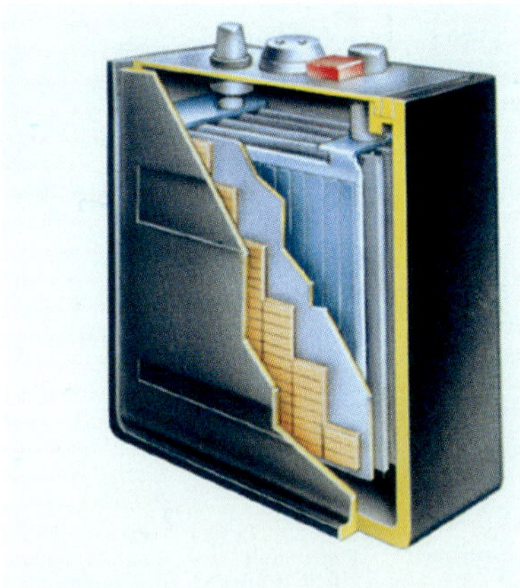

Fig. 22.11
A car battery is a lead acid accumulator. It is used to start the engine and power other electrical components in the car when the engine is not running.

THE THERMOCOUPLE

This is a source of emf. It is discussed on page 155.

THE MAINS

Electricity supplied to your home is called **mains electricity**. It is supplied at an emf of 230 volts.

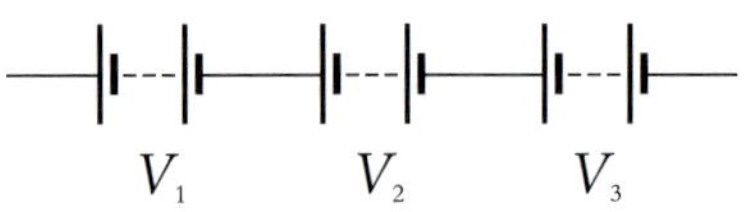

Fig. 22.12
Total voltage of battery $=V_1+V_2+V_3$.

BATTERIES CONNECTED IN SERIES

If a number of batteries (or other sources of emf) are connected in series with the positive terminal of one connected to the negative terminal of the next as in FIG. 22.12 then the total voltage of the combination is the sum of the individual voltages. Many everyday applications make use of this fact, e.g., in the torch in FIG. 22.13 three batteries, each of voltage 1.5 V are connected in series giving an overall voltage of 4.5 V.

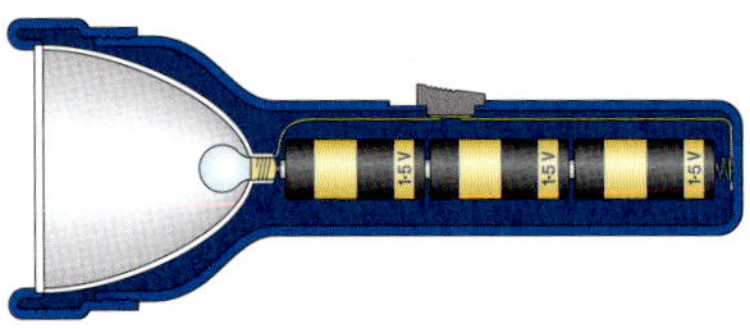

Fig. 22.13

Problem 5: What is the combined emf of the batteries in FIG. 22.14?

Solution: Since the batteries are connected + to – in series, the overall emf is the sum of the individual emfs i.e. Total emf = 4 + 6 + 2 = 12 volts.

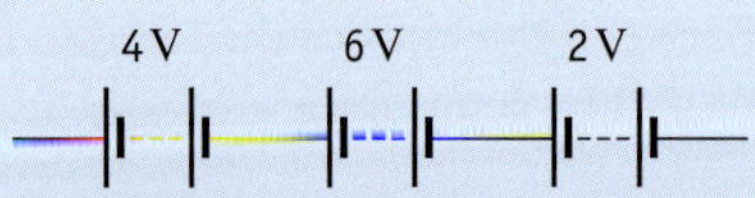

Fig. 22.14

EXERCISE 22.1

1. What is the combined emf of the batteries shown in FIG. 22.15?

Fig. 22.15

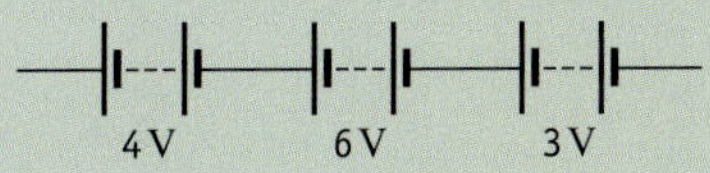

2. The p.d. across a bulb is 10 V. How much heat and light energy is given out from the bulb when:

   (a) 1 C, (b) 6 C, (c) 1 μC,

   passes through it?

3. 200 J of heat are produced in a wire when a charge of 50 C passes through it. What is the voltage across the wire?

4. The p.d. between two points A and B in a wire is 60 volts. In a certain time, 20 C of charge passes A. How much heat energy is produced between A and B?

5. 4000 J of heat are given off from a wire when charge of 80 C passes a point in the wire. What is the p.d. across the wire?

6. What current flows through a 100 W bulb when connected to a 230 V supply?

7. A 12 V car battery sends a current of 6.67 A through a car headlight bulb. What is the power developed in the bulb?

8. An electric motor operating at 220 volts draws a current of 4 A. Calculate:

   (i) the power of the motor,

   (ii) the electrical energy used by the motor in 1 hour.

9. Heat and light energy are dissipated at the rate of 100 W in a domestic light bulb. If the voltage across the bulb is 230 volts find the current flowing in the bulb.

10. A 220 volt electric generator is connected to 50 identical bulbs connected in series. If the power dissipated in each bulb is 5 watts, find the current in the circuit.

11. The p.d. between two points in a circuit is 10 volts. How much energy is released between the two points when:

    (i) a current of 5 A flows for 5 minutes,

    (ii) 1 C passes one of the points?

12. Three batteries of emfs 2 V, 3 V and 12 V are connected in series with a bulb. If the + terminal of one battery is connected to the – of the next and if the current in the bulb is 2 A, find the power dissipated in the bulb.

13. In FIG. 22.16 V_1 reads 6 volts, V reads 20 volts and A reads 3 A, how much heat and light energy is produced in Bulb 2 in 2 hours?

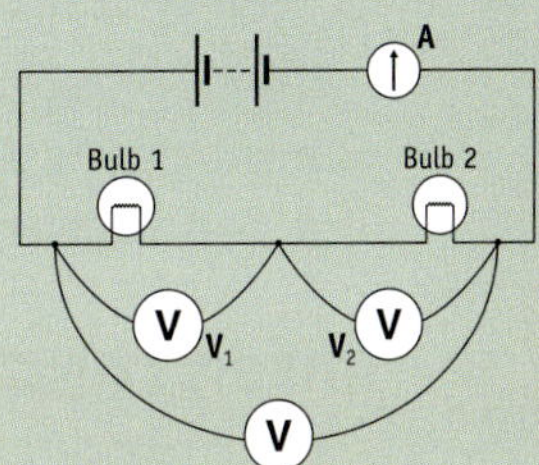

Fig. 22.16

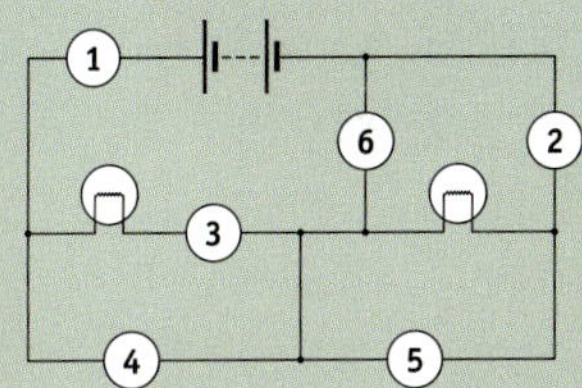

Fig. 22.17

14. Which of the meters 1, 2, 3, 4, 5, 6 in FIG. 22.17 are ammeters and which are voltmeters?

15. In FIG. 22.18 the bulbs are identical and the emfs of the batteries are as indicated. Which bulb lights the brightest?

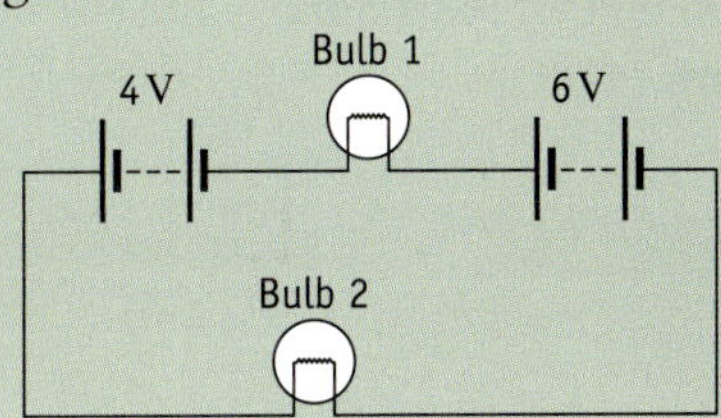

Fig. 22.18

CHAPTER CHECKLIST

- **State**: The unit of p.d.; The unit of emf; The unit of power.
- **Define**: Potential difference; Voltage; Electromotive force (emf); Power; The volt; The watt.
- **Recall** that: Potential difference is called voltage; A voltage when applied to a circuit is called an emf; Voltages (and emfs) in series add up; Voltages in parallel are the same; A voltmeter is connected in parallel.
- **Describe** a simple cell.
- **List** five sources of emf.
- **Recall** and **use** the formulae: $V = \frac{W}{Q}$; $P = IV$ to solve problems.

CHAPTER 23

Resistance

CONDUCTORS OPPOSE THE FLOW OF ELECTRIC CHARGE

As electrons move through a metal they continually collide with atoms and lose energy. This energy appears as heat. **In this way the material of the metal resists the movement of the electric charge through it.** Different conductors oppose current by different amounts.

- For most conductors if you increase the p.d. across them the current flowing through them increases.
- If a conductor offers a lot of opposition, a large p.d. causes only a small current and the ratio of V to I, i.e. V/I, is large.
- If a conductor offers a little opposition, a large p.d. causes a large current and the ratio V/I is small.

The last two facts are used to define a new quantity called the **resistance of a conductor**.

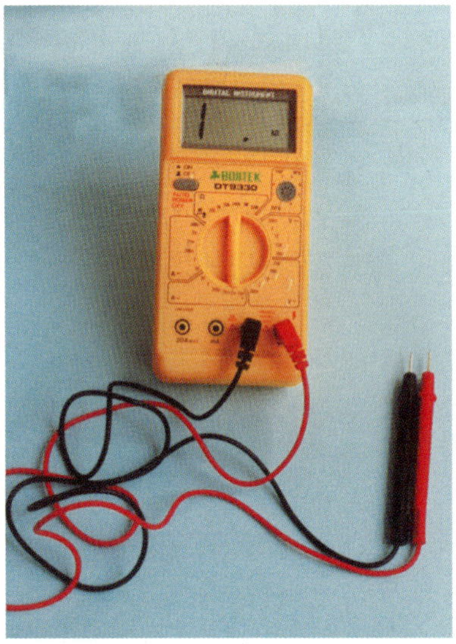

Fig. 23.1
A digital ohmmeter.

RESISTANCE
The ***resistance (R)*** of a conductor is the ratio of the p.d. across it to the current flowing through it, *i.e.*

$$R = \frac{V}{I}$$

Resistance is a **scalar** quantity. The unit of resistance is the **ohm** (Ω).

UNIT OF RESISTANCE
The unit of resistance is the ***ohm*** **(Ω).**

THE OHM
A conductor has a resistance of 1 ***ohm*** if the current through it is 1 ampere when the p.d. across it is 1 volt.

Problem 1: Find the resistance of a conductor if it carries a current of 4 A when the p.d. across it is 20 V.

Solution: $R = \frac{V}{I} = \frac{20}{4} = 5\ \Omega$

MEASURING RESISTANCE

USING AN AMMETER AND A VOLTMETER

A simple way to measure the resistance of a conductor is to pass an electric current through it, then measure the p.d. across it with a voltmeter and the current through it with an ammeter. Divide the voltage by the current and the result is its resistance ($R = V/I$).

USING AN OHMMETER

The most common way of measuring resistance is with an ohmmeter. This instrument is connected across the conductor and reads its resistance directly. You are likely to meet two types of ohmmeter. One type – called a digital ohmmeter (FIG. 23.1) – is often part of a digital multimeter. By selecting one of the resistance-measuring scales it reads the resistance of whatever is connected across the ends of its probes.

The other type is a **moving coil ohmmeter** or an **analogue ohmmeter** and is often part of an analogue multimeter (FIG. 23.2). To use such an instrument you:

- choose a suitable resistance scale on the instrument,
- connect the two probes together firmly and adjust the knob on the meter until it reads zero, i.e. zero the ohmmeter,
- connect the two probes across the resistance to be measured and read the instrument.

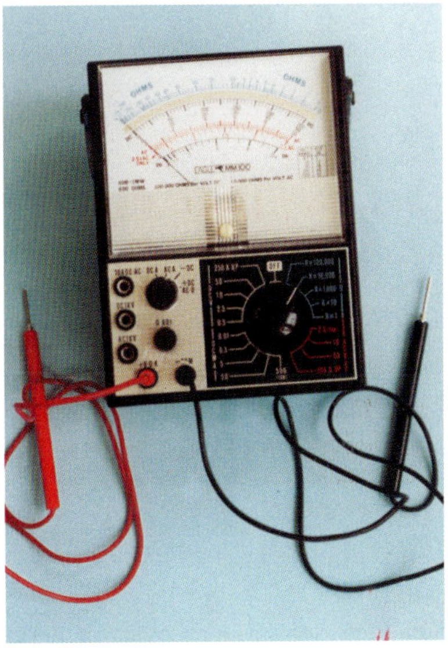

Fig. 23.2
An analogue ohmmeter.

OHM'S LAW

The resistance of some conductors changes as the p.d. across them (or the current through them) changes. For other conductors – mainly metals and some liquids – the resistance remains the same as the voltage across them changes. Such conductors are called **Ohmic conductors** and obey **Ohm's Law** which states:

> For certain conductors (mainly metals), at constant temperature the current flowing through them is directly proportional to the p.d. across them. i.e.
>
> $$V \propto I \Rightarrow \frac{V}{I} = \text{a constant}$$

The **constant of proportionality is the resistance (*R*) of the conductor**. Ohm's Law tells us that for certain conductors at constant temperature, their resistance remains constant as the size of the current through them (or the p.d. across them) changes. This fact was discovered by George Simon Ohm in 1826.

CIRCUIT DIAGRAMS AND CIRCUIT SYMBOLS

To illustrate circuits clearly we use circuit diagrams. These are diagrams where each electrical component is represented by a special symbol. FIG. 23.3 shows the circuit symbols used in this part of the course.

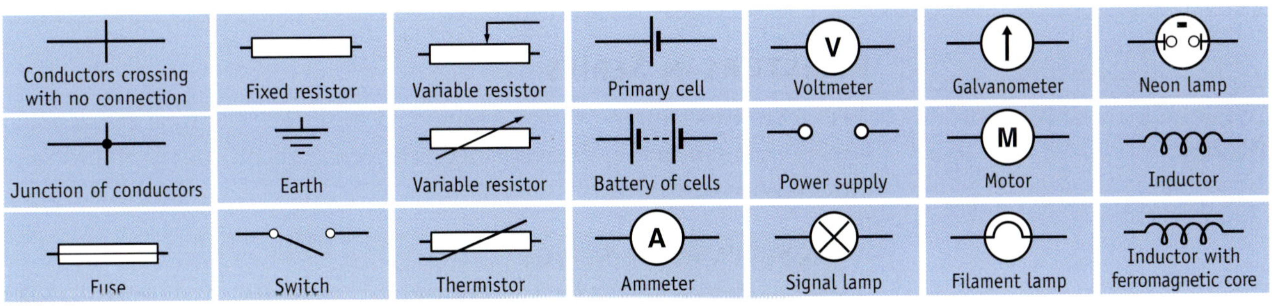

Fig. 23.3

PRACTICAL RESISTORS

Some conductors are specially made to have a resistance of a specific value.

VARIABLE RESISTORS

FIG. 23.4(A) on the next page shows one type of variable resistor. It is called a **rheostat**. Current flows through the rheostat as shown. By moving the sliding contact along the bar we can cause the current to pass through all of the wire, some of the wire or none of the wire in the coil – thus changing the total resistance of the circuit in which the rheostat is connected.

You should set up the circuit shown in FIG. 23.5 and see how changing the resistance of the rheostat causes the current to change. FIG. 23.4(B) shows another type of variable resistor found in radios (usually controlling volume). It operates similarly to the rheostat except that the current flows along a carbon track rather than a coil of wire.

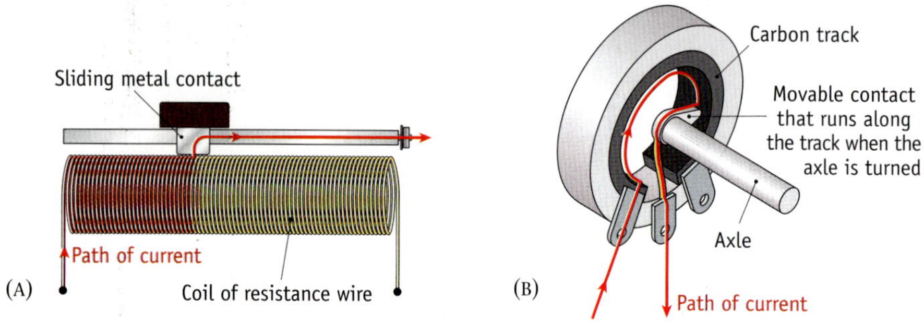

Fig. 23.4
Variable resistors – (A) A rheostat. (B) A potentiometer.

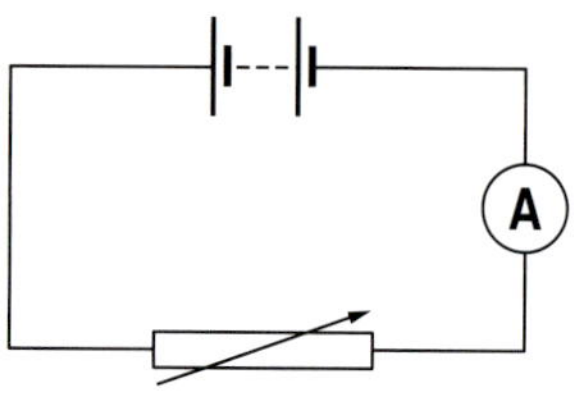

Fig. 23.5

EXPERIMENT

TO DEMONSTRATE OHM'S LAW.

Use the equipment in FIG. 23.6.

- By varying the rheostat, the current and p.d. across the metal conductor are varied.
- Measure a series of values of I and corresponding values of V.
- Plot a graph of I against V. It will be **a straight line through the origin,** (FIG. 23.7) **verifying that** $I \propto V$, i.e. verifying Ohm's Law.

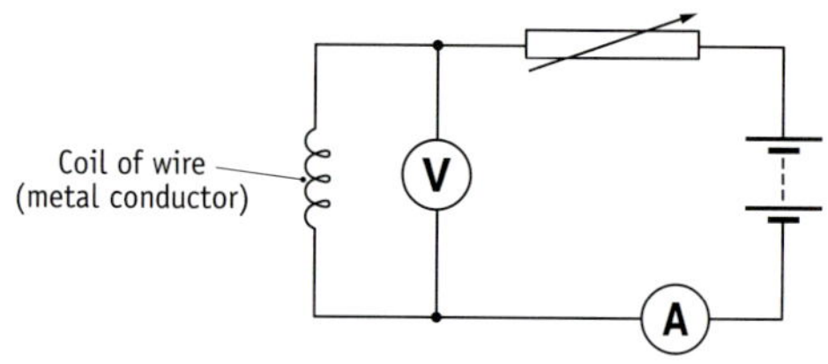

Fig. 23.6

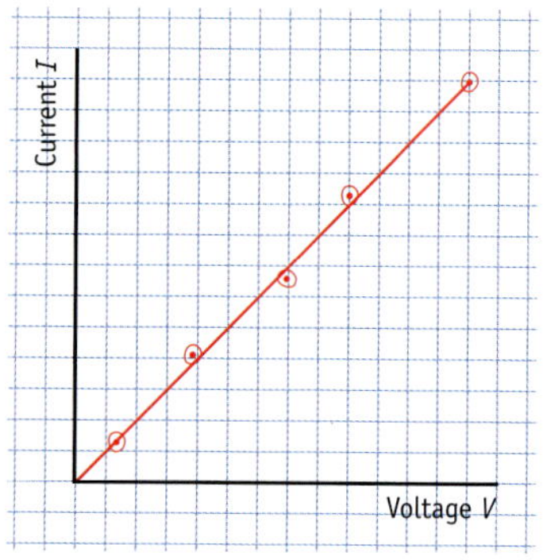

Fig. 23.7
The current voltage characteristic of a metal, i.e. of a conductor that obeys Ohm's Law.

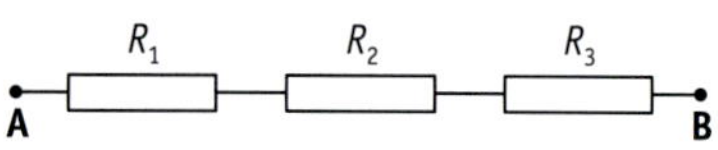

Fig. 23.8

RESISTORS IN SERIES

FIG. 23.8 shows three resistors connected in series. The total resistance of the combination (i.e. the resistance between A and B) can be found using the following formula:

RESISTORS IN SERIES

If two or more resistors are connected in series the total resistance of the combination is equal to the sum of the resistances of each resistor, i.e. in FIG. 23.8 the resistance (R) between A and B is given by:

$$R = R_1 + R_2 + R_3$$

Problem 2: Find the resistance between A and B in FIG. 23.9.

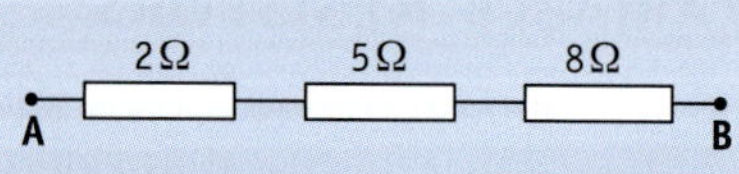

Fig. 23.9

Solution: Total resistance

$$R = R_1 + R_2 + R_3$$
$$= 2 + 5 + 8 = 15\ \Omega$$

Proof of Formula

In Fig. 23.8 let V_1, V_2 and V_3 be the voltages across each resistor. Let the current through each be I.

If V is the voltage and R is the total resistance between A and B, Ohm's Law gives: $V = IR$ **(1)**

Apply Ohm's Law to each resistor:

$V_1 = IR_1$ and $V_2 = IR_2$ and $V_3 = IR_3$

Now $V = V_1 + V_2 + V_3 \Rightarrow V = IR_1 + IR_2 + IR_3$

$\Rightarrow V = I(R_1 + R_2 + R_3)$ **(2)**

(1) and (2) $\Rightarrow IR = I(R_1 + R_2 + R_3) \Rightarrow R = R_1 + R_2 + R_3$

Resistors in Parallel

Fig. 23.10 shows three resistors connected in parallel. The overall resistance of the combination (i.e. the resistance between A and B) can be found from the following formula:

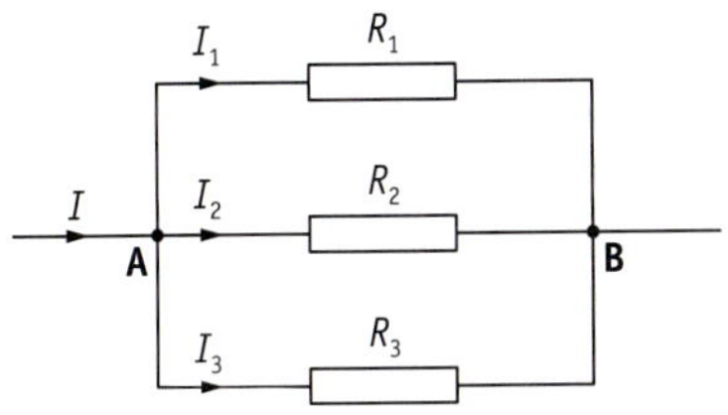

Fig. 23.10

RESISTORS IN PARALLEL

If resistors of resistances R_1, R_2 and R_3 ohms are connected in parallel with each other, the resistance of the combination can be calculated from the formula:

$$\frac{1}{R} = \frac{1}{R_1} + \frac{1}{R_2} + \frac{1}{R_3}$$

Problem 3: Find the total resistance R of the combination shown in Fig. 23.11.

Solution:

$$\frac{1}{R} = \frac{1}{R_1} + \frac{1}{R_2} + \frac{1}{R_3} = \frac{1}{4} + \frac{1}{5} + \frac{1}{10}$$

$$= \frac{11}{20} \Rightarrow R = \frac{20}{11} = 1.82\ \Omega$$

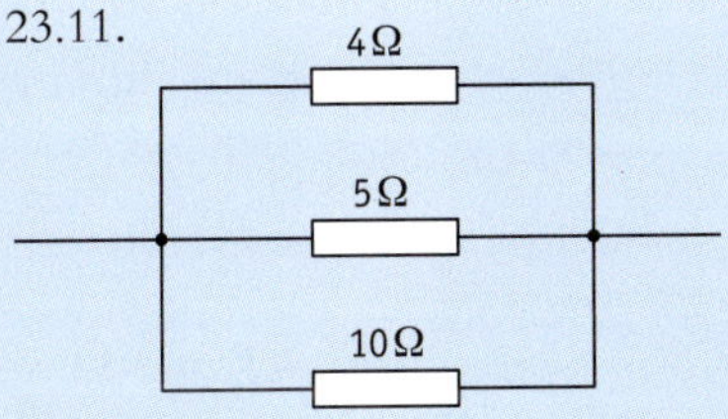

Fig. 23.11

Proof of Formula

In Fig. 23.10 let current I flow into the combination. Since the resistors are in parallel, the voltage (V) across each is the same. Let the currents in R_1, R_2 and R_3 be I_1, I_2 and I_3 respectively.

Apply Ohm's Law to each resistor: $I_1 = \frac{V}{R_1}$ $I_2 = \frac{V}{R_2}$ and $I_3 = \frac{V}{R_3}$

But $I = I_1 + I_2 + I_3 \Rightarrow I = \frac{V}{R_1} + \frac{V}{R_2} + \frac{V}{R_3} = V\left(\frac{1}{R_1} + \frac{1}{R_2} + \frac{1}{R_3}\right)$

If R is the total resistance, then by Ohm's Law: $I = \frac{V}{R}$

Thus: $\frac{V}{R} = V\left(\frac{1}{R_1} + \frac{1}{R_2} + \frac{1}{R_3}\right) \Rightarrow \frac{1}{R} = \frac{1}{R_1} + \frac{1}{R_2} + \frac{1}{R_3}$

Problem 4: Find the total resistance of the circuit shown in FIG. 23.12. Then, assuming the resistance of the ammeter and the battery are negligible, find the reading on the ammeter.

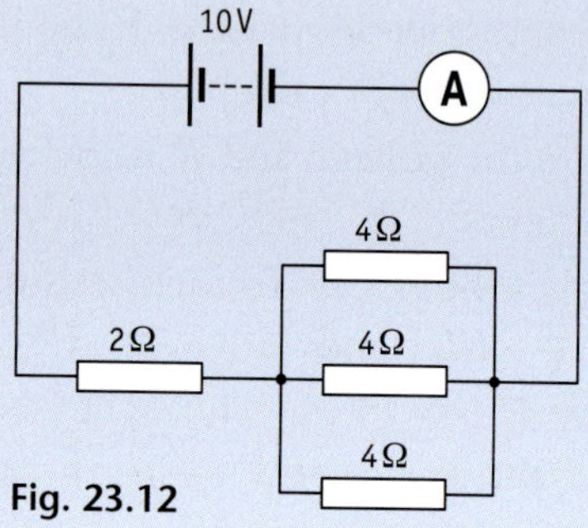

Fig. 23.12

Solution: Find the resistance of the parallel combination first:

$$\frac{1}{R} = \frac{1}{R_1} + \frac{1}{R_2} + \frac{1}{R_3}$$

$$= \frac{1}{4} + \frac{1}{4} + \frac{1}{4} = \frac{3}{4} \Rightarrow R = \frac{4}{3}\ \Omega$$

Total resistance of circuit (R_t) equals the resistance of 2 Ω and 4/3 Ω in series, which is = 10/3 Ω

Current in ammeter: $I = \dfrac{V}{R_t} = \dfrac{10}{(10/3)} = 3\text{ A}$

Problem 5: Find the current flowing through each resistor in FIG. 23.13.

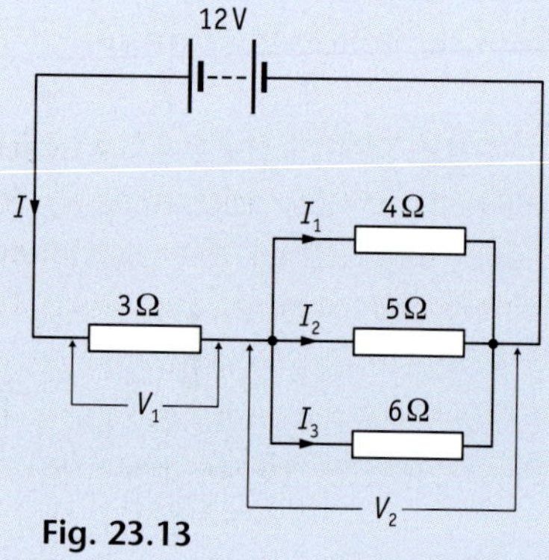

Fig. 23.13

Solution: Let the currents be I, I_1, I_2 and I_3 as indicated.

Resistance of three resistors in parallel:

$$\frac{1}{R_p} = \frac{1}{R_1} + \frac{1}{R_2} + \frac{1}{R_3} = \frac{1}{4} + \frac{1}{5} + \frac{1}{6}$$

$$\Rightarrow R_p = 1.62\ \Omega$$

Total resistance of circuit: $R = 3 + 1.62 = 4.62\ \Omega$

Calculate current from battery:

$$I = \frac{E}{R} = \frac{12}{4.62} = 2.6\text{ A} = \text{Current through 3 } \Omega \text{ resistor } R$$

p.d. across 3 Ω resistor: $V_1 = IR = (2.6)(3) = 7.8\text{ V}$

p.d. across parallel combination: $V_2 = 12 - V_1 = 12 - 7.8 = 4.2\text{ V}$

Calculate the remaining currents using $I = V/R$:

$$I_1 = 4.2 / 4 = 1.05\text{ A}$$

$$I_2 = 4.2 / 5 = 0.84\text{ A}$$

$$I_3 = 4.2 / 6 = 0.70\text{ A}$$

To check answers confirm that: $I = I_1 + I_2 + I_3$

EXERCISE 23.1

1. The current through a conductor is 4 A when the p.d. across it is 20 V. Calculate the resistance of the conductor.
2. What p.d. will produce a current of 5 A in a 12 ohm resistor?
3. What current flows through a resistance of 100 Ω when it is connected to a 230 volt supply?
4. A current of 4 A flows through the filament of a car headlight bulb when it is connected to a 12 volt battery. Find the resistance of the filament.
5. Find the potential difference across a 6 Ω resistor when it is carrying a current of 5 amperes.

6. At a certain temperature, the current through a conductor is 3 A when the p.d. across it is 24 V. Find the resistance of the conductor. When the temperature of the conductor is raised, the same p.d. causes a current of 2 A to flow through it. Find the increase in its resistance.

7. An ammeter and voltmeter are used as shown in FIG. 23.14 to measure the resistance of a coil of wire. If the readings on the instruments are 10 V and 2 A, find the resistance of the coil.

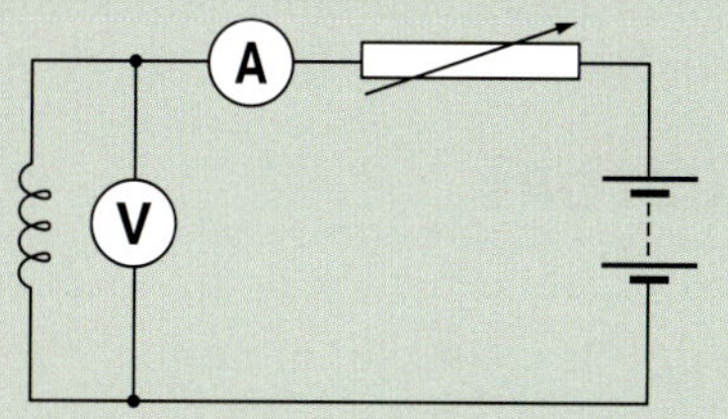

Fig. 23.14

8. Two resistors are connected in series as shown in FIG. 23.15. Calculate the total resistance of the combination. If the p.d. across the 2 Ω resistor is 12 V, find the current in each resistor. Find also the potential difference across the combination.

Fig. 23.15

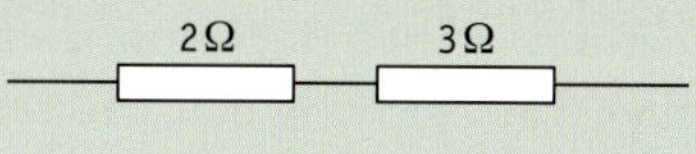

9. A bulb and a resistor of 4 Ω are connected in series with a 10 V battery. If the current flowing in the circuit is 1 A, find the resistance of the filament of the bulb.

10. Find the potential difference across each of the resistors in FIG. 23.16. Find also the potential difference across the combination.

Fig. 23.16

11. Find the total resistance between A and B in each of the arrangements shown in FIG. 23.17.

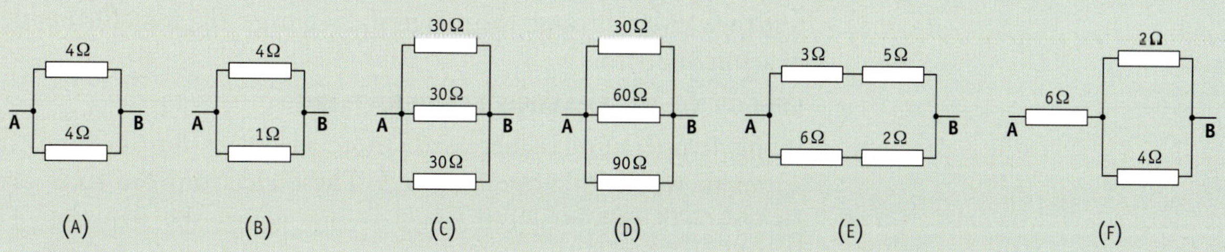

Fig. 23.17

12. Find the resistance of:

(i) two 20 Ω resistors connected in parallel,
(ii) two 20 Ω resistors connected in series,
(iii) a 2 Ω and a 3 Ω resistor connected in parallel.

13. Find the current through each resistor in FIG. 23.18.

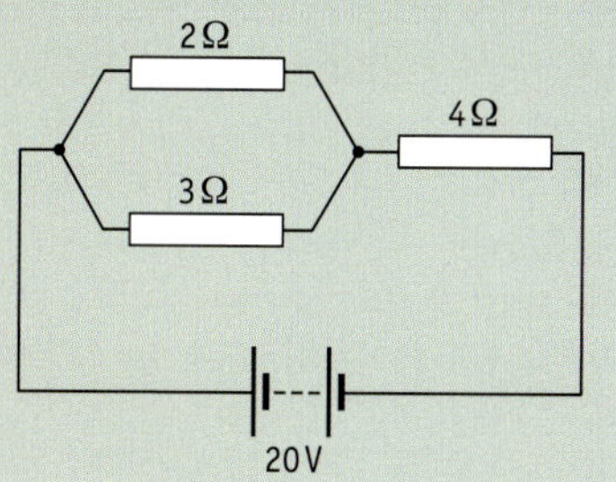

Fig. 23.18

14. In the circuit shown in FIG. 23.19 R is a variable resistor of maximum resistance 400 Ω. Find:

(i) the effective resistance of the circuit if R is set at maximum value,
(ii) the current in the 2 k Ω resistor.

If the value of R is reduced, what effect would this have on:

(i) the current flowing in the 600 Ω resistor,
(ii) the potential at A,
(iii) the current in the 2 k Ω resistor?

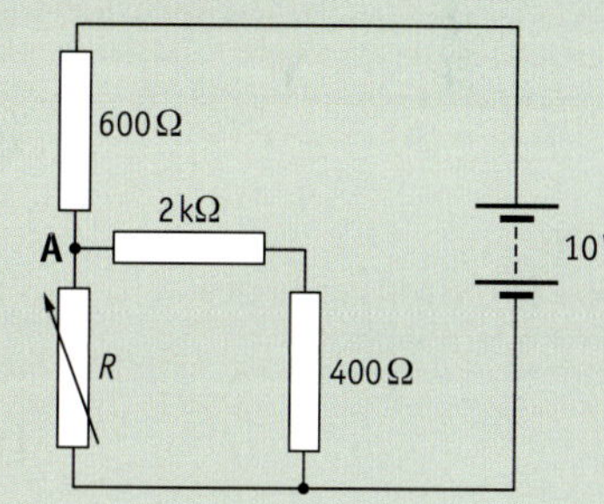

Fig. 23.19

Factors Affecting the Resistance of a Conductor

The **resistance** of a conductor depends on:

- its **temperature**,
- its **length**,
- its **cross-sectional area**,
- the **material** from which it is made.

Dependence of Resistance on Temperature

Experimentally it is found that:

- the resistance of a metallic conductor increases as the temperature increases,
- the resistance of most other substances (including carbon and semiconductors) decreases as the temperature increases.

Metallic Conductors

In a metal, some electrons are not stuck to any particular atom and are free to wander in the metal. For example, in copper there is usually one such 'free electron' from every copper atom. The movement of these electrons through the metal is an electric current. As the electrons move, they collide with the atoms and thus meet resistance to their motion. The greater the number of collisions the greater the resistance.

Effect of Increasing Temperature

As the temperature of the metal is increased, the metal atoms vibrate at a greater rate. Thus the electrons, in trying to move through, collide more frequently with the atoms. The electrons therefore meet increased resistance. This is another way of saying that **the resistance of a metallic conductor increases as the temperature increases**.

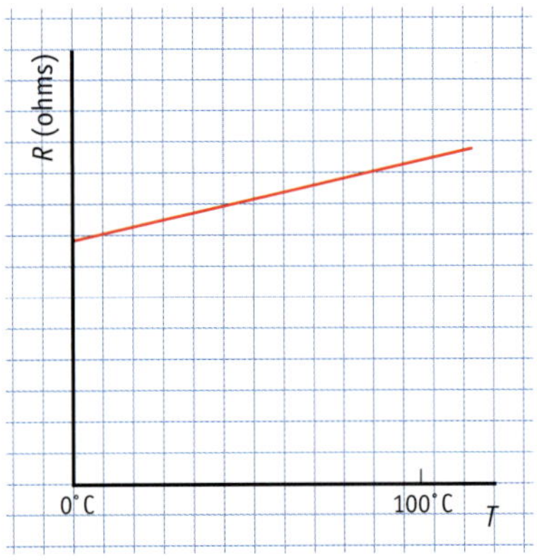

Fig. 23.20
Variation of the resistance of a **metal** with temperature.

For moderate changes in temperature (less than 100 °C) it is found that the change in resistance is proportional to the change in temperature. Fig. 23.20 shows graphically how the resistance of a metal changes with temperature. Note that the graph is a straight line, but it does not pass through the origin. We say **the resistance changes linearly with temperature**. Note also that the increase is not very great, e.g. a piece of copper that has a resistance of 10 Ω at 20 °C might have a resistance of 13 Ω at 100 °C.

Insulators and Semiconductors

In an insulator or semiconductor, most electrons are attached to particular atoms. Only a few (in semiconductors) and almost none (in insulators) are free to wander through the material. (See page 285 for a further discussion of semiconductors.)

Effect of Increasing Temperature

If the temperature is increased, more electrons 'break loose' from the atoms to which they are attached. These electrons can form an electric current if a potential difference is put across the material. Thus **the resistance of an insulator or a semiconductor decreases as the temperature increases**. As the temperature increases, the atoms in these materials vibrate more and thus tend to increase the resistance. However, the increase in the number of conduction electrons totally masks this effect and the resistance decreases. In a metal, virtually no extra conduction electrons are produced when the metal's temperature increases.

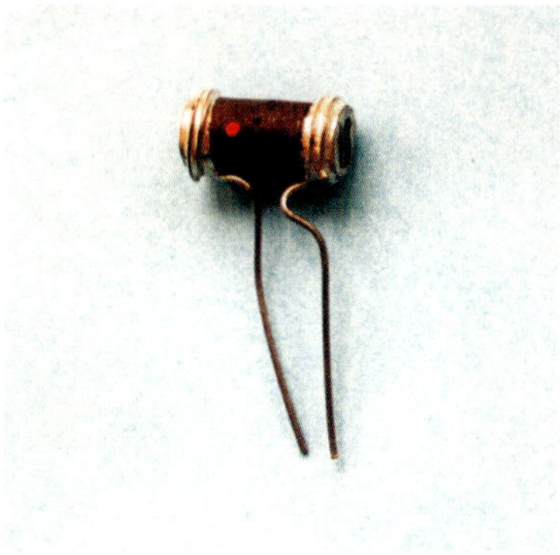

Fig. 23.21
A thermistor.

THE THERMISTOR

'Thermistor' is an abbreviation for the word 'thermal resistor'. **A thermistor** (FIG. 23.21) **is a semiconductor whose resistance decreases rapidly with increasing temperature.** A typical thermistor is made up of a mixture of oxides of nickel, cobalt, iron and small quantities of other substances. Such a thermistor might have a resistance of 800 Ω at 0 °C and a resistance of 130 Ω at 50 °C. FIG. 23.22 shows how the resistance of a thermistor varies with temperature. Note that the graph is not a straight line.

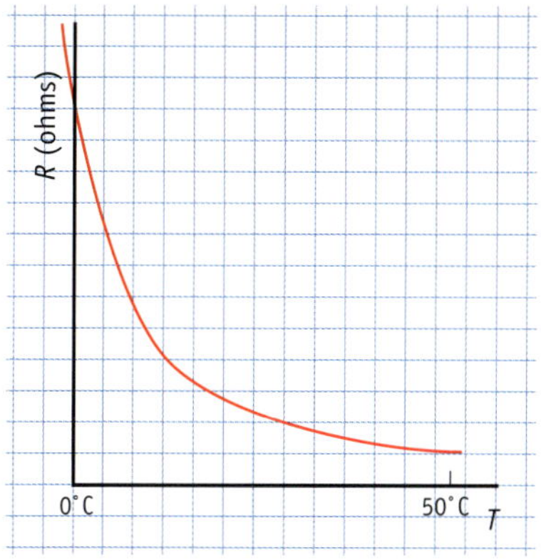

Fig. 23.22
Variation of the resistance of a thermistor with temperature.

EXPERIMENT

ELECTRICITY 3

TO INVESTIGATE THE VARIATION OF THE RESISTANCE OF A METALLIC CONDUCTOR WITH TEMPERATURE.

Summary of Method

In this experiment you will measure the resistance of a coil of wire a number of times with an ohmmeter. Each time the wire will be at a different temperature which you will also measure. You will then plot a graph of resistance against temperature.

Equipment Needed

- A coil of wire
- A thermometer (0 °C to 100 °C)
- Some liquid paraffin or some glycerol
- A Bunsen burner, tripod and gauze
- A large test tube
- A beaker (500 ml)
- An ohmmeter
- A retort stand and clamp

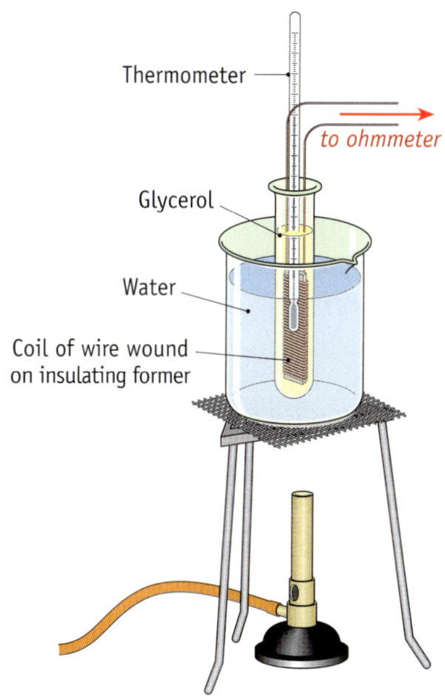

Fig. 23.23

Method

1. Using the ohmmeter, determine approximately the resistance of the coil. Then choose a suitable range on the ohmmeter for the rest of the experiment.
2. Connect the probes of the ohmmeter firmly together. If it is an analogue meter, turn the relevant knob on the meter until the scale reads zero. If it is digital, see if the leads have any noticeable resistance. If they have it must be subtracted from the readings taken below. Connect the probes of the ohmmeter firmly to the coil.
3. Place cold tap water in the beaker. Then set up the equipment as shown in FIG. 23.23.
4. Allow the coil and thermometer to cool down. When the thermometer gives a steady reading, measure the resistance of the coil with the ohmmeter and the temperature of the coil with the thermometer. Record these values.
5. Turn on the Bunsen and gently heat the water. When the temperature of the glycerol has risen by about 8 °C, remove the Bunsen, wait for the thermometer to stop rising then repeat the two measurements as in step 4.

   Warning! Keep the ohmmeter and connecting wires well away from the hot Bunsen.
6. Using the Bunsen burner to slowly increase the temperature, repeat step 5 for a series of values of temperature about 10 °C apart until the temperature is almost 100 °C. Record all measurements.

7. On graph paper plot a graph of resistance (on the *y*-axis) against temperature (on the *x*-axis.). It should look like that in FIG. 23.20 (page 262).

Temperature of coil $\theta/^\circ C$	Resistance of coil R/Ω

EXPERIMENTAL NOTES

Heat the water very slowly, otherwise the temperature of the coil will continue to rise a lot after you remove the Bunsen. Before reading the thermometer, wait a while until the coil is at the same temperature as the glycerol.

Questions

1. Why must the probes of the ohmmeter be connected firmly to the coil?
2. Why is glycerol (or paraffin) used in this experiment rather than water?
3. Why is it important to heat the water very slowly?

EXPERIMENT ELECTRICITY 4

TO INVESTIGATE THE VARIATION OF THE RESISTANCE OF A THERMISTOR WITH TEMPERATURE.

Summary of Method

In this experiment you will measure the resistance of a thermistor a number of times using an ohmmeter. Each time the thermistor will be at a different temperature which you will also measure. You will then plot a graph of resistance against temperature.

Equipment Needed

- A thermistor
- A thermometer (0 °C to 100 °C)
- Some liquid paraffin or glycerol
- An ohmmeter
- A large test tube (into which the thermistor can fit)
- A Bunsen burner, tripod and gauze
- A beaker (500 ml)
- A retort stand and clamp

Method

1. Using the ohmmeter, determine approximately the resistance of the thermistor. Then choose a suitable range on the ohmmeter. You may need to change this resistance range during the experiment.
2. Connect the probes of the ohmmeter firmly together. If it is an analogue meter, turn the relevant knob on the meter until the scale reads zero. If it is digital, see if the leads have any noticeable resistance. If they have it must be subtracted from the readings taken below. Connect the probes of the ohmmeter firmly to the thermistor.
3. Place cold tap water in the beaker and set up the equipment as in FIG. 23.24.
4. Allow the thermistor and thermometer to cool. When the thermometer gives a steady reading, measure the resistance of the thermistor and its temperature. Record these.
5. Turn on the Bunsen and gently heat the water. When the temperature has risen by about 8 °C, remove the Bunsen, wait for the thermometer to stop rising then repeat the two measurements as in step 4.

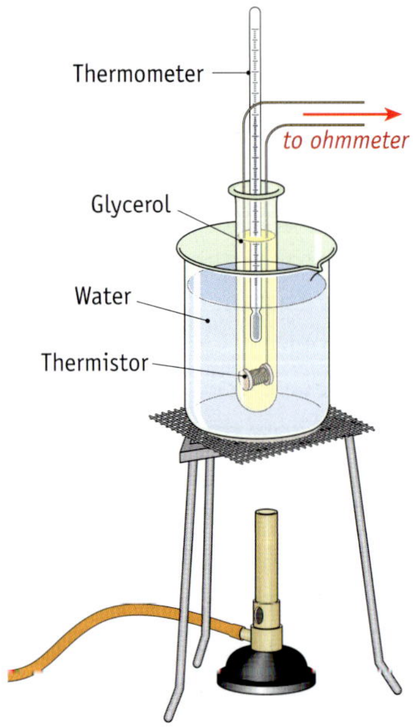

Fig. 23.24

Warning! Keep the ohmmeter and connecting wires well away from the hot Bunsen.

6. Using the Bunsen to slowly increase the temperature, repeat the measurements in step 4 for a series of values of temperature about 10 °C apart until the temperature is almost 100 °C. Record all measurements.
7. On graph paper plot a graph of resistance (on the y-axis) against temperature (on the x-axis). It should look like FIG. 23.22 (page 263).

Temperature of thermistor θ/°C	Resistance of thermistor R/Ω

EXPERIMENTAL NOTES

Heat the water very slowly, otherwise the temperature of the thermistor will continue to rise a lot after you remove the Bunsen. Before reading the thermometer wait a while until the thermistor is at the same temperature as the glycerol.

Questions

1. Why must the probes of the ohmmeter be connected firmly to the thermistor?
2. Why is glycerol (or paraffin) used in this experiment rather than water?
3. Why is it important to heat the water very slowly?

RESISTIVITY

As well as temperature, the resistance of a conductor also depends on:

(i) its **length**,
(ii) its **cross-sectional area**,
(iii) the **material** from which it is made.

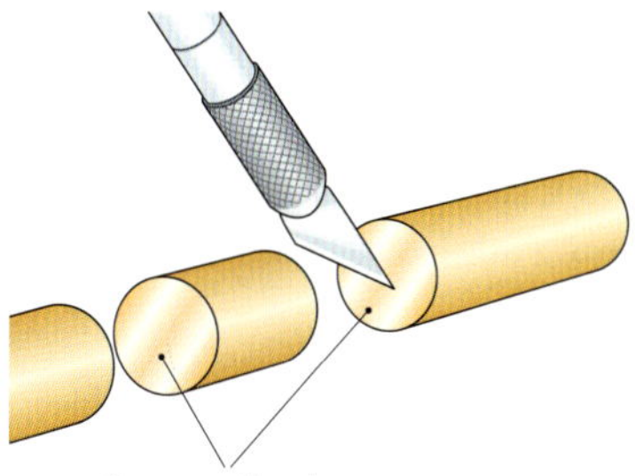

Fig. 23.25

FIG. 23.25 shows a wire conductor. If the wire is cut at right angles to its length as shown the exposed ends are circular. All these circles have the same area. We say the wire has **uniform cross-sectional area**.

RESISTANCE AND LENGTH

FIG. 23.26(A) shows a piece of wire of uniform circular cross section and of length l metres. Suppose its resistance is R ohms. FIG. 23.26(B) shows two such pieces of wire connected in series. The resistance of this piece must be 2 R ohms since it is made of two pieces of resistance R connected in series. Clearly if we add on any other number of pieces of the same wire the resistance would increase proportionally. This can be easily verified with an ohmmeter. Thus:

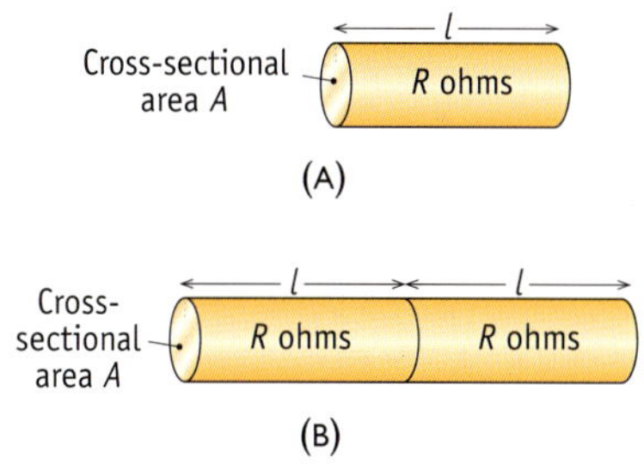

Fig. 23.26

The ***resistance*** of a uniform conductor is ***directly proportional to its length***

i.e. $R \propto l$ (for a fixed cross-sectional area A) (1)

RESISTANCE AND CROSS-SECTIONAL AREA

By experiment it is found that the resistance of a uniform conductor of fixed length l is inversely proportional to its cross-sectional area, i.e. provided the length remains the same:

- if the cross-sectional area is doubled the resistance is halved,
- if the cross-sectional area is trebled the resistance is three times smaller,
- if the cross-sectional area is quartered the resistance becomes four times as large, etc.

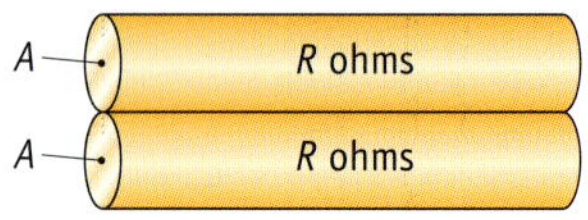

Fig. 23.27
The cross-sectional area is doubled and the resistance is halved.

This can be seen from the following. If two identical pieces of wire are connected in parallel, the resistance of the combination is ½ the resistance of each. But from FIG. 23.27 we see that connecting them in parallel is effectively doubling the cross-sectional area through which the electric current can flow.

The ***resistance*** of a uniform conductor is ***inversely proportional to its cross-sectional area***.

i.e. $$R \propto \frac{1}{A} \qquad \text{(for a fixed length } l\text{)} \qquad (2)$$

From (1) and (2) it follows that:

$$R \propto \frac{l}{A} \Rightarrow R = \frac{\rho l}{A} \quad \text{where } \rho \text{ is a constant} \qquad (3)$$

Resistivities at 20 ˚C of some Common Materials. (Unit: Ω m)	
Silver	1.6×10^{-8}
Copper	1.7×10^{-8}
Iron	9.8×10^{-8}
Manganin	4.2×10^{-7}
Nichrome	1.12×10^{-6}
Silicon	1×10^{3} approx.
Distilled water	5×10^{3}
Glass	10^{12} approx.

Fig. 23.28

The constant of proportionality ρ is only a constant for the particular material in the conductor. For different materials, the value of ρ is different. If the material is a good conductor ρ is small, if it is a bad conductor ρ is large. Thus ρ takes account of the fact that the resistance of a conductor depends on the material from which it is made.

ρ is called the **resistivity** of the material in the conductor.

Solving (3) for ρ gives: $$\rho = \frac{RA}{l}$$

It follows that **the unit of resistivity is the ohm meter (Ω m)**. FIG. 23.28 shows the resistivities of some common conducting materials.

Resistivity is defined as follows:

RESISTIVITY

If a conductor of length l and cross-sectional area A has a resistance R, the constant ρ given by:

$\rho = \frac{RA}{l}$ is called the **resistivity** of the material in the conductor.

Clearly, resistivity of a material could also be defined as the resistance of a piece of that material of length 1 m and cross-sectional area 1 m^2.

Problem 6: A uniform wire of length 2 m has a resistance of 12 Ω. Find the resistance of a piece of identical wire of length 11.4 m

Solution: The length of the second wire is 5.7 times ($^{11.4}/_2 = 5.7$) that of the first wire. Since resistance is proportional to length, the resistance of the second wire will be 5.7 times that of the first.

Thus: Resistance of second wire = 12 × 5.7 = 68.4 Ω

Problem 7: What length of copper wire of cross-sectional area 2 mm^2 is needed to make a resistor of resistance 10 Ω? Resistivity of copper = 1.7 × 10^{-8} Ω m.

Solution:
$$R = \frac{\rho l}{A} \Rightarrow l = \frac{AR}{\rho} = \frac{(2 \times 10^{-6})(10)}{(1.7 \times 10^{-8})} = 1176.5 \text{ m}$$

Note that length must be expressed in metres and cross-sectional area in metres squared when using this formula. Recall that $1 \text{ m}^2 = 10^6 \text{ mm}^2 \Rightarrow 1 \text{ mm}^2 = 1 \times 10^{-6} \text{m}^2$

Problem 8: A coil of copper wire 20 m long has uniform composition and uniform cross-sectional area. The diameter of the wire is 0.055 mm. Taking the resistivity of copper to be 1.7 × 10^{-8} Ω m, calculate the resistance of the coil.

Solution: First find the cross-sectional area of the coil. $d = 0.055 \text{ mm} = 0.055 \times 10^{-3} \text{ m}$

$$r = \frac{d}{2} = \frac{(0.055 \times 10^{-3})}{2} = 2.75 \times 10^{-5} \text{ m}$$

$$\therefore A = \pi r^2 = \pi (2.75 \times 10^{-5})^2 = 2.376 \times 10^{-9} \text{ m}^2$$

$$R = \frac{\rho l}{A} = \frac{(1.7 \times 10^{-8})(20)}{(2.376 \times 10^{-9})} = 143 \ \Omega$$

Measuring Resistivity in the Laboratory

If a wire has circular cross-sectional area A, radius r and diameter d, then:

$$A = \pi r^2 = \pi \left(\frac{d}{2}\right)^2 = \frac{\pi d^2}{4}$$

Substituting this into $\rho = \frac{RA}{l}$ gives: $\rho = \frac{R\pi d^2}{4l}$

By measuring its length l, its resistance R and its diameter d its resistivity ρ can be found.

Experiment — Electricity 2

To Measure the Resistivity of the Material of a Wire.

Summary of Method

In this experiment you will measure the resistance R and the length l of a piece of wire using an ohmmeter and a metre stick. You will measure the diameter d of the wire using a micrometer. You will calculate the resistivity ρ from the formula:

$$\rho = \frac{R\pi d^2}{4l}$$

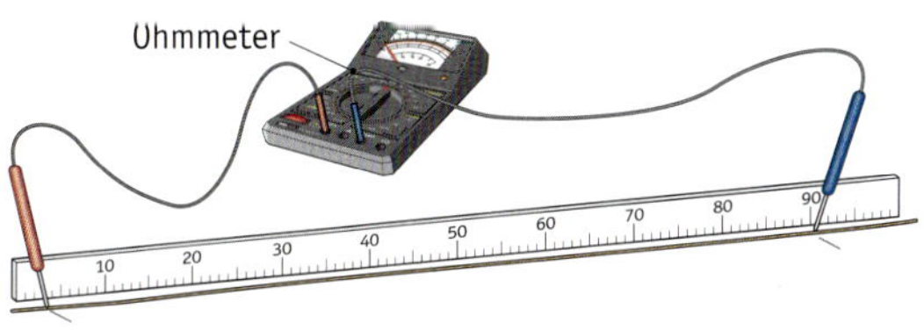

Fig. 23.29

Equipment Needed

- At least 1 m of wire
- 2 G-clamps (optional)
- A metre stick
- A micrometer
- An ohmmeter

Method

1. Ensure the wire is straight and free of any bends or kinks. Clamp the wire to the bench if clamps are available.
2. Switch on the ohmmeter, select a suitable resistance range and connect the probes of the ohmmeter firmly together. Set the resistance reading at zero or allow for the resistance of the leads.
3. Measure and record the resistance R of the wire between two points marked on it (FIG. 23.29 page 267).
4. With the metre stick measure the distance l between the two points marked on the wire. Do not measure the points on the wire that were in contact with the ohmmeter. Record this value.
5. Check the micrometer screw gauge for zero error (see Appendix 2) and record its value, if any. Measure the diameter of the wire at a number of different points along its length (at least six). Record these. Correct these readings for the zero error in the micrometer and calculate the average diameter d of the wire.
6. From the formula: $\rho = \dfrac{R\pi d^2}{4l}$ calculate the resistivity ρ.
7. If time permits, repeat the experiment for different values of l and calculate an average value of ρ.

<table>
<tr><td colspan="2">Resistance of wire R =</td><td>Ω</td></tr>
<tr><td colspan="2">length of wire l =</td><td>m</td></tr>
<tr><td rowspan="6">zero error on
micrometer = _____ mm

(to be added/to be subtracted)</td><td>Diameter of wire</td><td>Diameter of wire d
(corrected for zero error)</td></tr>
<tr><td></td><td></td></tr>
<tr><td></td><td></td></tr>
<tr><td></td><td></td></tr>
<tr><td></td><td></td></tr>
<tr><td></td><td></td></tr>
</table>

Average diameter of wire = ____________ mm

= ____________ m

Resistivity $\rho = \dfrac{R\pi d^2}{4l}$ = ____________ Ω m

Questions

1. Why must the wire be free of any bends or kinks before measuring its length?
2. Why should a reasonably long length of wire be used in this experiment?
3. Why is the diameter of the wire measured at a number of different points?
4. What would the effect on the final result be if the length of the wire were underestimated?
5. If the reading on the circular scale of the micrometer when it was fully closed were 0.03 mm but you ignored this zero error, how would the value of the final result have been affected?
6. If the reading on the circular scale of the micrometer when it was fully closed was 0.98 mm but you ignored this zero error, how would the value of the final result have been affected?
7. List two precautions which should be taken when determining the length of the wire to ensure a more accurate result.
8. List two precautions which should be taken when determining the resistance of the wire to ensure a more accurate result.
9. Outline carefully the precise steps taken when using the micrometer to measure the diameter of the wire.

EXERCISE 23.2

1. A piece of wire of uniform circular cross-section has a resistance 28.2 Ω, a length of 89.2 cm and a diameter 0.22 mm. Calculate its resistivity.
2. The length, diameter and resistance of a piece of nichrome wire were found to be 85.6 cm, 0.22 mm and 27.9 Ω. Calculate the resistivity of nichrome.
3. What length of wire of cross-sectional area 0.16 mm^2 and of resistivity 4.2×10^{-7} Ω m is needed to make a resistance of 4 Ω?
4. What length of wire of circular cross-section of radius 0.25 mm and of resistivity 1.7×10^{-8} Ω m is needed to make a resistance of 12 Ω?
5. What is the reading on each micrometer in FIG. 23.30?
6. A resistor of 400 Ω consists of a piece of wire of uniform cross-section of length 1.5 m. If the resistivity of the material of the wire is 1.2×10^{-6} Ω m, find the diameter of the wire.
7. A 20 Ω resistor consists of a piece of wire of uniform cross-sectional area and of length 130 cm. If the resistivity of the material in the wire is 1.3×10^{-6} Ω m, calculate the diameter of the wire.

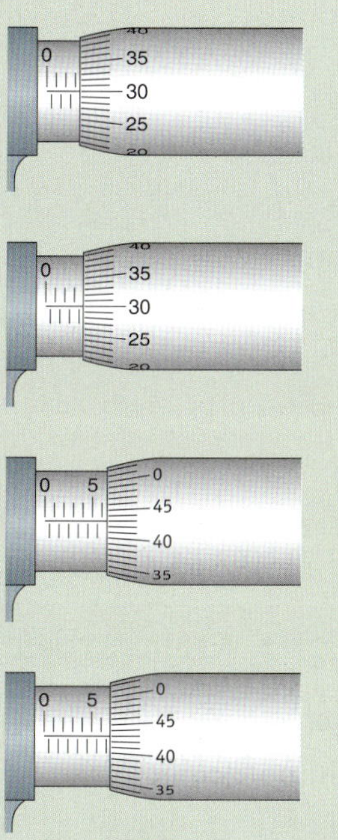

Fig. 23.30

8. A uniform wire of length 1.2 m has a resistance of 4 Ω. Find the resistance of an identical wire if its length is 7.2 m. What length of similar wire would have a resistance of 40 Ω?
9. The cross-sectional area of a piece of uniform wire is 1.5 mm^2. Its resistance is 2.6 Ω. Another wire of the same composition and length has a cross-sectional area of 6 mm^2. Find its resistance. What length should the second wire be if it is to have the same resistance as the first?
10. A uniform wire of length l, diameter d and resistivity ρ has circular cross-section. Prove that its resistivity ρ is given by:

$$\rho = \frac{R\pi d^2}{4l}$$

11. The resistance of a uniform wire is 10 Ω. Another piece of wire of the same uniform material has half the length and twice the cross-sectional area as the first piece. What is its resistance?
12. The resistance of a piece of wire of length 78.4 cm was found to be 6 Ω using an ohmmeter. A micrometer was used to measure the diameter of the wire at a number of places along its length. The measured values were: 0.45 mm, 0.44 mm, 0.46 mm 0.44 mm and 0.43 mm. If the reading on the micrometer when it was fully closed was 0.02 mm calculate:
    (i) the average diameter of the wire,
    (ii) the resistivity of the material in the wire.

MEASURING RESISTANCE USING A WHEATSTONE BRIDGE

The circuit in FIG. 23.31 is called a **wheatstone bridge circuit**. Suppose the values of the four resistors are arranged – by trial and error – so that no current flows in the galvanometer. The bridge is then said to be **balanced**. When this happens suppose current I_1 flows in the top path and I_2 flows in the bottom path of the bridge. Since the galvanometer reads zero, the potential difference between B and D = 0, *i.e.* $V_{BD} = 0$

$$\Rightarrow \quad V_{AB} = V_{AD} \quad \text{and} \quad V_{BC} = V_{DC}$$

Ohm's Law $\Rightarrow \quad I_1R_1 = I_2R_3 \quad \text{and} \quad I_1R_2 = I_2R_4$

Dividing gives: $\dfrac{I_1R_1}{I_1R_2} = \dfrac{I_2R_3}{I_2R_4}$

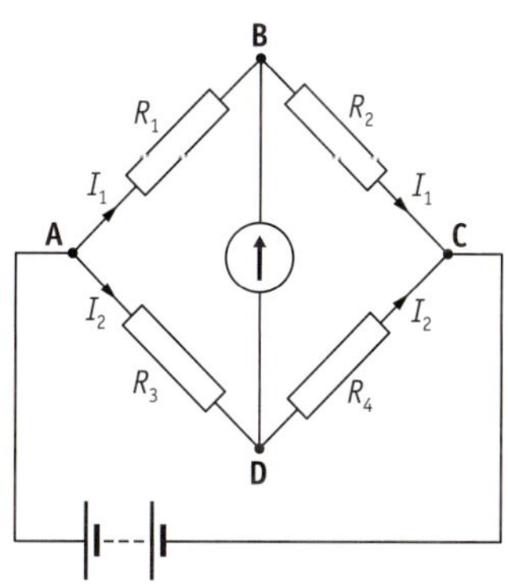

Fig. 23.31

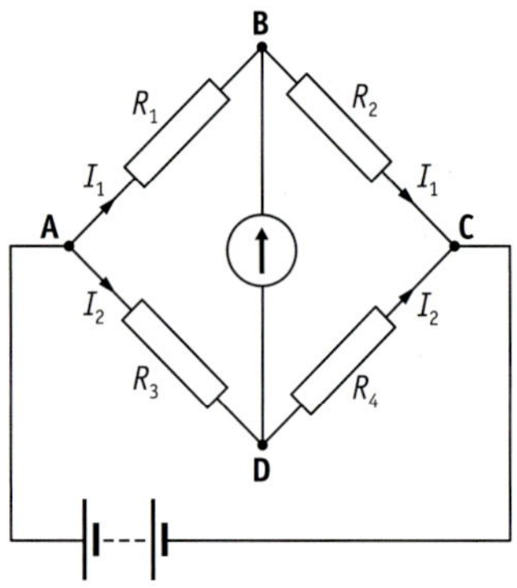

Fig. 23.32

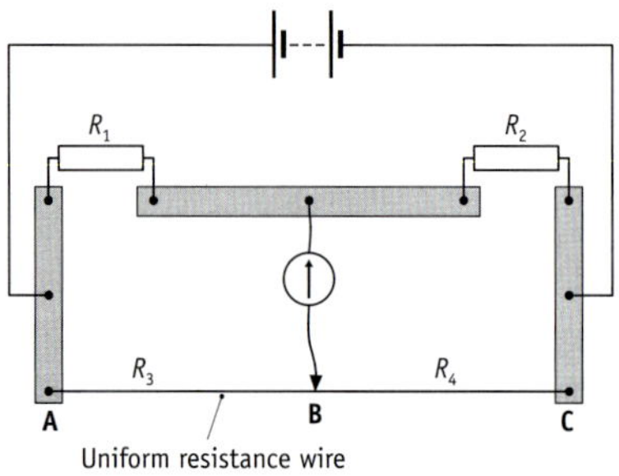

Fig. 23.33

Thus when a wheatstone bridge is balanced: $\frac{R_1}{R_2} = \frac{R_3}{R_4}$

Be careful when using this formula that R_1, R_2, R_3 and R_4 are in the positions shown in FIG. 23.32.

This simple circuit gives an accurate method of measuring resistance. Suppose R_1 was a resistance of unknown value. By inserting it in the bridge and adjusting some or all of R_2, R_3, R_4 until the bridge is balanced R_1 can be found. This method is accurate because it does not depend on the accuracy of the galvanometer but only on its sensitivity, i.e. on its ability to detect a small current. For this reason it is sometimes called a **null method**.

THE METRE BRIDGE

A simple and easy to use form of the Wheatstone bridge is the Metre bridge. It consists of a metre of uniform resistance wire and two resistors as in FIG. 23.33. Compare FIG. 23.32 with FIG. 23.33. Note that R_1 is the unknown resistance. R_2 is the known resistance. Resistance of the uniform resistance wire between A and B = R_3. Resistance of wire between B and C = R_4.

Wire uniform $\Rightarrow$ $R_3 = R_{AB} = k|AB|$ and $R_4 = R_{BC} = k|BC|$ where k is a constant.

The balance point is located by trial and error by touching the moving contact (sometimes called a **jockey**) to various points on the resistance wire until the galvanometer reads zero.

We then have: $R_1 = \left(\frac{R_2 R_3}{R_4}\right) = \left(\frac{R_2 k|AB|}{k|BC|}\right) = R_2\left(\frac{|AB|}{|BC|}\right)$

$|AB|$ and $|BC|$ are measured and R_2 is known, thus R_1 can be calculated.

PRACTICAL USES OF A WHEATSTONE BRIDGE

Temperature Control

Suppose a wheatstone bridge is balanced. If the value of one of the resistors now changes (e.g. due to a change in temperature) current will flow in the galvanometer. If the resistance increases, the current flows one way, if the resistance decreases, current flows the other way. The size of the current changes linearly with the change in resistance. Thus the direction and size of this current tells us whether the temperature has fallen or risen and by how much. The current can be used to control a heater/cooler to restore the temperature to its original value.

Fail-Safe Device

In a gas or oil flame boiler, a pilot flame should continuously be lighting. If this flame goes out, the fuel supply should automatically shut off. A Wheatstone bridge can be used for this purpose as follows: The pilot flame has a thermistor near it. This thermistor is one of the resistors in the bridge. If the flame goes out, the resistance of the thermistor increases and unbalances the bridge. The unbalanced current can be used to switch off the fuel. The Wheatstone bridge thus acts as a **fail-safe device**.

EXERCISE 23.3

1. In FIG. 23.34(A) find the value of R_4 if the bridge is balanced.

2. In FIG. 23.34(B) find the value of R_3 if the bridge is balanced.

3. In FIG. 23.34(C) $R_1 = 10R_2$ and $R_3 = 4\ \Omega$. Find the value of R_4.

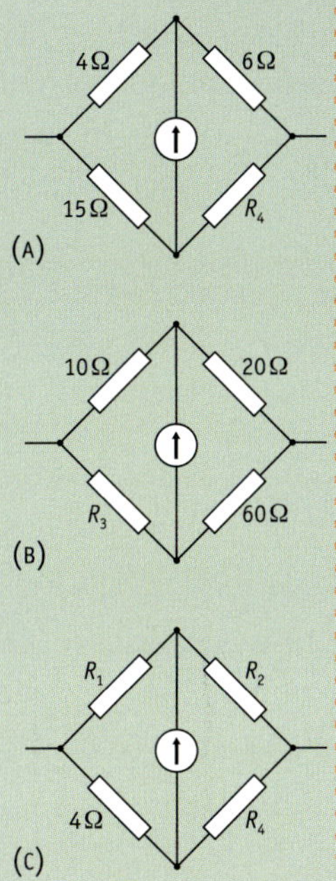

Fig. 23.34

4. In FIG. 23.35 the bridge is balanced.
   (i) If R_1 is doubled what change must be made to R_2 to re-balance the bridge?
   (ii) If R_1 is doubled, what change must be made to R_4 to balance the re-bridge?

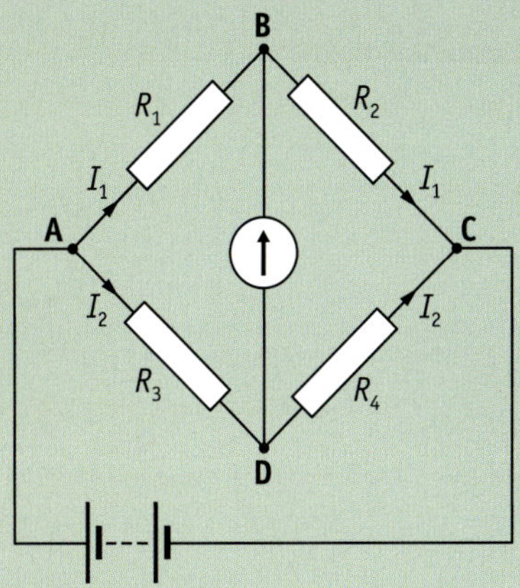

Fig. 23.35

THE POTENTIAL DIVIDER CIRCUIT

Problem 9: Calculate the p.d. across each resistor on FIG. 23.36.

Solution: Total resistance of circuit $= 2\ \Omega + 4\ \Omega = 6\ \Omega$

Current flowing $I = \dfrac{E}{R} = \dfrac{12}{6} = 2\ \text{A}$

p.d. across top resistor $V_1 = IR = (2)(2) = 4\ \text{V}$

p.d. across bottom resistor $V_2 = IR = (2)(4) = 8\ \text{V}$

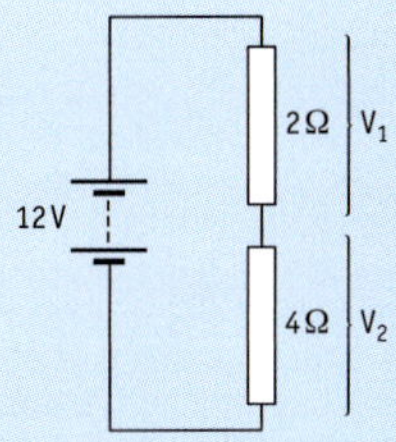

Fig. 23.36

Problem 10: Calculate the p.d. across each resistor on FIG. 23.37.

Solution: Total resistance of circuit $= 1000\ \Omega + 2\ \Omega = 1002\ \Omega$

Current flowing $I = \dfrac{E}{R} = \dfrac{12}{1002} = 0.011976\ \text{A}$

p.d. across $1000\ \Omega$ resistor $V_1 = IR$

$= (0.011976)(1000) = 11.976\ \text{V}$

p.d. across $2\ \Omega$ resistor $V_2 = IR$

$= (0.011976)(2) = 0.023952\ \text{V}$

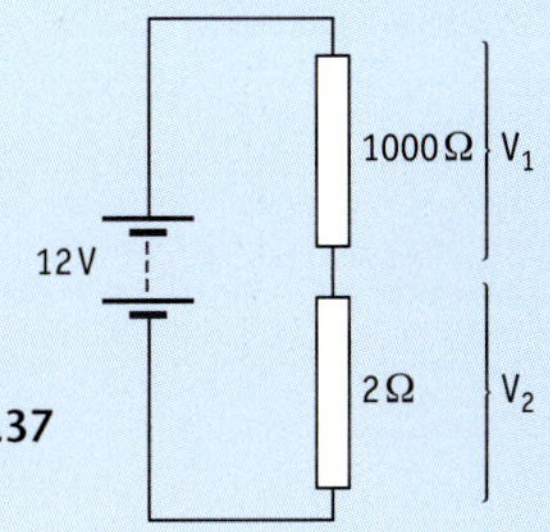

Fig. 23.37

We see from the last two problems that if two resistors are connected across a fixed voltage supply:

- the greater voltage is across the greater resistor,
- the sum of the voltages is the supply voltage,
- if one resistance is much larger than the other, the voltage across the small resistor is approximately zero and the voltage across the larger resistance is almost the same as the supply voltage. The circuits in the last two problems are examples of **potential divider circuits**.

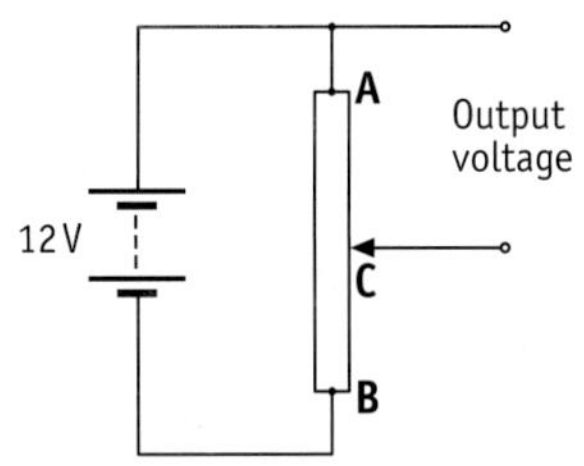

Fig. 23.38
A variable potential divider.

The circuit in FIG. 23.38 is called a **variable potential divider circuit**. As the moving contact C moves from A to B, the resistance of AC increases and so does the voltage across it. The voltage increases from zero volts when the contact is at A, to 12 volts when it is at B. Set up this circuit in the laboratory and measure the output voltage with a voltmeter. See how the voltage varies as C is moved. In many practical circuits the variable resistor would be a potentiometer rather than a rheostat.

CHAPTER CHECKLIST

- **Define:** Resistance; The Ohm; Resistivity
- **State:** The unit of resistance; Ohm's Law; The four factors on which the resistance of a conductor depends; The unit of resistivity; What a potential divider circuit does.
- **Recall** the formulae:

$$R \propto l \qquad R \propto \frac{1}{A}$$

$$R = \frac{V}{I} \qquad R = R_1 + R_2 + R_3$$

$$\frac{1}{R} = \frac{1}{R_1} + \frac{1}{R_2} + \frac{1}{R_3} \qquad R = \frac{\rho l}{A}$$

$$\frac{R_1}{R_2} = \frac{R_3}{R_4}$$

and use them to solve problems.

- **Describe** an experiment to: Demonstrate Ohm's Law; Investigate the variation of the resistance of a metallic conductor with temperature; Investigate the variation of the resistance of a thermistor with temperature; Measure the resistivity of the material of a wire.
- **Recall** that: The resistance of a metal increases as the temperature increases; The resistance of a semiconductor or an insulator decreases as the temperature increases; The resistance of a uniform conductor is directly proportional to its length; The resistance of a uniform conductor is inversely proportional to its cross-sectional area.
- **Draw**: A graph of resistance temperature for (a) a metal and (b) a thermistor; A graph of I/V for an ohmic conductor. A wheatstone bridge circuit; A metre bridge and explain how they can be used to measure resistance;
- **List** two practical applications of a wheatstone bridge.
- **Derive**: $R = R_1 + R_2 + R_3$; $\frac{1}{R} = \frac{1}{R_1} + \frac{1}{R_2} + \frac{1}{R_3}$

Effects of an Electric Current and Domestic Circuits

CHAPTER 24

THE HEATING EFFECT OF AN ELECTRIC CURRENT

The heating effect of an electric current has countless everyday applications such as in electric heaters, cookers, hairdryers, kettles etc. The amount of heat produced by an electric current depends on the size of the current I, the resistance R of the conductor through which it flows and the time t for which it flows. This can easily be demonstrated in the laboratory.

EXPERIMENT

TO SHOW THE HEATING EFFECT OF AN ELECTRIC CURRENT.

1. Set up the equipment as shown in FIG. 24.1.
2. Send a known current through the heating coil for three minutes and note the rise in temperature on the thermometer. This shows that there is indeed a heating effect.
3. Let the current flow for a longer time and observe that the rise in temperature is greater.
4. Repeat step 2 with a larger current. The rise in temperature produced in the same time will be greater.
5. Repeat step 2 with a heating coil of higher resistance. Keep the current and the time the same. The rise in temperature will be greater.

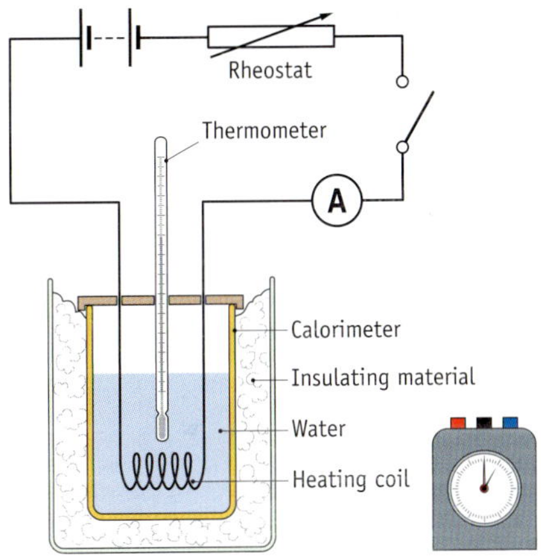

Fig. 24.1

Conclusion

The amount of heat produced by an electric current depends on the current, the resistance and the time for which the current flows.

James Joule (1818–89) experimentally investigated the factors that determined the amount of heat W given out from a current-carrying wire. He discovered that:

$W \propto I^2$ if t and R are fixed

$W \propto R$ if I and t are fixed

$W \propto t$ if I and R are fixed

It follows from this that:

$W \propto I^2Rt \Rightarrow W = kI^2Rt$ where k is a constant.

In our units the constant of proportionality k is 1, thus:

$$W = I^2Rt$$

> **JOULE'S LAW** states that the rate at which heat is produced in a conducor is directly proportional to the square of the current provided its resistance is constant,
>
> i.e. $P \propto I^2$

Since $W \propto I^2$, it follows that if a certain amount of heat is produced in a given time by a given current in a wire, in the same wire in the same time:
Twice the current will produce **four** times as much heat,
Three times the current will produce **nine** times as much heat etc.

$$W = I^2Rt \Rightarrow \frac{W}{t} = I^2R$$

Now $\frac{W}{t}$ = rate at which heat is produced
= the power (P) developed in the wire.

Thus we have: $$P = I^2R$$

If R is constant, this tells us that the rate at which heat is produced is directly proportional to the square of the current. This fact is known as **Joule's Law**.

Problem 1: Find the heat produced in a 20 Ω resistor by a current of 3 A flowing for 40 s.

Solution: $W = I^2Rt = (3^2)(20)(40) = 7200\text{ J}$

Problem 2: Find the rate at which heat is produced (i.e. the power dissipated) by a current of 80 mA in a 2 kΩ resistor.

Solution: $P = I^2R = (80 \times 10^{-3})^2(2000) = 12.8\text{ W}$

Problem 3: When a current of 3 A flows in a wire, heat is produced at the rate of 60 W. What current would produce heat at the rate of 540 W in the same wire?

Solution: $P = I^2R \Rightarrow 60 = 3^2R \Rightarrow R = \frac{60}{9} = 6.6667\ \Omega$

$P = I^2R \Rightarrow I = \sqrt{\frac{P}{R}} = \sqrt{\frac{540}{6.6667}} = 9\text{A}$

Alternative solution:

Since $P \propto I^2$ and 540 is 9 times 60, then the current must be three times larger, i.e. 9 A.

EXERCISE 24.1

1. Find the heat produced in a 20 Ω resistor by a 5 A current flowing for 3 s.
2. Find the heat produced in a 1 kΩ resistor by a 12 mA current flowing for 4 minutes.
3. Find the heat produced in a 400 Ω resistor by a 0.5 A current flowing for 1 hour.
4. Find the heat produced per second in a 10 Ω resistor by a current of:
   (i) 1 A (ii) 2 A
   (iii) 3 A (iv) 4 A
5. Find the heat produced per second in:
   (a) a 1 Ω, (b) a 2 Ω,
   (c) a 10 Ω, (d) a 100 Ω,
   resistor by a current of 2 A.
6. An electric heating coil of resistance 40 Ω takes a current of 6 A. Find the power dissipated in the heater. Find also the heat energy produced in 1 hour.
7. A 100 W bulb takes a current of 0.5 A. Calculate the resistance of the filament of the bulb.
8. A 3 kW electric fire takes a current of 10 A. Find the resistance of the heater.
9. A 2 kW heater operates on a 230 volt supply. Find:
   (a) the current flowing in the heater,
   (b) the resistance of the heater,
   (c) the time taken to produce 100 MJ of heat.
10. A potential difference of 230 V is maintained across a resistance of 40 Ω. How much heat energy is produced in the resistance in 30 s? How much charge passes through the resistance in this time?
11. When the current through a resistor doubles, what happens to the power developed in the resistor?

12. A heating coil carrying 20 A raises the temperature of 0.5 m^3 of water from 4 °C to 44 °C in 5 hours. If 80% of the electrical energy supplied appears as heat in the water, find the resistance of the heating coil.
(Specific heat capacity of water = 4180 J kg^{-1} K^{-1} and density of water = 1×10^3 kg m^{-3})

13. Two kilograms of water at 10 °C are placed in an electric kettle which operates on a 230 volt supply. If the current that flows in the kettle when it is switched on is 9 A, how long does it take to bring the water to the boil? How much longer does it take for three quarters of the water to boil off? (Specific heat capacity of water = 4180 J kg^{-1} K^{-1}. Specific latent heat of vaporisation of water = 2.3×10^6 J kg^{-1}.)

EXPERIMENT — ELECRICITY I

TO VERIFY JOULE'S LAW.

THAT IS: TO VERIFY THAT THE RISE IN TEMPERATURE PRODUCED BY A CURRENT IN A GIVEN TIME IS DIRECTLY PROPORTIONAL TO THE SQUARE OF THE CURRENT

i.e. TO SHOW THAT: $\Delta\theta \propto I^2$

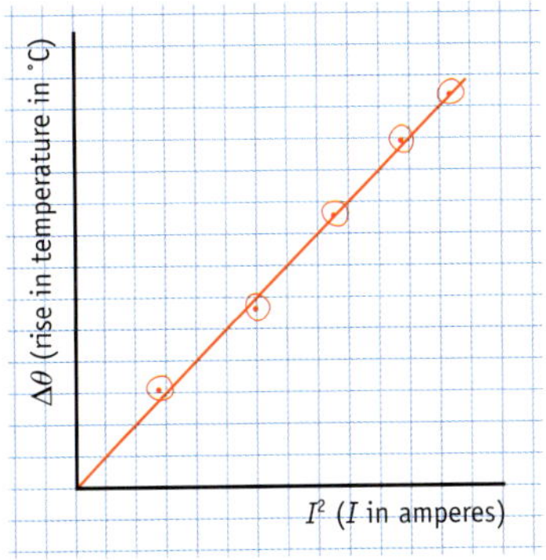

Fig. 24.2

Summary of Method

In this experiment you will send a known steady current through a coil of wire of constant resistance which is immersed in a calorimeter of water. You will measure the rise in temperature produced in a given time. You will repeat the procedure for a number of different values of the current, **the time interval being the same** and the water in the calorimeter being replaced by **an equal volume of cold water** each time. From your results you will plot a graph of rise in temperature $\Delta\theta$ against I^2 (current)^2.
A straight line through the origin will result (FIG. 24.2), showing that: $\Delta\theta \propto I^2$, hence verifying Joule's Law.

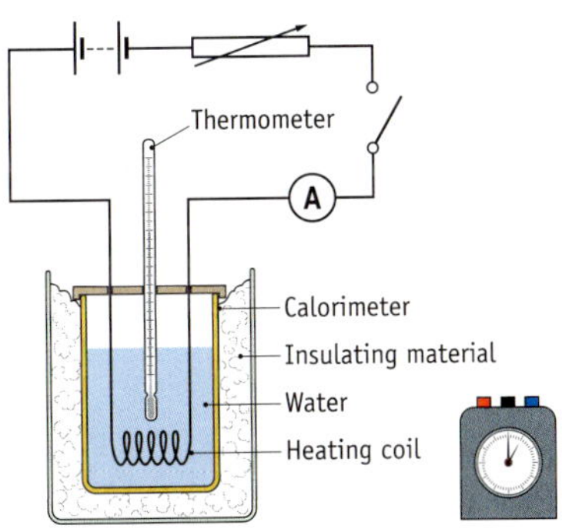

Fig. 24.3

Equipment Needed

- A low voltage d.c. power supply (0–25 V)
- An ammeter (0–6 A) and a variable resistor
- A thermometer (0–50 °C in steps of 0.1 °C)
- A heating coil (whose resistance changes very little with temperature)
- A calorimeter, insulating material, a lid and a stirrer
- A container (e.g. a large beaker) to hold the calorimeter and insulating material
- A clock or stopwatch, some connecting leads and a graduated cylinder

Method

1. Fill the calorimeter with enough cold water to completely cover the heating coil and then transfer the water into the empty graduated cylinder and note its volume. Transfer the water back into the calorimeter.
2. Set up the circuit shown in FIG. 24.3.
3. Turn on the current and adjust the rheostat so that a current of about 0.5 A flows. Turn off the current.
4. Stir the water, wait a minute or two, then measure and record its temperature on the thermometer θ_1.
5. Turn on the current and start the clock.
6. Take the reading on the ammeter I and record it. Allow the current to flow for a known time (say 5 minutes), stirring the water regularly.

7. When the current is flowing, watch the ammeter. If the current changes, immediately restore it to its original value by suitably adjusting the rheostat.
8. At the end of the time interval stop the clock and switch off the current. Stir the water, wait a while and note the highest temperature θ_2 reached.
9. Empty the warm water from the calorimeter and refill it with an equal volume of cold water.
10. Repeat steps 4 to 9 with larger values of current obtained by either reducing the resistance of the rheostat or increasing the voltage of the power supply, **the same time interval being used on each occasion**. Increase the current in steps of about 0.5 A.
11. Complete the Table calculating $\Delta\theta$, I^2, and $\frac{\Delta\theta}{I^2}$ for each set of readings. Find the average value of $\frac{\Delta\theta}{I^2}$.
12. On graph paper plot a graph of $\Delta\theta$ against I^2 with $\Delta\theta$ along the *y*-axis. Find the slope of the graph.

Result

Within the limits of experimental error, $\frac{\Delta\theta}{I^2}$ will be a constant verifying that $\Delta\theta \propto I^2$.

The graph of $\Delta\theta$ against I^2 will be a straight line passing through the origin, again verifying that $\Delta\theta \propto I^2$.

The slope of the graph will be almost the same value as the average value of $\frac{\Delta\theta}{I^2}$ calculated above.

Temperature before θ_1 / °C	Temperature after θ_2 / °C	Rise in Temp. $\Delta\theta$ / °C	Current I / A	Current squared I^2	$\frac{\Delta\theta}{I^2}$

EXPERIMENTAL NOTES

This experiment verify's Joule's Law because:
Heat theory $\Rightarrow$ Heat produced $\propto$ Rise in temperature, i.e. $\Delta H \propto \Delta\theta$
The graph (FIG 24.4) tells us that: $\Delta\theta \propto I^2$
It follows that: $\Delta H \propto I^2$ i.e. the heat produced in a given time is directly proportional to I^2 $\Rightarrow$ Heat produced per second $\propto$ I^2 i.e. Joule's Law.
A coil of wire whose resistance changes only very little with temperature (such as one made from Constantan or Manganin) must be used.

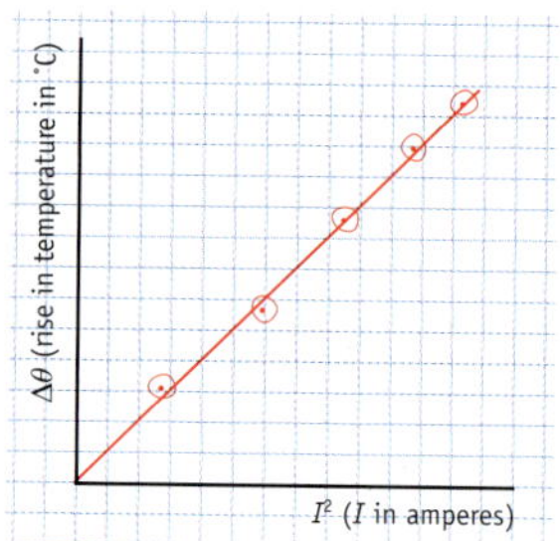

Fig. 24.4

Questions

1. Why must the heating coil be completely covered by the water?
2. Why is the same volume of water used each time the procedure is repeated?
3. Why must the current be kept constant? If the current were found to have increased, should you increase or decrease the resistance of the rheostat?
4. Why should the warm water be replaced by an equal volume of cold water each time the procedure is repeated?
5. Why must the same time interval be used each time the procedure is repeated?
6. What is the disadvantage in allowing the current to flow for a long time, say 15 minutes?
7. What is the disadvantage in allowing the current to flow for only a very short time, say 30 seconds?
8. Why do you plot a graph of $\Delta\theta$ against I^2 instead of against I?
9. List three precautions you would take in the experiment to ensure an accurate result.

Advantage of Use of High Voltage in the Transmission of Electrical Energy

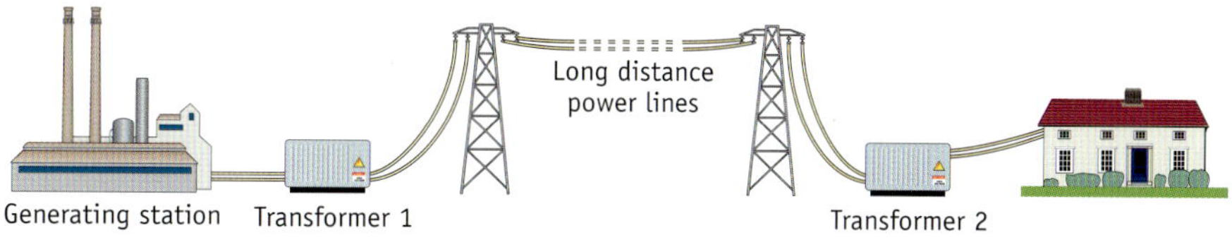

Fig. 24.5

Suppose that in Fig. 24.5 the generating station sends electrical energy to the house through wires of total resistance R ohms. If the current flowing in the wires is I then the heat produced (wasted) in the wires per second is given by Joule's Law as $P = I^2R$. Thus the larger the current, the larger the heat energy wasted. **Transformers** are devices that can change the voltage of the supply (see Chapter 28). Since power = current × voltage, i.e. $P = IV$, the bigger the voltage the smaller the current for a given power. Transformer 1 increases the voltage to several kilovolts and so the current in the wires is small. Thus the heat losses in the wires are small. The second transformer lowers the voltage to a safer value to use in a house. The high voltage used is called **extra high tension (EHT)**.

The Chemical Effect of an Electric Current

In Chapter 21 you saw that **an electric current may cause a chemical reaction when it passes through a liquid**, i.e. an electric current has a chemical effect. This phenomenon is called **electrolysis**.

- The liquid in which the chemical reaction takes place is called an **electrolyte**.
- The rods or plates that dip into the electrolyte are called **electrodes**.
- The electrode connected to the positive of the power supply is called the **anode**.
- The electrode connected to the negative of the power supply is called the **cathode**.
- The container, electrolyte and electrodes together are called a **voltameter**.
- If the electrodes do not take part in the chemical reaction they are called **inactive electrodes**.
- If the electrodes do take part in the chemical reaction they are called **active electrodes**.

Examples of Electrolytes

- A solution of an acid, base or salt in water.
- An ionic compound in its molten state.

Experiment

To Demonstrate the Chemical Effect of an Electric Current.

1. Set up the equipment as in Fig. 24.6.
2. Turn on the current.
3. Observe that the anode gradually gets eaten away and the cathode becomes coated with copper.

The following reactions have taken place:

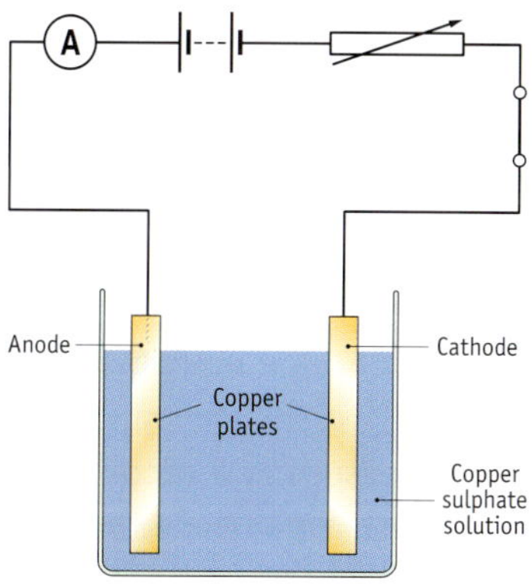

Fig. 24.6

At the Anode

Copper atoms at the surface of the anode lose two electrons and go into the electrolyte as copper ions. The anode slowly dissolves into the electrolyte as the electrolysis proceeds. In chemical symbols: $\mathbf{Cu \rightarrow Cu^{++} + 2\,e^-}$

At the Cathode

Positive copper ions from the electrolyte pick up electrons from the negative cathode and become copper atoms. These atoms become plated onto the cathode and it becomes coated in copper. In chemical symbols: $\mathbf{Cu^{++} + 2\,e^- \rightarrow Cu}$

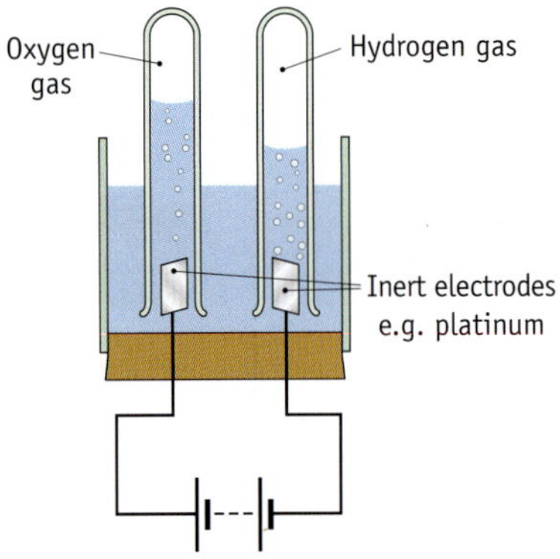

Fig. 24.7
The electrolysis of water – breaking water up into hydrogen and oxygen by passing an electric current through it.

ION
An atom or a molecule that has lost or gained one or more electrons is called an ***ion.***

CHARGE CARRIERS IN AN ELECTROLYTE
In an electrolyte the charge carriers are positive and negative ***ions***.

ELECTROLYSIS OF WATER

FIG. 24.7 shows the electrolysis of water with inert electrodes. The water breaks up into hydrogen and oxygen. Hydrogen gas bubbles off at the cathode and oxygen gas bubbles off at the anode. This can easily be set up in the laboratory.

APPLICATIONS OF THE CHEMICAL EFFECT

- Electroplating, i.e. covering one metal with a thin layer of another, usually to protect the first one from corrosion and make it look better. The metal object to be electroplated must be the cathode.
- Extracting metals from their ores.
- Purifying metals.
- In the manufacture of one particular type of capacitor called electrolytic capacitors. The dielectric is produced by electrolysis and is very thin. The value of the capacitance can then be very large (see Chapter 20).

EXPERIMENT — ELECTRICITY 5(C)

To Investigate the Variation of Current (I) with p.d. (V) for Copper Sulphate Solution with Copper Electrodes

Summary of Method

In this experiment you will measure the current flowing through the electrolyte for a series of different values of the voltage, which you will also measure. You will plot a graph of I against V. A straight line through the origin will result.

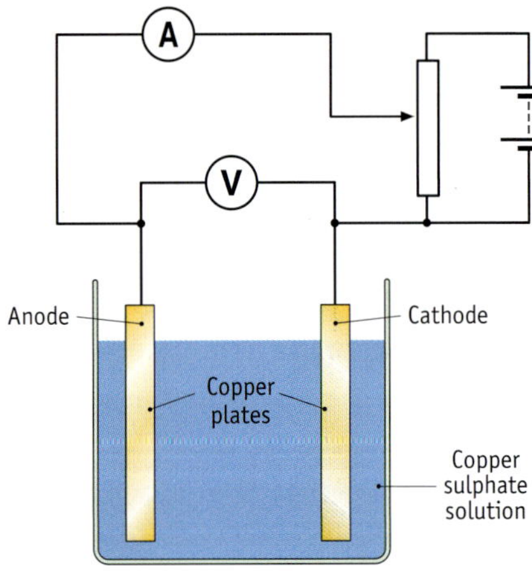

Fig. 24.8

Equipment Needed

- A beaker, two copper electrodes and copper sulphate solution
- A d.c. power supply (0–12 V)
- A rheostat, ammeter and voltmeter

Method

1. Set up the equipment as shown in FIG. 24.8.
2. Adjust the rheostat until the p.d. across the electrolyte is small (e.g. 1 V). Measure and record the p.d. (V) across and the current I through the electrolyte.
3. Adjust the rheostat increasing the p.d. and the current. Measure and record V and I again.
4. Repeat step 3 a number of times, recording all values.
5. On graph paper, plot a graph of I (on the y-axis) against V.

V / V	I / A

Result

A straight line through the origin will result.

Relationship between Current and Voltage for Different Conductors

A metallic conductor

Provided the temperature remains fairly constant, the resistance of a metal conductor does not change as the current through it increases. The voltage and the current are thus directly proportional. The metal obeys Ohm's Law. Fig. 24.9 shows the I–V graph.

In a **metal** the charge carriers are **negative electrons**.

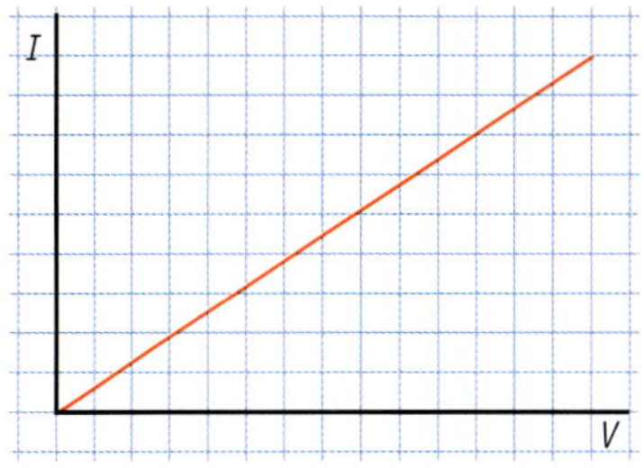

Fig. 24.9
The I/V graph of a metal at constant temperature.

A filament bulb

As the p.d. across a filament bulb is increased, the current increases. As it does so, the filament gets significantly hotter and its resistance increases. Thus, when the filament is hot, a given increase in V does not produce as much of an increase in I as when it is cooler. The I–V graph thus becomes less steep (Fig. 24.10).

In a **filament bulb** the charge carriers are **negative electrons**.

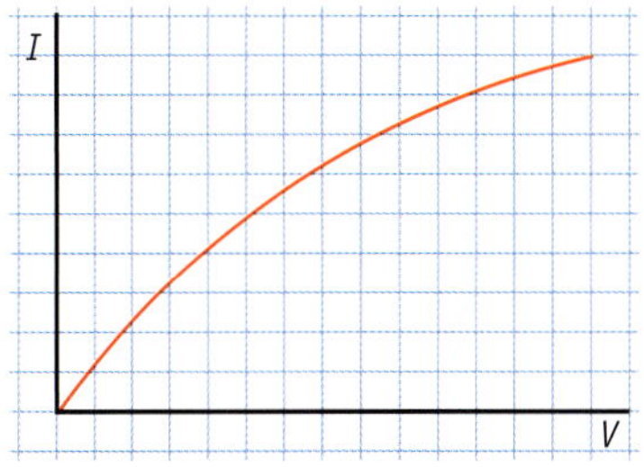

Fig. 24.10
The I/V graph of a filament bulb.

A semiconductor e.g. a thermistor

As the p.d. across a semiconductor is increased, the current increases. As it does so, the semiconductor gets hotter. This produces many more holes* and electrons which are available for conduction and its resistance drops. A further increase in V produces a much larger increase in I than when it was cold. The I–V graph thus gets much steeper (Fig. 24.11).

In a **semiconductor** the charge carriers are **negative electrons** and **positive holes**.

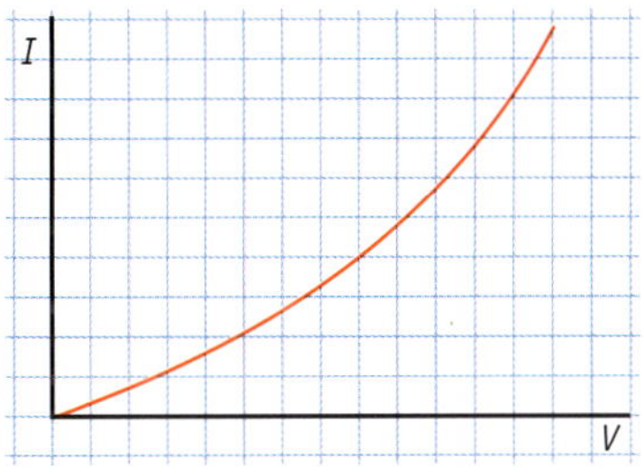

Fig. 24.11
The I/V graph of a semiconductor.

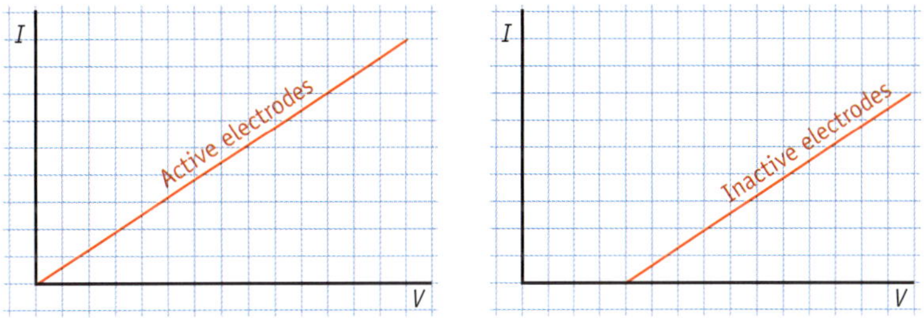

(A)
The I/V graph of an electrolyte with active electrodes.

(B)
The I/V graph of an electrolyte with inert electrodes.

Fig. 24.12

Ionic solutions i.e. electrolytes

As the p.d. increases so does the current. The resistance remains constant and thus the I/ V graph is a straight line. If the electrodes are **active** (i.e. they take part in the chemical reactions), the electrolyte obeys Ohm's Law and the graph passes through the origin (Fig. 24.12(A)).

If the electrodes are **inactive**, the voltameter behaves as a cell and produces an emf across its plates. The applied voltage must be greater than this emf before the current will flow. The I/ V graph is as shown in Fig. 24.12(B).

In an **electrolyte** the charge carriers are **positive ions** and **negative ions**.

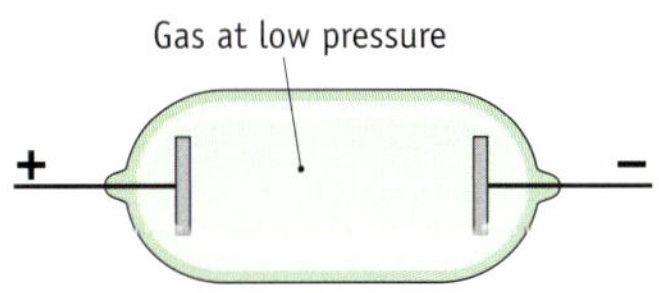

Fig. 24.13
A gas discharge tube.

A gas

Fig. 24.13 shows a sealed container with two electrodes in it. It contains a gas at low pressure. It is called a discharge tube. Due to background radioactivity and cosmic rays there are always some ions being formed in the gas. These ions recombine with their electrons after a while. If a p.d.

* See Chapter 25.

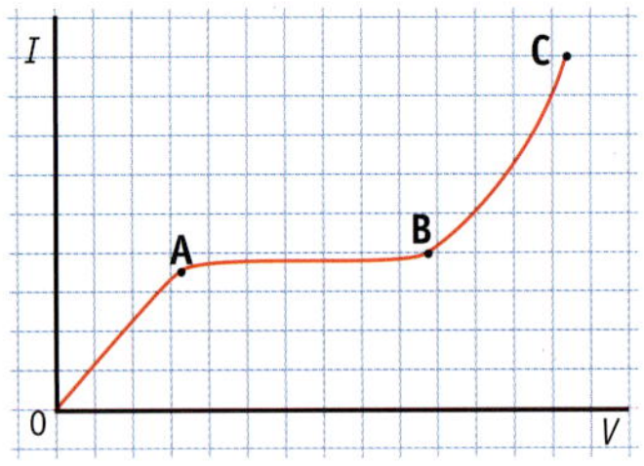

Fig. 24.14
The I/V graph of a gas.

is put across the tube, the positive ions move towards the negative electrode and the electrons move towards the positive electrode. A current thus flows.

As the p.d. is increased, the number of ions crossing the tube increases, as does the current. This corresponds to the region OA in Fig. 24.14. At a certain p.d. all the ions produced in the tube cross the tube before recombining. Further increasing the p.d. produces no further increase in the current – the graph levels out as in region AB.

As the voltage is further increased, a stage is reached where the ions and electrons reach sufficient speed to produce further ions by collision. The current then increases with the p.d. (region BC on the graph).

In a gas, the charge carriers are **positive ions**, **negative electrons** and a few **negative ions**.

The sodium vapour lamp (page 208), sodium vapour street lamps (yellow/ orange) and neon lamps are examples of gas discharge tubes, where conduction takes place in a gas.

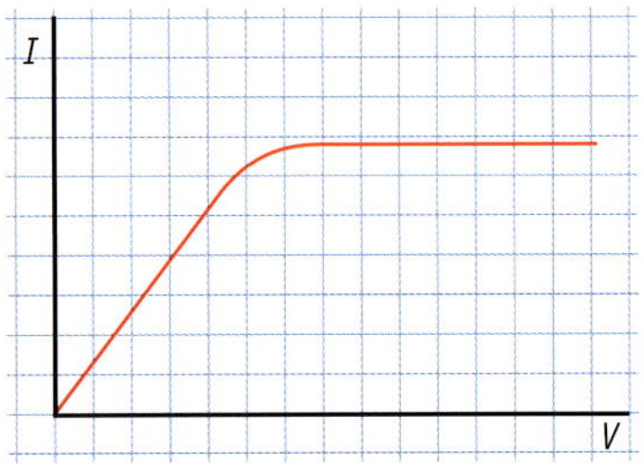

Fig. 24.15
The I/V curve of a vacuum, if thermionic emission occurs at the cathode.

A VACUUM

A vacuum will not conduct electricity because there are no charge carriers present. If, however, the cathode (negative electrode) is heated sufficiently, thermionic emission will occur at the cathode (page 328). Electrons get sufficient energy to leave the cathode. As the p.d. across the tube is increased, the current increases until all the electrons emitted from the cathode are carried across the tube. Further increases in p.d. cause no further increases in the current and the I/V curve flattens out (Fig. 24.15).

EXPERIMENT **ELECTRICITY 5**

To Investigate the Variation of Current (I) with p.d. (V) for:

(a) A Metallic Conductor (b) A Filament Bulb
(c) Copper Sulphate Solution with Copper Electrodes
(d) A Semiconductor Diode.

For (a) see the experiment to verify Ohm's Law on page 258 or alternatively use a coil of wire instead of the filament lamp below.
For (c) see page 278. For (d) see page 289.

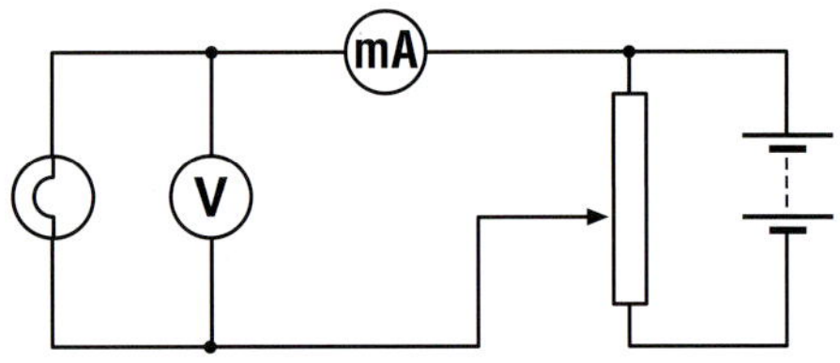

Fig. 24.16

Equipment Needed

Equipment needed to investigate the I/V relationship for a filament bulb:

- A d.c. power supply (0–12 V) and a 12 V filament bulb
- A rheostat, ammeter and voltmeter

Method

1. Set up the equipment as shown in Fig. 24.16.
2. Adjust the rheostat until the p.d. across the bulb is small (e.g. 1 V). Measure and record the p.d. V across and the current I through the bulb.
3. Adjust the rheostat increasing the p.d. and the current. Measure and record V and I again.
4. Repeat step 3 a number of times, recording all values.
5. On graph paper plot a graph of I (on the y-axis) against V.

V / V	I / A

Result

A graph similar to Fig. 24.10 (page 279) will result.

Domestic Electric Circuits

Mains electricity is supplied at 230 V a.c. in Ireland and the EU. Fig. 24.17 shows a simplified picture of the electric wiring found in a house.

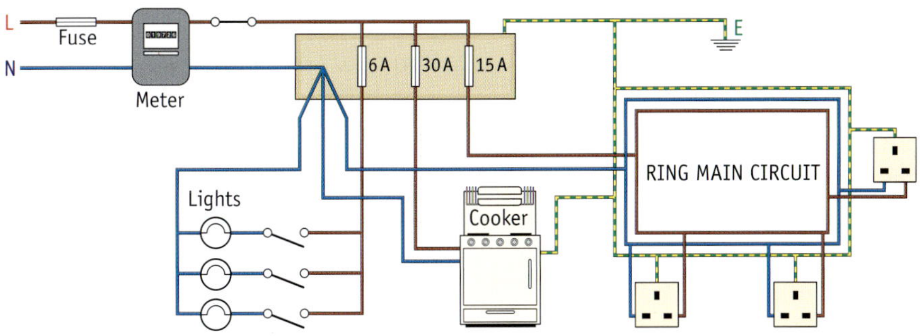

Fig. 24.17

Two wires enter the house from the mains. One of these is the **live**. Its voltage varies from about +325 V to −325 V. **It is very dangerous. Contact with this wire could be fatal.** The other wire is called the **neutral** and should be at or very near zero volts. When the live enters the house, it is first interrupted by the house's main fuse and possibly a main switch. It then passes through a meter (similar to a joulemeter) which can measure the amount of electrical energy used. It then goes to the **distribution box**. The neutral wire follows a similar route.

To supply power to the various appliances in the house a live wire and a neutral wire must be connected to each appliance. These wires come from the distribution box.

APPLIANCES THAT TAKE A LARGE CURRENT

Appliances that take a large current, such as an electric cooker, an immersion heater or an electric shower have a separate live and neutral wire coming from the distribution box. Such a circuit is called a **radial circuit**. Each radial circuit has its own fuse.

CONNECTIONS TO LIGHTS

Since lights do not take much current, a number of them may be connected to the same fuse. Each light has a separate switch in the live wire. The lights are connected in parallel with each other, so that if one bulb fails the others are not affected.

CONNECTIONS TO THE SOCKETS – THE RING CIRCUIT

In a **ring circuit**, the live terminals of each socket are connected together. Power is thus fed along both sides of the ring to each socket. The neutrals are also connected together, which is then connected back to the neutral at the distribution box. Each ring circuit has a fuse in its live.

SWITCHES

A switch should always be connected in the live wire, so that when the switch is open the appliance is disconnected from the live.

FUSES AND MINIATURE CIRCUIT BREAKERS (MCBs)

Fuses

If an excessive current passes through an appliance it may cause it to overheat or, worse still, cause a fire. A **fuse** is a piece of wire that will melt when a current of a certain size passes through it. In this way, the fuse protects an appliance should a fault occur whereby an excessive current is flowing. **A fuse should be connected in the live wire**, so that if it blows, the appliance is disconnected from the live. Fig. 24.18 shows the circuit symbol for a fuse.

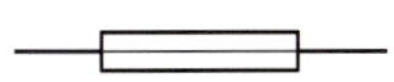

Fig. 24.18 Circuit symbol for a fuse.

Miniature circuit breakers (MCBs)

Miniature circuit breakers (MCBs) are usually used instead of fuses in the distribution box. They contain a bimetallic strip and an electromagnet. When the current is larger than a pre-set value, one of these pulls two contacts apart and breaks the circuit. The bimetallic strip causes the switch to trip for small currents and the electromagnet does it for larger currents. They operate faster than a fuse and can be reset by flicking a switch.

As well as a fuse or an MCB, the circuits connected to the sockets in a house are protected by **residual current devices (RCDs)**. They operate by detecting a difference between the current in the live and the neutral – which could arise if someone comes in contact with the live and current flows through them to earth. When the difference in the value of the current between the live and the neutral reaches a pre-set value (usually 30 mA), the RCD trips very quickly, disconnecting the circuit from the live. They thus protect against electrocution where a fuse or MCB would not blow fast enough to do so.

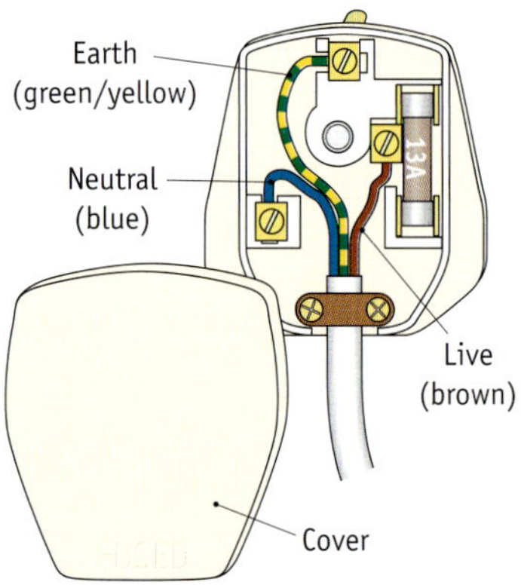

Fig. 24.19

BONDING

All metal water pipes, metal taps, metal water tanks etc. in a house must be connected to earth (bonded to earth). Bonding to earth is a safety precaution. If the metal pipes etc. become accidentally connected to live they will not be at 230 V but will remain at zero potential and will not pose the danger of electrocuting someone.

PLUGS

FIG. 24.19 shows a plug. Take note of the colours of the wires.

Live is **BROWN**,

Neutral is **BLUE** and

Earth is **GREEN/YELLOW**.

When you are wiring a plug these must be connected correctly, with the live connected to the pin with the fuse as shown in the diagram. Make sure the chord grip is tightened properly, holding the cable in place.

Fig. 24.20

EARTHING

Look at FIG. 24.20, suppose a fault develops whereby the live wire comes in contact with the metal of the kettle. If someone were to then touch the kettle, an electric current would pass through them, possibly killing them.

If, however, the metal of the kettle is connected to earth, current from the live flows through the kettle to earth. Since this path has low resistance, a large current will flow. This will blow the fuse and disconnect the kettle from the live. The fuse should not be replaced until the fault in the kettle is fixed. If the current is not large enough to blow the fuse, it is still safe to touch the kettle, since being connected to earth its potential is zero. All electrical equipment with exposed metal parts should therefore be earthed.

Problem 4: A lamp has a power rating of 100 W. It operates on 230 V mains. What current does it draw? What size fuse should be put in its plug, a 3 A or a 13 A?

Solution: $P = IV \Rightarrow I = \frac{P}{V} = \frac{100}{230} = 0.43 \text{ A}$

The 3 A fuse is much more suitable.

Problem 5: An electric cooker has four 500 W plates, a 2 kW grill and a 3 kW oven. It operates on 230 V mains. Is a 40 A fuse suitable for this cooker?

Solution: Total power $P = (4)(500) + 2000 + 3000 = 7000 \text{ W}$

$P = IV \Rightarrow I = \frac{P}{V} = \frac{7000}{230} = 30.4 \text{ A}$

i.e. the maximum current the cooker takes is 30.4 A, therefore a 40 A fuse is suitable.

THE KILOWATT-HOUR

The joule is a very small unit of energy to use when dealing with domestic electricity. A **unit of energy** called the **kilowatt-hour (kW h)** is used instead. It is the amount of energy used by a 1000 W appliance in one hour. The number of 'units' recorded in an electricity bill is the number of kilowatt-hours used.

KILOWATT-HOUR

The ***kilowatt-hour* (kW h)** is the amount of energy used by a 1000 W appliance in one hour.

Problem 6: How many joules in a kilowatt-hour?

Solution: 1000 W = 1000 joules per second. 1 hour = 60 × 60 seconds

Energy = Power × Time = $(1000)(60)(60) = 3.6 \times 10^6 \text{ J} = 3.6 \text{ MJ}$

EXERCISE 24.2

1. A heater has a power rating of 1000 W. It operates on 230 V mains. What current does it draw? What size fuse should be put in its plug, a 3 A or a 13 A?
2. An electric cooker has two 500 W plates, a 1 kW grill and a 2 kW oven. It operates on 230 V mains. Is a 30 A fuse suitable for this cooker?
3. A 2000 W appliance operates for 3 hours. How many kilowatt-hours of energy does it use?
4. A 75 W lamp is operating for 40 minutes. How many kilowatt-hours of energy does it use?

CHAPTER CHECKLIST

- **State**: The three factors on which the amount of heat produced in a current-carrying conductor depend; Joule's Law; What the charge carriers in an electrolyte are; The colour of live, earth and neutral; The purpose of a fuse; The purpose of an MCB; The Purpose of an RCD.
- **Describe** how to wire a plug.
- **Define**: Chemical effect of an electric current; Electrolyte; Cathode; Anode; Active electrode; Inactive electrode; Ion; Kilowatt-hour.
- **Explain**: The advantage in using high voltage in the transmission of electrical energy; What each of the following mean in the context of domestic electric supplies: Live, Earth, Neutral, Ring circuit, Radial circuit, Bonding, Earthing.
- **Recall**: That an electric current can cause a chemical reaction in a liquid; Examples of electrolytes; Where in a mains circuit the switch and fuse should be put.
- **Describe** and carry out an experiment to: Show the heating effect of an electric current; Verify Joule's Law; Show the chemical effect of an electric current; Investigate the variation of current (I) with p.d. (V) for a metallic conductor, a filament bulb and copper sulphate solution with copper electrodes.
- **Recall** and use the formulae: $W = I^2 Rt$; $P = I^2 R$ to solve problems.
- **Draw**: the I-V graph for a metal, a filament bulb, a semiconductor, an ionic solution, a gas, a vacuum and state what particles are the charge carriers in each case.
- **List**: Two examples of electrolytes; Four practical uses of the chemical effect of an electric current.

CHAPTER 25

Semiconductors

SEMICONDUCTORS

Virtually every electronic device now available is based on the conduction of electricity by materials known as semiconductors. Computers televisions, sound and communication systems contain many different types of semiconductors. Semiconductors, called light emitting diodes (LEDs), can produce light (FIG. 25.1). You are probably familiar with LEDs that are used as indicator lamps on much electrical equipment to show whether it is turned on or not. They are also used on the displays of large calculators, televisions, videos, digital clocks etc. More powerful LEDs are used as high intensity rear brake lights on cars and as normal lights on bicycles.

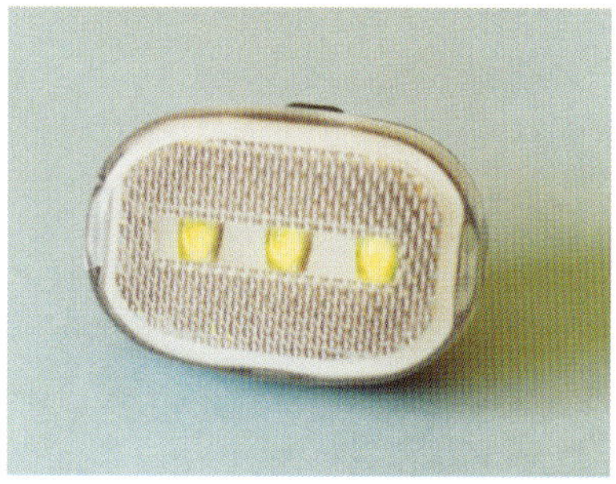

Fig. 25.1
Three LEDs on a bicycle light.

> A ***semiconductor*** is a substance whose resistivity is between that of a good conductor and a good insulator. The resistivity of a semiconductor decreases as its temperature increases.

Examples of semiconductors are: silicon, germanium, cadmium sulphide.

CONDUCTION IN SEMICONDUCTORS

FIG. 25.2 shows the outer electron arrangement in the semiconductor silicon at a temperature very near zero kelvin. Each silicon atom has its four outer electrons (called **valence electrons**) involved in covalent bonds with four other silicon atoms. These electrons are not free to wander from atom to atom. **Silicon is thus an insulator near zero kelvin.**

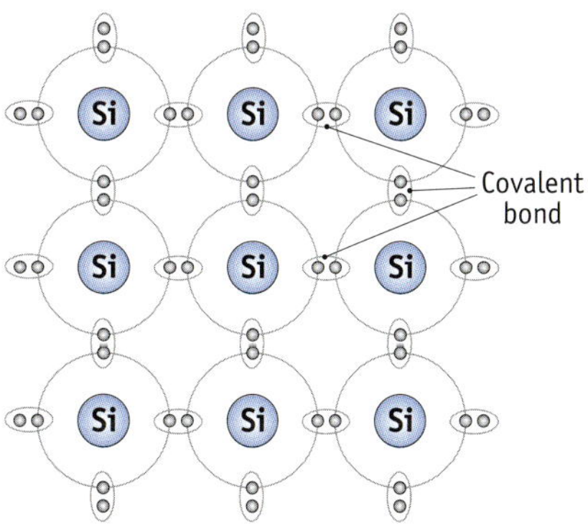

Fig. 25.2
The outer electron arrangement in pure silicon at a temperature very near zero kelvin. The grey dots are the outer electrons. Silicon is an insulator.

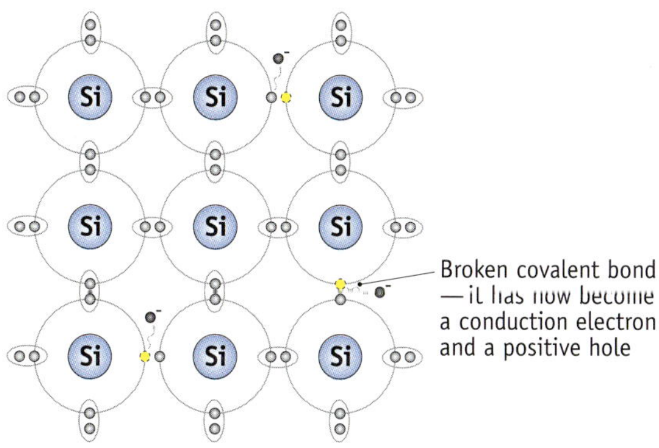

Fig. 25.3
The outer electron arrangement in silicon at room temperature.

FIG. 25.3 shows the outer electron arrangement in silicon at room temperature. At this temperature some of the valence electrons have enough thermal energy to break their covalent bonds and are free to wander from atom to atom. Such electrons are called **conduction electrons**. The energy of a conduction electron is greater than the energy of a valence electron.

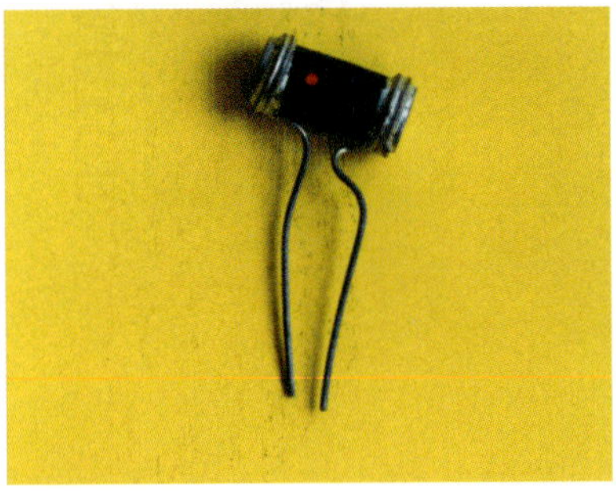

Fig. 25.4
A thermistor and its circuit symbol.

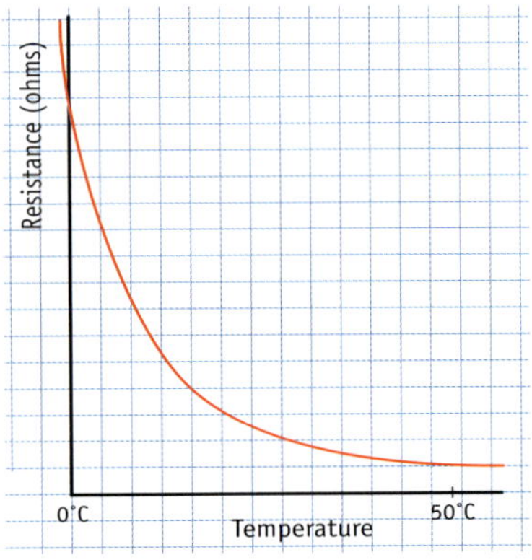

Fig. 25.5
Variation of the resistance of a thermistor with temperature.

THE THERMISTOR

A ***thermistor*** is a semiconductor device whose resistance decreases rapidly with increasing temperature.

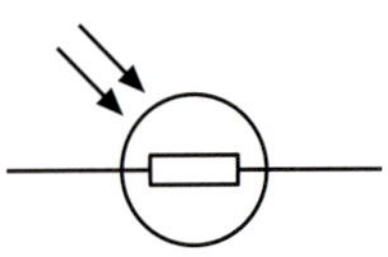

Fig. 25.6
A light dependent resistor (LDR) and its circuit symbol.

A LIGHT DEPENDENT RESISTOR

A light dependent resistor (LDR) is a semiconductor whose conductivity is increased when light shines on it.

HOLES

When an electron breaks free from a covalent bond, it leaves behind a gap in the atom from which it came. This gap is called a ***positive hole*** or simply a ***hole***.

A hole is positively charged since the atom in which it occurs has lost an electron. A valence electron from a neighbouring atom can move into this hole, producing another hole in the atom from which it came. Thus as an electron moves into one hole it produces another. This process can continue and can be viewed as a movement of positive holes through the silicon.

INTRINSIC CONDUCTION

Conduction in a pure semiconductor due to electrons moving from negative to positive (– to +) and an ***equal number*** of holes moving in the opposite direction is called ***intrinsic conduction***. The semiconductor is called an ***intrinsic semiconductor***.

The resistivity of an intrinsic semiconductor is quite large, e.g. a piece of pure silicon of diameter 1 mm and length 1 cm has a resistance of about 8 million ohms at 0°C. To increase the conductivity (i.e. to reduce the resistivity) of a semiconductor you must increase the number of mobile charge carriers present in it. This can be done in a number of ways.

INCREASING THE TEMPERATURE OF A SEMICONDUCTOR INCREASES ITS CONDUCTIVITY

By increasing the temperature of a semiconductor, more electrons get enough energy to break out of their covalent bonds. Thus more free electrons and holes are produced. These are charge carriers and therefore the conductivity of the semiconductor is increased. The thermistor (FIG. 25.4) is a good example of a semiconductor whose resistance decreases rapidly with increasing temperature.

The variation of the resistance of a thermistor can easily be seen by connecting the thermistor to an ohmmeter and seeing how the resistance changes as you heat the thermistor. An experiment to see accurately how its resistance varies with temperature and produce the graph in FIG. 25.5 is described on page 264.

SHINING LIGHT ON SOME SEMICONDUCTORS INCREASES THEIR CONDUCTIVITY

The conductivity of some semiconductors can be increased by shining light on them. For example, cadmium sulphide shows a notable increase in conductivity when light shines on it. Valence electrons in the material get enough energy from the light to become conduction electrons, thus producing more free electrons and holes and reducing the resistivity of the material. Typically its resistance changes from several megaohms in darkness to a few hundred ohms in daylight. A semiconductor that behaves like this is called a light dependent resistor (LDR).

FIG. 25.6 shows one type of LDR and its circuit symbol. Being made of cadmium sulphide it is sometimes called a cadmium sulphide (CdS) cell. The variation of its resistance with brightness can be demonstrated by connecting the LDR to an ohmmeter to read its resistance. This can be measured when the LDR is in the dark and in varying degrees of brightness (by varying its distance from a bulb which is lighting).

THE CONDUCTIVITY OF A SEMICONDUCTOR CAN BE INCREASED BY THE PRESENCE OF CERTAIN IMPURITIES

The conductivity of a semiconductor may be increased by adding small amounts of certain impurities to it. This procedure is called **doping**.

DOPING

The adding of small controlled amounts of certain impurities to a pure semiconductor to increase its conductivity is called ***doping***.

In the case of silicon, small amounts of substances with a valency of either 5 or 3 are added when it is in a molten state at manufacture. The semiconductors so formed are called n-type or p-type semiconductors. The more impurity added the greater the increase in conductivity.

N-TYPE SEMICONDUCTOR

An ***n-type semiconductor*** is one in which the impurity added produces more free electrons available for conduction, e.g. phosphorous in silicon.

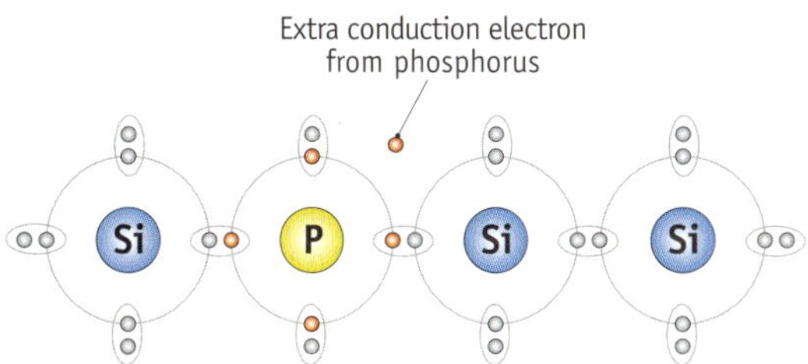

Fig. 25.7
An n-type semiconductor. Each phosphorous atom introduces an extra electron for conduction.

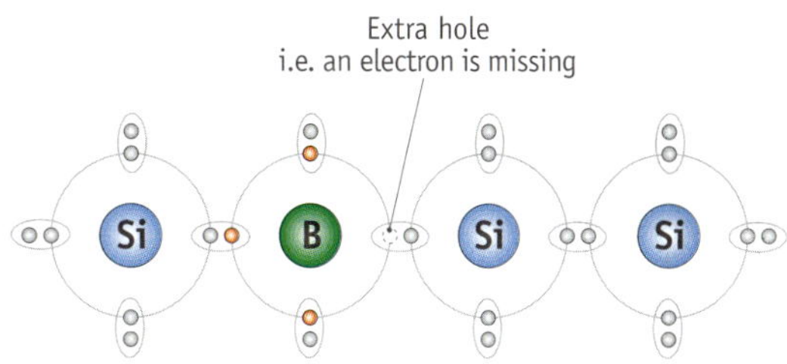

Fig. 25.8
A p-type semiconductor. Each boron atom introduces an extra hole.

Fig. 25.7 shows silicon where some of the silicon atoms have been replaced with atoms of phosphorous. A phosphorous atom has 5 valence electrons. Phosphorous atoms are roughly the same size as silicon atoms and fit into the crystal structure of silicon. Only 4 of the 5 valence electrons are needed to form the 4 covalent bonds keeping the phosphorous in the lattice. The other electron becomes free to wander throughout the crystal, i.e. it becomes a conduction electron. By having sufficient phosphorous atoms present, the conductivity of the material can be greatly increased due to these extra conduction electrons. **The majority charge carriers are thus negatively charged electrons** and the material is called an n-type semiconductor. There is also some intrinsic conduction taking place with some electrons and an equal number of holes moving in the material. **The holes are the minority charge carriers**. Note that a piece of n-type semiconductor does not contain excess negative charge. It is still neutral.

P-TYPE SEMICONDUCTOR

A ***p-type semiconductor*** is a semiconductor in which the impurity added produces extra holes which are available for conduction, e.g. boron in silicon.

Fig. 25.8 shows silicon with some of the silicon atoms replaced with atoms of boron. Boron atoms only have 3 valence electrons. Boron atoms are similar in size to silicon atoms and fit into the crystal structure of silicon. Since boron atoms have only 3 valence electrons there is a vacancy or gap – called a **positive hole** – in the crystal lattice at every boron atom.

When a p.d. is put across this material, a valence electron from a nearby atom may move into the hole, causing a hole to appear in the atom from which it came. It is as if the hole moved from the boron atom to the other atom. This process will continue and the hole appears to move through the silicon. In such a semiconductor **the majority charge carriers are positively charged holes**. The material is therefore called a p-type semiconductor. There is also some intrinsic conduction taking place with some electrons and an equal number of holes moving in the material. Obviously **the electrons are the minority charge carriers**.

EXTRINSIC CONDUCTION
Increased conduction in a semiconductor due to the addition of impurities is called ***extrinsic conduction***. The semiconductor formed is called an ***extrinsic semiconductor***.

THE P-N JUNCTION

P-N JUNCTION
A ***p-n junction*** is a piece of semiconductor with part of it doped p-type and the rest doped n-type.

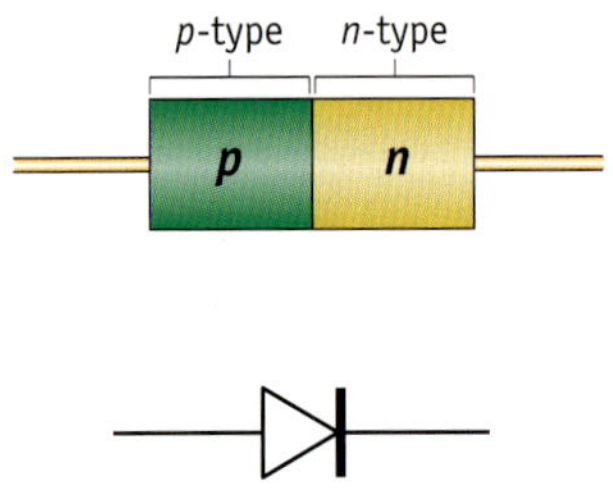

Fig. 25.9
A p-n junction and circuit symbol

FIG. 25.9 shows a piece of semiconductor which was formed by doping one side of a piece of pure semiconductor p-type and the rest of it n-type. The resulting semiconductor is called a p-n junction. It is also called a **p-n diode** or a **semiconductor diode**.

Recall that in a p-type semiconductor there are a number of holes which behave as mobile positive (+) charge carriers. The material is electrically neutral since it has no excess charge. In an n-type semiconductor there are a number of electrons not involved in bonding and are mobile negative (-) charge carriers. The material is electrically neutral. When the two types join together the following happens:

- Free electrons in the n-type material wander into the p-type. They do this since the mobile electron concentration is higher in the n-type than the p-type. Here they meet holes. When an electron meets a hole it fills the hole. Thus both the electron and the hole are no longer available for conduction.
- Similarly holes in the p-type material wander into the n-type. They do this since the concentration of holes is higher in the p-type than in the n-type. Here they meet electrons. When a hole meets an electron the electron fills the hole. Thus again both the electron and the hole are no longer available for conduction.
- In this way a region forms near the junction containing virtually no free majority charge carriers (FIG. 25.10). It is called the **depletion layer** and behaves as an insulator.

DEPLETION LAYER
The ***depletion layer*** is the region at both sides of a p-n junction that contains no free majority charge carriers. It thus behaves as an insulator.

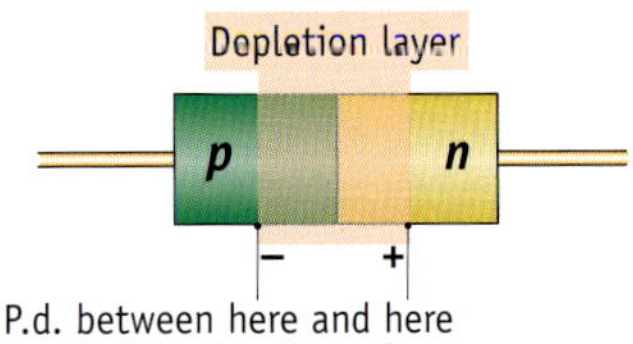

P.d. between here and here is called the *junction voltage*

Fig. 25.10

Neither all the electrons from the n-type nor all the holes from the p-type material move towards the junction. This is because as electrons leave the n-type material it becomes positively charged. Likewise holes entering it increase its positive charge. When enough positive charge accumulates in the n-type material no more electrons leave it – since they are attracted by the positive charge. In a similar way, a negative charge accumulates in the p-type material as electrons enter it and holes leave it. The charge that accumulates across the junction causes a small voltage to appear across the junction called the **junction voltage** (FIG. 25.10).

JUNCTION VOLTAGE
The p.d. that exists across a p-n junction caused by holes and electrons moving across the junction when it was formed is called the ***junction voltage***.

For silicon its value is about 0.6 volts and for germanium it is about 0.2 volts.

Reverse Biased P-N Junction

If the **positive terminal of a battery is connected to the n-type and the negative terminal is connected to the p-type, the diode is said to be reverse biased** (Fig. 25.11). A reverse biased p-n junction will not conduct current. This is because the voltage of the battery increases the width of the depletion layer since the positive of the battery attracts electrons from the n-type and the negative of the battery attracts holes from the p-type. Since the depletion layer is an insulator, no current can flow.

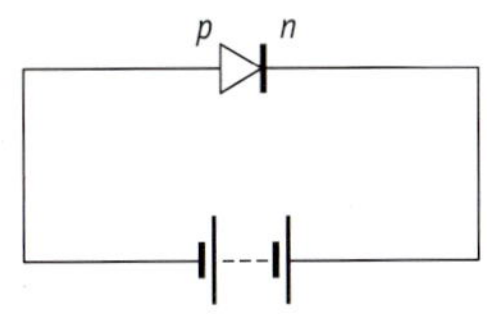

Fig. 25.11
A reverse biased p-n junction.

Forward Biased P-N Junction

If the battery is connected with its **positive terminal to the p-type and the negative terminal to the n-type** the diode is said to be **forward biased** (Fig. 25.12). A forward biased p-n junction will conduct electricity provided the voltage of the battery is greater than the junction voltage of the diode. This is because the negative of the battery forces electrons into the depletion layer and the positive of the battery forces holes into the depletion layer thus reducing its width. If the battery voltage is greater than the junction voltage the depletion layer is eliminated and current can flow across the junction. This can easily be demonstrated with the equipment shown in Fig. 25.13. When it is forward biased the bulb lights but when it is reverse biased the bulb does not light.

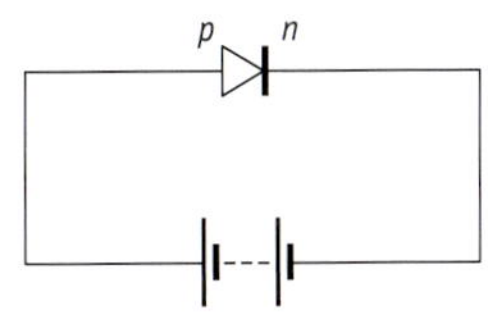

Fig. 25.12
A forward biased p-n junction.

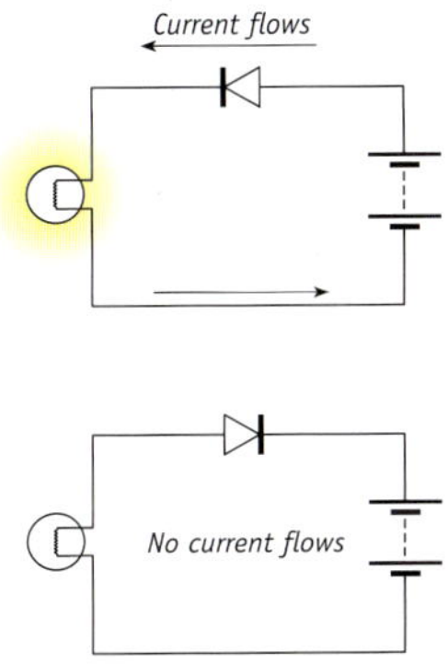

Fig. 25.13
A forward biased p-n junction conducts current. A reverse biased p-n junction does not conduct current.

> A ***forward biased*** p-n junction conducts current.
> A ***reverse biased*** p-n junction does not conduct current.

Experiment

ELECTRICITY 5(D)

To Investigate the Variation of Current (I) with p.d. (V) for a Semiconductor Diode.

Summary of Method

In this experiment you will measure the voltage V across and the current I flowing through a semiconductor diode for a number of different values of the voltage. You will do this when the diode is forward biased and then reverse biased. You will then plot a graph of I against V.

Equipment Needed

- A semiconductor diode
- A low voltage d.c. power supply (0–20 V)
- A variable resistor (0–2 kΩ)
- A milliammeter (0–20 mA)
- A microammeter (0–10 μA)
- A high-resistance voltmeter (0–20 V)

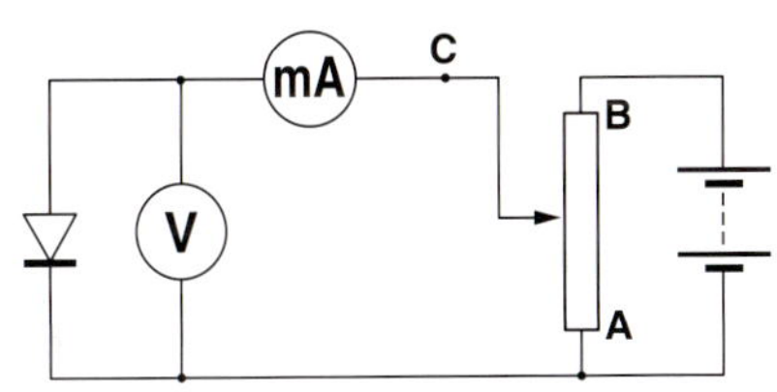

Fig. 25.14

Method

1. Set up the circuit shown in Fig. 25.14. With this circuit the diode is **forward biased**.
2. With the movable terminal of the variable resistor at A read the value of the p.d. across the diode from the voltmeter and the corresponding current on the milliammeter. Record these values.
3. By adjusting the variable resistor towards B, increase the voltage across the diode by about 0.2 volt. Measure and record the new values of V and of I.

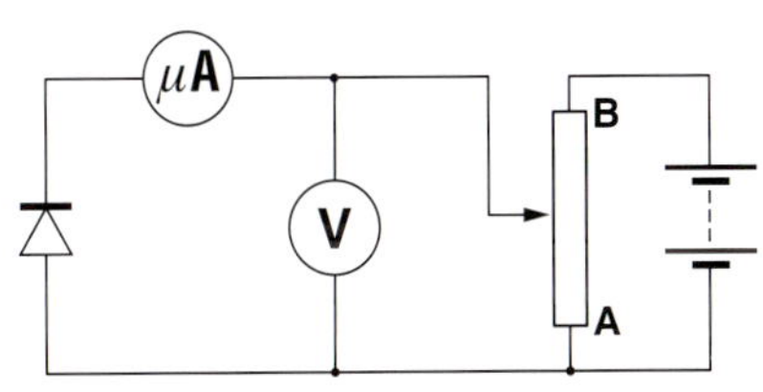

Fig. 25.15

4. Repeat step 3 until the current or the voltage is just less than the maximum recommended by the manufacturer of the diode (typically 50 mA).
5. Reverse the connections on the diode. Remove the milliammeter and connect the microammeter in the circuit as shown in FIG. 25.15 (page 289). Note that the microammeter is not in the same position as the milliammeter. In this circuit the diode is reverse biased.
6. Starting with the movable terminal of the variable resistor at A again, measure and record a series of values for I and V. This time V can be increased in larger steps than before.
7. On graph paper plot a graph of I (on the y-axis) against values of V (on the x-axis). A graph similar to FIG. 25.16 will result. Note that the scales on the y-axis are different for forward and reverse bias.

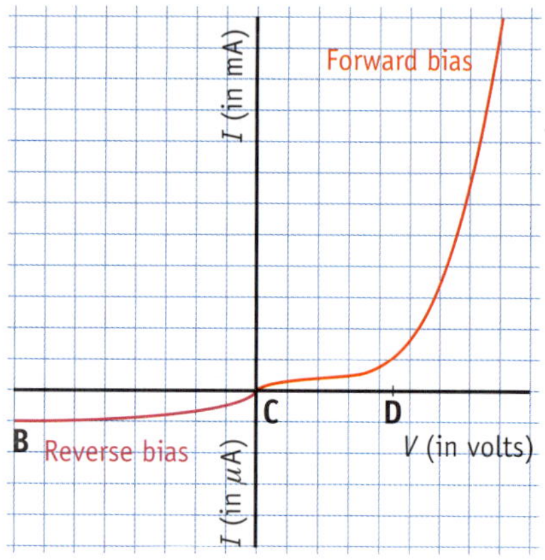

Fig. 25.16
The variation of current (I) with p.d. (V) for a semiconductor diode

Forward Biased		Reverse Biased	
Voltage V / V	Current I / mA	Voltage V / V	Current I / µA

EXPERIMENTAL NOTES

- When the diode is forward biased you may connect a resistor in series with the milliammeter at C. The value of the resistor is chosen so that if the full supply voltage is put across the diode the size of the current in the circuit will not exceed the safe maximum for the diode.
- A forward biased diode has a very low resistance and a reverse biased diode has a very high resistance. The position of the voltmeter in relation to the current reading meter must be arranged to allow for this. In FIG. 25.14 the diode is forward biased. The voltmeter reads the p.d. across the diode. The milliammeter reads the sum of the currents through the voltmeter and the diode. The resistance of the voltmeter is huge and thus takes almost no current. The reading on the milliammeter is thus the current through the diode to a high degree of accuracy. If the diode is reverse biased it has a very high resistance. Thus, if the same circuit is used, the current through the diode and voltmeter may be similar in size.
  The reading on the microammeter will not then be the current through the diode. With the circuit in FIG. 25.15 the microammeter reads the current flowing through the diode correctly. The voltmeter reads the sum of the p.d.'s across the diode and microammeter. The resistance of the microammeter is tiny compared with the resistance of the reverse biased diode, thus virtually all the p.d. is across the diode and the reading on the voltmeter is the p.d. across the diode to a high degree of accuracy.

Questions

1. Why is a milliammeter used when the diode is forward biased but a microammeter is used when it is reverse biased?
2. Why is the position of the current reading meter changed during the experiment?
3. When the diode is forward biased why is the current small until the voltage reaches about 0.6 V for a silicon diode or about 0.2 volts for a germanium diode?
4. Does a forward biased diode obey Ohm's Law?
5. What will happen if the forward current is allowed to get too large?
6. Why might a resistor be connected in series with the milliammeter in the first part of the experiment?

Explanation of the Graph (Fig 25.16)

When forward biased

- In the region C → D the applied voltage is less than the junction p.d. and the depletion layer still exists. The only current flowing is the leakage current which is very small.
- When the applied voltage exceeds the junction voltage (about 0.6 V for a silicon diode) a significant forward current flows. This current increases rapidly with the forward voltage. If the voltage is increased too much the current will be large enough to overheat and permanently damage the diode. Check with the manufacturer of the diode to find out the maximum safe forward current.

When reverse biased

In the region B → C the diode only allows a very small current to flow. As we have seen, this current is called the leakage current and is typically less than 1 μA. Its value remains fairly steady over a large range of voltage. If the voltage is increased too much, suddenly the insulating properties of the diode break down and a large current flows. This current will destroy the diode. The voltage at which this occurs is called the **breakdown voltage**.

Rectification of a.c.

In Fig. 25.17 an alternating voltage source is connected in the circuit with the diode and the resistor. When the diode is forward biased it conducts and current flows in the resistor. When it is reversed biased no current flows. Thus the input and output voltages are as shown. A diode can be used to convert alternating current (a.c.) to direct current (d.c.). This is called **rectification**. Note, however, that the current in the resistor is not a steady d.c.

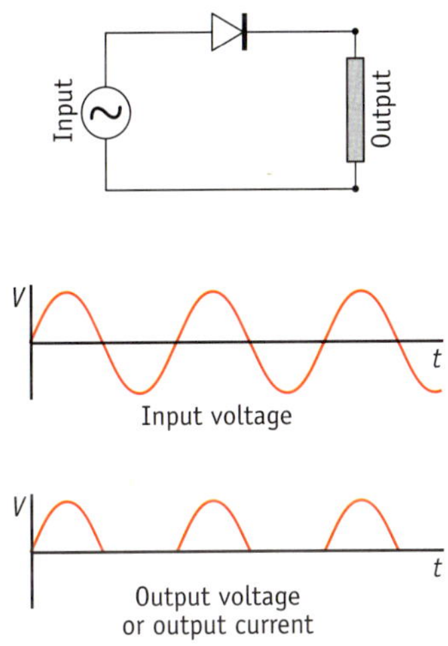

Fig. 25.17
Rectification of a.c.

Integrated Circuits (*I*Cs)

An integrated circuit is a circuit containing some or all of the components of transistors, diodes, resistors and capacitors on one small chip of silicon. Fig. 25.18(a) shows a number of ICs in a personal computer from the 1970s. Each IC contains 3500 transistors. Connections to the chip are made via the pins. Fig. 25.18(b) shows an IC from a personal computer of today. This chip contains more than 5 000 000 transistors.

(A)
Main logic assembly board of Commodore PET 2001 microcomputer, 1979.

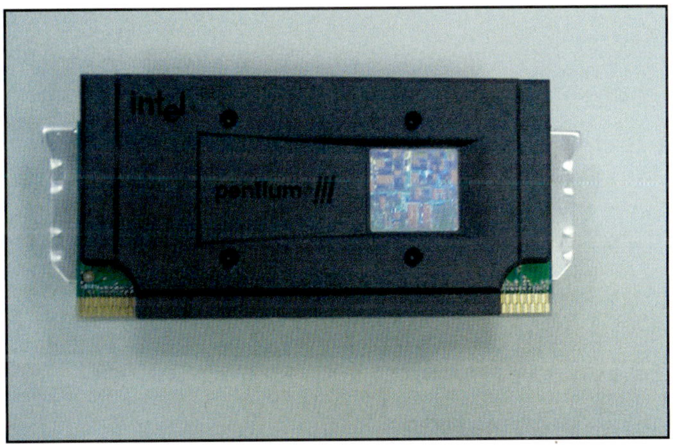

(B)

Fig. 25.18

CHAPTER CHECKLIST

- **Define**: Semiconductor, Thermistor, Light Dependent Resistor (LDR).
- **State**: Two materials that are semiconductors; What is meant by forward and reverse biased; What is meant by rectification.
- **Recall** that: A reverse biased p-n junction does not conduct current; A forward biased p-n junction conducts current.
- **Explain** what is meant by: Valence electron; Conduction electron; Intrinsic conduction; Positive hole; Doping; P-type semiconductor; N-type semiconductor; Extrinsic conduction; P-N junction; Depletion layer; Junction voltage.
- **Describe** and carry out an experiment to: Investigate the variation of current (I) with p.d. (V) for a semiconductor diode; Demonstrate the action of an LDR and a thermistor.
- **Draw** the I-V graph for a semiconductor diode and explain its features.

Magnets and Magnetic Fields

CHAPTER 26

INTRODUCTION

Over 2400 years ago the Greeks knew that a certain ore of iron was able to attract small pieces of iron to it. Today this ore is called lodestone or magnetite. They also knew that if a piece of lodestone was suspended by a string it would turn until a particular end of it was pointing north. Today we explain these properties of lodestone by saying that the piece of lodestone is **magnetised**. We can investigate magnetic properties like those of lodestone using bar magnets.

EXPERIMENT

TO DEMONSTRATE THE MAIN PROPERTIES OF MAGNETS.

A magnet attracts certain materials to it

Such materials are called **ferromagnetic** materials. Test a number of different materials with a bar magnet by bringing it near them. The magnet will attract **iron**, **steel**, **nickel**, **cobalt** and some alloys of these. The magnet has no noticeable effect on other materials.

A bar magnet is strongest at each end

Dip the bar magnet into a jar of iron filings or a box of pins. It attracts the filings or the pins to it and most cling on at each end of the magnet (FIG. 26.1). The regions of greatest strength at each end are called **magnetic poles**.

If a bar magnet is suspended by a piece of thread it will line up approximately north-south

Make a paper stirrup and place the magnet in it. Suspend it from a wooden retort stand and note that it always comes to rest pointing roughly north-south. The pole of the magnet that always points north is called the **north seeking pole** or simply the north pole. The pole that points south is called the **south seeking pole** or simply the south pole. Experiments have always shown that:

- For every north pole there is always a south pole, i.e. **magnetic poles exist in pairs**.
- The strength of the north pole is exactly the same as the strength of the south pole.

Like poles repel and unlike poles attract

The north pole of one magnet repels the north pole of another. The south pole of one magnet repels the south pole of another. The north pole of one magnet is attracted to the south pole of another. Place the poles of different magnets near each other and see that **like poles repel and unlike poles attract**. Note also that the force of repulsion or attraction between two magnets increases the nearer the magnets are to each other and decreases as the magnets are placed further apart.

A magnet causes ferromagnetic material brought near or touching it to become magnetised

This magnetism is called **induced magnetism**. If the original magnet is then taken away some materials (called permanent magnets) hold on to their magnetism but others (called temporary magnets) lose most of it.

(A)
Pins attached to a bar magnet.

(B)
Iron filings attracted to a bar magnet.

Fig. 26.1

Get some pins and notice that they do not stick to each other, i.e. they are not magnets. Place a strong bar magnet into the heap of pins and see that as well as being attracted to the magnet the pins are also attracted to **each other** (FIG. 26.1(A)). A similar effect can be seen with iron filings. See also that the pins hold onto some of their magnetism when the bar magnet is removed. Hard steel holds onto magnetism very well whereas soft iron does not. A hacksaw blade, a knitting needle or steel from a clock spring are made of hard steel. Ordinary nails are usually made from soft iron.

MAGNETIC FIELDS

MAGNETIC FIELD

A ***magnetic field*** is any region of space where magnetic forces can be felt. The direction of the magnetic field at a point is the direction of the force on a north pole if it were placed at that point.

Fig. 26.2
A plotting compass.

An easy way to show the presence of a magnetic field is with a **plotting compass**. This is a very small light magnet suspended so that it can freely rotate about a vertical axis (FIG. 26.2). If no other magnets are nearby it will line up north-south. If there is another magnetic field present this will cause the compass needle to deflect from its north-south position. If the other field is strong enough, the compass needle will line up almost parallel to the field rather than north-south. Magnetic fields can be visually represented by lines called **magnetic field lines**.

MAGNETIC FIELD LINE

A line drawn in a magnetic field so that the tangent to it at any point shows the direction of the magnetic field at that point is called a ***magnetic field line***.

Magnetic field lines around a magnet can be drawn using a plotting compass.

EXPERIMENT

TO PLOT THE MAGNETIC FIELD DUE TO A BAR MAGNET.

- Place a bar magnet on a sheet of paper.
- Place a plotting compass next to one pole of the magnet and mark with dots on the paper both ends of the compass needle (FIG. 26.3).
- Move the compass (as in FIG. 26.3). Mark the other end of the needle.
- Repeat this step until you end up at the other pole of the magnet. Join the dots with a smooth curved line.
- Repeat this process drawing a number of lines at both sides of the magnet.
- Mark each line with an arrowhead showing the direction of the magnetic field (pointing from north to south).
- A field pattern similar to that in FIG. 26.4 will result.

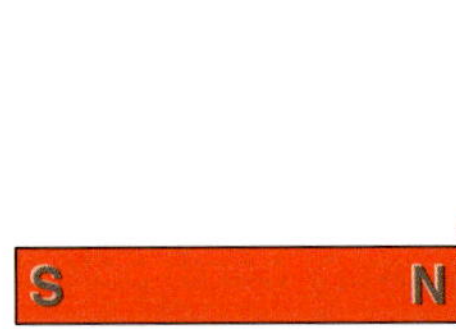

Fig. 26.3

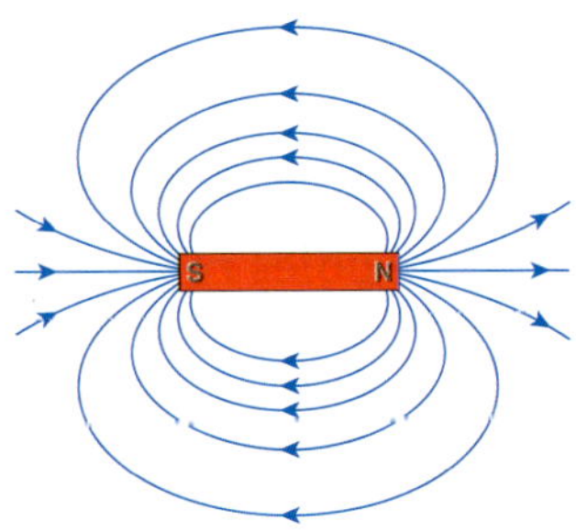

Fig. 26.4
The magnetic field around a bar magnet.

NOTE
- The field lines in the space around a bar magnet start at the north pole and end at the south pole of the magnet.
- Near the poles – where the magnetic field is strongest – the lines are close together. Further away where the field is weaker the lines are far apart.

The magnetic field around a bar magnet may also be plotted with iron filings. A sheet of paper or cardboard is placed over the magnet and its outline is drawn in. Iron filings are sprinkled on the paper and it is shaken gently. The filings line up as in Fig. 26.5. It must, however, be realised that the magnetic field around a magnet is, in reality, three-dimensional.

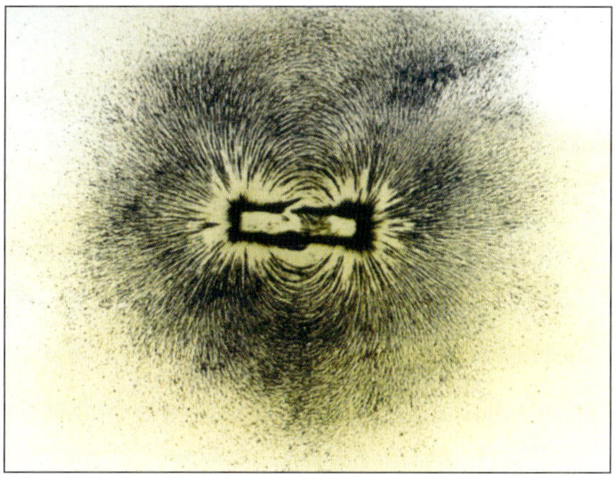

Fig. 26.5 Iron filings showing the shape of the magnetic field around a bar magnet.

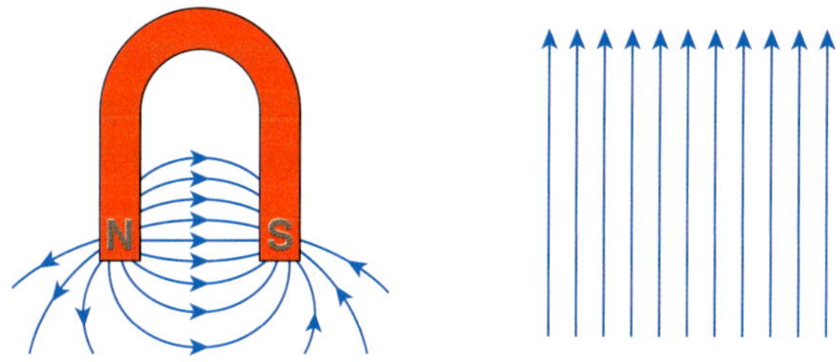

Fig. 26.6 Magnetic field of a horseshoe magnet and the Earth's magnetic field.

Fig. 26.6 shows the magnetic field due to (i) a horseshoe magnet, (ii) the Earth. They can easily be plotted using a plotting compass.

The Magnetic Effect of an Electric Current

In 1819 in Copenhagen, Hans Christian Oersted discovered that any current-carrying conductor has a magnetic field around it as long as the current is flowing. When the current stops flowing the magnetic field disappears. This can easily be demonstrated in the laboratory as follows:

Experiment

To Show the Magnetic Effect of an Electric Current

- Align a piece of wire north-south and place a plotting compass underneath it (Fig. 26.7(a)). The compass needle also lines up north-south due to the Earth's magnetic field acting on it.
- Send a steady current (e.g. 2 A) through the wire and the compass needle will deflect (Fig. 26.7(b)). The direction in which it deflects depends on the direction of the current. Reverse the direction of the current and the needle deflects in the opposite direction.
- Switch off the current, the magnetic field due to the current disappears and the needle again lines up north-south. Thus we conclude the following:

> Every current-carrying conductor has a magnetic field around it caused by the current.

- The magnetic field due to a single current-carrying wire is weak unless the current is very large.

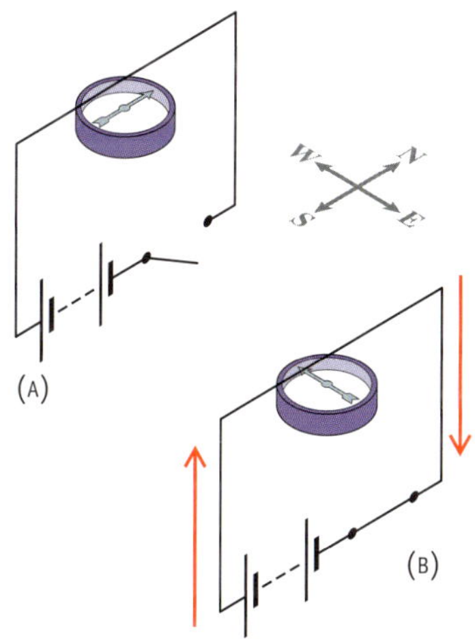

Fig. 26.7

EXPERIMENT

To Plot the Magnetic Field Due to the Current in a Long Straight Wire

- Use the equipment in FIG. 26.8 and send a current through the wire (2 A is suitable).
- Place a plotting compass near the wire and mark with dots on the paper both ends of the compass needle.
- Move the compass so that its south pole is at the dot that marked the north pole. Mark the other end of the needle with a dot.
- Repeat this until you end up back at the point where you started. Join the dots with a smooth curved line. It will be found to be a circle.
- Continue this process drawing a number of circles around the wire.
- Mark each circle with an arrowhead showing its direction.

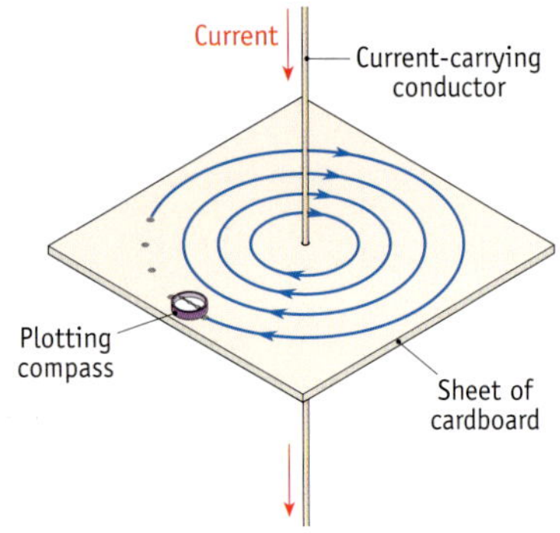

Fig. 26.8
Magnetic field lines due to a long straight current-carrying conductor.

The Right-Hand Grip Rule

This is a useful rule that relates the direction of the current flowing in a conductor and the direction of the magnetic field around that conductor.

THE RIGHT-HAND GRIP RULE states that if the right-hand clasps a conductor with the thumb pointing in the direction of the current, then the fingers give the direction of the magnetic field around the conductor (FIG. 26.9).

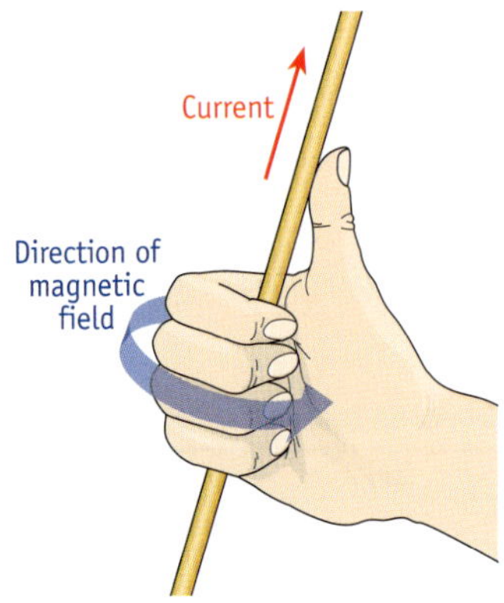

Fig. 26.9
The right-hand grip rule.

Note that the thumb points in the direction of conventional current, i.e. from + to – .

Magnetic Field Due to Current in a Circular Loop and Due to Current in a Coil

In FIG. 26.10 a circular loop of wire is carrying a current in the direction shown. Using the right-hand grip rule at a number of points on the wire gives the shape of the magnetic field around the loop. The side of the loop facing us behaves like a south pole (the magnetic field lines are going into it) and the other side like a north pole (the magnetic field lines are coming out of it). A number of loops wound closely together is called a coil. FIG. 26.11 shows a current-carrying coil and the magnetic field around it. The magnetic field due to a coil is stronger than that of a similar sized loop carrying the same current. The shapes of the magnetic fields are similar.

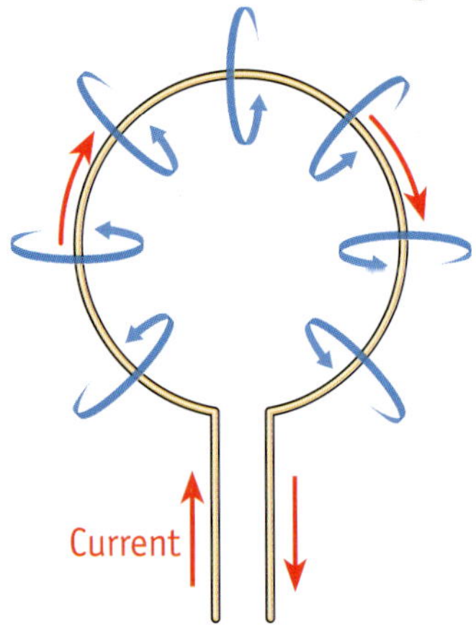

Fig. 26.10
The magnetic field due to a current-carrying loop.

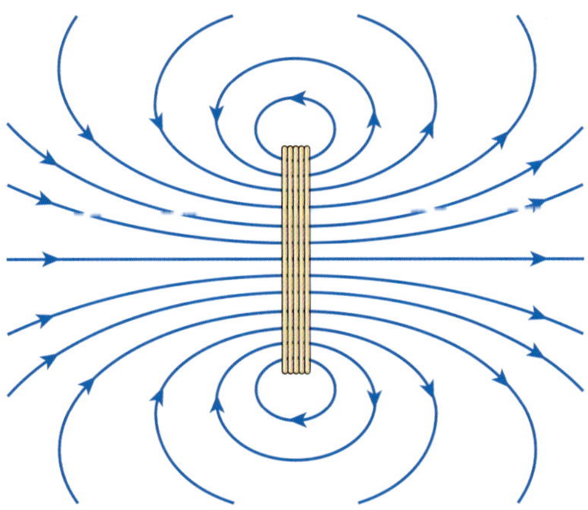

Fig. 26.11
The magnetic field due to a current-carrying coil.

Magnetic Field Due to Current in a Solenoid

Fig. 26.12 shows a **solenoid** and the magnet field around it due to current flowing in it. **A solenoid is a coil whose length is much longer than its radius.** The right-hand grip rule enables us to see why the magnetic field around a solenoid is the shape that it is. Consider the current going in the top of the solenoid. The right-hand grip rule gives the direction of the magnetic field as shown. Between the wires on top, the magnetic field cancels out, giving the overall field as shown. By the same reasoning the field inside the solenoid is along the axis of the solenoid. Note that the field due to a solenoid is very like that of a bar magnet. In fact in a bar magnet the magnetic field lines actually run through the magnet just like in Fig. 26.12.

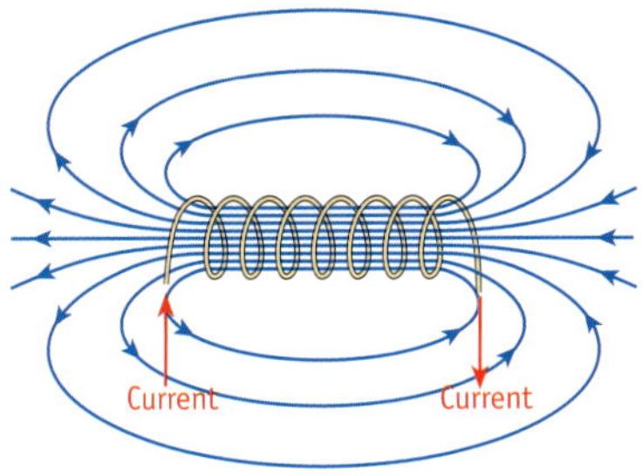

Fig. 26.12
The magnetic field due to a current-carrying solenoid.

Experiment

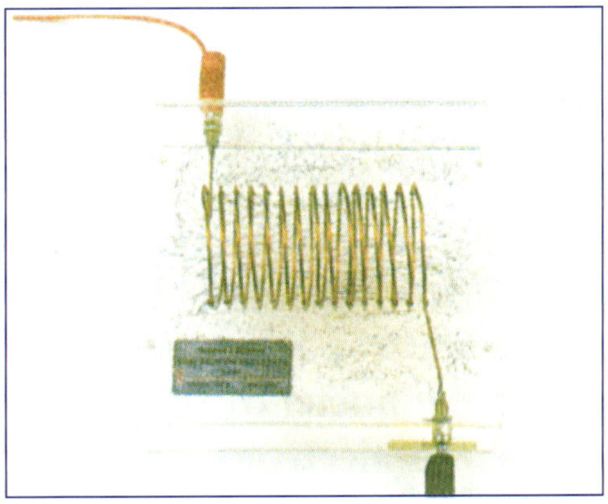

Fig. 26.13

To Plot the Magnetic Field Due to a Current in a Loop, a Coil or a Solenoid.

Using either of the methods used for the bar magnet, the magnetic field around a loop, coil or solenoid may be plotted. Fig. 26.13 shows a solenoid wound on a perspex plate specially for doing this.

An easy way to remember which side of a loop, coil or solenoid behaves like a north pole and which like a south pole is to notice the following:

When looking into the loop, coil or solenoid, if the current is moving in a clockwise direction the end facing you is a south pole, if the current is moving in an anticlockwise direction it is a north pole (Fig. 26.14).

Fig. 26.14

Electromagnet

If a soft iron core (e.g. some large nails) is placed in a solenoid and current passed through the solenoid, the core becomes magnetised. When the current is switched off the core loses its magnetism. The solenoid and core together is called an **electromagnet**. If there is a large number of turns and a large current in the coil the magnet will be quite powerful.

Uses of Electromagnets

Electromagnets have many practical uses. Fig. 26.15 shows a powerful electromagnet lifting scrap iron and steel. Electromagnets are used in most electric motors (page 301) and electromagnetic relays (page 379).

Fig. 26.15

THE EARTH'S MAGNETIC FIELD

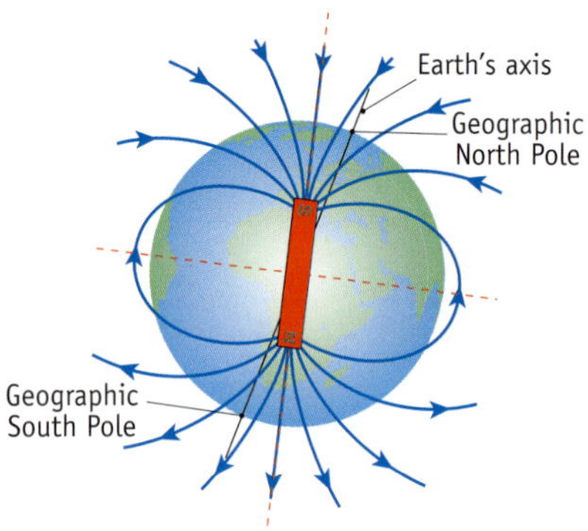

Fig. 26.16

You saw on page 295 that the Earth has a magnetic field around it. It is thought to be caused by circulating electric currents in the Earth's core. The shape of the Earth's magnetic field is as shown in FIG. 26.16. It is as if there is a huge bar magnet at the centre of the Earth with the north pole of this imaginary magnet in the southern hemisphere and the south pole in the northern hemisphere. This imaginary magnet is not quite aligned north-south. Thus the direction that a magnetic compass needle lines up is not exactly north-south. There is thus a difference between true north and magnetic north as given by a compass. The angle between true north and magnetic north at a point on the Earth is called the **magnetic declination** or the **magnetic variation** at that point.

Fig. 26.17

The size of the declination varies from place to place and at a given place it varies slowly with time. In Cobh in 1992 the value in was 8° and decreasing by about 5' every year. In Donegal it was 10° and decreasing by about 5' annually.

The magnetic compass (FIG. 26.17) has been used for hundreds of years in marine navigation, since it always enables you to know the direction in which you are travelling. The value of magnetic variation (declination) is also of importance. Charts and maps needed for marine navigation have its value in the locality of the chart noted since navigators must allow for it in their calculations.

CHAPTER CHECKLIST

- **Define**: Magnetic field; Magnetic field line; Electromagnet.
- **Recall that**: Magnetic poles exist in pairs; Each pole of a bar magnet is the same strength; Like poles repel; unlike poles attract; The Earth has a magnetic field around it that is used in navigation; Every current-carrying conductor has a magnetic field around it due to the current.
- **State**: The right-hand grip rule.
- **Draw** the magnetic field due to: A bar magnet; A horseshoe magnet; A loop; A coil; A solenoid; The Earth.
- **Describe** an experiment to plot the magnetic field due to: A bar magnet; A horseshoe magnet; A long straight current-carrying wire; A current-carrying loop, A current-carrying solenoid.
- **List** four practical uses of electromagnets.
- **List** a practical use of the Earth's magnetic field.

Current in a Magnetic Field

CHAPTER 27

Force on a Current-Carrying Conductor in a Magnetic Field

A current-carrying conductor has a magnetic field around it due to the current. If the current-carrying conductor is placed in **another** magnetic field, the magnetic field due to the current interacts with the other magnetic field and causes a force on the current-carrying conductor. In simple terms, we can imagine that the two magnetic fields push off each other. If the conductor is free to move it will do so under the influence of this force.

Experiment

To show the Force on a Current-Carrying Conductor in a Magnetic Field

Method

- Set up the equipment shown in Fig. 27.1.
- Send a current through the tinfoil (2 A is suitable).
- The foil will be seen to move forwards or backwards depending on which direction the current is flowing.

Conclusion:

A current-carrying conductor in a magnetic field experiences a force.

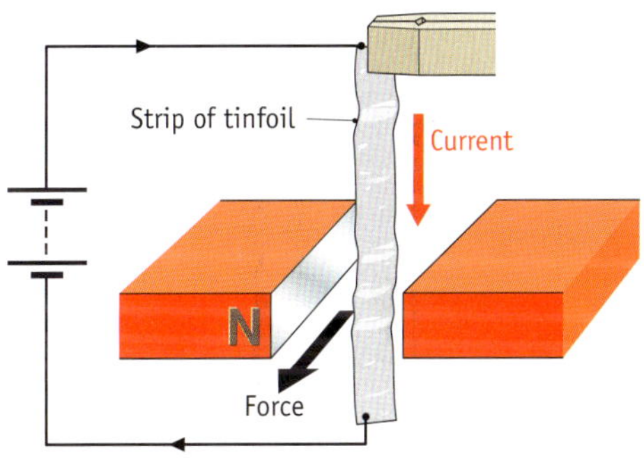

Fig. 27.1
The force on a current-carrying conductor in a magnetic field.

Experimentally it is found that the direction of the force on the conductor is perpendicular to the current and perpendicular to the magnetic field. If the current-carrying conductor is placed parallel to the magnetic field it experiences no force.

A current-carrying conductor in a magnetic field will always experience a **force** unless the conductor is parallel to the magnetic field. The direction of the force is always:

- **perpendicular to the current** and
- **perpendicular to the magnetic field**

This is the principle on which the **d.c. electric motor**, the **moving coil loud-speaker**, **moving coil galvanometer**, **moving coil voltmeter** and **moving coil ohmmeter** are based.

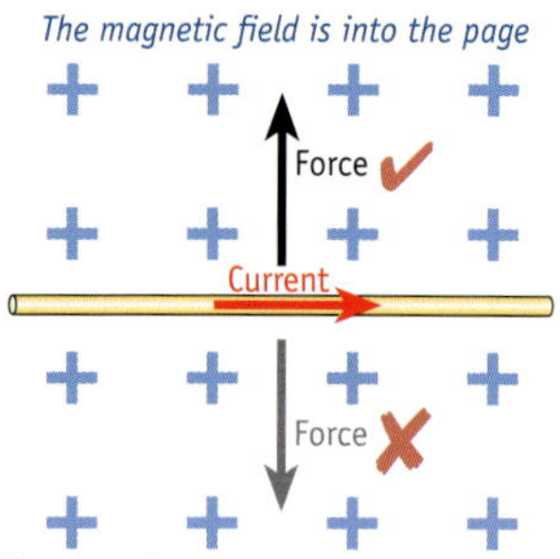

Fig. 27.2

Direction of the Force on a Current-Carrying Conductor in a Magnetic Field

The direction of the force on the conductor is perpendicular to both the current and the magnetic field. There are, however, two such directions, e.g. in Fig. 27.2 we have a magnetic field going into the page (signified by the +) and the wire is carrying the current in the direction shown. The force then is either up the page or down the page. A simple rule – called **Fleming's left-hand rule** – tells us which of these directions it is.

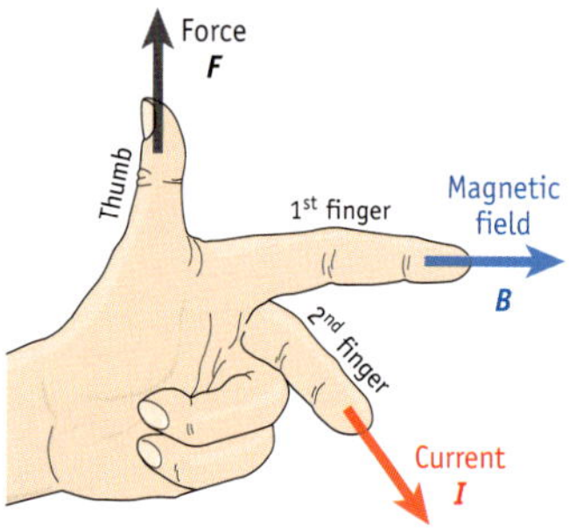

Fig. 27.3
Fleming's left-hand rule.

> **FLEMING'S LEFT-HAND RULE** states that if the thumb, first finger and second finger of the left hand are held at right angles (Fig. 27.3), with the first finger pointing in the direction of the magnetic field and the second finger pointing in the direction of the current, then the thumb points in the direction of the force.

Applying this in Fig. 27.2 we see that the force is upwards. You must learn to apply this rule where and when it is needed. Note that the direction of the current is the direction of conventional currents, i.e. flowing from + to –.

Problem 1: Fig. 27.4 shows a magnetic field and a wire carrying an electric current into the page. In what direction is the force on the wire?

Solution: Applying Fleming's left-hand rule:

First finger is in the direction of the magnetic field.

Second finger is in the direction of the current (into page).

The thumb gives direction of force which is therefore downwards as shown.

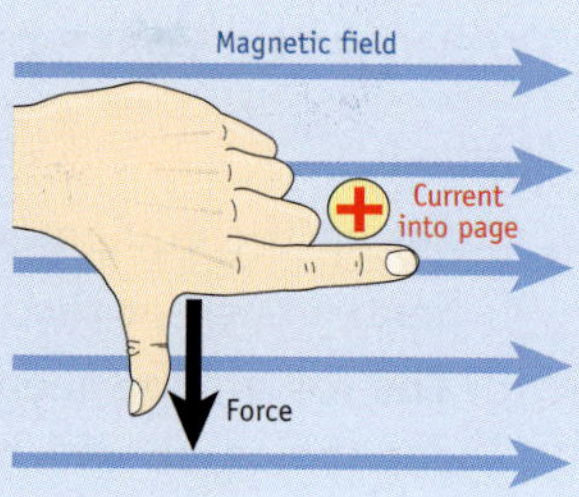

Fig. 27.4

Force on a Current-carrying Coil in a Magnetic Field

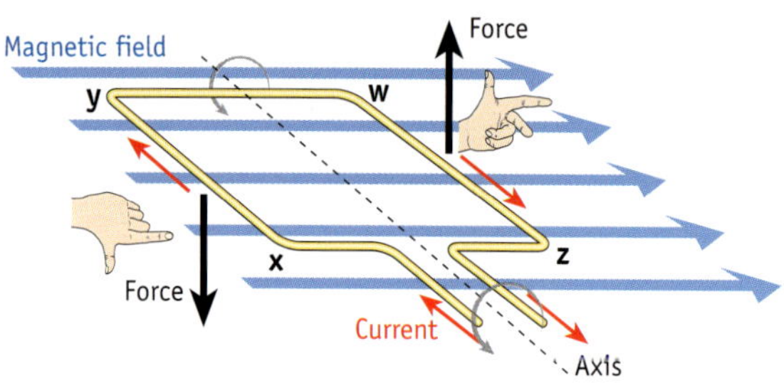

Fig. 27.5

Fig. 27.5 shows a coil in a magnetic field. The coil is free to rotate about the axis shown. Suppose the coil is carrying an electric current (conventional current) in the direction shown. Then by Fleming's left-hand rule the sides *xy* and *wz* experience forces in the direction shown. These forces tend to cause the coil to rotate in the direction indicated. Since the coil is free to rotate it does so.

Fig. 27.6 (which is a front view of Fig. 27.5) shows the directions of the forces on the coil as it rotates. When it reaches the vertical position, the forces no longer tend to rotate the coil (Fig. 27.6(c)). If the coil is sufficiently free, its momentum will carry it beyond the vertical position to that shown in Fig. 27.6(d). However, the forces which act on the coil in this position tend to rotate it back again to the vertical position. Hence the coil comes to rest in the vertical position - perhaps having oscillated for a while. If we could reverse the direction of the current in the coil as it passes through the vertical position, the forces acting on the coil would tend to keep the coil rotating in the same direction (Fig. 27.6(e)). If this is done every time the coil passes through the vertical position, the coil will rotate continuously and we have **a simple d.c. motor** (see Chapter 33 for more detail).

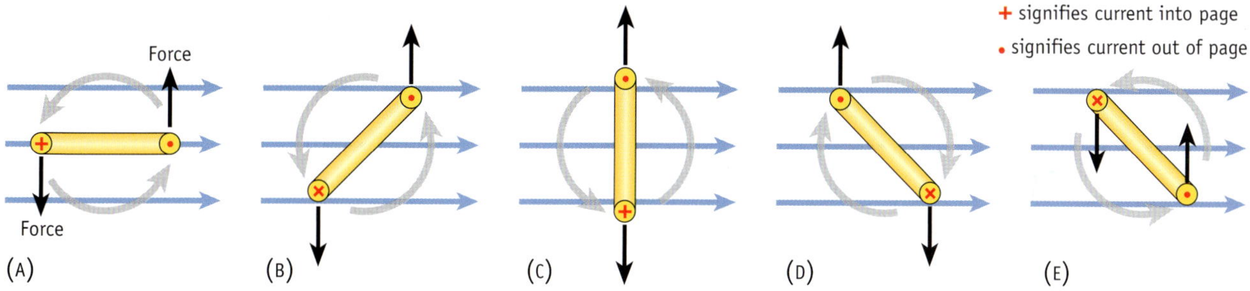

Fig. 27.6
As the coil rotates the turning moment on the coil changes.

The force on a current-carrying coil in a magnetic field can easily be demonstrated in the laboratory with suitable equipment; like that in FIG. 27.5.

SIZE OF THE FORCE ON A CURRENT-CARRYING CONDUCTOR IN A MAGNETIC FIELD

Further experiments with equipment like that in FIG. 27.1 (page 299) show that the size of the force on a current-carrying conductor in a magnetic field depends on:

- the size of the current I,
- the length of the conductor l,
- the strength of the magnetic field.

We all understand the difference between a weak and a strong magnetic field. However, in Physics we need to be able to put a numerical value on the strength of the field. To do this we introduce a new quantity called **Magnetic Flux Density (*B*)**. Magnetic flux density is a **vector**. At any point **the direction of *B*** is the direction of the force on a north pole placed at that point, i.e. its direction is the direction of the magnetic field. Its **magnitude** is defined in terms of the size of the force on a current-carrying conductor placed at that point in the magnetic field.

Further accurate experiments show that if a conductor of length l, carrying a current I, is placed at right angles to a uniform magnetic field it experiences a force F where: $F \propto I$ and $F \propto l$

It follows that: $F \propto Il \Rightarrow F = IlB$ where B is a constant.

The value of B depends on the strength of the magnetic field. In a strong magnetic field, B is large and in a weak field, B is small. Thus B is a measure of the strength of the magnetic field.

If a conductor of **length *l***, carrying a **current *I*** is placed at right angles to a magnetic field of **flux density *B*** it experiences a **force *F*** given by:

$$F = IlB$$

MAGNETIC FLUX DENSITY

At a point in a magnetic field, the ***magnetic flux density (B)*** is a vector whose:

- direction is the direction of the force on a north pole placed at that point,
- magnitude is the value of B from the equation $F = IlB$.

The ***magnetic flux density (B)*** at a point in a magnetic field, is a vector whose magnitude is equal to the force that would be experienced by a conductor of length 1 m carrying a current of 1 A at right angles to the field at that point and whose direction is the direction of the force on a north pole placed at that point.

UNIT OF MAGNETIC FLUX DENSITY
The unit of magnetic flux density is the ***tesla* (T)**.

One tesla is defined as follows:

THE TESLA
The magnetic flux density at a point is 1 ***tesla*** (T) if a conductor of length 1 m carrying a current 1 A experiences a force of 1 N when placed perpendicular to the field.

Problem 2: A straight piece of wire of length 3 m carrying a current of 2 A experiences a force of 12 N when placed perpendicular to a uniform magnetic field. Calculate the value of the magnetic flux density.

Solution: $F = I\,l\,B \Rightarrow B = \frac{F}{I\,l} = \frac{12}{(2)(3)} = 2\text{T}$

Problem 3: Calculate the force acting on a conductor of length 40 cm which is carrying a current of 3 A and which is placed perpendicular to a uniform magnetic field of flux density 5.2 T.

Solution: $F = I\,l\,B = (3)(0.4)(5.2) = 6.24\text{ N}$

Problem 4: Fig. 27.7 shows a rectangular loop of wire which is free to rotate about the axis shown. The plane of the loop is parallel to a uniform magnetic field of flux density 0.6 T. A current of 4 A flows in the coil and its dimensions are 20 cm × 12 cm. Draw a diagram showing the directions of the forces on the 20 cm sides of the coil.

(i) Find the magnitude of the force acting on one of the 20 cm sides of the loop.

(ii) Find the moment of this force about the axis.

(iii) Why does the moment of the force decrease as the coil rotates?

(iv) Will the moment of the force ever be zero?

(v) Calculate the couple on the coil when it is in the position shown in Fig. 27.7.

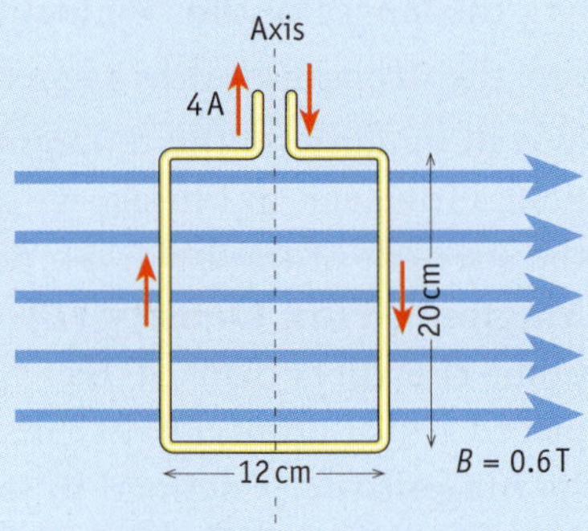

Fig. 27.7

Solution: Fig. 27.8 shows the forces acting on the 20 cm sides of the coil.

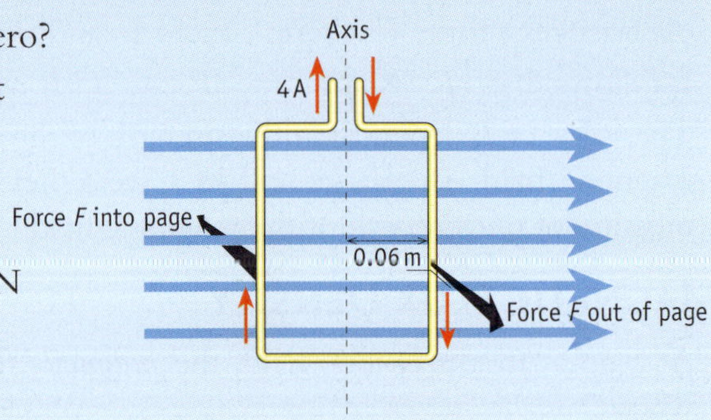

Fig. 27.8

(i) $F = I\,l\,B = (4)(0.2)(0.6) = 0.48\text{ N}$

(ii) Moment = force × perpendicular distance from axis = $(0.48)(0.06)$
$= 0.0288\text{ N m}$

(iii) As the coil rotates, the perpendicular distance between the force and the axis decreases, thus the moment of the force also decreases.

(iv) Yes, the moment will be zero when the coil rotates through $90°$ from the position shown in Fig. 27.7 as the perpendicular distance between the force and the axis will then be zero.

(v) Moment of couple = (Force)(Perp. Dist. between forces)
$= (0.48)(0.12) = 0.0576\text{ N m}$

What if the Current-carrying Conductoris not Perpendicular to the Field?

If the conductor is not perpendicular to the field, resolve the magnetic flux density B into two perpendicular components – one parallel to the conductor and the other at right angles to the conductor. **It is the component of B that is perpendicular to the conductor that causes the force on it.** The parallel component has no effect on it. Recall that a current-carrying conductor placed parallel to a magnetic field experiences no force.

Problem 5: A straight piece of wire carrying a current of 4 A is placed at an angle of 30° to a magnetic field of flux density 2 T. The wire is 2 m long. Resolve the flux density into components parallel and perpendicular to the wire. Which component causes the force on the wire? What is the magnitude of the force? In what direction does it act?

Solution: From Fig. 27.9:

Component parallel to wire = $B \text{ Cos } 30° = (2)(0.866) = 1.73$ T

Component perpendicular to wire = $B \text{ Sin } 30° = (2)(0.5) = 1$ T

The component perpendicular to the wire causes the force on it.

Magnitude of force $= I\,l\,B_{perp} = (4)(2)(1) = 8$ N

By Fleming's left-hand rule this force acts into the page.

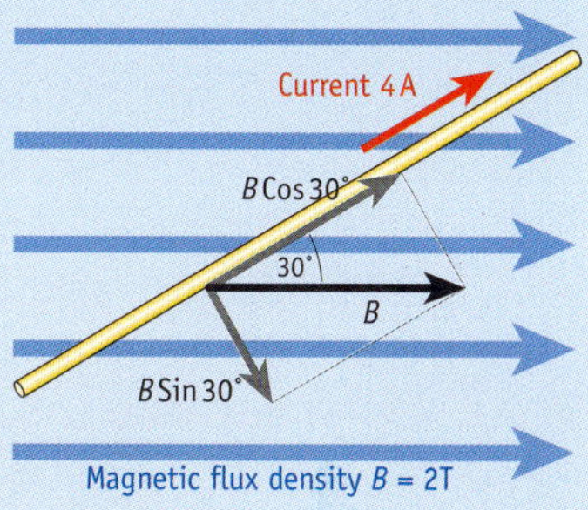

Fig. 27.9

EXERCISE 27.1

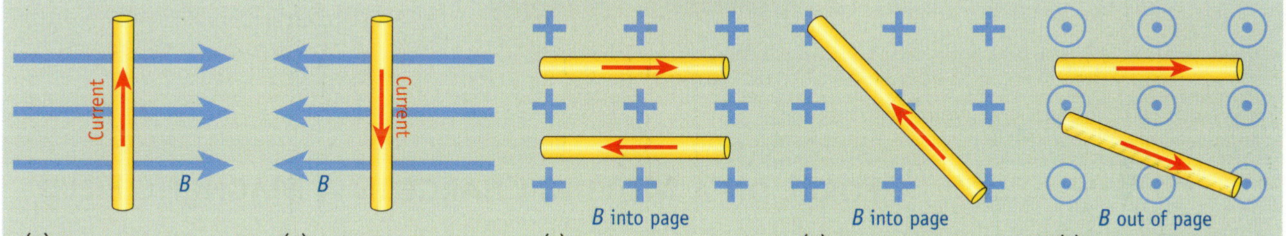

Fig. 27.10

1. Use Fleming's left-hand rule to find the direction of the force on each conductor in Fig. 27.10.
2. A straight piece of wire of length 0.5 m and carrying a current of 3 A experiences a force of 2 N when placed perpendicular to a uniform magnetic field of magnetic flux density B. Find the value of B.
3. A straight wire of length 2 m carrying a current of 4 A is placed perpendicular to a magnetic field of magnetic flux density 2.5 T. What is the force on the wire? In what direction does the force act?
4. A straight wire of length 1 m carrying a current of 3 A experiences a force of 4 N when placed perpendicular to a uniform magnetic field. What is the magnetic flux density at the wire?
5. In what direction is the force on a wire carrying a current vertically upward when placed in the Earth's magnetic field in Ireland?
6. In what direction is the force on a horizontal straight piece of wire carrying current from West to East in Ireland?

7. FIG. 27.11 shows a rectangular coil of wire in a uniform magnetic field. The coil is carrying current in the direction shown. Use Fleming's left-hand rule to find the direction of the force on:

(i) the side AB,

(ii) the side CD of the coil.

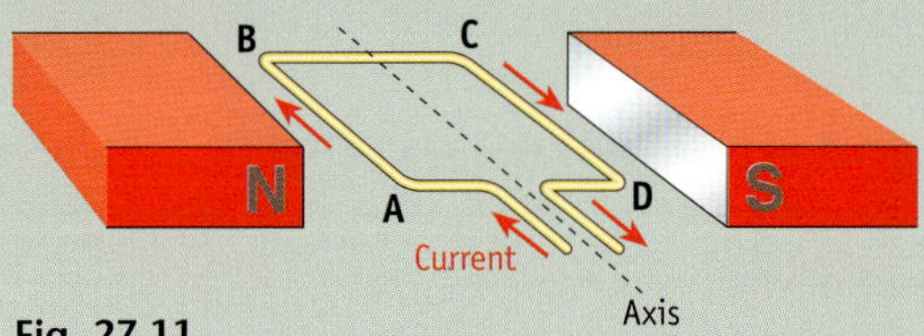

Fig. 27.11

If the coil was free to rotate about the axle shown what would you expect it to do?

8. FIG. 27.12 shows a rectangular loop of wire which is free to rotate about the axis shown. The plane of the loop is parallel to a uniform magnetic field of flux density 0.4 T. A current of 10 A flows in the coil and its dimensions are 30 cm x 16 cm. Draw a diagram showing the directions of the forces on the 30 cm sides of the coil.

(i) Find the magnitude of the force acting on the 30 cm side of the loop.

(ii) Find the moment of this force about the axis.

(iii) Why does the moment of the force decrease as the coil rotates?

(iv) Will the moment of the force ever be zero?

(v) Calculate the couple on the coil when it is in the position shown in FIG. 27.12.

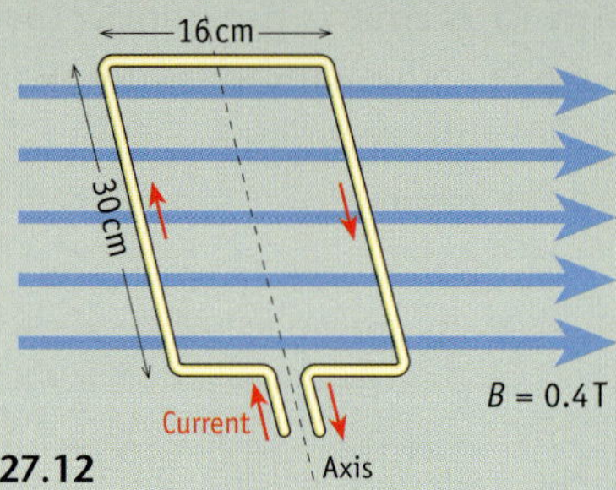

Fig. 27.12

9. A straight piece of wire of length 3 m is carrying a current of 3 A in a magnetic field of magnetic flux density 2 T. What is the force on the wire if it makes 30° with the magnetic field? At what angle would the force on the wire be zero?

10. A straight piece of wire carrying a 2.5 A current is placed at an angle of 60° to a magnetic field of magnetic flux density 3 T. The wire is of length 50 cm. Resolve the magnetic flux density into components parallel and perpendicular to the wire. Which component causes the force on the wire? What is the magnitude of the force? In what direction does it act?

FORCE ON A MOVING CHARGE IN A MAGNETIC FIELD

Fig. 27.13
The beam of electrons in the cathode ray tube is an electric current.

On page 295 you saw that around any current-carrying conductor there is a magnetic field that is caused by the current, i.e. the magnetic field is caused by the moving charges in the conductor. You would therefore expect that if we had a stream of moving charges without a conducting wire, there would still be a magnetic field around it. This is found to be the case.

In FIG. 27.13 a beam of electrons in a cathode ray tube (page 328) moves in a vacuum. The beam of electrons passing close to the fluorescent screen shows up as a beam of light. The moving electrons have negative charge and thus are an electric current. They therefore have a magnetic field around them. This magnetic field will interact with any other magnetic field placed near it. FIG. 27.14 shows the beam of electrons deflecting due to the presence of a bar magnet. FIG. 27.15 shows a proton (a positively charged particle) moving with a speed of v m s^{-1} perpendicular to a uniform magnetic field of flux density B. The moving proton creates a magnetic field around itself. This field interacts with the uniform field B causing a force on the moving proton and it deflects from its original path.

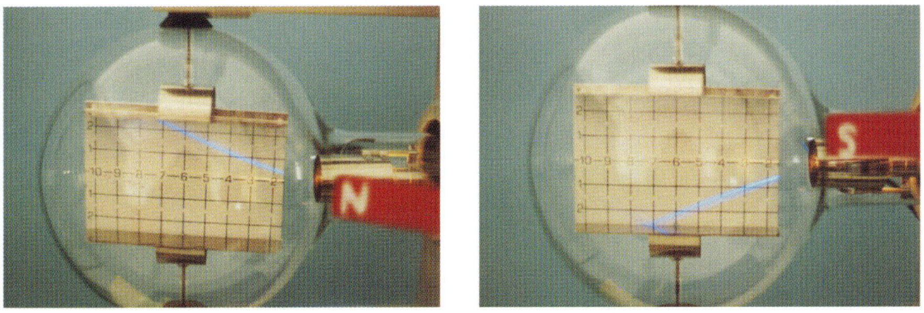

Fig. 27.14
The beam of electrons deflects in the magnetic field. The N pole deflects the beam one way and the S pole the other obeying Fleming's left-hand rule.

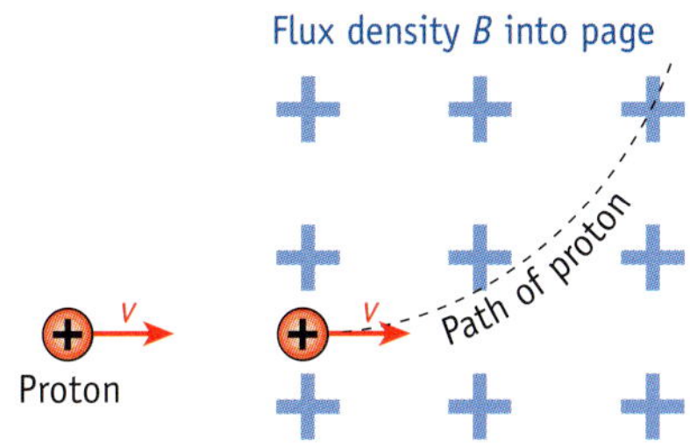

Fig. 27.15

The following can be proved:

> If a charge q coulombs is moving with a speed of v metres per second at right angles to a magnetic field of flux density B teslas then there is a force F on it given by:
> $$F = qvB$$

The direction of F is perpendicular to that of v and B. Fleming's left-hand rule tells us which of the two directions it is. Note that the direction in which the second finger points – the direction of the current – is the direction in which a positive charge is moving. If the moving charge is negative (e.g. an electron) the second finger must point in the opposite direction to the motion of the charge. FIG. 27.16 shows a positive charge and a negative charge entering a magnetic field. By using Fleming's left-hand rule convince yourself that they move in the directions shown in the diagram.

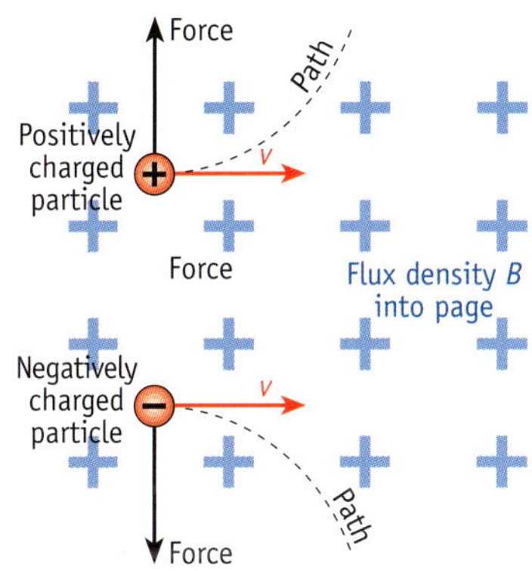

Fig. 27.16

Problem 6:	A particle of charge of 2×10^{-3} C moving at 100 m s^{-1} moves at right angles to a uniform magnetic field of flux density 3 T. What is the force on the charge?
Solution:	$F = qvB = (2 \times 10^{-3})(100)(3) = 0.6$ N

Charged Particle Moving in a Circle

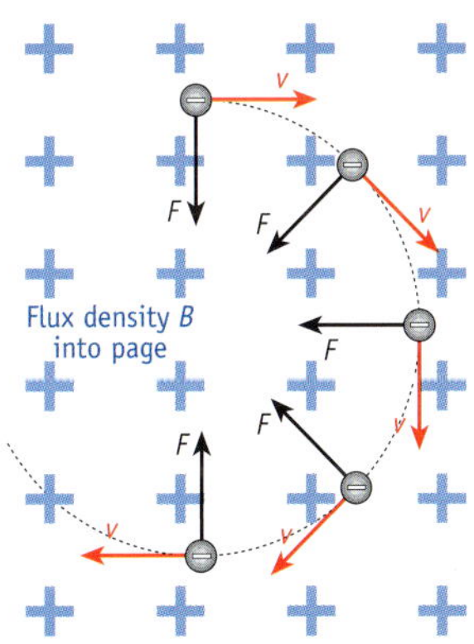

Fig. 27.17

When a charged particle moving at a constant speed enters a uniform magnetic field and moves at right angles to the field it is found that the particle moves in a circular path. In FIG. 27.17 as the charged particle moves in the magnetic field the force on it is at right angles to its direction of motion and of constant magnitude $F = qvB$. The speed of the particle does not change but its direction of motion does. It turns. As it turns **the force always remains at right angles to the direction of motion**. Thus the particle moves in a circular path*.

NOTE
If a charged particle moving at constant speed enters a uniform magnetic field and moves at right angles to the field, the particle moves in a circle.

* Recall from page 140 that a moving particle will move in a circle at a steady speed if the resultant force on it is of fixed magnitude and always acts at right angles to the direction of motion. The force is always directed towards a fixed point, i.e. towards the centre of the circle.

Problem 7: An electron of charge 1.6×10^{-19} C and mass 9.1×10^{-31} kg enters a uniform magnetic field of flux density 2×10^{-2} T travelling at a speed of 400 m s^{-1}. Find the radius of the path it follows in the field.

Solution: Centripetal Force = Force on electron due to B field

$$\frac{mv^2}{r} = qvB \Rightarrow \frac{(9.1 \times 10^{-31})(400)^2}{r} = (1.6 \times 10^{-19})(400)(2 \times 10^{-2})$$

$$\Rightarrow r = \frac{(9.1 \times 10^{-31})(400)^2}{(2 \times 10^{-2})(1.6 \times 10^{-19})(400)} = 1.14 \times 10^{-7}\text{m}$$

Derivation of F = qvB

Consider a conductor of length l containing n charges per unit length each moving with speed v (Fig. 27.18). Suppose the size of each charge is q.

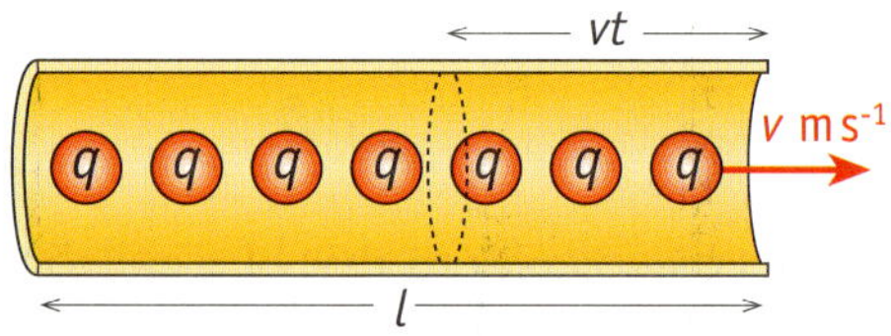

Fig. 27.18

In a time t, the amount of charge passing any point in the conductor is the amount of charge in a length vt of the conductor.

i.e. amount of charge passing in time t is $= qnvt$

$$\text{The current } I = \frac{\text{Charge passing}}{\text{Time taken}} = \frac{qnvt}{t} = nvq$$

Force on length l of the conductor $= IlB = nvqlB$

Force per unit length (i.e. on 1 metre) $= nvqB$

This is the force on n moving charges, therefore the force on one moving charge is n times less, i.e. $F = qvB$.

EXERCISE 27.2

1. A charge of 2 C moves at a speed of 10 m s^{-1} at right angles to a magnetic field of flux density 2 T. What is the force on the charge?
2. A particle of charge of 3×10^{-6} C moving at 200 m s^{-1} enters a uniform magnetic field of flux density 4 T and moves at right angles to the field. Calculate the force on the particle.
3. What is the force on an electron travelling at 6×10^{7} m s^{-1} at right angles to a magnetic field of flux density 4 T? (Charge on electron $e = 1.6 \times 10^{-19}$ C).
4. An electron of charge 1.6×10^{-19} C enters a uniform magnetic field of flux density 2 T and moves at right angles to the field. If the force on the electron is 2×10^{-18} N calculate the speed of the electron.

5. FIG. 27.19 shows a uniform magnetic field directed into the page. It also shows the paths of three particles *A*, *B* and *C* as they move through the field at constant speeds. What can be concluded about the three particles?

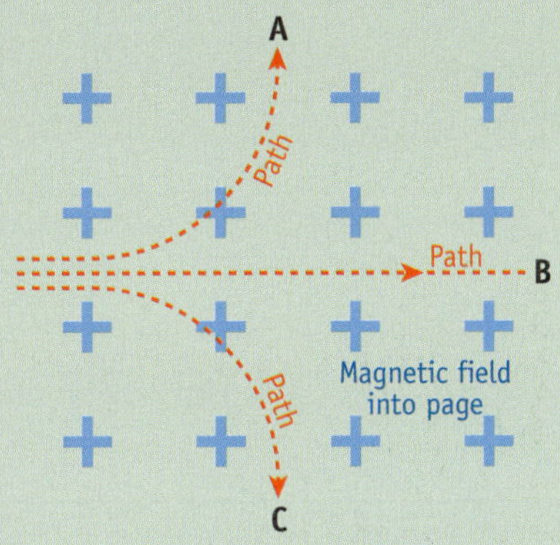

Fig. 27.19

6. A proton of mass 1.67×10^{-27} kg travelling at a speed of 2×10^{7} m s^{-1} enters a uniform magnetic field of flux density 0.03 T travelling in a plane perpendicular to the field. Explain why the proton travels in a circular path and find its radius.
(Charge on proton = 1.6×10^{-19} C)

7. An electron of charge 1.6×10^{-19} C and mass 9.1×10^{-31} kg enters a uniform magnetic field of flux density 3×10^{-2} T travelling at a speed of 2000 m s^{-1}. Find the radius of the path it follows in the field.

8. A beam of cathode rays is bent into a circular path or radius 10 cm by a magnetic field of flux density 2×10^{-3} T. With what speed are the electrons travelling?
(Charge on electron $e = 1.6 \times 10^{-19}$ C, mass of electron = 9.1×10^{-31} kg.)

9. Prove that the period *T* of the circular orbit of an electron of mass *m* and charge *e* when moving at speed *v* perpendicular to a magnetic field of flux density *B* is given by:
$T = 2\pi m / Be$

10. Identical particles each carrying a charge of 2×10^{-6} C are travelling at a speed of 0.1 m s^{-1} through a conductor. There are 104 particles in each metre of the conductor.
(i) How many particles pass any point in the conductor in one second?
(ii) How much charge passes any point in the conductor in one second?
(iii) What is the size of the current flowing in the conductor?

11. Identical charges of charge 1.6×10^{-19} C each are travelling at a speed of 0.02 cm s^{-1} through a conductor. There are 10^{12} charges per metre of conductor. What is the size of the current flowing in the conductor?

The Magnetic Force Between Two Current-Carrying Conductors

FIG. 27.20 shows two parallel conductors carrying current in opposite directions. Each has a magnetic field around it. These magnetic fields interact with each other and cause a force of repulsion on each wire, pushing the wires apart. If the wires carry current in the same direction it is found that there is a force on each pulling it towards the other, i.e. a force of attraction.

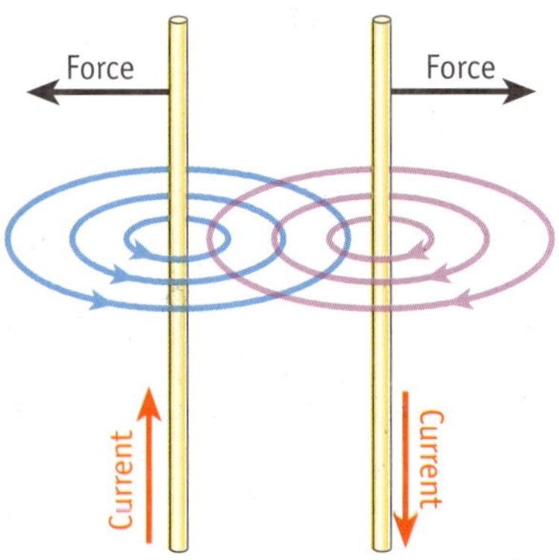

Fig. 27.20
The force between two parallel current-carrying conductors is due to their magnetic fields.

This can easily be demonstrated in the laboratory as follows:

EXPERIMENT

TO SHOW THE MAGNETIC FORCES BETWEEN TWO CURRENT-CARRYING CONDUCTORS.

Method

1. Set up the equipment in FIG. 27.21.
2. Pass a current of about 4 A through the parallel strips of tinfoil.
3. The foil strips will be seen to move away from each other.

Conclusion

There is a force between current-carrying conductors due to their magnetic fields.

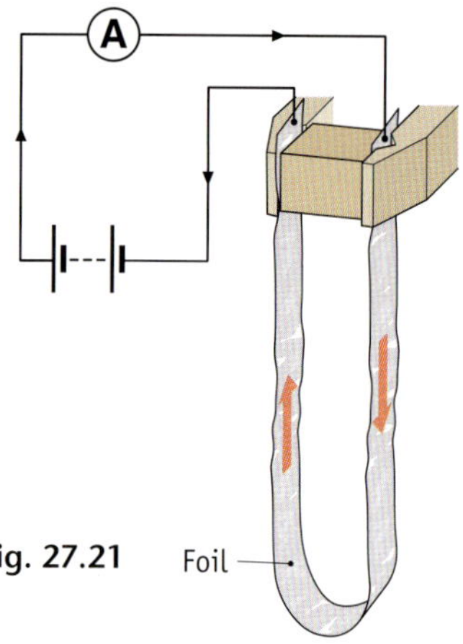

Fig. 27.21

Fig. 27.22

DEFINITION OF THE AMPERE

You saw on page 246 that the size of an electric current is the amount of charge passing any point in a circuit per second. Therefore in a metal, the greater the number of electrons passing per second, the greater the current. Since we cannot count electrons directly, we use the fact that the size of the magnetic effect increases with the size of the current to measure the amount of current. In particular, it is found that the force of attraction or repulsion between two parallel current-carrying conductors increases with the size of the current. This force can be measured and can be used to indicate the size of the current.

THE AMPERE

***The ampere* (A)** is that constant current which, if maintained in two straight parallel conductors of infinite length, of negligible cross section and placed 1 metre apart in a vacuum, would produce a force on each conductor of 2×10^{-7} newtons per metre of length.

The last experiment above is also **an experiment to show the principle on which the definition of the ampere is based**.

THE UNIT OF ELECTRIC CHARGE

Recall from page 222 and 246 that the unit of electric charge is the coulomb (C). The coulomb is defined as follows:

THE COULOMB

The ***coulomb* (C)** is the amount of charge that passes any point in a circuit when a current of 1 ampere flows for 1 second.

CHAPTER CHECKLIST

- **Define**: Magnetic flux density; The tesla;
  The ampere; The Coulomb.
- **State**: Fleming's left-hand rule; The unit of magnetic flux density; The factors on which the size of the force on a current-carrying conductor in a magnetic field depends.
- **Recall** that: A current-carrying conductor experiences a force when placed in a magnetic field (unless parallel to field); The direction of the force is perpendicular to the current and perpendicular to the magnetic field; The size of the force is directly proportional to the current (I), the length of the conductor (l) and the strength of the magnetic field; Current-carrying conductors exert forces on each other due to their magnetic fields; A charged particle moving in a magnetic field experiences a force;
  A charged particle moving perpendicular to a magnetic field moves in a circle.
- **Describe** an experiment to show: The force on a current-carrying conductor in a magnetic field; The forces on a current-carrying coil in a magnetic field; That current-carrying conductors exert forces on each other;
  The principle on which the definition of the ampere is based.
- **List** three practical applications of the force on a current-carrying coil in a magnetic field.
- **Recall** and use the formulae: $F = I\,l\,B$;
  $F = qvB$; $F = \dfrac{mv^2}{r}$
  to solve problems.
- **Derive** the formula: $F = qvB$

CHAPTER 28

Electromagnetic Induction

In Chapter 26 you saw that an electric current causes a magnetic field in the space around it. In this chapter we look at how **a changing magnetic field** can cause an electric current to flow, i.e. we shall study electromagnetic induction.

ELECTROMAGNETIC INDUCTION
Whenever the magnetic field passing through a coil changes, an emf appears in the coil. This phenomenon is called ***electromagnetic induction***.

EXPERIMENT

EXPERIMENTS TO SHOW ELECTROMAGNETIC INDUCTION.

(i)
- Use the equipment in FIG. 28.1.
- Move the north pole of the magnet towards the coil. The galvanometer deflects, indicating that a current flows in the circuit.
- Stop the magnet moving and the meter reads zero, i.e. the current stops.
- Move the north pole away from the coil and the meter deflects in the opposite direction, indicating that current flows in the opposite direction.
- Turn the magnet so that the south pole is nearest the coil and repeat the above. Current will flow as long as the magnet is moving. The current is in the opposite direction to that caused by the moving north pole.

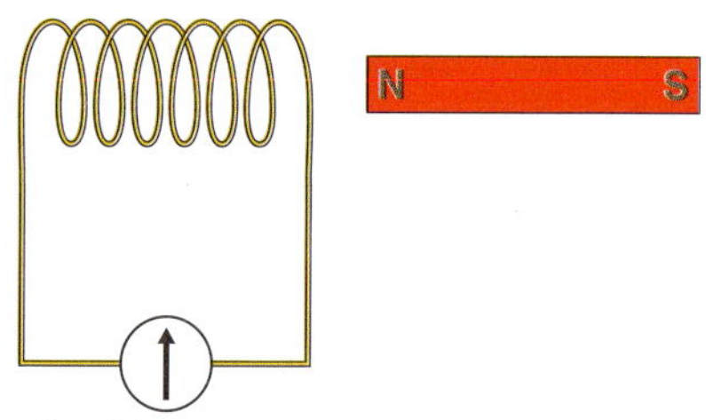

Fig. 28.1

(ii)
- Hold the magnet stationary and move the coil away from or towards the magnet.
- Again the meter deflects showing that current flows.
- Thus any relative motion between the coil and the magnet causes a current to flow. This current is called an **induced current**.

(iii)
- Use the equipment in FIG. 28.2.
- At the instant that the switch S is closed, the meter in circuit B gives a deflection. When the switch remains closed, causing a steady current in circuit A, no current flows in circuit B.
- If the switch is now opened, the meter again deflects, but this time in the opposite direction. In each case it is found that the size of the deflection would be significantly greater if both coils were wound on (but electrically insulated from) a soft iron core.

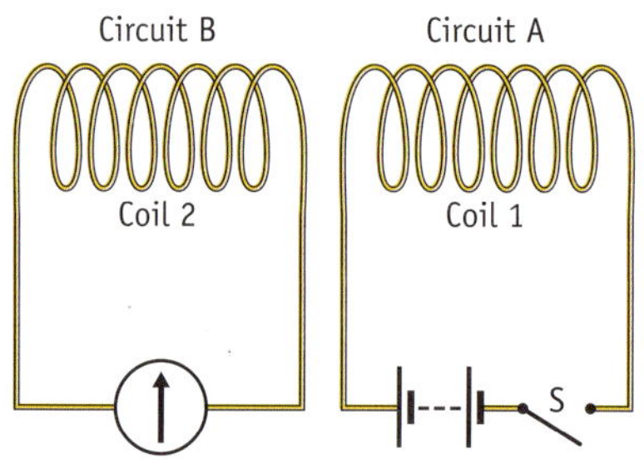

Fig. 28.2

(iv)
Set up the circuit in FIG. 28.3 where the meter is one that reads a.c.
The meter will indicate that a current always flows in the second circuit.

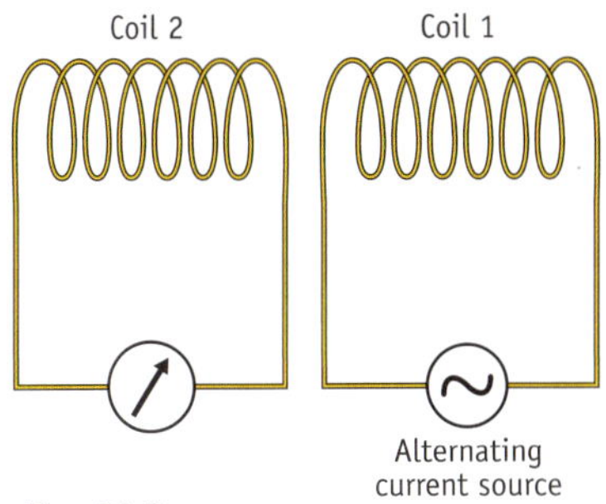

Fig. 28.3

Conclusion

From the above and from many similar experiments we see the following:
Whenever the magnetic field passing through a coil **is changing**, an emf appears in that coil. Such an emf is called an **induced emf.** This emf causes the induced current. If the magnetic field is not changing there is no induced emf and hence no induced current, no matter how strong the magnetic field.
Note that **induced current** will only flow if the circuit is a complete circuit.
The emf is induced whether the circuit is complete or not.

NOTE

REMEMBER:
There is no induced emf unless the magnetic field is CHANGING!

MAGNETIC FLUX

To calculate the size of the induced emf we use the quantity **magnetic flux Φ**. Suppose a plane loop of wire of area A has a magnetic field at right angles to it (FIG. 28.4). Suppose the value of the magnetic flux density at any point on the loop is B, then the **magnetic flux** passing through the area A is defined by:

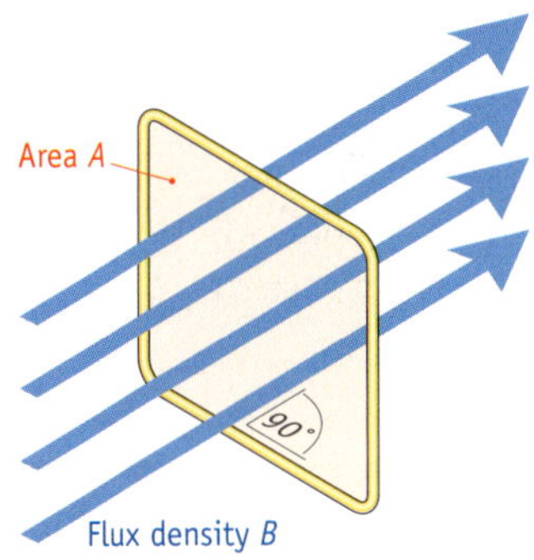

Fig. 28.4
Magnetic Flux.

$$\begin{pmatrix}\textbf{Magnetic flux}\\ \textbf{through Area } A\end{pmatrix} = \textbf{Magnetic flux density} \times \textbf{Area}$$

$$\textbf{i.e} \quad \Phi = BA$$

Magnetic flux is a **scalar quantity**.

UNIT OF MAGNETIC FLUX
The unit of magnetic flux is the ***weber (Wb)***.

$\Phi = BA \quad \Rightarrow$ Unit of Flux
$=$ Unit of magnetic flux density $\times$ Unit of area
$=$ (tesla)(metre squared) i.e. $\mathbf{1\ Wb = 1\ T\ m^2}$

THE WEBER
If the magnetic flux density over an area of 1 m^2 is 1 tesla then the flux through the area is 1 ***weber***.

If the magnetic flux density is not perpendicular to the area, the flux through A is the component of B perpendicular to A multiplied by the area.

Problem 1: What is the magnetic flux through a loop of area 0.4 m^2 placed at right angles to a magnetic field of 2 T?

Solution: $\Phi = BA = (2)(0.4) = 0.8$ Wb

Problem 2: What is the magnetic flux through a loop of area 0.4 m^2 in a magnetic field of 2 T if the flux density makes 30° with the loop?

Solution: Fig. 28.5 shows the situation.

Component of B perpendicular to the coil

$= B \text{ Sin}30° = 2 \text{ Sin}30° = 1$ T

Flux through coil = (comp. of B perp to coil)(area)

$= (1)(0.4) = 0.4$ Wb

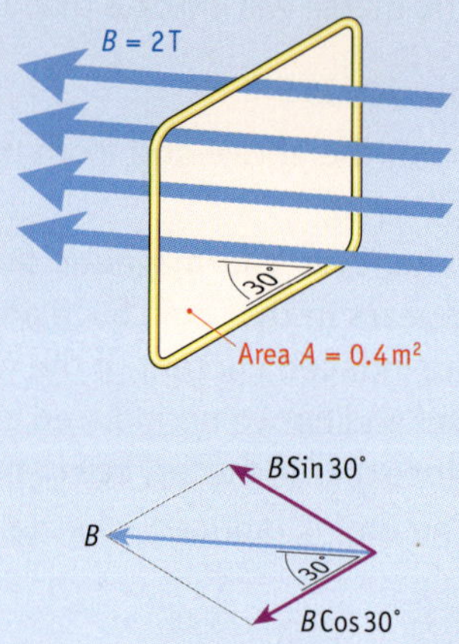

Fig. 28.5

EXERCISE 28.1

1. What is the magnetic flux through a loop of area 0.3 m^2 when placed at right angles to a magnetic field of flux density 2 T?
2. The magnetic flux passing through a coil is 0.4 Wb when placed at right angles to a magnetic field flux density 0.5 T. What is the area of the coil?
3. A planar loop of wire of area 100 cm^2 is perpendicular to a uniform magnetic field of flux density 2 T. What is the magnetic flux passing through the loop?
4. The magnetic flux passing through a coil of wire is 2×10^{-2} Wb. The field is uniform and perpendicular to the plane of the coil. The area of the coil is 200 cm^2. Find the flux density at any point in the coil.
5. The magnetic flux passing through a one turn circular coil when placed perpendicular to a uniform magnetic field of flux density 3×10^{-3} T is 2×10^{-2} Wb. Find the radius of the coil.
6. The flux density of a uniform magnetic field changes in value from 1.2 T to 2.4 T. Find the change in the flux passing through a rectangular coil of dimensions 10 cm × 6 cm whose plane is perpendicular to the field.
7. The plane of a coil of area 0.2 m^2 makes an angle of 30° with a magnetic field of flux density 4 T. Calculate the flux passing through the coil.
8. A circular coil of radius 20 cm is placed in a uniform magnetic field, the angle between the plane of the coil and the field being 40° (Fig. 28.6). If the flux density is 2.5×10^{-3} T, find the flux through the coil.

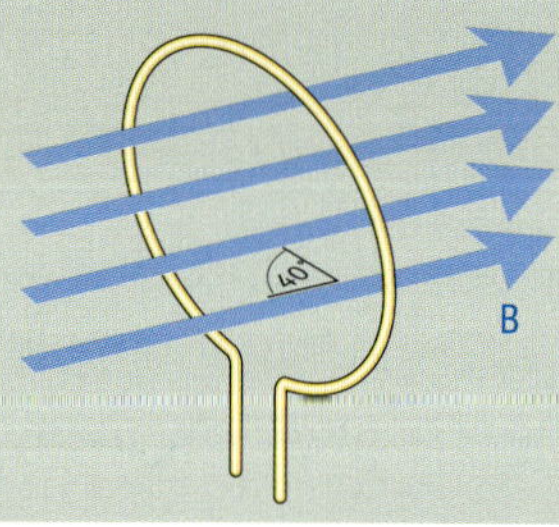

Fig. 28.6

Faraday's Law of Electromagnetic Induction

The experiments on page 310 indicate that the size of an induced emf depends on how quickly the magnetic flux is changing. If the flux is changing quickly the induced emf is large, if the flux is changing slowly the induced emf is small. The relationship between the size of the induced emf and the changing magnetic flux is given by **Faraday's Law*** **of Electromagnetic Induction**.

* Faraday's Law and Lenz's Law (below) are sometimes known as The Laws of Electromagnetic Induction.

FARADAY'S LAW OF ELECTROMAGNETIC INDUCTION
states that the size of the induced emf is directly proportional to the rate of change of flux.

From Faraday's Law: Induced emf $\propto$ Rate of change of flux

$$\text{i.e.} \quad E \propto \frac{\text{Change in } \Phi}{\text{Time taken}} \quad \text{or} \quad E \propto \frac{\text{Final flux} - \text{Initial flux}}{\text{Time taken}}$$

$$\Rightarrow \quad E = \frac{k(\text{Final flux} - \text{Initial flux})}{\text{Time taken}} \quad \text{where } k \text{ is constant.}$$

The value of the constant of proportionality k depends on the units in which the various quantities are measured. In SI units its value is 1 (don't worry why). Thus we have:

$$\textbf{Induced emf } E = \frac{(\textbf{Final flux} - \textbf{Initial flux})}{\textbf{Time taken}}$$

This is the formula you will use in most numerical problems. Using calculus notation it is written as:

$$\textbf{Induced emf } E = -\frac{d\Phi}{dt}$$

The minus sign indicates the direction of the induced emf and will be explained below under Lenz's Law. You can ignore the minus sign in all numerical calculations.

Problem 3: The magnetic flux through a one turn coil changes from 2 Wb to 8 Wb in 4 seconds. Find the average emf induced in the coil. If the coil has a resistance 10 Ω and is a complete circuit, find the induced current.

Solution:

$$\text{Induced emf } E = \frac{(\text{Final } \Phi - \text{Initial } \Phi)}{\text{Time taken}} = \frac{(8-2)}{4} = 1.5 \text{ V}$$

$$\text{Induced Current } I = \frac{\text{Induced emf}}{\text{Resistance}} = \frac{E}{R} = \frac{1.5}{10} = 0.15 \text{ A}$$

Problem 4: Find the emf induced in a 100 turn coil if the flux through it changes from 0 Wb to 6 Wb in 0.2 s.

Solution:

$$\text{Emf induced in one turn of the coil} = \frac{(\text{Final } \Phi - \text{Initial } \Phi)}{\text{Time taken}} = \frac{6-0}{0.2} = 30 \text{ V}$$

The coil has 100 turns connected in series. An emf of 30 volts is induced in each one. Therefore:

Total emf induced = 100 × 30 = 3000 volts

Note that if a coil has a number of turns and if the flux through the coil is changing, then the total emf induced in the coil is the sum of the emfs induced in each turn, because the turns are in series. Thus if a coil has N turns and the rate at which the flux through the coil is changing is $d\Phi/dt$, then the induced emf in the coil is given by:

$$E = -N\frac{d\Phi}{dt}$$

EXPERIMENT

TO DEMONSTRATE FARADAY'S LAW OF ELECTROMAGNETIC INDUCTION.

Method

- Set up the equipment in FIG. 28.7.
- Move the magnet away from or towards the coil slowly (the flux through the coil thus changes slowly).
- The galvanometer gives a small deflection (indicating a small induced emf).
- Move the magnet away from or towards the coil very rapidly (the flux through the coil thus changes rapidly).
- The galvanometer gives a large deflection (indicating a large induced emf).

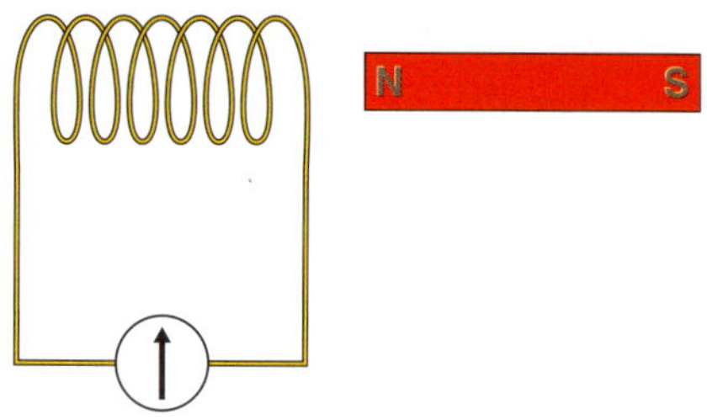

Fig. 28.7

Conclusion

This shows approximately that the size of the induced emf is proportional to the rate of change of flux through the coil.

Problem 5: A 200 turn rectangular coil of area 0.05 m^2 is placed perpendicular to a uniform magnetic field of flux density 3 T. If the flux density increases to 10 T in 0.4 s, find the average induced emf in the coil.

Solution:

$\Phi = BA \Rightarrow$ Initial flux through coil $= BA = (3)(0.05) = 0.15$ Wb

Final flux through coil $= BA = (10)(0.05) = 0.5$ Wb

$$\text{Average induced emf } E = \frac{N(\text{Final } \Phi - \text{ initial } \Phi)}{\text{Time taken}} = \frac{(200)(0.5 - 0.15)}{0.4} = 175 \text{ V}$$

Problem 6: A rectangular coil of 1 turn and dimensions 2 cm × 4 cm enters a magnetic field of flux density 3 T. If the plane of the coil is perpendicular to the field and it moves at 6 m s^{-1} parallel to its 4 cm side (FIG. 28.8), find the emf induced in the coil.

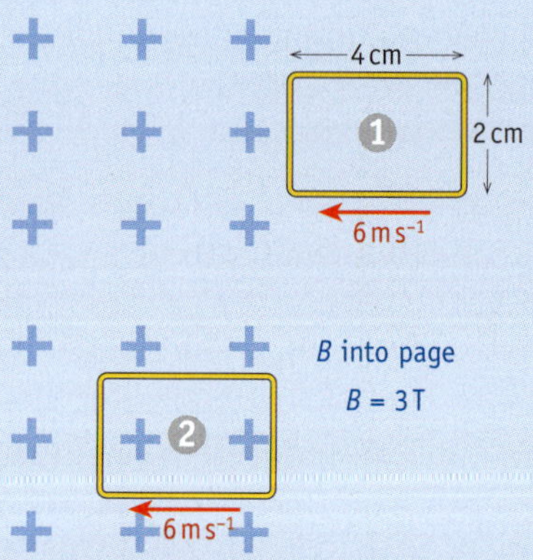

Fig. 28.8

Solution:

$$\text{Induced emf} = \frac{(\text{Final flux} - \text{Initial flux})}{\text{Time taken}}$$

From FIG. 28.8 initial Φ through coil $= 0$

From FIG. 28.8 final Φ through coil $= BA = 3(2 \times 10^{-2})(4 \times 10^{-2}) = 2.4 \times 10^{-3}$ Wb

Time to go from position 1 to position 2 is time to travel 4 cm at 6 m s^{-1}

$$= \frac{(4 \times 10^{-2})}{6} = 6.67 \times 10^{-3} \text{ seconds}$$

$$\text{Average induced emf} = \frac{(\text{Final } \Phi - \text{ initial } \Phi)}{\text{Time taken}} = \frac{(2.4 \times 10^{-3} - 0)}{6.67 \times 10^{-3}} = 0.36 \text{ V}$$

Problem 7: A 100 turn rectangular coil of dimensions 6 cm × 8 cm rotates at constant angular speed in a uniform magnetic field of flux density 2 T (FIG. 28.9). If the coil undergoes 5 rotations per second, find the average induced emf in the coil in going from the position in FIG. 28.9(A) to that in FIG. 28.9(B). Sketch a graph to show how the induced emf varies during this time.

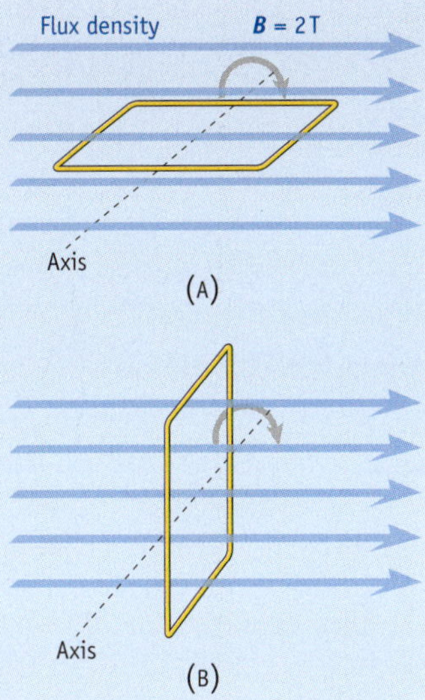

Fig. 28.9

Solution: In FIG. 28.9(A) flux through coil is zero.

In FIG. 28.9(B) flux through coil = BA

i.e. $\Phi = (2)(6 \times 10^{-2} \times 8 \times 10^{-2}) = 0.0096$ Wb

Time taken to go from FIG. 28.9(A) to FIG. 28.9(B) is the time to undergo a ¼ rotation.

Coil does 5 rotations per second

⇒ Time for one rotation = ⅕ s

⇒ Time for quarter of a rotation = 1/20 s

$$\text{Average induced emf: } = N\frac{(\text{Final } \Phi - \text{ initial } \Phi)}{\text{Time taken}} = \frac{(100)(0.0096-0)}{(1/20)} = 19.2 \text{ V}$$

In position (A) the flux through the coil is zero but a further small rotation causes a large change in flux. The rate of change of flux is greatest at (A) and so is the induced emf. In (B) a further small rotation causes only a tiny change in flux. In fact the instantaneous rate of change of flux at (B) is zero and the induced emf is instantaneously zero there. The graph is shown in FIG. 28.10.

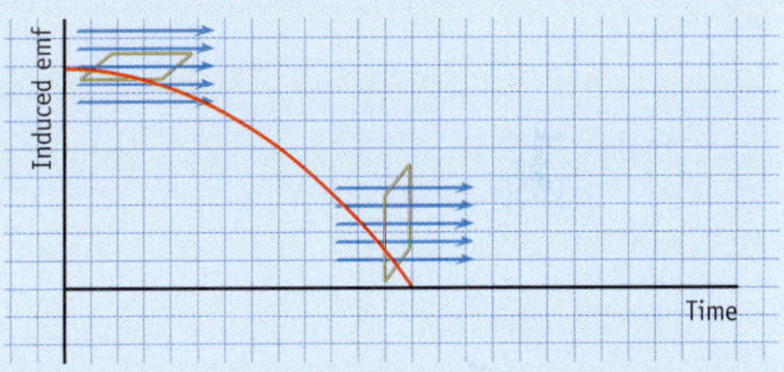

Fig. 28.10

EXERCISE 28.2

1. The magnetic flux through a loop of wire changes from 1 Wb to 5 Wb in 2 s. Calculate the average induced emf in the loop.
2. The magnetic flux through a coil is reduced from 0.4 Wb to zero in 0.2 s. The coil has 200 turns. Find the emf induced in one turn of the coil. Find the total emf induced in the coil.
3. Find the emf induced in a 600 turn coil if the magnetic flux through it changes from 0 Wb to 2.4 Wb in 0.6 s.
4. The magnetic flux through a coil of 200 turns changes uniformly from 2 Wb to 4 Wb in 0.3 seconds. Find the magnitude of the emf induced in the coil. If the coil is a complete circuit of resistance 4 Ω, find the induced current.
5. The flux density of a uniform magnetic field changes in value from 1.2 T to 2.4 T. Find the change in the flux passing through a rectangular coil of dimensions 10 cm × 6 cm whose plane is perpendicular to the field.
6. A 100 turn coil of area 0.08 m^2 is placed perpendicular to a uniform magnetic field of flux density 2 T. If the flux density increases to 6 T in 0.5 s, find the average induced emf in the coil.
7. A rectangular coil of 1 turn and dimensions 6 cm × 8 cm enters a magnetic field of flux density 2 T. If the plane of the coil is perpendicular to the field and it moves at 3 m s^{-1} parallel to its 8 cm side (FIG. 28.11), find the emf induced in the coil.

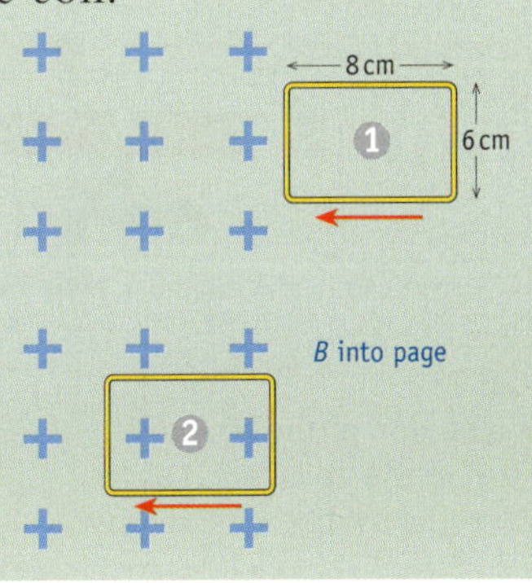

Fig. 28.11

8. A 200 turn rectangular coil of dimensions 4 cm × 6 cm rotates at constant angular speed in a uniform magnetic field of flux density 4 T (FIG. 28.12). If the coil undergoes 10 rotations per second find the average induced emf in the coil in going from the position in FIG. 28.12(A) to that in FIG. 28.12(B).

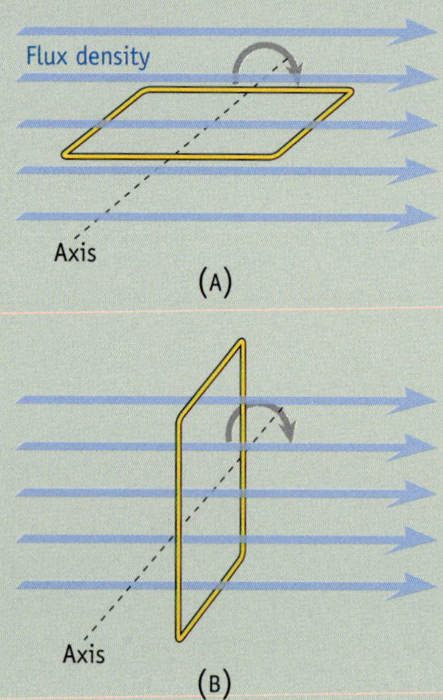

Fig. 28.12

9. A 10 000 turn square coil of side length 3 cm is withdrawn from a region where the flux density is 2×10^{-2} T to one where the flux density is zero. If the coil is withdrawn in 0.1 seconds find the average induced emf in the coil.

10. A planar 200 turn coil of area 500 cm² is rotating at 40 revolutions per second about an axis through its centre in the plane of the coil. The axis of rotation is perpendicular to a uniform magnetic field. If at a given instant the flux through the coil is zero and the average emf induced in the coil as it rotates through 90° from this position is 4 V, find the flux density of the field.

11. When the flux through a one turn coil of resistance 12 Ω changes from 0 to 0.25 T a charge of 4 mC passes through the coil. What is the area of the coil?

12. A 500 turn coil of area 400 cm² has its plane perpendicular to a magnetic field of flux density 1.8×10^{-5} T. The coil is rotated through 90°. If the resistance of the coil is 10 Ω, what charge passes through the coil?

LENZ'S LAW

On page 310 you saw that the direction of the induced current (and emf) depends on whether the magnetic flux through the coil is increasing or decreasing. **Lenz's Law** tells us the direction of the induced current caused by a changing magnetic field.

LENZ'S LAW states that the direction of an induced current is always such as to oppose the change producing it.

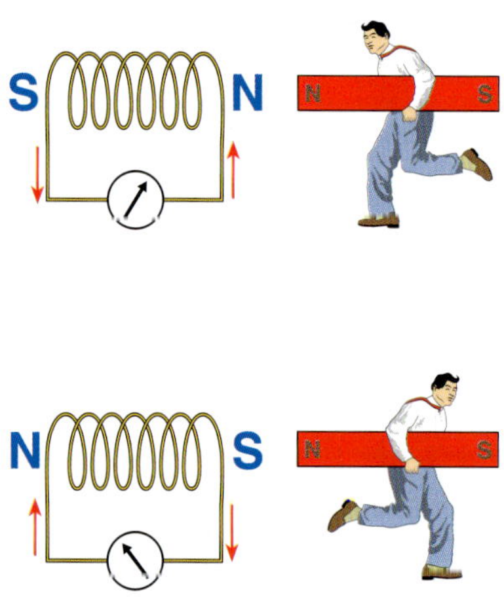

Fig. 28.13
The direction of the induced current in the coil is such as to oppose the motion of a magnet.

For example, if the changing flux in a coil is produced by an approaching north pole, Lenz's Law says the induced current will flow in such a direction as to oppose the approaching north pole.

FIG. 28.13 shows a man running towards a coil with a magnet. The magnetic flux passing through the coil is increasing as the north pole approaches and an emf is induced in the coil.

By Lenz's Law the induced current in the coil must flow in a direction that opposes the north pole of the magnet coming towards it.

The induced current in the coil must therefore flow in such a direction that the end of the coil facing the approaching magnet behaves like a north pole, thus opposing the approaching north pole.

Because of this opposition the man must do work in bringing the magnet towards the coil (north repels north). The work he does appears as electrical energy in the coil.

If the man runs away from the coil the direction of the induced current must change so that his motion is still opposed. The direction of the induced current changes so that a south pole appears at the right-hand side of the coil. The work the man does in pulling a north pole away from a south again appears as electrical energy in the coil.

Consider what would happen if Lenz's Law were not true. In FIG. 28.13, as the north pole of the magnet approached the coil, suppose the induced current flowed so that a south pole appeared facing the approaching north pole. The south pole would attract the north pole and it would move faster towards the coil causing an increased induced emf and an increased induced current. Work would be done on the magnet AND electrical energy would be expended in the circuit. Thus energy would be coming from nowhere thereby violating the Principle of Conservation of Energy. Thus Lenz's Law follows from the Principle of Conservation of Energy.

In the above circuit the man does work and loses energy. The energy he loses becomes electrical energy in the circuit.

CHANGING MECHANICAL ENERGY TO ELECTRICAL ENERGY – GENERATORS

Electromagnetic induction is the principle on which the electric generator is based and on which modern large-scale production of electrical energy is based. In an electricity generating station some form of energy, e.g. chemical energy from coal or oil is used to produce steam which causes a turbine to rotate – i.e. the turbine is given kinetic energy. The turbine rotates a coil in a magnetic field thereby causing the magnetic flux through it to change and an emf is induced in it. Thus the kinetic energy is converted to electrical energy.

Fig. 28.14
An alternator in a car. This particular alternator has a maximum power output of 1 kW.

Everyday examples of electricity generators are:

- **Electricity power stations** which generate huge quantities of electricity.
- The **alternator in a car** (FIG. 28.14) is turned by the engine and generates electricity to supply power to the car's electrical system and keep the car battery charged.
- The **dynamo** on a bicycle generates electricity to operate the bicycle's lights.

ELECTRIC GENERATOR
An ***electric generator*** is a device that converts mechanical energy to electrical energy.

Problem 8: If the resistance of the coil in FIG. 28.13 is 5 Ω and the induced current is 0.2 A find the force on the magnet if it moves towards the coil at 10 m s^{-1}.

Solution: Let F = the force on the magnet

In one second work done by man = Force × Distance = $F \times 10$ joules

In one second electrical energy dissipated in coil = $I^2R = (0.2)^2(5)$ (by Joule's Law)

By the principle of conservation of energy these must be equal.

i.e. $10F = (0.2)^2(5) \Rightarrow F = 0.02$ N

EXPERIMENT

TO DEMONSTRATE LENZ'S LAW.

Method

- Suspend a light aluminium ring from a piece of thread (FIG. 28.15).
- Move one pole of a bar magnet quickly towards it.
- The ring moves away from the approaching magnet.
- Move the magnet quickly away from the ring and the ring follows the magnet.

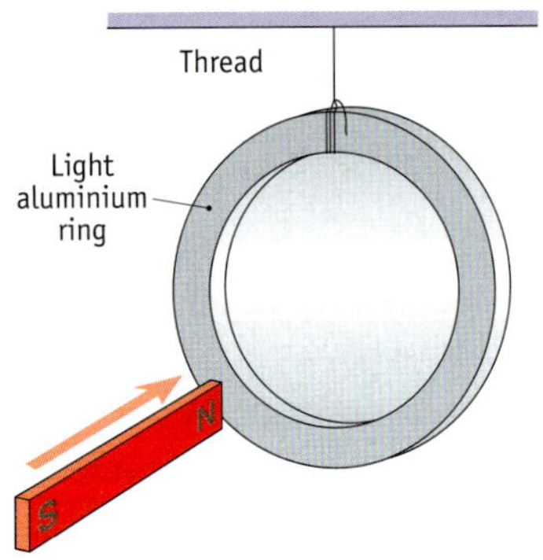

Fig. 28.15

Conclusion

As the magnet approaches the ring, current is induced in the ring, a pole the same as the approaching one appears at the side of the ring nearest the magnet. This repels the magnet and the magnet repels it, with the result that the ring is seen to move away from the approaching magnet. When the magnet is taken away from the ring the direction of the induced current changes causing an opposite pole to appear at the side of the ring facing the magnet. Thus the ring is attracted to the magnet and follows it. This follows from Lenz's Law and thus the law is demonstrated.

Alternative Method

FIG. 28.16 shows an **alternative way of demonstrating Lenz's Law**. A cylindrical piece of metal is dropped through a copper pipe. How long it takes to fall through the pipe is noted. A strong cylindrical magnet (the same size and weight as the piece of metal) is then dropped through the same pipe. It takes much longer to fall through the pipe.

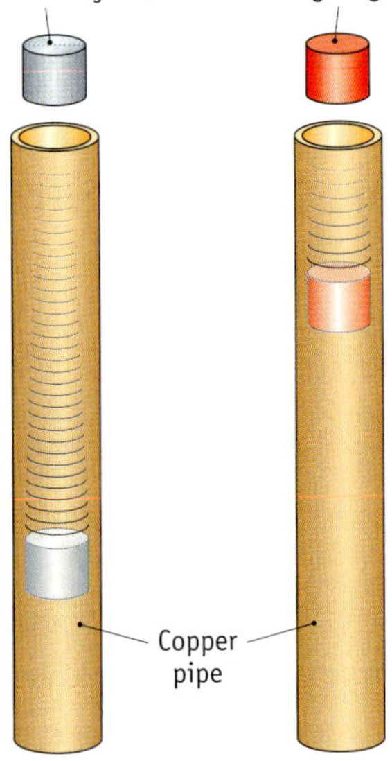

Fig. 28.16

Conclusion

As the magnet falls through the copper pipe its changing magnetic field induces currents in the pipe. By Lenz's Law these currents flow in such a direction as to oppose the change producing them; i.e. the moving magnet. They therefore exert forces on the magnet slowing it down. The unmagnetised metal cylinder does not experience these forces. Thus Lenz's Law is demonstrated.

EXERCISE 28.3

1. If the resistance of the coil in FIG. 28.13 (page 316) is 20 Ω and the induced current is 0.6 A, find the force on the magnet if it moves towards the coil at 30 m s^{-1}.

2. A rectangular coil of one turn and dimensions 5 cm × 12 cm enters a uniform magnetic field of flux density 2 T which is perpendicular to the plane of the coil. If the coil moves at 4 m s^{-1} parallel to the 12 cm side and has a resistance of 5 Ω, find:
   (i) the emf induced in the coil and
   (ii) the force that must be exerted on the coil to keep it moving at that speed.

3. A square coil of total resistance R has side L and contains N turns (FIG. 28.17). It is travelling in a direction parallel to the side EH, as shown in the diagram, when it enters a magnetic field which is perpendicular to the plane of the coil and is of uniform magnetic flux density B. Given that the two terminals of the coil are connected together, explain why the coil slows down as it is entering the field.

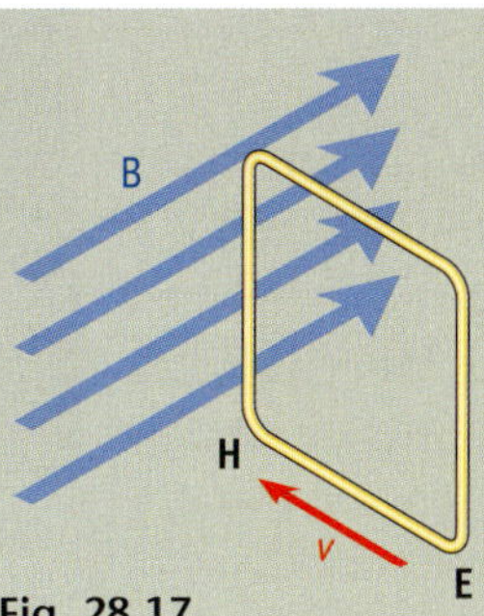

Fig. 28.17

Use Faraday's law to show that the emf induced in the coil at any instant while it is entering the field is given by $E = NBLv$, where v is the speed of the coil at that instant. Hence, derive an expression, in terms of v, for:

(i) the current in the coil,

(ii) the force on the coil.

ALTERNATING CURRENT

You saw on page 248 that when direct current (d.c.) flows in a wire, the electrons move in the same direction through the wire. You also saw that when alternating current (a.c.) flows, the electrons change their direction of motion regularly. For example, in the circuit shown in FIG. 28.18, an alternating current of frequency 50 Hz is flowing. This means that:

for the first 1/100 of a second the electrons flow from A towards B,

in the next 1/100 of a second they flow from B towards A,

in the next 1/100 of a second they flow from A towards B again.

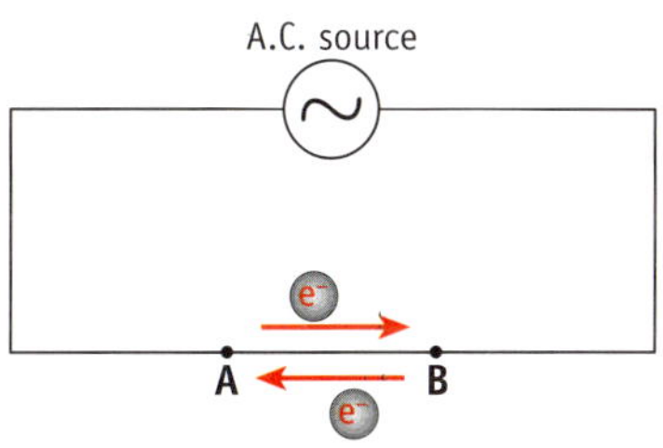

Fig. 28.18

This process repeats. Each electron undergoes a full cycle (also called an oscillation) in 1/50 of a second*. FIG. 28.19 is a graph of the size of the current plotted against time.

+ values signify that the current is flowing in one direction, – values signify that it is flowing in the opposite direction. (You may have seen a graph like this in your Maths course. It is the same graph as that of y = Sin x).

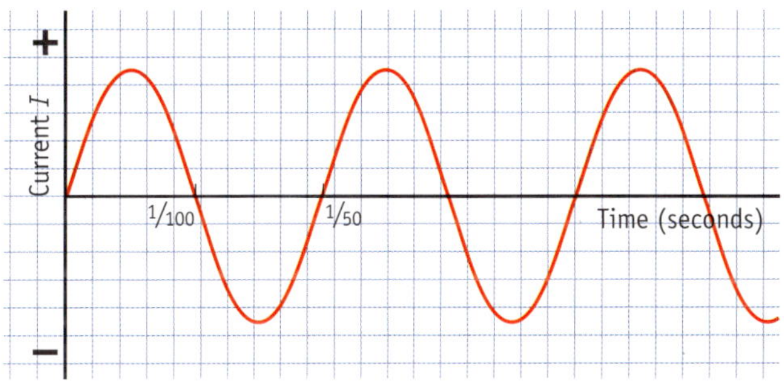

Fig. 28.19
Graph of a.c. against time.

A.C. VOLTAGE

To produce an alternating current like this an alternating voltage is necessary. A graph of the a.c. voltage against time is very similar to that in FIG. 28.19. The y-axis would be labelled 'voltage'. If the a.c. voltage is applied across a pure resistor (i.e. one that has no induction or capacitance), the current flowing at any instant is found from Ohm's Law, i.e.

$$i = \frac{v}{R} \qquad \textbf{Current at any instant} = \frac{\textbf{Voltage at that instant}}{\textbf{Resistance}}$$

Thus when v is large so is i, when v is small so is i and when v changes direction so does i.

* One complete forward and backward movement of an electron is a cycle or an oscillation.

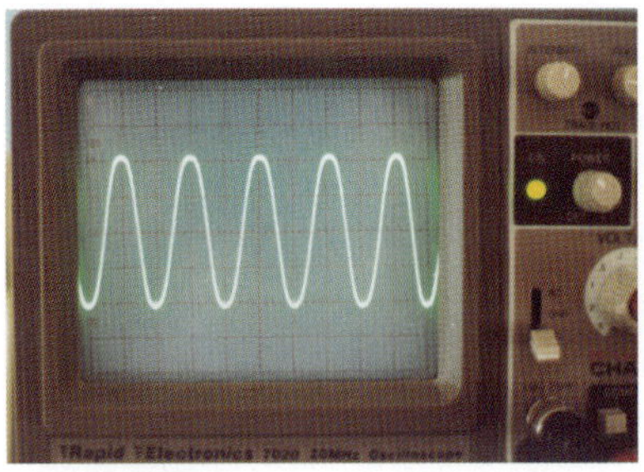

Fig. 28.20
Using an oscilloscope to show an a.c. voltage.

Using an Oscilloscopeto Show a.c.

If an alternating p.d. is placed across the y-plates of an oscilloscope and if a suitable time base voltage is placed across the x-plates a display similar to Fig. 28.20 will result. Note how similar this is to the graph in Fig. 28.19. If a d.c. voltage is placed across the y-plates the display will be like that in Fig. 28.21.

Fig. 28.21
Using an oscilloscope to show a d.c. voltage.

The Heating Effect of an Alternating Current

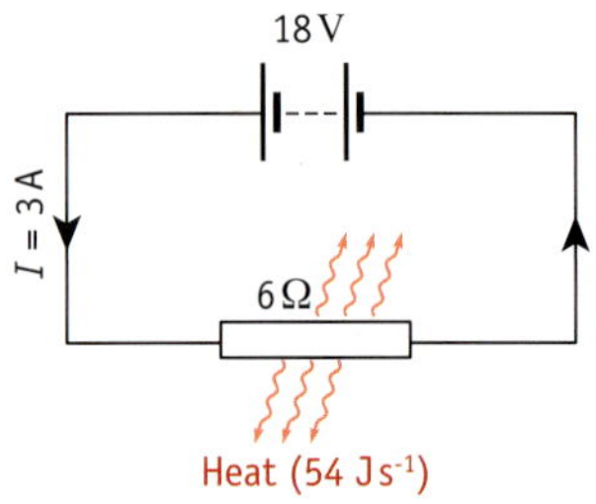

Fig. 28.22

In Fig. 28.22 a current of 3 A flows through the 6 Ω resistor. Heat is produced in the resistor at a rate given by Joule's Law,

i.e. $P = I^2R = (3)^2(6) = 54$ joules per second.

Fig. 28.23 shows an alternating current flowing through the same resistor. What is the maximum value of the a.c. in either direction if the a.c. produces heat at the same rate as the 3 A direct current?

It should be obvious that if the a.c. only reaches 3 A in each direction, it will not produce heat at the rate of 54 J s^{-1}, because the current is less than 3 A at all other times in each cycle. Thus if the a.c. is to produce heat at the same rate as the 3 A direct current its maximum value in each direction must be bigger than 3 A.

It can be proved (and you need not prove it) that the maximum value of the a.c. (I_o) needed to have the same heating effect as the 3 A d.c. is given by: $I_o = 3 \times \sqrt{2} = 3 \times 1.414 = 4.24$ A

RMS Values of a.c.

When we give the value of an alternating current as a certain value (say 5 A) we mean that this alternating current has the same heating effect as a 5 A direct current. Since alternating current varies with time, to have the same heating effect as a 5 A d.c. it must have a maximum value in each direction which is greater than 5 A. It is usual to give the value of an alternating current in this way. It is called its **rms value**. The following results can be shown to hold:

$$I_{rms} = \frac{I_0}{\sqrt{2}} \qquad I_0 = I_{rms} \times \sqrt{2}$$

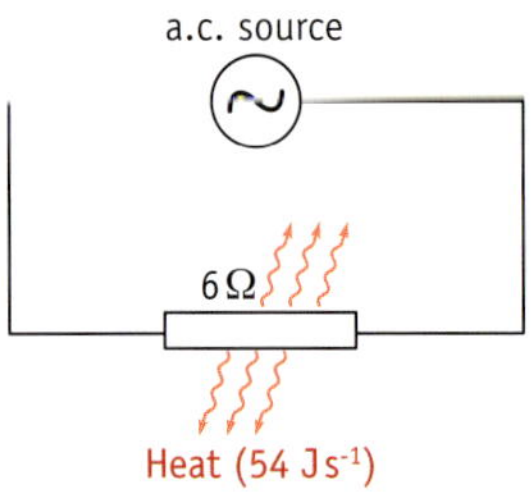

Fig. 28.23

The same is done for alternating voltages:

$$V_{rms} = \frac{V_0}{\sqrt{2}} \qquad V_0 = V_{rms} \times \sqrt{2}$$

It should also be clear that when a.c. flows in a pure resistance, the power P (the rate at which heat is produced) is given by the following:

$$P = I_{rms} \times V_{rms} \quad \text{or} \quad P = I^2_{rms}R$$

Problem 9: Domestic electricity is supplied at an rms voltage of 230 volts. Find the maximum value of the voltage in any one cycle.

Solution: $V_0 = V_{rms} \times \sqrt{2} = 230 \times 1.414 = 325.2$ V

Problem 10: The peak value of an a.c. current is 10 A. Find its rms value.

Solution: $I_{rms} = \frac{I_0}{\sqrt{2}} = \frac{10}{\sqrt{2}} = 7.07$A

Problem 11: An alternating current flowing in a wire of resistance 10 Ω produces heat at the rate of 60 W. Find:

(i) the rms value of the current,

(ii) the rms voltage,

(iii) the peak value of the voltage across the wire.

Solution:

(i) $P = I^2R \Rightarrow 60 = I^2_{rms}(10) \Rightarrow I^2_{rms} = \sqrt{\frac{60}{10}} = 2.45$ A

(ii) $V_{rms} = I_{rms} \times R = (2.45)(10) = 24.5$ V

(iii) $V_0 = V_{rms} \times \sqrt{2} = 24.5 \times \sqrt{2} = 34.65$ V

EXPERIMENT

To Compare Peak Values and rms Values of a.c.

Method

- Set up the equipment as shown in Fig. 28.24. The Oscilloscope is being used as a voltmeter.
- Connect the filament bulb to the a.c. source and note how bright it is.
- Measure the peak value of the a.c. voltage with the oscilloscope.
- Connect the filament bulb to the d.c. source and adjust the rheostat until it shines with the same brightness as it did when connected to the a.c. You may need to switch from one to the other a number of times to achieve this.
- When this is happening the heating effect of both the a.c. and the d.c. are the same.
- Measure the p.d. across the bulb due to the d.c. source.
- Multiply the d.c. voltage by $\sqrt{2}$. It will be approximately equal to the peak value of the a.c. voltage.

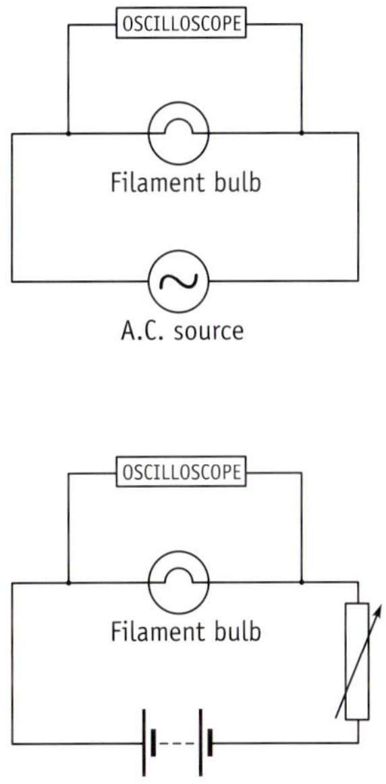

Fig. 28.24

EXERCISE 28.4

1. The peak value of an a.c. voltage is 20 V. Calculate its rms voltage.
2. The rms value of an a.c. voltage is 20 V. Calculate its peak value.
3. Domestic a.c. is supplied at 230 V. What is the maximum value that this voltage has in a cycle?
4. Alternating current passes through a resistor, where $I_{rms} = 2$ A and $V_{rms} = 110$ V. Calculate the power dissipated in the resistor.

5. An alternating current flowing in a wire of resistance 20 Ω causes heat to be produced in the wire at the rate of 500 J s^{-1}. Find the rms value of the current (i.e. the equivalent d.c. value). Find also:
   (i) the rms voltage,
   (ii) the peak voltage of the a.c. source.
6. The instantaneous maximum value of an alternating current is 3 A. Find:
   (i) the rms current,
   (ii) the heat produced per second in a 200 Ω resistor by the current,
   (iii) the rms voltage across the resistor,
   (iv) the peak value of the voltage.
7. The peak value of the emf from an a.c. source is 520 volts. The maximum value of the current is 3 A. How long does the source take to expend 2 kJ of electrical energy in a resistor?
8. A 400 turn coil has a resistance of 200 Ω and is connected to an a.c. supply. Over a 1 ms time interval the flux through the coil increases by 5×10^{-4} Wb. Find the average induced emf over the 1 ms interval. Find also the average current in the coil if the average applied voltage over the 1 ms time interval is 300 V.

Mutual Induction

A changing electric current in a coil produces a changing magnetic field in the space around it. If another coil is placed in this changing magnetic field an emf is induced in this coil. If this happens we say there is **mutual induction** between the two coils.

MUTUAL INDUCTION

If a changing magnetic field in one coil causes an induced emf to appear in a nearby coil there is said to be ***mutual induction*** between the two coils.

The bigger the induced emf for a given **rate of change of magnetic flux** in the first coil the greater the mutual induction. The mutual induction between two coils depends on how well the magnetic field from one coil links the other coil.

The **size of the induced emf and hence the mutual induction may be increased by**:

- having the coils nearer each other,
- winding the coils on the same soft iron core,
- increasing the number of turns on either or both of the coils.

Experiment

To show Mutual Induction.

Method

- Use the equipment in Fig. 28.2 (page 310).
- When the switch is either opened or closed the current in coil 1 changes and thus the magnetic field around it changes. This changing magnetic field passes through coil 2.
- As this happens, the galvanometer in coil 2 gives a deflection showing that an emf is induced in coil 2.
- There is thus mutual induction between the two coils.

If a steady current flows in coil 1, no emf is induced in coil 2. It is only when the current in coil 1 **changes** that an emf is induced in coil 2. If both coils are wound on the same soft iron core, the size of the induced emfs are much larger due to the increased mutual induction.

Mutual induction occurs in the **transformer** (page 325) and the **induction coil** (page 384).

SELF INDUCTION

EXPERIMENT

TO SHOW SELF INDUCTION.

Method

- Set up the equipment shown in FIG. 28.25.
- Close the switch.
- After the switch is closed, it will be found that the bulb does not light immediately. It takes a number of seconds for the bulb to reach full brightness. This is due to the self induction occuring in the coil which is explained below.

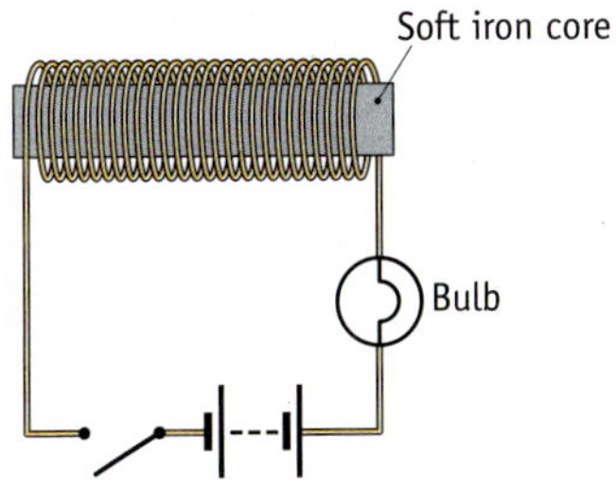

Fig. 28.25

Explanation

- When the switch is closed the current starts to flow and immediately produces a magnetic field around the coil. This field is increasing.
- Since the coil now has a changing magnetic field in it, by Faraday's Law an emf will be induced in the coil.
- By Lenz's Law the direction of the emf is such as to oppose the change producing it, i.e. it opposes the increasing current.
- The induced emf opposes but does not succeed in preventing the current from increasing. Such an emf is called a **back emf.** The result is that it delays the build up of the current. This phenomenon is called **self induction**.

SELF INDUCTION
Whenever the current passing through a coil changes, the magnetic field surrounding that coil changes. This changing magnetic field induces an emf in the coil that opposes the changing current (i.e. a back emf). This phenomenon is called ***self induction***.

The size of the back emf may be increased by winding the coil on a soft iron core. Thus the amount of self induction in a coil is increased if it is wound on a soft iron core. A coil which has the property of self induction is often called an **inductor**.

Another effect of self induction may be shown with the circuit in FIG. 28.26, which consists of a 12 V battery, a switch, a coil with a soft iron core and a neon lamp. The neon lamp will only light when a p.d. of at least 90 V is put across it.

When the switch is opened the current flowing through the coil is suddenly reduced to zero and the magnetic field around it rapidly disappears. As it does, an emf is induced in the coil. This emf is large enough to cause the neon bulb to flash because:

- the magnetic field decreases to zero very rapidly,
- the coil has a large number of turns,
- the coil has a soft iron core.

If the core were removed from the coil, the emf induced in it may not be large enough to light the neon bulb.

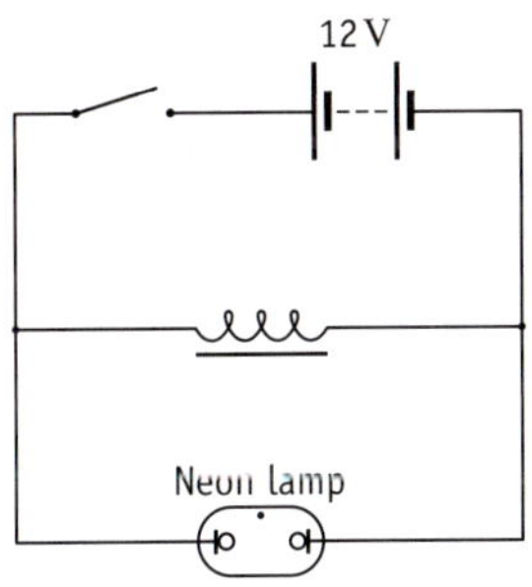

Fig. 28.26

Problem 12: A coil of 400 turns is carrying a direct current which causes a flux of 2.3×10^{-2} Wb to pass through the coil. The current is suddenly switched off causing the flux to become zero in 0.01 seconds. Find the emf induced in the coil.

Solution: Induced emf $= N(\text{Rate of change of flux}) = \dfrac{(400)(0 - 2.3 \times 10^{-2})}{0.01} = 920$ V

a.c. AND INDUCTORS

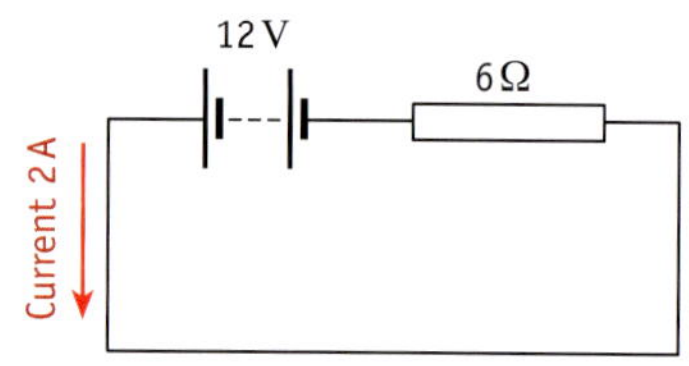

Fig. 28.27

In FIG. 28.27 the steady d.c. flowing is found from Ohm's Law. In FIG. 28.28 the battery is replaced by an equivalent 12 volt a.c. source (i.e. $V_{rms} = 12$ V) and the resistor by a coil of resistance 6 Ω with a soft iron core through it. It is now found that a current flows, but it is much less than 2 A. The reason for this is as follows:

An alternating current is continually changing in size. As it changes, the magnetic field around the coil also changes and thus by Faraday's and Lenz's Laws **an emf is induced in the coil which always opposes the changing current**. It is this back emf that causes the coil to offer more opposition to a.c. than to d.c.

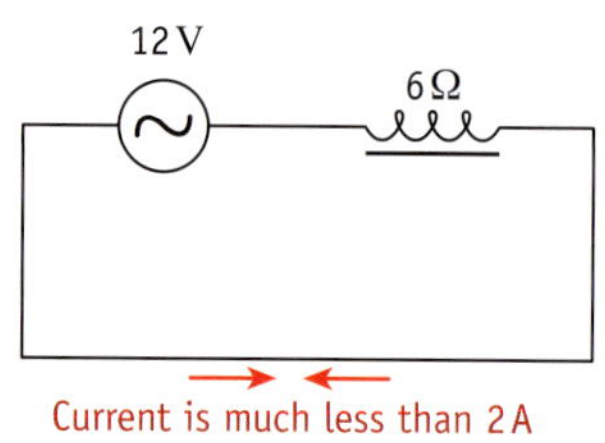

Fig. 28.28

A coil (an inductor) ***opposes the flow of direct current (d.c.)*** with its ohmic resistance.
A coil (an inductor) ***opposes the flow of alternating current (a.c.)*** with its ohmic resistance and the back emf induced in it.

The larger the amount of self induction in the coil, the greater the opposition it offers to a.c. This is because the increased self induction causes a larger rate of change of magnetic flux in the coil and hence a larger back emf, which decreases the current. In FIG. 28.29 the lamp is lighting. If a soft iron core is inserted in the coil, the brightness of the lamp decreases or the lamp stops lighting due to the decreased current.

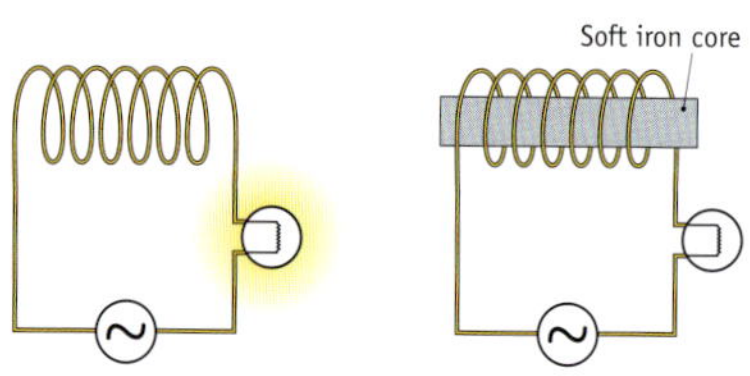

Fig. 28.29

It should also be clear that the greater the frequency of the a.c., the greater the opposition an inductor has to it, since the greater the frequency the greater the rate of change of magnetic flux.

USES OF INDUCTORS

Inductors are used:

- to smooth out slight variations in d.c. in power supply units,
- in the tuning circuits of radios to tune to different stations,
- in dimmer switches used in stage lighting.

EXERCISE 28.5

1. A filament lamp and a coil with an iron core are connected in series with an a.c. power supply of variable frequency and fixed voltage. If the frequency is increased, what happens to the brightness of the lamp? Why does this happen?
2. A coil of wire of resistance 10 Ω has a d.c. voltage of 20 V across its ends. What steady d.c. current flows in the coil? If the d.c. source is removed and replaced with a 20 V a.c. source what changes if any will take place in the current? Why?
3. Explain the terms:
   (i) self induction, (ii) mutual induction.

4. The circuit in Fig. 28.30 is set up. The value of the voltage is adjusted so that the bulb just lights up. Explain the effect on the brightness of the bulb of:
   (i) increasing the p.d. of the source,
   (ii) closing switch S_1,
   (iii) placing a soft iron core in the coil,
   (iv) increasing the frequency of the a.c. source,
   (v) replacing the a.c. source with a d.c. source of the same voltage.

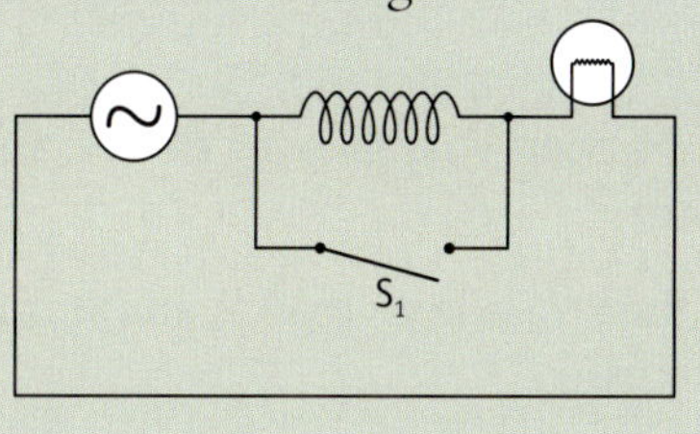

Fig. 28.30

5. In Fig. 28.31 why does the bulb flash briefly if S_1 is opened after having been closed? Where else in the circuit might the bulb be connected to bring about the same effect?

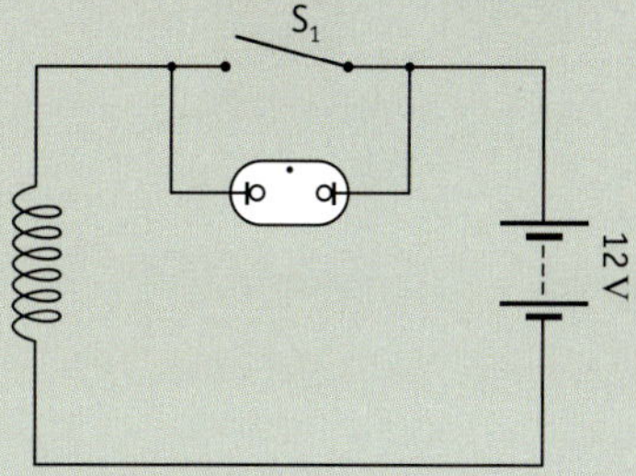

Fig. 28.31

6. The resistance of the lamp in Fig. 28.32 is 40 Ω. The resistance of the rest of the circuit is negligible. If the current flowing in the coil is 100 mA, find the emf induced in the coil.

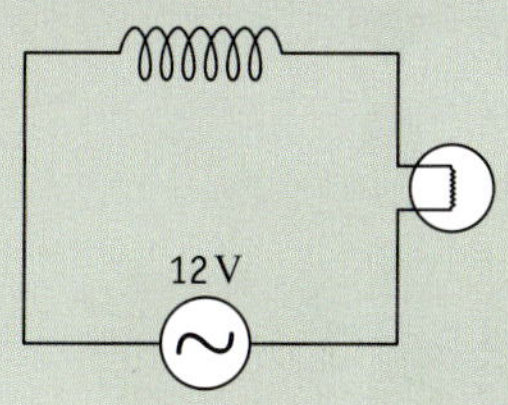

Fig. 28.32

Capacitors and a.c.

If the circuit in Fig. 28.33(A) is set up and the switch closed, current flows for a short while and the capacitor charges up. When the capacitor is charged no more current flows. Thus **a charged capacitor blocks d.c.**

If the circuit in Fig. 28.33(B) is set up, alternating current will flow. This is because as the a.c. changes direction, the capacitor continually charges and discharges. It can be shown that the greater the capacitance of the capacitor the less opposition it offers to a.c.

> A ***charged capacitor blocks d.c.*** A ***capacitor conducts a.c.*** since it charges and discharges as the a.c. changes direction.

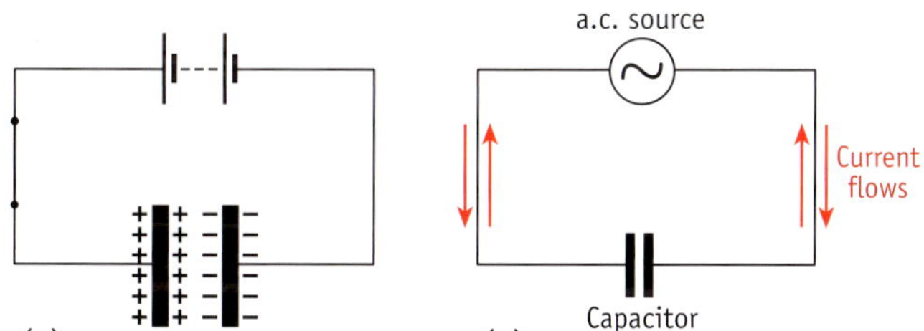

Fig. 28.33

(A) Capacitor charges and then no further current flows.

(B) The capacitor allows a.c. to flow.

The Transformer

You saw on page 248 that mains electricity is a.c. You saw on page 277 that the voltage of mains electricity is increased to a very high value when the electricity is being transported around the country on the national grid. This is because at high voltages the I^2R heat losses in the wires are minimised. a.c. is used rather than d.c. because it is very easy to change the value of an alternating voltage with a device called a **transformer**.

> **THE TRANSFORMER**
> A ***transformer*** is a device used to change the value of an alternating voltage.

Fig. 28.34 on the next page shows the structure of a transformer. It consists of two coils of wire wound on a soft iron core. One coil, called the **primary coil** has an alternating voltage applied to it. The transformer causes a different voltage to appear across the other coil, called the **secondary coil**. The voltage across the secondary coil is the output voltage.

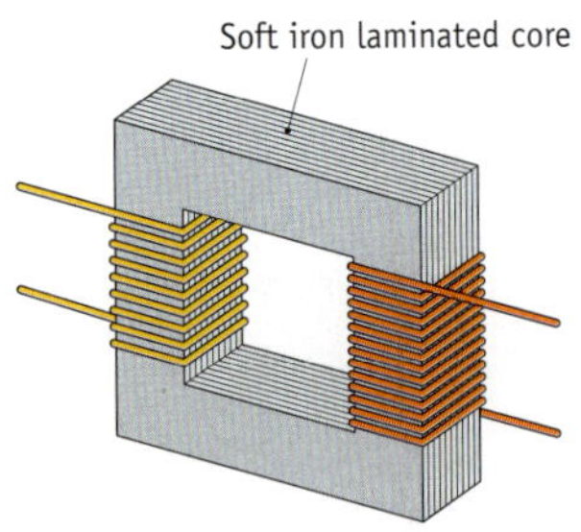

Primary coil
No. of turns = N_p
Input voltage = V_i

Secondary coil
No. of turns = N_s
Output voltage = V_o

(A)
A transformer.

(B)
A demonstration transformer for a school laboratory.

Fig. 28.34

The transformer operates as follows:

- The input voltage V_i across the primary coil causes alternating current to flow in the primary coil.
- This current causes an alternating magnetic flux in the iron core.
- This alternating flux passes through the secondary coil and induces an emf in it. This is the output voltage V_o.
- The size of the emf in the secondary V_o depends on the number of turns in the secondary N_s.
  Recall that induced emf $=$ N_s × (Rate of change of flux)
- Depending on the number of turns on the secondary, the output voltage can be less than, equal to, or greater than the input voltage.

Suppose N_p = Number of turns on the primary coil, then the following is true:

If N_s is greater than N_p then V_o is greater than V_i and it is called a **step up transformer**.

If N_s is less than N_p then V_o is less than V_i and it is called a **step down transformer**.

It can be shown that: $\dfrac{V_i}{V_o} = \dfrac{N_p}{N_s}$

If there are no energy losses in the transformer the power in must equal the power out thus:

$$V_i I_p = V_o I_s$$

Problem 13: A transformer with 100 turns on the primary and 5000 turns on the secondary has its primary connected across a 220 volt a.c. supply. Find the voltage across the secondary assuming no energy losses in the transformer.

Solution: $\dfrac{V_i}{V_o} = \dfrac{N_p}{N_s} \Rightarrow = V_o = \dfrac{V_i\,N_s}{N_p} = \dfrac{(220)(5000)}{100} = 11\ 000\text{ V}$

EXPERIMENT

To Demonstrate the Action of a Transformer

Method

- Set up the equipment as in FIG. 28.35.
- With an a.c. voltmeter, measure the input voltage across the primary and the output voltage across the secondary coil.
- It will readily be seen that if $N_s > N_p$ then $V_i > V_o$ and vice versa.
- By noting the number of turns on each coil the truth of the formula $\dfrac{V_i}{V_o} = \dfrac{N_p}{N_s}$ can be verified.

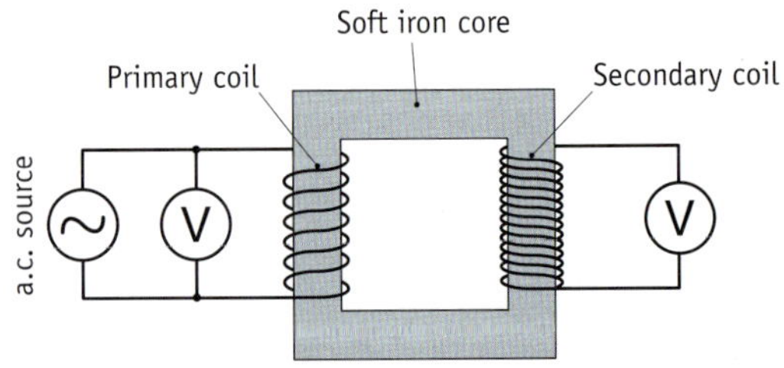

Fig. 28.35

USES OF TRANSFORMERS

- Generating stations usually generate electricity at between 20 kV and 30 kV. This value is stepped up to 220 kV or 400 kV for distribution on the national grid. Its value is reduced to 230 V at substations, which is the value it is in houses. All changes in voltages are made with transformers.
- Computers, radios, televisions and similar equipment all contain transformers to supply the necessary voltages to their various parts. A television has both step up (to provide the necessary high voltage for the cathode ray tube – page 328) and step down (to power other electrical components in it) transformers.

Fig. 28.36
A charger for a mobile phone contains a small transformer and a rectifier.

EXERCISE 28.6

1. A transformer has 500 turns of wire in its primary coil and 100 turns in the secondary coil. Calculate the output voltage across the secondary if the primary is connected to an a.c. power supply of 230 V.
2. The primary coil of a transformer has 2000 turns and the secondary has 100 turns. If the voltage across the secondary is 4 V, what is the voltage connected across the primary?
3. A step down transformer converts 3000 V to 220 V. If the number of turns on the secondary is 60, find the number on the primary.
4. A step down transformer converts 4000 V to 220 V. If the number of turns on the primary is 10 000, find the number on the secondary. If the current drawn from the secondary is 5 A, find the primary current if:
   (i) the energy losses in the transformer are negligible,
   (ii) 90% of the energy input is available at the output.
5. What is likely to happen if a transformer designed to operate on 220 V a.c. has its primary coil connected to a 220 V d.c. supply?

CHAPTER CHECKLIST

- **Define**: Electromagnetic Induction; Magnetic Flux;
  Mutual Induction; Self Induction.
- **State**: The unit of magnetic flux; Faraday's Law of Electromagnetic Induction; Lenz's Law; What an electric generator is; What a transformer is.
- **Recall** that: A coil (an inductor) opposes a.c. with both its ohmic resistance and the back emf induced in it; A capacitor blocks d.c. but conducts a.c; Electromagnetic induction occurs in a generator; Mains electricity (from the national grid) is a.c.
- **Draw**: A graph of an a.c. voltage or current against time; A graph of d.c. voltage or current against time; A labelled diagram of a transformer.
- **Explain**: Induced emf; Induced current; Alternating current; Step up transformer; Step down transformer; Primary coil; Secondary coil.
- **Describe** an experiment to: Show electromagnetic induction; Demonstrate Faraday's Law; Demonstrate Lenz's Law; Show a.c. using an oscilloscope; Show the effect of an inductor on a.c; Show that a capacitor conducts a.c. but not d.c; Demonstrate the action of a transformer; Compare peak and rms values of a.c;
  Show mutual induction; Show self induction.
- **Recall** and use the following formulae to solve problems: $V_{rms} = \dfrac{I_0}{\sqrt{2}}$; $I_{rms} = \dfrac{V_0}{\sqrt{2}}$

  $\dfrac{V_i}{V_o} = \dfrac{N_p}{N_s}$; $\Phi = BA$; $\text{Average induced emf} = \dfrac{\text{Final flux} - \text{Initial flux}}{\text{Time taken}}$; $E = -\dfrac{d\Phi}{dt}$
- **List:** Practical applications of generators; Practical uses of inductors; Practical uses of transformers.

CHAPTER 29

The Electron

PROPERTIES OF THE ELECTRON

On page 222 you saw that every atom has a dense central part called a nucleus around which electrons orbit. Electrons have the following properties:

- An electron is a **particle that orbits the nucleus** of an atom.
- An electron has a **very small mass**. Its mass is 9.1×10^{-31} kg.
- An electron is **negatively charged**. The charge on the electron is 1.6×10^{-19} coulombs. This charge is usually represented by *e*. Thus $e = 1.6 \times 10^{-19}$ C.
- The charge on the electron is the **smallest amount of charge found in nature**. It is the indivisible quantity of charge. All other amounts of electric charge are a whole number of times the charge on the electron.

The word electron was first used by the Irish physicist **GJ Stoney** for the fundamental amount of charge found in electrolysis. The value of the charge on the electron was first measured by the American **Robert Millikan** in 1911 in his famous Oil Drop Experiment. He found that the charge on an electrostatically charged oil drop was 1.6×10^{-19} C or some whole number of times this value. He concluded that this was the amount of charge on the electron.

THERMIONIC EMISSION
Thermionic Emission **is the giving off of electrons from the surface of a hot metal.**

THERMIONIC EMISSION

In a metal, some electrons are not stuck to any particular atom but are free to wander throughout the metal. They move in random directions with various speeds. The higher the temperature the faster their average speed. At room temperature these electrons do not have enough speed to escape from the attractive forces holding them in the metal. (If an electron left the metal there would be a positively charged atom left behind which would attract it back.) If the temperature is raised high enough by heating the metal (usually to around 800°C), some electrons get enough energy to escape from the surface of the metal. This phenomenon – the emission of electrons from the surface of a hot metal – is called **thermionic emission.**

Thermionic emission can be used to produce a beam of electrons in an evacuated glass tube. Such a tube is called a cathode ray tube.

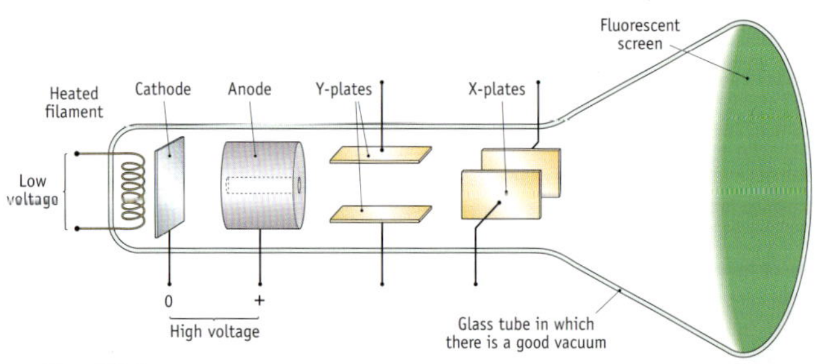

Fig. 29.1
A cathode ray tube.

THE CATHODE RAY TUBE

FIG. 29.1 shows a simple cathode ray tube. It consists of:

- a glass tube from which most of the air has been removed, i.e. there is a good **vacuum** in the tube,
- a thin wire called the **filament** through which a small electric current flows. The filament acts as a **heater** and heats the cathode,

- two electrodes called the **cathode** and the **anode**. The anode has a hole in its centre. The anode is positive (+) with respect to the cathode,
- a fluorescent **screen**.
- two sets of parallel plates which can control the position where the beam of electrons strike the screen (see page 330).

The cathode ray tube operates as follows:

- A low voltage (e.g. 6 V) is placed across the filament. Current flows through it and it becomes white hot. It heats the cathode.
- **Thermionic emission** occurs at the cathode, i.e. electrons are emitted from the cathode.
- There is a large voltage (called the **anode voltage**) between the cathode and the anode. The anode is positive with respect to the cathode. A typical value would be 2000 V. The electrons emitted from the cathode accelerate to high speed as they move towards the anode due to this voltage. Since there is a high vacuum in the tube there are no gas molecules into which the electrons would collide. Thus the motion of the electrons is not opposed.
- Many of the electrons that reach the anode pass through the hole in it and travel along to the end of the tube, i.e. a **beam of electrons** travels from the anode to the end of the tube.
- The end of the tube is coated with a layer of **fluorescent material** (also called phosphors). This is the **screen**. When an electron strikes the screen most of its kinetic energy is converted into light and a **bright spot of light** is produced on the screen.

CATHODE RAYS
Streams of high speed electrons moving from the cathode are called ***cathode rays***.

The streams of high-speed electrons moving from the cathode are called **cathode rays**.

Cathode rays have the following properties:

- They travel from the cathode in fairly straight lines.
- They cause certain substances to give out light when struck by them e.g. zinc sulphide. Such substances are used as the fluorescent coating on the screen in a cathode ray tube.
- They have kinetic energy.
- They can be deflected in electric and magnetic fields. This shows that they are charged particles.
- They can produce X-rays when they strike a metal target (page 340).
- They are invisible but their presence can be detected in a tube by making them strike fluorescent material. In Fig. 29.2 the beam can be seen as it passes the fluorescent sheet because some of the electrons in the beam strike the sheet and cause light to be emitted.

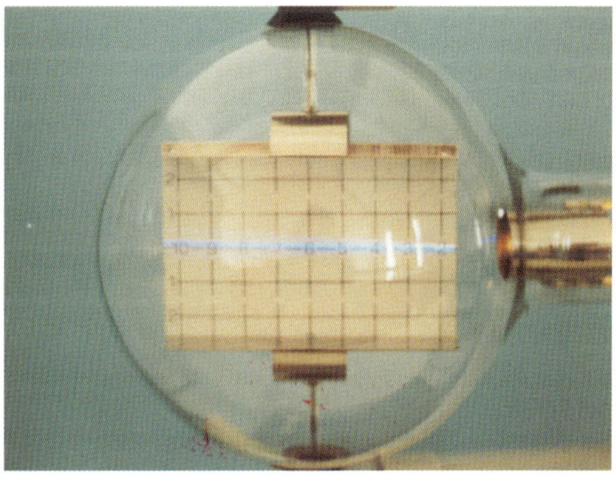

Fig. 29.2
A beam of electrons in a cathode ray tube.

Deflection of a Beam of Electrons in an Electric Field

In Fig. 29.3 a beam of electrons passes between two parallel plates. The plates carry opposite charge. Since the electrons are negatively charged they are attracted to the positive plate and repelled from the negative plate. Thus the beam deflects towards the positive plate. This can be

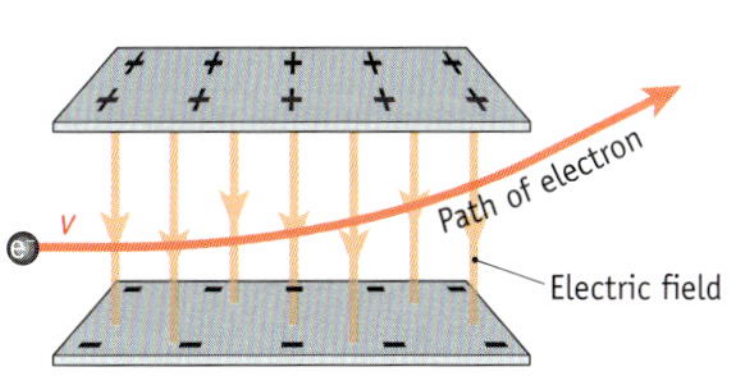

Fig. 29.3

seen in the laboratory with a suitable cathode ray tube (FIG. 29.4). Here a high voltage is placed across the parallel plates and the beam deflects. If the p.d. across the plates is reversed, the beam will deflect upwards.

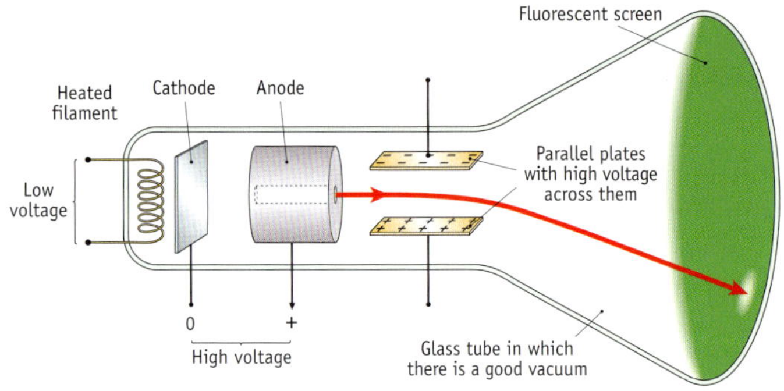

Fig. 29.4
Deflection of a beam of electrons in an electric field.

The larger the p.d. the more the beam deflects.

In the cathode ray tube in FIG. 29.1, the position of the spot on the screen is controlled by the voltages put on the X and Y plates. The Y-plates control the vertical positioning and the X-plates the horizontal positioning.

ENERGY OF ELECTRONS IN AN ELECTRIC FIELD

FIG. 29.5 shows an electric field. If an electron is released at A, the field exerts a force on it. The electron accelerates in the direction shown. As it moves it loses potential energy and gains kinetic energy, the loss being equal to the gain.

If the electron is projected against the field (i.e. from A to C) it loses kinetic energy and gains potential energy as it slows down. Again the loss is equal to the gain. You saw on page 235 that when a charge Q moves through a voltage V the work W done (i.e. the energy lost or gained) is given by: $W = QV$. For an electron this becomes $W = eV$. Thus we have the following:

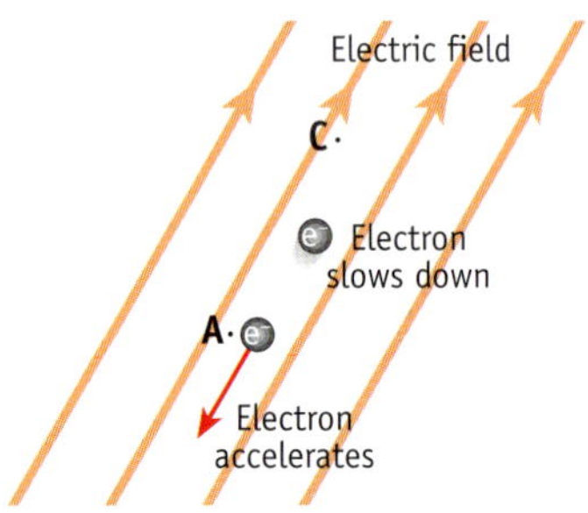

Fig. 29.5

	Loss in potential energy	=	Gain in kinetic energy
or	Gain in potential energy	=	Loss in kinetic energy
	in either case ***eV***	=	$\frac{1}{2}mv^2$

where: e = charge on electron = 1.6×10^{-19} C and
m = Mass of electron = 9.1×10^{-31} kg.

Use these values in the following problems.

Problem 1: An electron is accelerated from rest through a voltage of 3000 V. Find:
(i) the electrical potential energy it loses,
(ii) the kinetic energy it gains,
(iii) the speed it acquires.

Solution:
(i) E_p lost $= eV = (1.6 \times 10^{-19})(3000) = 4.8 \times 10^{-16}$ J
(ii) E_k gained $= E_p$ lost $= 4.8 \times 10^{-16}$ J
(iii) $E_k = \frac{1}{2}mv^2 = \frac{1}{2}(9.1 \times 10^{-31})v^2 = 4.8 \times 10^{-16}$

$$\Rightarrow v^2 = \frac{(2)(4.8 \times 10^{-16})}{(9.1 \times 10^{-31})} \Rightarrow v = \sqrt{1.0549 \times 10^{15}} \text{ i.e. } v = 3.2 \times 10^7 \text{m s}^{-1}$$

Problem 2: An electron strikes the screen of a cathode ray tube with a speed of 2×10^7 m s^{-1}. What is the voltage across the tube?

Solution:. Let V be the voltage across the tube. Then:

$$eV = \tfrac{1}{2}mv^2 \Rightarrow 1.6 \times 10^{-19} V = \tfrac{1}{2}(9.1 \times 10^{-31})(2 \times 10^7)^2 \Rightarrow V = 1137.5 \text{ V}$$

THE ELECTRONVOLT

When measuring the energy of electrons or other small particles, the joule is awkwardly large. Another unit, called the **electronvolt**, is often used instead. Its definition is:

THE ELECTRONVOLT
The ***electronvolt* (eV)** is the amount of energy gained or lost by an electron when it moves through a potential difference of one volt.

When a charge Q moves through a voltage V, the work done (i.e. energy lost or gained) is given by $W = QV$. Thus if an electron of charge e moves through a p.d. of 1 volt then:

Work done: $W = QV = (e)(1) = (1.6 \times 10^{-19})(1) = 1.6 \times 10^{-19}$ J.

Thus:

1 electronvolt = 1.6×10^{-19} joules   1 eV = 1.6×10^{-19} J

Larger amounts of energy are expressed in: keV, MeV and GeV, where:

1 kiloelectronvolt	=	**1 keV**	=	1000 eV	=	10^3 eV
1 megaelectronvolt	=	**1 MeV**	=	1 000 000 eV	=	10^6 eV
1 gigaelectronvolt	=	**1 GeV**	=	1 000 000 000 eV	=	10^9 eV

See your Maths tables page 5.

Problem 3: A particle has a kinetic energy of 4 keV. Express this energy in joules.

Solution: 4 keV = 4000 eV = $(4000)(1.6 \times 10^{-19}) = 6.4 \times 10^{-16}$ J

Problem 4: An electron has an energy of 6×10^{-18} J. What is its energy in electronvolts?

Solution: $1 \text{ eV} = 1.6 \times 10^{-19} \text{ J} \Rightarrow 1 \text{ J} = \dfrac{1}{(1.6 \times 10^{-19})} \text{ eV}$

$\therefore\ 6 \times 10^{-18} \text{ J} = \dfrac{(6 \times 10^{-18})}{(1.6 \times 10^{-19})} \text{ eV} = 37.5 \text{ eV}$

DEFLECTING A BEAM OF ELECTRONS IN A MAGNETIC FIELD

A beam of electrons is a flow of charge and is thus an electric current. It will therefore experience a force when placed in a magnetic field (page 304). This can easily be demonstrated by placing a bar magnet near the cathode ray tube shown in FIG. 29.2 (page 329). FIG. 29.6 is the result and the beam of electrons deflects according to Fleming's left-hand rule.

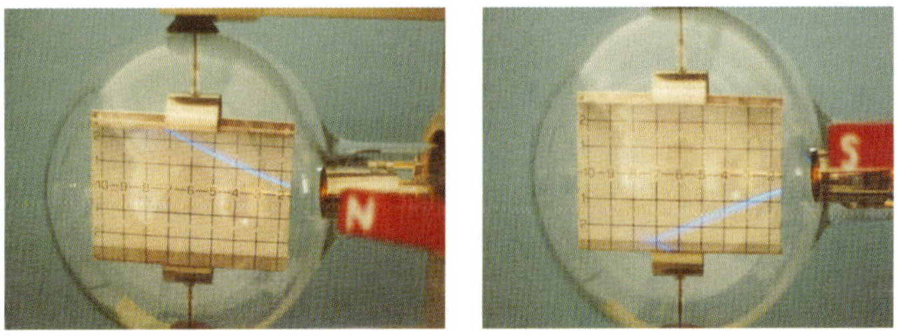

(A) (B)

Fig. 29.6
Deflecting a beam of electrons in a magnetic field.

A BEAM OF ELECTRONS IN A UNIFORM MAGNETIC FIELD MOVES IN A CIRCLE

FIG. 29.7 shows a beam of electrons moving at right angles to a uniform magnetic field. By Fleming's left-hand rule the force on the electrons is as shown. The force is perpendicular to their direction of motion and they deflect. As they move you can see that the force is always perpendicular to the direction of motion. Thus the speed of the electrons does not change.

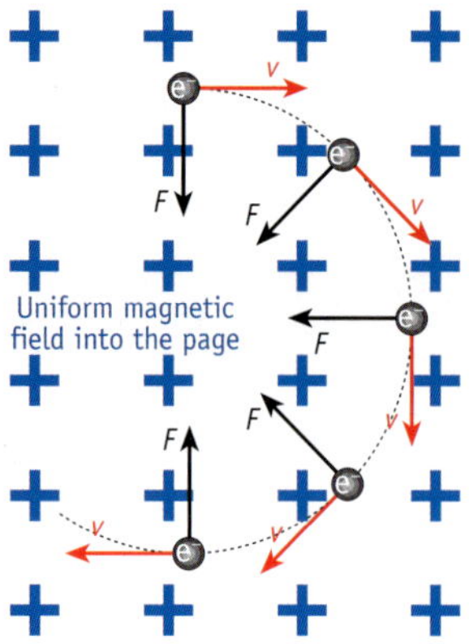

Fig. 29.7
A beam of electrons moving at right angles to a magnetic field moves in a circle.

When a charged particle of charge q moves at speed v at right angles to a magnetic field of flux density B it experiences a force F given by: $F = qvB$ (page 305). This formula gives us the force on an electron in a magnetic field. Since the speed of the electrons does not change, the force on them is of constant magnitude.

A force of constant magnitude always acting at right angles to the direction of motion is exactly the condition needed for an object to move in a circle (page 140). Thus:

A beam of electrons moving at right angles to a magnetic field moves in a circle (FIG. 29.7).

In the following problems $e = 1.6 \times 10^{-19}$ C *and mass of electron* $= 9.1 \times 10^{-31}$ kg.

Problem 5: What is the force on an electron moving with a speed of 4×10^6 m s^{-1} at right angles to a magnetic field of flux density 5 T?

Solution: $F = qvB = (1.6 \times 10^{-19})(4 \times 10^6)(5) = 3.2 \times 10^{-12}$ N

Problem 6: FIG. 29.8 shows an electron beam moving into a uniform magnetic field. In what direction will the electron beam deflect in the field?

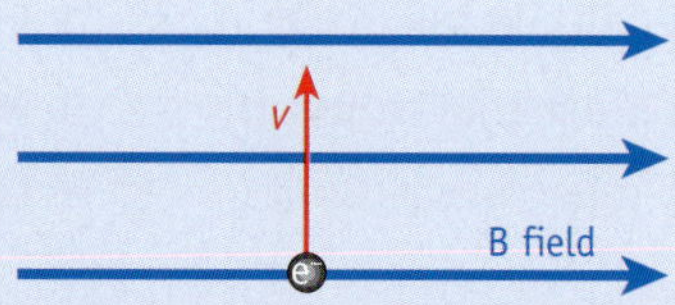

Fig. 29.8

Solution: Negative charges moving up the page correspond to a conventional current moving down the page. Using Fleming's left-hand rule we see that the beam of electrons deflects out of the page towards us.

Problem 7: An electron moving at a speed of 4×10^7 m s^{-1} enters a uniform magnetic field moving at right angles to the field. The electron then moves in a circular path. If the flux density of the field is 1.5×10^{-2} T, calculate the radius of the path.

Solution: Centripetal force on electron = Force on electron due to magnetic field

$$\frac{mv^2}{r} = evB \Rightarrow r = \frac{mv}{Be} = \frac{(9.1 \times 10^{-31})(4 \times 10^7)}{(1.5 \times 10^{-2})(1.6 \times 10^{-19})} = 1.52 \times 10^{-2}\text{m}$$

APPLICATIONS OF A CATHODE RAY TUBE

Fig. 29.9
Cathode ray tubes for televisions being manufactured.

One of the most common applications of a cathode ray tube is the tube of a **television** or a **computer monitor** (FIG. 29.9). The picture you see is produced by the beam of electrons hitting the fluorescent screen at the front of the tube. The beam of electrons rapidly scans the screen causing light of various intensities to be emitted thus producing a picture. In a colour TV there are three beams of electrons and three different types of phosphor on the screen to produce the three primary colours (page 217). In a TV, the beam of electrons is deflected by magnetic fields produced by current-carrying coils.

The cathode ray oscilloscope (CRO) is a cathode ray tube which is used to display electrical signals. The deflection of the beam is usually done by X and Y-plates. In such a tube the spot of light produced by the electrons scans the screen horizontally at regular time intervals. By placing a varying voltage on the Y-plates its shape can be seen on the screen.

In medicine the CRO is used in an **ECG (electrocardiogram)** to display electrical signals in the heart. The regular contractions of the heart's muscle are controlled and preceded by electrical impulses that spread through the heart muscle. These signals can be displayed in an ECG (Fig. 29.10). This can be used to help diagnose various heart disorders.

A CRO is also used in medicine in an **EEG (electroencephalogram)**. The small varying electrical signals produced in the brain can be displayed on an EEG. Certain features of the activity and health of the brain can be studied by this means.

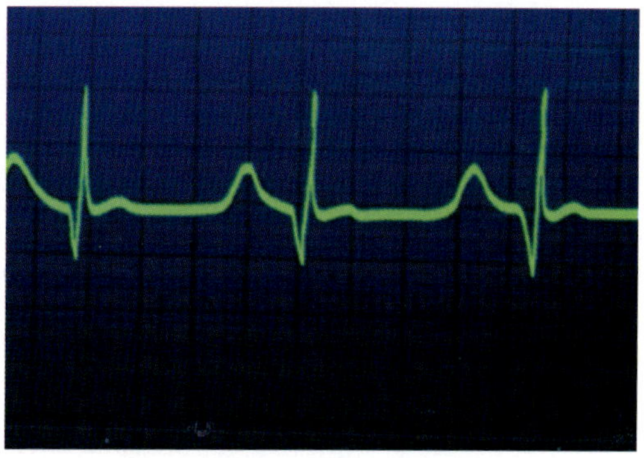

Fig. 29.10
An ECG (electrocardiogram) displays electric signals produced in the heart.

EXERCISE 29.1

In the following exercises e = 1.6×10^{-19} C *and mass of electron* = 9.1×10^{-31} kg

1. An electron is accelerated from rest through a voltage of 8000 volts. Find:
   (i) the electrical potential energy it loses,
   (ii) the kinetic energy is gains,
   (iii) the speed it reaches.
2. An electron is accelerated from rest through a voltage of 10 000 V. Find the speed it reaches.
3. An electron on striking the screen of a cathode ray tube has a speed of 1.5×10^7 m s^{-1}. What is the voltage across the tube?
4. Express each of the following in joules:
   (i) 5 eV (ii) 200 eV
   (iii) 40 keV (iv) 5 MeV
   (v) 40 GeV (vi) 4.2 eV
5. Express each of the following in eV:
   (i) 3 keV (ii) 6 MeV
   (iii) 2.5 GeV (iv) 1 J
   (v) 2×10^{-15} J (vi) 6.4×10^{-16} J
   (vii) 5.6×10^{-19} J
6. In Fig. 29.11 charged particles move in the paths shown in an electric field. What is the sign of the charge on each particle?

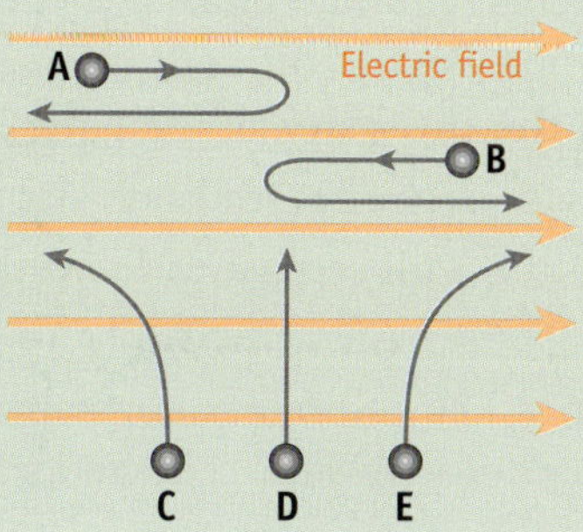

Fig. 29.11

7. What is the force on an electron travelling with a velocity of 2.1×10^6 m s^{-1} at right angles to a magnetic field of magnetic flux density 4.2 T?
8. An electron experiences a force of 2×10^{-12} N when moving with a uniform speed at right angles to a uniform magnetic field of flux density 3 T. Calculate the speed of the electron.
9. An electron travelling with a speed of 5.6×10^7 m s^{-1} enters a magnetic field of flux density 3×10^{-2} T moving perpendicular to the field. Find the radius of the circular path the electron follows in the field.
10. In Fig. 29.12 charged particles move in the paths shown in a magnetic field. What is the sign of the charge on each particle?

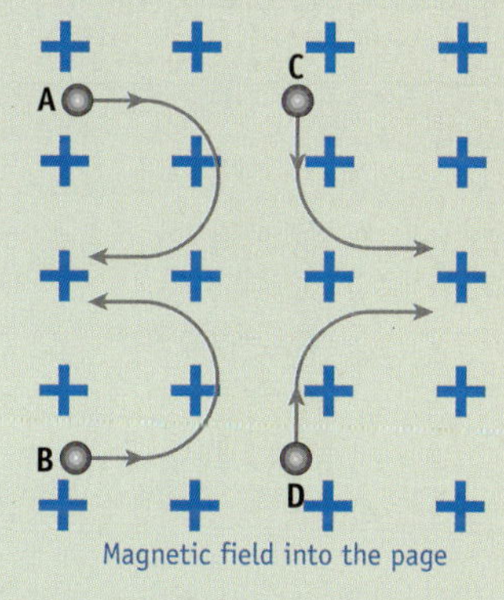

Fig. 29.12

11. Given that the charge on the electron is 1.6×10^{-19} C and the charge to mass ratio for the electron is 1.76×10^{11} C kg^{-1}, find the mass of the electron.

The Photoelectric Effect

THE PHOTOELECTRIC EFFECT
The ***photoelectric effect*** is the emission of electrons from the surface of a metal by electromagnetic radiation of a suitable frequency.

In 1887 Heinrich Hertz discovered that an electric spark discharge between two electrodes occurred more readily when ultravoilet light falls on one of the electrodes. In 1888 Hallwachs discovered that if ultraviolet light is shone onto a negatively-charged zinc plate, the plate loses its charge rapidly. Later it was shown that when the UV light falls on the negatively charged zinc plate, it causes electrons to be emitted from the zinc plate. This phenomenon is known as the **photoelectric effect**.

Experiment

To Show the Photoelectric Effect.

Method

- Use the equipment in Fig. 29.13.
- Place the zinc plate on the cap of the electroscope and charge the electroscope negatively (e.g. by induction).
- Shine UV light onto the zinc plate and notice that the gold leaf collapses quite quickly. If a sheet of glass is placed between the UV lamp and the zinc, the leaves do not collapse. UV light cannot get through the glass.
- Replace the UV source with visible light sources of various frequencies. The leaves will not collapse.

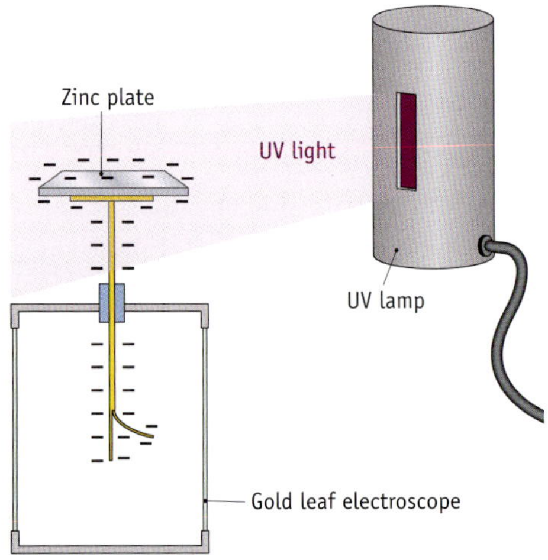

Fig. 29.13

Conclusion

Ultraviolet light causes electrons to be emitted from the zinc.

- Charge the electroscope positively and shine UV light on the zinc. The leaves will not collapse.

They do not collapse because any electrons emitted by the UV light are immediately attracted back by the positive charge on the zinc.

Photoemission occurs in zinc for UV light or X-rays but not for visible light. Photoemission will occur in sodium for some wavelengths of visible light as well as UV light and X-rays. Some other materials will show photoemission even for infra-red light.

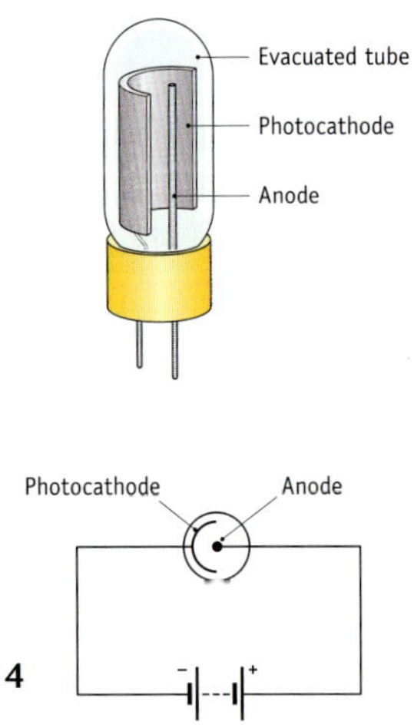

Fig. 29.14

The Photocell

Fig. 29.14 shows a photocell and its circuit symbol.

- A photocell (also called a **photoelectric cell**) is a device that conducts electric current when light of a suitable frequency shines on it. The size of the electric **current** it conducts is directly **proportional to the intensity** of the light falling on it.
- A photocell has two electrodes (cathode and anode), in a glass tube. The anode is positive with respect to the cathode.
- The cathode is called the **photocathode** and is usually semi-cylindrical in shape. The cathode is normally coated with material that will undergo photoemission. The anode is a rod running along the centre of the tube.
- The tube contains a **vacuum**, i.e. it is evacuated. Some tubes contain inert gas at very low pressure.

- If light of a suitable frequency strikes the photocathode, electrons (called **photoelectrons**) are emitted from it. These are attracted across the tube to the positive anode and a small current (a few microamperes) flows in the circuit. This current is called the **photocurrent**. Switch off the light and the current stops.

EXPERIMENT

To Demonstrate the Action of a Photocell.

Method

- Use the equipment shown in Fig. 29.15.
- Measure the distance d from the light source to the photocathode and measure the current I.
- Change the distance a number of times and measure both the distance and the current each time.
- Plot a graph of photocurrent against $\frac{1}{(\text{Distance})^2}$
- A graph like that in Fig. 29.16 will result showing that the photocurrent is directly proportional to $\frac{1}{d^2}$.
- It is a fact that the intensity of the light is inversely proportional to the square of the distance from the light source (i.e. doubling the distance makes the intensity one quarter of what it was etc.). Thus we have:
- photocurrent $\propto \frac{1}{d^2}$ and Intensity of light $\propto \frac{1}{d^2}$

  $\Rightarrow$ Photocurrent $\propto$ Intensity of light

The ***photocurrent*** is directly proportional to the ***intensity*** of the light.

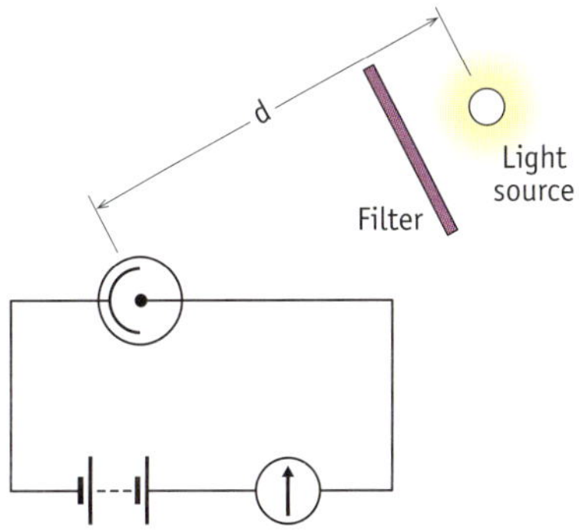

Fig. 29.15

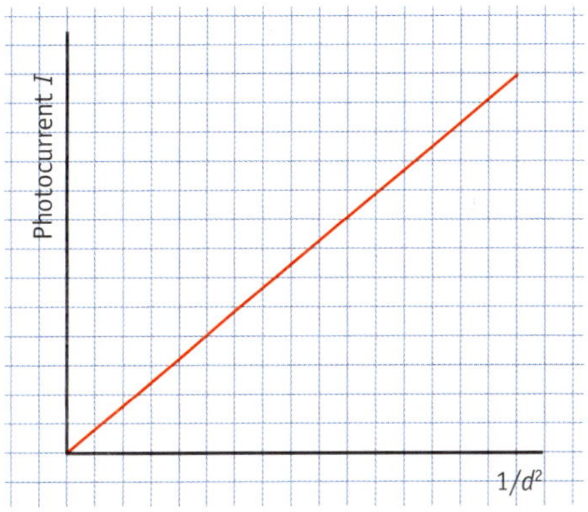

Fig. 29.16

A white light source emits light of a continuous range of frequencies. The frequency of the light striking the cathode can be controlled placing a coloured filter in front of the white light source. This blocks all but a very narrow band of frequencies. By changing the filter, the frequency of the light striking the cathode can be changed. Using equipment similar to that in Fig. 29.15 the following can also be shown:

For each metal there is a definite frequency (called the *threshold frequency*) below which no photoemission occurs no matter how intense the light.

For zinc this frequency is in the ultraviolet range. For some of the alkali metals it is in the visible and infra-red range.

Increasing the frequency above the threshold frequency does not change the size of the photocurrent. This means that the number of electrons being emitted per second from the photocathode does not depend on the frequency of the light. It depends only on the intensity as we saw above.

THRESHOLD FREQUENCY

For a given metal the frequency below which photoemission will not occur is called the ***threshold frequency***. Light of frequency above the threshold frequency will cause photoemission.

Around the end of the nineteenth century scientists could not explain the above aspects of the photoelectric effect. They knew that electrons in a metal are held there by certain forces and that to remove an electron, energy must be supplied to overcome these forces. The energy needed to remove the loosest electron is called the **work function** of the metal.

WORK FUNCTION
The ***work function*** ***(Φ)*** of a metal is the minimum energy needed to remove the loosest electron from the surface of that metal.

At the time, scientists assumed light was a continuous wave, its energy being proportional to its intensity. Thus if the light were bright enough or if it shone for long enough, it should supply enough energy to an electron to overcome the work function and cause the electron to be emitted. This does not happen. They could not explain why there was a threshold frequency.

EINSTEIN OFFERED THE FOLLOWING EXPLANATION:

- Light must be considered as a stream of 'packets of energy'. Each packet of energy is called a **photon** or a **quantum of energy**.
- The energy E of a photon depends only on the **frequency** f of the light. It is proportional to the frequency. The energy is given by the formula:

$E = hf$ where h is a constant called Planck's constant.

PHOTON
A ***photon*** is a packet of electromagnetic energy.
The energy E of a photon is given by $E = hf$, where f is its frequency and h is Planck's constant.

- **The brighter a light source the more photons it gives out per second.** Thus the greater the intensity of the light the more photons pass per second.
- Electrons in a metal are held there by certain forces. The energy needed to remove the loosest electron from the surface of a metal is the **work function** W of that metal.
- When light strikes the metal **an electron can only pick up the energy from one photon.**
- If the energy of each photon is less than the work function no electrons are emitted.
- If the energy of the photon is greater than the work function electrons are emitted.
- The amount by which the energy of the photon exceeds the energy needed to remove the electron appears as the kinetic energy of the emitted electron.

Using these ideas it is easy to see why, for a given metal, light of a certain frequency or higher is needed to cause photoemission and why the greater the intensity of the light the more electrons emitted per second and hence the greater the photocurrent.

Einstein's Photoelectric Law

It is also an experimental fact that:

- The velocities of the photoelectrons emitted by light of a frequency above the threshold frequency range from zero up to a definite maximum value.
- The maximum velocity (and hence the maximum kinetic energy) of the emitted electrons increases with the frequency of the light, but does not depend on the intensity.

EINSTEIN EXPLAINED THIS AS FOLLOWS:

The kinetic energy of the fastest electron emitted is the difference between the energy of the photon and the work function of the metal, i.e. $\frac{1}{2}mv^2_{max} = hf - \Phi$. Rearranging gives the following:

$$hf = \Phi + \tfrac{1}{2}mv^2_{max}$$

This equation is known as **Einstein's Photoelectric Law**.

- Electrons which are more tightly held in the metal will be emitted with a lesser speed. Also from the equation we see that as f increases so does v.
- If light of the threshold frequency f_o strikes the metal, an electron is just emitted, but it has no extra kinetic energy to move away from the metal. Obviously then the work function Φ is given by the following:

$$\Phi = hf_o$$

i.e. Work function = (Planck's constant)(Threshold frequency)

In the following problems take: $e = 1.6 \times 10^{-19}$ C; $h = 6.6 \times 10^{-34}$ J s; $c = 3 \times 10^8$ m s^{-1}

Problem 8: Calculate the energy of a photon of blue light of frequency 7.5×10^{14} Hz.

Solution: $E = hf = (6.6 \times 10^{-34})(7.5 \times 10^{14}) = 4.95 \times 10^{-19}$ J

Problem 9: Derive a formula for the energy of a photon in terms of its wavelength.

Solution: $c = \lambda f \Rightarrow f = \dfrac{c}{\lambda} \qquad E = hf \Rightarrow E = \dfrac{hc}{\lambda}$

Problem 10: What is the energy of a photon of red light of wavelength 6×10^{-7} m in: (i) joules (ii) eV?

Solution: $c = \lambda f \Rightarrow f = \dfrac{c}{\lambda} = \dfrac{3 \times 10^8}{(6 \times 10^{-7})} = 5 \times 10^{14}$ Hz

(i) $E = hf = (6.6 \times 10^{-34})(5 \times 10^{14}) = 3.3 \times 10^{-19}$ J

(ii) $3.3 \times 10^{-19}\text{ J} = \dfrac{(3.3 \times 10^{-19})}{(1.6 \times 10^{-19})}\text{ eV} = 2.0625$ eV

Problem 11: What is: (i) the frequency and (ii) the wavelength of a photon of energy 4 eV?

Solution: $4\text{ eV} = 4 \times 1.6 \times 10^{-19}\text{ J} = 6.4 \times 10^{-19}$ J

(i) $E = hf \Rightarrow f = \dfrac{E}{h} = \dfrac{(6.4 \times 10^{-19})}{(6.6 \times 10^{-34})} = 9.697 \times 10^{14}$ Hz

(ii) $c = \lambda f \Rightarrow \lambda = \dfrac{c}{f} = \dfrac{(3 \times 10^8)}{(9.697 \times 10^{14})} = 3.09 \times 10^{-7}$ m

Problem 12: A light source of frequency 4×10^{14} Hz has a power of 2 watts. Find the number of photons emitted per second by the source.

Solution: Energy of one photon $E = hf = (6.6 \times 10^{-34})(4 \times 10^{14}) = 2.64 \times 10^{-19}$ J

Number of photons emitted per second $= \dfrac{\text{Energy given out per second}}{\text{Energy of one photon}}$

$= \dfrac{2}{2.64 \times 10^{-19}} = 7.58 \times 10^{18}$ photons

Problem 13: The work function of a metal is 2 eV. What is the threshold frequency of this metal?

Solution: $2\ \text{eV} = (2)(1.6 \times 10^{-19})\ \text{joules} = 3.2 \times 10^{-19}\ \text{J} = \text{work function}$

$$\Phi = hf_o \Rightarrow f_o = \frac{\Phi}{h} = \frac{(3.2 \times 10^{-19})}{(6.6 \times 10^{-34})} = 4.85 \times 10^{14}\ \text{Hz}$$

Problem 14: The threshold frequency for a certain metal is 5.2×10^{14} Hz. What is the maximum kinetic energy of an electron emitted by light of frequency 9.6×10^{14} Hz?

Solution: $\frac{1}{2}mv^2_{\text{max}} = hf - hf_o$

$$= (6.6 \times 10^{-34})(9.6 \times 10^{14}) - (6.6 \times 10^{-34})(5.2 \times 10^{14}) = 2.9 \times 10^{-19}\ \text{J}$$

Problem 15: When light of wavelength 150 nm falls on a metal surface, the maximum kinetic energy of the emitted electrons is 5 eV. What is the work function of the metal in joules?

Solution: $\lambda = 150\ \text{nm} = 150 \times 10^{-9}\ \text{m} = 1.5 \times 10^{-7}\ \text{m}$

$\frac{1}{2}mv^2_{\text{max}} = 5\ \text{eV} = (5)(1.6 \times 10^{-19})\ \text{J} = 8 \times 10^{-19}\ \text{J}$

Frequency $f = \frac{c}{\lambda} = \frac{(3 \times 10^{8})}{(1.5 \times 10^{-7})} = 2 \times 10^{15}\ \text{Hz}$

Now: $\frac{1}{2}mv^2_{\text{max}} = hf - \Phi$ i.e. $8 \times 10^{-19} = (6.6 \times 10^{-34})(2 \times 10^{15}) - \Phi$

$\Rightarrow$ Work function $\Phi = 5.2 \times 10^{-19}\ \text{J}$

Problem 16: When light of frequency 3.2×10^{16} Hz falls on a certain photoelectric cell, the photocurrent is 0.22 mA. If each incident photon causes photoemission and all of the emitted electrons cross the cell, find:

(i) the number of photons striking the cathode per second,

(ii) the light energy falling on the cathode per second.

Solution: (i) $0.22\ \text{mA} = 0.22 \times 10^{-3}\ \text{A} = 0.22 \times 10^{-3}$ coulombs per second.

$$\text{Number of electrons crossing tube per second} = \frac{\text{Charge crossing per second}}{\text{Charge on one electron}}$$

$$= \frac{0.22 \times 10^{-3}}{1.6 \times 10^{-19}} = 1.375 \times 10^{15}\ \text{electrons}$$

One photon emits one electron $\Rightarrow$ number of photons striking cathode per sec $= 1.375 \times 10^{15}$

(ii) Energy of one photon $E = hf = (6.6 \times 10^{-34})(3.2 \times 10^{16}) = 2.112 \times 10^{-17}\ \text{J}$

Energy falling on cathode per second = (energy of one photon) × (number of photons)

$$= (2.112 \times 10^{-17})(1.375 \times 10^{15}) = 0.02904\ \text{J s}^{-1}$$

Fig. 29.17

Applications of Photoelectric Sensing Devices

When a beam of light falls on a photoelectric cell connected in a circuit, like Fig. 29.14 (page 334), a small current flows. If the beam of light is broken, the current stops. This current stopping can be used to trigger other circuits which perform some useful purpose.

Examples are:

- some types of **burglar alarm**.
- **automatic doors**.
- **counting items** on a conveyor belt.
- monitoring and controlling the flame in a central heating burner.

Some film projectors reproduce the **sound-track in a film** by means of a photocell. Fig. 29.17 shows a piece of a reel of film and the sound-track down the left-hand side of the film. As the film is shown, a small beam of light passes through the sound track and falls onto a photoelectric cell. As the brightness of the light varies due to variations in the width of the track, so does the current in the photoelectric cell. This current varies in the same way as the current in the microphone that recorded the sound originally. The current is amplified and fed to a loudspeaker producing the sound that was originally recorded. The brightness of the flame in a central heating burner can also be monitored by a photoelectric cell. The photocurrent produced can be used to control the supply of fuel.

EXERCISE 29.2

$h = 6.6 \times 10^{-34}$ J s; $c = 3 \times 10^{8}$ m s^{-1}; $e = 1.6 \times 10^{-19}$ C; mass of electron $= 9.1 \times 10^{-31}$ kg

1. Red light has a frequency of 4×10^{14} Hz. Find the energy of a photon of red light in:
(i) joules, (ii) electron volts.

2. Blue light has a frequency of 8×10^{14} Hz. Find the energy of a photon of blue light in:
(i) joules, (ii) electron volts.

3. Find the energy of a photon of light of wavelength 5×10^{-7} m.

4. What is the energy of a photon of wavelength 600 nm?

5. A sodium light source has a power of 10 W. If the wavelength of the light emitted is 590 nm, find:
(i) the frequency of sodium light,
(ii) the energy of a photon of sodium light,
(iii) the number of photons emitted by this source per second.

6. What is:
(i) the frequency and
(ii) the wavelength
of a photon of energy 2.2 eV?

7. The current in a photoelectric cell when illuminated by a monochromatic light source is 2 μA. How many photons strike the photocathode per second? Assume all incident photons cause photoemission and all the emitted electrons cross the cell.

8. The work function of a certain metal is 4 eV. Express the work function in joules. What is the threshold frequency of this metal?

9. Light of frequency 2×10^{15} Hz just causes photoemission on a certain metal. What is the work function of the metal:
(i) in joules, (ii) in electronvolts.

10. The work function of a certain metal is 1.2 eV. Express the work function in joules. What is the threshold frequency of this metal? What is the largest wavelength that causes photoemission from this metal?

11. The threshold frequency for a certain metal is 8.8×10^{14} Hz. What is the maximum kinetic energy of the electrons emitted from it by light of frequency 9.2×10^{14} Hz? Give your answer in (i) joules, (ii) in electron volts.

12. When light of wavelength 350 nm falls on a metal surface the maximum kinetic energy of the emitted electrons is 2 eV. What is the work function of the metal in joules?

13. When light of frequency 2×10^{16} Hz falls on a certain photoelectric cell, the photocurrent is 0.2 mA. If each incident photon causes photoemission and all of the emitted electrons cross the cell, find:
(i) the number of photons striking the cathode per second,
(ii) the light energy falling on the cathode per second.

14. Light of wavelength 250 nm falls on a metal which has a work function of 1.6 eV. What is the maximum kinetic energy of the emitted electrons? What is the maximum speed of an emitted electron?

15. Radio station 2 FM operates on a frequency of 92 MHz. What is the energy of a photon of electromagnetic radiation of this frequency? If the transmitter has a power of 2 MW, how many photons are emitted per second?

> **X-RAYS**
> ***X- rays*** are high frequency electromagnetic radiation produced when high speed electrons in a cathode ray tube strike a metal target that has a high melting point.

X-RAYS

DISCOVERY OF X-RAYS

X-rays were discovered accidentally in 1895 by **Wilhelm Röntgen**. While working with a gas discharge tube (a type of cathode ray tube) he found that a fluorescent screen some distance away gave out light when the tube was operating. He covered the discharge tube with black cardboard. This would not allow any visible light or UV light to pass from the discharge to the fluorescent screen. Light was still produced on the fluorescent screen when the tube was operating. He concluded that some form of radiation was produced where the cathode rays struck the glass of the discharge tube. This radiation passed through the cardboard and the surrounding space and struck the fluorescent screen causing it to give out light.

The radiation was found to be extremely penetrating and could pass through many metals, but was stopped by a sheet of thick lead. The nature of the radiation was not known at the time – hence it was called **X-radiation** or simply **X-rays**. X-rays were later shown to be very high frequency (very short wavelength) **electromagnetic radiation** produced when high-speed electrons collide with an anode or the glass walls of a cathode ray tube. Nowadays X-rays are produced by causing high speed electrons in a cathode ray tube to strike a metal target that has a high melting point.

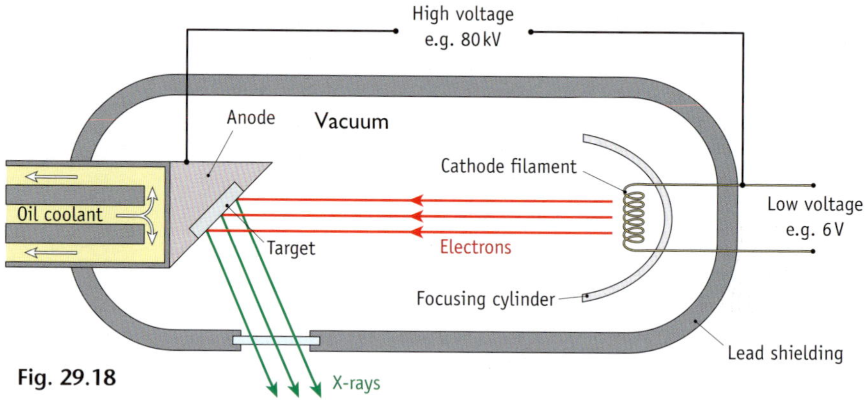

Fig. 29.18

THE HOT CATHODE X-RAY TUBE

FIG. 29.18 shows a modern hot cathode X-ray tube. It operates as follows:

- At the cathode thermionic emission occurs and a beam of electrons is produced.
- The very high voltage across the tube (typically 80 000 V) accelerates the electrons to very high speeds as they move across the tube towards the anode.
- When the electrons strike the metal target in the anode the kinetic energy of a small percentage of the electrons (< 1%) is converted into X-rays.
- The rest of the electrons' energy appears as heat in the anode and must be removed. This is the function of the circulating coolant. The target itself must also have a high melting point – for this reason tungsten is often used.
- People operating the tube must be protected from the X-rays. The tube is therefore surrounded with a lead shield which does not allow X-rays pass through it. It has a small window in it through which the X-rays can pass.

X-RAY PRODUCTION AS THE INVERSE OF THE PHOTOELECTRIC EFFECT

In the photoelectric effect (page 334), electromagnetic radiation strikes a metal and its energy is given to electrons. The electrons are then emitted from the metal. In X-ray production the opposite happens; high speed electrons strike a metal target and lose their energy. The energy is given off in the form of electromagnetic radiation.

PENETRATING POWER OF X-RAYS

The penetrating power of an X-ray depends on its frequency. The higher the frequency, the greater the penetrating power. The frequency in turn depends on the voltage across the tube. The higher the voltage the higher the frequency. Thus the penetrating power depends on the voltage across the tube. Very penetrating X-rays are known as **hard X-rays**. Less penetrating X-rays are known as **soft X-rays**.

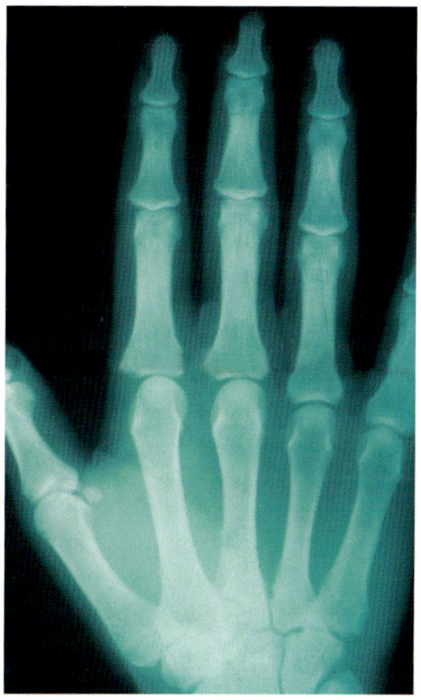

Fig. 29.19

PROPERTIES OF X-RAYS

- X-rays are **electromagnetic radiation** with wavelengths between 10^{-9} m and 10^{-15} m.
- X-rays **ionise** the material through which they pass. This means that they can knock electrons from the atoms of the material through which they pass. Atoms that have lost electrons are called ions.
- X-rays **penetrate** through materials. The denser the material the more it absorbs X-rays and the less X-rays get through it.
- X-rays are **not deflected** in electric or magnetic fields.
- X-rays produce **fluorescence** in certain materials if they strike them. For example, zinc sulphide or barium platinocyanide.
- X-rays affect **photographic emulsions**.
- X-rays can produce **interference patterns** and undergo **diffraction**.
- X-rays can cause **photoemission** (i.e. they can cause the photoelectric effect).

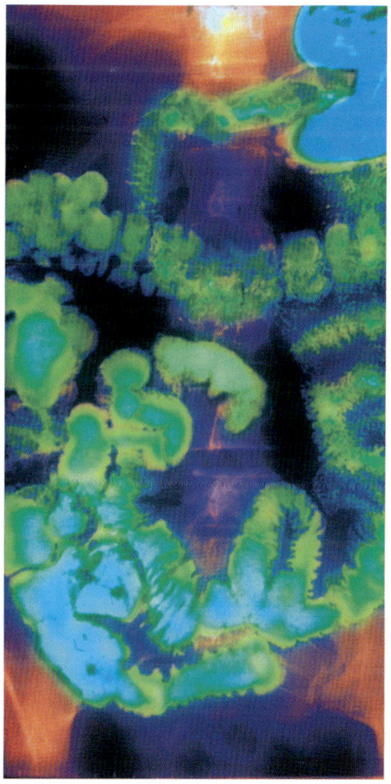

Fig. 29.20

USES OF X-RAYS IN MEDICINE

X-ray Photographs

The denser a substance, the more it absorbs X-rays. Use is made of this fact in medicine to take X-ray photographs. Bones are denser than the rest of the body. Diseased tissue is a different density to similar healthy tissue. Thus bones and damaged tissue (such as cancerous growths) will show up on X-ray photographs (Fig. 29.19).

The stomach and the intestines can also be made visible by getting the patient to eat food containing barium sulphate (a 'barium meal'). The barium shows up on the X-ray photograph showing the shape of the digestive organs (Fig. 29.20).

X-rays can be used to Destroy Cancerous Cells

Diseased tissue, such as a cancerous growth, is more easily damaged by X-rays than healthy tissue. Thus a beam of X-rays can be used to destroy some cancers.

USES OF X-RAYS IN INDUSTRY

X-rays are used in industry to detect cracks and flaws in metals, welds and castings. X-rays are used to photograph inside machines without taking them apart. X-rays can also be used to determine the thickness of things and to find out how full packages are.

HAZARDS OF X-RAYS

X-rays are ionising radiation and thus have a damaging effect on human tissue. This is discussed more fully on page 365.

CHAPTER CHECKLIST

- **Define:** Thermionic emission; Cathode rays; The electronvolt; The photoelectric effect; Threshold frequency; Work function; Photon; X-rays.
- **State:** The relationship between the energy of a photon and its frequency; Who discovered X-rays; The relationship between the eV, the keV, the MeV and the GeV;
  Einstein's Photoelectric Law.
- **Recall** that: The charge on the electron is the indivisible quantity of charge; The quantity of charge on the electron was first measured by Millikan; 1 electronvolt = 1.6×10^{-19} joules; A beam of electrons can be deflected by an electric or a magnetic field; The relationship between the photocurrent and the intensity of the light; Einstein's explanation of the photoelectric effect; X-rays can ionise material through which they pass; X-ray production is the inverse of the photoelectric effect; X-rays can be dangerous to humans.
- **Explain:** How a cathode ray tube works; How a photocell operates; How a hot cathode X-ray tube works;
  Why a beam of electrons moving at right angles to a magnetic field moves in a circle.
- **Describe** an experiment to: Demonstrate the photoelectric effect; Demonstrate the action of a photocell; Demonstrate the production of a beam of electrons in a cathode ray tube and to demonstrate their deflection in electric and magnetic fields.
- **Recall** and use the formulae:
  $W = QV$; $eV = \frac{1}{2}mv^2$; $E = hf$;
  $F = qvB$; $hf = \Phi + \frac{1}{2}mv^2_{max}$
- **List:** Three properties of the electron; Four properties of the cathode rays; Four applications of a cathode ray tube; Four applications of photoelectric sensing devices; Six properties of X-rays; Three uses of X-rays.
- **Draw:** A labelled diagram of a cathode ray tube; A labelled diagram of a photocell; A labelled diagram of a hot cathode X-ray tube.

The Atom, the Nucleus and Radioactivity

CHAPTER 30

ATOMS

Atoms are so small that they cannot be seen with the most powerful light microscope. About 2 000 000 atoms would cover one of the full stops on this page. There are 104 different kinds of atom, 92 of these occur naturally. The others are made artificially in nuclear reactors (page 360). Each kind of atom has one or two letters as its symbol (its **atomic** or **chemical symbol**), e.g. H is the symbol for a hydrogen atom; He is the symbol for a helium atom. An element is a substance which is made up of atoms of the same kind, e.g. hydrogen, helium, oxygen, copper, gold, zinc, carbon and uranium. Since there are 104 different kinds of atom there are 104 different elements. The **Periodic Table** (page 44 of your Maths tables) is a list of the elements.

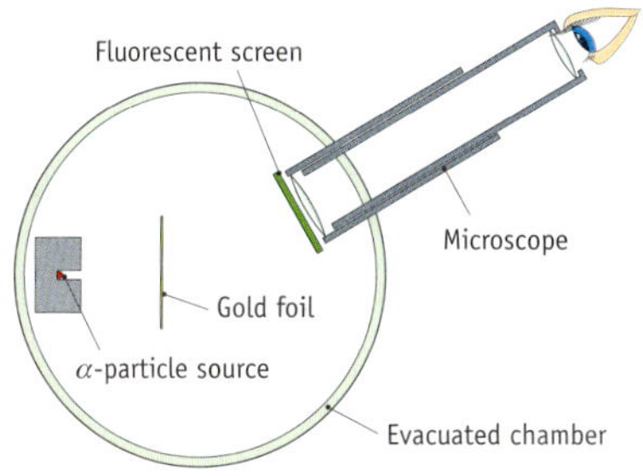

Fig. 30.1

Around 1900, it was known that atoms contained negatively charged particles called **electrons**. It was also known that since atoms were electrically neutral they contained an equal amount of positive charge. The location and arrangement of the electrons and positive charge within an atom were unknown. In 1911 Ernest Rutherford performed an experiment which showed that the positive charge in an atom was contained at its centre in what became known as the **nucleus**. Rutherford thus proposed the **nuclear model of the atom**.

RUTHERFORD'S EXPERIMENT

Rutherford bombarded a very thin piece of gold foil (about 2000 atoms thick) with particles known as **alpha-particles (α-particles)**. α-particles are actually the nuclei of helium atoms. At the time they were known to be positively charged particles with twice the charge of an electron. The α-particles could be detected by small flashes of light (called scintillations) that they produced on a fluorescent screen (FIG. 30.1). He found that:

- most α-particles were undeflected and passed straight through the gold foil,
- some were deflected through small angles,
- a very small number were turned back through angles greater than 90°.

He explained this by using the nuclear model of the atom, i.e. by assuming that each gold atom had a small positively charged nucleus at its centre containing most of the mass of the atom (FIG. 30.2). The explanation is as follows:

- The nucleus is very small compared to the size of the atom. Thus an atom is mainly empty space. This is why most of the α-particles passed straight through.
- All the positive charge of the atom is concentrated in the nucleus. If an α-particle passed near the nucleus it was deflected since positive charge repels positive charge.

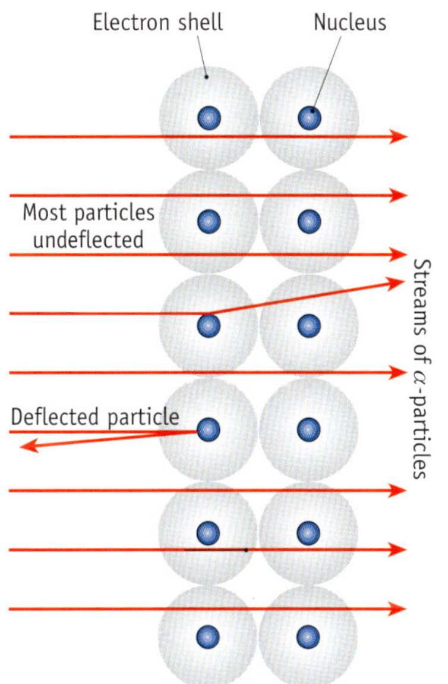

Fig. 30.2

- If an α-particle is about to, or nearly about to collide head on with a nucleus it is deflected back through an angle greater than 90°.
- The relatively light electrons which are negatively charged orbited the nucleus in various orbits.

From the number of alpha-particles deflected through different angles, Rutherford was able to estimate the radius of the nucleus.

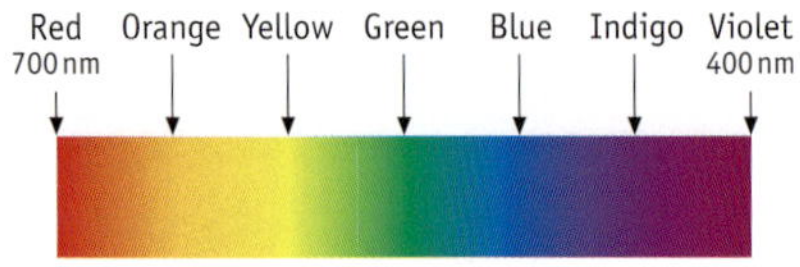

Fig. 30.3
A continuous spectrum.

THE RADIUS OF THE NUCLEUS

The radius of the nucleus is in the order of 10^{-15} m. The radius of the atom is in the order of 10^{-10} m. Thus most of the volume of an atom is empty space. If the nucleus were the size of a tennis ball the outer electrons would be about three km away.

The Arrangement of Electrons in an Atom

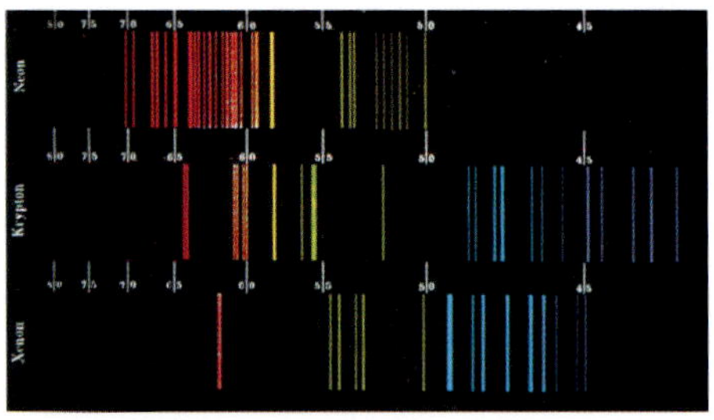

Fig. 30.4

In 1913 the Danish scientist Niels Bohr proposed a model – based on Plank's quantum theory – describing how electrons are arranged in certain orbits around the nucleus. Evidence for this model came from his explanation of emission spectra.

Emission Spectra

If sufficient energy is supplied to the atoms of a solid, liquid or gas, the atoms may give out light. The energy may be supplied by heating or by passing an electric current through the substance. For example:

- The filament of a bulb is a heated solid that gives out light.
- White hot iron is a liquid that gives out light.
- Most yellow street lights are tubes of sodium vapour. This is a gas that gives out light.

> **EMISSION SPECTRUM**
> When light from a luminous source undergoes dispersion the resulting pattern is called an ***emission spectrum***.

If this light is passed through a prism (or a diffraction grating) dispersion occurs and an **emission spectrum** is formed.

We shall look at two types of emission spectra. They are a **continuous spectrum** and a **line spectrum**.

CONTINUOUS SPECTRUM

A **continuous spectrum** is produced by an **incandescent solid or liquid**. It consists of a continuous spread of changing colours going from red to violet (Fig. 30.3). All visible wavelengths are emitted. Continuous spectra are not characteristic of the material (solid or liquid) from which they come. For example, white hot iron produces the same spectrum as white hot tungsten.

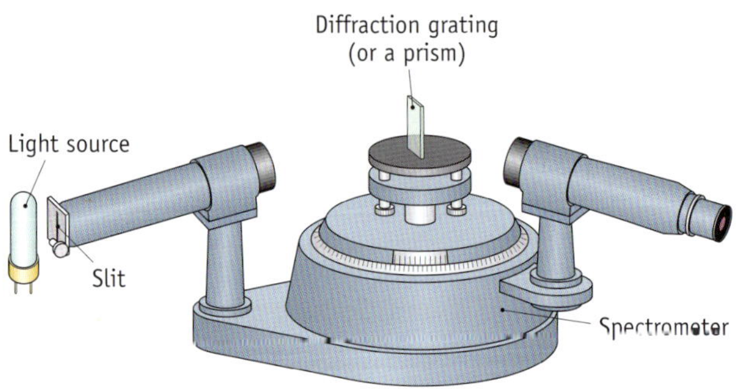

Fig. 30.5
Demonstrating spectra in the laboratory.

LINE SPECTRUM

If the atoms of a **gaseous element** are given sufficient energy, they give out coloured light. The colour of the light given out depends on the element in question. If this light is passed through a prism (or a diffraction grating) a series of bright lines on a dark background called a **line emission spectrum** is formed. Each element has a different line spectrum that is characteristic of that element. Fig. 30.4 shows the line spectrum of xenon, krypton and neon.

Both continuous and line spectra may be conveniently demonstrated in the laboratory using a spectrometer as in Fig. 30.5. The light source is an incandescent filament bulb for the continuous spectrum and a gas discharge tube for the line spectrum.

Spectroscopy as a Tool in Science

Since no two elements have the same line spectrum, light from a vapour of an element can be used to identify that element accurately. If a mixture of different elements is vaporised and made to give out light, the individual elements present can be identified by their spectra.

Also, the relative amounts of each element can be determined from the brightness of the spectra. Spectroscopic analysis like this can be used where only minute amounts of the substance in question are present.

Line spectra give us information on how electrons are arranged in an atom.

Electrons and Emission Spectra – The Bohr Model

- Electrons in an atom can only move in certain allowed orbits. While in a particular orbit, an electron will not emit electromagnetic radiation (light). Thus the energy of an electron in a given orbit is a fixed value.
- When an atom is supplied with energy, an electron in a particular orbit may absorb some of this energy and move from its existing orbit (of energy E_1) to an orbit of higher energy (E_2). This atom is then said to be in an excited state (Fig. 30.6(a)).
- After a very short time, the electron falls back to its original energy level, giving out an amount of energy equal to the difference between the two levels (i.e. $E_2 - E_1$) (Fig. 30.6(b)). This energy is given out as a photon of electromagnetic radiation of frequency f given by:

$$hf = E_2 - E_1, \quad \text{where } h = \text{Planck's constant.}$$

- Light of a given frequency is a given colour and appears as a line if dispersed by a prism or a diffraction grating.
- Because there are definite orbits in an atom between which an electron may move, there are definite frequencies emitted and thus definite coloured lines formed.
- Different atoms contain different numbers of electrons in different orbits, thus each atom gives out its own particular set of frequencies, i.e. different elements have their own characteristic line spectra.

Bohr's model was successful in explaining the hydrogen spectrum. It ran into problems with other atoms and today more complicated mathematical theories are used to explain spectra. None the less, the basic idea of electrons jumping between energy levels is correct.

ENERGY LEVEL
An ***energy level*** is a fixed energy value that an electron can have in an atom.

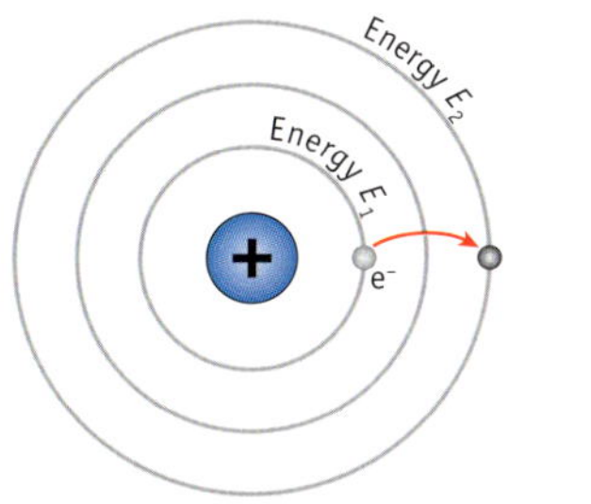

(A) Supply energy and electron moves to a higher energy level

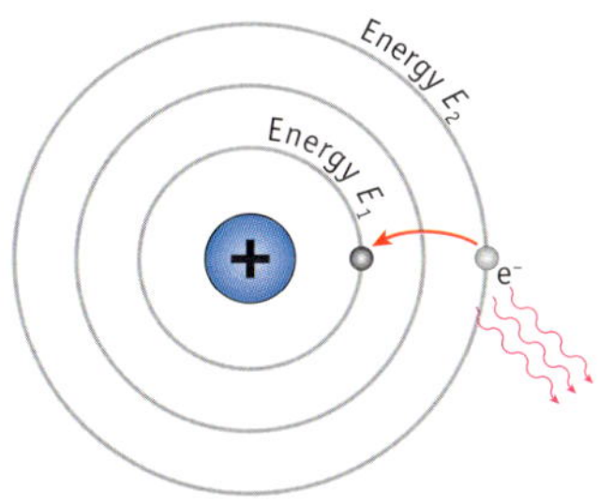

(B) Electron drops back to a lower energy level emitting a photon of energy, hf where $E_2 - E_1 = hf$

Fig. 30.6

Lasers

When an electron is excited into a higher energy level in an atom it will eventually drop back to its original energy level with the emission of a photon of light. If a photon of light of the same frequency as that which the atom is about to emit strikes the atom it may cause that atom to emit its photon. Thus two identical photons are produced. In a laser, many atoms have their electrons excited to a higher energy level. All these atoms are stimulated (by light of the same frequency) to emit these photons together, producing an intense beam of coherent light – a laser beam. 'Laser' is an acronym of Light Amplification by Stimulated Emission of Radiation.

Lasers are used in:

- Telecommunications, where laser light is used to send digital signals along optic fibres.
- Medicine, where they are used to burn away cancer cells, remove birthmarks, stop bleeding, cut tissue and fix certain eye problems.
- Industry, where they are used for cutting and welding.

Lasers are also found in CD players and supermarket checkouts to read bar codes.

STRUCTURE OF THE NUCLEUS

By 1932 it had been discovered that the nucleus of an atom was itself made up of at least two more particles called **protons and neutrons**. FIG. 30.7 shows the main properties of the subatomic particles: electrons, protons and neutrons.

Particle	Symbol	Location	Charge in coulombs	Relative Charge	Mass (kg)	Relative mass
Proton	p	nucleus	$+1.6 \times 10^{-19}$	+1	1.67×10^{-27}	1
Neutron	n	nucleus	0	0	1.68×10^{-27}	1
Electron	e^-	orbiting nucleus	-1.6×10^{-19}	−1	9.1×10^{-31}	1/2000

Fig. 30.7

ATOMIC NUMBER

The ***atomic number*** **(*Z*)** of an element is the number of ***protons*** in the nucleus of an atom of that element.

The number of protons in the nucleus of an atom tells us what element it is. A uranium atom **always** has 92 protons in its nucleus – it would not be uranium otherwise. If an atom has 92 protons in its nucleus it is a uranium atom. The **periodic table** (Maths tables page 44) is a list of the elements in order of increasing atomic number.

MASS NUMBER

The total number of protons and neutrons in the nucleus of an atom is called the ***mass number*** **(*A*)** of that atom.

When writing the symbol for an atom the mass number is usually written on top, for example:

Mass number

$^{1}_{1}H \quad ^{4}_{2}He \quad ^{12}_{6}C \quad ^{235}_{92}U$

Atomic number

It should be obvious that:

Number of neutrons in nucleus = Mass Number – Atomic Number

i.e. Number of neutrons = $A - Z$

ISOTOPES

Atoms of an element that have the same number of protons but different numbers of neutrons are called ***isotopes*** of that element.

For example, naturally occurring hydrogen is approximately 99.98% ^1_1H and 0.02% ^2_1H (called deuterium). All elements have more than one isotope, though some of the isotopes occur in very small quantities. Many more isotopes can be produced artificially in nuclear reactors.

RADIOACTIVITY

In 1896 Becquerel noticed that a uranium salt caused a nearby photographic plate to go black. The plate was in a dark paper wrapper which prevented any light from falling on it. He concluded that the uranium was giving out some kind of radiation that passed through the wrapper and caused the photographic plate to go black, i.e. it caused the plate to be exposed. Today we say the uranium is **radioactive**. Within a few years scientists discovered that some isotopes of other elements were also radioactive.

RADIATION FROM THE NUCLEUS

It is now known that the nuclei of certain isotopes are unstable – i.e. they contain excess energy. These nuclei can become stable by getting rid of this energy. The process of getting rid of this energy is the radioactivity discovered by Becquerel. For this reason it is also called **nuclear radiation**. The nucleus is said to undergo **radioactive decay** or **radioactive disintegration**.

RADIOACTIVITY

Radioactivity is the disintegration or decay of the nuclei of certain atoms with the emission of one or more types of radiation.

THREE KINDS OF NUCLEAR RADIATION

Scientists soon discovered that there are three kinds of nuclear radiation. They are called **alpha**-radiation (α), **beta**-radiation (β) and **gamma**-radiation (γ). Some radioactive isotopes emit one of these radiations. Some isotopes emit two and others emit all three.

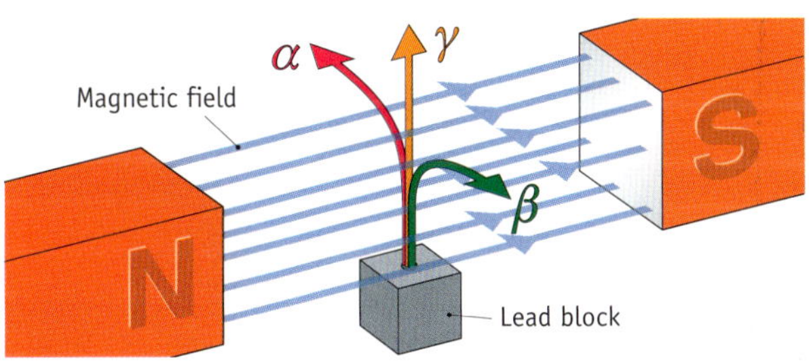

Fig. 30.8 (A)

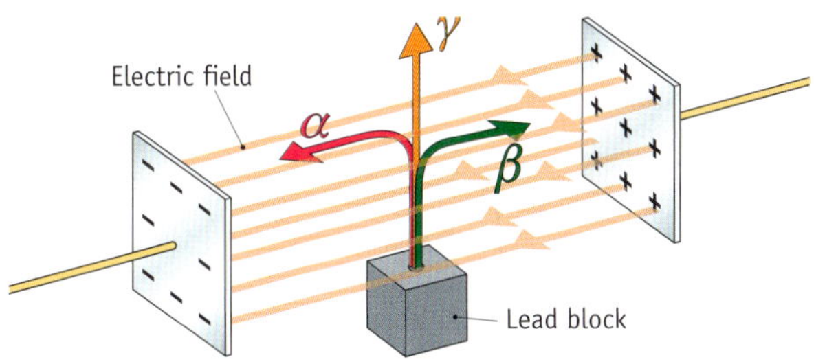

Fig. 30.8 (B)

EXPERIMENTAL EVIDENCE FOR THE THREE KINDS OF RADIATION

DEFLECTION IN ELECTRIC OR MAGNETIC FIELDS

When a narrow beam of the radiation from certain radioactive nuclei is passed at right angles through a magnetic field (FIG. 30.8(A)) or an electric field (FIG. 30.8(B)) it is found to split into three components. Some deflect in the same way as a positively charged particle (the α-rays), some deflect like a negatively charged particle (the β-rays) and the others are not deflected (the γ-rays).

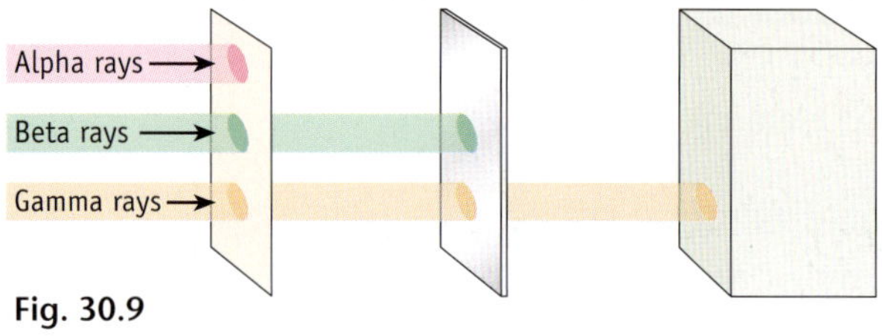

Fig. 30.9

PENETRATING POWER

The penetrating power of the radiations varies considerably. Some are stopped by a sheet of paper (the α-rays). Some will be stopped by a thin sheet (a few mm) of aluminium (the β-rays), while the others (the γ-rays) are very penetrating and will only be stopped by a thick block of lead or a few feet of concrete (FIG. 30.9).

IONISING ABILITY

Nuclear radiation was found to cause a charged electroscope to lose its charge. This is because it ionises any matter through which it passes, i.e. it knocks off electrons from the atoms in the matter thus producing positive ions. Again, one type (the α-rays) are found to produce a lot of ionisation and cause the electroscope to lose its charge rapidly. Some (the β-rays) cause the electroscope to lose its charge more slowly, since they produce less ionisation. Some (the γ -rays) produce very little ionisation and hardly affect the charged electroscope.

EXPERIMENT

TO DEMONSTRATE THE PENETRATING POWER OF α, β AND γ RAYS.

Method

- Set up the equipment as shown in FIG. 30.10 without the source in place. Record the number of counts registered in two minutes and calculate the background count rate in counts per second.
- Place an α-source about 1 cm from the GM tube*. Record the number of counts in a one-minute interval and calculate the count rate.
- Place a sheet of paper between the source and the GM tube and measure the count rate. It will have gone back to the background count rate.
- Remove the sheet of paper and slowly move the GM tube back from the source. As you do, the count will fall and eventually reach the background count.

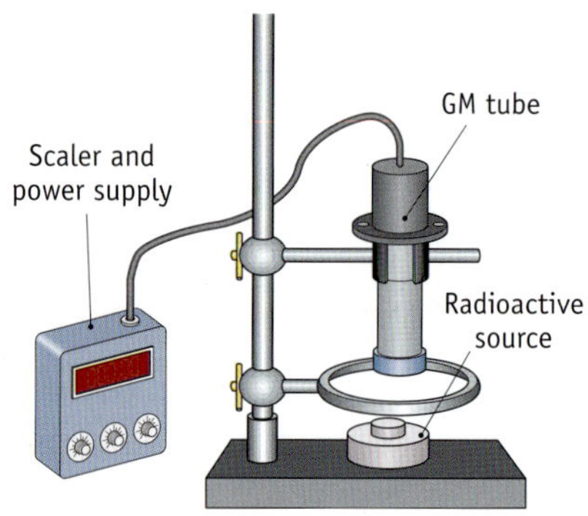

Fig. 30.10

Conclusion

α-rays are stopped by a sheet of paper or a few cm of air.

- Repeat the above steps using a β-source. This time use aluminium sheets and the paper as stoppers.

β-rays are not stopped by air or a sheet of paper, but will be stopped by a sheet of aluminium a few mm thick.

- Repeat the above steps using a γ-ray source.

It takes a sheet of lead a few cm thick to completely block the γ-rays.

* The GM tube detects the presence of nuclear radiation. Its operation is explained on page 353.

EXPERIMENT

TO DEMONSTRATE THE IONISING EFFECT OF NUCLEAR RADIATION.

Method

- Charge an electroscope. It should remain charged for at least a few minutes with the leaf diverged.
- Bring a radioactive source near it as in FIG. 30.11. The leaf will initially collapse a little due to the change in the capacitance of the system.
- The leaf will then slowly collapse due to the nearby radioactive source.

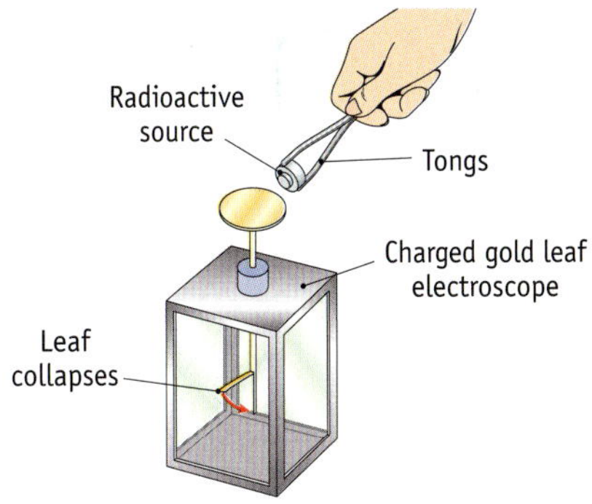

Fig. 30.11

Explanation

The radiation ionises air molecules above the cap of the electroscope. Ions with opposite charge to that on the cap are attracted to it and neutralise the charge on it. Thus the leaf collapses.

THE NATURE OF α, β AND γ RADIATION

Alpha (α) radiation is fast moving helium nuclei ejected from the nuclei of radioactive atoms.

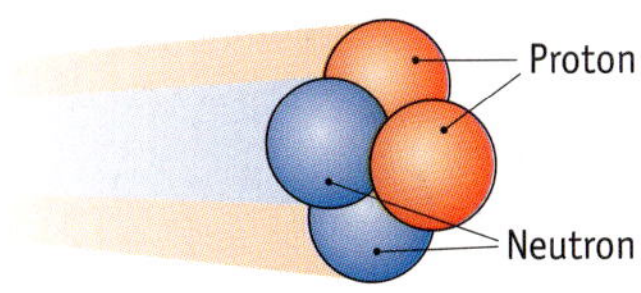

Fig. 30.12
An α-particle is a helium nucleus

A helium nucleus is a bundle of two protons and two neutrons stuck together (FIG. 30.12). Each bundle is called an **alpha-particle**.

Beta (β) radiation is high-speed electrons ejected from the nuclei of radioactive atoms.

Each electron is called a **beta-particle**.

Gamma (γ) radiation is high-frequency electromagnetic radiation (with frequencies above those of normal X-rays) emitted from the nucleus of a radioactive atom.

Gamma radiation is usually called **gamma-rays**.

When a nucleus emits an alpha or a beta-particle, the number of protons in it changes. Thus it becomes a **different nucleus**. The nucleus that emits the particle is called the **parent nucleus** and the new nucleus formed is called the **daughter nucleus**. The daughter nucleus may also be radioactive.

ALPHA EMISSION

If a nucleus emits an alpha-particle ($^{4}_{2}He$) its atomic number decreases by 2 (it has lost 2 protons) and its mass number decreases by 4 (it has lost 2 protons and 2 neutrons). Thus the **daughter nucleus is two places to the left of the parent nucleus on the periodic table**, e.g.:

$$^{226}_{88}Ra \rightarrow {}^{222}_{86}Rn + {}^{4}_{2}He$$

Radium → Radon + α-particle

In general, for **alpha emission:**

Parent nucleus → Daughter nucleus + Alpha-particle

$$^{A}_{Z}\mathbf{X} \rightarrow {}^{A-4}_{Z-2}\mathbf{Y} + {}^{4}_{2}\mathbf{He}$$

Fig. 30.13

In general, for **beta emission**				
Parent nucleus	→	Daughter nucleus	+	Beta particle
$^{A}_{Z}X$	→	$^{A}_{Z+1}Y$	+	$^{0}_{-1}e$

Fig. 30.14

Beta Emission

When a nucleus emits a beta-particle, a neutron in the nucleus splits up and becomes a proton and an electron. The proton remains in the nucleus and the electron is ejected out at high velocity (the β -particle).

$$^{1}_{0}n \rightarrow {}^{1}_{+1}p + {}^{0}_{-1}e$$

Neutron → Proton + Electron (β -particle)

Since the mass of the beta-particle is very small, there is almost no change in the mass of the parent nucleus. Since there is now one more proton in the nucleus, **the atomic number of the daughter nucleus is one greater than that of the parent** and the daughter is one place to the right of the parent on the periodic table, e.g.:

$$^{228}_{88}Ra \rightarrow {}^{228}_{89}Ac + {}^{0}_{-1}e$$

Radium → Actinium + β -particle

Gamma Rays

Gamma rays are very high frequency electromagnetic radiation with frequencies above those of normal X-rays. When gamma rays are emitted from the nucleus, the structure of the nucleus remains the same. However, it will have lost energy and become more stable. Gamma rays are normally emitted from a nucleus that has already emitted an alpha or beta-particle.

Fig. 30.15 summarises the main properties of the three types of nuclear radiation.

	Nature	Ionising Ability	Penetrating Power	Range	Charge	Relative Mass	Deflection in electric and magnetic fields.
α -Particle	helium nucleus	greatest	least	few cm of air thin paper sheet	+2	4	as + charged particle
β -Particle	electron	less than alpha	more than alpha	few mm of Al	-1	≈ 0	as - charged particle
γ -Rays	em radiation	least	most	many cm of lead few feet of concrete			undeflected

Fig. 30.15

Problem 1: How many protons, neutrons and electrons are there in each of the following?

$^{1}_{1}H$; $^{4}_{2}He$; $^{235}_{92}U$

Solution: In each one the lower number (the atomic number) is the number of protons. Since each atom is overall neutral, the atomic number is also the number of electrons.

Number of neutrons = Mass number − Atomic number. Thus:

For Hydrogen: Number of neutrons = 1 - 1 = 0 neutrons

For Helium: Number of neutrons = 4 - 2 = 2 neutrons

For Uranium: Number of neutrons = 235 - 92 = 143 neutrons

Problem 2: What is X and what is Y in each of the following nuclear equations?

(i) $^{220}_{86}\text{Rn} \rightarrow X + ^{4}_{2}\text{He}$ (ii) $^{228}_{88}\text{Ra} \rightarrow Y + ^{0}_{-1}\text{e}$

Solution: (i) α-emission $\Rightarrow$ mass number decreases by 4 to 216 and atomic number decreases by 2 to 84. Periodic table says element with atomic number 84 is polonium (Po)

$\Rightarrow$ X is $^{216}_{84}\text{Po}$

(ii) β-emission $\Rightarrow$ mass number is the same and atomic number increases by 1 to 89 which is actinium (Ac).

Thus Y is $^{228}_{89}\text{Ac}$

Problem 3: When an isotope undergoes radioactive decay the daughter nucleus is often radioactive. This in turn decays and the process continues until a stable isotope is formed. The series of isotopes formed is called a **radioactive decay series**.

Construct a radioactive decay series beginning with $^{226}_{88}\text{Ra}$ and ending with $^{214}_{83}\text{Bi}$

Consider only isotopes of elements with atomic numbers 82, 84 and 86.

Solution: Decrease in mass no. is caused only by alpha-emission.

Decrease in mass no. = 226 − 214 = 12 $\Rightarrow$ 3 alpha-particles emitted

3 α-particles emitted $\Rightarrow$ atomic no. decreased by 6 (i.e. 3 × 2) giving: 88 − 6 = 82

To convert this to $^{214}_{83}\text{Bi}$ the atomic no. must increase by 1, i.e. one β-particle is emitted.

Thus using the periodic table the series is: $^{226}_{88}\text{Ra} \xrightarrow{\alpha} ^{222}_{86}\text{Rn} \xrightarrow{\alpha} ^{218}_{84}\text{Po} \xrightarrow{\alpha} ^{214}_{82}\text{Pb} \xrightarrow{\beta} ^{214}_{83}\text{Bi}$

EXERCISE 30.1

1. What is the order of magnitude of the diameter of a typical atom? What is the order of magnitude of the diameter of a typical nucleus? Define:

   (i) atomic number, (ii) mass number.

   What are isotopes?

2. How many protons, neutrons and electrons are there in a neutral atom of each of the following?

   $^{1}_{1}\text{H}$ $^{2}_{1}\text{H}$ $^{4}_{2}\text{He}$ $^{12}_{6}\text{C}$ $^{14}_{6}\text{C}$ $^{235}_{92}\text{U}$

3. The atomic number of an atom is Z, its mass number is A. How many neutrons are there in its nucleus?

4. Radium decays to Radon with the emission of an α-particle. Complete the reaction:

   $^{226}_{88}\text{Ra} \rightarrow ^{\square}_{\square}\text{Rn} + ^{4}_{2}\text{He}$

5. Complete each of the following nuclear equations:

   (i) $^{220}_{86}\text{Rn} \rightarrow \alpha + \square$

   (ii) $^{232}_{90}\text{Th} \rightarrow \alpha + \square$

   (iii) $^{A}_{Z}\text{X} \rightarrow ^{\square}_{\square}\text{Y} + ^{4}_{2}\text{He}$

   (iv) $^{238}_{92}\text{U} \rightarrow \square + ^{4}_{2}\text{He}$

   (v) $^{14}_{6}\text{C} \rightarrow \square + \beta$

   (vi) $^{40}_{19}\text{K} \rightarrow ^{40}_{20}\text{Ca} + \square$

   (vii) $^{A}_{Z}\text{X} \rightarrow ^{\square}_{\square}\text{Y} + ^{0}_{-1}\text{e}$

   (viii) $^{1}_{0}\text{n} \rightarrow ^{1}_{+1}\text{p} + \square$

6. Calculate the number of α-particles and the number of β-particles emitted in the decay of: $^{238}_{92}U$ to $^{230}_{90}Th$

7. Calculate the number of α-particles and the number of β -particles emitted in the decay of: $^{238}_{92}U$ to $^{226}_{88}Ra$

8. Construct a radioactive decay series beginning with $^{238}_{92}U$ and ending with $^{226}_{88}Ra$ considering only atoms of atomic numbers 90,91,92 (see Maths tables page 44 and 46).

9. The following represents part of the natural radioactive disintegration of thorium ($^{232}_{90}Th$). Name the elements X, Y, Z and W.

$$^{232}_{90}Th \xrightarrow{\alpha} X \xrightarrow{\beta} Y \xrightarrow{\beta} Z \xrightarrow{\alpha} W$$

10. A sample of radioactive gas in a container is found to emit α-particles. If the daughter nuclei formed are radon-219, write an equation to represent the reaction taking place in the container.

11. Polonium-218 is an α-emitter. Write down an equation to represent this decay and name the products formed.

12. A radioactive nucleus emits 4 α-particles and 3 β-particles. By how much does its atomic number and mass number change?

13. Radium-226 decays to polonium-218 in two stages, with the same particle emitted in each stage. Name this particle and give an equation for the process.

THE ACTIVITY OF A RADIOACTIVE NUCLEUS

UNIT OF ACTIVITY

The unit of activity is the ***becquerel* (Bq)**.

1 becquerel is equal to 1 radioactive disintegration per second.

The number of nuclei of a radioactive sample decaying per second is called its **activity (*A*)**. Its unit is the **becquerel (Bq)**.

ACTIVITY

The ***activity (A)*** of a radioactive substance is the number of nuclei of that substance decaying per second.

THE LAW OF RADIOACTIVE DECAY

Given a sample of radioactive material we cannot predict which nuclei in it will decay next. Neither can we predict when a given nucleus will decay. Radioactive decay is a **random process**. Because the nuclei decay at random it follows that at any particular instant the number decaying per second is directly proportional to the number undecayed at that instant. For example, if there are 2 grams of undecayed radioactive uranium present in a sample and the number decaying per second is 5×10^6 then in a 4 gram sample of the same material, 10×10^6 nuclei would be decaying per second.

LAW OF RADIOACTIVE DECAY

The number of nuclei decaying per second (i.e. the activity) is directly proportional to the number of nuclei undecayed.

i.e. Rate of decay $\propto N \Rightarrow$ **Rate of decay** $= \lambda N$

Where N is the number of atoms undecayed and λ is a constant called the **radioactive decay constant** or simply the **decay constant**, note that:

- λ has different values for different radioactive isotopes. It is a constant for a given isotope,
- since $\lambda = \dfrac{\text{Rate of decay}}{N}$ it follows that the **unit of λ** is

$$\frac{\text{Number of atoms per second}}{\text{Number of atoms}} = \textbf{per second} = \mathbf{s^{-1}}$$

HALF-LIFE

Suppose we have 4 grams of a radioactive isotope at a given instant. Wait until ½ of it undergoes decay. Suppose it takes a time of $T_{½}$ for this to happen. We now have 2 grams left undecayed.

Wait again until ½ of this 2 grams undergoes decay, i.e., until 1 gram remains undecayed. This takes exactly the same time, namely $T_{½}$. In general, the time taken for the number of undecayed atoms to halve is always the same and does not depend on the size of the sample. This time is called the **half-life** of the isotope. Its symbol is $T_{½}$. The value of a half-life may be very large (5×10^5 years for plutonium-242), small (54.5 seconds for radon-220) or very small (3×10^{-7} seconds for polonium-212). FIG. 30.16 shows a graph of number of atoms undecayed against time for a radioactive isotope. N_o is the number of atoms undecayed at time $t = 0$.

Since Rate of decay = λN, it follows that if N is halved in a given time, the value of the rate of decay is also halved in this time.

This means that the number of atoms disintegrating per second also decreases by half in one half-life. In other words, the activity is also halved in a half-life.

If a graph of the activity against time is drawn for a radioactive isotope it will look very similar to that in FIG. 30.16. FIG. 30.17 is such a graph.

HALF-LIFE
The ***half-life*** $T_{½}$ of a radioactive isotope is the time taken for half of the undecayed atoms to undergo decay.

HALF-LIFE
The ***half-life*** of a radioactive element is also the time taken for its activity to decrease by half.

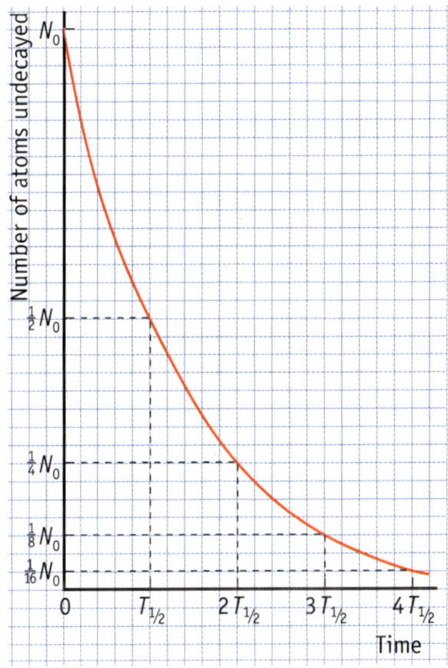

Fig. 30.16

RELATIONSHIP BETWEEN HALF-LIFE AND THE DECAY CONSTANT

It can be proved that the half-life and the decay constant of a radioactive isotope are related by the equation

$$T_{½} = \frac{\ln 2}{\lambda} \qquad \text{i.e. } T_{½} = \frac{0.693}{\lambda}$$

You do not need to be able to prove these formulae. Simply remember and be able to use them. Note that the logs used are natural logs – **not logs to base 10!** Thus: $\ln 2 = \log_e 2$

DETECTING NUCLEAR RADIATION

The presence of nuclear radiation can be detected by the effect it has on other matter. We shall look at two types of detector here. One is the **Geiger-Müeller** tube which detects the ionisation produced by the radiation. The other is called a **solid state detector** and relies on the electron hole pairs produced when radiation strikes certain semiconducting materials. You need to know the structure and operation of one of these detectors.

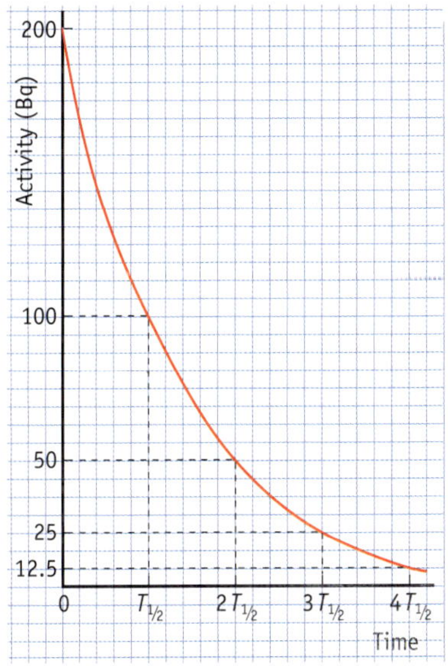

Fig. 30.17

THE GEIGER-MÜELLER TUBE

The GM tube detects the presence of radioactivity by the ionisation it produces. It can also be used to measure the activity of a radioactive sample. The GM tube (FIG. 30.18) operates as follows:

- Radiation passes through the thin mica window into the argon gas at low pressure.
- It ionises some of the argon atoms there producing positive argon ions and negatively charged electrons.

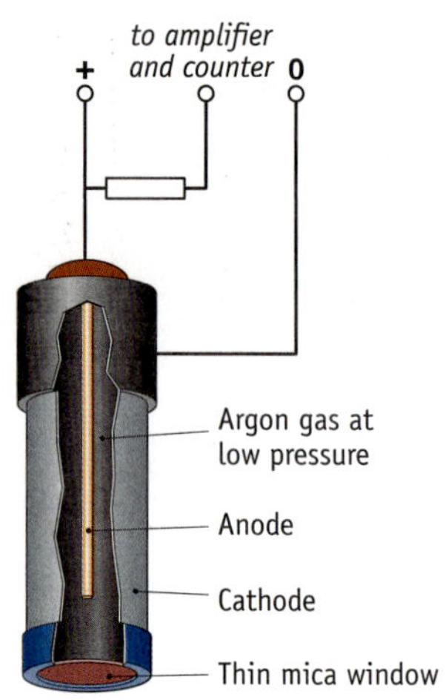

Fig. 30.18
A GM tube.

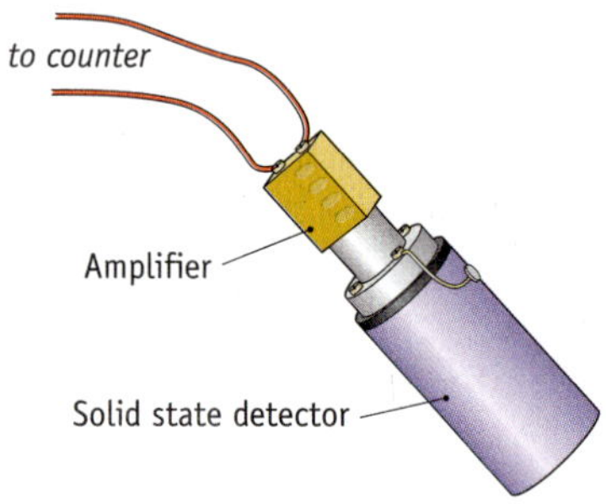

Fig. 30.19
A solid state detector.

- The electrons pick up high speed in the very strong electric field near the wire anode. These produce further ions and electrons by collision with other argon atoms. In this way an **avalanche of electrons** is produced.
- The electrons reach the anode and a pulse of current flows in the external circuit.
- This number of pulses can be counted on some form of **electronic counter such as a scaler or a ratemeter**.

The Solid State Detector

Fig. 30.19 shows a solid state detector.

- It consists of a reverse biased p-n junction connected to some form of counting device such as a ratemeter or a scaler.
- When nuclear radiation strikes the depletion layer, some electron-hole pairs are formed there.
- These charge carriers move under the influence of the voltage across it and so a pulse of current is formed.
- This is amplified before being passed to a pulse counter.

Artificial Radioactivity

Most non-radioactive (i.e. stable) isotopes can be made radioactive by bombarding them with neutrons. The neutrons are captured by the nuclei of the atoms. This is usually done in a nuclear reactor (page 360). Such isotopes are called artificial radioactive isotopes. Many of the isotopes used in medicine and industry are produced this way.

Uses of Radioisotopes

MEDICAL IMAGING

Small amounts of short-lived isotopes are placed in a particular organ of the body. An image of the organ can be seen from the radiation given off.

MEDICAL THERAPY

Radiation will kill cancer cells more easily than healthy cells.

FOOD IRRADIATION

Gamma rays can be used to sterilise food.

AS RADIOACTIVE TRACERS

In medicine and agriculture they are used to trace the movements of various substances in living matter.

CARBON DATING

The age of archaeological specimens can be determined by the activity of the isotope $^{14}_{6}C$ contained in them.

IN INDUSTRY

They are used to check the thickness of objects, the fullness of containers, to find leaks and to detect wear in components.

SMOKE DETECTORS

In an ionisation smoke detector a radioactive source ionises the air between two electrodes. Thus a small electric current can flow between them. If smoke particles enter this space they stick to the ionised molecules and the current is reduced. The decrease in current triggers the alarm.

Problem 4: A radioactive isotope has a half-life of 5 years. What fraction of the isotope will remain after 20 years? What fraction of the original sample will have decayed in 20 years?

Solution:

After 1 half-life,	i.e. 5 years	½ of sample remains
After 2 half-lives,	i.e. 10 years	¼ of sample remains
After 3 half-lives,	i.e. 15 years	⅛ of sample remains
After 4 half-lives,	i.e. 20 years	1/16 of original sample remains.

$\Rightarrow$ 15/16 of original sample will have decayed in 20 years.

Problem 5: The half-life of a certain radioactive isotope is 30 minutes.

(i) How many half-lives in 4 hours?
(ii) What fraction remains undecayed after 1.5 hours?
(iii) What fraction has decayed in 3 hours?

Solution:

(i) 4 hours = 8 × 30 minutes = 8 half-lives.

Number of half-lives	Fraction undecayed	Number of half-lives	Fraction undecayed
1	½	4	1/16
2	¼	5	1/32
3	⅛	6	1/64

(ii) 1.5 hours = 3 half-lives $\Rightarrow$ ⅛ of original sample remains.
(iii) 3 hours = 6 half-lives $\Rightarrow$ 1/64 of original remains $\Rightarrow$ 63/64 of sample has decayed.

Note that in general: **after *n* half-lives $\frac{1}{2^n}$ of the original sample remains.**

Problem 6: The activity of a sample of a radioactive isotope decreases to 1/32 of its original value after 250 years. What is its half-life?

Solution:

$$\frac{1}{2^n} = \frac{1}{32} \Rightarrow 2^n = 32 \qquad \text{i.e. } 2^n = 2^5 \Rightarrow n = 5$$

i.e. 250 years is 5 half-lives $\Rightarrow$ half-life = 50 years.

Problem 7: The isotope ${}^{90}_{38}\text{Sr}$ decays by beta-emission and has a decay constant of $8 \times 10^{-10}\ \text{s}^{-1}$.

Calculate the number of atoms present in a sample of this isotope which emits 2.4×10^4 beta-particles per second.

Solution:

$$\text{Rate of decay} = \lambda N \Rightarrow N = \frac{\text{Rate of decay}}{\lambda}$$

$$\text{i.e. Number of particles present} = \frac{(2.4 \times 10^4)}{8 \times 10^{-10}} = 3 \times 10^{13} \text{ beta-particles}$$

Problem 8: The half-life of a certain radioactive isotope is 10 hours. What is its decay constant?

Solution:

$$T_{½} = \frac{0.693}{\lambda} \Rightarrow \lambda = \frac{0.693}{T_{½}} = \frac{(0.693)}{(10)(60)(60)} \quad \text{i.e. } \lambda = 1.925 \times 10^{-5}\ \text{s}^{-1}$$

Problem 9: The half-life of U 235, which is an alpha emitter, is 8×10^8 years. Find the number of alpha-particles emitted per second from a sample containing 2.6×10^{24} atoms.

Solution:

$$8 \times 10^8 \text{ years} = (8 \times 10^8)(365)(24)(60)(60)\text{ s} = 2.5228 \times 10^{16}\text{ s}$$

$$T_{½} = \frac{0.693}{\lambda} \Rightarrow \lambda = \frac{0.693}{T_{½}} = \frac{0.693}{2.5228 \times 10^{16}} = 2.7469 \times 10^{-17}\ \text{s}^{-1}$$

$$\begin{pmatrix}\text{Number of particles}\\ \text{emitted per second}\end{pmatrix} = \text{Rate of decay} = \lambda N = (2.7469 \times 10^{-17})(2.6 \times 10^{24}) = 7.1 \times 10^7$$

EXERCISE 30.2

1. FIG. 30.20 shows a graph of the activity of a radioactive isotope plotted against time. From the graph find the half-life of the isotope.

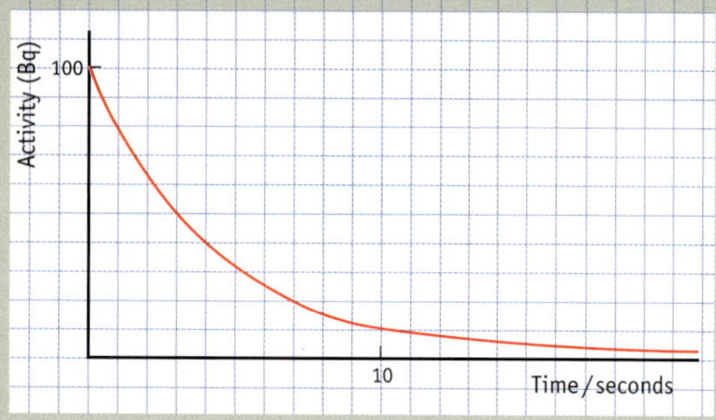

Fig. 30.20

2. FIG. 30.21 shows a graph of the number of undecayed atoms of a radioactive isotope plotted against time. From the graph find the half-life of the isotope.

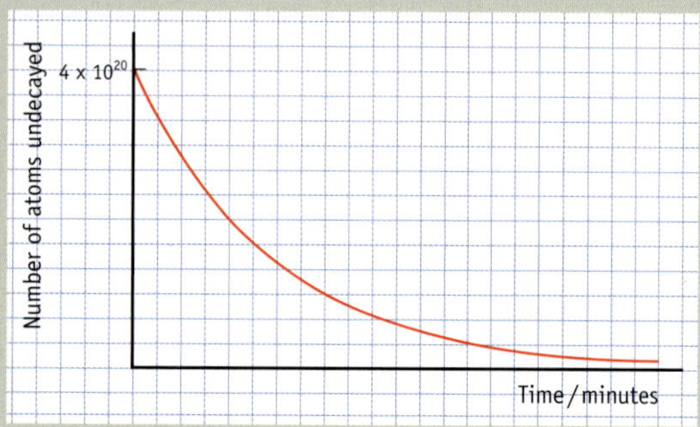

Fig. 30.21

3. The half-life of a certain radioactive isotope is 3 years. What fraction of the isotope remains undecayed after:
   (i) 3 years, (ii) 6 years,
   (iii) 9 years?
4. The half-life of a certain radioactive isotope is 10 years. What fraction remains undecayed after 40 years? What fraction has decayed in this time?
5. The half-life of a certain isotope is 20 minutes.
   (i) How many half-lives in 5 hours?
   (ii) What fraction remains undecayed after 3 hours?
   (iii) What fraction has decayed in 40 minutes?
   (iv) What fraction remains undecayed after n half-lives?
6. A radioactive isotope has a half-life of 120 s. Calculate its decay constant.
7. A radioactive isotope has a half-life of 5.5 minutes. Calculate its decay constant.
8. A radioactive isotope has a half-life of 2.4 years. Calculate its decay constant.
9. The decay constant of a radioactive isotope is 5×10^{3} s^{-1}. Find its half-life.
10. The decay constant of a radioactive isotope is 2×10^{-3} s^{-1}. Find its half-life.
11. The decay constant of a certain radioactive isotope is 9.627×10^{-5} s^{-1}. What fraction remains undecayed after 6 hours?
12. The decay constant of a certain radioactive isotope is 8×10^{-8} s^{-1}. A sample of this isotope undergoes 3×10^{3} disintegrations per second. What is the number of undecayed atoms in this sample?
13. An alpha-emitter has 2×10^{15} atoms undecayed at a given instant. If $\lambda = 8 \times 10^{-3}$ s^{-1}, find the activity (i.e. the number of atoms undergoing decay per second) at that instant.
14. A beta-emitter has 6×10^{20} undecayed atoms present in a sample at a given instant. If its half-life is 4 minutes, find its activity at that instant.
15. The half-life of U 235, which is an alpha-emitter, is 8×10^{18} years. Find the number of alpha-particles emitted per second from a sample containing 6.2×10^{16} U 235 atoms.
16. If a sample of radium contains 2.6×10^{21} radium-226 nuclei and is emitting 3.5×10^{10} particles per second calculate:
   (i) the decay constant and
   (ii) the half-life of radium-226.

Atomic Mass

Since hydrogen is the simplest atom (containing only one proton in its nucleus), scientists used the mass of one hydrogen atom as a unit of mass and expressed the mass of other atoms relative to it. Thus since an oxygen atom has eight protons and eight neutrons it has approximately sixteen times the mass of the hydrogen atom. Similarly, the approximate relative atomic mass of carbon–12 ($^{12}_{6}C$) is twelve. Since the proton and neutron have approximately the same mass, the relative atomic mass of an element is approximately equal to its mass number.

The Unified Atomic Mass Unit

In 1960 scientists slightly changed the size of the unit of relative atomic mass. They took 1⁄12 the mass of the $^{12}_{6}C$ as the new unit of atomic mass. They called this new unit the **unified atomic mass unit (u)**.

The Mole

THE MOLE

A ***mole*** of any substance is the amount of that substance that contains as many particles as there are atoms in exactly 12 grams $^{12}_{6}C$. This number is 6.02×10^{23}, and is called ***Avogadro's number***.

Thus: a mole of electrons is 6.02×10^{23} electrons,
a mole of Iron atoms is 6.02×10^{23} Iron atoms.

Important fact:
An atomic mass of any element expressed in grams contains 6.02×10^{23} atoms.

Using this fact we can find the number of atoms in a given mass of an element or mass in grams of any number of atoms of a given element.

Problem 10: How many atoms are there in 10 kg of Lead–207, ($^{207}_{82}Pb$)?

Solution: Atomic mass of $^{207}_{82}Pb = 207 \Rightarrow 207$ grams of $^{207}_{82}Pb$ has 6.02×10^{23} atoms

$$\Rightarrow 1 \text{ gram of } ^{207}_{82}Pb \text{ has } \frac{6.02 \times 10^{23}}{207} \text{ atoms}$$

$$\Rightarrow 10 \text{ kg of } ^{207}_{82}Pb \text{ has } \frac{(10\,000)(6.02 \times 10^{23})}{207} = 2.9 \times 10^{25} \text{ atoms}$$

Problem 11: The decay constant of U 235 which is an alpha-emitter is $2.75 \times 10^{-17} s^{-1}$. Find the number of alpha-particles emitted per second from a 1 kg sample of this substance.

Solution: 235 grams of $^{235}_{92}U$ has 6.02×10^{23} atoms $\Rightarrow$ 1 gram of $^{235}_{92}U$ has $\frac{6.02 \times 10^{23}}{235}$ atoms

$$\Rightarrow 1 \text{ kg of } ^{235}_{92}U \text{ has } \frac{(1000)(6.02 \times 10^{23})}{235} = 2.56 \times 10^{24} \text{ atoms}$$

Rate of decay $= \lambda N \Rightarrow$ Number of α-particles emitted per second

$$= \lambda N = (2.75 \times 10^{-17})(2.56 \times 10^{24}) = 7.0 \times 10^{7}$$

EXERCISE 30.3

Take Avogadro's constant = 6.02×10^{23} mol^{-1}

1. How many atoms are there in 2 kg of copper? 1 mole of copper = 64 grams.

2. How many atoms in 3.5 μg of Po–218? 1 mole of Po–218 = 218 grams.

3. A sample of Po–218, which is an α-emitter, has a mass of 2.4 μg. Calculate the number of α-particles emitted per second from the sample, if its half-life is 3.1 minutes.

4. A certain radioactive substance decays by α-emission with a half-life of 22 hours. At a certain time a sample of the substance is found to be emitting 150 α-particles per second. If one mole of the substance has a mass of 228 grams, calculate the mass of the substance present.

CHAPTER CHECKLIST

- **State**: The names of the three types of nuclear radiation; The nature and properties of α, β and γ radiation; The unit of activity of a radioactive source;

  The Law of Radioactive Decay; The unit in which decay constant is measured.
- **Work out** the daughter nucleus formed when a given nucleus undergoes α-decay, β-decay or γ-decay.
- **Define**: Emission spectrum; Energy level; Atomic number; Mass number; Radioactivity; Activity of a radioactive source; The becquerel; Half-life;

  Decay constant.
- **Recall**: The principle of Rutherford's Experiment and the conclusion it leads to; That continuous spectra are produced by incandescent solids or liquids and are not characteristic of the material producing them; That a line spectrum is produced by an elemental gas and is characteristic of the element producing it; That spectra can be used to identify minute amounts of a substance.
- **Describe**: The location of the proton, electron and neutron in an atom; The principle of operation of a detector of ionising radiation; The deflection of α, β and γ rays in electric and magnetic fields.
- **Explain** line spectra in terms of electron transitions between different energy levels.
- **Describe** an experiment to: Demonstrate continuous spectra and line spectra; Demonstrate the ionisation produced by and the penetrating power of α, β and γ radiation; Demonstrate the G-M tube or the solid-state detector.
- **Recall** and use the formulae:

  Number of neutrons = $A - Z$; $\quad hf = E_2 - E_1$

  Rate of decay = λN; $\quad T_{1/2} = \dfrac{\ln 2}{\lambda}$
- **List** six uses of radioactive isotopes.

Fission, Fusion and Nuclear Energy

CHAPTER 31

Nuclear Fission

In 1939 Hahn and Strassmann discovered that when uranium is bombarded with neutrons, a new type of nuclear reaction occurs. They called this type of reaction **nuclear fission**. Nuclear fission is the splitting up of a large nucleus into two smaller nuclei of roughly the same size.

- Fission is produced in a large nucleus by **bombarding it with neutrons**. The neutrons act as a trigger for the reaction.
- During fission very large amounts of **energy are given out**, about 200 MeV per nucleus.
- More **neutrons are produced** in the fission reaction. These can produce further fission.

NUCLEAR FISSION
Nuclear Fission is the splitting up of a large nucleus into two smaller nuclei of roughly the same size.

Fission of Uranium

Natural uranium is a mixture of two main isotopes. These are ^{235}U (0.7%) and ^{238}U (99.3%).

Uranium 235

If ^{235}U is bombarded with fast or slow neutrons it undergoes fission. It is much more likely to undergo fission with slow neutrons, i.e. those moving with kinetic energies equal to the average kinetic energy of the surrounding atoms (called **thermal neutrons**). The two nuclei formed during fission vary from one fission to the next, but their masses are roughly similar. FIG. 31.1 shows a typical reaction.

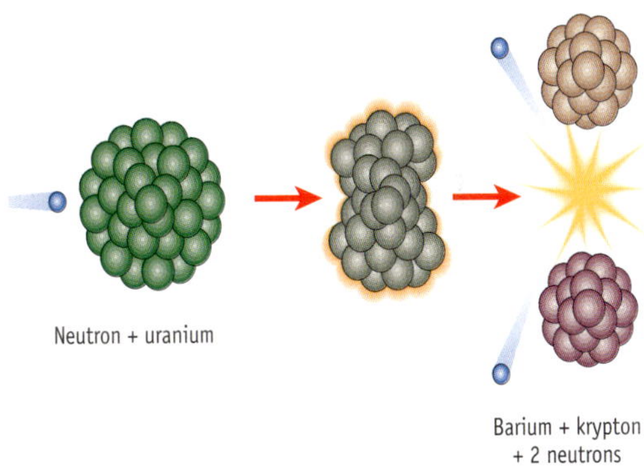

$$^{235}_{92}U + ^{1}_{0}n \rightarrow ^{144}_{56}Ba + ^{90}_{36}Kr + 2^{1}_{0}n$$

Fig. 31.1
Nuclear fission.

The products formed are called the **fission fragments**. Most of the energy released goes to the kinetic energies of the fission fragments. These are very often themselves radioactive. The neutrons released are fast neutrons and may trigger further fission. If at least one neutron from each atom that undergoes fission produces further fission we have a **chain reaction** (FIG. 31.2). In a small sample of fissionable material, many neutrons escape and are not available for further fission. If the size of the sample is increased, a size is reached – called the **critical size** – whereby a chain reaction can occur. For ^{235}U, this is about the size of a tennis ball and has a mass of about 10 kg. Plutonium-239 (^{239}Pu) also undergoes fission with both fast and slow neutrons. ^{239}Pu and ^{235}U are said to be **fissile**.

Fig. 31.2
A fission chain reaction in ^{235}U.

Fig. 31.3
The devastation produced in Hiroshima.

Atomic Bomb (A Fission Bomb)

In a fission bomb at least two pieces of fissile material of sub-critical mass are very suddenly brought together and an uncontrolled chain reaction occurs with an enormous release in energy (Fig. 31.3). The material used is either plutonium-239 or uranium-235.

Nuclear Reactor (Fission Reactor)

In a nuclear reactor nuclear energy is released at a slow controllable rate. The energy may be used to drive a turbine to generate electricity.

THE THERMAL NUCLEAR REACTOR

It is an experimental fact that ^{238}U absorbs fast neutrons without undergoing fission. It only absorbs slow neutrons to a small extent. If, in a sample of natural uranium, a ^{235}U isotope undergoes fission, the neutrons produced are fast. Since the majority of the atoms in natural uranium are ^{238}U, the neutrons get captured by the ^{238}U and do not produce further fission in ^{235}U. Thus in pure uranium a chain reaction will not occur.

However, **if the neutrons were slowed down they would produce further fission in the ^{235}U instead of being captured by the ^{238}U.**

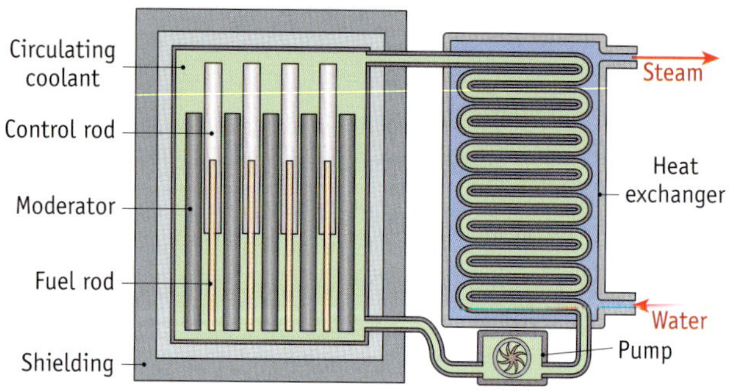

Fig. 31.4
A nuclear reactor.

Thus a chain reaction can be made to occur in a large enough sample of natural uranium. This happens in the thermal reactor (Fig. 31.4). Such a reactor operates as follows:

- The **fuel** is natural uranium or uranium slightly enriched with ^{235}U.
- The **moderator** is graphite or heavy water (D_2O). It slows down neutrons so that they produce further fission in ^{235}U rather than be absorbed by the ^{238}U. (The neutrons collide with the nuclei of the atoms in the moderator; this slows them down.)
- The **control rods** are usually made from steel with cadmium or boron. They absorb neutrons. Placing them into the core of the reactor slows the reaction. Placing them in fully stops the reaction. As they are removed the rate at which the reaction occurs increases.
- The **shielding** stops radiation escaping, since the fission fragments are highly radioactive.
- The **coolant** takes heat from the core to the heat exchanger.
- The **heat exchanger** uses the heat to produce steam. The steam drives a turbine to generate electricity.

ENVIRONMENTAL IMPACT OF FISSION REACTORS

- **Mining Uranium ore.** The mining of uranium ore releases radon gas, which can cause lung cancer in miners. The area around the mine may contain radioactive materials.
- **Containment of radioactive materials within the reactor.** There have been accidents in reactors with the release of radioactivity into the atmosphere or leakage from the cooling system. Major accidents are rare, but the consequences can be very serious, with large amounts of radioactivity being released into the atmosphere, e.g. the Chernobyl accident in 1986.
- **The removal and treatment of spent fuel rods, i.e. nuclear reprocessing.** These are withdrawn from the core and transferred to a cooling pond to cool down. They are then transported to a reprocessing plant to separate uranium and plutonium from the fission products. There are problems associated with the transport of these materials.
- **Radioactive waste.** The remaining waste products must be stored securely for a very long time. This is likely to be a very big problem for future generations.

NUCLEAR FUSION

Another kind of nuclear reaction that can occur is the joining together of two small nuclei to form a larger nucleus. Such a reaction is called nuclear fusion.

An important example is the fusion of two heavy hydrogen atoms (Deuterium) to form Helium.

$$^{2}_{1}H + ^{2}_{1}H \rightarrow ^{3}_{2}He + ^{1}_{0}n$$

Another is the fusion of deuterium and tritium:

$$^{2}_{1}H + ^{3}_{1}H \rightarrow ^{4}_{2}He + ^{1}_{0}n$$

- Fusion can only occur if the two reacting nuclei are forced together with sufficient force to overcome the coulomb repulsion between them. This is done by heating them to extremely high temperatures, typically greater than 10^8 K.
- When fusion starts, energy is released which can help keep the reaction going.
- No one has yet managed to achieve a sustained controlled fusion reaction. A great deal of effort is currently being put into this project.
- The hydrogen bomb is an uncontrolled fusion reaction. The initial high temperatures are produced by a small fission bomb exploding in the deuterium.
- Nuclear fusion in the interior of the Sun is the principal source of the Sun's energy. In a series of reactions hydrogen fuses to form helium, releasing energy in the process.

NUCLEAR FUSION
Nuclear Fusion is the joining of two small nuclei to form a larger nucleus.

ADVANTAGES OF FUSION OVER FISSION AS A SOURCE OF POWER

- There is less radioactive waste produced.
- There is no possibility of an uncontrolled runaway reaction occurring.
- The fuel, deuterium, is readily available in the oceans and can be extracted cheaply.

The Equivalence of Mass and Energy

In 1905, Einstein, in his Special Theory of Relativity, first proposed that mass is a form of energy. Thus mass and energy are not independent quantities but are related to each other. He further stated that **mass can be converted into energy and energy can be converted into mass**.

If some petrol and oxygen are placed in a completely sealed container and burned, energy is released. If the container is allowed to cool down this energy has left the container. According to Einstein, the mass of the container and contents must have decreased to account for this energy loss. However, the loss in mass is incredibly small. It can be calculated from the now famous equation:

$$E = mc^2$$

Change in Energy = Change in Mass × (Speed of Light)^2

> **NOTE** Mass must be expressed in kilograms when using the formula $E = mc^2$

Because the value of c, the speed of light, is so large, a tiny decrease in mass causes an enormous release of energy. The changes in mass that occur in normal physical and chemical reactions (like burning petrol) are so small that they cannot be detected. In nuclear reactions it is different. It has been estimated that about 1 gram of matter was converted into energy in the atomic bomb that was dropped on Hiroshima in 1945.

Problem 1: When 10 litres of petrol are completely burned in air 350 000 000 J of energy are released. Find the decrease in mass.

Solution: $E = mc^2 \Rightarrow m = \frac{E}{c^2} = \frac{350\,000\,000}{(3 \times 10^8)^2} = 3.9 \times 10^{-9}$ kg

This is a very small mass and cannot be measured on any currently available balance.

Problem 2: How much mass does a 100 W bulb dissipate in a year?

Solution: 1 year = (365)(24)(60)(60) seconds = 3.15×10^7 s

Energy released in 1 year = power × time = $(100)(3.15 \times 10^7) = 3.15 \times 10^9$ joules

$E = mc^2 \Rightarrow m = \frac{E}{c^2} = \frac{3.15 \times 10^9}{(3 \times 10^8)^2} = 3.5 \times 10^{-8}$ kg

Problem 3: The mass of the Sun decreases by 4×10^6 kg every second due to the emission of electromagnetic radiation (sunlight). Find the energy emitted as electromagnetic radiation by the sun in one second.

Solution: $E = mc^2 = (4 \times 10^6)(3 \times 10^8)^2 = 3.6 \times 10^{23}$ joules per second

Mass-energy Conservation in Nuclear Reactions

In nuclear reactions the size of the energy changes occurring are much larger and changes in mass can be detected. In a nuclear reaction the mass of the products is often noticeably different from the masses of the reactants. The energy corresponding to the change in mass is either given out or taken in during the reaction.

Mass of products > Mass of reactants ⇒ Energy must be supplied
Mass of reactants > Mass of Products ⇒ Energy is given out.

Energy released is in the form of the kinetic energies of the products or as gamma rays or both.

Problem 4: In a nuclear reaction the difference in mass between the reactants and the products is 5×10^{-30} kg. The mass of the reactants is greater. Calculate the energy liberated in this reaction.

Solution: $E = mc^2 = (5 \times 10^{-30})(3 \times 10^8)^2 = 4.5 \times 10^{-13}$ J

Problem 5: When lithium is bombarded with a proton two alpha-particles are produced. How much energy is released in this reaction? Give your answer in both joules and MeV. The mass of a lithium nucleus is $1.165\,007 \times 10^{-26}$ kg, the mass of a proton is $1.673\,493 \times 10^{-27}$ kg and the mass of an alpha particle $6.646\,322 \times 10^{-27}$ kg.

Solution: The reaction is: ${}^{7}_{3}\text{Li} + {}^{1}_{1}\text{H} \rightarrow {}^{4}_{2}\text{He} + {}^{4}_{2}\text{He} +$ Energy

Total mass of reactants $= 1.165\,007 \times 10^{-26}\text{ kg} + 1.673\,493 \times 10^{-27}\text{ kg}$

$= 1.332\,356 \times 10^{-26}$ kg

Total mass of products $= 2 \times 6.646\,322 \times 10^{-27}\text{ kg} = 1.329\,264 \times 10^{-26}$ kg

Decrease in mass $= 1.332\,356 \times 10^{-26} - 1.329\,264 \times 10^{-26} = 3.092 \times 10^{-29}$ kg

Energy released $E = mc^2 = (3.092 \times 10^{-29})(3 \times 10^8)^2 = 2.7828 \times 10^{-12}$ J

Recall that 1 eV $= 1.6 \times 10^{-19}$ J

$\therefore$ Energy released $= \dfrac{2.7828 \times 10^{-12}}{1.6 \times 10^{-19}}\text{ eV} = 1.739 \times 10^{7}\text{ eV} = 17.39$ MeV

Problem 6 A radioactive nucleus, initially at rest, emits an alpha-particle of mass 6.68×10^{-27} kg to produce a new nucleus of mass 3.67×10^{-25} kg. Explain how the principle of conservation of momentum and energy apply to this reaction.

Calculate the ratio of the speed of the alpha-particle to the speed of the new nucleus. Given that their total mass when at rest is 9.50×10^{-30} kg less than the mass of the original nucleus, calculate their total initial kinetic energy and hence the speed of each.

Solution: FIG. 31.5 shows the situation. The particles move in opposite directions so that the total momentum before emission = Total momentum after emission. The loss in rest mass during the emission = The kinetic energy of the particles after the emission

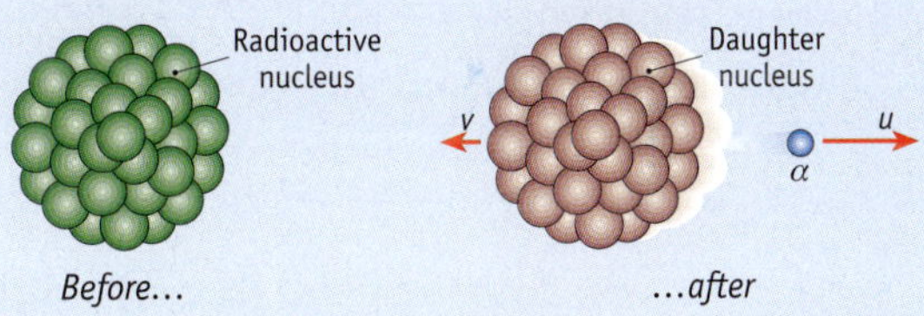

Fig. 31.5

Conservation of momentum: $0 = -3.67 \times 10^{-25} V + 6.68 \times 10^{-27} U \Rightarrow U/V = 54.94$

Loss in rest mass $= 9.5 \times 10^{-30}$ kg. This appears as the total kinetic energy E of the particles.

$E = mc^2 = (9.5 \times 10^{-30})(3 \times 10^8)^2 = 8.55 \times 10^{-13}$ J

$(½)(3.67 \times 10^{25})V^2 + (½)(6.68 \times 10^{-27})U^2 - 8.55 \times 10^{-13}$

Also $U = 54.94\,V$.

Solving these two simultaneous equations for U and V gives:

$U = 1.58 \times 10^{7}\text{ m s}^{-1}$ and $V = 2.866 \times 10^{5}\text{ m s}^{-1}$

Ernest Walton with his daughter.

The reaction in problem 5 (page 363) was carried out by Cockroft and the Irish scientist Walton in 1932. It was the first nuclear reaction produced by artificially accelerated particles, i.e. the **first transmutation produced by artificially accelerated particles**. It was also the first experimental verification of Einstein's equation $E = mc^2$, since the masses and speeds of all the particles could be measured. They won a Nobel Prize in 1951. (See also Chapter 32.)

EXERCISES 31.1

Charge on electron $e = 1.6 \times 10^{-19}$ C *Speed of light* $c = 3.00 \times 10^8$ m s^{-1}

1. When 40 litres of petrol (a typical full car tank) are completely burned in air 1.4×10^9 J of energy are released. Find the amount of mass converted to energy.

2. The mass of the Sun decreases by 4×10^6 kg every second due to the emission of electromagnetic radiation (sunlight). Find the energy emitted as electromagnetic radiation by the sun in one hour.

3. The sun emits 3.6×10^{23} joules of energy every second. Calculate the decrease in mass of the Sun in one year.

4. In a nuclear reaction the difference in mass between the reactants and the products is 8×10^{-27} kg. The mass of the reactants is greater. Calculate the energy liberated in this reaction.

5. In 1919 Rutherford bombarded Nitrogen with α-particles and the following nuclear reaction took place:

$${}^{14}_{7}\text{N} + {}^{4}_{2}\text{He} \rightarrow {}^{17}_{8}\text{O} + {}^{1}_{1}\text{H}$$

This was the **first artificial transmutation of an element**, i.e. the converting of one element into another. It also showed that **the nuclei of elements contain protons**. The mass of ^{14}N is $2.325\,210 \times 10^{-26}$ kg, the mass of ^{4}He is $6.646\,322 \times 10^{-27}$ kg, the mass of ^{17}O is $2.822\,706 \times 10^{-26}$ kg, and the mass of ^{1}H is $1.672\,623 \times 10^{-27}$ kg. If the energy of the incident α-particle is 7.68 MeV, calculate the kinetic energy of the proton assuming it gets 17/18 of the available kinetic energy.

6. How much energy is released in the following fusion reaction?

$${}^{2}_{1}\text{H} + {}^{2}_{1}\text{H} \rightarrow {}^{4}_{2}\text{He}$$

The mass of each deuterium nucleus is 3.344×10^{-27} kg and the mass of the helium nucleus is 6.646×10^{-27} kg.

7. Complete the following nuclear reactions and comment on their historical significance:

$${}^{7}_{3}\text{Li} + \square \rightarrow {}^{4}_{2}\text{He} + {}^{4}_{2}\text{He}$$

$${}^{14}_{7}\text{N} + \square \rightarrow {}^{17}_{8}\text{O} + {}^{1}_{1}\text{H}$$

8. When cobalt-59 is irradiated with neutrons a radioactive isotope of cobalt is formed and a gamma ray photon is emitted. Write an equation to represent this reaction. The mass of the cobalt-59 nucleus is 9.7859×10^{-26} kg and the mass of the nucleus produced in this reaction is 9.9520×10^{-26} kg. Given that the mass of the neutron is 1.6749×10^{-27} kg, find the energy of the photon produced in the reaction. (Neglect the kinetic energies of the particles involved and take the speed of light to be 3.0×10^8 m s^{-1}.)

9. The fission of one nucleus of uranium-235 releases 200 MeV of energy. Calculate the decrease in mass for this reaction.

10. A radioactive nucleus, initially at rest, emits an alpha-particle of mass 6.68×10^{-27} kg to produce a new nucleus of mass 4.24×10^{-25} kg. Using the principle of conservation of momentum, calculate the ratio of the speed of the alpha-particle to the speed of the new nucleus. Given that their total mass, when at rest, is 1.20×10^{-29} kg less than the mass of the original nucleus, calculate their total initial kinetic energy, and hence the speed of each.

11. With the exception of ${}^{1}_{1}$H, all nuclei contain neutrons. An isolated neutron is not a stable particle but decays into a proton. Its half-life is 10.6 minutes. Write an equation to show this reaction. Calculate the energy released in this reaction.

Mass of neutron = $1.674\,929 \times 10^{-27}$ kg
Mass of proton = $1.672\,623 \times 10^{-27}$ kg
Mass of electron = $9.109\,390 \times 10^{-31}$ kg

Health Hazards of Ionising Radiation

Ionising radiation is a general word for any form of radiation that will knock off outer electrons of atoms, thus forming ions. Such radiations will cause ionisation when they are absorbed by human tissue.

α-radiation, β-radiation, γ-radiation, X-rays and neutrons are thus examples of ionising radiation.

All ionising radiations are harmful to the human body. The amount of damage done depends on:

- the type of radiation,
- the activity of the source producing it,
- the duration of the exposure,
- the type of tissue irradiated (i.e. the part of the body).

Ionising radiation can cause:

- skin burns similar to intense sunburn,
- cataracts, leukaemia and other cancers,
- genetic defects in children of parents exposed to the radiation,
- death.

The radiation produces highly reactive ions or radicals within cells. These disrupt the normal operation of the cell. Chemical reactions can occur that break bonds in proteins and other vital molecules. The radiation may also ionise vital molecules. In large doses, damage to sufficient molecules in a cell may kill it. The death of many cells may irreversibly damage the organism. Cells that do survive the radiation may become defective. These may produce more defective cells when they divide, causing cancer. If the reproductive cells are affected, we say genetic damage has occurred. Damage to genes can lead to defective offspring.

The effects of short-term exposure to high levels of radiation are well known, e.g. survivors of the atomic bombs dropped in Hiroshima and Nagasaki suffered what is called radiation sickness within a few hours of exposure. This consists of vomiting, diarrhoea, fever and dehydration. Recovery is possible if the dose is not too great. Larger doses will kill within a few days or weeks. Those who survive will have weakened defences to other diseases. Leukaemia and other cancers often appear after a number of years. Long-term exposure to lower levels is more difficult to quantify, but damage to DNA in cells leads to genetic changes and some forms of cancer, particularly leukaemia. **It is therefore important to minimise one's exposure to any form of ionising radiation.**

We are all exposed all the time to some radiation, called **background radiation**. Background radiation comes from the following:

- **Outer space.** Radiation coming from space is known as **cosmic rays**.
- **Rocks in the Earth's crust.** Rocks in the Earth's crust contain traces of uranium and its decay products, one of which is **radon gas**. In Ireland, regions of granite rock release radon gas, which can accumulate in houses to levels that increase the risk of lung cancer (Fig. 31.22).
- **Man-made radioactive materials.**

Natural sources account for about 87% of background radiation.

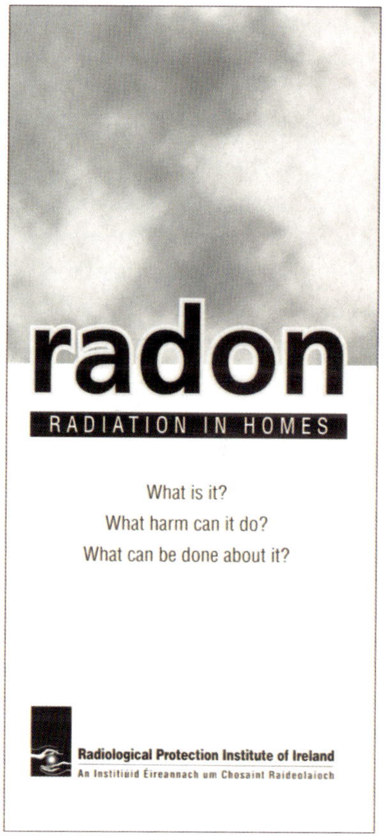

Fig. 31.22
An information leaflet on radon gas available from the Radiological Protection Institute of Ireland.

PRECAUTIONS WHEN USING IONISING RADIATIONS

- Estimate dose rate before using any source of radiation and measure dose rate while proceeding.
- Minimise the time spent using sources of radiation.
- Use proper protective clothing, e.g. gloves, glasses, coat etc.
- Make sure sources are properly shielded from you.
- Keep as far away from the sources as possible.
- Do not eat, drink or smoke, i.e. make sure unsealed sources do not get ingested.
- Use tongs for handling sources.

Keep sources securely locked away when not in use and far away from people.

EXPERIMENT

TO MEASURE BACKGROUND RADIATION.

Method

1. Set up the equipment shown in FIG 31.7.
2. Switch on the scaler or ratemeter.
3. Set the high voltage supply to its lowest value and allow the tube to warm up for a few minutes.
4. Set the tube to its operating voltage.
5. Measure the background count rate. If a scaler is used, count the number of counts registered in 100 s. If a ratemeter is used, take an integrating time of 25 s.
6. Note how the count rate varies with time, showing the random nature of background radiation.

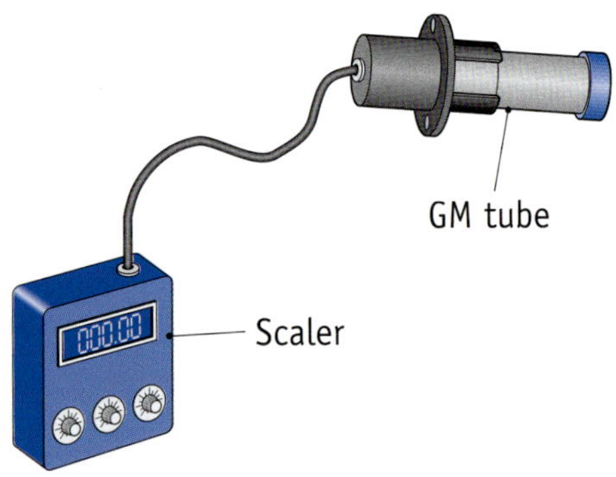

Fig 31.7

CHAPTER CHECKLIST

- **State**: The two main isotopes of uranium; The advantages of fusion over fission as a source of power; The factors that determine how much damage ionising radiation causes in human tissue.
- **Define**: Nuclear fission; Nuclear fusion.
- **Explain**: The principles of nuclear fission and nuclear fusion; Chain reaction; Critical size; Moderator; Control rods; What is meant by mass-energy conservation in nuclear reactions.
- **Recall**: That energy can be converted into mass and vice versa; That nuclear fusion is the source of the Sun's energy.
- **Describe**: the operation of a nuclear reactor.
- **List**: Three environmentally negative aspects of fission reactors; Three health hazards of ionising radiation; Five precautions to be taken when using sources of ionising radiation.
- **Describe an experiment** to measure background radiation.
- **Use the formula**: $E = mc^2$ in problems on mass-energy conservation in nuclear reaction.

Option 1 (Honours Only) Particle Physics

CHAPTER 32

Conservation of Energy and Momentum in Nuclear Reactions

In Chapter 30 you saw that many elements are naturally radioactive and their nuclei undergo changes when they undergo radioactive decay. In α and β emission, the parent nucleus decays into the daughter nucleus and an α or a β-particle is emitted.

In reactions such as these and in all other nuclear reactions the principle of conservation of energy (mass-energy) and the principle of conservation of momentum apply. Also the nett amount of electric charge before the reaction is the same as the nett amount of charge after the reaction, i.e. electric charge is conserved. When a radioactive disintegration takes place spontaneously, energy is released. This energy is called the **disintegration energy**. Its symbol is Q.

IN A NUCLEAR REACTION:

Mass-Energy is conserved;
Momentum is conserved;
Electric charge is conserved.

In the following problems take: $c = 3.00 \times 10^{8}$ m s^{-1}; $e = 1.60 \times 10^{-19}$ C

Problem 1: ${}^{238}_{92}\text{U}$ is an alpha-emitter and decays spontaneously as follows:

$${}^{238}_{92}\text{U} \rightarrow {}^{234}_{90}\text{Th} + {}^{4}_{2}\text{He} + \text{Q}$$

The energy released Q is shared between the daughter nucleus (${}^{234}_{90}\text{Th}$) and the α-particle (${}^{4}_{2}\text{He}$) as kinetic energy. Given that the masses are:

^{238}U: 3.952833×10^{-25} kg
^{234}Th: 3.886294×10^{-25} kg
^{4}He: 6.646322×10^{-27} kg

find the energy released. Give the answer in both joules and MeV.

Solution: The masses are given correct to 6 decimal places. We must use all of these until we have calculated the change in mass.

$$\begin{aligned}\text{Loss in mass} &= \text{mass of reactants} - \text{mass of products} \\ &= (3.952833 \times 10^{-25}) - (3.886294 \times 10^{-25} + 6.646322 \times 10^{-27}) \\ &= 7.578 \times 10^{-30}\text{ kg}\end{aligned}$$

Energy released $E = mc^2 = (7.578 \times 10^{-30})(3.00 \times 10^{8})^2 = 6.8202 \times 10^{-13}$ J

1 eV $= 1.60 \times 10^{-19}$ J, $\Rightarrow$ energy released $= \dfrac{6.8202 \times 10^{-13}}{1.60 \times 10^{-19}}$ eV

$= 4.26 \times 10^{6}$ eV

$= 4.26$ MeV

Problem 2: In the reaction $^{238}_{92}\text{U} \rightarrow {}^{234}_{90}\text{Th} + {}^{4}_{2}\text{He}$, 6.8×10^{-13} J of energy are released as the kinetic energies of the products. If the ratio of the masses of $^{234}_{90}\text{Th}$ to $^{4}_{2}\text{He}$ is 234:4, find the kinetic energy of the α-particle.

Solution: FIG. 32.1 shows the situation

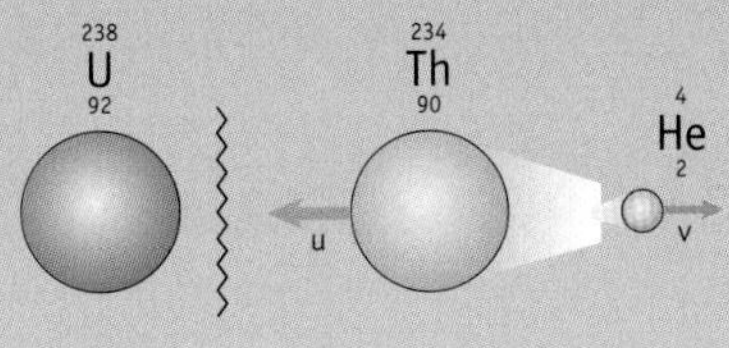

Fig. 32.1

Let M_T be the mass of the thorium atom and M_H the mass of the alpha-particle.

Conservation of momentum

$$\Rightarrow \quad M_T u = M_H v \Rightarrow \frac{v}{u} = \frac{M_T}{M_H} = \frac{234}{4} = 58.5$$

$$\text{Ratio of kinetic energies} = \frac{\frac{1}{2}M_T u^2}{\frac{1}{2}M_H v^2} = \left(\frac{M_T}{M_H}\right)\left(\frac{u^2}{v^2}\right) = \left(\frac{v}{u}\right)\left(\frac{u^2}{v^2}\right) = \frac{u}{v} = \frac{1}{58.5}$$

$\Rightarrow$ Kinetic energy of α-particle is 58.5 times the kinetic energy of the Thorium nucleus

$$\Rightarrow \quad \text{Kinetic energy of } \alpha\text{-particle} = \left(\frac{58.5}{58.5+1}\right)(6.8 \times 10^{-13}) = 6.69 \times 10^{-13}\ \text{J}$$

Note that as is typical in reactions like this the lighter particle, the α-particle, gets most of the kinetic energy.

THE NEUTRINO

In any nuclear decay process, the energy of the products is determined by the change in mass of the system as in Problem 1. When this kind of analysis was applied to β-decay a problem arose. The calculated and the experimentally determined values for energies of the β-particle did not always agree. Both energy and momentum appeared not to be conserved.

In 1931, the Austrian physicist Wolfgang Pauli proposed that a third particle must also be emitted to carry away the missing energy and momentum. This particle became known as the **neutrino**. The neutrino (ν) only interacts very weakly with matter and was not detected experimentally until 1956 by Cowan and Reines.

EXERCISE 32.1

In the following exercises $c = 3.00 \times 10^8\ \text{ms}^{-1}$; $e = 1.6 \times 10^{-19}$C

1. ^{226}Ra undergoes α-decay according to the equation:

$$^{226}_{88}\text{Ra} \rightarrow {}^{222}_{86}\text{Rn} + {}^{4}_{2}\text{He} + Q$$

If the masses of Ra, Rn and He are:

^{226}Ra: 3.753152×10^{-25} kg,
^{222}Rn: 3.686602×10^{-25} kg,
^{4}He: 6.646322×10^{-27} kg,

calculate the disintegration energy Q in both joules and MeV.

2. In the reaction:

$$^{226}_{88}\text{Ra} \rightarrow {}^{222}_{86}\text{Rn} + {}^{4}_{2}\text{He} + Q$$

7.8×10^{-13} J of energy are released as the kinetic energies of the products. If the ratio of the masses of $^{222}_{86}\text{Rn}$ to $^{4}_{2}\text{He}$ is 222:4, find the kinetic energy of the α-particle.

3. If ^{14}C undergoes β-decay according to:

$$^{14}_{6}\text{C} \rightarrow {}^{14}_{7}\text{N} + {}^{0}_{-1}\text{e}$$

and the loss in mass is 2.77×10^{-31} kg calculate the kinetic energy of the β-particle.

The First Splitting of a Nucleus by Artificially Accelerated Particles

In 1932, the English physicist John Cockroft and the Irish physicist Ernest Walton produced a nuclear disintegration by bombarding Lithium with artificially accelerated protons. The following reaction took place:

$$^{7}_{3}\text{Li} + {}^{1}_{1}\text{H} \rightarrow {}^{4}_{2}\text{He} + {}^{4}_{2}\text{He} + \text{Energy}$$

This was the first artificial splitting of a nucleus. It was also the first transmutation* using artificially accelerated particles. The age of artificially induced nuclear reactions was born. It continues today, and continues to give us a deeper understanding of the fundamental structure of matter and of the universe.

Fig. 32.2 shows a simplified arrangement of the equipment used by Cockroft and Walton.

- They used transformers, rectifiers and capacitors to produce the necessary **high d.c. voltage** to accelerate the protons.
- Protons produced in a Hydrogen discharge tube were injected into the accelerating tube. Here they were accelerated by the high voltage.
- The protons struck a Lithium target placed at an angle of 45° to the beam.
- The products of the reaction were emitted at right angles to the proton beam and struck Zinc sulphide screens producing small flashes of light called scintillations which could be seen with a microscope.
- A number of different tests showed that the products were Helium nuclei, i.e. α-particles. If momentum was to be conserved, two Helium nuclei should be emitted in opposite directions with the same speed. This was found to be the case.

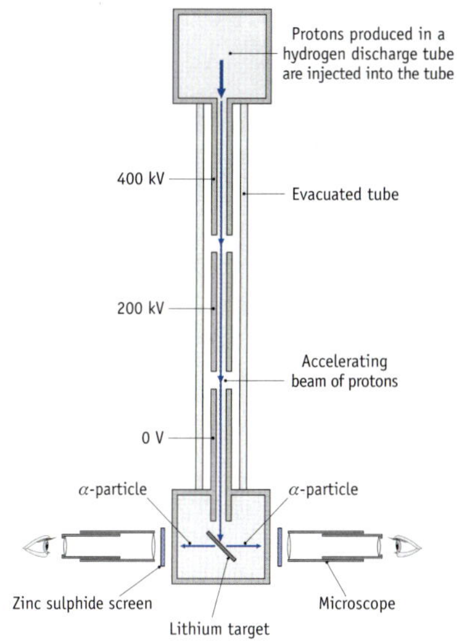

Fig. 32.2

Fig. 32.3 The Cockroft and Walton accelerator.

Converting Mass Into Energy

In Cockroft and Walton's experiment, the incident proton had an energy of about 1 MeV. The combined kinetic energies of the Helium nuclei produced was 17.3 MeV. Thus there was a **gain in energy** in this experiment. This came from the loss in mass that occurred during the experiment. In problem 3 below $E = mc^2$ will be used to calculate the energy that should be liberated in this reaction. In 1932 the experimental values and the calculated values for energies agreed with each other. This experiment was thus the **first experimental verification of Einstein's equation $E = mc^2$** in the laboratory. Cockroft and Walton won a Nobel Prize in physics in 1951.

---

* A transmutation is the changing of a nucleus of one atom into a nucleus with a different atomic number, i.e. the changing of an atom of one element into an atom of another element. In 1919 Rutherford had bombarded Nitrogen atoms with alpha-particles producing Oxygen according to the equation: ${}^{4}_{2}\text{He} + {}^{14}_{7}\text{N} \rightarrow {}^{17}_{8}\text{O} + {}^{1}_{1}\text{H}$.

This was **the first artificial transmutation**.

Problem 3: When Lithium is bombarded with a proton two alpha-particles are produced. How much energy is released in this reaction? Give your answer in both joules and MeV. The mass of a Lithium nucleus is 1.165007×10^{-26} kg, the mass of a proton is 1.673493×10^{-27} kg and the mass of an alpha-particle 6.646322×10^{-27} kg.

Solution: The reaction is: ${}^{7}_{3}\text{Li} + {}^{1}_{1}\text{H} \rightarrow {}^{4}_{2}\text{He} + {}^{4}_{2}\text{He} + \text{Energy}$

Total mass of reactants $= 1.165007 \times 10^{-26}\text{ kg} + 1.673493 \times 10^{-27}\text{ kg}$

$= 1.332356 \times 10^{-26}\text{ kg}$

Total mass of products $= 2 \times 6.646322 \times 10^{-27}\text{ kg} = 1.329264 \times 10^{-26}\text{ kg}$

Decrease in mass $= 1.332356 \times 10^{-26} - 1.329264 \times 10^{-26} = 3.092 \times 10^{-29}\text{ kg}$

$E = mc^2 = (3.092 \times 10^{-29})(3 \times 10^{8})^2 = 2.7828 \times 10^{-12}\text{ J}$

i.e. Energy released $= 2.7828 \times 10^{-12}\text{ J} = \dfrac{2.7828 \times 10^{-12}}{1.6 \times 10^{-19}}\text{ eV} = 17.39\text{ MeV}$

Which agrees with the experimentally measured values.

The Unified Atomic Mass Unit (u)

In particle physics it is more usual to express masses of particles in unified atomic mass units rather than in kilograms. On page 357 you saw that the symbol for the unified atomic mass unit was u.

It can be shown that $1\text{ u} = 1.66 \times 10^{-27}$ kg.

If a problem is given to you with masses expressed in terms of atomic mass units, calculate the change in mass in u, convert to kg and then use $E = mc^2$. For example:

Problem 4: When Lithium is bombarded with a proton two alpha-particles are produced. How much energy is released, given that the atomic masses of Lithium, a proton and an α-particle are: 7.01600 u, 1.00783 u and 4.00260 u respectively ($1\text{ u} = 1.66 \times 10^{-27}$ kg)?

Solution: The reaction is ${}^{7}_{3}\text{Li} + {}^{1}_{1}\text{H} \rightarrow {}^{4}_{2}\text{He} + {}^{4}_{2}\text{He} + \text{Energy}$

Total mass of reactants $= 1.00783 + 7.01600 = 8.02383$ u

Total mass of products $= 4.00260 + 4.00260 = 8.00520$ u

Decrease in mass $= 8.02383 - 8.00520 = 0.01863\text{ u} = (0.01863)(1.66 \times 10^{-27}\text{ kg})$

$= 3.0925 \times 10^{-29}$ kg

Energy released $E = mc^2 = (3.0925 \times 10^{-29})(3 \times 10^{8})^2 = 2.78 \times 10^{-12}\text{ J}$

EXERCISE 32.2

1. Calculate the energy in eV that is equivalent to 1 u.
2. Calculate the energy (in both joules and MeV) released in the following reaction:
   ${}^{238}_{92}\text{U} \rightarrow {}^{234}_{90}\text{Th} + {}^{4}_{2}\text{He}$, given that the atomic masses of Uranium, Thorium and Helium are: 238.050784 u, 234.043593 u and 4.002603 u respectively.

CONVERTING ENERGY INTO MASS

ANTIMATTER

The Positron

In 1932, the American physicist Carl David Anderson in California was observing the tracks produced by cosmic rays in a cloud chamber. (A cloud chamber is a chamber where moving charged particles produce tracks that can be seen). Some of the tracks produced were similar to those that would be produced by electrons, but when a magnetic field was placed in the chamber the particles deflected in the direction that a positively charged particle would. Thus the particles had the same mass and charge as the electron but the charge was of opposite sign. This particle was called a **positron**. Anderson won a Nobel prize in 1936. The positron is the **antiparticle** of the electron.

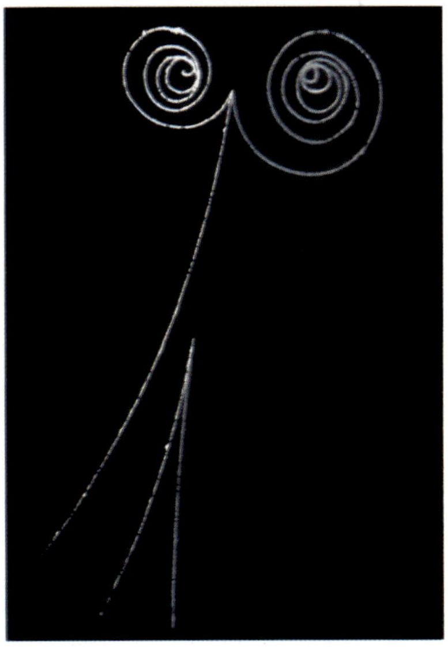

Fig. 32.4
An electron positron pair produced when a high energy gamma ray interacts with matter.

Pair Production

Since Anderson's discovery, the positron has been observed in other experiments. If a lead plate is placed in a cloud chamber and bombarded with high energy gamma rays, two similar tracks originating from the same point are produced. In a magnetic field, the tracks curve in opposite directions (FIG. 32.4). The particles produced have the same mass and the same amount of charge, but the charges are of opposite sign. They are a positron and an electron (FIG. 32.5).

The production of an electron and a positron like this is called **pair production**. It is an example of the conversion of energy into matter. Pair production occurs when a high energy γ-ray photon loses its energy hf when it collides with a nucleus. Some of the energy is converted into the mass of the electron and positron. The rest is the kinetic energy of the electron and positron.

PAIR PRODUCTION
Pair Production is the creating of two particles from energy.
In the process a particle an its anti-particle are produced.
Momentum and electric charge are conserved.

$$\text{i.e. } hf = 2mc^2 + E_{k1} + E_{k2}$$

In the reaction charge is conserved. The γ-ray has no charge, thus an equal amount of + and – charge is produced, i.e. there is no net charge after the interaction. In the reaction momentum is also conserved. (In this context the incident gamma ray has some momentum. That is why the electron-positron pair do not move in opposite directions. We need not worry about the mathematical details of momentum here).

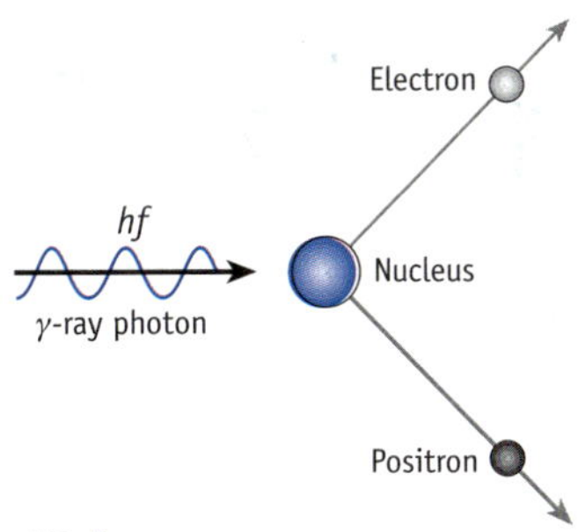

Fig. 32.5

Problem 5: If the mass of an electron is 9.109×10^{-31} kg, find the minimum energy that a γ-ray photon can have if it is to cause pair production.

Solution: The energy of the photon must be equivalent to the energy of the positron and electron

$$E = mc^2 = (2)(9.109 \times 10^{-31})(3 \times 10^8)^2 = 1.64 \times 10^{-13}\text{ J}$$

$$= \frac{1.64 \times 10^{-13}}{1.6 \times 10^{-19}}\text{ eV} = 1.02 \times 10^6\text{ eV}$$

$$= 1.02\text{ MeV}$$

If the energy of the γ-ray photon is less than $2mc^2$, pair production will not occur.

Other pairs of particles, such as a proton and antiproton can be produced if the incident photon has enough energy.

We now know that each particle has an antiparticle. **Antiparticles** have the same mass as their corresponding particles. If a particle is charged, its antiparticle has the same amount of charge but of the opposite sign. We usually denote an antiparticle with the same symbol as the particle with a bar over it. Thus the antineutrino as $\overline{\nu}$, the antiproton is $\overline{\text{p}}$. The positron is symbolised as $\overline{\text{e}}$ or more usually e^+.

The English physicist Paul Dirac predicted mathematically the existence of positrons and other antiparticles in the late 1920s. The existence of antiprotons and antineutrons was confirmed experimentally in 1955. Since then the full range of antiparticles has been confirmed. In 1995 Physcists at CERN produced nine antihydrogen atoms. They only existed for 40 billionths of a second before meeting ordinary atoms and turning back into energy.

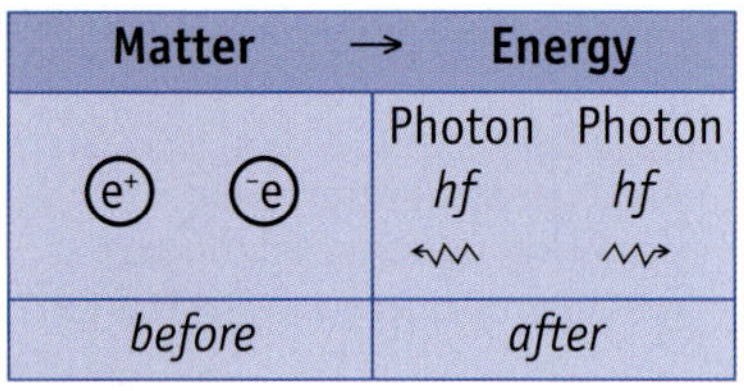

Fig. 32.6
Pair annihilation producing two photons.

Pair Annihilation

An electron and a positron which are almost at rest and near each other will join together and be annihilated. Matter disappears and energy is produced. The initial momentum of the system is zero, therefore the momentum after must be zero. Thus one photon cannot be produced, but two photons of equal energy moving in opposite directions (FIG. 32.6) are produced. The total charge before the interaction is zero as it is after the interaction.

Pair Annihilation: $\text{e}^+ + \text{e}^- \rightarrow 2\,hf$

Similarly, any particle will annihilate its antiparticle (producing energy) if they meet.

PARTICLE ACCELERATORS

Fig. 32.7

In particle accelerators such as that of Cockroft and Walton, the energy of the incident protons is quite small (about 1 MeV). It disrupts the nucleus with which it collides, and some mass is converted into energy as the kinetic energy of the products.

When particles of much higher energies collide, it is found that some of this extra energy is itself converted into matter in the form of **new particles**. Thus **energy is being converted into matter**.

However, to accelerate particles to the required high energies much better particle accelerators were required. Between 1929 and 1939, Ernest O. Lawrence in the US developed the cyclotron. This was the first circular particle accelerator. In it, magnetic fields are used to confine and control the position of the particle beams. Electric fields are used repeatedly for accelerating purposes. FIG. 32.7 is an aerial photograph of the accelerator centre in CERN in Switzerland. The smaller accelerator has a circumference of 7 km, the larger one 27 km. The tunnels for both accelerators are underground.

Beginning in about 1945, improved circular particle accelerators were made. Scientists then discovered that by accelerating known particles, such as protons, to very high speed and causing them to collide with other protons, many new particles were produced. These particles were almost always very unstable with half-lives between 10^{-6} and 10^{-23} s. Over 400 unstable temporary particles have been found.

e.g. p + p + energy → p + p + additional particles

The higher the energy made available by better particle accelerators, the greater the mass and variety of the additional particle produced.

THE SEARCH FOR THE BASIC BUILDING BLOCKS OF NATURE

ANCIENT GREECE

The ancient Greeks saw nature as being made up of earth, fire, air and water. They had elaborate explanations as to how various phenomena were made up of combinations of these. The idea that all matter is made up of small indivisible particles – atoms – also comes from this time.

NINETEENTH CENTURY

In the nineteenth century the existence of atoms and the forces between them became reasonably well understood. All matter is made up of combinations of about 90 different kind of atoms. Atoms were, however, considered indivisible (*atomos* is Greek, meaning indivisible). However, towards the end of the nineteenth century, scientists suspected that atoms were divisible and had an internal structure. Atoms contained smaller particles.

TWENTIETH CENTURY

In 1911, Rutherford's Experiment showed that the atom consisted of a minute positively charged nucleus surrounded by a cloud of negatively charged particles called electrons. The nucleus contained particles called protons which were about 1800 times more massive than the electrons. A proton has the same amount of charge as an electron but of the opposite sign. The electrons were attracted to the nucleus by electrostatic forces. The atom itself was mainly empty space.

In 1932, the neutron was discovered. Thus the nucleus consisted of two particles, positively charged protons and uncharged neutrons. The mass of a proton is about the same as the mass of a neutron.

The question then arose as to what force kept the positively charged protons together in the nucleus? Surely their positive charge would repel them apart with considerable force since they are very close to each other? There must therefore be a very **strong nuclear force** binding the protons and neutrons together in the nucleus. This force must have a **very short range**. The range is less than the diameter of the nucleus. Two protons passing each other at a distance greater than this do not interact by this force.

THE FUNDAMENTAL FORCES OF NATURE

To understand the behaviour of particles, we must be able to describe the forces that act on them. Particles in nature are subject to one or more of the four fundamental forces discussed below.

1. THE GRAVITATIONAL FORCE

You should be very familiar with the **force of gravity** that acts between particles that have mass (Chapter 10). This force is always attractive and is very weak unless the masses involved are large. The distance over which it can act is infinite. Its size is inversely proportional to the square of the distance between the particles on which it acts. Even though this force is very important in our daily lives and is the force keeps the planets, stars and galaxies together, its effect on elementary particles and the structure of the nucleus is negligible.

2. THE ELECTROMAGNETIC FORCE

This is the force that binds electrons and protons together in atoms and binds atoms and molecules together in ordinary matter. Run into a wall and the force that stops you (with pain) is electromagnetic. It is the force on which all modern electric and electronic technology is based. You are also familiar with the electromagnetic force in the form of the electrostatic force that acts between charged particles and magnetic forces that act between moving charged particles (Chapters 19 and 26). This force can be attractive or repulsive. The electromagnetic force between two particles is about 10^{40} times stronger than the gravitational force between the same particles. The distance over which it can act is infinite. Its size is inversely proportional to the square of the distance between the particles on which it acts.

3. THE STRONG NUCLEAR FORCE

This is the strongest of the four forces. We do not, however, experience it in everyday life because its range is very small. It is negligible beyond 10^{-15} m from any elementary particle. At distances less than this it is very powerful and is responsible for binding the nucleus together*. It holds the nucleus together against the repulsive electrostatic force between the protons. Some particles, e.g. electrons, do not feel the strong force at all.

4. THE WEAK NUCLEAR FORCE

This force can be felt by all particles and when very near each other it is the force between particles that are not subject to the strong force. The distance over which this force acts, 10^{-18} m, is extremely small. It is much less powerful than the strong nuclear force. The β-decay of nuclei and the decay of a neutron into a proton occur via the weak force.

THE FOUR FUNDAMENTAL FORCES OF NATURE				
Force	**Relative Strength** (based on force between two protons just touching)	**Acts on**	**Occurrence**	**Range**
Strong nuclear	1	Protons, neutrons	Binds the nucleus	Short (10^{-15} m)
Electromagnetic	10^{-2}	Charged particles	Binds atoms and molecules	Infinite ($\propto 1/r^2$)
Weak nuclear	10^{-7}	All particles	β-decay	Short range (10^{-18} m)
Gravitational	10^{-38}	All particles	Keeps the universe together	Infinite ($\propto 1/r^2$)

FAMILIES OF PARTICLES

> **NOTE**
> The higher the energy of the colliding particles
> - the greater the variety of new particles produced,
> - the greater the mass of the particles.

When very high energy particles collide, some of their energy is converted into mass. The higher the energy E, the greater the masses m of the particles produced, since from $m = \frac{E}{c^2}$ we see that the bigger E, the bigger m. Also with more mass being produced, it is found that a greater variety of particles is produced. Until the 1960s the bewildering array of particles

* This is the force that strongly binds quarks together to form protons, neutrons and other particles called hadrons. A residual effect of the force then binds protons and neutrons together in the nucleus, much like the way the forces between molecules in matter is a residual effect of the electromagnetic forces between the charged particles in the molecules.

being produced by high energy collisions was likened to the animals in a zoo. There was no apparent relationship between them or any way of categorising them. These particles became known as the '**Particle Zoo**'.

Problem 6: Two protons of equal energy collide head on, producing two further particles (a pion π^+) and an antipion π^-) according to the following equation:

$$p + p \rightarrow p + p + \pi^+ + \pi^-$$

If the mass of each pion produced is 2.5×10^{-28} kg, find the minimum energy (in MeV) that each incident proton must have.

Solution: Mass 'created' $= (2)(2.5 \times 10^{-28})$ kg

Energy needed (from $E = mc^2$) $= (2)(2.5 \times 10^{-28})(3 \times 10^8)^2$ J

$$= \frac{(2)(2.5 \times 10^{-28})(3 \times 10^8)^2}{1.6 \times 10^{-19}} = 281 \text{ MeV}$$

Thus energy of each proton $= \frac{281}{2} = 140$ MeV

EXERCISE 32.3

1. Two protons of equal energy collide head on, producing a further particle, a pion, according to the following equation:
$$p + p \rightarrow p + p + \pi^+$$
If the mass of the pion produced is 2.5×10^{-28} kg, find the minimum energy (in MeV) that each incident proton must have.

2. Two protons of equal energy collide head on, producing further particles (a neutron and three pions) according to the following equation:
$$p + p \rightarrow p + n + \pi^+ + \pi^0 + \pi^0$$
If the mass of each pion produced is 2.5×10^{-28} kg, and the neutron and proton have the same mass, find the minimum energy (in MeV) that each incident proton must have.

Classification of the Particles – Leptons and Hadrons

After a while, physicists began to see some order in this apparent chaos. Particles were classified according to whether they felt the strong force or not. Thus two families of particles emerged, namely those that were affected by the weak force but not the strong force (called **leptons**) and those that were affected by both strong and weak forces (called **hadrons**)

An ***elementary particle*** is a particle that has no other particles inside it or has no subparts to it.

Leptons

Particles that feel the weak force* and not the strong force are called leptons. The leptons appear to have no internal structure and are now considered to be elementary particles. They are not made up of still smaller particles. Leptons are point particles with no finite dimensions.

At present six leptons and their antiparticles are known. They are the electron, the muon and the tau and the neutrinos associated with these; the electron neutrino, the muon neutrino and the tau neutrino and also the antiparticles of each of these (FIG. 32.8).

LEPTONS

A ***lepton*** is a particle that does not feel the strong force.

* Leptons are affected by the gravitational and electromagnetic force (if they are charged) as well.

THE LEPTON FAMILY					
Particle name	**Symbol**	**Relative mass**	**Charge**	**Mean life**	**Antiparticle symbol**
Electron	e^-	1	−1	stable	e^+
(Electron) neutrino	ν_e	≈ 0	0	stable	$\bar{\nu}_e$
Muon	μ^-	207	−1	2.2×10^{-6} s	μ^+
(Muon) neutrino	ν_μ	≈ 0	0	stable	$\bar{\nu}_\mu$
Tau	τ^-	3 500	−1	$< 4 \times 10^{-13}$ s	τ^+
(Tau) neutrino	ν_τ	≈ 0	0	stable	$\bar{\nu}_\tau$

Fig. 32.8

HADRONS

HADRON
A particle that feels the strong force is called a ***hadron***.

Particles that feel both the strong and the weak force are called **hadrons***. There are over one hundred types of hadron known. The hadrons can be divided into two families, the **baryons** and the **mesons**.

BARYONS

The **baryons** have masses greater than or equal to that of the proton (*baryon* means 'heavy' in Greek). Protons, neutrons and heavier particles are baryons.

MESONS

The **mesons** have masses between electrons and protons.

FIG. 32.9 shows some examples of each type of particle:

SOME MEMBERS OF THE HADRON FAMILY					
Particle name	**Symbol**	**Relative mass**	**Charge**	**Mean lifetime**	**Antiparticle**
Baryons					
Proton	p	1	+1	stable	$\bar{p}$
Neutron	n	1	0	9×10^2 s	$\bar{n}$
Lambda	Λ^o	1.2	0	2.6×10^{-10} s	$\bar{\Lambda}^o$
Sigma	Σ^+	1.3	+1	0.8×10^{-10} s	$\bar{\Sigma}^+$
Mesons					
Pion	π^+	≈ 0.1	+1	2.6×10^{-8} s	π^-
Kaon	K^+	≈ 0.5	+1	1.24×10^{-8} s	K^-

Fig. 32.9

THE QUARK MODEL

QUARKS
Quarks are elementary particles out of which baryons and mesons are made. They have a charge of ±⅓e or ± ⅔ e.

Leptons appear to have no internal structure, i.e. they are truly elementary particles. They have no measurable size or internal structure. They are limited in number and do not break down into anything simpler.

However, **baryons** and **mesons** are complex particles. They have size and an internal structure. There are over a hundred different baryons and mesons. In 1963, the American physicists Gell-Mann and Zweig independently proposed that mesons and baryons are themselves composed of smaller particles called **quarks** and their antiparticles **antiquarks**. Thus protons and neutrons are not elementary particles but are made up of simpler particles. The quark model has brought an order to the hadrons. Using the quark model, the existence of other particles has been predicted. Subsequently, many of these have been observed in experiments.

* Hadrons feel the gravitational and electromagnetic force also.

To account for all known baryons and mesons there are six different kinds of quarks (and their antiparticles – called antiquarks). The quarks are named up (u), down (d), strange (s), charmed (c), top (t) and bottom (b). At the moment we believe that quarks are fundamentally structureless particles – point-like objects with no internal parts. They are thus like leptons. However, leptons have either zero charge or one unit of charge. Quarks have either ⅓ or ⅔ of a unit of charge. The antiquarks have the same amount of charge as the corresponding quark but of opposite sign. The mass of a quark and the corresponding antiquark are the same. FIG. 32.10 lists the six quarks, the six antiquarks and their charges.

The name quark was used by Gell-Mann. Initially he had proposed only three quarks. It comes from a quotation in James Joyce's *Finnegan's Wake*, 'Three quarks for Muster Mark!'.

QUARKS		
Name	**Symbol**	**Charge**
Up	u	+ ⅔ e
Down	d	– ⅓ e
Strange	s	– ⅓ e
Charmed	c	+ ⅔ e
Bottom	b	– ⅓ e
Top	t	+ ⅔ e

ANTIQUARKS		
Name	**Symbol**	**Charge**
Anti-up	$\bar{u}$	– ⅔ e
Anti-down	$\bar{d}$	+ ⅓ e
Anti-strange	$\bar{s}$	+ ⅓ e
Anti-charmed	$\bar{c}$	– ⅔ e
Anti-bottom	$\bar{b}$	+ ⅓ e
Anti-top	$\bar{t}$	– ⅔ e

Fig. 32.10

MESONS

A meson is made up of any one quark and any one antiquark:

e.g. The pion π^+ consists of the up and anti-down (FIG. 32.11) i.e. $u\bar{d}$.

Its charge is therefore + ⅔ + ⅓ = +1

The Kaon K^+ consists of up and antistrange i.e. $u\bar{s}$.

Its charge is therefore + ⅔ + ⅓ = +1

A ***meson*** is made up of any one quark and any one antiquark.

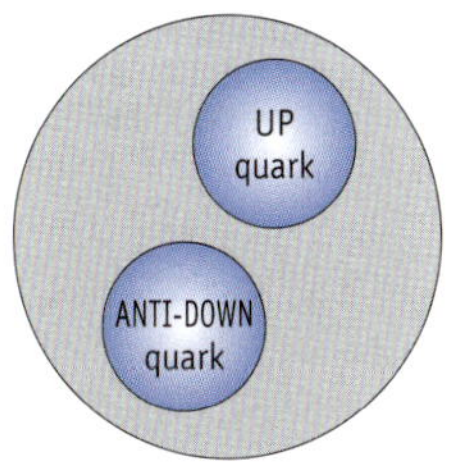

Fig. 32.11
A pion (π^+) consists of an up quark and an anti-down quark

BARYONS

Baryons are made up of any three quarks and antibaryons are made up of any three antiquarks:

e.g. The proton p = uud (FIG. 32.12). Its charge is: + ⅔ + ⅔ – ⅓ = +1

The neutron n = udd. Its charge is: + ⅔ – ⅓ – ⅓ = 0

The antiproton $\bar{p}$ = $\bar{u}\,\bar{u}\,\bar{d}$. Its charge is: – ⅔ – ⅔ + ⅓ = –1

The antineutron $\bar{n}$ = $\bar{u}\,\bar{d}\,\bar{d}$. Its charge is: – ⅔ + ⅓ + ⅓ = 0

A ***baryon*** is made up of any three quarks.
An ***antibaryon*** is made up of any three antiquarks.

SOME EXAMPLES OF QUARK COMPOSITION			
Particle name	**Symbol**	**Charge**	**Quark composition**
Baryons			
Proton	p	+1(= + ⅔ + ⅔ – ⅓)	uud
Neutron	n	0(= + ⅔ – ⅓ – ⅓)	udd
Lambda	Λ^0	0(= + ⅔ – ⅓ – ⅓)	uds
Sigma	Σ^+	+1(= + ⅔ + ⅔ – ⅓)	uus
Mesons			
Pion	π^+	+1(= + ⅔ + ⅓)	$u\bar{d}$
Kaon	K^+	+1(= + ⅔ + ⅓)	$u\bar{s}$

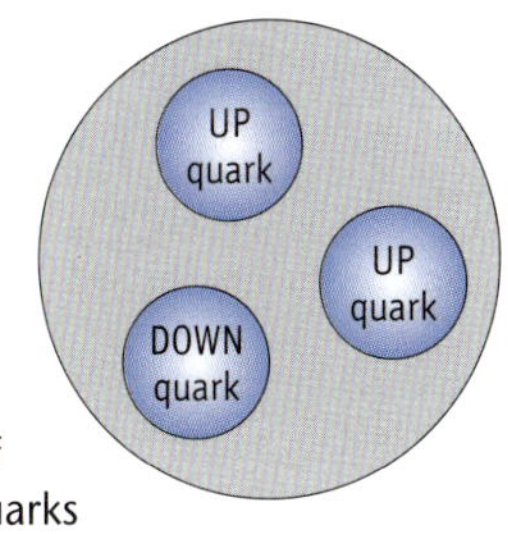

Fig. 32.12
A proton consists of two up quarks and a down quark.

EXERCISE 32.4

1. For the particle made up of each of the following quark combinations; state whether the particle is a meson or a baryon and state its charge:
(i) $u\bar{d}$ (ii) $\bar{u}d$ (iii) uud (iv) udd (v) $u\bar{s}$ (vi) $\bar{u}s$ (vii) uds (viii) uus (ix) dds (x) dss (xi) sss

Quarks feel all four forces. However, they principally interact via the strong force. Because this force is so strong, it would be extremely difficult to isolate a quark. Nonetheless, claims to have observed a quark in isolation have already been made.

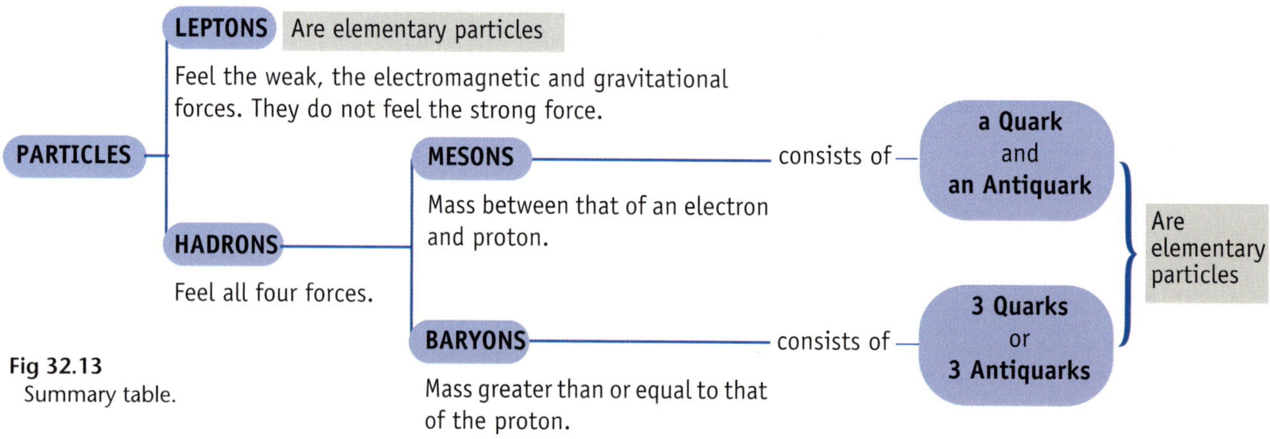

Fig 32.13
Summary table.

CHAPTER CHECKLIST

- **Define**: Antiparticle; Positron; Antimatter; Pair production; Pair annihilation; Lepton; Hadron; Baryon; Meson; Quark.
- **State**: Who produced the first splitting of a nucleus by artificially accelerated particles; The four fundamental forces of nature; The names of the six quarks and their charges; The names of the six antiquarks and their charges; What is meant by 'Particle Zoo'; The relative strengths and ranges of the four fundamental forces.
- **Describe**: Cockroft and Walton's experiment and recall its equation.
- **Recall** that: In a nuclear reaction mass-energy, momentum and electric charge is conserved; The neutrino was predicted to exist when conservation of momentum and energy was applied to β-decay; When particles collide in high energy accelerators, the larger the energy of the colliding particles, the greater the mass and variety of particles produced; Paul Dirac predicted anti-matter mathematically in the late 1920s; The word 'Quark' comes from a quotation in James Joyce's novel *Finnegan's Wake*.
- **Recall** and use the formula: $E = mc^2$ to solve problems.
- **Be able to**: Use conservation of mass-energy and conservation of momentum to solve numerical problems; Identify the nature and charge of a particle from its quark composition.

Option 2 (Honours Only) Applied Electricity

CHAPTER 33

1. Current in a Solenoid

ELECTROMAGNETIC RELAY

When you start a car on a cold morning, the starter motor may need a current of 100 A to operate it. This is a very large current which needs thick copper wire to carry it. There may be significant electrical sparking at the switch that turns on the current. We usually start a car by turning a key in the dashboard of the car. We do not want thick wires coming to the dashboard nor do we want sparking there. The usual way to overcome these problems is as follows:

When you turn the key, it closes a switch, completing a circuit containing a **coil** with a soft iron core, i.e. an **electromagnet** (Fig. 33.1). This electromagnet attracts a soft iron armature, which is free to pivot towards it. When one end of the armature moves towards the electromagnet the **contacts** at the other end of it close, completing the heavy duty circuit connecting the battery to the starter motor. The large current flows and the car engine starts. The electromagnet, pivoting armature and contacts together make up an **electromagnetic relay**. Fig. 33.2(a) shows the circuit symbol for an electromagnetic relay and Fig. 33.2(b) shows the structure of one particular type of relay. A relay is simply a switch that is operated by an electromagnet.

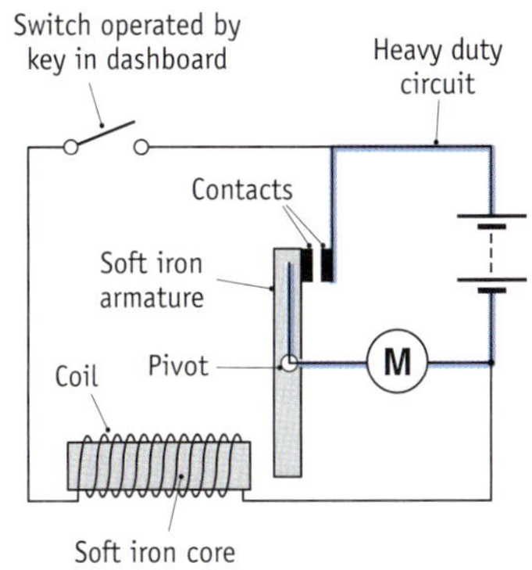

Fig. 33.1
An electromagnetic relay in a circuit.

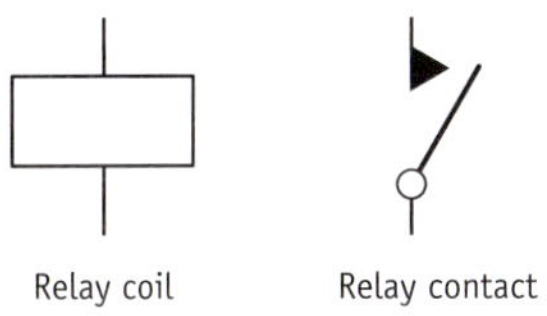

Fig. 33.2 (a)
Circuit symbol for an electromagnetic relay.

> **ELECTROMAGNETIC RELAY**
> An ***electromagnetic relay*** is a switch in an electric circuit that is turned on or off by means of an electromagnet.

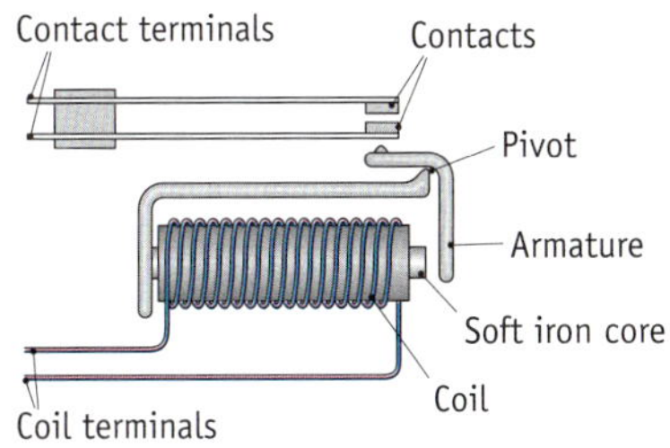

Fig. 33.2 (b)

USES OF A RELAY

Relays are used:

- in cars in the circuits for the starter motor, the heater fan, the horn and the heated rear window,
- for switching on and off most large electric motors,
- in residual current circuit breakers (residual current devices page 282). The current due to the difference in the size of the current between live and neutral operates a relay which breaks the circuit.

The action of a relay can easily be demonstrated in the laboratory with equipment like that in Fig. 33.3. When the terminals of the coil are connected to a battery or power supply the motor operates.

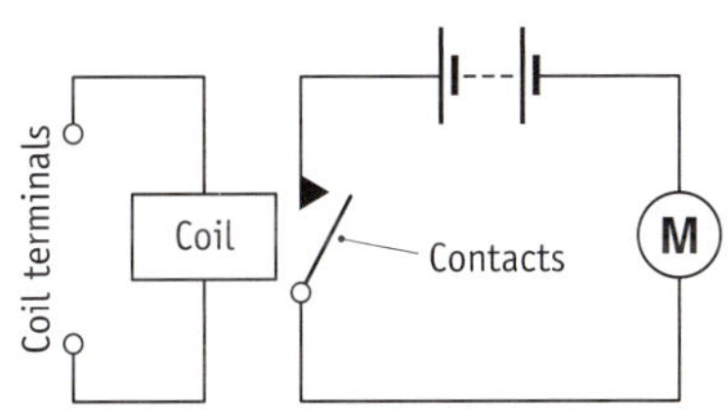

Fig. 33.3
Demonstrating the action of a relay.

2. Current in a Magnetic Field

FORCE ON A CURRENT-CARRYING COIL IN A MAGNETIC FIELD

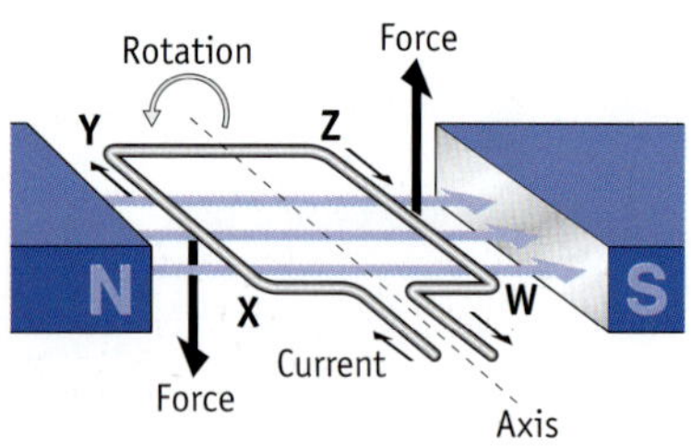

Fig. 33.4
The forces on a current-carrying coil in a uniform magnetic field.

Fig. 33.4 shows a coil in a uniform magnetic field. The coil is free to rotate about the axis shown. The coil is carrying a current (conventional current) in the indicated direction. By Fleming's left-hand rule, the sides XY and WZ experience forces in the direction shown. These forces cause the coil to rotate.

Fig. 33.5(A) is a front view of Fig. 33.4. The current flows in the left-hand side and flows out the right-hand side. Fig. 33.5(B) shows the forces on the coil after it has rotated a little. Fig. 33.5(C) shows the forces on the coil when it reaches the vertical position. Here the forces no longer tend to rotate the coil. If the coil is sufficiently free, its momentum will carry it beyond the vertical position to that shown in Fig. 33.4(D). However, the forces which now act on the coil tend to rotate it back to the vertical position. Hence the coil will come to rest in the vertical position – perhaps having oscillated for a while.

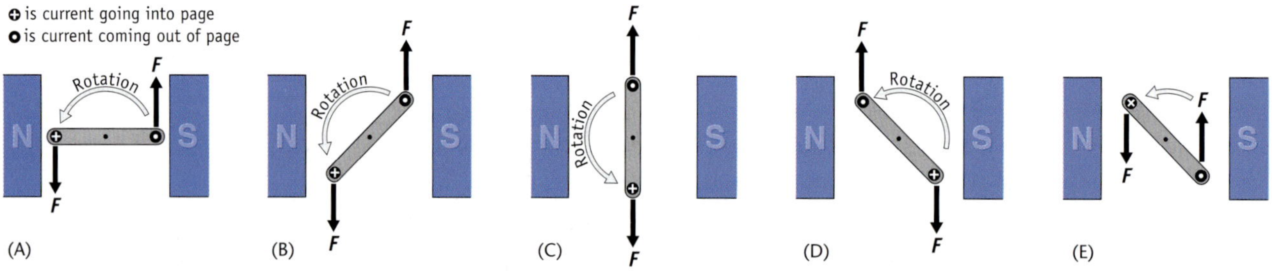

Fig. 33.5

If we could reverse the direction of the current in the coil as it passes through the vertical position, the forces acting on the coil would tend to keep the coil rotating in the same direction (Fig. 33.5(E)). If this is done every time the coil passes through the vertical position, the coil will rotate continuously. A coil that moves like this is called a **simple d.c. motor**.

SIMPLE d.c. MOTOR

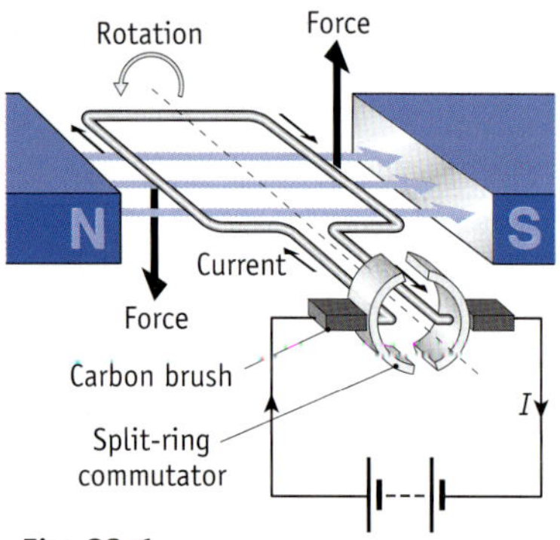

Fig. 33.6
A simple d.c. motor.

An easy way to make the coil rotate continuously is shown in Fig. 33.6. The ends of the coil are connected to a split circular ring of conducting material that rotates with the coil. This ring is called a **split-ring commutator**. Two pieces of carbon – called **carbon brushes** – touching the sides of the ring, allow current to pass from the battery through one half of the ring, into the coil and out through the other half of the moving ring. Convince yourself that this arrangement always causes current to flow in the left side of the coil and out the right side of the coil as shown in Fig. 33.6. We thus have a simple electric motor that operates on direct current, i.e. **a simple d.c. motor**. Note that **the simple d.c. motor is based on the principle that a current-carrying conductor in a magnetic field experiences a force.** You should not have any difficulty in naming ten common uses of electric motors.

The action of a simple d.c. motor can easily be demonstrated in the laboratory with simple equipment specially designed for this purpose.

It should be clear from Fig. 33.5 that the turning moment (i.e. the torque) on the coil is greatest when the coil is in the horizontal position.

FIG. 33.7 shows a way of keeping the value of the torque on the coil the same as the coil rotates. The magnet has semicircular pole pieces and the coil is wound around a cylindrical soft iron core.

The direction of the magnetic field in the air gap is along a radius of the cylinder. It is called a **radial magnetic field**. You can see from the diagram that the torque on this coil is always the same (except when in the vertical position), because the perpendicular distance between the forces remains the same. Smoother rotation is obtained. The split-ring commutator must still be used.

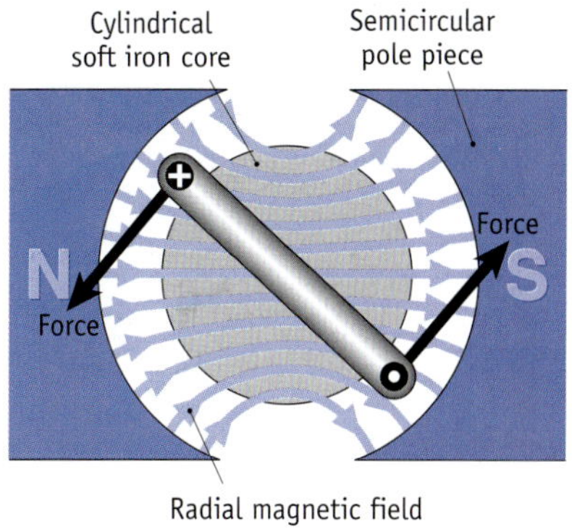

Fig. 33.7
The torque on a current-carrying coil in a radial magnetic field remains the same as the coil rotates.

THE MOVING COIL LOUD-SPEAKER

FIG. 33.8 is a simplified diagram of a moving coil loud-speaker. It is **based on the principle that a current-carrying conductor in a magnetic field experiences a force.** It consists of a **coil of wire** wound on a cardboard **tube** that can move along the central piece of a very strong **magnet**. The coil and tube is connected to a large **paper cone** that moves with it.

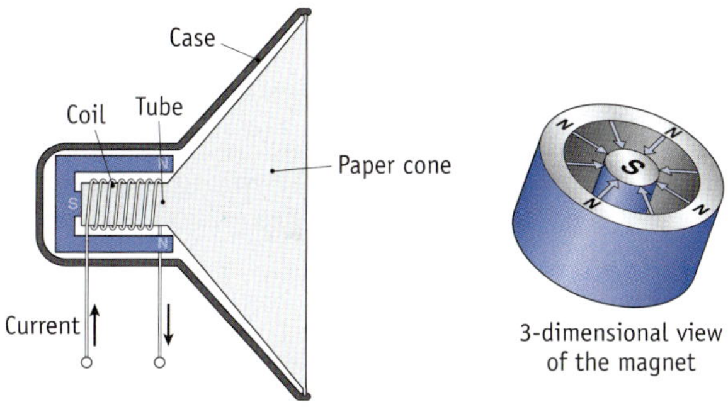

Fig. 33.8
A moving coil loud-speaker.

Fig. 33.9

When current flows in the coil in the direction shown, you can see – using Fleming's left-hand rule – that there is a force on the coil pushing it to the left. It moves slightly, pulling the cone with it. If the direction of the current is reversed, the force on the coil is to the right and the cone moves to the right. If an alternating current (a.c.) is passed into the coil, the coil and the cone repeatedly move in and out with a frequency which is the same as that of the a.c. If this frequency is within the frequency limits of audibility, the vibrating cone produces a sound wave of the same frequency in the surrounding air. FIG. 33.9 is a dissected loud-speaker showing its parts.

THE MOVING COIL GALVANOMETER

A galvanometer is an instrument that is used to measure the size of a small electric current. FIG. 33.10 shows the structure of a moving coil galvanometer. Its operation is **based on the principle that a current-carrying conductor in a magnetic field experiences a force.**

Construction

- The coil of wire is wound on an aluminium former.
- The coil and former together are mounted on bearings.
- The coil can rotate in the gap between the cylindrical poles of a magnet and a soft iron core.

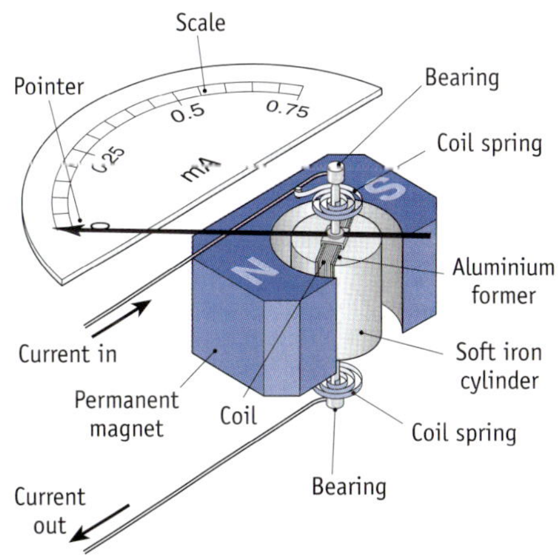

Fig. 33.10
A moving coil galvanometer.

- The angle through which the coil rotates is indicated by a pointer attached to it.
- Two coil springs are connected to the axle on which the coil rotates. These oppose the rotation of the coil.

Operation

- When current passes through the coil there are forces on the sides of the coil causing it to rotate.
- In the air gap between the poles and the core, the magnetic field is **radial**. Thus, as the coil is rotated, the perpendicular distance between the forces on the sides of the coil always remains the same (FIG. 33.7 page 381). The **moment of the couple** (i.e. the torque) on the coil due to the current is therefore constant as the coil rotates. The torque on the coil can be shown to be directly proportional to the current ($T_1 = kI$).
- As the coil rotates under the action of the current, the springs wind up and exert another couple on the coil. This couple opposes the couple due to the current.
- The torque the springs exert is proportional to the angle through which the coil rotates ($T_2 = c\theta$).
- The coil comes to rest when the torque due to the springs is equal to the torque due to the current.
  Thus $T_2 = T_1 \Rightarrow c\theta = kI \Rightarrow \theta \propto I$, i.e the angle through which the coil rotates is directly proportional to the current.
- The angle of rotation can be seen from a pointer connected to the coil. The scale over which the pointer runs is calibrated to read current. Thus the scale is **linear**.

When current flows in the coil, the pointer moves to a particular point on the scale. It would, however, overshoot that point and oscillate about it, taking some time to come to rest if the aluminium former were not present. The aluminium former stops this by a process called **eddy current damping**. The aluminium former cuts the lines of magnetic flux as the coil rotates and currents, called **eddy currents**, are induced in it. By Lenz's Law, the direction of these currents is to oppose the motion of the former. They thus exert a damping effect on the motion of the coil.

The moving coil galvanometer is a sensitive instrument in the sense that it can measure μA or mA. Any large current would cause the coil or suspension system to break.

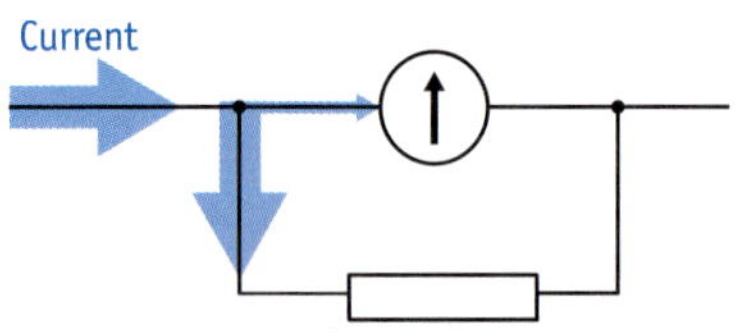

Fig. 33.11
Converting a galvanometer to an ammeter.

CONVERSION OF A GALVANOMETER TO AN AMMETER

A galvanometer may be converted to an ammeter by placing a low resistance called a shunt in parallel with it (FIG. 33.11). The value of this resistor is chosen so that most of the current flows through the resistor and a known fraction of it flows through the galvanometer. Since this fraction is known, the fraction through the shunt is also known. Thus the total current is known. Study Problem 1 on the next page carefully.

> **NOTE**
> A **galvanometer** may be converted to an **ammeter** by placing a **low resistance** – called a shunt – in **parallel** with it.

Problem 1: A galvanometer of full-scale deflection 4 mA and internal resistance 50 Ω is to be converted to an ammeter to read up to 6 A. Find the resistance of the shunt that must be connected in parallel with it.

Solution: For a full-scale deflection we want the situation shown in FIG. 33.12 to occur.

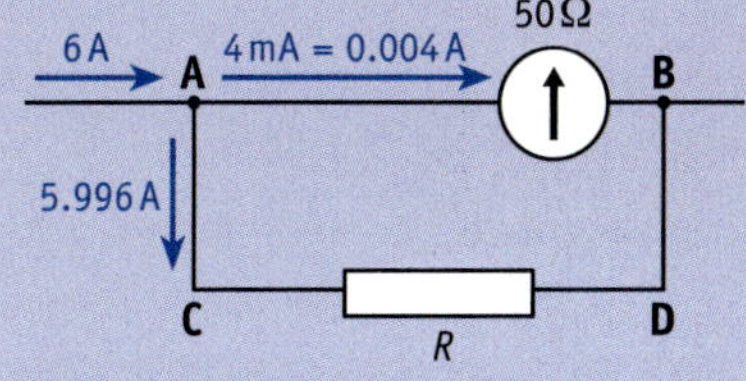

Fig. 33.12

Since AB and CD are in a parallel, the p.d. across each is the same. Thus by Ohm's law ($V = IR$) we have:

$V_{CD} = V_{AB} \Rightarrow 5.996R = (0.004)(50) \Rightarrow R = 0.0334\ \Omega$

Therefore, if a resistance of this value is connected in parallel with the meter, when 6 A flows in the whole circuit, the meter will give a full-scale deflection. You should also check that when 3 A flows in the circuit the meter gives a half-scale deflection. In practice, a moving coil ammeter found in a school laboratory is a galvanometer and a shunt enclosed in a plastic case.

CONVERSION OF A GALVANOMETER TO A VOLTMETER

A galvanometer may be converted to a voltmeter by connecting a high resistance – called a multiplier – in series with it. Study Problem 2 below to see how to find the value of the resistance.

NOTE
A **galvanometer** may be converted to a **voltmeter** by connecting a **high resistance** – called a multiplier – **in series** with it.

Problem 2: Find the value of the resistor that must be connected in series with a galvanometer of full-scale deflection 5 mA and internal resistance 4 Ω to convert it to a voltmeter with a full-scale deflection of 12 volts.

Solution: In the circuit in FIG. 33.13 it is required that when 12 volts is put across the combination, the galvanometer must give a full-scale deflection, i.e. 5 mA (0.005 A) must flow through the galvanometer.

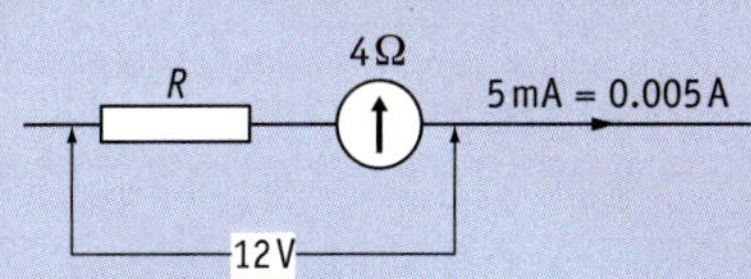

Fig. 33.13
Converting a galvanometer to a voltmeter.

p.d. across R + p.d. across 4 Ω = 12 V

i.e. $0.005 \times R + 0.005 \times 4 = 12 \Rightarrow R = 2396\ \Omega$

CONVERSION OF A GALVANOMETER TO AN OHMMETER

Consider the circuit shown in FIG. 33.14. If A is connected to B, the resistance between A and B is zero. The value of R is such that the galvanometer gives a full-scale deflection under these circumstances. If any other conductor is connected between A and B, the resistance between A and B is increased. Thus the size of the current flowing in the circuit is reduced. The position of the pointer on the galvanometer on the scale indicates this reduced current. The larger the resistance, the smaller the deflection on the galvanometer scale. Thus the position of the pointer indicates the value of the resistance. It is calibrated at manufacture to read the resistance directly. Note that the resistance scale runs in the opposite direction to the current scale. Note also that the resistance scale is very non-linear, since it goes from zero to infinity in the length of the scale.

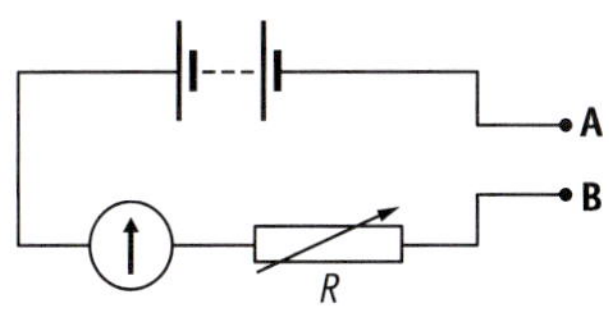

Fig. 33.14
Converting a galvanometer to an ohmmeter.

NOTE
A **galvanometer** may be converted to an **ohmmeter** by connecting a **battery** and a **variable resistor** in series with it.

EXERCISE 33.1

1. Find the value of the resistor that must be connected in parallel with a galvanometer of internal resistance 5 Ω and full-scale deflection 15 mA to convert it into an ammeter with a full-scale deflection of 1 A.
2. Find the value of the resistor that must be connected in series with a galvanometer of resistance 5 Ω and full-scale deflection 10 mA to convert it into a voltmeter with a full-scale deflection of 12 volts.
3. A galvanometer has a full-scale deflection of 6 mA and a resistance of 10 Ω.
   (i) How should a resistor be connected to it so that it can read currents up to 12 A? What should the value of this resistor be?
   (ii) How should a resistor be connected to it so that it can read voltages up to 20 volts? What should the value of this resistor be?
4. An ammeter consists of a galvanometer of resistance 4 Ω and a shunt in parallel with it of resistance 0.02 Ω. What current flows through the galvanometer when the current through the ammeter is 20 A?
5. A voltmeter consists of a galvanometer of resistance 5 Ω and a resistor of 3000 Ω connected in series with it. If the reading on the instrument is 40 V, what is the potential difference across the galvanometer?
6. A voltmeter has a full-scale deflection of 20 V and a resistance of 20 kΩ. How may it be converted to read up to 100 volts?
7. A galvanometer coil consists of 18 m of copper wire. The wire is of uniform circular cross-section and has a diameter of 0.085 mm. The resistivity of copper is 1.7×10^{-8} Ω m. Calculate the resistance of the coil. Given that the full-scale deflection of the galvanometer is 2 mA calculate:
   (i) the maximum voltage which should be applied between its terminals,
   (ii) the resistance of the resistor required to convert the galvanometer to a voltmeter of full-scale deflection 10 V.

3. Electromagnetic Induction

THE INDUCTION COIL

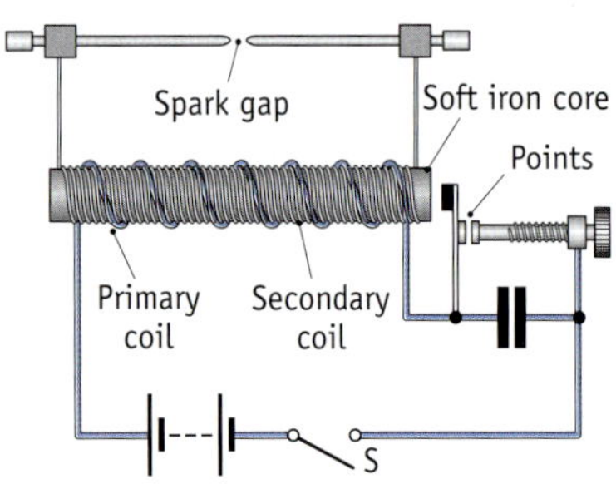

Fig. 33.15
An induction coil.

In 1836, in Maynooth, Dr. Nicolas Callan invented the **induction coil**. This is a device that produces a very high voltage from a low voltage source, such as a battery. Using an induction coil, he demonstrated how large electric sparks could be produced in air using a low voltage battery. Fig. 33.15 shows an induction coil. It consists of a coil of thick wire with a small number of turns wound around a soft iron core. This coil is called the **primary coil**. It is connected to the battery with a make-break mechanism similar to that found in an electric bell. Wound around the primary coil is another coil with a very large number (many thousands) of turns of wire. This coil is called the **secondary coil**.

The induction coil operates as follows:

- When the switch S is closed, current flows in the primary coil and the soft iron core becomes magnetised.
- The contact breaker is attracted to the core, the primary circuit is broken and the primary current stops flowing.
- The magnetic field due to the current in the primary coil changes quickly, i.e. it disappears. The contact breaker springs back and current flows again in the primary circuit.
- This process continues.

THE INDUCTION COIL
An *induction coil* is a device used to get a very high voltage from a low voltage d.c. source such as a battery.

- Every time the primary circuit is broken, the magnetic field becomes zero very quickly. This rapidly changing magnetic field passes through the secondary coil and because of its large number of turns, a very large emf is induced in the secondary coil.
- The very large induced emf is enough to break down the insulation of the air in the spark gap and large sparks are produced there (Fig. 33.16).
- An emf is also induced in the secondary every time the contact breaker points close, but because it takes longer for the magnetic field to build up as the current grows in the primary, the induced emf is much smaller.

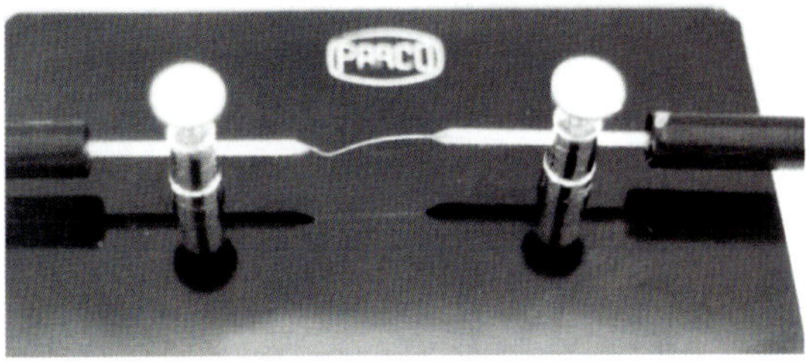

Fig. 33.16
Demonstrating the induction coil in a school laboratory.

At the break, an emf is also induced in the primary. This is large enough to produce some sparking at the contacts which would eventually burn out. To reduce this, a capacitor is connected as shown. This charges, absorbing the energy which would otherwise produce sparks.

Uses of an induction coil

- It is used in a petrol engine in a car to produce the necessary high voltage for the spark plugs.
- It is used in electric fences to produce the necessary high voltage (4 – 10 kV) from a 12 V battery.
- Historically it was used to produce a high voltage to operate a gas discharge tube and can still be used to do so.

4. Alternating Current

THE SIMPLE a.c. GENERATOR

Fig. 33.17(a) and (b) show a coil in a uniform magnetic field.

- If the coil is made to rotate about the axis, the magnetic flux passing through the coil changes.
- As the coil goes from A to B the flux passing through it increases, as it moves from B to C the flux passing through it decreases.
- An emf is thus induced in the coil.
- The emf is in one direction when the flux is increasing and in the opposite direction when the flux is decreasing. Thus **if a coil is rotated in a uniform magnetic field and alternating, emf is induced in the coil.**
- If the coil is connected to a complete circuit, current will flow. We say current is generated in the coil and the arrangement is called a **generator**.
- In Fig. 33.17(b) the coil is connected to the external circuit with **slip rings**. As the coil spins so do the rings. The carbon brushes do not move. Thus the same side of the coil is always connected to X and the other side always to Y. Because of this, **alternating current** flows in the external circuit. The magnet, coil and slip rings together are called a **simple a.c. generator**.

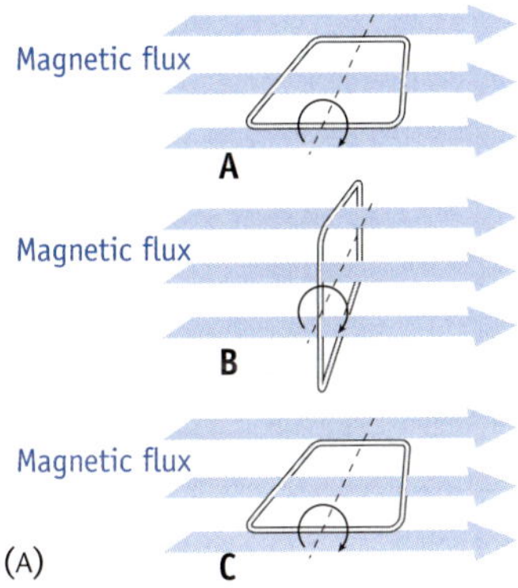

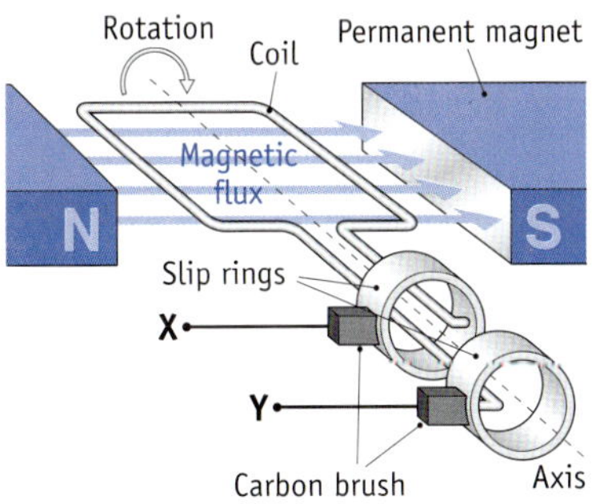

Fig. 33.17
Simple a.c. generator.

GENERATOR
An ***electrical generator*** is a device that converts kinetic energy to electrical energy in the form of an electric current.

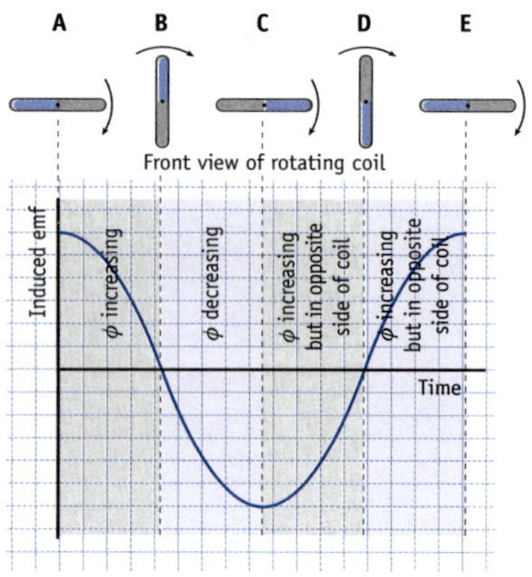

Fig. 33.18
Variation of the induced emf with time and position of the coil in an a.c. generator.

If the field is uniform and the coil spins at a constant rate, the a.c. generated varies in the same manner as the graph of $y = Sin\ x$. It is called a sinusoidal a.c. FIG. 33.18 shows a graph of emf (or current) against time. Also shown is the position of the coil at a number of instants. The emf is greatest when the coil is in positions A, C and E because the rate of change of flux is greatest there. The emf is instantaneously zero at B and D as the flux is not changing at each of these instants.

The operation of a simple a.c. generator can easily be shown in the laboratory. If a centre zero voltmeter is connected to the output, the pointer will be seen to move back and forth showing the a.c. produced.

The alternator in a car and a dynamo on a bicycle are two common examples of a.c. generators.

> **NOTE**
>
> ***Alternating current* (a.c.)** is generated in a coil spinning in a uniform magnetic field.
>
> If the coil is connected to the external circuit with ***slip rings*** we have a simple a.c. generator.

THE INDUCTION MOTOR

FIG. 33.19 shows a light aluminium disk that is free to rotate about an axle. If a strong magnet is rotated quickly around the disc as shown, the disc will be seen to follow the moving magnet. This is the **principle on which the induction motor is based**. It is explained as follows:

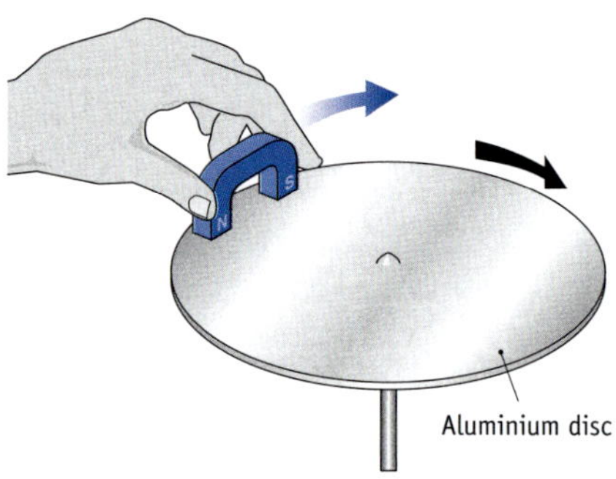

Fig. 33.19

- The **rotating magnetic field** causes a changing magnetic field in the aluminium disc.
- **Induced currents** flow in the aluminium since it is a conductor.
- By Lenz's Law the direction of these currents is such as to oppose the change producing them, i.e. oppose the magnet moving. The currents thus exert a **force on the magnet** trying to stop it from moving away.
- By Newton's 3rd Law an **equal but opposite force acts on the aluminium disc**.
- This force causes the **disc to rotate** in the same direction of the magnet.

In a real induction motor the rotating magnetic field is produced by coils connected to a.c. The rotating magnetic field produced causes a metallic cylinder to rotate. We need not worry about the exact details. Induction motors have no carbon brushes that can wear out. They are rugged, efficient and relatively cheap. The vast majority of electric motors used in industry are induction motors. They are used in pumps, fans and compressors, where reliability is important.

FACTORS AFFECTING THE EFFICIENCY OF A TRANSFORMER

If a transformer was 100% efficient, the useful energy output would equal the energy input. There are, however, a number of energy losses in a transformer which reduce its efficiency. Most of the energy lost appears as heat in the transformer. Very large transformers are oil cooled to remove this heat. A well designed transformer could be 90% efficient or better.

The main energy losses in a transformer are as follows:

- **I^2R heat losses in the coils.** This can be reduced by using thick wire in the low voltage coil.
- **Eddy current losses in the core,** i.e. induced electric currents in the core itself. These can be reduced by laminating the core (FIG. 33.20).
- **Hysteresis losses.** Energy is needed to repeatedly magnetise, demagnetise then magnetise the core in the opposite direction. These losses – called hysteresis losses – appear as heat in the core.
- **Leakage of magnetic flux.** Not all the flux from the primary may link the secondary.

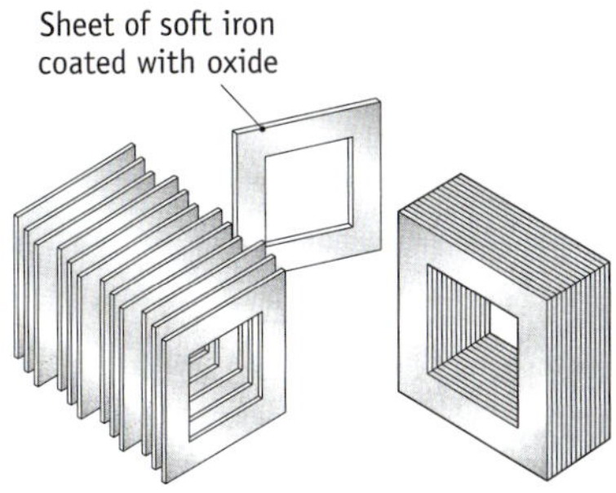

Fig. 33.20
A laminated core consists of thin layers of soft iron separated from each other by thin layers of oxide which is an insulator. This reduces eddy currents.

5. APPLICATIONS OF A DIODE

RECTIFICATION OF a.c.

Mains electricity is a.c. Much electrical and electronic equipment needs d.c. on which to operate. To rectify a.c. is to convert it to d.c. This can be done using a semiconductor diode.

Half-Wave Rectifier

In FIG. 33.21 an alternating voltage source is connected in the circuit with the diode. When the diode is forward biased, it conducts and current flows. When it is reversed biased no current flows. The current through the load (represented as a resistor) thus flows in one direction only, i.e. it is d.c. However, the current is not steady. It is pulsating d.c. The input and output voltages are as shown. Since the diode conducts for a half of each cycle it is called a **half-wave rectifier**.

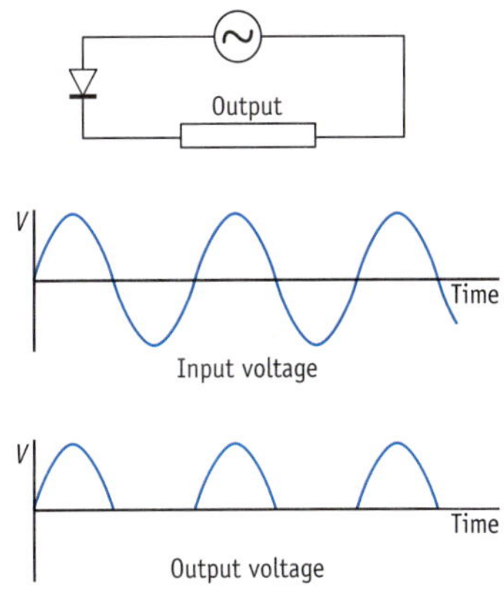

Fig. 33.21
Half-wave rectification.

Full-Wave Rectification – The Bridge Rectifier

FIG. 33.22 shows an a.c. source connected to four diodes connected in a bridge network. This arrangement is called a **bridge rectifier**. It produces d.c. as follows:

- When A is + with respect to B, current follows the path shown in FIG. 33.22(A).
- When the direction of the a.c. reverses, i.e. when B is + with respect to A, current follows the path shown in FIG. 33.22(B).
- In either case current always flows in the same direction through the resistor R. Thus d.c. is produced.
- Since d.c. flows on each half of the a.c. cycle, it is called **full-wave rectification**. FIG. 33.23 shows how the input and output voltages vary with time.

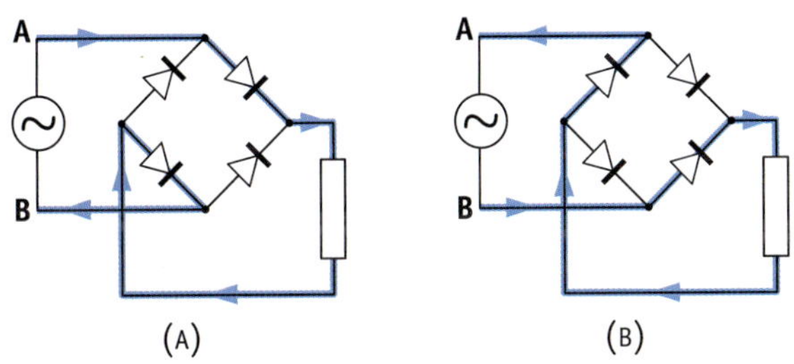

Fig. 33.22
Full-wave rectification using a bridge rectifier.

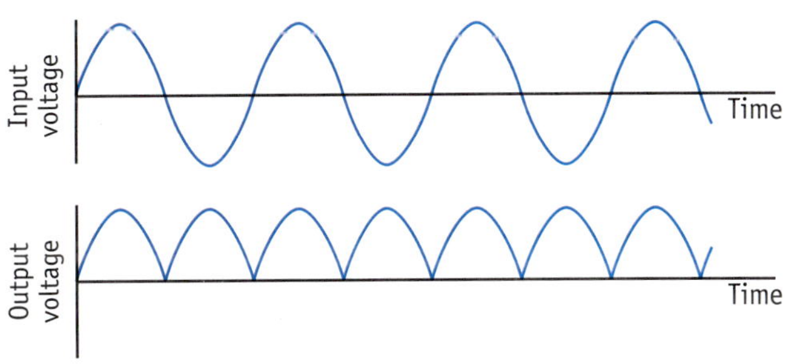

Fig. 33.23
Unsmoothed d.c. after full-wave rectification.

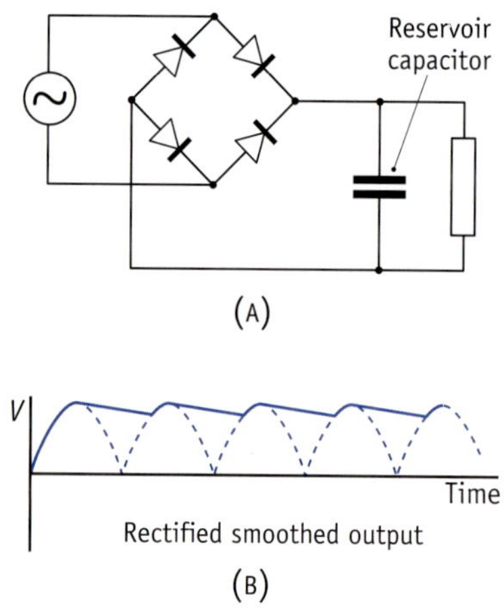

Fig. 33.24
Smoothing d.c. after full wave rectification.

If the circuit is modified, as in FIG. 33.24(A), by connecting in a large capacitor (e.g. 20 μF), much smoother d.c. is produced. The capacitor is called a **reservoir capacitor**. When the p.d. rises on the first half-cycle, current flows and the capacitor charges up. When the p.d. begins to drop, the capacitor starts to discharge. It cannot send current back through the diodes, so it discharges through the load, keeping the current near its maximum value. When the p.d. increases again, the capacitor tops up its charge and the process continues. The resulting output voltage is shown in FIG. 33.24(B). It is not perfectly smooth. The effect of half-wave and full-wave rectification and the action of the smoothing capacitor can be demonstrated in the laboratory with the circuits shown above. A CRO can be used to view the output.

THE LIGHT-EMITTING DIODE (LED)

A light-emitting diode (LED) is a p-n diode which when forward biased, and thus conducting current, gives out light (FIG. 33.25).

It is usually made of the semiconductor galium arsenide phosphide and has its junction very near the surface. When current flows through it, electrons and holes recombine at the junction. When an electron drops into a hole it loses energy. This energy is given out as light.

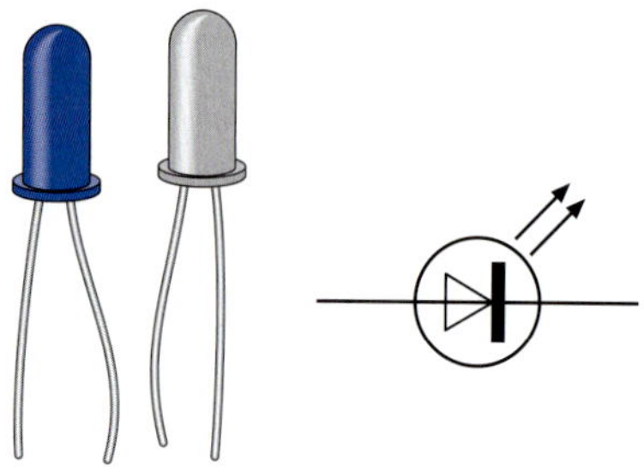

Fig. 33.25
Light-emitting diodes (LEDs) and their circuit symbol.

> **LIGHT-EMITTING DIODE**
> A ***light-emitting diode*** is a p-n junction which when forward biased gives out light.

LED's are often used as indicator lamps to show when a particular piece of electrical equipment is turned on. They are also used in the displays of large calculators, televisions, videos, digital clocks etc. More powerful and brighter LED's are used as bicycle lights and high brake lights on cars.

The operation of an LED can easily be shown in the laboratory. A resistor is normally connected in series with the LED to prevent large currents flowing which could damage the diode.

PHOTODIODE

The photodiode (FIG. 33.26) is a reverse biased p-n junction which creates electron-hole pairs in the depletion layer and the diode conducts when light shines on the junction.

In a photodiode, the size of the current flowing is proportional to the intensity of the light. Photodiodes are used in light meters and in some burglar alarms. In telecommunications they are used as fibre optic receivers. Telecommunications signals sent along optical fibres as pulses of light can be converted back into electrical pulses by photodiodes at the receiving end.

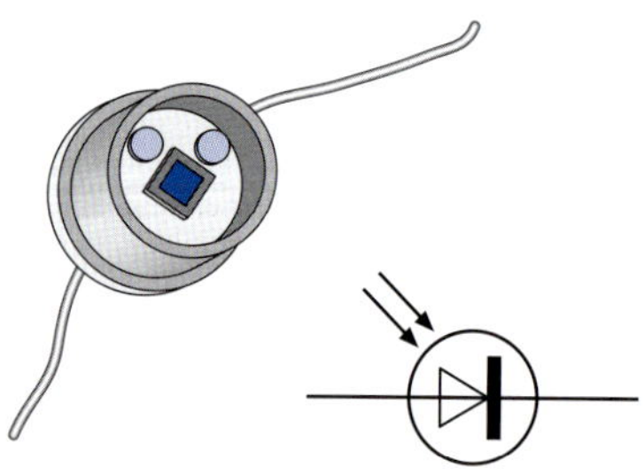

Fig. 33.26
A photodiode and its circuit symbol.

> **PHOTODIODE**
> A ***photodiode*** is a reverse biased p-n junction that conducts electricity when light shines on it.
>
> The ***current*** flowing is ***directly proportional*** to the ***intensity of the light***.

6. The Transistor

At the heart of almost any piece of modern electronic equipment is a transistor or a number of transistors. In Leaving Certificate Physics we study the action of a bi-polar transistor. There are two types of bi-polar transistor: the n p n and the p n p. We shall look here at the n p n transistor.

The N P N Bi-Polar Transistor

An n p n bi-polar transistor consists of a lightly doped piece of p-type semiconductor (called the **base**) sandwiched between two other thicker more heavily doped n-type pieces (called the **collector** and the **emitter**). The three are formed on the one crystal of semiconductor material. There are three connections to a transistor, one to the collector, one to the base and the other to the emitter. Fig. 33.27 shows an n p n bi-polar transistor, its circuit symbol and a simplified picture of its construction.

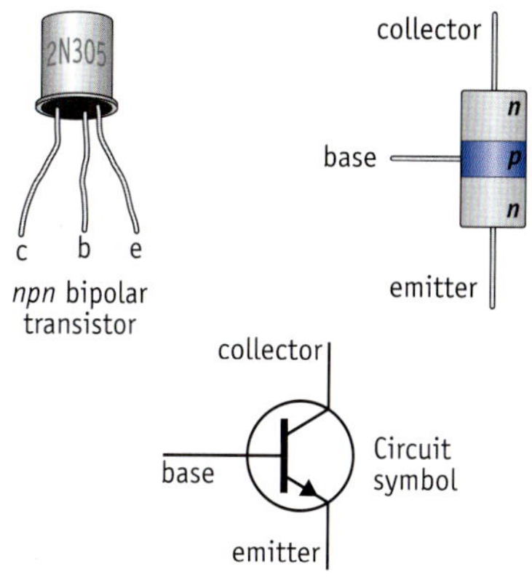

Fig. 33.27
The n p n bi-polar transistor.

N P N Transistor in Common Emitter Configuration

Fig. 33.28 shows an n p n bi-polar transistor connected in common emitter configuration. It is so called because the emitter is connected to both the collector and the base via the batteries. In Fig. 33.28 conventional current can flow in the paths shown.

- The current flowing in through the collector is the **collector current (I_c)**.
- The current flowing out through the emitter is the **emitter current (I_e)**.
- The current flowing in through the base is the **base current (I_b)**. Clearly:

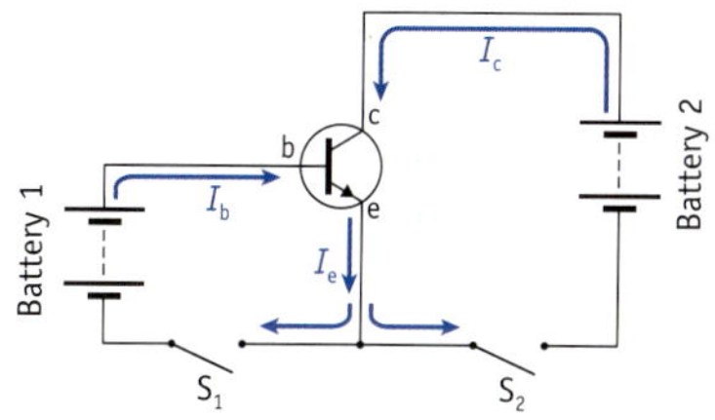

Fig. 33.28

$$I_e = I_c + I_b$$

Emitter Current = Collector Current + Base Current

Transistor Action

What is it that makes a transistor so important and of such great use? Very simply the answer is as follows:

A small current flowing in the base turns on or off and controls the size of a much larger current in the collector. With the circuit in Fig. 33.28 the following is found:

- With S_1 open and S_2 closed no current flows in either path, i.e. the collector current is zero if the base current is zero even if there is a large voltage between the collector and the emitter (battery 2).

  i.e. $I_b = 0 \Rightarrow I_c = 0$
- When S_1 is closed and if the value of the voltage between the emitter and the base is about 0.6 volts (for a silicon transistor) a small current (I_b) will flow in through the base. (The base emitter junction is a forward biased p n junction.)
- When base current flows, the collector current will also flow provided S_2 is closed.
- **I_c is a lot bigger than I_b.** Typically I_c is between 10 and 1000 times I_b depending on the transistor. I_c is proportional to I_b (Fig. 33.29).

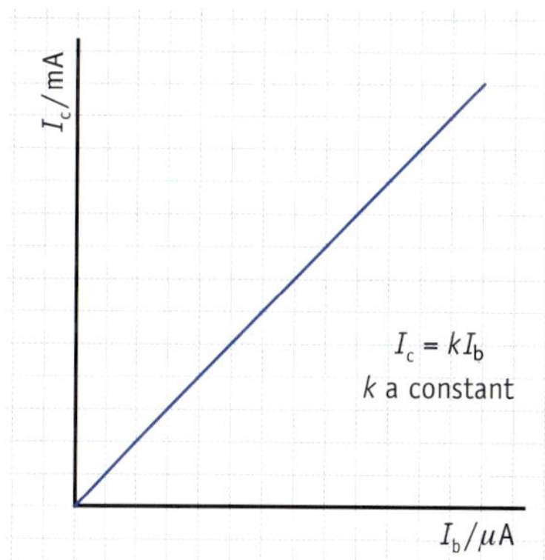

Fig. 33.29
The collector current is much bigger than and proportional to the base current.

We shall accept these results as experimental facts that can easily be demonstrated in the laboratory. If the base current exceeds a certain value, the transistor will be permanently damaged. A high resistance should be connected in the base lead to limit the current should too high a p.d. be accidentally put between the base and the emitter.

> **THE TRANSISTOR AS A SWITCH**
> No collector current flows unless a base current flows; i.e. the base current turns on (or off) the larger collector current. In this sense the transistor acts as a switch.

Demonstration of the Transistor as a Switch

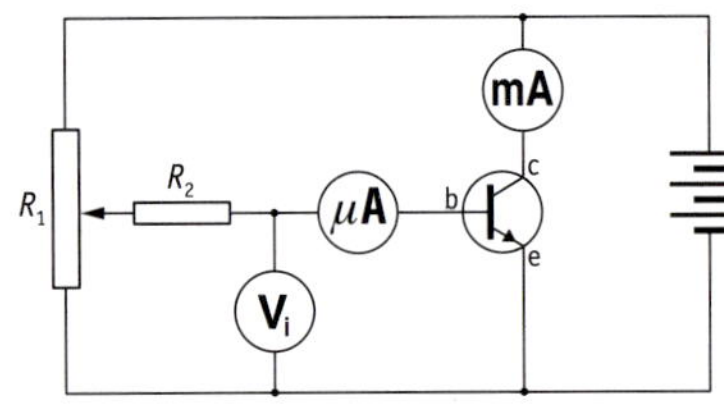

Fig. 33.30

Set up the equipment as in Fig. 33.30. Instead of using a separate battery to cause the base current, the variable resistor R_1 acts as a potential divider across the battery. The base emitter voltage then depends on the p.d. across the bottom piece of R_1 and can be varied. The resistor R_2 is the protective resistor mentioned above. By varying R_1, the input voltage V_i can be increased. For values of V_i less than about 0.6 V no base current flows and no collector current will be seen to flow either (this is because the base emitter junction voltage is not yet overcome). Increasing V_i above 0.6 V causes I_b to flow. As I_b increases so does I_c. Note that I_c is much bigger than I_b. I_b is in the order of microamperes whereas I_c is in milliamperes. If V_i is decreased below 0.6 V, I_b becomes zero and so does I_c. **Thus the base current switches on or off the larger collector current.**

Applications of Switching

In the following circuits a sensor allows the base current to flow, turning on the collector current which then turns on a bulb in the collector circuit. Other devices, such as an LED, a bell or a buzzer could replace the bulb.

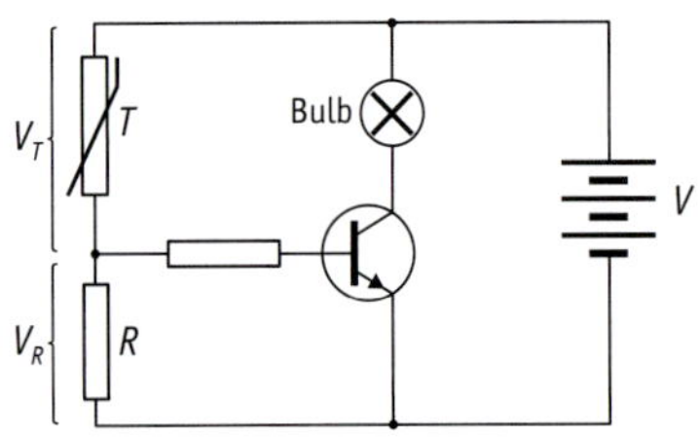

Fig. 33.31
A rise in temperature causes the bulb to light.

A Temperature-Controlled Switch

In Fig. 33.31 the thermistor T and the resistor R form a potential divider across the battery. The value of the resistance of T and R are chosen so that at low temperatures most of the voltage is dropped across the thermistor. The voltage across R is too small to cause a base current.

As the temperature is increased, the resistance of the thermistor decreases and so does the voltage (V_T) across it. Hence the voltage across R (V_R) increases. When V_R is large enough the base emitter voltage becomes large enough to cause I_b to flow. The transistor is turned on and I_c flows. **The bulb comes on indicating that the temperature has risen.** The value of R determines the temperature at which the bulb comes on.

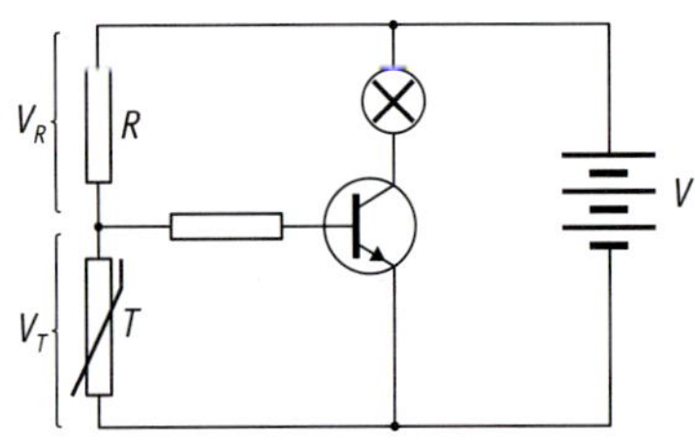

Fig. 33.32
A fall in temperature causes the bulb to light.

In Fig. 33.32 the positions of R and the thermistor are interchanged (the value of R being suitably adjusted). The circuit is now a detector of a fall in temperature (e.g. a frost detector in a greenhouse). When the temperature drops, the resistance of the thermistor increases, increasing V_T to a value that turns on the base current.

A Light-Controlled Switch

In Fig. 33.33(a) the thermistor is replaced with a light-dependent resistor (LDR). This circuit will automatically turn on a light in the dark. The value of R is chosen so that in daylight its resistance is much larger than that of the LDR. Hence most of the voltage is dropped across R and no base current flows. When it gets dark the resistance of the LDR increases and so does the p.d. across it, thus increasing the p.d. across the base emitter junction. When this value is about 0.6 V the base current flows and thus turns on the transistor, the collector current flows, and the bulb lights. Change the position of the LDR and R (FIG. 33.33(B)) and we have a circuit that comes on in brightness.

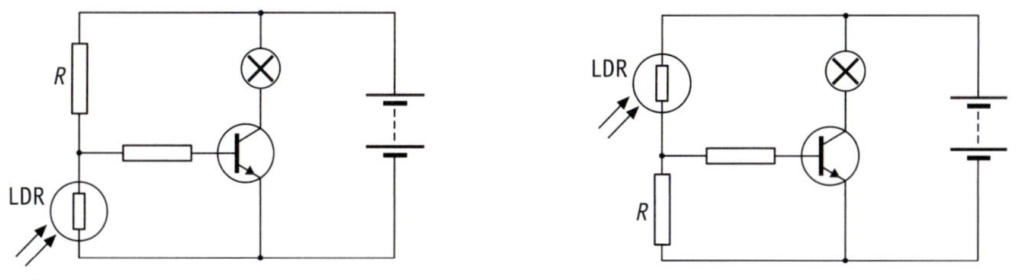

(A) The bulb comes on in the dark.

(B) The bulb comes on in brightness.

Fig. 33.33

USING A TRANSISTOR TO TURN ON A RELAY SWITCH

When using a transistor as a switch, the size of the collector current may be too small to operate whatever must be switched on (such as a bell or a light). We can, however, use the collector current to operate a relay switch (page 379). The relay then connects the warning device directly to the power supply. The relay is such that the collector current is large enough to operate it.

FIG. 33.34 shows a suitable circuit.

Fig. 33.34
A transistor switching on a relay.

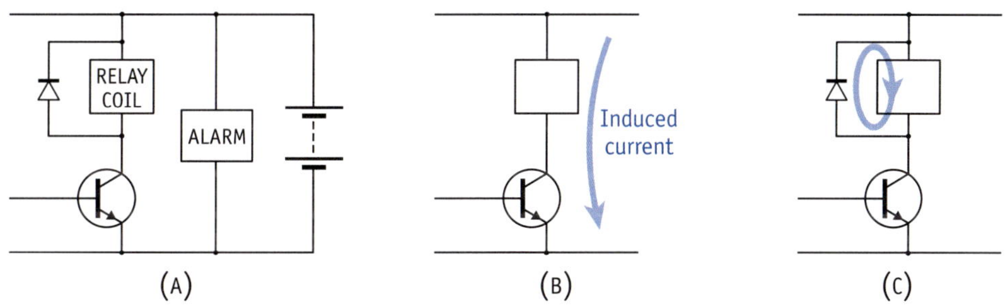

Purpose of the Diode in a Relay Circuit

In the circuit in FIG. 33.34 the collector current I_c becomes zero when the base current becomes zero, and hence the current through the coil of the relay becomes zero also. Since the coil has self inductance, (page 323) an induced emf appears in it causing an induced current to flow. By Lenz's Law the direction of this current is such that it opposes the decrease in I_c, i.e. it flows in the direction of the original collector current. This emf may be large and the current could be large enough to permanently damage the transistor. With the diode connected as in FIG. 33.34(C), it is forward biased and gives a low resistance path through which this current flows harmlessly.

THE TRANSISTOR AS A VOLTAGE INVERTER

If the input voltage V_i is sufficient to cause a base current in FIG. 33.35, the collector current flows. It flows through the resistor R and causes a p.d. between its ends. Since R and the transistor form a potential divider across the battery and the value of R is much larger than the resistance of the transistor, it follows that the voltage across the transistor V_o is almost zero.

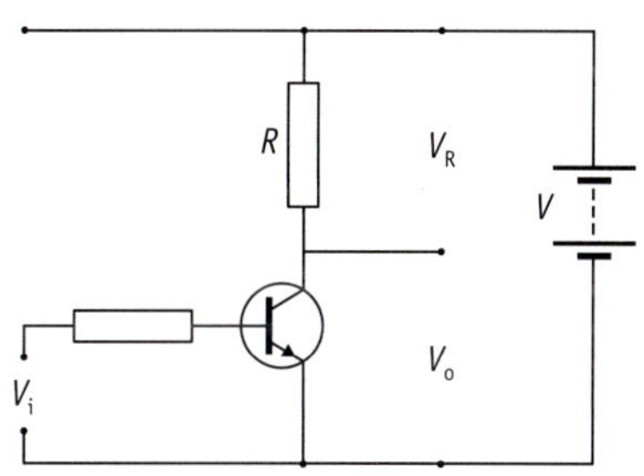

Fig. 33.35
The transistor as a voltage inverter.

Furthermore as V_i increases, I_c increases and hence V_o decreases, i.e.:

as the input voltage increases the output voltage decreases (1)

Likewise as V_i decreases so does I_b and I_c and hence so does V_R. But $V_R + V_o = V$ which is fixed. Therefore V_o increases. Thus:

as the input voltage decreases the output voltage increases. (2)

From (1) and (2) we see that the output voltage behaves in the opposite way to the input voltage. In this sense the **transistor acts as a voltage inverter**. This can be demonstrated easily in the laboratory. The output voltage can be measured with a high resistance (preferably digital) voltmeter.

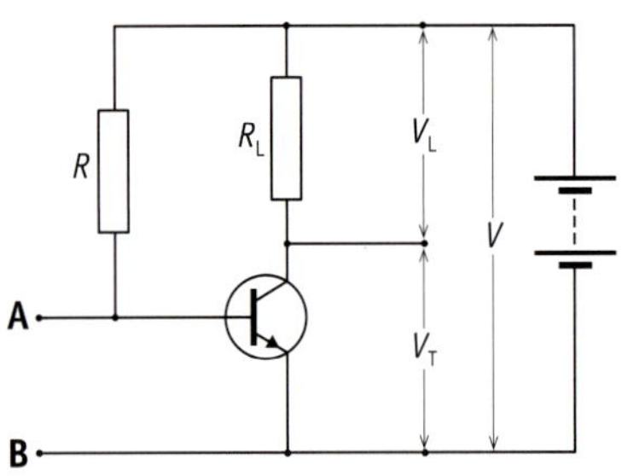

Fig. 33.36
A transistor as a voltage amplifier.

VOLTAGE AMPLIFIERS

A **voltage amplifier** is a circuit in which a small change in the voltage across its input causes a corresponding larger change in the value of the voltage across its output. Such a circuit will also produce across its output a magnified copy of a small alternating voltage across its input.

THE TRANSISTOR AS A VOLTAGE AMPLIFIER

FIG. 33.36 shows a transistor in a circuit that will behave as a voltage amplifier. The value of R is chosen so that the base emitter junction is always forward biased (even if the input voltage goes negative) and a suitable steady base current flows and turns on the collector current. R is called the **base bias resistor**. The input voltage is the voltage put across AB.

Suppose we place a small voltage across AB, with A positive with respect to B. The following happens:

- The voltage increase causes a small increase in the base current I_b.
- The small increase in the base current causes a corresponding **larger** increase in the collector current I_c.
- I_c passes through the resistor R_L and causes an increase in the voltage drop across it. R_L is called the **load resistor**.
- The voltage increase across the load resistor is much bigger than the corresponding increase in the input voltage.
- In FIG. 33.36 $V_L + V_T = V$ (the battery voltage). Since V is fixed, it follows that as V_L increases V_T decreases and vice versa. Thus the small **increase in the input voltage** causes a corresponding **larger decrease** in the voltage across the transistor (and vice versa).
- Either V_L or V_T can be taken as the output voltage.
- Thus voltage amplification is brought about.
- A small decrease in the input voltage, causes a decrease in I_b, a corresponding larger decrease in I_c and hence a large decrease in the voltage across R_L.

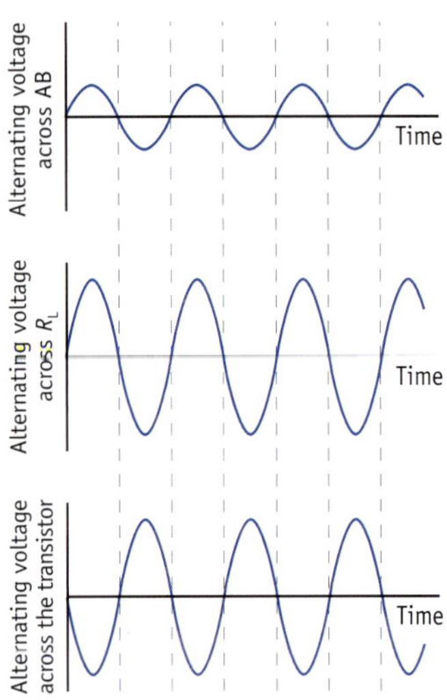

Fig. 33.37

Note that voltage amplification is brought about by both the transistor and the **load resistor** R_L. Without R_L there would be no voltage amplification.

If a small alternating voltage is placed across the input, it follows from the above that a corresponding larger alternating voltage will appear across R_L and also across the transistor. Note that the alternating voltage across the transistor is **inverted** compared with the input (FIG. 33.37).

The action of a transistor as a voltage amplifier can be demonstrated in the laboratory with the above circuit. The input and output voltages can be measured on a CRO.

> **NOTE**
>
> The ***base bias resistor R*** ensures that the base emitter junction is always forward biased and thus the collector current flows (even if the input voltage goes negative).
>
> The ***load resistor R_L*** converts large changes in the collector current I_c to large changes in the voltage across it.

Fig. 33.38
George Boole (1815–64).

7. Logic Gates

A **logic gate** is an electronic circuit that has an input and an output and whose output voltage depends on the input voltage in a definite manner. In this course you study three different logic gates called an **AND gate**, an **OR gate** and a **NOT gate**. These three and other logic gates are used in computers. The logical rules determining the ways they combine – called Boolean Algebra – were first derived by George Boole (Fig 33.38) in 1854 when he was professor of mathematics in what is now UCC.

AND GATE

Fig. 33.39 shows the circuit symbol for an **AND gate**. It has two input connections and one output connection. When in use:

- a voltage of zero volts or else a definite voltage (called the supply voltage) is applied to either of the 2 inputs,
- if the supply voltage is applied to an input, that input is said to be **high** (represented by a 1),
- if the voltage applied to an input is zero that input is said to be **low** (represented by a 0),
- the circuit is called an AND gate because it is **only when inputs A and B are high that the output is high**. Otherwise the output is low.

The table in Fig. 33.40 is called a **truth table** and summarises the way an AND gate behaves. In most practical applications (e.g. computers) logic gates are parts of integrated circuits.

OR GATE

Fig. 33.41 shows the circuit symbol and truth table for an **OR gate**. The circuit is called an OR gate because the **output C is high when either A or B or both are high.** If both inputs are low the output is low.

2 inputs 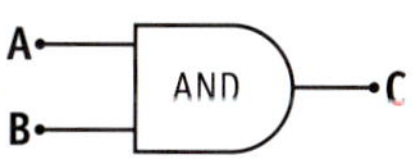 1 output

Fig. 33.39

2 inputs 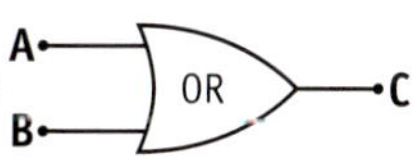 1 output

Input A	Input B	Output C (A and B)
1	1	1
1	0	0
0	1	0
0	0	0

Fig. 33.40
Truth Table for an **AND** gate.

Input A	Input B	Output C (A or B)
1	1	1
1	0	1
0	1	1
0	0	0

Fig. 33.41
Circuit symbol and Truth Table for an **OR** gate.

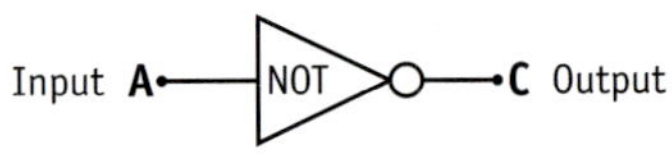

Fig 33.42

Input A	Output C (not A)
1	0
0	1

Fig 33.43
The Truth Table for a **NOT** gate.

NOT GATE

FIG. 33.42 shows the circuit symbol for a **NOT gate**. It has one input and one output. The circuit is called a NOT gate because **if the input is high the output is low** (not high) and **if the input is low the output is high** (not low). FIG. 33.43 is the truth table for a NOT gate.

A circuit consisting of a transistor and two resistors (FIG. 33.44) will behave as a NOT gate. The p.d. between A and Y is the input voltage and the p.d. between C and Y is the output voltage.

Input High

If A is connected to X the input is **high**. In this case, base current flows, the transistor gets turned on and collector current I_c flows. I_c passes through the 1 kΩ resistor and the transistor. Since the resistance of the transistor is small compared with 1 kΩ, most of the voltage is dropped across the 1 kΩ resistor. The potential of P is +6 volts, thus the potential of C is almost zero, i.e. the output is **low.**

Input Low

If A is connected to Y the input is **low**. No base current flows and therefore no collector current flows. There is thus no p.d. across the 1 kΩ resistor. The potential at each end of it is, therefore, the same. But the potential of P is +6 V. Therefore, the potential of C is +6 V, i.e. the output is **high**.

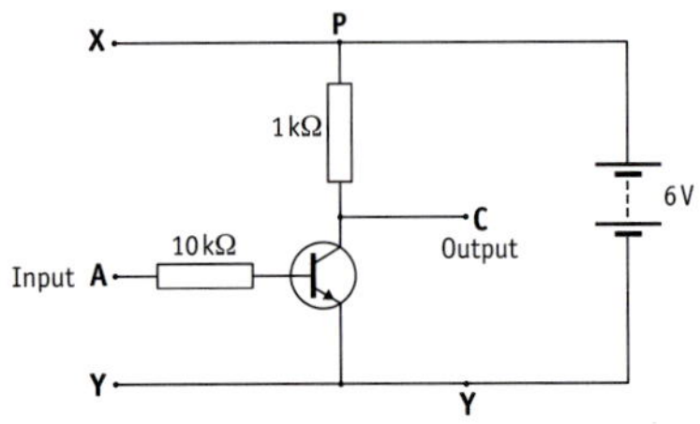

Fig 33.44

ESTABLISHING THE TRUTH TABLES EXPERIMENTALLY

Integrated circuits (ICs) are available that contain a number of logic gates on one chip. Such ICs are available mounted on plastic bases with only one of the gates connected to terminals on the base. These can be used in the laboratory to establish the truth tables.

EXPERIMENT

TO ESTABLISH THE TRUTH TABLES FOR AND, OR AND NOT GATES.

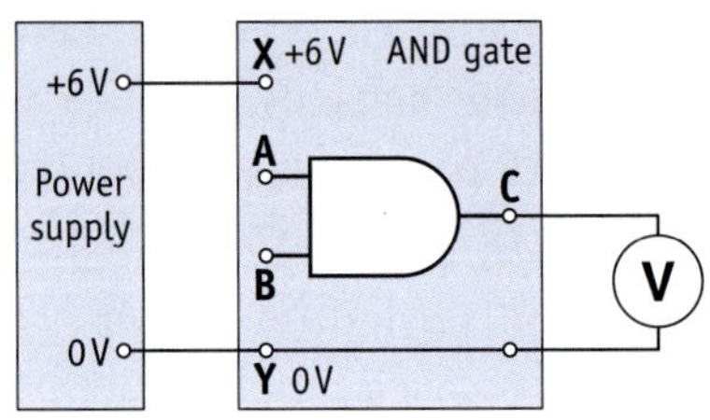

Fig 33.45

For the AND gate

1. Set up the circuit shown in FIG 33.45 for the AND gate.
2. When using the voltmeter do the following:
   - If the reading is zero or approximately so the output is low. Record this as a 0.
   - If the reading is 6 V or approximately so the output is high. Record this as a 1.
3. Connect both inputs to *X* (both are high) note the reading on the voltmeter.
4. Connect *A* to *X* and *B* to *Y* (*A* is high and *B* is low) note the reading on the voltmeter.
5. Connect *B* to *X* and *A* to *Y* (*A* is low and *B* is high) note the reading on the voltmeter.
6. Connect both inputs to *Y* (*A* is low and *B* is low) note the reading on the voltmeter.
7. Complete the truth table.

For the OR gate

Repeat steps 1 to 7 above using the OR gate.

For the NOT gate

Connect the input to +6 V and then connect it to 0 V. Note the reading on the voltmeter each time and complete the truth table.

Truth Tables

Input A	Input B	Output C
1	1	
1	0	
0	1	
0	0	

AND gate

Input A	Input B	Output C
1	1	
1	0	
0	1	
0	0	

OR gate

Input A	Output C
1	
0	

NOT gate

EXPERIMENTAL NOTES

If a digital voltmeter is not available, a light emitting diode and a resistor can be connected across the output in each gate. The diode lights when the output is high and not otherwise.

CHAPTER CHECKLIST

- **State** what each of the following is: An electromagnetic relay; An electric motor; A loud-speaker; A galvanometer; An induction coil; An a.c. generator; An induction motor; A half-wave rectifier; A bridge rectifier; A light-emitting diode (LED); A photodiode.
- **Recall**: The factors affecting the efficiency of a transformer; How a galvanometer may be converted to an ammeter, a voltmeter and an ohmmeter; How to calculate the value of the resistance of the resistor needed to convert a galvanometer to a voltmeter or an ammeter; That a simple d.c. motor, a moving coil loud-speaker and a moving coil galvanometer are based on the principle that a current-carrying conductor in a magnetic field experiences a force.
- **Explain** with the aid of a labelled diagram how each of the following operates: An electromagnetic relay; A simple d.c. motor; A moving coil loud-speaker; A moving coil galvanometer; An induction coil; A simple a.c. generator; A half-wave rectifier; A bridge rectifier; A transistor as a switch; A transistor as a voltage inverter; A transistor as a voltage amplifier.
- **Describe** an experiment to demonstrate: The action of a relay; The operation of a d.c. motor; The operation of an induction coil; The operation of a simple a.c. generator; The principle of the induction motor; The action of a transistor as a switch; A voltage inverter and a voltage amplifier.
- **Describe** an experiment to establish the truth tables for an AND, OR and NOT gate.
- **List**: three practical uses of a relay; Five uses of an electric motor; Two uses of an induction coil; Four uses of a transformer and a generator; Four practical applications of an LED.
- **Draw**: The circuit symbol for a diode, an LED, a photodiode and a transistor; The circuit symbol and truth table for an AND, OR and NOT gate.

APPENDIX 1

Proportionality

Direct Proportionality

Suppose a car is travelling at a steady speed of 27 m s^{-1} along a straight road. Then: in 1 second the car travels 27 m; in 2 seconds the car travels 54 m; in 10 seconds the car travels 270 m; in t seconds the car travels $27t$ m, i.e. the distance s travelled by the car in a time t seconds is given by: $s = 27t$

In this example you see that:

- if the time is doubled the distance travelled is doubled,
- if the time is trebled the distance travelled is trebled,
- if the time is increased by a factor of 4 the distance is also increased four times,
- if the time is halved the distance travelled is also halved, etc.

We say that the distance travelled is **directly proportional** to the time taken.

In general, if two quantities P and Q are related in such a manner that:

- if one is doubled so is the other,
- if one is trebled so is the other etc...,
- if one is halved so is the other,
- if one is quartered so is the other etc...,

the quantities are said to be **directly proportional** to each other, i.e. P is directly proportional to Q (we can just as well say that Q is directly proportional to P).

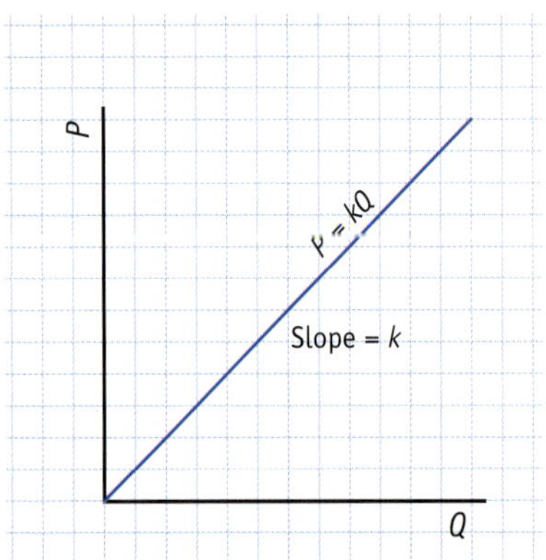

Fig. A1.1
P is directly proportional to Q.

DIRECTLY PROPORTIONAL
Two related quantities P and Q are ***directly proportional*** to each other if when one is increased n times so is the other; if one is decreased n times so is the other.

If P is directly proportional to Q we write: $\mathbf{P \propto Q}$

It follows mathematically that if $P \propto Q$ then $P = kQ$ where k is a constant number whose value depends on the units in which P and Q are measured. k is called the **constant of proportionality**.

IMPORTANT RESULT:

If two quantities P and Q are **directly proportional** to each other (i.e. if $P \propto Q$) then:

A **graph** of P against Q is a straight line through the origin (Fig. A1.1).

The **slope** of the graph = **the constant of proportionality**

Proof: If $P \propto Q$ then $P = kQ$ where k is a constant.

In the co-ordinate geometry section of your Maths course, you will find out that the graph of the equation $y = mx$ where m is a constant, is a straight line through the origin (0, 0) of slope m.

Comparing $P = kQ$ with $y = mx$, we see that a graph of P against Q is also a straight line through the origin (0, 0) of slope k.

Suppose we obtain from an **experiment** a series of corresponding values of two quantities P and Q. There are two ways to find out if $P \propto Q$:

1. Calculate the value of P / Q for each pair of values. Within the limits of experimental error the value of P / Q will always be the same if P is directly proportional to Q. This must be so since if $P / Q = k$ then $P = kQ \Rightarrow P \propto Q$.
2. Plot a graph of P against Q. If the plotted points appear to be in a more or less straight line that passes through the origin, we may conclude that $P \propto Q$.

Inverse Proportionality

If two quantities P and Q are related in such a manner that:

- if one is doubled the other is halved,
- if one is trebled the other is reduced by a factor of three etc...,
- if one is halved then the other is doubled,
- if one is quartered the other is increased four times etc...,

then the two quantities are said to **inversely proportional** to each other.

INVERSELY PROPORTIONAL

Two quantities P and Q are **inversely proportional** to each other if when one is increased n times the other becomes n times smaller; if one is made n times smaller the other becomes n times bigger.

It follows that if P is inversely proportional to Q then P is directly proportional to $1 / Q$.

i.e. P is inversely proportional to $Q \Rightarrow P \propto \frac{1}{Q} \Rightarrow P = k\frac{1}{Q} \Rightarrow PQ = k$ where k is a constant.

NOTE

IMPORTANT RESULT:

If two quantities P and Q are inversely proportional to each other, then a graph of P v $1 / Q$ is a straight line through the origin (Fig. A1.2).

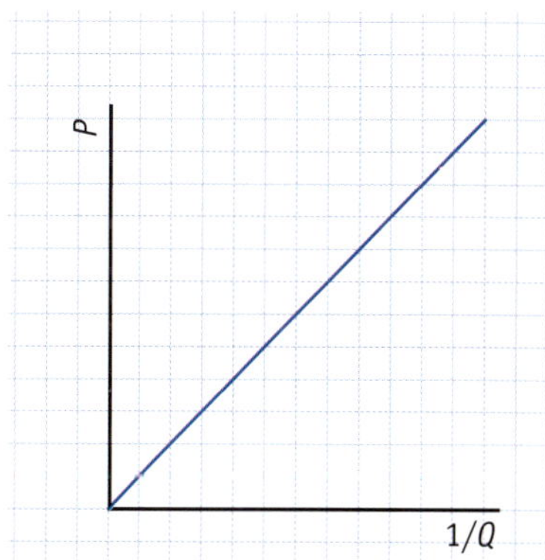

Fig. A1.2
P is inversely proportional to Q.

Note that a graph of P v Q is **NOT** a straight line.

Suppose we obtain from an **experiment** a series of corresponding values of two quantities P and Q. There are two ways to find out if P is inversely proportional to Q:

1. Calculate the value of the product PQ for each pair of values. If the value of PQ is more or less the same each time, P is inversely proportional to Q. This must be so since if $PQ = k$ then $P = k(1 / Q) \Rightarrow P \propto 1 / Q$.
2. Plot a graph of P against $1 / Q$. If the plotted points appear to be in more or less a straight line that passes through the origin, then we may conclude that P is inversely proportional to Q.

TWO OTHER IMPORTANT RESULTS:

Let P, Q and R be three related variable quantities:

(i) If $P \propto Q$ when R is constant and if $P \propto R$ when Q is constant then $P = kQR$ where k is a constant.

(ii) If $P \propto Q$ when R is constant and if $P \propto \frac{1}{R}$ when Q is constant then $P = \frac{kQ}{R}$ where k is a constant.

You need not worry about proving these facts.

EXERCISE A1.1

1. Two quantities P and Q are directly proportional to each other if:
   (a) when Q increases P decreases,
   (b) when P is doubled Q is halved,
   (c) when P is doubled Q is quadrupled,
   (d) when P is increased by two so is Q,
   (e) when Q is increased n times so is P,
   (f) none of the above.
2. If $P \propto Q$ then:
   (a) $P + Q = k$, (b) $PQ = k$,
   (c) $Q = P$, (d) $P / Q = k$,
   (e) none of the above.
   In each case k is a constant.
3. If a car has a constant velocity then:
   (a) the time taken to travel a distance is directly proportional to the distance,
   (b) the velocity is directly proportional to the distance,
   (c) the acceleration is inversely proportional to the velocity,
   (d) the velocity is directly proportional to the time.
4. If two related quantities P and Q are directly proportional to each other then:
   (a) a graph of P against Q passes through the origin,
   (b) a graph of P against Q is a straight line not passing through the origin,
   (c) a graph of P against Q has a constant slope,
   (d) (a), (b) and (c) are true,
   (e) (a) and (c) only are true.
5. Two related quantities P and Q are inversely proportional to each other. Then:
   (a) when P is doubled Q is halved,
   (b) when P is doubled Q is quadrupled,
   (c) when P is increased by two so is Q,
   (d) when Q is increased n times so is P.
   (e) none of the above.
6. If P is inversely proportional to Q:
   (a) $P + Q = k$, (b) $PQ = k$,
   (c) $Q = P$, (d) $P / Q = k$,
   (e) none of the above.
   In each case k is a constant.
7. If P is inversely proportional to Q then:
   (a) $P \propto Q$, (b) $Q \propto P$,
   (c) $1 / P \propto 1 / Q$, (d) $P \propto 1 / Q$,
   (e) none of the above.
8. If P is inversely proportional to Q then:
   (a) a graph of P against Q is a straight line through the origin,
   (b) a graph of P against $1 / Q$ is the same as a graph of P against Q,
   (c) a graph of P against $1 / Q$ is curved,
   (d) a graph of P against $1 / Q$ is a straight line through the origin,
   (e) none of the above.
9. If $P \propto Q$ when R is constant and if $P \propto R$ when Q is constant, then:
   (a) $P = kQR$, (b) $P = kQ / R$,
   (c) $R = kPQ$, (d) $P/Q = k R$,
   (e) none of the above.
10. $P \propto Q$, Q is increased to $9Q$, what is the new value of P,
    $P \propto 1 / Q$, Q is increased to $9Q$, what is the new value of P,
    $P \propto 1 / Q$, P is decreased to $P / 4$, what is the new value of Q.

Measuring Length

APPENDIX 2

The Vernier Scale

A vernier is a scale which, when used with another scale for measuring length (or angle), enables you to measure accurately to one more significant figure than the other scale alone would do. For example, when using a ruler graduated in millimetres, we can measure accurately the number of whole millimetres and make an estimate of the number of tenths of a millimetre. Using a suitable vernier scale with the millimetre scale allows us to measure both the number of whole millimetres and the number of tenths of a millimetre accurately. The vernier scale was invented by the French mathematician Pierre Vernier in 1631.

To Read a Vernier Scale

Proceed as follows:

1. Read the main scale as far as the zero on the vernier scale.
   In Fig. A2.1 this reading is: 4.9 ? cm.
2. Find a line on the vernier scale that is in line with a line on the main scale.
3. Read the vernier scale at this line.
   This gives the next significant figure.
   In Fig. A2.1 this is 7, thus the overall reading is 4.97 cm.

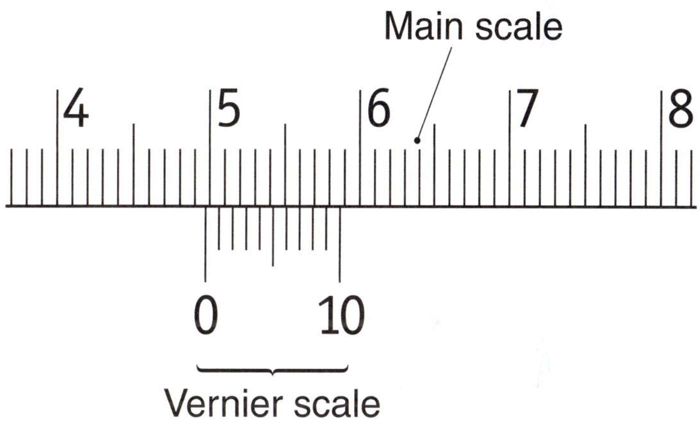

Fig. A2.1

EXERCISE 2.1

1. What are the readings on each scale in Fig. A2.2:

Fig. A2.2

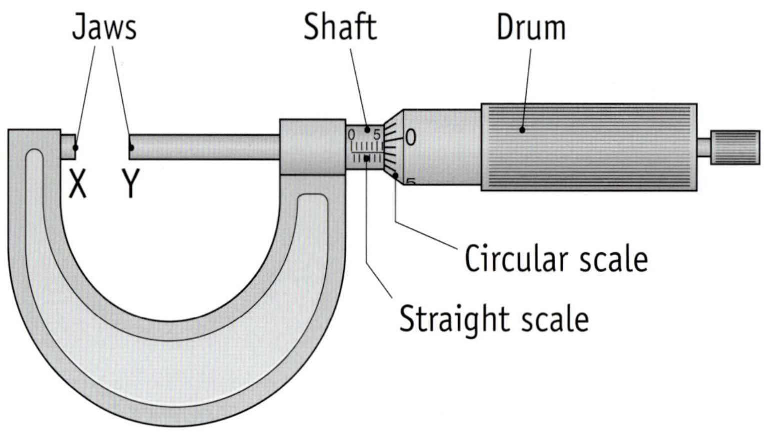

Fig. A2.3
A micrometer.

The Micrometer

The micrometer is an instrument which can be used to measure small distances, usually correct to 1⁄100 mm (i.e. 0.01 mm). Fig. A2.3 shows a typical micrometer.

There are two scales on a micrometer. They are the **straight scale** on the shaft and the **circular scale** on the drum. In one type of micrometer the straight scale is graduated in millimetres (mm) and the circular scale has 50 equal divisions on it (numbered 0 to 50). On such an instrument, turning the drum one full turn anticlockwise causes the distance between X and Y to increase by 0.5 mm. You must be very careful when using it as it is very easy to make a mistake. To read it (Fig. A2.4) you must proceed as follows:

Read the straight scale first. This gives either a whole number of mm or a whole number of mm plus 0.5 mm, e.g. The straight scale reading in Fig. A2.4(a) is 6.0 mm and the reading in Fig. A2.4(b) is 6.5 mm.

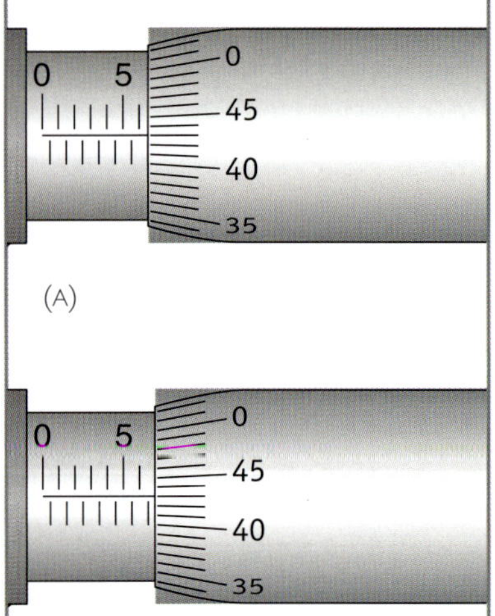

Fig. A2.4

Then read the circular scale. This gives two more figures which must be added to the reading from the straight scale. In Fig. A2.4 these figures are 0.43. The final readings then are:

For Fig. A2.4(a)	Fig. A2.4(b)
6.0 + 0.43 = 6.43 mm	6.5 + 0.43 = 6.93 mm

Zero Error on a Micrometer

When a micrometer is fully closed, with nothing between the jaws, the zero on the circular scale should be in line with the main line (the horizontal line) on the straight scale. If this is not the case, **the instrument has zero error** which must be taken into account. For example:

- Suppose the reading on the circular scale when fully closed is 0.02 mm. This means that all further reading will be **too big**. To correct for this, 0.02 mm must be **subtracted** from each reading.
- Suppose the reading on the circular scale when fully closed is 0.48 mm. This means that all further reading will be **too small**, by 0.02 mm. To correct for this, 0.02 mm must be **added** to each reading when using the instrument.

Useful Information

APPENDIX 3

Some Websites for Physics Students

Physics 2000	http://www.colorado.edu/physics/2000
Scoilnet	http://www.scoilnet.ie
Contemporary Physics Education Project (CPEP)	http:www.cpepweb.org
Institute of Physics	http://Physicsweb.org/resources
Physics of Everyday Life	http://howthingswork.virginia.edu/home.html
Particle Physics	http://pdg.lbl.gov/cpep/other sites.html
CERN	http://www.cern.ch/
Radiological Protection Institute of Ireland (RPPI)	http://www.rpii.ie

The Greek Alphabet

Upper case	Lower case	English name
Α	α	Alpha
Β	β	Beta
Γ	γ	Gamma
Δ	δ	Delta
Ε	ε	Epsilon
Ζ	ζ	Zeta
Η	η	Eta
Θ	θ	Theta
Ι	ι	Iota
Κ	κ	Kappa
Λ	λ	Lambda
Μ	μ	Mu
Ν	ν	Nu
Ξ	ξ	Zi
Ο	ο	Omicron
Π	π	Pi
Ρ	ρ	Rho
Σ	σ	Sigma
Τ	τ	Tau
Υ	υ	Upsilon
Φ	φ	Phi
Χ	χ	Chi
Ψ	ψ	Psi
Ω	ω	Omega

APPENDIX 4

How to Revise Leaving Certificate Physics

Pick a Chapter in Your Book:

1. ***Learn off by heart or be able to state in your own words the basic factual material in that chapter; so that you can:***
   - Recall every **definition**.
   - State every **Law**.
   - Recall the **unit** of each new quantity in that chapter and state whether it is a vector or scalar.
   - Define the unit of each quantity (except the second, the metre and the kilogram).
   - Recall each **formula** and state what quantity each letter in the formula stands for.
   - State what a particular piece of equipment is used for.
   - State who discovered what and in what century.

2. ***Be able to do straightforward numerical calculations with the formulae, that is:***
   - Recall what formula applies in a given situation.
   - Be able to put the right numbers in the right places in the formula.
   - Be able to calculate the answer and give it with the correct unit.

- Formula
- Substitution
- Calculation
- Answer and Unit

3. ***(a) Be able to describe any demonstration experiment as follows:***
   - Draw a **labelled diagram** of the equipment used.
   - Outline the **procedure** to be followed.
   - State what is observed and what **conclusion** is reached.

   (b) Be able to describe any of the mandatory experiments as follows:
   - Draw a **labelled diagram** of the equipment used in the experiment.
   - List the step-by-step **procedure** you would carry out.
   - State clearly **what quantities you measure** and with what piece of equipment.
   - Write down the **formula** or formulae relevant to the experiment.
   - State three **precautions** you would take to ensure a more accurate result.
   - Where appropriate, state what **graph** you would plot and what shape the graph would have.
   - Be able to use the graph to calculate the value of the quantity that is needed.

4. ***Be able to handle a set of data for each of the mandatory experiments as follows:***
   - Be able to calculate the value of the desired variable from the given data.
   - Be able to adjust the data where necessary so that you can draw a suitable graph.
   - Be able to measure the slope of the graph.
   - Be able to calculate the value of a relevant quantity from the slope.

5. ***Be able to explain the* physical principles *behind any everyday application of physics or piece of equipment described in the chapter.***

6. ***Be able to derive (prove) the formulae indicated in the chapter.***

> You now have a very good knowledge of the material in the chapter and should be able to handle any exam question on this topic, with the exception of the more difficult numerical problems.

7. ***Be able to handle more difficult problems.***

Practice as many examples as possible to improve this aspect of your learning.

Repeat **Steps 1 to 7** of this procedure for every chapter in your book. Do the chapters you like or are good at first, so that you get the most revision done in whatever time is available.

The more often you revise a chapter, the more quickly you will be able to revise it the next time and the more of it you will remember. The Leaving Certificate Physics course is not a short course. You will not be able to learn it all the week before the exam!

APPENDIX 5

Answers

CHAPTER 1

EXERCISE 1.1

2. kg m s^{-1} **3.** m s^{-2} **4.** kg m^{-3} **5.** N m^{-2}
6. (a) 10^4 **(b)** 10^6 **(c)** 10^3
7. (i) 5×10^{-4} m^2 **(ii)** 4×10^{-3} m^2 **(iii)** 1×10^{-6} m^3 **(iv)** 4.56×10^{-4} m^3 **(v)** 10^3 m^3
8. (i) 1.05×10^5 m **(ii)** 5.7×10^{-2} m **(iii)** 6.67×10^{-13} m **(iv)** 6×10^{24} kg **(v)** 9×10^3 kg m^{-3} **(vi)** 27.78 m s^{-1} **(vii)** 5×10^{-9} N **(viii)** 1×10^{-5} W **(ix)** 5×10^9 m
9. kg m s^{-2} **10.** N m **11.** kg m^2 s^{-2} **12.** J s^{-1}, kg m^2 s^{-3}

CHAPTER 2

EXERCISE 2.1

1. 1.27 s **4.** 2 m towards mirror, 3.5 m towards mirror, **5.** 1 m **6.** 4.2 years **7.** 50°, 50°, 40° **8.** 2 m s^{-1}
9. 0.9 m

CHAPTER 3

EXERCISE 3.1

1. 18.75 cm **2.** 60 cm **3.** 30 cm, real, 2, 4 cm
4. 20 cm, virtual, 2 **5.** 50 cm **6.** 26.67 cm, 13.33 cm
7. 75 cm, 25 cm **8.** 66.67 cm **9.** 40 cm, real
10. 160 cm from mirror

EXERCISE 3.2

1. 5.45 cm behind mirror, virtual, 0.545 **2.** 8.57 cm behind mirror, virtual, 0.29 **3.** 30 cm, 7.5 cm
4. 16.67 cm, **5.** 12 cm **6.** 40 cm from mirror
7. 40 cm from mirror **8.** At a distance from the mirror equal to its focal length

CHAPTER 4

EXERCISE 4.1

1. 1.5 **2.** 2.42 **3.** 22.1° **4.** 0.67 **5.** 0.88 **6.** 1.63
7. 6.1° **8.** 55.9°

EXERCISE 4.2

1. 1.5 **2.** 7.52 m **3.** 1.07 m **4.** 1.2 **5.** 9.99 cm

EXERCISE 4.3

1. 2×10^8 m s^{-1} **2.** 1.24×10^8 m s^{-1} **3.** 1.5 **4.** 2.4

EXERCISE 4.4

1. 1.56 **2.** 2.4 **3.** 56.4° **4.** 48.75° **5.** 37.04°
6. 7.7 cm **7.** 4.56 m **8.** 1.33 **9.** 1.414

CHAPTER 5

EXERCISE 5.1

1. 15.38 cm, real **2.** 42.86 cm, real, 1.71 cm **3.** 60 cm, virtual, 6 cm **4.** 75 cm, 25 cm **5.** 30 cm, 10 cm
6. 26 cm **7.** Convex, 25 cm **8.** 19.2 cm, 1.5, 0.67
9. 5.3 cm, 5 cm

EXERCISE 5.2

1. 12 cm, virtual **2.** 10 cm, virtual, 2.5 cm **3.** 120 cm, 40 cm

EXERCISE 5.3

1. + 2.5 m^{-1}, –1.67 m^{-1} **2.** 8.33 cm **3.** 40 m
4. + 16 m^{-1}, 6.25 cm **5.** –12 m^{-1}, 8.33 cm
6. + 0.03 m^{-1}, convex **7.** 60 cm, virtual, 6 cm **8.** 30 cm,
9. 20 cm **10.** 13.33 cm

CHAPTER 6

EXERCISE 6.1

1. 2×10^{-2} s, 0.5 s, 4 s, 1×10^{-6} s, 5×10^{-5} s, 1800 s, 172 800 s, 31 536 000 s **2.** 11.57 days **3.** 1000
4. 6.25×10^3 m, 1×10^{-2} m, 2×10^{-2} m, 4×10^{-9} m
5. 8.33 m s^{-1}, 10 m s^{-1} **6.** 8000 s **7.** 10 000 s
8. 2.09 m s^{-1} **9.** 50 m, 3600 m, 2×10^{-3} m, 2 *t* metres

EXERCISE 6.2

1. 6.67 m s^{-1} South East **2.** 300 m South **3.** 10 m, 100 m, 10 *t* metres, 2×10^{-5} m **4.** 2 m North, 2 m South, 2 m West, 2 m East, 2.83 m North West, 2.83 m South East **5.** 15.71 m, 10 m East, 1.571 m s^{-1}, 1 m s^{-1} East **7.** No, Yes, a particle moving in a circle at a steady speed **8.** Yes **9.** 5 m E 53.10° N, 1 m s^{-1} E 53.10° N

EXERCISE 6.3

1. Dog has a constant speed, 4 s, 22.5 m, 5, 5 m s^{-1}
2. Car stopped at pothole, Car stopped some distance from pothole, Car passes through pothole and moves away with steady speed **3.** 2.5 m s^{-1}, 0.625 m s^{-1}, 8 s

CHAPTER 7

EXERCISE 7.1

1. 6.67 m s^{-2} North **2.** 5 m s^{-2} West **3.** –1.5 m s^{-2}
4. 3 m s^{-1}, 12 m s^{-1}, 40.5 m s^{-1}, 3 *t* m s^{-1} **5.** 1.67 m s^{-2}
6. 5.6 m s^{-2} West

EXERCISE 7.2

1. 34 m s^{-1}, 264 m **2.** 0.8 m s^{-2}, 440 m **3.** 1.4 m s^{-2}, 7.14 s **4.** –10 m s^{-2}, 160 m **5.** 1.467 m s^{-2}, 94.02 m s^{-1}
6. –2 m s^{-2} **7.** 6.667 s **8.** 240 m s^{-1} in original direction of motion, Stopped **9.** 25 m s^{-1} in original direction, 15 m s^{-1} in opposite direction **10.** 12 s, 144 m from P

EXERCISE 7.3

1. (i) Steady speed of 6 m s^{-1} **(ii)** Starts from rest and moves with constant acceleration **(iii)** Starts at 5 m s^{-1} and moves with constant acceleration **(iv)** Starts at 9 m s^{-1} and decelerates to rest in 8 seconds **(v)** Object is stopped **(vi)** Starts from rest and moves with increasing acceleration **(vii)** Starts from rest, accelerates uniformly, then immediately decelerates uniformly to rest with a larger deceleration **(viii)** Starts from rest, accelerates uniformly, moves at a steady speed for a while then decelerates uniformly to rest
2. 7.3 m s^{-1}, 4 s, 0.75 m s^{-2}, 29.5 m
3. 0.49 m s^{-2}, 11.1 m, 7.2 m

EXERCISE 7.4

1. 4.453 m s^{-2} **2.** 2.025 m s^{-2}

EXERCISE 7.5

1. 3.5 s. 34.29 m s^{-1} **2.** 2040.8 m, 20.41 s **3.** 44.1 m **4.** 8.067 m s^{-2}, 2.73 s **5.** 326.5 m, 8.16 s, 15.02 s, 1.304 s **6.** 59.8 m s^{-1}, 182.45 m, 8.204 s **7.** 45.39 m, 5.493 s, 29.83 m s^{-1} **8.** 12 m **9.** 48.28 s, 2331.2 m **10.** 50.8 m s^{-1}, 168.4 m

CHAPTER 8

EXERCISE 8.1

1. 5 m in their common direction, 0 m, 2 m West, 8 N in their common direction, 2 N in the direction of the 8 N force **3.** 5 N in their common direction, 1 N in the direction of the 3 N force, 0 N, 5 N at 36.87° with 4 N force, 2.83 N at 45° with each, 13 N at 22.62° with 12 N force

EXERCISE 8.2

1. 7.83 N in direction of 5 N force, 10.24 N in direction of 6 N force, 4.14 N at 45° to each of the perpendicular 10 N forces, 3.66 N in opposite direction to 2 N force

EXERCISE 8.3

1. 68.4 N, 187.9 N **2.** 3172.1 N, 1479.2 N **3.** 50, 86.6 **4.** 128.6, 153.2 **5.** 518.5 s, 2383 m, 3111.1 m **6.** 20 N, 34.64 N **7.** 3.46 m s^{-2}, 2 m s^{-2} **8.** 28.19 N, 10.26 N **9.** WCosθ, WSinθ **10.** 1.73 m, 1.0 m, –2.5 m s^{-1}, 4.33 m s^{-1}, –2.6 m s^{-2}, –1.5 m s^{-2}

CHAPTER 9

EXERCISE 9.1

1. 100 N **2.** 0.4 m s^{-2} **3.** 200 N, **4.** 1333.3 kg **5.** 4 m s^{-2}, 54 m s^{-1}, 360 m **6.** 40 N **7.** 9.8 N, 0.0098 N, 1029 N, 9.8 *m* newtons **8.** 500 kg, 80 m s^{-1}, 800 m, 400 kN **9.** 1.5 m s^{-2}, 4 m s^{-2}, 0 m s^{-2}, 23 m s^{-1}, 28 m s^{-1}, 20 m s^{-1} **10.** 16.67 m s^{-2} **11.** No, 4630 N

EXERCISE 9.2

1. 600 N, 50.35 s **2.** 60 N **3.** 19 600 N, 15 600 N **4.** 98 N, 118 N, 78 N, 98 N, 0 N **5.** 980 N, 980 N, 980 N, 1280 N, 680 N, 0 N **6.** 45 N

EXERCISE 9.3

1. 16 000 kg m s^{-1} **2.** 36 000 kg m s^{-1}, 0 kg m s^{-1}, 33 333 kg m s^{-1}, 6000 kg m s^{-1} **3.** 33.33 m s^{-1}, 20 000 m s^{-1} **4.** 200 kg m s^{-1} in direction of 50 kg's motion **5.** 6.96 m s^{-1} **6.** 8 m s^{-1} in original direction of motion **7.** 7 m s^{-1} in original direction of motion of 600 kg **8.** 2 m s^{-1} **9.** 3.8 m s^{-1} in opposite direction to initial velocity **10.** 0.133 m s^{-1} **11.** 2 m s^{-1}, 4000 N **12.** 0.18 m s^{-1}, 900 N **13.** 18.87 m s^{-1} at 32° with direction of lorries original velocity **14.** 40.1995 m s^{-1} at 5.7° with original direction of motion of rocket **15.** 0.286 kg m s^{-1}, 0.286 kg m s^{-1}, 2.86 N

CHAPTER 10

EXERCISE 10.1

1. 0.2 kg, 0.004 kg, 200 kg, 2.4 × 10^{-5} kg **2.** 1 × 10^{-6} m^3, 1.2 × 10^{-4} m^3, 4 × 10^{-3} m^3, 2 m^3 **3.** 1 × 10^{-4} m^2, 2.2 × 10^{-2} m^2, 4 × 10^{-6} m^2, 3 m^2 **4.** 333.33 kg m^{-3} **5.** 11 180 kg m^{-3} **6.** 3.8 × 10^{-4} m^3, 31.5 grams **7.** 0.0136 kg **8.** 1.37 m^2

EXERCISE 10.2

1. 20 Pa **2.** 24 000 Pa **3.** 2.613 × 10^4 Pa, 1.452 × 10^4 Pa, 8.711 × 10^3 Pa **4.** 24 N **5.** 6210 N **6.** 15.7 N **7.** 2691 N, 25.79 kg m^{-3} **8.** 62 388.7 Pa

EXERCISE 10.3

1. 1960 Pa **2.** 323 400 Pa **3.** 1960 pa, 26 656 Pa **4.** 9800 Pa, 10 780 Pa, 588 N, 646.8N, It sinks

EXERCISE 10.4

1. 1 m^3 **2.** 25 000 Pa, 50 000 Pa, 4 × 10^6 Pa **3.** None, decreases, increases, increases **4.** 350 cm^3, 100 cm^3, 1400 cm^3 **5.** 20 Pa m^3

EXERCISE 10.5

1. 6.7 × 10^{-11} N **2.** 745.9 N **3.** 1.508 × 10^{-6} N **4.** 123.34 N **5.** 1.95 × 10^{20} N **6.** 3.39 × 10^{22} N

EXERCISE 10.6

1. 26 m s^{-2}, 2340 N **2.** 1.62 m s^{-2}, 97.2 N **3.** 263 m s^{-2} **6.** 9.51 m s^{-2} **7.** 2.7 × 10^6 m, 1.39 × 10^7 m **8.** 5.99 × 10^{24} kg **9.** 1.09 m s^{-2} **10.** No, Radius of Earth varies **11.** 2.86 m s^{-2} **12.** 3.8 × 10^7 m from moon

EXERCISE 10.7

1. 4 N m, 0 N m, 9 N m **2.** 1.1, 1.1, 1.1 **3.** 42 N, 8.33 cm, 38.33 cm, 35 **4.** 420 N **5.** 2 N **6.** One third of the weight of the beam at the 10 cm mark and two thirds of its weight at the 70 cm mark **7.** 280 N, 700 N, 1.667 m from centre **8.** 1.6 m from man

EXERCISE 10.8

1. 9000 N **2.** 20 N m **3.** 212.5 N **4.** 6000 N

CHAPTER 11

EXERCISE 11.1

1. 300 J **2.** 12 000 J **3.** 58.82 m **4.** 980 N, 58 800 J **5.** 196 J **6.** 4704 J **7.** 360 000 J **8.** 246 914 J **9.** 6614.28 J, 10 290 J

EXERCISE 11.2

1. 1440 J **2.** 392 000 J **3.** 1.024 × 10^{-16} J **4.** 800 kg **5.** 6.32 m s^{-1} **6.** 320 J, 313 600 J, 3.9 × 10^6 J 26 J **7.** 90 J, 9000 J, 8910 J, 742.5 m, 8910 J **8.** 160 N, 141.4 m s^{-1} **10.** 1:16

EXERCISE 11.3

1. 117 600 J **2.** 102 m **3.** 1764 J, 34.3 m s^{-1} **4.** 510.2 m **5.** 49 000 J, 19 600 J **6.** 3.13 m s^{-1} **7.** 3.028 m s^{-1} **8.** 12 m s^{-1}, 2700 J **9.** 0.073 cm

EXERCISE 11.4

1. 50 kW **2.** 10 W **3.** 1000 W **4.** 2778 W **5.** 1.08 × 10^6 J **6.** 3.9 × 10^7 J **7.** 2042 W **8.** 408 W **9.** 78.4 W **10.** 12.987 s **11.** 3.6 MJ

EXERCISE 11.5

1. 80 % **2.** 65.3 % **3.** 390 kW

CHAPTER 12

EXERCISE 12.1

1. π, $\pi/2$, $\pi/4$, $\pi/6$, $\pi/3$, 2π, $\pi/18$, 0.838 rad **2.** 36°, 25.7°, 57.3°, 240.6°, 60°, 30°, 45° **3.** 3 m, 9.42 m, 2.36 m, 6.44 m **4.** 0.857 cm **5.** 8.67 × 10^5 miles **6.** 0.384 m

EXERCISE 12.2

1. 0.333 m s^{-1}, 0.111 rad s^{-1}, 14.15 s, 9.43 s **2.** 25 s **3.** 7.5 rad s^{-1} **4.** 0 m s^{-1}, 1.8 m s^{-1} **5.** 83.78 m s^{-1} **6.** 7.27 × 10^{-5} rad s^{-1}, 465 m s^{-1}, 280 m s^{-1}

EXERCISE 12.3

1. 3750 N **2.** 16.2 N **3.** 144 m s^{-2}, 480 m s^{-2}, 720 N, 2400 N **4.** 10 rad s^{-1}, 40 m s^{-2}, 400 N **5.** 7330 N, **6.** 10 m s^{-1} **7.** 2834 m **8.** 3.05 m s^{-1}, 11.63 m s^{-2}, 2.79 N

EXERCISE 12.4

1. 2669.8 m s^{-1}, 36.87 hours **2.** 3.2×10^6 m, 11 208 m s^{-1} **3.** 11.9 years **4.** 29.28 years **5.** 6×10^{26} kg **6.** 4.24×10^7 m, 3079 m s^{-1} **7.** 465 m s^{-1}, 0.03385 m s^{-2}

CHAPTER 13

EXERCISES 13.1

1. 80 N, 0.25 m **2.** 2.67 N, 11.25 cm

EXERCISE 13.2

1. 0.2 s, 5 Hz **2.** 0.4 s **3.** 0.4 m s^{-2}, 0.125 m **4.** 0.5 m **5.** 25.27 N **6.** 2.565 s **7.** 0.673 Hz **8.** 4.39 m s^{-2}

EXERCISE 13.3

1. 2.84 s **2.** 9.75 m s^{-2} **3.** 99.29 cm **4.** 1.6 s

CHAPTER 14

EXERCISE 14.1

1. 273.15 K, 173.15 K, 293.15 K, 373.15 K **2.** –173.15°C, –0.15 °C, 99.85 °C, 226.85 °C

EXERCISE 14.2

1. 35.6 °C **2.** 30.3 °C, 30.0 °C, Because different thermometric properties vary differently with changing temperature.

CHAPTER 15

EXERCISE 15.1

1. 105 000 J **2.** 1667°C **3.** 400 J K^{-1} **4.** 568 480 J **5.** 514 800 J **6.** 385 J kg^{-1} K^{-1} **7.** 12.32 kg **8.** 22.39°C **9.** Copper **10.** 22 377.6 J **11.** 1881 s **12.** 38.5°C **13.** 29.25°C

EXERCISE 15.2

1. 3.3×10^6 J **2.** 165 kJ **3.** 3.03 kg **4.** 9.2×10^5 J **5.** 1.84×10^5 J **6.** 2.23 MJ **7.** 3.05 MJ **8.** 159 318 J **9.** 19.5 grams **10.** 34.6°C **11.** 25.6°C, 24.5°C **12.** 3.3×10^5 J kg^{-1} **13.** 4.54 min, 13.03 min **14.** 102.9 J, 3.19 grams

CHAPTER 16

EXERCISE 16.1

1. 10, 0.1 s **2.** 4 m **3.** 1 m **4.** 7.5 m s^{-1} **5.** 3×10^8 m s^{-1} **6.** 12 m **7.** 6×10^{16} Hz – 3×10^{19} Hz **8.** 480 m s^{-1}, remains the same, doubles **9.** 199 998 m **10.** 58.82 cm

EXERCISE 16.2

1. They are the same **2.** $\lambda/2$, $\lambda/2$, $\lambda/4$ **3.** 0.2 Hz **4.** 5 cm **5.** 680 Hz

EXERCISE 16.3

1. 2349.7 Hz **2.** 1740.9 Hz **3.** 56.67 m s^{-1} **4.** 241.5 Hz **5.** 238.9 Hz **6.** 16.19 m s^{-1}, 3818.2 Hz, 0.388 s, 0.194s **7.** 37.78 m s^{-1}, 888.89 Hz **8.** 12.92 m s^{-1}

CHAPTER 17

EXERCISE 17.1

1. 18.5 m **2.** 1360 m

EXERCISE 17.2

1. 44.2 Hz **2.** 7200 N **3.** 0.0625 kg m^{-1}, 25 Hz **4.** 25 Hz **5.** 4 times **6.** 520 Hz, 581 Hz **7.** 230 Hz, 184 Hz

CHAPTER 18

EXERCISE 18.1

1. 1.269×10^{-6} m **2.** 1.607×10^{-6} m **3.** 5×10^{-6} m **4.** 2×10^{-6} m **5.** 6.25×10^{-7} m **6.** 6.44×10^{-7} m **7.** 2.978×10^{-7} m **8.** 0.2999 m **9.** 4th **10.** 8th

EXERCISE 18.2

1. 32° 30', 126° 30' **2.** 5.914×10^{-7} m **3.** 243° 30', because the wavelength calculated using it differs from the others, 5.94×10^{-7} m

EXERCISE 18.3

1. 15.46° **2.** **(i)** Because in the higher orders n is bigger. Calculate the angle between red and violet when $n = 1$ and when $n = 3$ and you will see. **(ii)** Because red has a longer wavelength than violet **(iii)** d metres, $d/2$ metres **(iv)** The bigger d the more orders visible **(v)** In a grating red is diffracted through a bigger angle than violet. In the prism it is the opposite way around.

CHAPTER 19

EXERCISE 19.1

1. 1.958×10^{-11} F m^{-1} **2.** 4.49 **3.** 2.682×10^{10} N, the same **4.** 1.0058×10^{-2} N, attraction **5.** 0.397 N towards the –2 μC charge **6.** 5.53×10^{-3} N towards the 4 μC **7.** 14.3 N, 1.17×10^{-35} N **8.** 9.46×10^{-8} C **9.** $F/9$ **10.** + 2 μC

EXERCISE 19.2

1. 3×10^6 N C^{-1} **2.** 6×10^{-3} N **3.** 3.5 nC **4.** 8.94×10^3 N C^{-1} **5.** 1.789×10^{13} N C^{-1}, 1.789×10^{11} N C^{-1}, 1.789×10^7 N C^{-1}, in each case it is towards the negative charge **6.** $Q/(4\pi\varepsilon_o r^2)$ away from Q, $Q/(4\pi\varepsilon_o r^2)$ towards Q **7.** 5.33×10^{20} m s^{-2} **8.** 1.79×10^6 N C^{-1} towards the + 2 μC charge, 8.94 N towards the + 2 μC charge **9.** 2.24×10^6 N C^{-1} towards the negative charge, 4.47 N towards negative charge **10.** 0.4142 m from the 5 μC charge **11.** 0.5463 m from the 5 μC charge and not between the two charges

CHAPTER 20

EXERCISE 20.1

1. 3 V **2.** 10 V **3.** 80 J **4.** 9.6×10^{-5} J **5.** 3000 V **6.** 1.6×10^{-19} J, 4.8×10^{-17} J **7.** 4000 J, 4000 V **8.** 2×10^4 N, 2×10^4 N C^{-1}, 3.2×10^{-15} N, 6.4×10^{-17} J, 6.4×10^{-17} J, 1.2×10^7 m s^{-1}

EXERCISE 20.2

1. 2×10^{-10} F **2.** 3.33×10^{-7} F **3.** 8 μC **4.** 6 μC **5.** 1.33×10^6 V **7.** 4.167×10^{-7} F **8.** 5×10^{-3} C

EXERCISE 20.3

1. 1.78×10^{-10} F **2.** 1.335×10^{-10} F **3.** 1.12×10^8 m^2

4. 4.45×10^{-11} F, 11.57×10^{-11} F **5.** 7.79×10^{-8} C **6.** 120 J **7.** 4×10^{-6} J **8.** 1.472×10^{-4} J **9.** 1.05×10^{-4} J **10.** 3.32×10^{-4} C **11.** 1.39×10^{-4} F **12.** 40 V, 3.2×10^{-4} C, 1.6×10^{-4} C **13.** 13.33 V, 1.067×10^{-4} C, 26.67 V **14.** 36 V, 2.16×10^{-3} C, 3.33×10^{-8} J

CHAPTER 21

EXERCISE 21.1

1. 1×10^{-3} A, 5×10^{-5} A, 5×10^{-5} A, 1A, 2×10^{-7}A **2.** 1000 mA, 10^5 mA, 25 mA, 1×10^{-3} mA, 0.6 mA **3.** 3 A, 4 A, 2 A, 3 A, **4.** 3 C, 180 C, 10 800 C **5.** 1 A, 0.01667 A, 10 A **6.** 86 400 C **7.** 6.25×10^{18} **8.** 7.5×10^{20} **9.** 32 A **10.** 7200 s **11.** 2.25×10^{19} electrons **12. (i)** All read 2 A **(ii)** All read 2 A **(iii)** meter 1 and 2 read 2 A **(v)** meter 2 reads 4 A, meter 1 and 4 read 8 A **(iv)** meter 4 reads 10 A, meter 3 reads 8 A

CHAPTER 22

EXERCISE 22.1

1. 13 V **2.** 10 J, 60 J, 1×10^{-5} J **3.** 4 V **4.** 1200 J **5.** 50 V **6.** 0.435 A **7.** 80 W **8.** 880 W, 3.168 MJ **9.** 0.435 A **10.** 1.136 A **11.** 15 kJ, 10 J **12.** 34 W **13.** 3.024×10^5 J **14.** Meters 1, 2 and 3 are ammeters, Meters 4, 5 and 6 are voltmeters **15.** Both the same

CHAPTER 23

EXERCISE 23.1

1. 5 Ω **2.** 60 V **3.** 2.3 A **4.** 3 Ω **5.** 30 V **6.** 8 Ω, 4 Ω **7.** 5 Ω **8.** 5 Ω, 6 A, 30 V **9.** 6 Ω **10.** 10 V, 20 V, 40 V, 70 V **11.** 2 Ω, 0.8 Ω, 10 Ω, 16.36 Ω, 4 Ω, 7.33 Ω **12.** 10 Ω, 40 Ω, 1.2 Ω **13.** 3.85 A, 2.31 A, 1.54 A **14.** 942.9 Ω, 0.0015 A, Increases, Decreases, Decreases

EXERCISE 23.2

1. 1.2×10^{-6} Ω m **2.** 1.24×10^{-6} Ω m **3.** 1.52 m **4.** 138.6 m **5.** 3.30 mm, 3.80 mm, 6.43 mm, 6.93 mm **6.** 7.569×10^{-5} m **7.** 3.28×10^{-4} m **8.** 24 Ω, 12 m **9.** 0.65 Ω, 4 times longer **11.** 2.5 Ω **12.** 0.424 mm, 1.08×10^{-6} Ω m

EXERCISE 23.3

1. 22.5 Ω **2.** 30 Ω **3.** 0.4 Ω **4.** Double it, Half it

CHAPTER 24

EXERCISE 24.1

1. 1500 J **2.** 34.56 J **3.** 360 000 J **4.** 10 J, 40 J, 90 J, 160 J **5.** 4 J, 8 J, 40 J, 400 J **6.** 1440 J, 5 184 000 J **7.** 400 Ω **8.** 30 Ω **9.** 8.7 A, 26.42 Ω, 50 000 s **10.** 39 675 J, 172.5 C **11.** 4 times larger **12.** 14.5 Ω **13.** 363.5 s, 1666.7 s

EXERCISES 24.2

1. 4.35 A, 13 A **2.** Yes **3.** 6 kW h **4.** 0.05 kW h

CHAPTER 27

EXERCISE 27.1

1. Into page, Into page, Top wire up, Bottom wire down, Towards bottom left, Top down, Towards bottom left **2.** 1.33 T **3.** 20 N, perp. to both B and wire **4.** 1.33 T **5.** To the West **6.** Vertically upwards **7.** Down, Up, Rotate anticlockwise **8.** 1.2 N, 0.096 N m, Perpendicular distance to axis decreases, Yes, 0.192 N m **9.** 9 N, 0°, i.e. wire parallel to field **10.** 3.25 N

EXERCISE 27.2

1. 40 N **2.** 2.4×10^{-3} N **3.** 3.84×10^{-11} N **4.** 6.25 m s^{-1} **5.** A has positive charge, B is uncharged, C has Negative charge **6.** 6.9583 m **7.** 3.7916×10^{-7} m **8.** 3.52×10^7 m s^{-1} **10.** 10.4, 2.08×10^{-5} C, 2.08×10^{-5} A **11.** 3.2×10^{-11} A

CHAPTER 28

EXERCISE 28.1

1. 0.6 Wb **2.** 0.8 m^2 **3.** 0.02 Wb **4.** 1 T **5.** 1.46 m **6.** 7.2×10^{-3} Wb **7.** 0.4 Wb **8.** 2.02×10^{-4} Wb

EXERCISE 28.2

1. 2 V **2.** 2 V, 400 V **3.** 2400 V **4.** 1333.3 V, 333.3 A **5.** 7.2×10^{-3} Wb **6.** 64 V **7.** 0.36 V **8.** 76.8 V **9.** 1.8 V **10.** 2.5×10^{-3} T **11.** 0.192 m^2 **12.** 3.6×10^{-5} C

EXERCISE 28.3

1. 0.24 N **2.** 0.4 V, 8×10^{-3} N **3.** $(NBLv) / R$, $(N^2B^2L^2v) / R$

EXERCISE 28.4

1. 14.14 V **2.** 28.28 V **3.** 325.3 V **4.** 220 W **5.** 5 A, 100 V, 141.42 V **6.** 2.12 A, 900 W, 424.53 V, 600.37 V **7.** 2.56 s **8.** 200 V, 0.5 A

EXERCISE 28.5

1. Decreases because higher frequency ⇒ greater back emf **2.** 2 A, It will decrease due to back emf **4.** Gets brighter, Gets brighter, Goes out, Gets brighter **5.** Due to large emf induced in coil, Across the coil **6.** 8 V

EXERCISE 28.6

1. 46 V **2.** 80 V **3.** 818 turns **4.** 550 turns, 0.275 A, 0.306 A **5.** Coil will overheat and maybe burn out

CHAPTER 29

EXERCISE 29.1

1. 1.28×10^{-15} J, 1.28×10^{-15} J, 5.3×10^7 m s^{-1} **2.** 5.93×10^7 m s^{-1} **3.** 639.8 V **4.** 8×10^{-19} J, 3.2×10^{-17} J, 6.4×10^{-15} J, 8×10^{-13} J, 6.4×10^{-9} J, 6.72×10^{-19} J **5.** 3000 eV, 6 000 000 eV, 2.5×10^9 eV, 6.25×10^{18} eV, 12 500 eV, 4000 eV, 3.5 eV **6.** A is negative, B is positive, C is negative, D is uncharged, E is Positive **7.** 1.41×10^{-12} N **8.** 4.2×10^6 m s^{-1} **9.** 1.06×10^{-2} m **10.** A is negative, B is positive, C is positive, D is negative **11.** 9.09×10^{-31} kg

EXERCISE 29.2

1. 2.64×10^{-19} J, 1.65 eV **2.** 5.28×10^{-19} J, 3.3 eV **3.** 3.96×10^{-19} J **4.** 3.3×10^{-19} J **5.** 5.08×10^{14} Hz, 3.3528×10^{-19} J, 2.98×10^{19} photons **6.** 5.33×10^{14} Hz, 5.63×10^{-7} m **7.** 1.25×10^{13} **8.** 6.4×10^{-19} J, 9.70×10^{14} Hz **9.** 1.32×10^{-18} J, 8.25 eV **10.** 1.92×10^{-19} J, 2.91×10^{14} Hz, 1.03×10^{-6} m **11.** 2.64×10^{-20} J, 0.165 eV **12.** 2.46×10^{-19} J **13.** 1.25×10^{15}, 1.65×10^{-2} J **14.** 5.36×10^{-19} J, 1.09×10^6 m s^{-1} **15.** 6.072×10^{-26} J, 3.29×10^{31} photons

Chapter 30

EXERCISE 30.1

4. $^{222}_{86}Rn$ **5.** $^{216}_{84}Po$, $^{228}_{88}Ra$, $^{A-4}_{Z-2}Y$, $^{234}_{90}Th$, $^{14}_{7}N$,

$^{0}_{-1}\beta$, $^{A}_{Z+1}Y$, $^{0}_{-1}\beta$,

6. 2 α-particles, 2 β-particles **7.** 3α and 2β

8. $^{238}_{92}U \rightarrow {}^{234}_{90}Th \rightarrow {}^{234}_{91}Pa \rightarrow {}^{234}_{92}U \rightarrow {}^{230}_{90}Th \rightarrow$

$^{226}_{88}Ra$

9. Ra, Ac, Th, Ra

10. $^{223}_{88}Ra \rightarrow {}^{219}_{86}Rn + {}^{4}_{2}He$

11. $^{218}_{84}Po \rightarrow {}^{214}_{82}Pb + {}^{4}_{2}He$,

Lead and an alpha-particle

12. Atomic number decreases by 5, Mass number decreases by 16

13. Alpha,

$^{226}_{88}Ra \rightarrow {}^{222}_{86}Rn + {}^{4}_{2}He \rightarrow {}^{218}_{84}Po + {}^{4}_{2}He$,

EXERCISE 30.2

1. 3 s **2.** 4 minutes **3.** ½, ¼, ⅛ **4.** 1/16, 15/16 **5.** 15, 1/512, ¾, $1/2^n$ **6.** $5.776 \times 10^{-3}\ s^{-1}$ **7.** $2.1 \times 10^{-3}\ s^{-1}$
8. $9.158 \times 10^{-9}\ s^{-1}$ **9.** 1.386×10^{4} s **10.** 5.776 min
11. ⅛ **12.** 3.75×10^{10} **13.** 1.6×10^{13} Bq
14. 1.73×10^{18} Bq **15.** 1.7×10^{-10} α-particles per second
16. $1.346 \times 10^{-11}\ s^{-1}$, 1633 years

EXERCISE 30.3

1. 1.88×10^{25} atoms **2.** 9.665×10^{15} atoms
3. 2.47×10^{13} **4.** 6.5×10^{-15} grams

Chapter 31

EXERCISE 31.1

1. 1.56×10^{-8} kg **2.** 1.296×10^{27} J **3.** 1.26×10^{14} kg
4. 7.2×10^{-10} J **5.** 6.58 MeV **6.** 3.78×10^{-12} J
7. 1.25×10^{-12} J **8.** 3.56×10^{-28} kg **9.** 63.5:1, 1.08×10^{-12} J, $1.78 \times 10^{7}\ m\ s^{-1}$, $2.81 \times 10^{5}\ m\ s^{-1}$

10. $^{1}_{0}n \rightarrow {}^{1}_{1}p + {}^{0}_{-1}e$, 1.26×10^{-13} J

11. $^{1}_{1}H$, $^{4}_{2}He$

Chapter 32

EXERCISE 32.1

1. 7.81×10^{-13} J, 4.88 MeV **2.** 7.66×10^{-13} J
3. 2.493×10^{-14} J

EXERCISE 32.2

1. 931MeV **2.** 4.28 MeV

EXERCISE 32.3

1. 70 MeV **2.** 211 MeV

EXERCISE 32.4

(i) meson +1 **(ii)** meson -1 **(iii)** baryon +1
(iv) baryon 0 **(v)** meson +1 **(vi)** meson -1
(vii) baryon 0 **(viii)** baryon +1 **(ix)** baryon -1
(x) baryon -1 **(xi)** baryon -1

Chapter 33

EXERCISE 33.1

1. 0.07614 Ω **2.** 1195 Ω **3.** In parallel, 0.005 Ω, In series, 3323.3 Ω **4.** 0.0995 A **5.** 0.0666V **6.** Connect a resistance of 80 kΩ in series with it **7.** 53.93 Ω, 0.1079 V, 4946 Ω

Appendix 1

EXERCISE A1.1

1. e **2.** d **3.** a **4.** e **5.** a **6.** b **7.** d **8.** d **9.** a
10. 9*P*, *P*/9, 4*Q*

Appendix 2

EXERCISE A2.1

1. 1.66, 2.15

Index